Glass ... Current Issues

NATO ASI Series

Advanced Science Institutes Series

A Series presenting the results of activities sponsored by the NATO Science Committee, which aims at the dissemination of advanced scientific and technological knowledge, with a view to strengthening links between scientific communities.

The Series is published by an international board of publishers in conjunction with the NATO Scientific Affairs Division

A	Life Sciences	Plenum Publishing Corporation
B	Physics	London and New York
C	Mathematical and Physical Sciences	D. Reidel Publishing Company Dordrecht and Boston
D	Behavioural and Social Sciences	Martinus Nijhoff Publishers Dordrecht/Boston/Lancaster
E	Applied Sciences	
F	Computer and Systems Sciences	Springer-Verlag Berlin/Heidelberg/New York
G	Ecological Sciences	

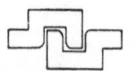

Series E: Applied Sciences – No. 92

Glass ... Current Issues

edited by

A.F. Wright
Institute Max von Laue-Paul Langevin
38042 Grenoble Cedex, France

J. Dupuy
Département de Physique des Matériaux
Université Claude Bernard Lyon I
69622 Villeurbanne Cedex, France

1985 **Martinus Nijhoff Publishers**
Dordrecht / Boston / Lancaster
Published in cooperation with NATO Scientific Affairs Division

Proceedings of the NATO Advanced Study Institute on Glass ... Current Issues,
Tenerife, Spain, April 2-13, 1984

Library of Congress Catalog in Publication Data

ISBN-13: 978-94-010-8758-2   e-ISBN-13: 978-94-009-5107-5
DOI: 10.1007/978-94-009-5107-5

Distributors for the United States and Canada: Kluwer Boston, Inc., 190 Old Derby
Street, Hingham, MA 02043, USA

Distributors for the UK and Ireland: Kluwer Academic Publishers, MTP Press Ltd,
Falcon House, Queen Square, Lancaster LA1 1RN, UK

Distributors for all other countries: Kluwer Academic Publishers Group, Distribution
Center, P.O. Box 322, 3300 AH Dordrecht, The Netherlands

PREFACE

      Glass... Current Issues is the proceedings of a NATO Advanced
Study Institute held in Puerto de la Cruz, Tenerife between the
2nd and 13th April 1984. The objectives of the School were twofold.
Firstly to inform participants of actual and developing technological applications of glassy materials in which fundamental science
makes a strong contribution, and secondly to bring together
scientists from the widely different backgrounds of glass science
and technology to promote mutual understanding and collaboration.

      The amorphous state has for more than a decade now been a
renaissance of scientific and technological activity extending
beyond traditional glass technology research. Striking developments
of amorphous materials have been made in fields such as metallurgy,
electronics and telecommunications and even in disciplines until
recently less concerned by materials science, such as colloid
chemistry, medicine and agriculture. The physical and chemical
properties brought into application here result from the
interaction between the glass composition and its non-crystalline
structure. One rôle of the basic research is to understand this
interaction, which in time through development, helps to extend
the range of properties and applications. In this meeting we hoped
to sensitize participants to the vast range of applications of
amorphous materials which exploit their unique properties, and
thus broaden future investigation.

      The program was organised around seven topics, signposts of
scientific and technological activity in the 1980's : optical
materials, amorphous metals, crystallisation phenomena, electronic
and electrical devices, sol-gel preparative methods, composite
materials and long-term applications. The latter included glass

for nuclear waste storage, and applications in medicine and biology. In addition active discussion sessions were held and participants were invited to contribute to a permanent poster exhibition.

An excellent example of the complementarity of basic and applied research can be found in the development of optical wave-guides. Early work of fibre optic communications centred on SiO_2 based glass because of its promising optical qualities and known fibre drawing behaviour. Processing experience gained with SiO_2 or SiO_2-GeO_2, has given fibres with a best loss figure of 0.2 dB.km^{-1} at 1.55 μm. This excellent value, close to the intrinsic minimum imposed by multiphonon absorption and Rayleigh scattering is sufficient for many applications, but still imposes a requirement for frequent repeater stations. The need to displace the multi-phonon edge to higher wavelength has led to further development of chalcogenide and fluoride glasses. The optical window in the former extends to 10 μm and these glasses find use in the optical components of IR cameras, night vision and thermal surveillance systems. Fluoride glasses (based on ZrF_4) exhibit a broad optical window from 0.3 to 7 μm, but more importantly have the ultra low loss potential of 10^{-3} dB.km^{-1} at 3 μm. Many processing problems must be overcome however, since these glasses have poor fibre drawing properties, are prone to crystallisation and are sensitive to water, but the enormous potential merits the development efforts.

The interdisciplinary nature of the meeting brought many physicists and materials scientists together for the first time with chemists working on sol-gel preparative techniques. This mixture of polymer chemistry, colloid science and ceramics technology is advancing extremely rapidly to produce novel amor-phous oxides with unique physical properties. The method opens up applications ranging from electrically conducting films to optical fibres as well as refractory monolithic materials which are impossible to make by classical melt-quench procedures. For many years a domain dominated by its chemistry, sol-gel methods of glass formation have now reached a stage where these new materials are readily available in modest quantity. We can expect in the next few years growing interest in measurement of the physical proper-ties and structure as a function of preparation variables. The aim here will be to improve reproducibility in the development programs.

Similar observations to these outlined above can equally well be drawn from the texts of the other topics which contributed to the 55 lectures delivered. We would encourage the reader to take account of the wide range of disciplines presented. This reflects the current broad scientific and technological base of amorphous materials, which motivated us to organise the meeting and suggests a bright future for glasses.

We address our sincere thanks firstly to Jean-Pierre Causse, who, in spite of his heavy work-load as Directeur Général Adjoint of Saint Gobain Research accepted the role of co-director of the A.S.I. and contributed in many ways and over a very long period to the success of the meeting.

We thank all the chairmen lecturers and participants who, cooperated so fully in the discussions and exchanges, on the one hand between different disciplines, and on the other between applied technologies and basic research. We ware only too aware of the difficulty to achieve this type of exchange. That many lively and spontaneous discussions continued until late in the evenings was a credit to the far-sightedness and understanding of the participants.

We thank all those organisations who generously supported the meeting :

- NATO, Scientific Affairs Division provided 50 % of the funds under the framework of the NATO double jump programme, the aim of which is to generate closer collaboration between industry and basic research organisations of the member countries.

- The industrial and national organisations below responded most admirably, ensuring the remaining support, either by direct financial aid or by providing one or more key lecturers according to our wishes and at their own expense for such an extended period : Compagnie Saint-Gobain, U.S. Air Force, (EAORD, London), D.R.E.T., Allied Corporation, Institut Laue-Langevin, C.N.E.T., C.E.A., Schott Glasswerke, Pilkington Brothers, C.N.R. (Italy), and N.S.F.

- We are also very grateful to Mr. Gobin for his assistance and support through the Ministère de l'Industrie et de la Recherche.

- Finally we express our personal gratitude for the continued devotion of the secretariat over many months : Danielle de Baere and France Parisot.

J. DUPUY, Lyon
A.F. WRIGHT, Grenoble.

May 1984.

TABLE OF CONTENTS

X

CHAPTER VII : LONG-RANGE APPLICATIONS OF GLASS

SHORT COMMUNICATIONS

'Dawn of glass technology'
From a eighteenth century encyclopedia
by Diderot d'Alembert

CHAPTER I: NUCLEATION AND CRYSTALLIZATION

NUCLEATION AND CRYSTALLIZATION IN GLASS-FORMING SYSTEMS

Donald R. Uhlmann

Department of Materials Science and Engineering
Massachusetts Institute of Technology
Cambridge, Massachusetts 02139

I. INTRODUCTION

The present paper will attempt in a brief space to summarize
the present state of knowledge and outstanding problems in the
area of nucleation and crystal growth in glass-forming liquids.
It is intended to provide background for the papers on crystallization phenomena to follow at this meeting. Specifically, it
will not address applications of controlled crystallization to
form glass-ceramic materials, which will be considered by Beall;
and it will not address the often-subtle effects of thermal history
near and below the glass transition, which will be considered by
Wright.

The combined effects of nucleation and crystal growth are
important in the formation of glasses as well as in the development
of glass-ceramic materials with desired combinations of properties;
and an attractive approach to treating glass formation will be
briefly discussed. This will be followed by separate considerations
of nucleation and crystal growth processes in both single-component
and multi-component systems, with emphasis on developments of the
past decade. Finally, consideration will be given to differences
expected in the nucleation and crystal growth behavior of metal
alloys, familiar oxide systems, and sol-gel derived oxides.

II. TRANSFORMATION KINETICS AND GLASS FORMATION

The kinetics of overall crystallization in an initially
amorphous material are generally described using a form of analysis
introduced by Avrami (1, e.g.). The most widely-used form of this
analysis represents the volume fraction crystallized, V_c/V, as a

function of time under isothermal conditions:

$$\frac{V_c}{V} = 1 - \exp(-\int_o^t I_v [\int_{t'}^t ud\tau]^3 dt')$$

(1)

Here I_v is the nucleation rate per unit volume and u is the crystal growth rate.

In evaluating V_c/V as a function of time, it is thus necessary to know the time dependence of both I_v and u. For the simple but frequently applicable case where the nucleation rate and growth rate are independent of time, Eqn. (1) reduces for isotropic growth to:

$$V_c/V \sim \pi/3 \ I_v u^3 t^4$$

(2)

This formalism can be used to describe volume crystallization and glass formation by constructing time-temperature transformation (TTT) curves (2). Such curves define the loci of times at various temperatures required to form a given volume fraction crystallized. To obtain a reasonable estimate of the cooling rate required to form a glass, it is useful to assume a minimum detectable V_c/V and construct a continuous cooling (CT) curve corresponding to that value of V_c/V. Such constructions are carried out following the approach of Grange and Kiefer (3, 4).

In considering glass formation, a minimum detectable V_c/V of 10^{-6} has been assumed (2). This seems reasonable as a definition of glass in light of the detection limits of established experimental techniques. From first principles, it may be useful to consider only glasses whose thermal stability is independent of the conditions (cooling rates) used in their formation. In such cases, the critical V_c/V would be about 10^{-10} (5). Use of 10^{-10} rather than 10^{-6} as the volume fraction crystallized would increase the critical cooling rates by about an order of magnitude.

Examples of TTT and CT curves corresponding to $V_c/V = 10^{-6}$ and 0.5-0.9 are shown in Fig. 1 for a lunar composition containing 35.5 wt. % SiO_2, 3.7% Al_2O_3, 14.3% TiO_2, 23.5% FeO, 11.6% MgO, 11.1% CaO and 0.2% Na_2O. The constant rate cooling curves shown in the figure indicate a critical cooling rate to form $V_c/V=10^{-6}$ of about 10^2 K min^{-1}.

A preferred description of glass formation and the development of partial crystallinity is provided by the analysis of crystallization statistics (6). The analysis introduces a crystal distribution function, $\psi(r,t,R)$, defined such that $\psi(r,t,R)dVdR$ is the number of crystals in a volume dV at position r with radii

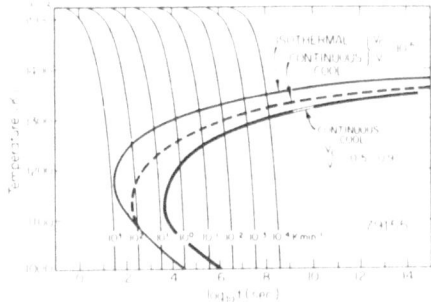

<u>Figure 1</u> - TTT and CT curves for lunar composition 79155

between R and R+dR. Using this approach, it is possible to
evaluate the detailed statistics of the number and size distributions of crystallites in a sample as a function of temperature and
thermal history. In estimating the critical cooling rate for glass
formation, one calculates the time and temperature where V_c/V
reaches the value of interest.

With this approach, the volume fraction crystallized is
expressed:

$$\frac{V_c}{V}(t_j) = 1-\exp \quad -\left\{\sum_{i=1}^{j} \frac{4\Pi}{3} R_i^3(t_j,t_i)I_{vi}(t_i)\Delta T\right\} \tag{3}$$

where $V_c/V(t_j)$ is the volume fraction crystallized at time t_j;
$I_{vi}(t_i)$ is the nucleation frequency at time t_i; and $R_i(t_i, t_j)$ is
the radius at time t_j of crystals nucleated at time t_i. This may
be represented:

$$R_i = R_i^* + \sum_{k=1}^{j} u_k(t_k)\Delta t \tag{4}$$

where R_i^* is the radius of the critical nucleus at t_i and u_k is the
crystal growth rate at time t_k.

Using this approach, it is possible to predict cooling rates
required to form glasses which are in excellent agreement with
measured values. Examples of this agreement are shown in Fig. 2
for a series of Na_2O-SiO_2 glasses. The minimum in critical cooling
rate corresponds closely with the eutectic seen in the phase
diagram for this system. The lower calculated curve in the figure
represents predictions of a simplified model of glass formation to
predict critical cooling rates (7).

Using the analysis of crystallization statistics, it is also
possible to describe the effects of complicated thermal histories
on crystallization behavior, including cooling a liquid to the
glassy state at a given rate and reheating it at the same or
different rates. It is also possible to describe the competition

4

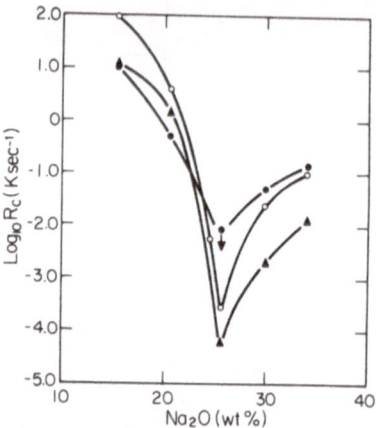

Figure 2 - Measured and calculated critical cooling rates
for forming glasses of Na_2O-SiO_2 composition. ● = measured
critical cooling rates; o = rates calculated using analysis
of crystallization statistics; ▲ = rates calculated using
simplified model of glass formation

between crystallization and viscous sintering under conditions of
continuous cooling or heating. Many of these applications have
been described in a recent review (8).

III. CRYSTAL NUCLEATION

Having discussed briefly the overall process of crystallization and its application to glass formation, let us treat the
individual kinetic processes of nucleation and crystal growth.
Consider first the homogeneous nucleation of crystals in liquids.
According to classical theory, the steady state rate of such
nucleation can be expressed (9, e.g.).

$$I_v \sim N_v^o \nu \exp[-K\sigma^3/kT(\Delta G_v)^2] \tag{5}$$

where N_v^o is the number of molecules per unit volume; ν is the
frequency factor for transport at the nucleus-matrix interface;
σ is the crystal-liquid surface energy; and ΔG_v is the difference
in Gibbs free energy per unit volume between the liquid and crystal.

In the frequently-encountered cases where free energy data
are not available, ΔG_v is approximated by either of two expressions
for congruently melting compounds:

$$\Delta G_v \sim \Delta H_{fv} \Delta T/T_E \tag{6a}$$

$$\Delta G_v \sim \Delta H_{fv} \Delta TT/T_E^2 \tag{6b}$$

where ΔH_{fv} is the heat of fusion per unit volume; ΔT is the undercooling; and T_E is the melting point. Eqn. (6a) is preferred for
metallic systems (10), and Eqn. (6b) for oxides.

In cases where transport at the interface involves similar molecular motions as transport in bulk liquid, ν can be related to the viscosity, η:

$$\nu = kT/3\pi a_o \eta \tag{7a}$$

or

$$\nu = kT/a_o \eta \tag{7b}$$

where a_o is the molecular diameter. Eqn. (7a) is appropriate for simple molecular liquids, while Eqn. (7b) seems preferred for oxides.

Combining Eqns. (5), (6b) and (7), one predicts that the log $I_v \eta$ vs. $1/T_r^3 \Delta T_r^2$ relation should be a straight line of negative slope. Here $T_r = T/T_E$ and $\Delta T_r = \Delta T/T_E$. Relations of this form have been observed for a few liquids, an example of which is shown in Fig. 3 for anorthite. The pre-exponential factor obtained from Fig. 3 is in good agreement with that predicted by classical theory.

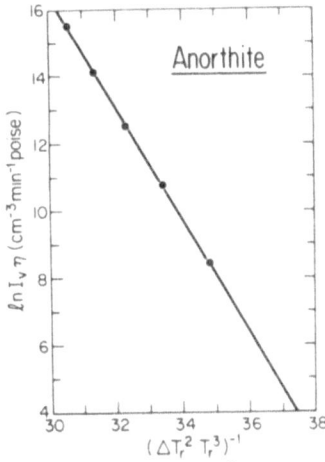

Figure 3 - Logarith $(I_v \eta)$ vs $1/T_r^3 \Delta T_r$ relation for anorthite, after Ref. 11

In contrast to the good agreement between experimental data and classical theory for anorthite, the data on crystal nucleation in $Li_2O \cdot 2SiO_2$ stand in sharp disagreement with theory (12-14); and more recent data on $Na_2O \cdot 2CaO \cdot 3SiO_2$ (15) also indicate a pronounced discrepancy between theory and experiment. In the case of $Li_2O \cdot 2SiO_2$, the measured (extrapolated) pre-exponential factor is larger by some 20 orders of magnitude than the predicted value. Previously-reported data on ΔG_v were used in obtaining the pre-exponential factor.

The origin of the discrepancy between theory and experiment is believed by the present author to lie in the temperature dependence of ΔG_v over the range of very large undercoolings where the nucleation data were obtained, rather than in a specific deficiency of the theory for describing silicate systems. At such large undercoolings, neither of the approximate relations of Eqn. (7) should provide a reliable estimate of ΔG_v. The change in ΔG_v from the reported data required for a pre-exponential factor in accord with classical theory is reasonable.

As further support for this view, it might be noted that classical nucleation theory has been used in predicting critical cooling rates for a variety of silicate liquids; and good agreement has been found between predicted and measured critical cooling rates. A discrepancy of 10^{20} in the pre-exponential factor in I_v would lead to an error of about 10^5 in critical cooling rate; and this is far beyond the observed differences between predictions and data. Whether or not the present suggestion is correct, it seems very important to resolve the issue posed by the data on $Li_2O \cdot 2SiO_2$.

In addition to the steady state nucleation rates discussed above, time-dependent nucleation rates can also be important in the range of high melt viscosities ($\sim 10^{11}$ poise or higher). Such nucleation is usually represented by the expression of Zeldovich:

$$I_v(t) = I_v^{ss} \exp(-\tau/t) \tag{8}$$

where $I_v(t)$ is the nucleation rate at time t; I_v^{ss} is the steady state nucleation rate; and τ is the incubation time. The last quantity is often represented:

$$\tau \approx (n^*)^2/n_s \nu \tag{9}$$

where n^* is the number of atoms in the critical nucleus; and n_s is the number of atoms on its surface.

It has been suggested (16) that Eqn. (9) be modified by an accommodation factor in treating crystal nucleation; and this suggestion is presently being explored by several investigators. More generally, it should be noted that Eqn. (8) itself represents an approximation which was derived to describe behavior at short times, and that more accurate representations can be provided by computer simulations.

It is widely recognized that nucleation often occurs heterogeneously rather than homogeneously. Anyone who has heat treated silicate liquids is familiar with the preferential formation of crystals at the external surfaces of bodies and their growth into the central regions, as well as nucleation on second-phase heterogeneities contained in the bodies. The present author has suggested (17) that liquid surfaces per se should not serve as preferred nucleating sites, and that the nucleation seen at exterior surfaces and interior bubble surfaces is associated with dirt and other second-phase heterogeneities. One can greatly reduce or eliminate nucleation at surfaces by fire polishing the surfaces, and can often increase such nucleation by handling the surfaces.

The key to many applications of controlled crystallization is the use of nucleating agents such as the noble metals or oxides such as TiO_2 and ZrO_2 in silicate melts. Despite the technological importance of such controlled nucleation, the process is not at all well characterized in any detail; and effective nucleating agents for non-silicate oxide melts remain largely unexplored.

Heterogeneous nucleation is often described using a spherical cap model (9), according to which the nucleation barrier should decrease as $(2+\cos\theta)(1-\cos\theta)^2$, where θ is the contact angle between nucleating heterogeneity and crystal being nucleated. In greater detail, the change in nucleation barrier has been represented in terms of the match in lattice parameter between heterogeneity and crystal; and useful descriptions are available for the case of cubic crystals. It should be noted, however, that most crystals of interest in ceramics or polymers are not cubic; and the description of interfacial strain effects requires modification for such cases.

It should also be noted that reliable data on the kinetics of heterogeneous nucleation are almost totally lacking, and that even differentiating unequivocally between homogeneous and heterogeneous nucleation is by no means trivial. With data over a sufficiently wide range of undercooling, such differentiation can readily be effected; but often data are obtained over only a modest range of temperature.

Crystal nucleation in multicomponent systems involving sizable changes in composition on crystallization is very poorly characterized. It is well known that the thermodynamic driving force for crystal nucleation differs in general from that for overall crystallization (18, e.g.), and that metastable rather than stable phases are often nucleated. The principal issues here are the lack of thermodynamic data on nearly all systems of interest, and the almost total absence of systematic experimental data on nucleation kinetics in such systems.

The nucleation of crystal melting exhibits a marked uniformity in behavior: For all materials investigated to date, liquid nucleated at the external surfaces of crystals at negligible superheats above the melting point. Internal nucleation of melting is almost never observed because of the sizable strain energy associated with the change in volume on melting. Silicate crystals have been superheated by hundreds of Centigrade degrees above their melting points without the occurrence of internal nucleation of melting (17). Such nucleation could most readily be investigated with crystals which exhibit a small volume change on melting coupled with slow melting kinetics (the former to minimize strain energy effects and the latter to preserve a superheated volume of crystal).

IV. CRYSTAL GROWTH

The area of crystal growth in glass-forming liquids is characterized by better models and much better and more abundant data than are available for nucleation behavior. The critical models of crystal growth and available experimental data have been summarized in a recent review (19). As noted there, the simple analytical models of 20 years ago have been clarified by computer simulations of growth (20, 21, e.g.). These simulations, particularly those using Monte Carlo techniques, have provided a more detailed and reliable description of the growth process than available previously, and have provided strong support for the view that the nature of the crystal-liquid interface--and particularly the distribution of step sites on the interface--has a decisive influence on the kinetics and morphology of growth.

The characteristic features of crystal growth in materials with small and large entropies of fusion are summarized in Table I.

Table I

Comparison of Crystallization and Melting Characteristics Between Small ΔS_f and Large ΔS_f Materials

	Small ΔS_f	Large ΔS_f
Interface Morphology	Nonfaceted in both crystallization and melting	Faceted in crystallization; nonfaced in melting
Anistropy	Largely isotropic; no strongly pre- ferred growth di- rection	Anisotropic
Interface Site Factor	Independent of under- cooling and super- heat; described in form by normal growth model	Increases with increasing undercooling; not well de- scribed by any standard kinetic model
Continuity at Melting Point	Crystallization and melting data con- tinuous with similar slope through T_E; melting and crystal- lization kinetics, corrected for vis- cosity, equal at equal small departures from equilibrium	Change in slope of kinetic data at melting point; melting more rapid than crystallization at given small departure from equilib- rium even after correction for viscosity

The classification shown in the Table was based on the original model of Jackson (23, 24), who considered the change in free energy on adding extra atoms to an initially-smooth interface. From the variation of free energy with occupied fraction of surface sites, it was suggested that for materials with molar entropies of fusion less than 2R, even the most closely packed interface planes should be rough on an atomic scale, and the growth rate anisotropy (differences in growth rate for different orientations) should be small. For such atomically rough interfaces, the normal growth (Wilson-Frenkel growth) model should provide a useful description of the kinetics:

$$u = \nu a_o [1 - \exp(-\Delta H_{fM} \Delta T / RTT_E)] \tag{10}$$

Here u is the growth rate and ΔH_{fM} is the molar heat of fusion.

In contrast, for materials with molar entropies of fusion greater than 4R, Jackson's model predicted that the most closely packed planes should be atomically smooth, while the less closely packed planes should be rough. On this basis, considerable growth rate anisotropy would be expected; and growth on the smooth interface planes should be sensitive to defects such as screw dislocations which provide sources of repeatable steps. For growth taking place exclusively at dislocation ledges, the growth rate should vary as:

$$u = f \nu a_o [1 - \exp(-\Delta H_{fM} \Delta T / RTT_E)] \tag{11}$$

where f, the fraction of preferred growth sites on the interface, can be approximated:

$$f \sim \Delta T / 2\pi T_E \tag{12}$$

For a smooth interface free of dislocations, growth can take place by the formation of two-dimensional nuclei on the interface. The kinetics of such growth have the form:

$$u = c\nu \exp(-B/T\Delta T) \tag{13}$$

where C is a slowly-varying function of temperature, and

$$B = \frac{\pi a_o V_M T_E \sigma_E^2}{3k\Delta H_{fM}} \tag{14}$$

Here V_M is the molar volume of the crystal and σ_E is the edge surface energy of the nucleus.

The mechanism of growth is conveniently explored by constructing the reduced growth rate, u_R, vs. undercooling relation.

10

Here

$$u_R \equiv \frac{u\,\eta}{[1-\exp(-\Delta H_{fM}\Delta T/RTT_E)]} \qquad (15)$$

Using Eqn. (7) to relate ν to η, the u_R vs. ΔT relation is a horizontal line for normal growth, a straight line of positive slope passing through the origin for screw dislocation growth and a curve with positive curvature passing through the origin for surface nucleation growth. For surface nucleation growth, Eqn. (13) indicates that the logarithm $(u\eta)$ vs. $1/T\Delta T$ relation should be a straight line of negative slope.

Experimental data on a variety of glass-forming liquids, summarized in Ref. 19, are in remarkable accord with the predictions outlined in Table I. These will be illustrated by GeO_2 $(\Delta S_{fM}\sim1.3R)$, $Na_2O\cdot3SiO_2(\Delta S_{fM}\sim4R)$ and anorthite $(\Delta S_{fM}\sim10.9R)$. For GeO_2, u_R is independent of ΔT (Fig. 4), and the crystal-liquid interface is non-faceted. Using the modified Stokes-Einstein relation of Eqn. (7b), the magnitude of the growth rate is in close accord with predictions of the normal growth model. For $Na_2O\cdot3SiO_2$, u_R increases linearly with ΔT (Fig. 5), and the interface morphology is faceted. Again using Eqn. (7b), the magnitude of the growth rate is predicted well by the screw dislocation growth model. In the case of anorthite, the u_R vs ΔT relation exhibits positive curvature; but the log $u\eta$ vs $1/T\Delta T$ relation (Fig. 6) is not a simple straight line of negative slope.

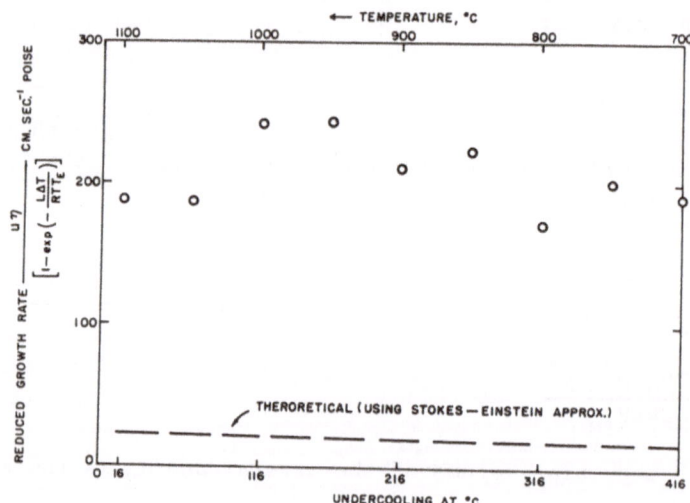

Figure 4 - Reduced growth rate vs. undercooling relation for GeO_2, after Ref. 22

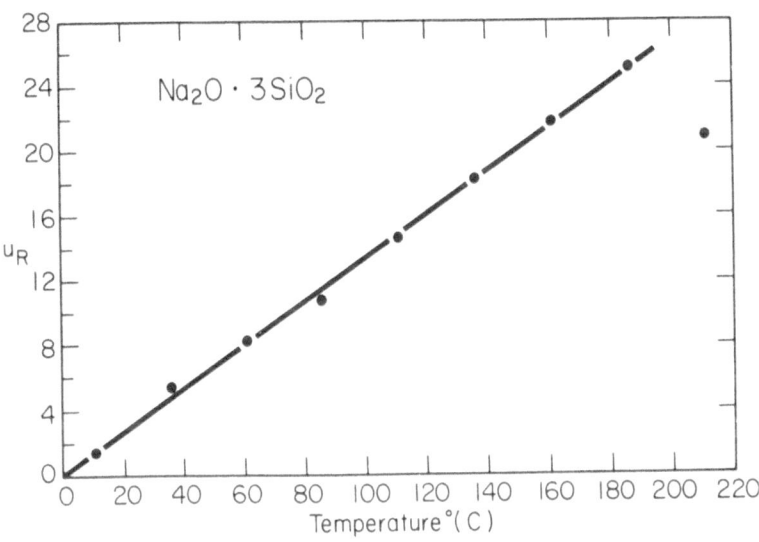

<u>Figure 5</u> – Reduced growth rate vs. undercooling relation for
Na$_2$O·3SiO$_2$, after Ref. 25

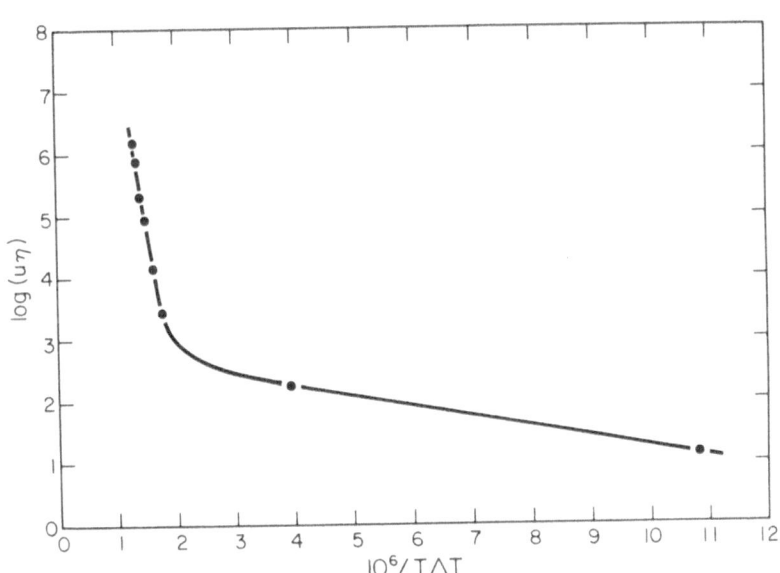

<u>Figure 6</u> – Logarithm (μη) vs. 1/TΔT relation for anorthite.
After Ref. 26

A reduced growth rate independent of undercooling was observed for the other classic low-ΔS_f material, SiO_2 (27). Relations of the form shown in Fig. 5 have been observed for several materials over limited ranges of temperature (28, e.g.); while relations of the form shown in Fig. 6 are more commonly observed. In particular, a similar relation was found for the purest high-ΔS_f material investigated--100^+ pass zone-refined o-terphenyl (29).

The failure of the surface nucleation growth model to describe the crystal growth behavior of materials like o-terphenyl and anorthite very likely reflects its neglect of growth at all sites save those at the perimeters of two-dimensional nuclei on the interface. Computer simulations such as those cited above suggest that the surface nucleation growth model represents a limiting case for materials with very high entropies of fusion crystallizing at modest undercoolings. For materials with ΔS_{fM}=5-10R, detailed consideration must be given to the departures of the interface from smoothness. Such consideration is readily provided by computer simulations of interfaces and growth dynamics.

In contrast to crystal growth, where a diversity of kinetic behavior is observed, the interfaces in melting are always rough on an atomic scale and kinetics of the Wilson-Frenkel type are observed. For materials with large entropies of fusion, melting takes place more rapidly than crystallization at equal small departures from equilibrium (even after correcting for the variation of mobility with temperature). This behavior (30, 31, e.g.) is illustrated by the data in Fig. 7 and reflects the difference in operative site distributions between melting and crystal growth (growth requires the stepwise formation of new layers of atoms on atomically smooth planes, while melting can always initiate at high energy corner sites).

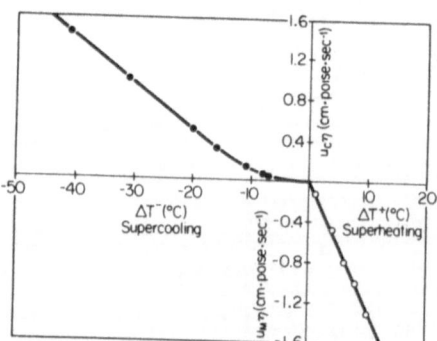

Figure 7 - Rates of crystal growth and melting, corrected for the variation of viscosity with temperature, for $Na_2O \cdot 2SiO_2$. After Ref. 30

In contrast, for materials with small entropies of fusion, the rates of melting and crystal growth are equal at equal small departures from equilibrium (after correcting for the variation of mobility with temperature). This is shown by the data in Fig. 8 for GeO_2, and reflects the similarity of operative site distributions in melting and growth for the atomically rough interfaces of such materials.

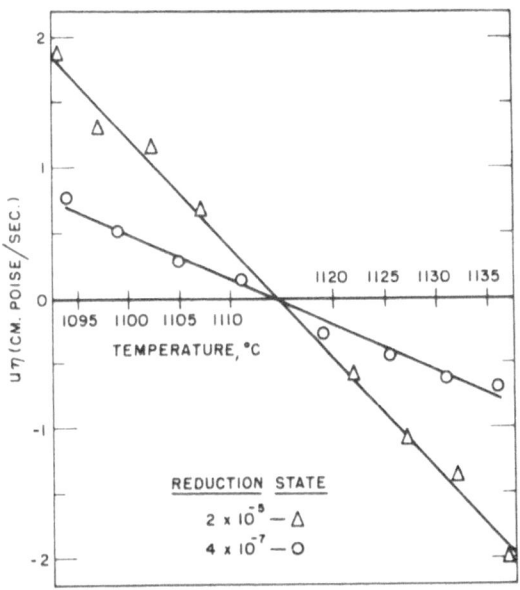

Figure 8 – Rates of crystal growth and melting, corrected for the variation of viscosity with temperature, for GeO_2. After Ref. 31

The discussion to here has been concerned with materials whose crystal growth and melting behaviors are limited by interface attachment (detachment) kinetics. For many materials, however, the rates of growth and melting are limited by diffusive processes (heat flow or mass diffusion) rather than by interface kinetic processes. Examples are provided by most multicomponent systems, and are illustrated by the data on Na_2O-SiO_2 melts shown in Fig. 9. The growth rates shown there were independent of time at all temperatures, similar to the findings of Swift (33, 34) in his classic study of devitrification in soda-lime-silica glasses.

With such large changes in composition as those involved in the studies of Swift or Scherer and Uhlmann, one might have expected the growth rate to decrease as the square root of time (as observed for many cases of diffusion-controlled growth). The origin of the growth rate being independent of time, even under conditions of diffusion control, lies in the dendritic growth morphology which is commonly observed. For such morphologies,

14

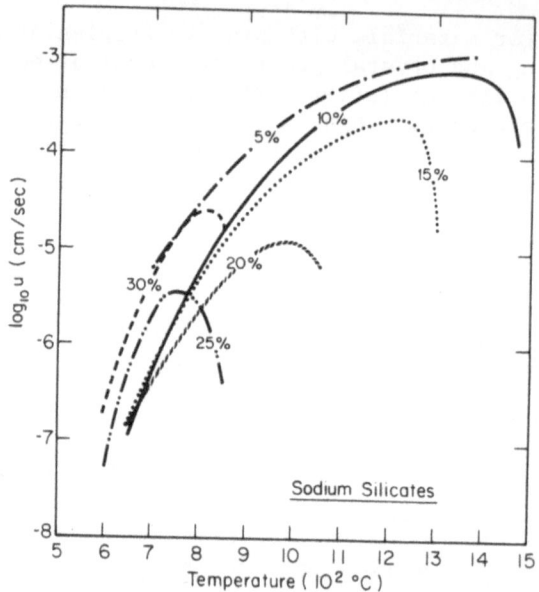

Figure 9 - Crystal growth rates for Na_2O-SiO_2 glasses (percentages are for Na_2O molar content). After Ref. 32.

the scale of the diffusion field can be unchanging with time; and hence the growth rate should be constant.

Several models have been advanced to describe diffusion-controlled growth. Among them three will be specifically noted here. The first, developed by Horvay & Cahn (35), describes the growth of an isolated dendrite having the shape of a paraboloid of revolution. A time-independent growth rate was obtained:

$$u = \frac{C_o - C_E}{C_s - C_E} = p\ exp(p) \int_p^\infty \frac{exp(-t)}{t}\ dr \qquad (16)$$

where C_o is the composition of bulk liquid; C_E is the liquidus composition; C_s is the composition of the solid; and $p = ur/2\tilde{D}$. Here r is the radius of curvature of the dendrite tip and $\tilde{D}$ is the interdiffusion coefficient.

The second model, introduced by Christensen, et. al. (36), considered growth as limited by diffusion through a boundary layer of thickness δ. According to this model, the growth rate can be expressed:

$$u = \frac{\tilde{D}}{\delta} \frac{(C_o - C_E)}{(1 - C_E)} \qquad (17)$$

A third approach (37) considered the growth of parallel arrays of dendrites with overlapping diffusion fields. Several forms of the composition gradient ahead of the interface were considered; and corresponding expressions for the growth rate were obtained. The results indicate that the growth rate is determined by the interdiffusion coefficient, the radius of curvature and spacing of the dendrites, and the difference of the interface concentration from that in bulk liquid and in the crystal.

The last approach can provide useful descriptions of the diffusion-controlled growth of dendrite arrays, at least for the very limited number of compositions on which a sufficient quantity of data are available. The model leaves as an outstanding problem the à priori determination of the radius of curvature of the dendrites. It is often assumed, but without proper justification, that the dendrite grows at the maximum velocity possible for a given departure from equilibrium.

Much work is needed, both theoretical and experimental, to clarify crystal growth behavior in systems undergoing large compositional changes in growth. Experimentally, there is considerable need for determinations of growth rate, radius of curvature and spacing of dendrites for a range of composition. In many systems, a change in interface morphology from faceted to dendritic is observed in going from small undercoolings to large undercoolings (38, 39). There remains a question at the present time if this change in morphology represents a change in mechanism from diffusion-controlled to interface-controlled growth, or if it reflects diffusion controlled growth over the entire range with the growth rate anisotropy occurring on only a local scale at small undercoolings.

There is also a question concerning the appropriate boundary condition at the interface during crystal growth with large changes in composition. The familiar condition suggests that the solid which forms at the interface is such that the free energy of the system is lowered to the greatest extent possible for a given amount crystallized. This condition leads to the familiar common tangent construction for describing the composition which crystallizes, and results in normal zoning profiles.

An alternative approach (40) considers the composition of the solid at the interface as that which is in local equilibrium with the liquid composition at the interface. Rather than the common tangent construction, this condition leads to a tangent from C_L (the liquid composition at the interface) to the solid free energy vs. composition curve. It also leads to a prediction of reverse zoning profiles. As in other cases, experimental data to clarify the situation is needed.

V. COMPARISON OF CRYSTALLIZATION BEHAVIOR OF OXIDES, METALS AND GEL-DERIVED GLASSES

In comparing the crystallization behaviors of oxides, metals and gel-derived glasses, a number of general observations can be made:

(1) In comparing the crystallization behavior of oxide glasses and metallic glasses, it is important to consider the similarities or differences between the molecular motions involved in liquid viscosity and those involved in the kinetic processes at the crystal-liquid interface. For the familiar oxide and organic liquids, transport at the interface is believed to involve similar molecular rearrangements as those taking place during viscous flow (reorientation of the molecules, breaking of directional bonds, etc.).

In the case of metals, reorientation and bond breaking is not required for crystal growth; and the relationship between ν and η may depart greatly from that given in Eqn. (7) above--if there is any relation at all. On this basis, it is well possible for metal glasses to exhibit nucleation and crystal growth at temperatures below the glass transition. In contrast, such crystallization (at temperatures below the glass transition) has never been observed for oxides in reasonable experimental times (as opposed to geological times) provided mineralizers such as water are not present.

In the case of metal glasses, it is often simple to obtain internal nucleation and form a metallic glass-ceramic body. In the case of metals, unlike oxides, such internal crystallization seems to impair rather than improve the mechanical properties. This difference is very likely related to the differences in mechanical behavior and the relevant molecular processes between metal and oxides.

(2) The use of sol-gel techniques offers the opportunity of forming novel glasses not obtainable by cooling from the melt. This opportunity is based on avoiding limitations imposed by the critical cooling rate required to form glasses on cooling from the liquid state. The use of sol-gel techniques introduces, however, a number of complications in considering crystallization behavior. These include:

(a) The presence of water and other mineralizers, which are almost invariably present during the early stages of processing. These "impurities" are at least partially driven off with increasing temperature; but their presence in various concentrations at various times lead to the properties changing with temperature, and with time at a given temperature, in a generally-unspecified manner.

(b) The effects of adventitious impurities and reactions with the environment, which can appreciably affect the observed crystallization behavior. A classic example of this behavior has been reported (41, 42) in the preparation of fused silica by wet chemical techniques. In that work, SiO_2 prepared from precursors containing sodium ions were observed to crystallize much faster than predicted by kinetic theory (or than SiO_2 prepared from sources free of sodium ions). The difference in behavior was found to be associated with sodium ions in the dried porous body (before densification) reacting with CO_2 in the atmosphere to form Na_2CO_3 crystals, which serve as heterogeneous nuclei for the crystallization process. The specific details of this system are essential; but the potential complication of interaction with the environment whenever highly porous bodies are processed seems to represent a matter of wide-ranging concern.

(c) The differences in properties between gel-derived and conventionally-melted glasses. It has been well documented by several investigators (43-46) that the properties of gel-derived glasses can differ considerably from those of conventionally melted glasses. Nielson and Weinberg (43) have demonstrated differences as large as 100°C in liquidus temperature between gel-derived and melted Na_2O-SiO_2 glasses--even after the gel-derived glasses were melted at temperatures hundreds of Centigrade degrees above the liquidus temperature. With such large differences in equilibrium properties such as liquidus temperature, it would not be surprising to find sizable differences in kinetic behavior; and such effects are presently being explored in our laboratory.

(d) The competition between crystallization and viscous sintering, which is critical in the formation of bulk glasses using sol-gel techniques. With these methods, it is generally simple to form an amorphous powder or dried gel (e.g., amorphous mullite or ZrO_2); but on heating, the materials often crystallize before sintering to a dense body.

The competition between viscous sintering and crystallization has been described using the analysis of crystallization statistics for assemblages of particles cooled from high temperatures (47); and the results were applied to the formation of brecciated rocks. Work is presently underway in our laboratory in developing the same type of analysis to describe the competition between crystallization and viscous sintering of powder assemblages and porous gels subject to arbitrary thermal histories. The results of this analysis should be useful in defining windows of temperature and time in processing--if such exist for a given material--where densified amorphous bodies can be prepared.

VI. CONCLUDING REMARKS

The present paper has attempted to provide a survey of
nucleation and crystal growth behavior as background for the papers
to follow in this area. It has been seen that the present state of
the art represents a curious mixture of detailed knowledge and
glaring gaps in knowledge, and a curious combination of simple
concepts and complicated phenomenology. The field has been
characterized by intense activity for a quarter-century or more;
and important advances in both scientific understanding and commercial applications have been developed during that period.

While much has been accomplished, much remains to be done, as
evidenced from our brief discussion of the crystallization behavior
or different classes of glasses. The present author is hopeful
that the coming decade will see important advances in the area,
and that the demands for physical insight will lead to important
improvements in our general understanding of kinetic behavior.

ACKNOWLEDGEMENTS

The author wishes to express appreciation to NATO and to the
Organizing Committee of this Symposium, and to the Air Force Office
of Scientific Research for financial support of the present work.

REFERENCES

1. M. Avrami, J. Chem. Phys. 9, 177 (1941).
2. D.R. Uhlmann, J. Non-Cryst. Solids, 7, 337 (1972).
3. R.A. Grange and J.M. Kiefer, Trans. ASM, 29, 85 (1941).
4. P.I.K. Onorato and D.R. Uhlmann, J. Non-Cryst. Solids, 22, 367 (1976).
5. P.I.K. Onorato, D.R. Uhlmann and R.W. Hopper, J. Non-Cryst. Solids, 41, 189 (1980).
6. R.W. Hopper, G.W. Scherer and D.R. Uhlmann, J. Non-Cryst. Solids, 15, 45 (1974).
7. P.I.K. Onorato, D.R. Uhlmann and G.W. Scherer, "A Kinetic Treatment of Glass Formation. VII. Simplified Model", submitted for publication, J. Non-Cryst. Solids.
8. D.R. Uhlmann and H. Yinnon, in D.R. Uhlmann and N.J. Kreidl, eds., Glass Forming Systems (Academic Press, New York, 1983).
9. D. Turnbull, in D. Turnbull and F. Seitz, eds., Solid State Physics, Vol. 3 (Academic Press, New York, 1956).
10. C.V. Thompson and F. Spaepen, Acta Met., 27, 1855 (1979).
11. D. Cranmer, R. Salomaa, H. Yinnon and D.R. Uhlmann, J. Non-Cryst. Solids, 45, 127 (1981).
12. E.G. Rowlands and P.F. James, Phys. Chem. Glasses, 20, 1 (1979).

13. E.G. Rowlands and P.F. James, Phys. Chem. Glasses, 20, 9 (1979).
14. G.F. Neilson and M.C. Weinberg, J. Non-Cryst. Solids, 34, 137 (1979).
15. C.J.R. Gonzalez-Oliver and P.F. James, J. Non-Cryst. Solids, 38-39, 699 (1980).
16. I. Gutzow and D. Kashchiev, in L.L. Hench and S.W. Freiman, eds., Advances in Nucleation and Crystallization in Glass (American Ceramic Society, Columbus, 1971).
17. D.R. Uhlmann, J. Non-Cryst. Solids, 41, 347 (1980).
18. J.W. Christian, Theory of Transformations in Metals and Alloys, 2nd ed. (Pergamon Press, New York, 1975).
19. D.R. Uhlmann, in Advances in Ceramics, 4 (American Ceramic Society, Columbus, 1982).
20. H.J. Leamy and G.H. Gilmer, J. Crystal Growth, 24-25, 499 (1974).
21. G.H. Gilmer, J. Crystal Growth, 35, 15 (1976).
22. P.J. Vergano and D.R. Uhlmann, Phys. Chem. Glasses, 11, 30 (1970).
23. K.A. Jackson, in Liquid Metals and Solidification (ASM, Cleveland, 1958).
24. K.A. Jackson, in Growth and Perfection of Crystals (Wiley, New York, 1958).
25. G.W. Scherer and D.R. Uhlmann, J. Crystal Growth 29, 12 (1975).
26. L.C. Klein and D.R. Uhlmann, J. Geophys. Res., 79, 486 (1974).
27. F.E. Wagstaff, J. Am. Ceram. Soc., 52, 650 (1969).
28. I. Gutzow, A. Razpopov, and R. Kaischew, Phys. Status Solidi, 1, 159 (1970).
29. G.W. Scherer, D.R. Uhlmann, C.E. Miller, and K.A. Jackson, J. Crystal Growth, 23, 323 (1974).
30. C.Y. Fang and D.R. Uhlmann, J. Non-Cryst. Solids, 64, 225 (1984).
31. P.J. Vergano and D.R. Uhlmann, Phys. Chem. Glasses, 11, 39 (1970).
32. G.W. Scherer, "Crystal Growth in Binary Silicate Glasses", Ph.D. Thesis, Massachusetts Institute of Technology (1974).
33. H.R. Swift, J. Am. Ceram. Soc., 30, 165 (1947).
34. H.R. Swift, J. Am. Ceram. Soc., 30, 170 (1947).
35. G. Horvay and J.W. Cahn, Acta Met., 9, 695 (1961).
36. N.H. Christensen, A.R. Cooper and B.S. Rawal, J. Am. Ceram. Soc., 56, 557 (1973).
37. G.W. Scherer and D.R. Uhlmann, J. Cryst. Growth, 30, 304 (1975).
38. R.J. Kirkpatrick, L. Klein, D.R. Uhlmann and J.F. Hays, J. Geophys. Res., 84, 3671 (1979).
39. C. Guillemet and J. Denoncin, in Advances in Ceramics, 4 (American Ceramic Society, Columbus, 1982).
40. R.W. Hopper and D.R. Uhlmann, J. Crystal Growth 21, 203 (1974).
41. J. Phalippou, M. Prassas and J. Zarzycki, J. Non-Cryst. Solids, 48, 79 (1982).

42. M. Prassas, J. Phalippou, L.L. Hench and J. Zarzycki, J. Non-Cryst. Solids, 48, 79 (1982).
43. M.C. Weinberg and G.F. Neilson, J. Am. Ceram. Soc., 66, 132 (1983).
44. B.E. Yoldas, J. Non-Cryst. Solids, 51, 105 (1982).
45. S.P. Mukherjee, J. Zarzycki and J.P. Traverse, J. Mater. Sci., 11, 341 (1976).
46. L.L. Hench, M. Prassas and J. Phalippou, Ceram. Engineering Sci. Proc., 3. 477 (1982).
47. D.R. Uhlmann, L. Klein and R.W. Hopper, The Moon, 13, 277 (1975).

MICROSTRUCTURE CONTROL BY THERMAL TREATMENT : REAL TIME STUDIES OF CRYSTALLISATION BY NEUTRON SCATTERING

A.F. Wright

Institut Laue-Langevin, 156X, 38042 Grenoble Cedex, France

INTRODUCTION

The classical heat treatment schedule for glass ceramics includes a holding temperature, usually slightly above T_g, for between several minutes and a few hours to promote nucleation of the crystallites (1). This is followed by a temperature ramp through crystallisation and sometimes phase transformation zones (2) to the final ceramming temperature. The cooling cycle is usually unimportant. This type of heat treatment tends to give rather reproducible results for the microstructure and properties since the critical phase is the nucleation holding temperature. A small positive temperature ramp may even be introduced into the holding temperature to reduce the total heat treatment time without significantly influencing the final microstructure. It is however difficult to study the optimisation of the nucleation treatment by classical (indirect) techniques, in order to obtain *a priori* a predetermined microstructure. One exception, which has been used over the last few years to enrich our knowledge of nucleation and growth processes is the technique of small angle neutron scattering (SANS) during actual thermal treatment of bulk glass (3,4,5).

The measurements are very easy to carry out in a number of neutron scattering centres in Europe, USA and Japan, being an evolution of X-ray small angle scattering with the considerable advantage of long wavelength (4-15 Å) neutrons which are capable of penetrating several millimeters of matter with very low absorption. Furthermore, in many favourable cases the data analysis is relatively simple and rapid. Direct information about the microstructure of a glass-ceramic can be obtained in real time during the heat treatment schedule, providing the neutron flux is sufficiently high. At the ILL in Grenoble, this technique has been pioneered to study the evolution of the microstructure during simplified growth heat-treatment schedules. It has also been extended to study to the more difficult and nebulous nucleation stage with somewhat surprising results.

SMALL ANGLE NEUTRON SCATTERING FROM PARTLY CRYSTALLISED MATERIALS

A simplified review of the technique and data analysis is inevitable before the finer points of the study can be appreciated. Small angle neutron scattering (6) requires a very highly collimated (parallel) neutron beam with a limited wavelength spread ($\Delta\lambda/\lambda \approx 10\%$), or energy analysis by time-of-flight techniques. We are interested in the intensity of the scattering of this beam at low angles, measured in terms of the scattering vector $Q = 4\pi \sin \theta/\lambda$ where 2θ is the full scattering angle. Scattering in the Q range 10^{-3} Å^{-1} to 10^{-1} Å^{-1} is caused by inhomogeneities in the sample ranging from 10 to 1000 Å, conveniently covering nucleation and growth of crystallites in glass. The scattered neutrons are recorded on a 2-dimensional multidetector placed at a distance within the range 2 to 20 metres from the sample (7). The direct unscattered beam is absorbed in a beam stop before reaching the detector. Radial integration of the neutron counts recorded in the multidetector enable a statistically significant spectrum to be obtained in a few minutes.

Fig. 1 shows a series of spectra collected from a cordierite glass ceramic over short time intervals at 900°C. The shape

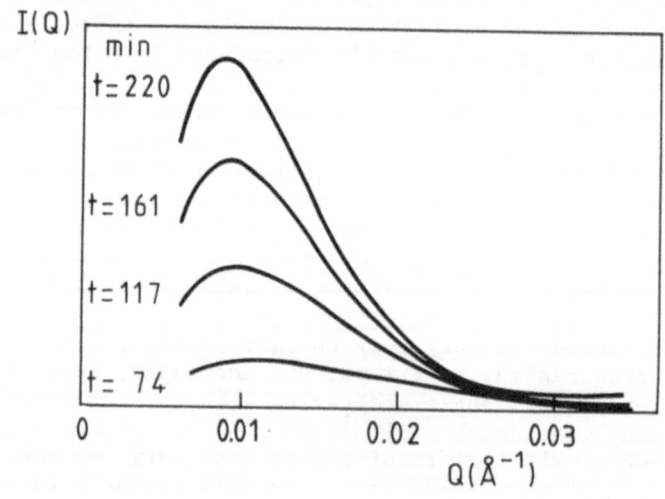

Fig. 1

Small angle neutron scattering from a cordierite glass during heat treatment at 900°C (ref. 12).

function of each curve, I(Q), may contain information on the mean crystallite size, the size distribution, and the spatial arrangement of the crystallites (8). The presence of a peak is clear evidence of a non-random spatial distribution of crystallites and of a narrow particle size distribution. In addition, the absolute intensity is a function of the particle number density and the contrast (from the mean neutron scattering length density) between the particles and the glassy matrix. A series of such time-related curves is therefore rich in qualitative and quantitative information about the physical-chemistry of the growth process, especially if it can be correlated with other physical measurements.

The full equation for the scattered intensity per atom as a function of Q is given by

$$I(Q) = K \frac{N_p V_p^2}{N} \cdot (\rho_p - \rho_m)^2 \, P(Q) \, S(Q) \tag{1}$$

where K is an experimental constant, N_p the number of particles of volume V_p, N the number of atoms in the sample and $(\rho_p - \rho_m)$ is the scattering length contrast between the particle and the matrix. $P(Q)$ is the form factor for scattering from a single particle normalised so that $|P(0)|^2 = 1$. It is identical to the form factor for X-ray scattering by atoms, decaying towards higher Q as a function of the particle size. $S(Q)$ is the interparticle inter-ference function which is the particle analogue of the liquid struc-ture factor, $S(Q)$, applied to the atomic structure of liquids. In the case of a random distribution of particles $S(Q)$ is unity, and the scattering curve is directly related to the particle size. This is not the case for nucleated glasses, and we must separate $S(Q)$ from $P(Q)$ by means of computer simulation to obtain both size and micro-structure information (8). The simulation procedure has been described elsewhere (9,10) and will not be repeated here.

We show in fig. 2 the deconvolution of the scattering curve $I(Q)$ into the particle and interparticle functions, with the approximation of a unique particle diameter based upon a spherically shaped particle. $S(Q)$ is the Fourier transformation of the particle pair distribution function g(r) which shows the probability of locating the centre of a neighbouring particle at a distance r (in units of the particle diameter) normalised to the mean probability over the whole sample. A series of these functions are shown in Fig. 3. One can see immediately that in this case, the crystallites have zero probability of contacting each other. They are in fact surrounded by an exclusion zone caused by depletion of the solute matter by the growing particle. This is not an artefact of the simulation procedure which may extend from a pure hard-sphere model (preferred contact) (11) through a liquid-type structure

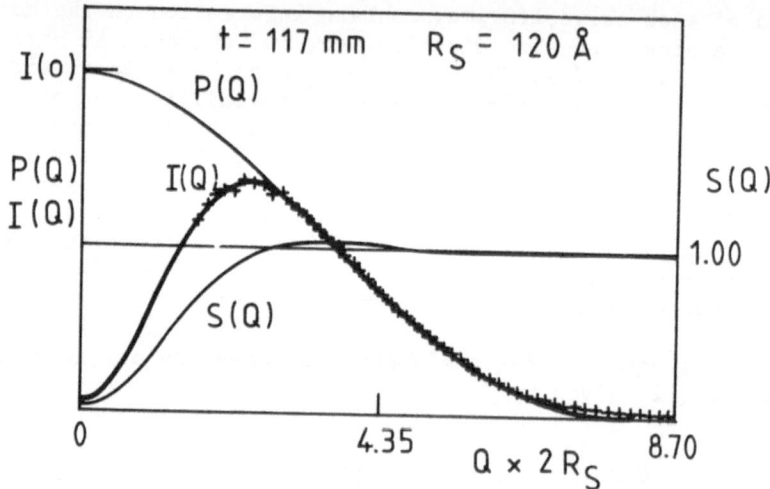

Fig. 2

Deconvolution of the experiment scattering curve into P(Q) and
S(Q).

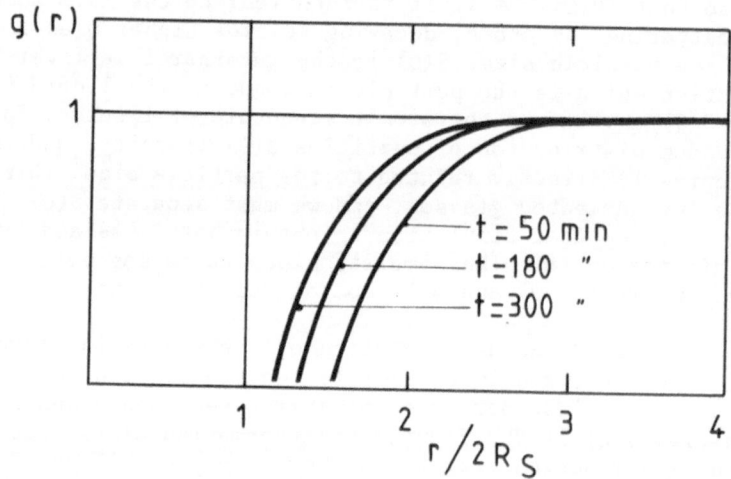

Fig. 3

The particle pair correlation functions g(r) for partially
crystallised cordierite glasses.

(present case) to a dilute gas type structure with insignificant interaction($g(r) = 1$ at all values of r).

In a series of spectra taken during heat treatment intended to develop the microstructure, we can also make use of the evolution of the particle size with respect to the intrinsic intensity and the time in order to study the nature of the growth process. The value of $P(Q)$ extrapolated to zero and scaled to the measured relative intensity, $I(0)$, has the merit of being simply related to the number and size of the particles and the contrast term.

$$I(0) = k' \frac{N_p V_p^2}{N} (\rho_p - \rho_m)^2$$

If the contrast term remains invariant, we can then expect, in the case of simple growth, the relationship $I(0) = k''r^6$ where r is the particle radius obtained from the simulation. Such a case is observed in the growth of cordierite glass nucleated by chromium dioxide (12) where $\Delta \rho$, which is essentially determined by the difference in atomic density between the matrix and the spinel crystallites, remains constant.

After very long heat treatments we may expect to observe the relationship $I(0) = k''r^3$ where the growth process is replaced by an Ostwald ripening process; elimination of smaller crystallites in favour of larger ones, within the constraint of a constant volume fraction of precipitate.

We have not yet observed this phenomenon in the case of crystallisation of glasses, but it is dominant during the phase separation of a silica rich droplet phase from lithium disilicate (13) glass (Fig. 4) and also during the growth of tetragonal ZrO_2 from a dehydrated $(Zr(OH)_4)_n$ gel at temperatures above 300°C (11). It is not easy to discriminate between growth and ripening from the raw data since both processes lead to an increase in the scattered intensity coupled with a shift of the maximum in $I(Q)$ to lower scattering vector although this shift is more pronounced during Ostwald ripening. The evolution of $g(r)$ during growth sometimes shows evidence of increasing spatial order as the treatment proceeds, particularly the formation of a maximum during the Ostwald ripening stage (8,11).

NUCLEATION

Consideration of equation (1) immediately reveals that the intensity of scattering from a nucleated glass will be many orders of magnitude lower than that of a crystallised sample, since $I(Q)$ varies as a function of $N_p V_p^2$, and we can expect the radius

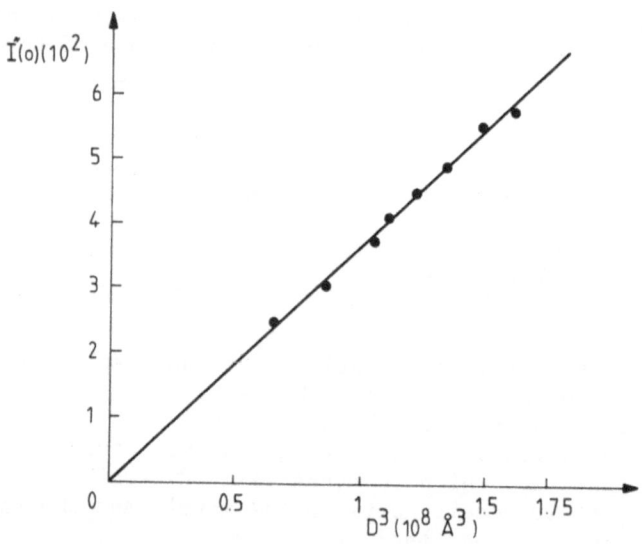

Fig. 4
Variation of I(0) with k"r³ for phase separation of a lithium
disilicate glass (ref. 13).

of the nucleus to be some ten or hundred times smaller than the
mature crystal. The nucleated glass then gives very little additional
scattering compared to that of the quenched glass, but during the
early stage of growth of the nuclei, the modification of the
scattering function caused by a large number of scattering centres is
immediately apparent (3). The existence of the peak in $I(Q)$ as well
as its position is directly related to the efficiency of the
nucleation treatment (4). This in turn depends upon temperature,
and the holding time, up to a limit, the duration of which increases
markedly as the holding temperature is reduced. As a general rule,
for a given growth treatment, the crystallite size decreases in
inverse proportion to the number density of nuclei, and we can
draw a nucleation response map in terms of particle size as a function of heat treatment. This is shown in Fig. 5 for a titania
nucleated glass. Note in particular the temperature dependent
saturation effects observed 740° and 760°C which limit the useful
nucleation time, but at the same time lead to reproducibility in
microstructure for a given nucleation temperature.

As we lower the nucleation temperature two phenomena become
important. (i) A long induction period during which nucleation is

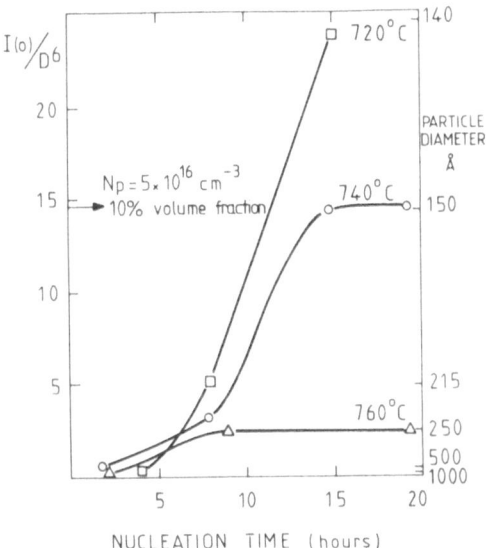

Fig. 5

Nucleation response map for a titania nucleated cordierite glass, showing particle size and number density as a function of heat treatment.

rather slow and the investigator may be misled into believing that no nucleation is occurring and (ii) the subsequent very high nucleation rate which leads to extremely fine stable microstructure.

At temperatures well below Tg, the very large number of nuclei created after days of heat treatment can compensate in some measure for the small particle size and enable direct measurement of the SANS without the need for a growth heat treatment. The measuring time is however increased, typically to a few hours in order to have modest statistical accuracy. We can see from the spectra shown in Fig. 6 the extra scattering observed after heat treatment of 36 hours at 720°C, which, being structureless, may be interpreted as coming from randomly spaced inhomogeneities of very small size. When this treatment time is doubled, the intensity strongly increases, and shows two effects : (i) at high Q, a negative slope from which we calculate a particle radius of about 5 Å, and (ii) a strong depression at low Q due to interparticle interference from nuclei which are no longer randomly spaced. Even at this small size, the creation of new nuclei becomes limited to regions beyond a minimum distance from the existing centres, presumably due to solute concentration profiles. The peak position

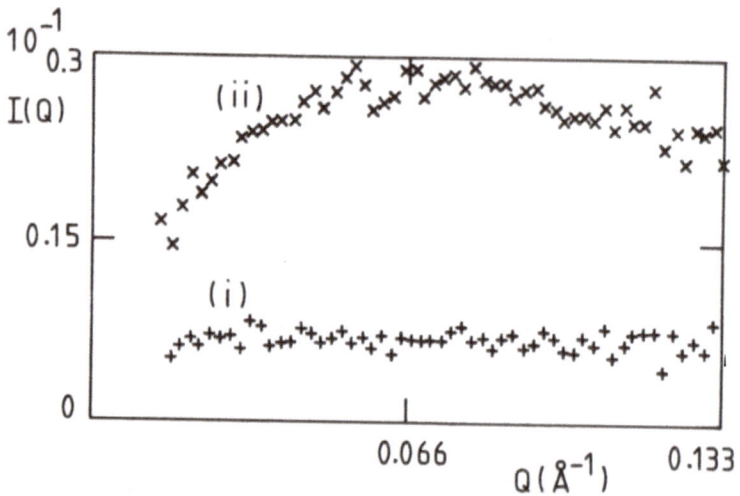

10^{-1}

0.3

$I(Q)$

(ii)

0.15

(i)

0

0.066 $Q(Å^{-1})$ 0.133

Fig. 6

Scattering data for titania glasses nucleated at 720°C
(i) 36 hours   (ii) 72 hours.

in $S(Q)$ becomes fixed at this stage, and the nucleation process
saturates.

    If we now heat this highly nucleated glass we can observe
the growth of these nuclei to form the crystallites (15). Somewhat
surprisingly however the microstructure is very resistant to
coarsening. Steady heating from 740 to 960° over 10 hours
only grows the mean radius to about 40 Å and the material remains
quite transparent. The relationship $I(0) = k'r^6$ is also maintained.
Furthermore the transition from the Al_2TiO_5 crystallite to the
β-quartz phase, normally observed around 900°C (2) is totally
suppressed. This result suggests that the particle size is a
parameter in the phase transformation, the variation of which also
modifies the stability of the glass-ceramic.

    The measurements plotted in Fig. 7 show a continuous evolution
in particle growth all the way from nuclei with a radius of 5 Å
to well defined crystallites with a radius of 40 Å. There is no
evidence to indicate a minimum (critical) radius for the particles.
One can reasonably conclude in this case that growth occurs via a
mechanism which allows continuous evolution from an amorphous
cluster (r < 5 Å) through to an ordered crystallite, employing both
chemical and structural variables to maintain minimum free energy
as the particle size increases.

29

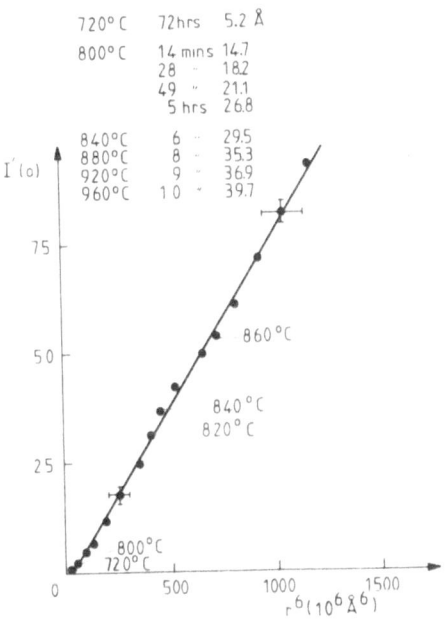

Fig. 7

Variation of I(0) with k'r⁶ for a titania nucleated glass heated
72 hours at 720°C and developed at temperatures up to 960°C.

CONCLUSION

Small angle neutron scattering studies can be used to survey
the nucleation and growth of crystallisation in glass ceramics real
time experiments. The measurements lead to optimisation of the heat
treatments for a given microstructure or can be used to explore the
time-dependent zones of heat treatment to obtain new ultramicro-
crystalline materials.

REFERENCES

1. McMillan P.W., "Glass Ceramics" Academic Press, London 1979.
2. Barry T.I., J.M. Cox, R. Morrell, J. Mat. Sci. 13(1978) 594-610.
3. Wright A.F., J. Talbot, B.E.F. Fender, Nature (1979) 227 (5695)
   366-368 (1979).
4. Wright A.F. Neutron scattering - 1981. Ed. J. Faber,
   359-367, American Institute of Physics (1982).
5. Bandyopadhyay, A.K., P. Labarbe, J. Zarzicki and A.F. Wright,
   J. Mat. Sci. 18 (1983) 709-716.
6. Schmatz W., T. Springer and J. Schelten, J. Appl. Crystallogr.
   7 (1974) 96.

30

7. Neutron Research Facilities at the ILL High Flux Reactor,
    Ed. B. Maier, Institut Laue-Langevin, Grenoble, 1983.
8. Wright A.F., P.W. McMillan and N.H. Brett, "Structure of Non-
    Crystalline Materials" - 1982" Ed. P.H. Gaskell, J.M. Parker,
    E.A. Davis, pp. 569-581. Society of Glass Technology,
    Taylor and Francis, London (1983).
9. Hayter J.B. and J. Penfold. Molecular Physics 42 (1981) 109,
    J. Chem. Soc., Faraday Transl. 77 (1981) 1851.
10. Hayter, J.B. and J.R. Hansen, ILL Report HA14T.
    Institut Laue-Langevin, Grenoble (1982).
11. Wright A.F., S. Nunn, N.H. Brett "Zirconia 83" Ed. N. Claussen,
    Advances in Ceramics, American Ceramic Society (1984).
12. Durville F., B. Champagnon, E. Duval, G. Boulon, F. Gaume,
    A.F. Wright and A.N. Fitch, To be published in Physics and
    Chemistry of Glasses, October 1984.
13. McMillan P.W. and A.F. Wright, unpublished work.
14. Fender B.E.F., A.N. Fitch and A.F. Wright, to be published.

PROPERTY AND PROCESS DEVELOPMENT IN GLASS-CERAMIC MATERIALS

G. H. Beall

Corning Glass Works, Corning, NY   14831

1.   INTRODUCTION

Glass-ceramics are polycrystalline solids produced by the controlled devitrification of glass. Glasses are melted, fabricated to shape, and then converted to a ceramic by a specific heat treatment. Glass-ceramic structure is characterized by fine-grained, randomly oriented crystals with some residual glass, but no voids, microcracks, or other porosity. The basis of controlled crystallization lies in efficient nucleation([1]). Except in rare cases, nucleation is heterogeneous; that is, the major crystalline phases normally silicates, precipitate upon precursor particles of relatively insoluble oxides, sulfides, fluorides, or metals. Often, amorphous phase separation on a very fine scale is the first stage in the nucleation event.

The properties of glass-ceramics are determined both by the inherent characteristics of the constituent phases and by the form of microstructure resulting from the nucleation and growth sequence. Glass-forming silicates can be divided into distinct structural groups: framework silicates, sheet silicates, and chain silicates, each predisposed to certain compositional and structural features. In the first two decades of research on glass-ceramics efforts naturally concentrated on relatively stable, highly polymerized glass-forming compositions where framework silicates with useful properties such as low thermal expansion behavior were the major constituents. The use of highly efficient nucleating agents has allowed the development of highly crystalline and transparent materials with grain sizes below 100 nm.

In the past decade, glasses of lower polymerization and stability have been studied in more detail, and new groups of glass-ceramics based upon anisotropic sheet and chain silicates developed. This has resulted in materials of enhanced mechanical strength and toughness, and in the case of certain fluormica glass ceramics, machinability and even flexibility. The structural polymerization of the major glass-ceramic silicate crystal phases is related to the viscosity and stability of the parent glass. Whereas framework silicates typically have a tetrahedral-ion-to-oxygen ratio (T/O) of 1:2, most sheet silicates have the characteristic 2:5 ratio, and chain silicates range from 1:3 to 2:5.

2. GLASS-CERAMICS BASED UPON FRAMEWORK SILICATES

Framework silicates form the basis of most glass-ceramic materials, and applications have relied primarily on their unique thermal properties like thermal shock resistance and dimensional stability over a wide range of temperatures. In terms of chemical composition, the SiO_2-Al_2O_3-Li_2O (MgO) system has produced glass-ceramics of lowest thermal expansion coefficient based upon either β-quartz or β-spodumene (keatite) solid solutions. The SiO_2-Al_2O_3-MgO system has produced refractory and thermal shock resistant materials of good strength and dielectric properties based on the ring-framework phase cordierite (indialite). Glass-ceramics based on sapphirine ($4MgO \cdot 5Al_2O_3 \cdot 2SiO_2$) are also interesting because of their hardness (KHn=1000). The SiO_2-Al_2O_3-Na_2O system has produced high expansion glass-ceramics ($\sim 100 \times 10^7/^{\circ}C$) which are easily glazed to give compressive skin strengthening required for strong table-ware[2]. Nepheline ($NaAlSiO_4$) is the major phase in these materials. High-alumina glass-ceramics containing mullite and/or pollucite are very refractory and chemically durable and are formed from SiO_2-Al_2O_3-(Cs_2O) glasses. These compositions have been suggested as hosts for radioactive waste[3]. All but the last system require the addition of a suitable nucleating agent to develop internal crystallization at sufficiently high viscosities to assure maintenance of the shape of the glass article.

The crystallization, structure, and properties of glass-ceramics based on the three most important of these framework silicates -- β-quartz, β-spodumene, and cordierite -- will be considered in more detail.

2.1 Glass-Ceramics Based on β-Quartz Solid Solution

Metastable glass-ceramics based on solid solutions of stuffed derivatives of β-quartz can be formed from a wide variety of glass formulations in the SiO_2-Al_2O_3-Li_2O-MgO-ZnO system[4]. Oxides TiO_2, ZrO_2, Ta_2O_5, and elemental silicon are all effective as nucleating agents for internal crystallization. The solid solution compositions are of the general type $(Li_2,R'')O.Al_2O_3.nSiO_2$, with n varying from 2 to 10. When n is in the range 6 to 8, birefringence in the quartz phase is minimized, and highly transparent glass-ceramics are most easily formed[5]. The choice of nucleation agent is critical in achieving very fine crystals (~1000 Å) which further serve to eliminate optical scattering and undesirable haze. β-quartz solid solutions are characterized by very low thermal expansion coefficients, generally near zero or even sometimes negative.

The effects of titania and zirconia on the crystallization of lithium aluminosilicates have been extensively described in the literature[6-13]. Doherty and Lee[6] found that certain compositions of the commercial type show phase separation on heating providing they contain greater than 2% titania. A large number of very small (about 50 Å) particles of an aluminum titanate compound form in the titania-rich amorphous phase and act as heterogeneous nuclei for the formation of β-quartz solid solution, which subsequently transforms to β-spodumene yielding a stable, opaque, fine-grained product. Tashiro and Wada[7] first described glass-ceramics nucleated effectively with zirconia. Beall et al[4] described the crystallization of β-quartz solid solutions over a wide range of ZrO_2-containing compositions. Tetragonal zirconia or cubic zirconia solid solution was identified as the nucleating oxide phase. Sack and Scheidler[8], and Stewart[9] both noted the effectiveness of mixtures of TiO_2 and ZrO_2 in nucleating aluminosilicate crystals. The latter showed a synergistic effect of the two oxides whereby the metastable β-quartz phase was nucleated at lower temperatures and higher viscosities, thus producing a finer grained and more transparent body with higher strength than could be produced by either agent alone.

Barry et al[12-13] studied a number of compositions at and near the SiO_2-$LiAlO_2$ join at the 1:1:4 and 1:1:6 $Li_2O:Al_2O_3:SiO_2$ compositions with 4 mole % titania added. The 1:1:6 stoichiometry, with excess alumina, is closely related to commercial formulations. With compositions at or near the join, in contrast to the subaluminous eutectic compositions, direct observation of glass-in-glass phase separation as a precursor to nucleation by titania was observed. In the 1:1:4 composition, electron microscopy revealed a minor phase of about 10% volume fraction and $10^{23}/m^3$ concentration of rounded 100 Å-diameter particles. This structure developed very rapidly when the glass was heated about $700°C$. Extremely high densities $(10^{23}m^{/3})$ of impinged metastable β-quartz solid solution were recorded as a result of further heating to $750°C$. Barry et al[13] noted a discrepancy between the high density and uniformity of the 1000 Å quartz crystals, the relatively long time (12×10^3 s) for crystallization measured at constant temperature ($735°C$) by differential thermal analysis, and their relatively quick growth rate when nucleated at the glass surface (3×10^{-10} m/s). They conclude that nucleation may be continuously triggered at the β-quartz crystallite growth fronts as growth proceeds, rather than occurring at each titania-rich nucleus simultaneously([12]).

Whether nucleation of β-quartz solid solution occurs simultaneously from countless titanate nuclei or from one or very few centers with subsequent nucleation triggered at growth fronts, may depend upon the solubility of titania in the original glass-in-glass phase separation. In compositions such as 1:1:4 β-spodumene with 4 mole % titania, the precipitation of crystalline titanates prior to β-quartz crystallization was not observed, and growth of quartz from one or very few nuclei (perhaps the first titanate crystals) appeared to trigger nuclei at the growth front, probably because titania was rejected from the β-quartz crystals. Anatase was observed by X-ray diffraction but only after substantial quartz crystallization. Thus these materials would crystallize fine-grained, but in a wave propagating from one or few scattered nuclei. This may explain the complex and intricate radial and concentric "spiderweb" fracturing observed in titania-nucleated glass-ceramics whose compositions lie close to the $LiAlO_2$-SiO_2 join (Fig. 1).

With commercial compositions, by contrast, phase separation and precipitation of titanates (or zirconates) are discrete events preceding the nucleation of β-quartz presumably by these oxides([6,14]). This sequence allows β-quartz to develop in countless scattered crystals throughout the body, thus eliminating the gross stresses and consequent fracturing associated with major growth fronts. The addition of excess Al_2O_3, or components such as MgO and ZnO, appears to allow the widespread precipitation of titanate (and zirconate) nuclei, perhaps because these oxides combine with TiO_2 (and ZrO_2) producing highly insoluble nucleating phases (e.g., $MgTi_2O_5$.Al_2TiO_5 s.s., $Al_2Ti_2O_7$, MgO.ZrO_2 s.s.). In the case of excess Al_2O_3, apparently the solubility of ZrO_2 and TiO_2 is decreased, which promotes their early formation.

A recent commercial β-quartz solid solution glass-ceramic, VISION®, has been developed as transparent cookware. This material is highly transparent because of the ultra-fine grain size and low birefringence of the β-quartz solid solution crystals. It has a coefficient of thermal expansion of $7x10^{-7}$/°C (0-500°C). Fig. 2 shows a transmission electron microphoto of this material depicting a highly crystalline microstructure of ultrafine (~600 Å) quartz crystallites, giving a high particle density of $5x10^{21}$/m^3. The nucleating phase, $ZrTiO_4$, is visible as small specks scattered throughout.

Current applications of β-quartz glass-ceramics requiring high transparency (or consequent polishability) include cookware, telescope mirror blanks, woodstove windows, infrared transmitting range tops, and heating coil sleeves.

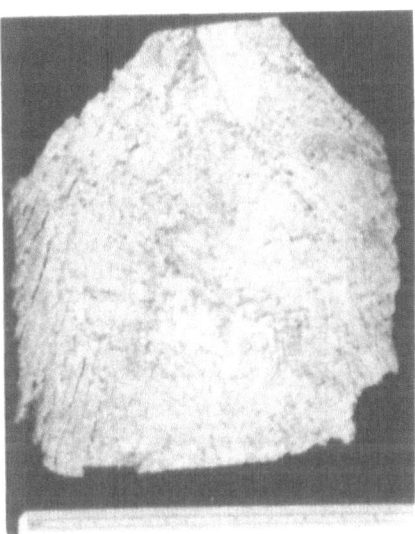

Fig. 1 Spiderweb fracturing
in TiO₂-nucleated 1:1:6 Li₂O-
Al₂O₃-SiO₂ glass-ceramic, probably a result of nucleation
at a growth front

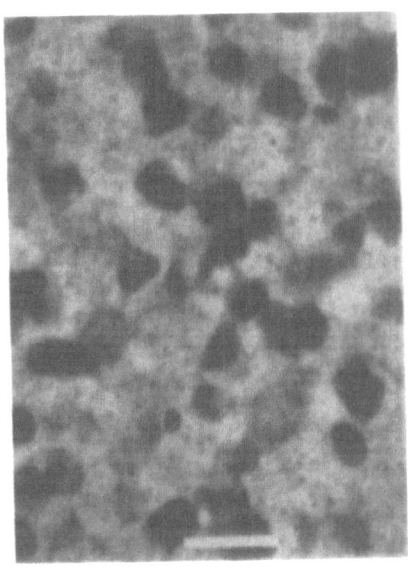

Fig. 2 Microstructure of
transparent β-quartz solid
solution glass-ceramic (white
bar = 0.1 μm)

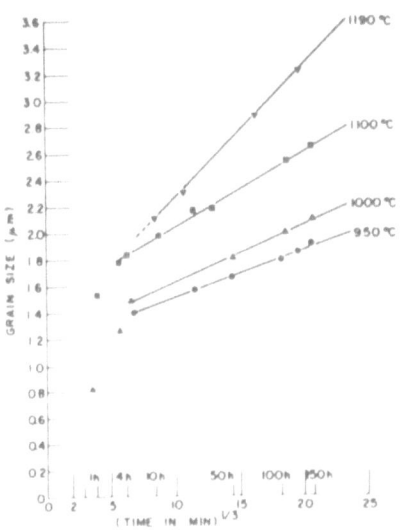

Fig. 3 Secondary grain growth
behavior in typical β-spodumene
solid solution glass-ceramic
(after Chyung[16])

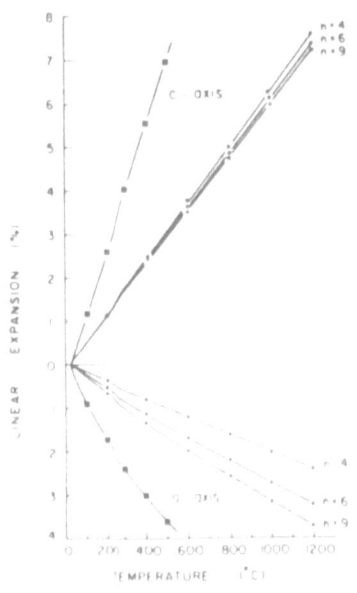

Fig. 4 Anisotropic thermal expansion in β-spodumene solid
solutions (after Ostertag etal[15])

2.2 Glass-Ceramics Based on β-Spodumene Solid Solution

Beta-spodumene solid solution is a stable low-expansion crystal phase of composition varying from $Li_2O.Al_2O_3.4SiO_2$ to $Li_2O.Al_2O_3.10SiO_2$[15]. It is a stuffed derivative of the polymorph of silica keatite. Considerable substitution of magnesium for lithium ($Mg^{2+} \rightleftarrows 2Li^+$) is permitted in the structure[4]. Beta-spodumene always forms from pre-existing β-quartz metastable solid solutions during glass-ceramic heat treatment, and thus the early stages of nucleation and growth of β-spodumene glass-ceramics is as described in the previous section. The transformation from β-quartz to β-spodumene solid solution usually occurs between 900° and 1000°C, and is irreversible -- β-spodumene being the stable phase. There is an increase in grain size of the silicate phase (typically by a factor of about 7) associated with this almost isochemical phase change. When TiO_2 is used as the nucleating agent, rutile development always accompanies the silicate transformation. Because of the high refractive index and birefringence of this phase, a high degree of opacity is developed.

The secondary growth behavior of β-spodumene crystals for a typical glass-ceramic as a function of temperature and time is illustrated in Fig. 3. The growth rate is slow and is linear with the cube root of time[17]. Grain size stability is one of three factors that provide unique thermal and dimensional stability in β-spodumene glass-ceramics. The others are the ultralow expansion of the solid solution crystals and the matching low thermal expansion of the siliceous residual glass. The resistance to grain growth is particularly important in view of the marked anisotropy in thermal expansion of β-spodumene crystals[15]. (Fig. 4)

Mechanical properties of β-spodumene glass-ceramics are not unusual. Abraded flexural strengths are typically around 100 MPa. Nevertheless, the extreme thermal shock resistance and dimensional stability afforded by the low expansion crystals allow a variety of applications such as cookware, hot plates, and ceramic regenerators for turbine engines.

2.3 Cordierite Glass-Ceramics

Glass-ceramics based on the ring-framework silicate cordierite, are characterized by high temperature stability, thermal and mechanical shock resistance, and excellent dielectric properties. They have been manufactured as radar-guided missile nose cones (radomes) for nearly 25 years[1]. Only recently have the excellent strength, toughness, and creep resistance of this material and cordierite glass-ceramics in general been documented, and their potential as substitutes for high quality alumina ceramics become clear[17,18].

Numerous studies of glass-ceramic formation in the $SiO_2-Al_2O_3-MgO-TiO_2$ system have been made[1,19-22]. The crystallization of glass in this system is extremely complex becuase of the large number of phases, many of them metastable, which can crystallize from glass. Barry et al[19] demonstrated that the sequence of crystallization can vary for near-stoichiometric cordierites depending upon the nucleation agents, i.e. titania or titania plus zirconia.

A typical sequence of crystallization for a near-stoichiometric cordierite glass-ceramic containing approximately 11 wt. % TiO_2 as the nucleation agent would proceed in the following manner: upon heating, the glass undergoes amorphous phase separation preceding the precipitation of the nucleating phase, $(Mg,Al)(Ti,Al)_2O_5$, a pseudobrookite structure[21,22]. Near $900^{\circ}C$, the first metastable silicate, a stuffed derivative of β-quartz solid solution, precipitates upon the oxide nuclei. With further heat treatment this solid solution phase breaks down to a very fine mixture of siliceous quartz and spinel, and at slightly higher temperatures, sapphirine, cordierite, and rutile appear. Cordierite and rutile form very rapidly near $1200^{\circ}C$ through a solid state reaction between quartz, sapphirine, spinel, and pseudobrookite, and they are the stable phases observed at temperatures above $1250^{\circ}C$[19,4].

Corning Code 9606 is a silica-rich cordierite glass-ceramic with the following phase assemblage as revealed by X-ray diffraction: cordierite, rutile, magnesium dititanate, and cristobalite. The cordierite is actually a subaluminous cordierite solid solution believed near stoichiometric $Mg_2Al_4Si_5O_{18}$, but with some $Mg^{+2}+Si^{+4}\rightleftarrows 2Al^{+3}$ substitution of the type described by Schreyer and Schairer[23]. The microstructure of this glass-ceramic is shown in Fig. 5. The grain size is about 1 μm, and the cordierite appears depressed relative to cristobalite and the titanates, because it is more easily dissolved in the HF etchant. A "coast and island" texture can be observed with resistant titanate-cristobalite intergrowth regions standing out in a background of cordierite. The grain boundaries are more easily seen in the titanate-cristobalite regions than in the cordierite matrix. Little or no evidence of glass is observed either in the microphotos or on X-ray diffraction traces.

The mechanical properties of this glass-ceramic have recently been extensively studied[17,18]. As is typical for brittle materials, the strength is variable and depends on processing parameters, particularly the heat treatment schedule and finishing techniques. Most samples tested for modulus of rupture fell within a reasonably narrow range, 250-300 MPa, and variations in strength within a single billet finished by surface grinding were small in comparison to those of most ceramic materials. Fig. 6 shows a typical cumulative strength distribution for production 9606 from bars surface ground in the transverse direction, normally a severe abrasion condition producing a large number of flaws aligned per-

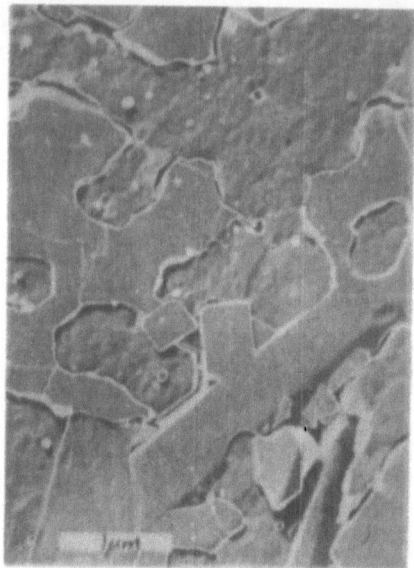

Fig. 5 Microstructure of Corning Code 9606 (replica electron microphoto; 1/2% HF etch)

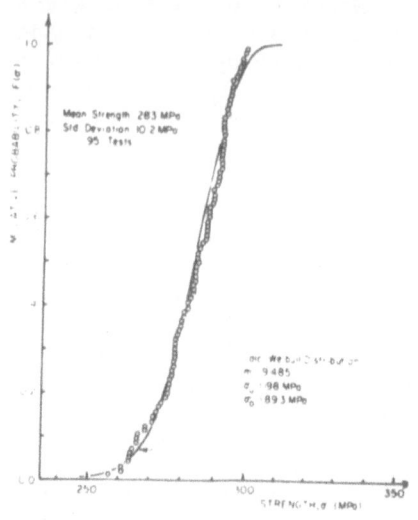

Fig. 6 Flexural strength distribution of Corning Code 9606 glass-ceramic (transverse surface grint) - after Lewis[17]

Fig. 7 Microstructure of highly crystalline alkaline earth mica glass-ceramic (white bar = 1 μm)

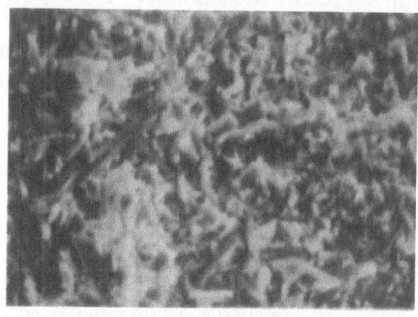

Fig. 8 Fracture surface of canasite glass-ceramic showing interlocking blade-like crystals (white bar = 1 μm)

pendicular to the bending stress.

The fracture toughness, K_{1c}, has been routinely measured at about 2.1 MPa $m^{\frac{1}{2}}$ for production 9606 which has a top crystallization hold near $1250^{\circ}C$ for 8 hrs. It has been reported[24] that this toughness can be increased to 2.5 MPa $m^{\frac{1}{2}}$ by lengthening this hold to 24 hrs. The interlocking and highly crystalline microstructure coupled with a high Young's modulus for a silicate material (~120 GPa) presumably accounts for such high toughness values.

Because of the unique properties of Corning 9606 -- moderately low thermal expansion, high strength, and excellent dielectric properties at microwave frequency -- it has long been the standard for high performance radomes.

3.   GLASS-CERAMICS BASED UPON SHEET SILICATES

Sheet silicates are based on the planar hexagonal network of linked silicon-oxygen tetrahedra. Aluminum may substitute for up to one-half of the silicon. Sheet silicates crystallize from glasses of intermediate polymerization -- the T/O ratio in the crystals typically being 2:5. Microstructures usually consist of tabular or lath-like crystals with substantial interlocking. Phases of this group which have been described in glass ceramics[26] include lithium silicate ($Li_2Si_2O_5$), sanbornite ($BaSi_2O_5$), fluormica ($X_{0-1}^{xii}Y_{2-3}^{vi}Z_4^{iv}O_{10}F_2$, where superscripts refer to coordination), and hexacelsian ($BaAl_2Si_2O_8$). Lithium disilicate and fluormica glass-ceramics have found wide application, and will be described in some detail.

3.1 Lithium Disilicate Glass-Ceramics

Glass-ceramic compositions based on lithium disilicate have been studied extensively[25], largely because they are relatively easy to melt, form a glass, and control subsequent nucleation and growth. Various nucleating agents including noble metals and P_2O_5 have been employed -- the latter being particularly effective[25]. Highly crystalline materials with lath-like lithium disilicate crystals can give good flexural strength (100-300 MPa) and toughness[27] (2-3 MPa $m^{\frac{1}{2}}$). They also show surprisingly high electrical resistivities (up to $3x10^9$ ohm-cm at $300^{\circ}C$) and low loss factors (0.002 at 1 MHz and $25^{\circ}C$) for such high-alkali glass-ceramics. Because of their relatively high thermal expansion coefficient, $\sim120x10^{-7}/^{\circ}C$, these materials can be sealed to several metals[25].

Lithium disilicate is also the major phase in Fotoceram®, a photosensitive glass-ceramic which is used for gas-discharge display panels, ink-jet printing plates, fluidic devices, and magnetic recording head pads. In this case, colloidal silver particles are formed in the parent glass by a photosensitive process involving ultraviolet radiation and a sensitizer cerous oxide. The Ce^{+3} ions donate electrons to silver ions on u-v exposure creating neutral silver atoms which agglomerate on thermal treatment. Colloidal silver then nucleates the chain silicate lithium metasilicate ($Li_2 SiO_3$), which forms a dendritic pattern easily etched and removed by dilute hydrofluoric acid[25,26]. Thus, the glass can be chemically machined according to the original ultraviolet exposure. Subsequent flood exposure and heat treatment can produce a strong (abraded M.O.R. ~150 MPa) dominantly lithium disilicate glass-ceramic with the high resolution etched or vacant pattern areas remaining.

3.2 Machinable Fluormica Glass-Ceramics

Machinable glass-ceramics based upon internally nucleated fluor-mica crystals in glass were originally described in the early 1970's[14,28]. At least one commercial material, Macor®, has been marketed for about ten years, and has found wide application in such diverse and specialty areas as precision electrical insulators, vacuum feedthroughs, windows for microwave tube parts, sample holders for field ion microscopes, seismograph bobbins, and boundary re-tainers on the space shuttle. The precision machinability of Macor® with conventional, metal-working tools is directly due to the fine-grained house-of-cards microstructure, where randomly-oriented flexible flakes tend to either arrest fractures or cause such deflection or branching that only local damage is done. High dielectric strength (~40 Kv/mm) and very low helium permeation rates are important in high vacuum applications.

While the early fluormica glass-ceramics were based on the fluorine phlogopite phase ($KMg_3AlSi_3O_{10}F_2$), recently other chemical formulations giving rise to different fluormicas have been developed. Thus subpotassic phlogopite ($K_{1-x}Mg_3Al_{1-x}Si_{3+x}O_{10}F$)[29,30], potassium tetrasilicic mica ($KMg_{2.5}Si_4O_{10}F_2$)[31], alkaline earth micas ($R_{0.5}Mg_3 AlSi_3O_{10}F_2$)[32] where R is Ba, Sr, or Ca, sodium phlogopite ($NaMg_3 AlSi_3O_{10}F_2$)[33], and biotite (K, Na) (Mg,Fe^{+2})$_3$ (Al,Fe^{+3}) $Si_3O_{10}F_2$[33] have been reported as major phases in glass-ceramics. The sub-alkali compositions tend to promote high-aspect-ratio flakes, with good thermal shock resistance and high fracture energies, up to 40 J/m^2, having been observed. The tetrasilicic mica glass-ceramics are characterized by fine grain-size (≤ 1 μm flakes), good strength (150 MPa), and translucency, which along with superior chemical durability, are suitable for dental restorations and similar pro-sthetic applications.

The alkaline earth mica glass-ceramics are currently the stron-gest of the machinables and also boast the highest thermal stability ($1100°C^+$) and some of the best dielectric properties including re-sistivities up to 10^{11} ohm-cm at $500°C$[32]. Abraded modulus of rupture values of 250 MPa have been measured on one highly crystal-line mixed barium-strontium phlogopite glass-ceramic whose highly

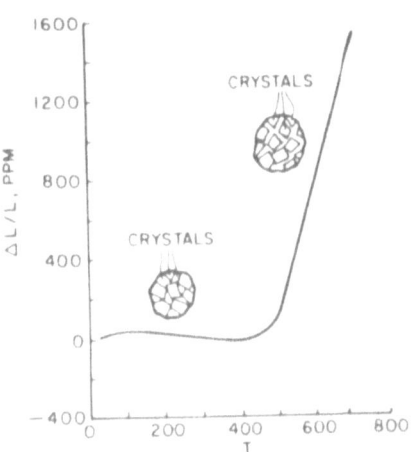

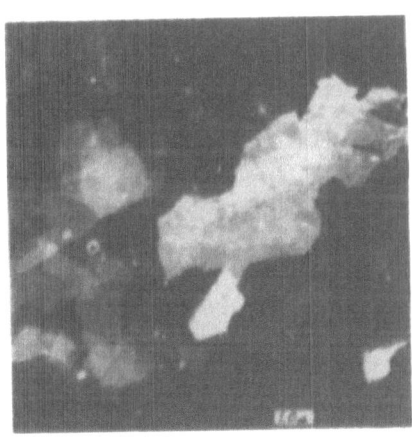

Fig. 10 Dark field T.E.M. photo of delaminated fluor-hectorite

Fig. 9 Thermal expansion curve of temperable β-quartz-borosilicate glass-ceramic showing crystal and glass dominated regimes

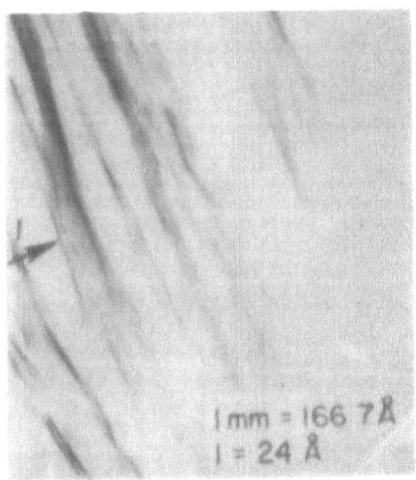

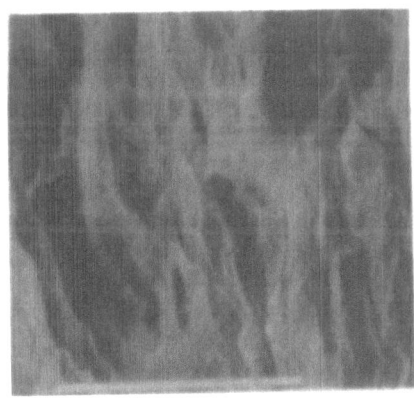

Fig. 12 S.E.M. photo of torn cross-section of mica film (arrows show strip crystallites along composite flake boundary; white bar = 1 μm)

Fig. 11 T.E.M. photo showing cross-section of thin hectorite flake (line l = double unit cell thickness)

interlocking microstructure is depicted in Fig. 7. Highly machinable glass-ceramics with both the "house-of-cards" structure and unusual spherulitic "cabbage head" development have recently been described in the soda phlogopite composition area([33]). Iron-containing biotite mica glass-ceramics, some of which show ferromagnetic behavior, have also been discovered([33]).

The crystallization cycle in mica glass-ceramics includes normal nucleation and growth temperature holds, but low-fluoride phases sometimes form on the surface if substantial fluoride species volatilize prior to crystallization. Thus a hard skin may have to be removed prior to normal etching. Machining resolution is limited only by the crystal size (10 μm-diameter flakes), and unlike other machinable ceramics, there is no porosity, and no post-machining firing is required.

4.   GLASS-CERAMICS BASED UPON CHAIN SILICATES

Because chain silicate compositions are usually marginal glass formers, most glass-ceramics containing these crystal-types also contain other phases like framework silicates or glass. This is especially true for single chain silicates, whose T/O ratio is only 1:3. Pyroxenes and pyroxenoids have been developed as the major phases in inexpensive glass-ceramics made from basalt or slag.

Double and multiple chain silicates correspond to more polymerized and stable glass compositions, and have recently been studied as glass-ceramic formers([34],[35]). Materials of considerable strength and toughness have resulted, especially with the alkali-lime fluorsilicate phases canasite and agrellite.

4.1   Inexpensive Glass-Ceramics Based on Pyroxenes and Pyroxenoids

Basalt glass-ceramics have been produced by reheating glasses made from natural basalts melted in an oxidizing atmosphere([36]). Magnetite (Fe_3O_4) nuclei separate from iron-rich amorphous clusters less than 100 Å in diameter, and upon further heating the clinopyroxine diopsidic-augite forms on the magnetite. Although the glass-ceramic is only about 50% crystalline, it has adequate strength (~100 MPa), good abrasion resistance (KHn ~900), and good chemical durability, particularly in alkaline environments.

Slag glass-ceramics have been manufactured for years in the U.S.S.R. and other eastern bloc nations([37]). Blast furnace slag is mixed with other inexpensive ingredients to form a glass which is then crystallized through internal nucleation by a reduced species, usually ZnS. Wollastonite and/or diopside is the major crystal phase. As with basalt, the major characteristics are abrasion resistance, good chemical durability, and body strength superior to glass. Building products (floor tile, cladding, bench tops) are the main application.

4.2 Fluoramphibole Glass-Ceramics

Amphiboles are a common natural group of chain silicate minerals characterized by the infinite double chain polyanion $(Si_4O_{11})^{-6}$ where aluminum can replace silicon up to 50%. The general formula of the fluoramphiboles is $W_{0-1}X_2Y_5(Z_4O_{11})F_2$ where the cationic co-ordination is $W = 12$, $X = 8$, $Y = 6$, and $Z = 4$. Several species have been crystallized from glass, among them tremolite $(Ca_2Mg_5Si_8O_{22}F_2)$, richterite $(Na_2CaMg_5Si_8O_{22}F_2)$, and proto-amphibole $(LiMg_{6.5}Si_8O_{22}F_2)$[34].

The microstructure of these materials varies from highly crystalline with blocky morphology to rodlike crystals with interstitial glass. Good abraded flexural strength (~150 MPa) and high dielectric strength (>50 Kv/mm) have been reported[34].

4.3 Multiple Chain Silicate Glass-Ceramics

Certain rare natural chain silicates contain crosslinked chains of tetrahedra where the T/O ratio may be as high as in sheet silicates, namely 2:5. Two such minerals, both of which are easily fused to a stable glass, are the alkali-lime fluorsilicates canasite[38] $(Na_4K_2Ca_5Si_{12}O_{30}F_4)$ and agrellite[39] $(NaCa_2Si_4O_{10}F)$. Glasses of these compositions are easily melted at $1300°C$ and can be internally well nucleated and fully crystallized by additions of excess fluoride[35].

Canasite is composed structurally of units of four parallel silicate chains crosslinked to make a long box-like backbone in which the potassium ions lie[38]. These box-like units are separated by networks composed of $Na(O,F)_6$ and $Ca(O,F)_6$ octahedra. The morphology of this phase in glass-ceramics reflects the crystal structure. The crystals are anisotropic and blade-like, and are highly interlocked (Fig. 8)[35]. The orientation is random, predetermined by nucleation on scattered centers of CaF_2 precipitated in the early stages of heating the glass above its annealing temperature. Canasite glass-ceramics are remarkable for both abraded body strength (MOR: 250-350 MPa) and toughness (3.5-5.0 MPa m$^{\frac{1}{2}}$)[40], especially considering their relatively low density (~2.7 g/cc) and elastic modulus (~80 GPa). The interlocking blade morphology and resultant high energy of fracture is presumed responsible for the strength and toughness.

Agrellite glass-ceramics are in many ways similar to the canasite materials[35]. Agrellite is also a tubular structure with double chains periodically branching into loops[39]. It is also nucleated by CaF_2 to form an interlocking lath-like structure. Abraded flexural strengths measured on agrellite glass-ceramics range up to 200 MPa.

5. ADVANCES IN PROCESSING OF GLASS-CERAMICS

The processing of glass-ceramics involves three steps: glass melting and forming, the thermal-crystallization cycle, and optional secondary forming, which may involve strengthening or reshaping in the crystallized state. Conventional glass melting and forming is generally used[41], although high fluoride glasses may require electrical or cold-crown melting. An optimum thermal-crystallization cycle is developed for each glass-ceramic composition type. This often involves one or more nucleation holds generally in the range 50 to 150°C above the glass annealing point, followed by a crystal growth hold at higher temperature to develop the desired microstructure. Care must be taken to avoid long periods of low viscosity lest the article deform or sag. Adequate time must also be allowed in temperature regions where densification or expansion due to phase transformation may occur[41]. For large articles, slow heating during early stages of crystallization generally serves to prevent thermal excursions due to the release and buildup of exothermic heat of crystallization. Other problems such as cracking due to nucleation and crystallization from a growth front as described in section 2.1, must be solved by compositional adjustment.

5.1 Secondary Processing: Strengthening

Secondary processing involves both reforming and strengthening techniques. The latter includes surface compression induced by ion exchange[2,414], differential surface crystallization[2542], glazing[2], and glass-ceramic lamination[42]. Frequently the increase in flexural strength induced by this surface compression processing is two or more times the original glass-ceramic strength. Unfortunately, processing involving molten salts, hot glazing, or the lamination of two glasses followed by differential densification during crystallization[42] involves increased operating or capital expense, and has not as yet been widely used.

5.1.1 Physical Tempering. Simple physical tempering has recently been applied with some success to low expansion and transparent β-quartz solid solution glass-ceramics containing a minor but significant proportion (20-25%) of residual potassium borosilicate glass[43]. The glass has a substantially higher thermal expansion coefficient than does the lithium aluminosilicate crystal phase, and tends to contract faster when the material is cooled. Thus a point can be reached where the continuous glass matrix is interrupted, with a continuous crystalline network developing and the glassy regions becoming largely isolated. This can result in a thermal expansion curve as shown in Fig. 9, where the ultra-low expansion β-quartz crystals dominate the expansion behavior below

d-spacing change from 9.7 to 12.5 Å, indicates the presence of a single layer of water in the interlayer of the mica.

In the case of fluorhectorite ($LiMg_2LiSi_4O_{10}F_2$) glass-ceramics, the water-swelling is even more dramatic[45]. Upon contact with water, this material delaminates to such an extent that the fine platelets of mica form a stable suspension. On drying this suspension or gel on a flat surface, a translucent film composed of oriented and overlapping hectorite flakes is observed[24]. This film is highly flexible but very hygroscopic. It can be easily re-dispersed in water to the gel-suspension state.

If the gel is extruded into an aqueous solution of common potassium salts, however, film or paper can be produced continuously by an extremely rapid ion exchange and flocculation step[45]. Typically a fluorhectorite gel is forced through a slot into salt solution of KCl, where $K^+ \rightleftharpoons Li^+$ exchange between the solution and the mica platelets causes immediate flocculation, producing a coherent and continuous film less than 25 μm thickness. This film, when washed and dried, is flexible and creasable. It is composed of oriented crystals of a stable potassium mica approaching $KMg_2LiSi_4O_{10}F_2$ in composition.

The morphology of the mica crystals both dispersed in the original aqueous suspension and in the ion exchanged and flocculated film is interesting. Very high aspect ratio platelets (Fig. 10) and narrow strips less than 100 Å thick, some approaching unit cell thickness (Fig. 11), are observed in the gel. In the ion-exchanged film, composite flakes composed of similar platelets and strips are present (Fig. 12). The high degree of orientation and overlapping of the mica crystallites is doubtless responsible for the flexibility and creasability of the film.

A thicker paper or even a board can be made by random flocculation of the gel, followed by wire-forming the floc on a conventional Fourdriner machine. Glass or other fibrous material may be introduced to impart strength and tear resistance.

These films, papers, and boards are about 50% porous, but the porosity is generally closed, allowing for controlled permeability. The papers are as strong as quality organic paper products, can withstand temperatures of over 500°C without embrittlement, and have dielectric strengths in excess of 20 Kv/mm. Additionally, they are durable over a wide range of pH.

Potential applications include fireproof paper and board for construction, capacitor film, electrical circuit boards, substrates, and flexible electrical and thermal insulation.

$500^{\circ}C$ and the borosilicate glassy phase dominates above this temperature. Tempering or chilling an article from the top crystallization temperature near $800^{\circ}C$ can therefore produce a rigid contracted surface below $500^{\circ}C$ before the interior cools. As the lagging interior glass-ceramic shrinks, the rigid surface is brought into a state of compression. In this way, near-zero-coefficient-of-expansion β-quartz glass-ceramics of marginal strength can be tempered to over twice their normal M.O.R., with abraded values of 150 MPa resulting. The thermal stability of such surface compression will, of course, deteriorate if the glass-ceramic article is reheated above $500^{\circ}C$ for long periods.

5.2 Secondary Processing: Reforming

It was discovered at Corning Glass Works that β-spodumene glass-ceramic sheet manufactured for laboratory benchtops could be vacuum formed or otherwise molded into sinks or other shapes at temperatures well below the initiation of melting, even though the crystallinity was over 90%. Raj and Chyung[44] have recently described the creep behavior of this type of material at sub-solidus temperatures and have attributed the high creep rates to solution-precipitation phenomena involving a minor (~5%) but somewhat fluid glassy phase. High strains (over 70%) were measured in bending samples at $950^{\circ}C$ for 10^5 seconds under 14 MPa stress. Superplastic creep occurs above $1000^{\circ}C$ as the original crystallization temperature (~$1100^{\circ}C$) is approached. A model for creep was developed where deformation occurs by species transport through the glass phase. The model assumes grain boundary islands of good fit separated by a thin glassy layer, consistent with the low dihedral angles at triple junctions observed in electron micrographs. The applied stress is supported at the islands where crystals meet while the glass expedites transport. Grain shape elongates in the direction of tensile stress. The results of the creep experiments are in agreement with interface reaction controlled creep.

More dramatic than reforming of glass-ceramics with stress is a recent technique of total reconstitution by an aqueous gelation technique applicable to certain sheet silicate glass-ceramics[45]. The result is oriented glass-ceramic film displaying mechanical flexibility.

5.3 Glass-Ceramic Paper

Some fluormica glass-ceramics were found to be spontaneously reactive with water and to disintegrate into a fine slurry of clay-like particles. Highly-crystalline and stoichiometric $Sr_{0.5}Mg_3AlSi_3O_{10}F_2$ glass-ceramic chunks can be observed to self-pulverize on contact with water at room temperature[32]. The individual fluormica crystals, about 2.5 μm in diameter, expand along the c-axes by some 25%. This expansion, seen on X-ray diffraction as an 001

6. SUMMARY

Considerable progress in the understanding and application of glass-ceramics has been made in recent years. Cordierite glass-ceramics are recognized as having strength and toughness rivaling alumina with the advantages of superior dielectric properties and lower thermal expansion. They can be made in large sizes with consistent and uniform microstructure. Transparent glass-ceramics based on the ultralow-expansion β-quartz phase is now manufactured with virtually no visible haze and can be strengthened by lamination or physical tempering techniques.

A variety of fluorsilicate glass-ceramics based on sheet and chain silicates has been recently developed. Some are machinable, and some, especially those based on the mineral canasite, have toughness-to-weight ratios comparable to transformation-toughened zirconia. Flexible inorganic films can be formed by aqueous delamination of certain mica glass-ceramics followed by reconstitution in oriented forms by a unique ion exchange and flocculation process.

REFERENCES

1. S.D. Stookey, Ind. Eng. Chem., 51 (7) 805, 1959.
2. D.A. Duke et al, J. Am. Ceram. Soc., 50, 2, 67, 1967.
3. G.H. Beall and H.L. Rittler, in Adv. in Ceramics 4, p. 301, Am. Ceram. Soc., 1982.
4. G.H. Beall et al, J. Am. Ceram. Soc., 50, 4, 181, 1967.
5. G.H. Beall and D.A. Duke, J. Mater. Sci., 4, 340, 1969.
6. P.E. Doherty et al, J. Am. Ceram. Soc., 50, 2, 77, 1967.
7. M. Tashiro and M. Wada, in Adv. in Glass Technology, Pr. Z, Plenum Press, NY 1963, p. 18.
8. W. Sack and H. Scheidler, Glastech. Ber. 39, 3, 126, 1966.
9. D.R. Stewart, in Adv. in Nucleation and Crystallization in Glasses, Ed. by L.L. Hench and S.W. Freeman, Am. Ceram. Soc. Sp. Pub. 5, 1971.
10. T.I. Barry et al, J. Mat. Sci. 4, 596, 1969.
11. T.I. Barry et al, J. Mat. Sci. 5, 117, 1970.
12. T.I. Barry et al, Discussions Faraday Soc. 50, 1970.
13. T.I. Barry et al, 7th I.S.R.S., Bristol, July, 1971.
14. G.H. Beall, in Adv. in Nucleation and Crystallization in Glasses, Am. Ceram. Soc. Sp. Pub. 5, 1971.
15. W. Ostertag et al, J. Am. Ceram. Soc., 51, 651, 1968.
16. C.K. Chyung, J. Am. Ceram. Soc., 52, 342, 1969.
17. D. Lewis III, Am. Ceram. Soc. Bull., 61, 11, 1208, 1982.
18. G.K. Bansal et al, Am. Ceram. Soc. Bull., 55, 3, 289, 1976.
19. T.I. Barry et al, J. Mater. Sci., 13, 594, 1978.
20. C.R. Grostelow and J.E. Restall, in Special Ceramics 6, Ed. by P. Popzer, Br. Ceram. Res. Assoc., July, 1974, p. 23.

21. R.C. Devekez and A.J. Majumdar, Glass Tech. 15 (1974).
22. D.G. Grossman, 14th Symposium on Electromagnetic Windows, Georgia Inst. of Tech., Atlanta, June, 1978.
23. W. Schreyer and J.F. Schairer, J. Petrology, 2, 324, 1961.
24. S. Bhaduri and D.P.H. Hasselman, Am. Ceram. Soc. Ann. Meeting, Chicago, April, 1983.
25. P.W. McMillan, Glass-Ceramics, 2nd. Ed., Academic Press, NY, 1979.
26. S.D. Stookey, U.S. Patent 2,684,911, 1954.
27. J.J. Mecholsky, in Adv. in Ceramics 4, p. 261, Am. Ceram. Soc., 1982.
28. G.H. Beall, U.S. Patent 3,689,293, 1972.
29. C.K. Chyung et al, 10th Int. Cong. Glass, 14, 33, 1974.
30. K. Chyung, in Fracture Mechanics of Ceramics, p. 495, Ed. R.C. Bradt, Plenum Press, NY, 1974.
31. D.G. Grossman, J. Am. Ceram. Soc., 55, 9, 446, 1972.
32. S.N. Hoda and G.H. Beall, in Adv. in Ceramics 4, p. 287, Am. Ceram. Soc., 1982.
33. W. Vogel and W. Holand, in Adv. in Ceramics 4, p. 125, Am. Ceram. Soc., 1982.
34. D.G. Grossman, U.S. Patent 3,839,056, 1974.
35. G.H. Beall, U.S. Patent 4,386,162, 1983.
36. G.H. Beall and H.L. Rittler, Bull. Am. Ceram. Soc., 55, 579, 1976.
37. A.I. Berezhnoi, Glass-Ceramics and Photo-Sitalls, Plenum Press, NY 1970.
38. M.I. Chigarov et al, Dokl. Akad. Nauk S.S.R., 185, 672, 1969.
39. S. Ghose and C. Wan, Am. Mineral., 64, 563, 1979.
40. K. Chyung and G.H. Beall, to be presented at the Am. Ceram. Soc. Ann. Mtg., 1984.
41. G.H. Beall and D.A. Duke, in Glass: Science and Technology, Vol. 1, Acad. Press, NY, 1983.
42. K. Chyung, in Adv. in Ceramics 4, Ed. by J.H. Simmons et al, p. 341, Am. Ceram. Soc., 1982.
43. G.H. Beall, U.S. Patent 4,391,914, 1983.
44. R. Raj and C.K. Chyung, Acta Met., 29, 159, 1981.
45. G.H. Beall et al, U.S. Patent 4,239,519, 1980.

ROUND TABLE ON
CRYSTALLIZATION OF NON-CONVENTIONAL GLASSES

C.A. ANGELL, chairman*
Panel members : C.T. MOYNIHAN, A.F. WRIGHT, J. DUPUY
F. LUBORSKY, M.G. SCOTT, B. CHAMPAGNON

This Round Table was set up to examine what, if anything, new could be learned about the phenomenon of nucleation and crystallization by studying what we might call unconventional glasses. By the term "unconventional glasses" we have in mind primarily non-oxide glasses but, since the study of metalic glasses has become so popular, we had the further restriction that metallic glass phenomenology would be treated as a separate subject. Thus, among unconventional glasses for this purpose would be considered the fluoride glasses of optical communications interest, and indeed any of the highly ionic molten salt type glasses, glasses formed by dissolution of ionic materials in molecular liquids such as the aqueous solution glasses, and finally glasses formed from purely molecular systems.

The chairman introduced the subject by drawing attention to two aspects of the study of nucleation in unconventional systems which were important to consider. The first of these was the possibility, using non-conventional systems of examining new states of the system. By new states one had in mind principally the possibility of forming microdispersed states of the system, and thereby eliminating heterogeneous nucleation from consideration as a phenomenon. However, it is possible also to go further and create truly nanoscopic systems in which the sample size is reduced to the order of some 10's of Å in diam. These are the conditions in microemulsion systems which have been recently getting a lot of attention for reasons quite distinct from their ability to help elucidate the crystallization phenomenon. The special advantage of microemulsion for our purposes is the fact that the individual systems are not much greater than the critical nuclei themselves. In such systems, indeed, the most rapidly crystillazing liquids seem to be capable of forming glasses - even benzene having now been obtained easily in the glassy state.

The second aspect of unconventional glass studies is the fact that frequently the viscosity factors controlling the growth of crystals already nucleated have very different temperature dependences from those in conventional oxide glasses. Thus, somewhat different phenomenology might be expected, and from the study of such differences new insight can be gained.

The particular case of the behavior of mixtures of isomers of xylene studied in emulsion form was considered. It was shown how both heterogeneously and homogeneously nucleated crystallization peaks could be identified during cooling of small samples in the differential scanning calorimeter environment, and how these two characteristic temperatures could be shifted by changes in binary solution composition. The condition for bulk glass-forming capability was found to be the condition that the heterogeneous nucleation temperature be depressed to within some 20 % of T_g.

Following these remarks the panel was assailed with a variety of questions from the audience as follows. Professor J. LUCAS asked what was being done at the moment in the way of development of techniques which can distinguish between the conditions for nucleation in fluoride glasses to occur and conditions under which these nuclei would grow to macroscopic sizes. C.T. MOYNIHAN responded with a display of recent results obtained by differential scanning calorimetry of small samples of some of the better known fluoride glasses. He first drew attention to problems associated with the study of unconventional glasses of the fluoride type where the composition lay close to binary, ternary or quaternary eutectics. The consequence is the possibility of occurence of many alternative crystallization sequences, the choice depending on a delicate balance of composition and thermal history. Thus, reproducability becomes a problem. After describing some elements of the confusion which can follow in nucleation studies of such systems he described, for one of the better behaved systems, some studies involving annealing of samples in the vicinity of T_g. Annealing temperatures separated from one another by several degrees dead to different positions of the crystallization peak during the subsequent reheating. These experiments revealed that nucleation was in fact proceeding at the glass transition and was producing different densities of nuclei for different time-temperature annealing histories. For a given annealing time, there was always a most effective nucleation temperature. One striking observation made here was that in fluoride glasses the nucleation and growth regimes seemed to overlap much less than in the conventional glasses.

N.J. KREIDL wanted to know if the classical method of detecting the optimum nucleation temperature by use of gradient furnace heat treatment was being applied to crystallization studies in non-conventional glasses. MOYNIHAN replied that as far as he was aware this technique was not yet, in fact, being applied. A new question from Dr. INGRAM brought attention to possible differences between weakly glass-forming ionic and metallic glasses. Dr. INGRAM was concerned about the meaning of the observation that fast ion conductor glasses which could only be successfully vitrified by splat or spray quenching could then be extensively manipulated in the vicinity of T_g without any crystallization occuring, whereas this apprently was not possible with the metallic splat-quenched glasses. What, INGRAM whished to know, was responsible for this difference. Dr. LUBORSKY observed that the crystallization in metallic glasses occured due to the fast diffusion of one or more species in the flass matrix, which Dr. INGRAM pointed out was also a feature of the fast ion conducting glasses in which he was interested. Evidently fast diffusion by itself is not sufficient to promote nucleation of the fast diffusion component. Dr. SCOTT said that in his experience of crystallization of metallic glasses, no nucleation stage was need because as far as their measurements could tell the nuclei were always already there. D. KÖSTER then noted that in the case of many of the metallic glasses, the phase being crystallized was characterized by extremely small unit cells ; for instance the case of iron-boron in which it is pure iron which is crystallizing. Pure iron has only two atoms in the unit cell so that the facility of creating a nucleus is such that it barely makes sense to talk about the need for critical nucleus anymore. Extending the subject, KÖSTER observed that they observed two cases : (a) crystallization via nuclei or protonuclei which had formed during the cooling and (b) crystallization starting from quenched-in nucleation "sites" which became activated on annealing. The nature of these "sites", however, remains unknown. Field ion microscopy studies, he noted, has shown the presence of large fluctuations which might be related.

Dr. R. ALMEIDA then raised the question of the extent of which nucleation in non-conventional glasses might be promoted by prior liquid-liquid phase separation in unconventional glasses. He noted that in ZrF_4 + ThF_4 binary glasses that he had studied, holding the glass significantly below T_g for a long time was sufficient to cause complete crystallization. C.T. MOYNIHAN said that there was no reason to expect any liquid-liquid phase separation to occur in any of the fluoride compositions currently under study, hence that nucleation in these glasses was unlikely to be related to this sort of phenomenon. The question was take up in a broader sense by A.F. WRIGHT who elaborated on the promotion of nucleation by concentration changes brought about by an initial liquid-liquid phase separation in the well studied case of $Li_2O.2SiO_2$ glass. Dr. WRIGHT then went on to pose the quasi-philosophic question of when was a phase separation distinguishably liquid-liquid rather than crystal-liquid, given the fact that "crystallization" could be carried out under conditions where no product particles larger in dimension than 10 - 15 Å was ever produced. Such "crystals" can barely be considered as crystals per se since the unit cells themselves are only of the order of 10 Å on a side in the cases in question. This difficulty thus implies that the initial stages of crystallization indeed involve a separation of amorphous clusters or protocrystals from the also amorphous matrix. This challenge led Dr.W.JOHNSON to raise the question of how the nucleation events should be properly viewed. He suggested that composition and topology should be regarded as independent dimensions in a configuration space in which the system moves along a curved trajectory during crystallization, i.e. the process should not be viewed independently as a composition change or a topology change in one sequence or another, but something more subtle and complex requiring at least two variables for its description.v variables for its descriptio

The question of sensitivity of experimental techniques for detection of the subtle changes involved in embryo crystal formation was the raised by Professor HENCH. He recalled investigations in which the initial stages of $Li_2O.2SiO_2$ glass had been studied by X-ray diffraction at small angles, dielectric relaxation and conductimetric analysis, and concluded that dielectric loss was the most sensitive. Due to the precipitation of a fast conducting crystalline phase, a Maxwell-Wagner-Sillars interfacial polarization could occur and loss peaks as large as $\tan \delta$ = 5-20 could arise allowing one to detect 0.01 vol% of crystallization. He wondered whether there were any significant developments in this respect in the last decade. The chairman responded that to the best of his knowledge, electrical, i.e. mass-transport related, detection was still the most sensitive, but, as Dr. HENCH had noted, it suffers from lack of direct structural information content. A technique which has the potential to be sensitive to extremely small quantities of crystalline material, and at the same time give direct information on local structure and its change during nucleation and growth is fluorescence spectroscopy, particularly with time resolution. Such measurements are actually in progress in the laboratory of Dr. CHAMPAGNON in Lyon. Dr. CHAMPAGNON then described some of his findings for the crystallization of Cr^{3+} containing glasses using the fluorescence method. He described the energy levels involved and the effects on fluorescence life time of different ordering processes, and then showed how the observations on nucleating chromium silicate glass implied that the prenucleation stage involved pairing or clustering of Cr^{3+} centers. The Round Table was brought to a close by some comments by Dr. CALLAS on EXAFS studies of nucleation in which changes in the environment of a dilute species in a crystallizing silicate glass can be monitored. Although both these latter studies were carried out on silicate glasses, hence where not strictly within the scope of the Round Table, they did illustrate the use of techniques of potential interest to those involved in refining our understanding of nucleation phenomenology.

* Purdue University, W-Lafayette IN 4707 (U.S.A.)

CHAPTER II: AMORPHOUS METALS

WHAT DO WE NEED TO KNOW ABOUT THE STRUCTURE OF AMORPHOUS METALS?

P.H.GASKELL

University of Cambridge,
Cavendish Laboratory,
Madingley Road,
Cambridge CB3OHE, U.K.

1. INTRODUCTION

The question posed in the title could be emphasised in several ways.

What do we need to know to really understand the structure of metallic glasses ("because it is there")?
Or, what must we know of the structure in order to control physical and chemical properties, devise new types of behaviour, predict new compositions?
A third and more comprehensive question could demand a definition of those essential structural characteristics of metallic glasses which offer an insight into the nature of the vitreous state itself.

Plainly, the questions become progressively more searching, and require a detailed analysis of individual experimental investigations and a synthesis of their conclusions. In what follows, a partial answer under each heading is attempted.

1.1 Is there a problem?

Metallic glasses are almost invariably disordered alloys, i.e. they are essentially mixtures of at least two different elements. If the structure can be adequately described as a random collection of atoms of two or more elements, then the problem largely disappears. The immediate neighbourhood of each atom - number of atoms of each element, the shape and symmetry of the nearest shell - is then determined principally by composition and radius ratio. We can say that neither chemical nor topological short-range ordering at the level of the local structure is significant. As will soon emerge, there is now adequate evidence that for certain glasses, the converse holds and chemical ordering in alloys of transition metals and metalloids - such as Ni-B, Pd-Si, etc. - is almost perfect. In Ni-B alloys, there is a strong preference for B atoms to be coordinated by a shell of Ni atoms, with B-B nearest-neighbours only at atomic concentrations greater than about 25%. Moreover, there is persuasive evidence for well-defined topological organisation of the nearest-neighbour Ni shell around B. These statements are not true in general though - certain alloys of two transition metals have been shown to have only slight chemical and (perhaps) no topological organisation.

Similar differences in behaviour may exist in the medium and long-range descriptions of structure, although here the evidence is less secure. Again, in certain TM-metalloid glasses, organisation appears to extend over distances of, perhaps, 1 - 2nm. - as shown by oscillations in reduced radial distribution functions for $Ni_{81}B_{19}$ (1) and $Mg_{85.5}Cu_{14.5}$ (2), for example. While such data appear to point towards a structure with more similarity to a micro-crystallite model than random packing, there is no reason to believe that this behaviour is general, as the absence of local organisation mentioned above reveals.

Adequate description of the structure of a metallic glass thus requires specification of the Local Structure - the presence or otherwise of well-defined first-neighbour coordination polyhedra - 'local structural units'- and the rules which govern packing of such units over distances in the range, say, 0.5 - 1.5 nm. - the Medium-Range Structure. Surprisingly rarely in metallic glasses (until recently, at least) has it been found necessary to define the extent of liquid-liquid immiscibility, i.e. to define the Long-Range Structure.

1.2 Structure and Properties

Forceful arguments could be made that there is no "need-to-know" the structure of amorphous materials in order to exploit their properties. Development of materials proceeds rapidly by experimental screening of materials and ad-hoc rules only are needed; structural work rarely has predictive value.

Although a good case can be made along these lines for oxide glasses, the argument is less easy to sustain for amorphous alloys. Partly this is due to the greater diversity of properties - electronic conductivity, amorphous magnetism, pseudo ductility and brittle behaviour in the same material under different conditions - and the fact that such macroscopic properties present severe challenges to the depth of our understanding.

Broadly, the electrical conductivity of a metallic glass, while clearly composition-dependent and affected by the presence of disorder, is only weakly structure-dependent. A detailed understanding may require structural data - for example an explanation of the temperature coefficient of resistance or the thermopower expressed in terms of the Faber-Ziman theory involves knowledge of the temperature-dependence of the first peak in the structure factor (3) but, as yet, a microscopic structure-based theory has not been devised.

Mechanical properties require structural information at all levels from local to long-range. For example, Fe-B bonding is clearly linked to the high hardness values observed in iron-boron glasses (4) indicating the importance of the local structure. Medium- and long-range structures reveal themselves in changes in ductility associated with annealing as measured, for example, by field ion microscopy (5). Finally, magnetic properties are so strongly coupled to details of the structure - at each level - that measurements of macroscopic properties represent a valuable source of structural data - a point to be elaborated in these Proceedings by Durand.

Atomic diffusion in the amorphous transition metal-metalloid glasses appears to reflect the strong metal-metalloid bonding and (perhaps) the well-defined local structure. Greer (6) has recently pointed out that the earlier ideas of fast metalloid diffusion by an interstitial mechanism and slow substitutional diffusion of the metal atoms are not supported by later data. Indeed the metal and metalloid appear to exhibit almost identical diffusion coefficients and activation energies suggesting some form of correlated motion.

Thermal properties represent a special case. Thermal expansion, conductivity and heat capacity are, like electrical resistivity, relatively insensitive to the detailed structure. In the region of the glass transition or crystallisation temperatures these quantities and derived thermodynamic parameters give sensible clues to the degree of disorder and indeed can be used to quantify amorphicity, but a detailed microscopic understanding of the collective motion involved in, say, the glass transition seems some way off. It should perhaps be said that the glass transition and the associated temperature and pressure-dependent changes in viscosity in amorphous metals are most simply expressed in terms of a free volume model in which atomic migration into adjacent vacant (sub-atomic) sites is explicitly considered. It is not however clear that this notion and others which derive from it -

pseudo-ductile behaviour due to dilatation in shear-bands for instance - necessarily contain a distance scale: such theories would appear to be valid if the moving unit is a cluster of atoms. Indeed similar models are envisaged for polymers in which the mobile species is a monomer segment (7). Consequently, the presence of any local structural units may in the future impinge via some suitably renormalized model.

2. LOCAL STRUCTURE

In this section an outline is given of the present state of knowledge of the local structure of certain amorphous alloys - chiefly the transition metal-metalloid glasses for which information is most complete. Specifically, we try to measure the local geometry around each atom site and to picture the local symmetry. Compositional ordering is considered implicitly.

2.1 Partial pair distribution functions

Structural information derives largely from measurements of the scattering of X-rays, neutrons or electrons and is conventionally represented by a so-called interference function $I(Q)$ or a structure factor $S(Q)$, Q being the modulus of a scattering vector $\underline{Q}$ ($Q = 4\pi\sin\theta/\lambda$ where 2θ is the scattering angle). Such data are difficult to interpret directly and may be transformed to a reduced pair distribution function:

$$G(r) = 4\pi r(\rho(r) - \rho_0) = \frac{2}{\pi} \int_0^\infty Q(S(Q) - 1) \sin QrdQ$$

Fluctuations in $G(r)$ reveal positive and negative deviations from the average atomic density, ρ_0, and the radial distribution function $J(r) = 4\pi r^2\rho(r)$ represents the probability of finding an atom at distance r from an origin atom and $\rho(r)$ is the atom density.

In a polyatomic amorphous solid, complete information requires the specification of the probability of finding an atom of species β around an origin atom α, that is, partial pair distri- bution functions (PPDFs) are needed: $G_{\alpha\beta}(r) = 4\pi r(\rho_{\alpha\beta}(r)/c_\beta - \rho_0)$.
The total function is then given by

$$G(r) = \sum_{\alpha,\beta} c_\alpha c_\beta \frac{f_\alpha(Q)f_\beta(Q)}{|<f(Q)>|} \cdot G_{\alpha\beta}(r)$$

Here c_α, $f_\alpha(Q)$ are the concentration and atomic scattering factors for atom α, $\rho_{\alpha\beta}(r)$ is the atomic density of β atoms at distance r from an origin α atom, ρ_0 is the mean atomic density and $|<f(Q)>| = c_\alpha f_\alpha(Q) + c_\beta f_\beta(Q)$.

Associated with each partial pair distribution function $G_{\alpha\beta}(r)$ is a partial structure factor $S_{\alpha\beta}(Q)$. Given a set of functions $G_{\alpha\beta}(r)$ or $S_{\alpha\beta}(Q)$, we have much of the information required to specify the local structure. However most scattering measurements only provide <u>total</u> functions $S(Q)$ as a weighted sum of partials and separation into three unknowns (the partials) from one set of data is not possible. Procedures for obtaining 'partials' are involved and will not be dealt with here - we merely point out that several measurements (three for a diatomic) allow an adequate separation in certain favourable cases. (See, for example, references 8 & 9). We proceed to consider the information content of those partial pair functions which have been adequately determined.

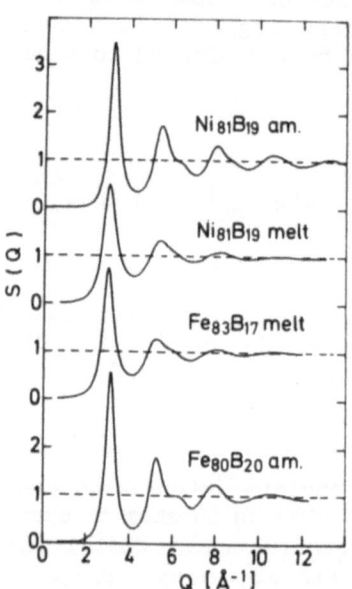 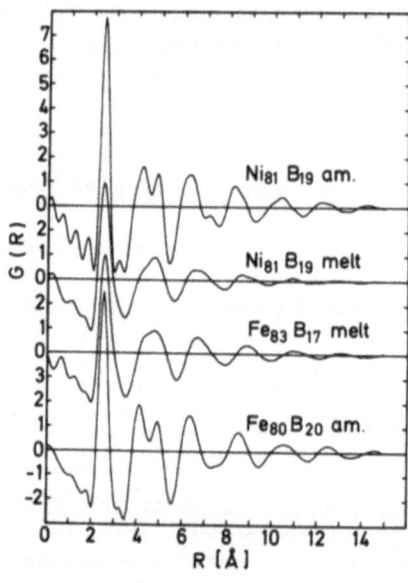

Figure 1. a) X-ray structure factors for Ni- and Fe-B melts and glasses and (b) corresponding total distribution functions (10)

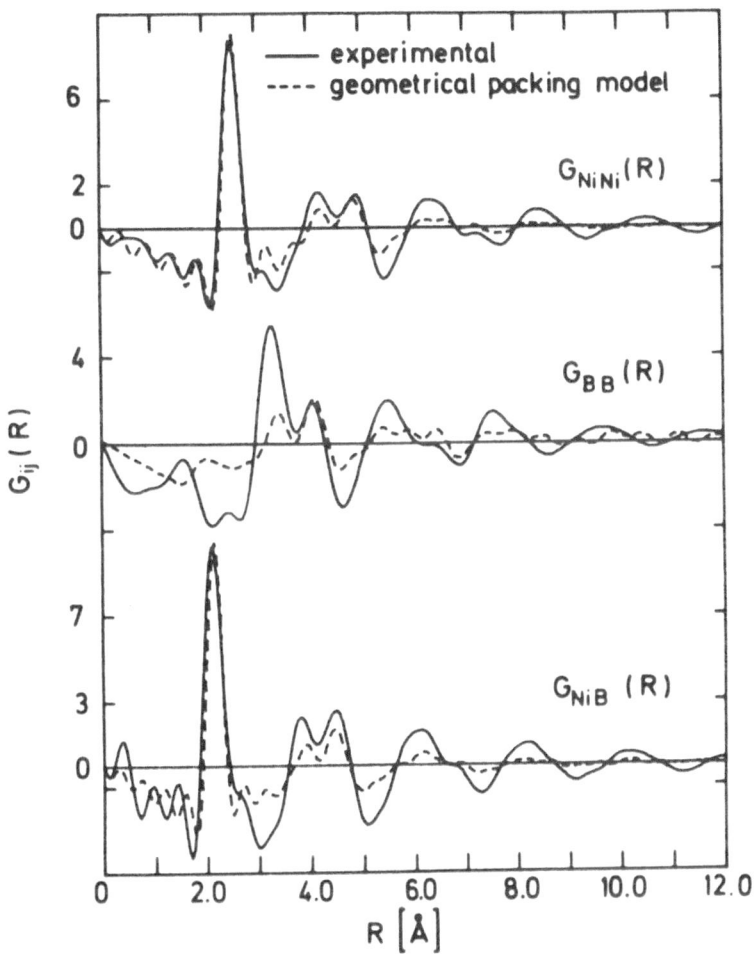

Figure 2. Partial pair distribution functions (PPDFs) for a-Ni$_{81}$B$_{19}$ obtained by neutron scattering from isotopically substituted samples (1). The dotted line shows PPDFs for a dense random packed model for the alloy.

2.2 Nickel-Boron

The total (Ashcroft-Langreth) structure factors for liquid and amorphous Ni- and Fe-B alloys (10) are shown in figure 1a and the corresponding <u>total</u> pair distribution functions in figure 1b. Note the characteristic differences in the second peak of G(r) between liquid and glass, which are not just manifestations of changes in amplitude of thermal vibrations but appear to reflect structural changes. PPDFs for a-$Ni_{81}B_{19}$ are shown in figure 2 and demonstrate several important features. The first B-B peak which would be expected at 0.17 nm (using values for the observed Ni-B and Ni-Ni distances - table 1) is clearly absent in $G_{BB}(r)$. The first B-B distance is found at about 0.3 nm. - that is, at somewhat larger distances than either the first neighbour Ni-B or Ni-Ni distances. This is a clear indication of the 'chemical' preference of boron for unlike neighbours in these alloys and appears to be a general rule (at least in dilute alloys) as indicated earlier by the classic work of Sadoc and Dixmier (11) on Co-P glasses.

The spatial extent of structural organisation in amorphous alloys is reflected, qualitatively, by the amplitude of oscillations in G(r) at high values of r. A quantitative guide can be obtained by a modification of the Scherrer formula for line-broadening in crystals to yield a correlation length, $\xi = 2\pi/\Delta Q$ where ΔQ is the breadth of the first peak of the structure factor, figure 1a. Values of ξ for molten and glassy forms of $Ni_{81}B_{19}$ are 0.78 and 1.19 nm. respectively (2). Values of the correlation length show considerable variations: for example for a-$Mg_{85.5}Cu_{14.5}$, $\xi \simeq 1.8$ nm, and similar values are found in $Mg_{84}Ni_{16}$ and $Mg_{30}Ca_{70}$.

Note also that the information content of partial distribution functions is much higher than that of the total G(r). Whereas the total function can be represented adequately by a number of qualitatively and quantitatively different atomic models, simulation of the partial functions represents a searching test of a model and to date the structure of $Ni_{81}B_{19}$ has not been adequately reproduced by any atomic model. The reason for this is not just that there is more information in the partials - three functions rather than one - but that the total function, being a weighted sum of partials, is degraded since peaks of one partial may overlap valleys of another.

This leads to a further general conclusion - the most secure structural interpretation must be based on experimental data of high information content and any 'successful' model should reproduce those data to the limits of accuracy both of experiment and theory.

This is a demanding constraint: in order to obtain generally valid conclusions we are forced either to extrapolate from about four satisfactory sets of experimental data or collect information in the way an insect builds up pictures of the world - by a compound, mosaic eye which views and integrates results of several structural techniques on, perhaps, different materials. (The message for young experimentalists is clear.)

2 3 Local structural information from PPDFs and other measurements

Table 1 collects a number of local structural parameters derived for amorphous TM-metalloid alloys by several structural techniques. The list is restricted to experimental results which allow the determination of PPDFs.

Alloy	Atom Pair	$\bar{r}_1$ (nm)	$\sigma_S(r_1)$ (nm)	N	Ref
$Co_{81}P_{19}$	P-Co	0.232	0.0105	8.9 ± 0.6	(11)
	Co-Co	0.254	0.0155	10.0 ± 0.4	
$Fe_{80}B_{20}$	B-Fe	0.214	0.0096	8.6	(24)
	Fe-Fe	0.257	0.0165	12.4	
$Ni_{81}B_{19}$	B-Ni	0.211	0.0142	8.9	(1)
	Ni-Ni	0.252	0.0147	10.5	
$Fe_{75}P_{25}$	P-Fe	0.238	0.018	8.1	(25)
	Fe-Fe	0.261	-ve	10.7	
$Pd_{84}Si_{16}$	Si-Pd	0.240	0.0106	9.0 ± 0.9	(26)
	Pd-Pd	0.276	0.0154	11.0 ± 0.7	

Table 1   Experimental mean, $\bar{r}_1$, standard deviation $\sigma_S(\bar{r}_1)$ and coordination numbers for nearest-neighbour metalloid-TM and TM-TM distributions in several amorphous alloys. Note, $\sigma_S(r_1)$ represents static and thermal broadening - termination broadening has been subtracted.

Several features deserve comment. (For a more detailed analysis see reference (9))

a) The coordination shell around the metalloid is approximately constant in the range 8.5 - 9.0.
b) The breadth of the distribution is represented by σ_S, the standard deviation of the distribution function (after correcting for artefacts due to truncation of the transform from scattering data) and is very narrow, suggesting a well-defined environment.

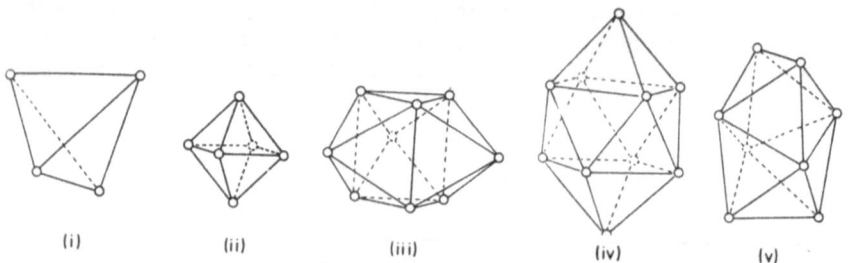

Figure 3. Polyhedra formed by packing of equal spheres. Note particularly the capped trigonal prism (iii) and Archimedean antiprism (iv) which form coordination shells for the metalloid in certain crystalline transition metal - metalloid alloys.

These facts suggest a parallel with oxide and other (partially) covalent glasses, in which the nearest-neighbour coordination shell is well-defined - a particular local structure is preferred for energetic or space-filling reasons. The structure is then described in terms of different ways of linking such units, e.g. corner-linked SiO_4 tetrahedra, and by the network statistics - the relative proportion of closed loops containing n such units. Moreover, the local units, if energetically preferred, would also be expected to form the basis for crystalline structures.

The suspicion that local structural units equivalent to those found in crystals form the basis for TM-M glasses is heightened by the fact that the crystalline local structural motif - predominantly a trigonal prism (figure 3) for a wide range of compositions and concentrations - contains nine metal atoms around the metalloid. Moreover attempts to reproduce the relatively constant coordination number by models with a random packing of atoms of two types have largely failed. Figure 4 shows the computed metalloid coordination numbers for models published up to mid-1982 (9) which follow closely a so-called staircase function giving the probability of packing N atoms around a central atom of radius ratio p. Random models thus produce a coordination number largely dictated by geometry alone. The situation has now changed somewhat with the recent publication (12) of a molecular dynamics simulation in which the coordination number of boron in Ni-B is found to be 8.6. It is not yet clear whether this is a general result or follows from the constraints imposed by containment of the fluid in a periodic box of limited size (2.8 nm.) under a hydrostatic pressure of some 'kilo- or mega-bars'.

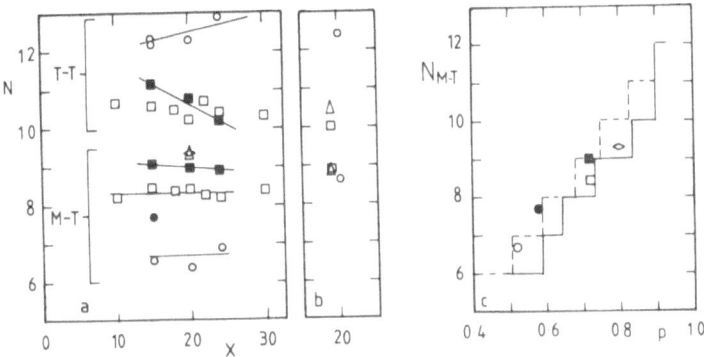

Figure 4. Calculated (a) and experimental (b) coordination numbers around the metalloid (M-T) and around the transition metal (T-T) in models for amorphous alloys. Lines merely connect points for compositionally-related systems. Note the relative constancy of the experimental coordination number in (b). (c) Comparison of the M-T coordination number for DRP models and the "staircase" curve giving the number of spheres surrounding a central sphere of radius ratio p. For further details, see reference (9).

2.4 Symmetry

Pair distribution functions indicate the <u>size</u> of the nearest-neighbour shell – coordination number and interatomic distance distribution – but contain no direct evidence on the <u>shape</u> of the polyhedra. Information on symmetry is, however, offered by some resonance techniques. Specifically, NMR or Mossbauer spectroscopy can provide a description of the symmetry of the environment surrounding the probe ion through the interaction between the nuclear quadrupole moment of the probe atom and the local <u>electric field gradient,(EFG).</u> The symmetry of the EFG reflects that of the local surroundings and can be represented in terms of the mean and standard deviation of the quadrupolar frequency v_Q, $\sigma(v_Q)$ and a so-called asymmetry parameter, η. The parameter v_Q is a measure of the departure from spherical or cubic symmetry whereas η measures the deviation from uniaxial local symmetry, for which $\eta = 0$.

Pannisod et al.(13) have shown that the local symmetry around the ^{11}B nucleus in $Mo_{70}B_{30}$, $Mo_{48}Ru_{32}B_{20}$ and $Ni_{78}P_{14}B_8$ is clearly related to the local structure found in crystaline phases of comparable composition. Moreover the distribution of local sites as measured by $\sigma(v_Q)$ was found to be comparatively narrow. More recent work by Pannisod, Bakonyi and Hasegawa (14) has thrown further light on the symmetry of the boron environment. NMR spectra were obtained for a series of ^{11}B-containing $Ni_{100-x}B_x$ glasses with x between 18.5 and 40% and were analysed by comparison with results obtained for $c-Ni_3B$ and orthorhombic Ni_4B_3 in which the nearest-neighbour polyhedron around B is a trigonal prism, and $c-Ni_2B$ composed of Archimedean antiprisms (fig.3). The uniaxial symmetry characteristic of $c-Ni_2B$ is not observed in the spectra of any glass; moreover it is possible to represent the composition-dependence in terms of a weighted sum of the environments of B in Ni_3B and $o-Ni_4B_3$. Further evidence pointing to the structural similarity of Ni-B glasses and that of the Ni_3B and Ni_4B_3 structures is obtained from the composition-dependence of the spin-lattice relaxation time. Data for glasses and crystals lie on a smooth curve with the exception of $c-Ni_2B$.

2.5 Summary

The conclusion stemming from the work to date is thus that the environment around B in Ni-B alloys is relatively well-defined. There is good evidence that the nearest-neighbour shell consists of nine Ni atoms (and no B atoms) and that the symmetry is strongly reminiscent of the trigonal prismatic environment found in certain crystalline borides. In this respect the picture of the structure of these amorphous metals strongly resembles that for the oxide and chalcogenide glasses, amorphous semiconductors, etc., that is, of a common local structural unit with differences in modes of packing distinguishing both the various crystalline polymorphs and differentiating the ordered and disordered states. These findings confirm earlier speculative suggestions on which a physical model for TM-metalloid glasses was built (15).

3. LOCAL STRUCTURE IN OTHER TYPES OF METALLIC GLASS

The picture is far more indistinct for metallic glasses other than amorphous TM-metalloid alloys. This is partly due to the relative paucity of data of sufficient quality - i.e. accurate partials, good NMR and Mossbauer data. Several clear guide-lines are emerging however.

Three recent papers list partial pair distribution functions for amorphous alloys of two transition metals. Lamparter et al. (16) have obtained data for $Cu_{57}Zr_{43}$ by isotopic substitution (of Cu), Fukunaga et al.(17) have studied $Ni_{40}Ti_{60}$ using ^{60}Ni and Lee et al.(18) have used combinations of X-ray and neutron scattering to extract partials for $Ni_{35}Zr_{65}$. In the Cu-Zr alloy the errors associated with separation of the partials were too great to allow any definite conclusions to be reached other than a strong tendency towards chemical ordering of the two species. $Ni_{40}Ti_{60}$ and $Ni_{35}Zr_{65}$

alloys were however considered to show similar local structure to those of crystalline phases NiTi$_2$ and NiZr$_2$ respectively. Similar conclusions were deduced from <u>total</u> pair distribution functions of Mg$_{85.5}$Cu$_{14.5}$ and Mg$_{84}$N$_{16}$ (2). There is adequate evidence in EXAFS work too for chemical ordering and for similarity of local structure in amorphous and crystalline phases - as shown in the work of the Orsay group (see for example reference (19)).

4. LOCAL ORDER AND THE INFLUENCE OF MEDIUM-RANGE STRUCTURE

It certainly cannnot be claimed that the evidence suggests any general equivalence in the local structures of crystals and glasses comparable to that found in the specific case of the TM-M glasses (indeed there is a greater weight of uncommitted and some opposed data). There is nonetheless a strong indication that in <u>certain</u> glasses (of all types) forces which produce locally-ordered structures in periodic assemblies of atoms are at work in aperiodic structures as well. In those cases where it is observed, the equivalence of local structures in crystals and glasses offers a simplifying principle. However it also raises what might perhaps be the <u>largest</u> structural question - why such preferred local structures <u>are preferred</u> and why the equivalence only applies in a limited number of cases, e.g. equivalence between a-Ni-B alloys and lattices containing trigonal prisms but not Archimedean antiprisms. A second question must be: does not the equivalence of local structures in crystal and glass suggest the presence of medium-range order and, by extension, that glasses are merely microcrystalline solids?

At present, there are no compelling reasons for believing that trigonal prismatic units - or indeed, nine-coordinated shells - are necessarily preferred either for energetic or space-filling reasons - unlike SiO$_4$ or BO$_3$ polyhedra where strong directional 'covalent' forces are assumed to stabilize the local cluster. It is difficult to see how the B-Ni bond, although undoubtedly strong, could be directional to any degree. Boron has no available d orbitals to form the basis for suitable 'hybrid' orbitals. Nonetheless, trigonal prisms represent the dominant motif in crystalline borides, phosphides, etc. over a wide range of composition, radius ratio, etc. In certain borides, both boron and metal atoms are found at the vertices of trigonal prisms.

The only satisfactory rationalisation appears to be that in crystals, at least, trigonal prisms occur because metalloid-centred trigonal prisms are (almost) large enough to accommodate the metalloid and that correlated arrangements of prisms allow a high fraction of close TM-TM contacts. In the Ni$_3$B structure for example, boron atoms lie in planes — so-called chemical twinning planes - and between the planes Ni atoms are close-packed, Figure 5. The lattice thus allows a high TM-TM coordination number with short average TM-TM bond length <u>and</u> a high coordination number for the boron atom

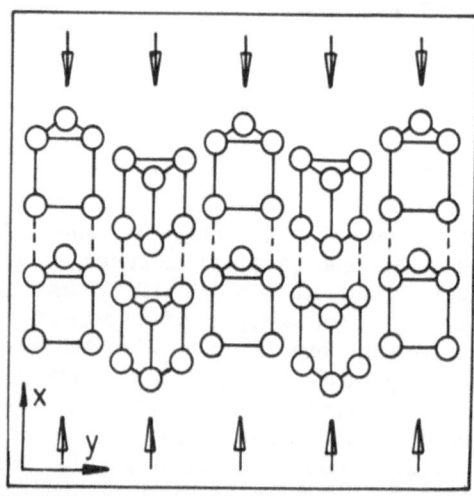

Figure 5. Trigonal prisms of transition metal atoms in the Fe_3C (cementite) lattice. Arrows show chemical twinning planes along which the metalloid atoms (not drawn) are located.

(even though boron is somewhat too large to fit within an undistorted prism). In crystals, therefore, trigonal prisms seem to owe their widespread occurrence to a mutually advantageous compromise between the separate bonding requirements of the transition metal and metalloid elements leading to a correlated arrangement of atoms of both species.

If trigonal prismatic coordination really is the dominant atomic arrangement in glasses, then we are forced to consider the possibility that the rules which guide this preference in the amorphous state may be closely similar, if not identical, to the rules appertaining in crystals.

Such speculation leads to the view that <u>local</u> trigonal prismatic coordination implies a <u>correlated</u> <u>medium-range</u> structure also.

There is a certain amount of direct experimental support for this assertion. High voltage, high resolution electron microscopic investigations (20) have shown the presence of domains of about 2nm in diameter within which atom sites appear to be correlated. Specifically, well-defined lattice fringe images were observed in the thinner parts of a specimen of an amorphous Pd-Si alloy. In certain regions, the structure appeared to be similar to images of multiply-twinned particles. More secure information (since it evades any possible suspicion of artefacts resulting from the relative thinness of the specimens necessary for high resolution imaging) comes from the neutron-scattering data on $Ni_{81}B_{19}$ (1) - particularly the oscillations in the B–B partial which extend to at least 1.2nm. If we assume that this structure is composed of local structural units then $G_{BB}(r)$ describes the correlation between the centres of such units. A random array of trigonal prisms certainly does not produce oscillation at high r values (1,15)(see also Figure 2). Consequently, the B-B partial indicates some relatively extensive organisation, again on a scale of perhaps 1-2 nm.

Corroborative evidence comes from the recent investigations of Dubois and co-workers (21). Models have been built for amorphous transition metal-metalloid alloys which contain a number of 1.2-2.0 nm. domains. Within each domain, atomic positions are correlated - specifically, chemical twinning planes have a common orientation. At the boundaries, atoms conform to the structural rules associated with two or more chemical twinning planes. The resulting structure consists of a series of topologically ordered domains with a local structure centred on the metalloid which is essentially equivalent to that of a crystal but with a medium-range structure which need not be that of any crystalline lattice. The medium-range structure of Dubois et al.'s model for 80:20 glasses is based on correlated domains of the same composition even though the crystalline TM_4M phase has not (yet) been identified experimentally. It is important to note that the positional constraints placed on atoms forming the boundaries between domains in these models are even more restrictive than those for atoms in the centre of domains. This represents an important contrast to microcrystallite models which, as usually defined, are built on the assumption of randomly-oriented domains.

Physical models built according to these principles were energy-minimised using parameters scaled for Ni-B and the resulting pair distribution functions computed. Experimental and calculated data for $Ni_{81}B_{19}$ and $Pd_{80}Si_{20}$ are shown in Figures 6 and 7.

Agreement is, on the whole, good - particularly for the Ni-Ni and Ni-B partials. However, a fit to these data does not provide a sufficiently discriminating test of a model and, paradoxically, the somewhat poorer agreement of the B–B partial provides more concrete evidence in favour of a correlated structure for this alloy. It should perhaps be emphasized that a good fit to the same experimental data has also been achieved by the molecular dynamics simulation of Beyer and Hoheisel (12) mentioned earlier.

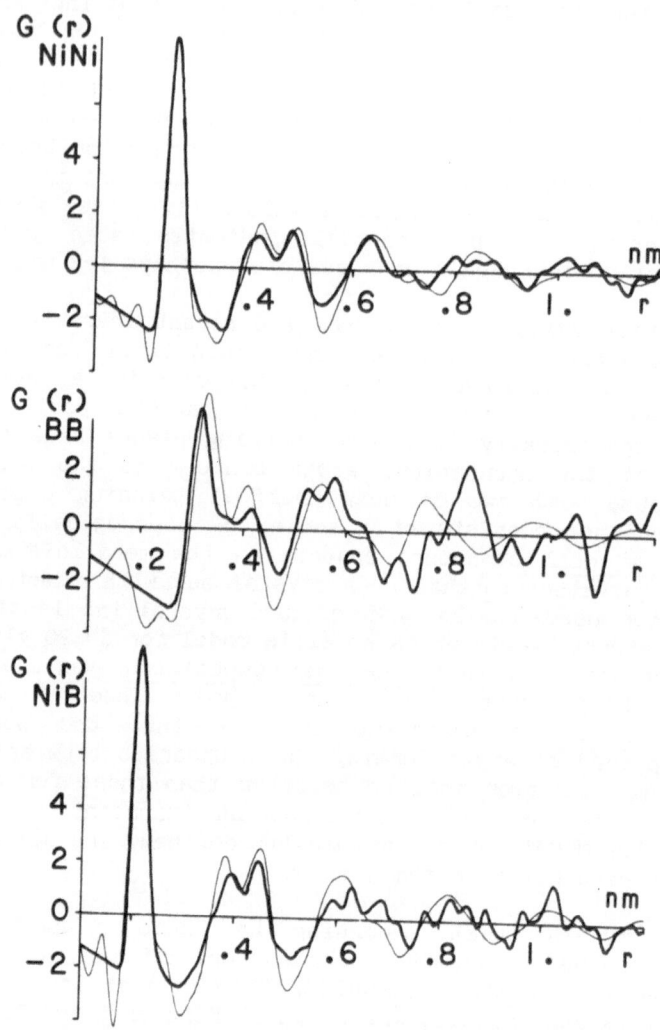

Figure 6. Experimental partial pair distribution functions (thin line) for a-Ni$_{81}$B$_{19}$ (1) and PPDfs calculated using a model constructed by Dubois et.al. (21) consisting of domains of size 1.2 nm.

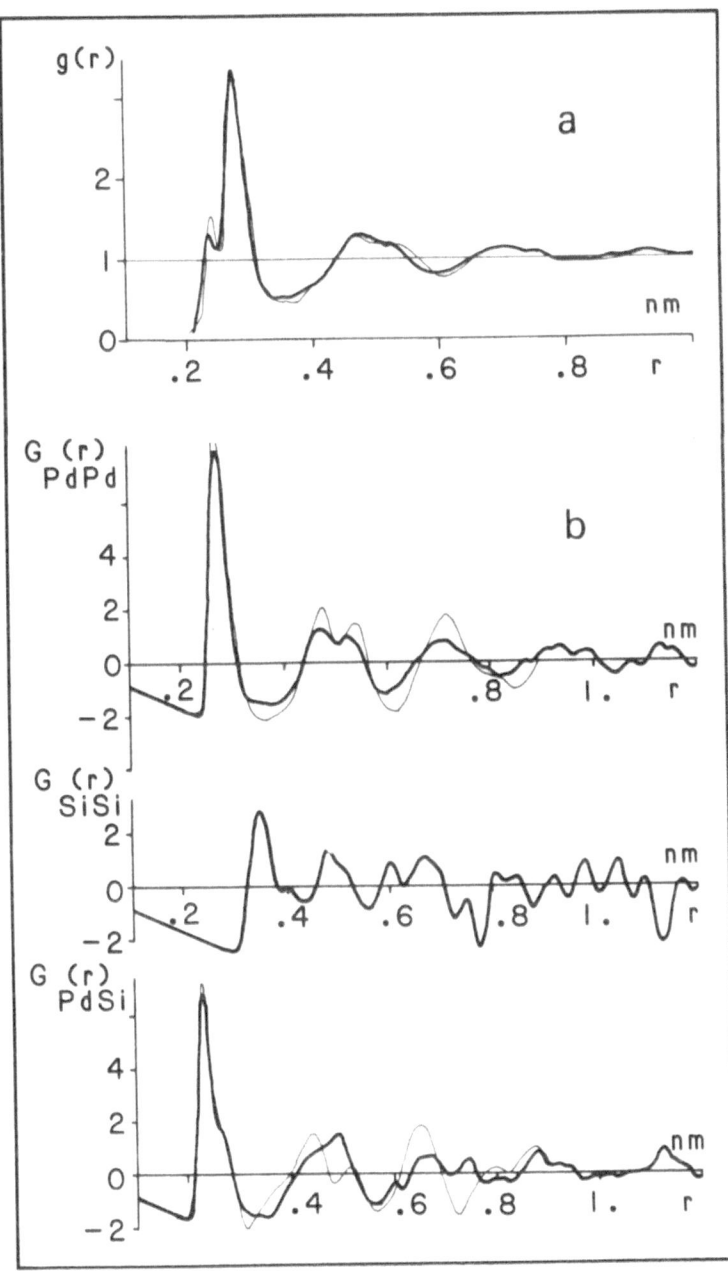

Figure 7. Total (a) and partial (b) distribution functions for a-$Pd_{80}Si_{20}$ calculated for the model of Dubois et. al. (21) with domains of size 1.5 nm. Experimental data of Sadoc and Dixmier (26) is shown by the thin line.

5   SUMMARY

The evidence surveyed in this review suggests that for many metallic glasses the structure is represented by some non-random arrangement of atoms. Certainly "chemical short-range ordering" is common. There is also an increasing body of evidence suggesting that there may be some similarity between the local structure of glasses and appropriate crystalline phases. In a very small number of cases, sufficient evidence exists to support the notion of a well-defined local structure which is also characteristic of the crystalline state. A more speculative argument leads to the proposition that the occurrence of a particular local structure may reveal a more pervasive pattern of structural organisation which extends into medium-range order. Surprisingly, perhaps, some direct experimental data (and a greater amount of circumstantial evidence not quoted here) support this viewpoint.

In conclusion, it should be mentioned that metallic glasses are not the only representatives of the class of close-packed amorphous solids. Fluoride glasses appear to be close-packed also and even in the alkali and alkaline earth silicates, which might be considered to be prototypical network structures, the environment of the 'network-modifying' cations closely approximates to dense-packing and it is arguable that the space-filling requirements of these atoms determines the configuration of the network (22,23) and not the reverse as is commonly assumed.

ACKNOWLEDGEMENT

Generous   support   from   Pilkington   PLC   is   gratefully acknowledged.

REFERENCES

1) Lamparter,P., W.Sperl, S.Steeb and J.Bletry Z.Naturforsch. 37a (1982) 1223
2) Nassif, E., P.Lamparter and S.Steeb ibid 38a (1983) 1206
3) Altounian, Z., C.L.Foiles, W.B.Muir and J.O.Strom-Olsen Phys. Rev. B 27 (1983) 1955
4) Joyner, D. J., O.Johnson, D.M.Hercules, D.W.Bullett and J. H. Weaver. Phys. Rev. B24 (1981) 3132
5) Piller, J. and P.Haasen Acta Met. 30 (1982) 1
6) Greer, A.L. J.Non-Cryst. Solids 61 & 62 (1984) 737
7) Adam, G. and J.H.Gibbs J.Chem.Phys. 43 (1965) 139
8) Steeb, S. in "Structure of Non-crystalline Materials 1982" eds. P.H.Gaskell E.A.Davis and J.M.Parker, London: Taylor and Francis, pp 441 1982.

9) Gaskell, P. H. in "Glassy Metals II " eds. H.Beck and H.-J.Guntherodt Berlin: Springer-Verlag 53 5 (1983)

10) Nassif, E., P.Lamparter, B.Sedelmeyer and S.Steeb Z.Naturforsch . 38 (1983) 1098

11) Sadoc, J. F. and J Dixmier Mater. Sci Eng. 23 (1976) 187

12) Beyer, O. and C.Hoheisel Z. Naturforsch. 38a (1983) 859

13) Pannisod, P., D.Aliaga Guerra, A.Amamou, J.Durand, W.L.Johnson, W.L.Carter and S.J.Poon. Phys. Rev. Lett. 44 (1980) 1465

14) Pannisod, P., I.Bakonyi and R.Hasegawa Phys. Rev. B28 (1983) 2374

15) Gaskell, P. H. J.Non-Cryst. Solids 32 (1979) 207

16) Lamparter, P., S.Steeb and E.Grallath Z.Naturforsch. 38a (1983) 1210

17) Fukunaga, T., N.Watanabe and K.Suzuki J. Non-Cryst. Solids 61 & 62 (1984) 343

18) Lee, A., G.Etherington and C.N.J.Wagner ibid 61 & 62 (1984) 349

19) Sadoc, A., D.Raoux, P.Lagarde and A.Fontaine ibid 50 (1982) 331

20) Gaskell, P.H. and D.J.Smith J. of Microscopy 119 (1980) 63

21) Dubois, J.M., P.H.Gaskell and G.LeCaer to be submitted

22) Dent-Glasser, L. S. Z. Krist. 149 (1979) 291

23) Gaskell, P.H. J.de Physique C9 43 (1982) 101

24) Nold, E., P. Lamparter, H. Olbrick, A. Rainer-Harbach and S. Steeb, Z. Naturforsch. 36a (1981) 1032

25) Waseda, Y., H. Okazaki and T. Masumoto in 'Structure of Non-Crystalline Materials', ed. by P. H. Gaskell (London: Taylor & Francis 1977) p 89

26) Sadoc, J. F. and J. Dixmier, ibid. p 85 (1977)

DIFFRACTION AND X-RAY SPECTROSCOPY FOR THE EVALUATION OF STRUCTURE OF AMORPHOUS BINARIES

S. Steeb

MAX-PLANCK-INSTITUT FUER METALLFORSCHUNG, SEESTR. 92, 7000 STUTTGART 1

The evaluation of partial structure factors by means of diffraction methods is shown especially by presenting the example of amorphous $Ni_{81}B_{19}$. Three models up to now were presented to reproduce the experimental data. Since there is a lack of information concerning the symmetry of nearest neighbour's arrangement, the use of X-ray absorption methods, namely EXAFS and furthermore XANES became more and more important. Using an electron microprobe, XANES spectra can be obtained in the laboratory. The sensitivity of the method is shown presenting the example of amorphous $Fe_{90}Zr_{10}$ once without and once with 10 % hydrogen.

1.   DIFFRACTION METHODS

1.1 Small angle scattering

Small angle scattering using X-rays and neutrons allows to reveal the medium range structure with dimensions up to 10000 Å. Using neutrons, magnetic domains also can be revealed; normally, however, features in the structure which differ in scattering length density $b \cdot \rho$ from their surroundings can be recognized. Using X-rays in the small scattering region, domains with an electron density different from the surroundings are detected. Fig. 1 shows an overview on the scattering behaviour as known up to date for amorphous $Fe_{80}B_{20}$. As usual, the intensity is plotted versus the momentum transfer $Q = 4\pi(\sin\theta)/\lambda$.

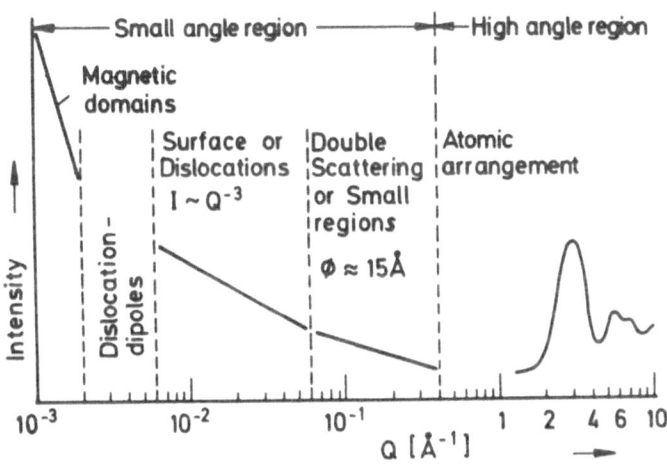

Fig. 1. Amorphous $Fe_{80}B_{20}$: Scattering behaviour

1.2 High angle scattering

Using high angle scattering, the atomic arrangement, the atomic
distances, the nearest neighbour numbers, and the short range
order parameters can be obtained. Most important are the topological arrangement, i. e. the occupation of distinct points by any
atoms regardless to which kind they belong and the chemical
arrangement, i. e. the question by which kinds of atoms are the
topological points occupied. From the experimental intensity after
corrections for background, multiple, and incoherent scattering
as well as normalization to absolute units one obtains finally the
so called total structure factor $S(Q)$. This function yields the
total pair correlation function $G(R)$ and the total number density
$\rho(R)$ by Fourier transform. From $\rho(R)$ one easily obtains the radial
distribution function $4\pi R^2 \rho(R)$ from which distances and coordination numbers can be evaluated.

1.3 Partial functions

The structure of amorphous binaries only can be revealed using the
partial functions, i. e. partial structure factors, partial pair
correlation functions, partial coordination numbers etc. This can
be done according to a more chemical description following Faber
Ziman or to a more physical description following Bhatia Thornton
according to the following scheme:

FABER ZIMAN $S = [1 - c_A c_B (\Delta b)^2] / <b>^2$

$$S = \frac{c_A^2 b_A^2}{<b>^2} \cdot S_{AA} + \frac{c_B^2 b_B^2}{<b>^2} \cdot S_{BB} + \frac{2 c_A c_B b_A b_B}{<b>^2} \cdot S_{AB}$$

$$= W_{AA} \cdot S_{AA} + W_{BB} \cdot S_{BB} + W_{AB} \cdot S_{AB}$$

BHATIA THORNTON $S' = 1 / <b^2>$

$$S' = \frac{<b>}{<b^2>} S_{NN} + \frac{(\Delta b)^2}{<b^2>} \cdot S_{CC} + \frac{2 <b> \Delta b}{<b^2>} \cdot S_{NC}$$

$$= W_{NN} \cdot S_{NN} + W_{CC} \cdot S_{CC} + W_{NC} \cdot S_{NC}$$

Normalized determinant of the coefficients W_{ij} : $D = 0 \ldots 1 (D_{max})$

METHOD	RADIATION	SCATTERING FACTORS b	D
→ Isotopic substitution	3 x N	$^{nat}Ni : 1.03$, $^{62}Ni : -0.87$, $^{60}Ni : 0.28$	good (0.5) ←
Magnetic scattering	3 x N	Co : 0.25, 0.61, -0.11 (magnetic field)	good
Three beam experiment	N + X + E	$b(N) \neq b(X) \neq b(E)$, but $b(X) \approx b(E)$	bad
Two beam experiment	N + X	$b(N) \neq b(X)$ (e.g. $b(N) < 0$, $b(X) > 0$)	good
Anomalous dispersion	3 x X	$b(Mo-K_\alpha) \neq b(Cu-K_\alpha) \neq b(Co-K_\alpha)$	very bad (10^{-4})
Isomorphous substitution	3 x N, 3 x X	A replaced by A^* (e.g. Zr ⟷ Hf)	good

For the determination of the three partial structure factors one needs three scattering experiments.

1.4 Example: Amorphous $Ni_{81}B_{19}$

The partial structure factors with this alloy were determined using three neutron diffraction experiments and the method of isotopic substitution. As one knows, the scattering length of an isotopic mixture can be calculated by a weighted sum of the scattering lengths of the single isotopes. Since also negative scattering lengths occur in nature one can for example produce zero scattering nickel (ØNi). Using this ØNi and B one obtains an alloy which yields directly from one neutron scattering experiment the B-B-correlation which was discussed in (1).

1.5 Computer modelling of the G_{ij}

The partial correlation functions G_{ij} with amorphous $Ni_{81}B_{19}$ up to now were modelled three times, namely by the Blétry model using single atoms (1), by the Dubois model using trigonal prisms (2), and by a molecular dynamic calculation from Beyer (3). It is to be expected that further approaches will be published in future and we feel the one dimensional result in R-space which one obtains from diffraction experiments not to be as comprehensive at is should be for the description of the atomic arrangement. There is a lack of information concerning the symmetry, i. e. the triplet correlation functions and we think that this could be overcome using X-ray spectroscopic methods.

2. X-RAY-SPECTROSCOPIC METHODS

2.1 Determination of the density of unoccupied and the density of occupied states

If an X-ray quantum is absorbed, normally an electron is emitted

from the K-shell and it occupies a normally unoccupied state. Thus
by X-ray absorption spectroscopy (XAS) we obtain the density of
unoccupied states. Normally the hole on the K-shell is then filled
up by an electron emerging from a normally occupied state. Thus by
X-ray emission spectrosopy (XES) we obtain the density of occupied
states. To become independent from various time consuming procedures
you must be able to do your experimental work in your own laboratory.
This is the reason why we perform the XAS- and XES-work by means
of an fully automatized electron microprobe (Type JXA 733; Jeol,
Tokyo). The emission spectra obtained with two different primary
electron energies yield according to Fig. 2 the absorption spectrum.

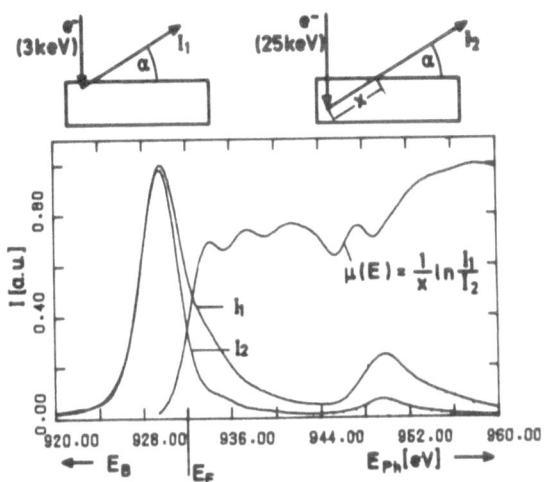

Fig. 2. Upper diagrams: schematic representation of beam geometry.
Lower diagram: Cu-$L_{\alpha\beta}$-spectrum; Emission line at 928 eV and
absorption spectrum μ(E)

2.2 X-ray Emisssion spectrum

From the emission spectrum one obtains according to (4) the charge
transfer which represents a quantitative measure for the strength
of binding. Furthermore the position and shape of the emission line
allows conclusions on the changes in chemical binding during transitions for example from the amorphous to the crystalline state. In
all cases investigated so far a decrease of the binding energy of
the electrons was observed (5,6), going from amorphous to crystalline state.

2.3 X-ray absorption spectrum

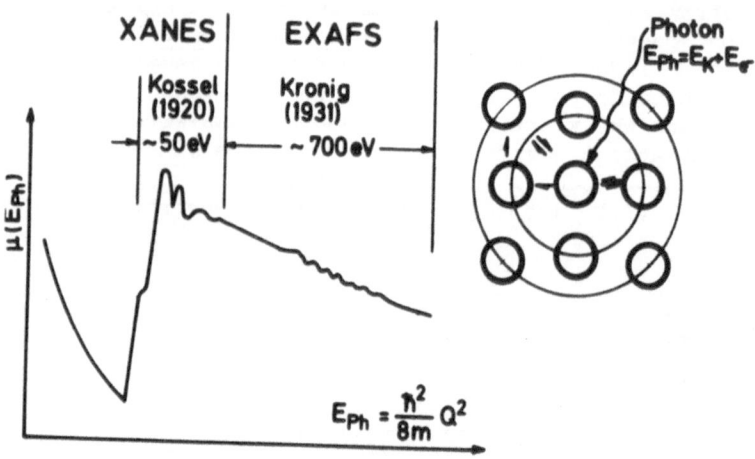

Fig. 3. a) Energy dependency of the X-ray absorption coefficient μ.
b) Mechanisms for the explanation of EXAFS and XANES

Fig. 3 shows on the left hand side in a schematic presentation the
run of the X-ray absorption coefficient μ versus the quantum energy
E_{ph} in the surrounding of one of the well known absorption edges.
The fine structure in the run can be subdivided in a more or less
arbitrary way in a far edge region and a near edge region. The
first one was treated by Kronig and is nowadays called Extended
X-ray Absorption Fine structure (EXAFS), the second one was treated
by Kossel and is now named X-ray Absorption Near Edge Spectroscopy
(XANES). The right hand side of Fig. 3 shows a central atom
surrounded by two coordination spheres. EXAFS can be explained by
a pair-wise interaction of nearest neighbour pairs. Thus, in principle, cannot yield more information than a diffraction experiment.
XANES, which occurs at lower Q's must have its origin in an interaction between atoms with larger distance and thus works in such
a way that the original K-electron which is pushed away during the
X-ray absorption process is scattered at least twice before inter-
acting with the original electron wave. Thus for the explanation
of XANES at least triplets play an important role and thus by this
method we should be able to learn more concerning the atomic arrangement as is the case by diffraction methods merely.

2.4 Example: Amorphous $Fe_{90}Zr_{10}$ with and without hydrogen

Fig. 4 shows the XANES-spectrum obtained recently (5) using an electron microprobe.

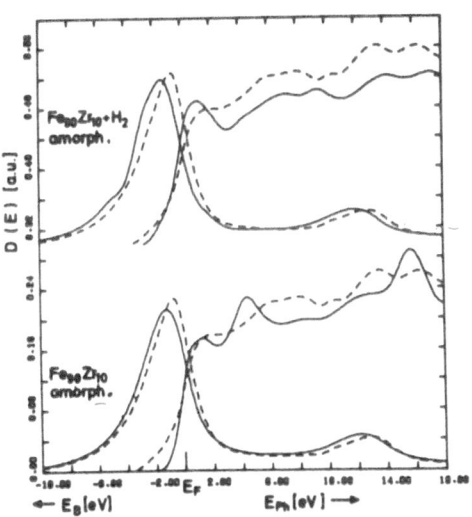

Fig. 4. Amorphous $Fe_{90}Zr_{10}$ (+10 a/o H_2) Fe-L-spectrum
--- cubic body centered (ferromagnetic) Fe

In the lower part we recognize in the left the density of occupied states (d-band) and in the right the density of unoccupied states. For comparison the spectrum of cubic body centered iron is presented as dashed line. The upper part shows the corresponding curves for a-$Fe_{90}Zr_{10}$+10 a/o H_2. Apparently the amount of electrons with larger binding energies E_B increases. In the right hand side of the spectrum we feel that the shape of the absorption spectrum has become more similar to that of b_{cc} Fe which is ferromagnetic. Thus one could understand why the Curie temperature of a-$Fe_{90}Zr_{10}$ increases drastically by the hydrogen treatment. And this only by the fact that the hydrogen atoms bring with them electrons. At the same time we learn by diffraction measurements that there is no change in the intensity curves during the hydrogen treatment.

3. SUMMARY

Concerning the structure of amorphous binaries, the following
items can be stated:

i)   The binding energy of the electrons normally is stronger
     than within the corresponding components.
ii) The chemical short range order parameter with amorphous bina-
     ries up to now always indicated compound formation and often
     reaches 100 %, i. e. perfect ordering.
iv) Regions with diameters of 15 to 20 Å seem to form structural
     units which are packed together as dense as possible.
v)   Using the method of isotopic substitution and neutron diffrac-
     tion it is possible to obtain the partial functions.
vi) Since the partial functions given by diffraction methods only
     yield pair (and not triplet) correlations, a variety of models
     describes the experimental results well.
vii) To limit the number of convenient models, triplet correlations
     must be introduced which also should be verified by experi-
     mental methods, i. e. XANES.

1.   Lamparter, P., Sperl, W., Steeb, S., and J. Blétry.
     Z. Naturforschung 37a (1983) 325-328

2.   Dubois, J.M, and P. Gaskell. Private communication

3.   Beyer, O, and C. Hoheisel. Z. Naturforschung 38a (1983)
     859-863

4.   Wenger, A., and S. Steinemann, Helvetia Physica Acta 47
     (1974) 321-325

5.   Falch, S., Thesis work, University Stuttgart, 1983

6.   Falch, S., Rainer-Harbach, G., Schmückle, F., and S. Steeb,
     Z. Naturforschung 36a (1981) 937

INFLUENCE OF SOLIDIFICATION PARAMETERS AND RELAXATION
ON PROPERTIES OF METALLIC GLASSES

Uwe Köster

Dept. Chem. Eng., University Dortmund
D-4600 Dortmund 50, F.R.Germany

From silicate glasses it is well known that the glass transition temperature T_g is influenced by the quenching conditions as well as annealing thus changing the glass structure and a number of properties. Higher quenching rates for example result in a higher glass transition temperature, but e.g. smaller density. Some controversy about actual numbers of physical properties of metallic glasses may be just due to differences in solidification parameter.

Amorphous materials can be obtained by quenching from their melts or by one of the techniques of deposition onto cooled substrates, for example by vacuum evaporation, cathodic sputtering, electrodepostion, or electron beam and laser surface melting. Amorphous solids of the same material prepared by different techniques or under different conditions show significant differences in many properties. But it seems doubtful that there is any fundamental microscopic distinctions between amorphous solids prepared by different techniques.

Whichever way it is prepared, an amorphous solid is not in configurational equilibrium, but is slowly relaxing by a homogeneous process toward an "ideal" metastable amorphous state. Relaxation toward this ideal amorphous state is distinct from heterogeneous crystallization which results in the stable crystalline state of the material. Real amorphous solids can be assumed to include structural defects with a so far unknown structure and compositional fluctuations. Apart from the type, distributions and the concentrations of these defects, the structure of amorphous solids prepared by different techniques or conditions appear to be essentially similar. In another concept such a microscopic state of a real glass can be characterized by excess free volume or in more detail by the atomic short-range order /1/, which includes both chemical and topological short-range order of the glass.

INFLUENCE OF SOLIDIFICATION PARAMETERS:

Metallic glasses usually form only by rapid solidification techniques with quenching rates of 10^5 K/s or more. Whereas planar flow casting /2/ is the method for industrial large scale production of wide strips of metallic glasses, most labs on the other hand can effort only small amounts of materials and prepare ribbons by melt spinning (see fig. 1).

As important parameter for melt spinning which may influence not only the geometry, but also a number of properties, are known: The melt temperature T, ejection pressure p_e, orifice diameter $\emptyset$, orifice-substrate spacing a, angle of inclination α, surface speed u of the quenching wheel, as well as temperature, material and diameter of the wheel 2R, ambient atmosphere and pressure.

The width and thickness of glassy ribbons have been observed to increase with superheating of the melt and decreases with increasing ambient pressure and substrate velocity, the thickness always behaves the contrary /3,4,5/. Liebermann /3/ investigated the effect of the ambient gas on the dimensions of the glassy ribbon and found that there is a critical gas boundary layer Reynolds number below which a smooth edge profile is obtained.

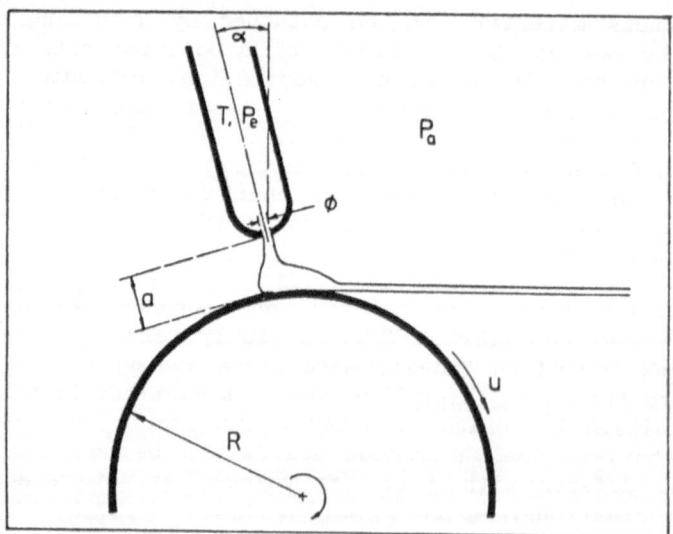

Fig. 1: Schematic sketch of the melt-spinning process indicating the variety of casting parameters.

Ribbons produced by melt spinning exhibit smooth top surfaces (free surfaces), but the contact sides are characterized by very typical microtopological asperities (see fig. 2); e.g. large lift-off (non-contact) areas due to gas entrapment under the solidifying material. Reducing the pressure as well as the substrate velocity, both will lead to smoother contact surfaces reducing the the non-contact areas and the mount of whirl streets along grooves. These whirl streets may be caused by liquid flow instabilities such as capillarity wave or Marangoni instability.

The best surface conditions for (Fe,Ni)B-glasses have been found after casting at 200 mbar He and 20 m/s substrate velocity. Slower velocities may produce ribbons with further improved surfaces, but result in partially crystallized glasses.

Surface asperities as described above are assumed not only to influence mechanical properties but also a number of other properties, e.g. pinning of domain walls in magnetic applications.

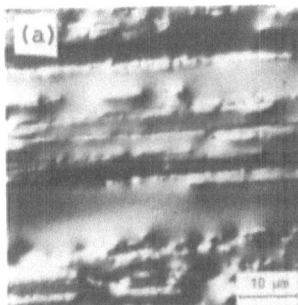 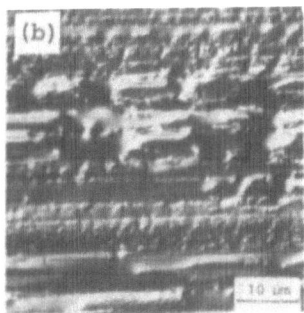 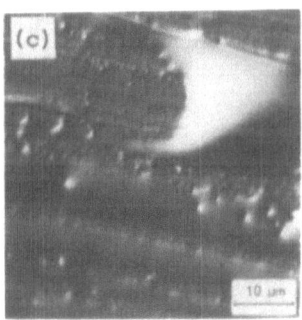

Fig. 2: Influence of substrate velocity u and pressure conditions on the topography of the contact side for glassy $Fe_{40}Ni_{40}B_{20}$ ribbons /5/: (a) 20 m/s; 200 mbar He; (b) 40 m/s; 200 mbar He; (c) 40 m/s; 1 bar air.

As shown in fig. 3a tensile strength R_m and ductility have been found to be very sensitive against casting conditions. These changes in strength may be due to structural changes reflecting different quenching rates and surface irregularities even in edge-polished ribbons. Higher cooling rates due to faster substrate velocities and/or lower melt temperatures (due to the smaller amount of heat which has to be removed) result in a larger amount of free volume which leads to better ductility. On the other hand, very fast substrate velocities will increase surface roughness and serration of the edges thus resulting in a higher number of stress raisers.

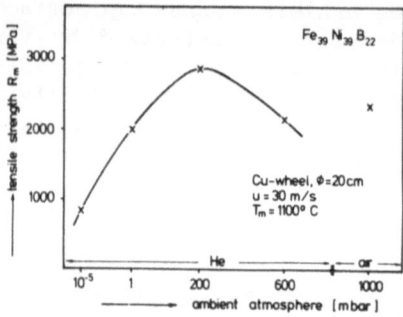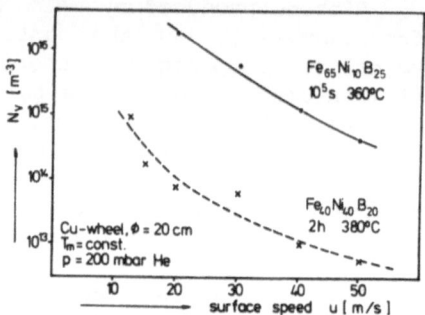

Fig. 3: Influence of solidification parameters on properties of metallic glasses: (a) tensile strength R_m versus ambient atmosphere; (b) number of quenched-in nucleation sites N_v versus substrate velocity u.

With the helium pressure the quenching rate after leaving the wheel increases due to the increasing thermal transport by convection thus improving ductility; at relatively high pressures, however, an increasing number of surface irregularities will overcompensate this effect in the ductility.

In addition, the duration of contact with the quenching wheel is of large importance and is assumed to be influenced by the wheel diameter; on loss of contact the cooling rate decreases catastrophically, depending on thermal conductivity, pressure and relative velocity of the ambient atmosphere. The contact time will determine the state of quench, since a glass that leaves the wheel at higher temperature will tend to undergo more relaxation during the low cooling rate afterwards.

Due to the high cooling rates required to form metallic glasses the effective quench rate decreases even over the thickness of the ribbon, leading to local structural changes, local increase of the number of quenched-in nucleation sites, etc. Very recently, M.Atzmon et al. /6/ reported changes of the amorphous structure in $(Au,Cu)_{91}La_9$ glasses as a function of the distance from the casting wheel which have been interpreted as an effect due to a decreasing cooling rate with increasing distance.

Crystallization of metallic glasses has been observed to proceed by heterogeneous nucleation at quenched-in nucleation sites in the bulk or at the surface /7/. The number of these nucleation sites decreases significantly with increasing cooling rate resulting from increasing surface velocity of the quenching wheel. Air pockets are expected to reduce locally the cooling rate thus leading to an significant increase of quenched-in nucleation sites as one can see in fig. 4a.

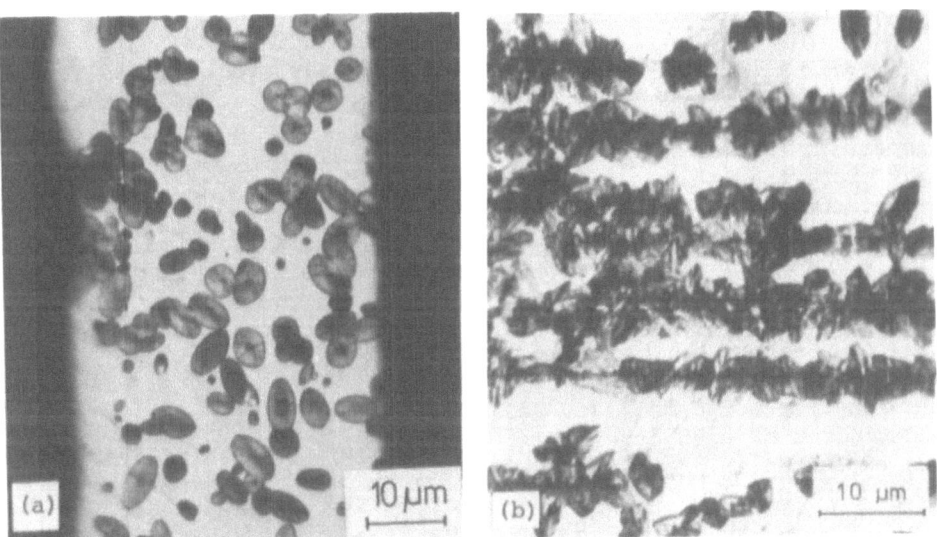

Fig. 4: Crystallization of metallic glasses:
(a) $Fe_{78}Mo_2B_{20}$ (cross section) - 4 h at 435°C: see the higher density of crystals on the left hand side near the dent due to an air pocket.
(b) $Fe_{39}Ni_{39}B_{22}$ (contact side) - 20 min at 380°C: preferred nucleation sites are along the grooves, i.e. areas which had the best contact during the casting process.

Allia et al. /8/ observed a dependence of the electrical properties in a number of iron based metal-metalloid glasses on the cooling rate.

The influence of casting conditions on the magnetic properties have been investigated by a number of authors: Takayana et al. /9/ showed for $Fe_{40}Ni_{40}B_{20}$ glassses that over a limited range of wheel speeds the coercivity H_c decreases, whereas magnetization M and permeability μ increases with increasing surface velocity. Luborsky et al. /10/ showed that a decrease in coercivity of $Fe_{81.5}B_{14.5}Si_4$ alloys with increasing speed is only due to a decrease in the degree of crystallinity. Above a critical speed the quench rate through the thickness of the sample is great enough to produce a completely amorphous ribbon. At this critical speed, the coercivity is minimum and with further increase in wheel speed, H_c increases again.

INFLUENCE OF RELAXATION TREATMENTS:

Since changes in solidification parameter are known to result in large differences in a number of properties, relaxation treatments are assumed to be of utmost importance for comparing physical properties of metallic glasses. On annealing metallic glasses, many properties (e.g., density, ductility, internal friction, Curie temperature, superconducting transition temperature, etc.) have been observed to change significantly /1,11/: Whereas most properties exhibit an irreversible relaxation towards a constant value which is most likely characteristic for the equilibrium amorphous state as far as this property is concerned, a number of properties are known for reversible relaxation.

There are two modes of relaxation which may take place nearly independent of each other /11/: Irreversible structural relaxation was explained in terms of the relaxation of the atomic level density fluctuations, which affects e.g. the density, and can be described by annealing out of free volume or topological short range ordering. The reversible relaxation below T_g has been attributed to the chemical short range ordering and could occur by minor cooperative atomic rearrangement. Very recently, Morito et al. /12/ assumed that most likely shear transformations are the common mechanism among all the reversible structural relaxation effects observed for electrical resistivity, superconductive transition temperature and Curie temperature.

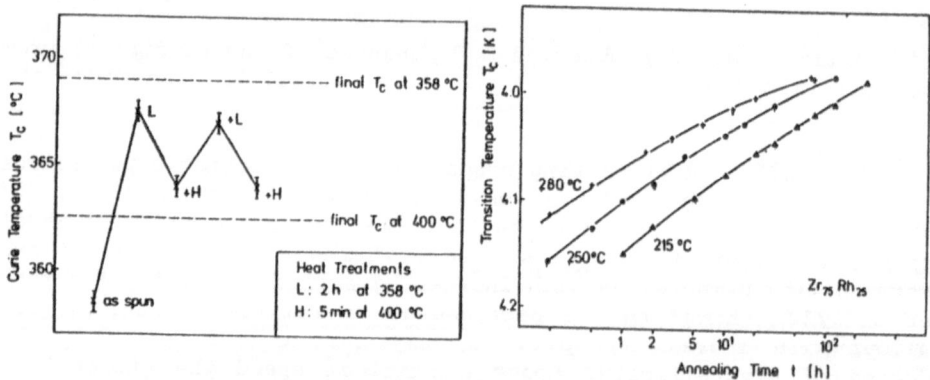

Fig. 5: Influence of relaxation on physical properties:
   (a) effect of annealing on the superconducting transition temperature /13/;
   (b) effect of successive heat treatments on the Curie temperature /14/.

The Curie temperature depends essentially on the nearest neighbour interactions and is sensitive to the chemical short range order, superconductivity depends on the bulk phonon frequency as well as the short range order of the glass. The influence of relaxation treatments on these two properties are shown in fig. 5.

Since the annealing for relaxation are carried out usually in a temperature range near the glass transition temperature, one has to ensure that no crystallization occurs during the annealing, in particular at surfaces where preferred crystallization /7/ has been observed even at temperatures far below the so called "crystallization temperature.

REFERENCES:

1. T.Egami, "Atomic short-range ordering in amorphous metal alloys", in: "Amorphous Metallic Alloys", ed. F.E.Luborsky, Butterworths, London 1983, p.100-113

2. M.C.Narasimhan, U.S.Patent 4,142,571 (1979)

3. H.H.Liebermann, Proc. Rapidly Quenched Metals III, ed. B.Cantor (London 1978), Vol. I, p.34

4. F.E.Luborsky, H.H.Liebermann, Mat.Sci.Eng. 49 (1981),257

5. U.Köster, U.Herold, H.-G.Hillenbrand, Script.Met. 17 (1983), 867

6. M.Atzmon, W.L.Johnson, Caltech Report CALT-822-138 (1982)

7. U.Köster, Z.Metallkunde 75 (1984), to be published

8. P.Allia, R.S.Turtulli, F.Vinai, G.Riontino, Solid State Comm. 43 (1982), 821

9. S.Takayama, T.Oi, J.Appl.Phys. 50 (1979), 1595

10. F.E.Luborsky, H.H.Liebermann, J.L.Walter, Proc. Conf. Metallic Glasses: Science and Technology, Budapest 1980, Vol.I, p.203

11. A.L.Greer, J.Non-Cryst.Sol. 61&62 (1984), 737-748

12. N.Morito, T.Egami, Acta Met. 32 (1984), 603-613

13. S.J.Poon, priv. comm. 1982

14. A.L.Greer, J.A.Leake, Proc. Rapidly Quenched Metals III, ed. B.Cantor (London 1978), Vol. I, p.299

HYDROGEN IN METALLIC GLASSES

Uwe Köster, Hans-Werner Schroeder

Dept. Chem. Eng., University Dortmund
D-4600 Dortmund 50, F.R.Germany

INTRODUCTION:

In recent years much attention has been paid to metallic hydrides, in particular because of their application in hydrogen storage systems. Such systems will be of crucial importance in a possible future "hydrogen economy". Earlier studies on glassy and crystalline Ti-Cu and Zr-Cu showed, that under similar charging conditions, the metallic glasses exhibited larger hydrogen absorption capacities than their crystalline counterparts /1/ as shown in table 1.

It is well known that hydrogen embrittles a large number of metallic alloys; for example the hydride forming crystalline FeTi disintegrates into fine particles after few charging cycles. The absence of hydrides in metallic glasses /2/ combined with their high yield stress and ductility is expected to result in less desintegration during the charging process. In addition, there exists a more scientific point of interest, since hydrogen can be used as a probe to study structure and properties of metallic glasses in more detail.

HYDROGEN CHARGING TECHNIQUES:

Prior to any hydriding metallic glass ribbons are usually surface cleaned by abrasion with emery paper and then ultrasonically cleaned in acetone and/or ether to remove surface oxide layers. Activation by annealing in a hydrogen atmosphere at elevated temperatures is used in the case of crystalline materials. For metallic glasses, however, such a heat treatment may result in partial crystallization in the bulk as well as at the surfaces of the ribbon. Since solidification parameters, e.g. quenching rate, are known to influence the excess free volume in a metallic glass, relaxation treatments are of utmost importance to avoid differences only due to different solidification parameters.

Table 1:   Hydrogen solubility in Ti-Cu alloys at room temperature and atmospheric pressure of hydrogen /3/:

	amorphous		crystalline	
$Ti_{35}Cu_{65}$	H/M = 0.37	-		-
$Ti_{50}Cu_{50}$	0.68	TiCu	H/M = 0.47
$Ti_{60}Cu_{40}$	0.96	-		-
$Ti_{65}Cu_{35}$	1.15	Ti_2Cu	0.92

Hydrogen can be introduced from the gas phase at room temperature or at elevated temperatures by exposing the glass to a hydrogen atmosphere; hydrogen charging can also occur cathodically in a e.g. H_2SO_4-solution containing an arsenic compound as recombination poison. Both methods are shown schematically in fig. 1. For comparison care has to be taken to ensure comparable charging conditions, i.e. constant pressure and temperature during charging in a hydrogen gas atmosphere or overvoltage and temperature during cathodical charging.

HYDROGEN SOLUBILITY:

The amount of hydrogen absorbed from a hydrogen atmosphere can be calculated, using the ideal gas law, from the weight of the sample, the known volume of the system and the change in hydrogen pressure. Hydrogen concentrations have been measured by an electrochemical permeation technique /4/. High hydrogen concentrations as found in intertransition metal glasses can be determined by careful weighing before and after hydriding with a very sensitive microbalance. Other more sophisticated techniques include atomic mass spectroscopy, NMR, lithium nuclear microprobe method, etc.

So far, there are only very few informations on hydrogen solubility in metallic glasses available. Most metal-metalloid glasses are known to solve only minor hydrogen contents (H/M < 0.05). In

88

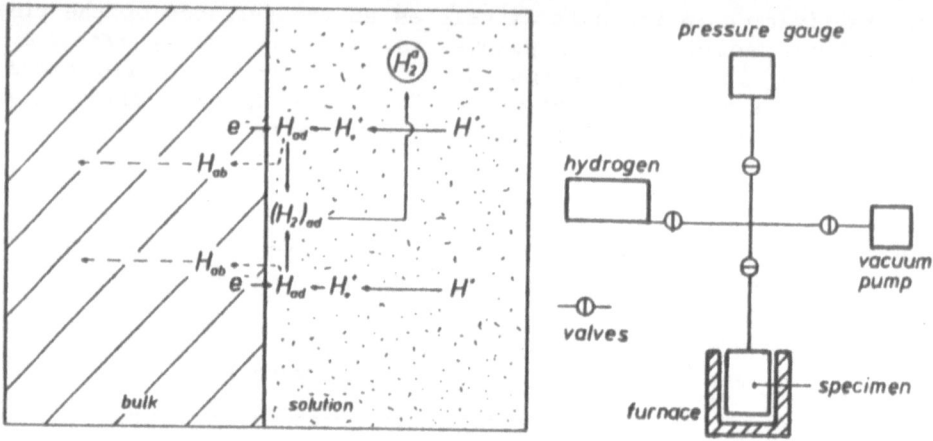

Fig. 1: Hydrogen charging techniques:
   (a) principles of electrolytical absorption of hydrogen;
   (b) schematic diagram for gaseous charging with hydrogen.

Pd-Si glasses /5/ hydrogen solubility increases dramatically with increasing silicon content, e.g. at a hydrogen pressure of 130 mbar from H/M = 0.05 in $Pd_{86}Si_{14}$ to H/M = 0.003 in $Pd_{78}Si_{22}$.

   Intertransition metal glasses, on the other hand, have been observed to store hydrogen up to H/M = 2.0 and even more. As shown in fig. 2a maximum hydrogen content in Ni-Zr glasses has been found to increase with the zirconium concentration, since hydrogen is preferential located in the tetrahedral like sites formed by the zirconium atoms /6/. At higher zirconium contents some of these

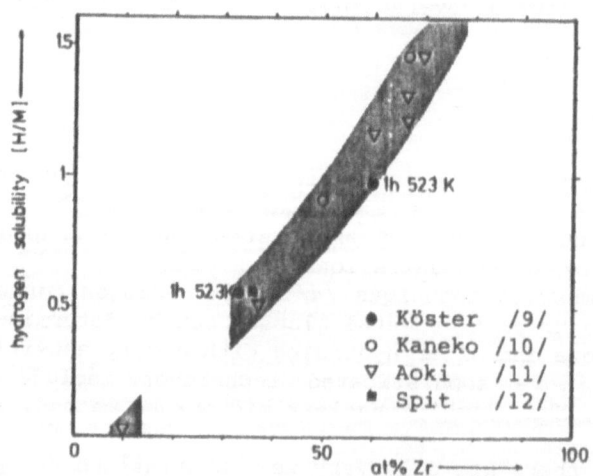

Fig. 2a: Hydrogen solubility in Ni-Zr glasses.

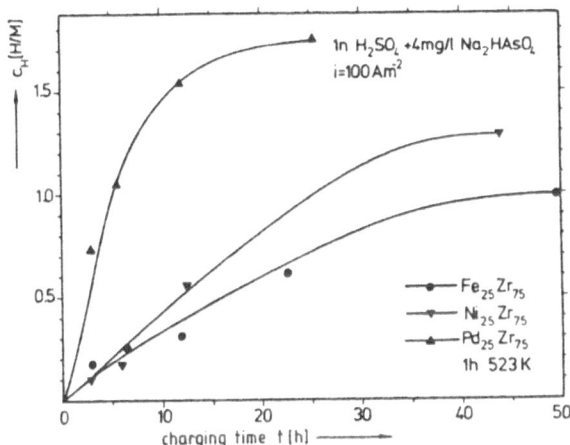

Fig. 2b: Influence of charging time on the hydrogen content
of zirconium based intertransition metal glasses.

tetrahedra may be excluded from occupation by the Switendick-criterion /7/, which demands H-H distances of at least 0.21 nm. Such an assumption would be in accordance with the asymmetric character of the hydrogen sites /8/ as observed in internal friction experiments. The late transition metal has been observed to exhibit some influence on hydrogen solubility in a number of zirconium-based glasses as shown in fig. 2b. The very high solubility in the Pd-Zr glass could be explained assuming that not only zirconium tetrahedra but also due to the size and electronic structure of the palladium atom mixed or even pure palladium tetrahedra can be occupied

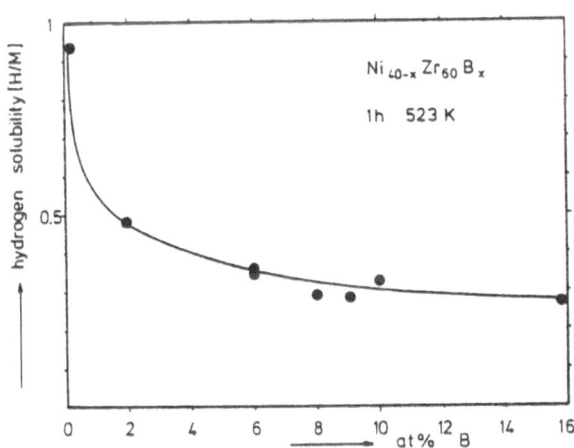

Fig. 2c: Hydrogen solubility of Ni-Zr-B glasses.

by the hydrogen. Even small boron additions to $Ni_{40}Zr_{60}$ result in a dramatic decrease in hydrogen solubility (fig. 2c) indicating a phase transition of the metallic glass /9/.

HYDROGEN DIFFUSION:

Diffusion of hydrogen in metallic glasses has been measured by a number of techniques, e.g. electrochemical permeation technique, NMR /13/, internal friction, Gorsky effect /14/, etc. Whereas internal friction data indicate a spectrum of activation energies for hydrogen diffusion, the Gorsky effect gives rise to only one value thus averaging out the distribution of local jumps into a single-valued activation energy. The temperature dependence of the diffusivity can be described by an Arrhenius-type equation. Examples of typical activation energies and preexponentials are given in table 2.

In a number of metallic glasses hydrogen diffusivity has been observed to increase with the hydrogen content /8,15,16,17/. In a $Ni_{64}Zr_{36}$ glass /8/, for example, the activation energy has been found to decrease from about 40 kJ/mol to 33 kJ/mol at H/M = 0.05 and 0.25, respectively. Such a behaviour was mentioned first by Kirchheim /15,16/, who gave a very plausible explanation: Metallic glasses are assumed to contain a spectrum of "defects" or sites suitable for hydrogen occupation corresponding to different energy

Table 2: Temperature dependence of hydrogen diffusivity in metallic glasses:

glass		D_o $[m^2/s]$	Q $[kJ/mol]$	cathodic current density i $[A/m^2]$
$Fe_{78}B_{13}Si_9$	/17/	$8.9 \cdot 10^{-9}$	37.6	100
$Fe_{40}Ni_{40}B_{20}$	/17/	$3.7 \cdot 10^{-7}$	42.6	100
$Pd_{738}Cu_{82}Si_{18}$	/17/	$5.9 \cdot 10^{-9}$	24.0	10
		$5.6 \cdot 10^{-10}$	15.9	100
$Pd_{80}Si_{20}$	/14/	$3.0 \cdot 10^{-7}$	24	H/M = unknown
$Pd_{775}Cu_6Si_{165}$	/15/	$1.4 \cdot 10^{-4}$	48	H/M = $3.2 \cdot 10^{-5}$
		$4.4 \cdot 10^{-6}$	37.9	H/M = $1.3 \cdot 10^{-4}$
		$1.0 \cdot 10^{-6}$	32.6	H/M = $2.1 \cdot 10^{-3}$

levels. At low hydrogen concentrations sites in the lower energy tail of the distribution are still free of hydrogen and can act as trap sites. With further increase of the hydrogen content these trap sites are more and more filled until they possess only little or no effect on hydrogen mobility, thus leading to a faster diffusion process.

PROPERTIES OF HYDROGENATED METALLIC GLASSES:

As in the case of crystalline materials the structure of metallic glasses as well as most of their properties have been found to be influenced even by minor hydrogen contents. In $Ni_{64}Zr_{36}$ glasses /18/, for example, the value of the electrical resistivity at 4.2 K is remarkable increased by hydrogen and roughly proportional to the hydrogen content with the rate of $3.8 \cdot 10^{-8}$ Ωm/at.%; the trend of originally negative temperature coefficient of resistivity is strongly intensified by the hydrogen. In $Ni_{33}Zr_{67}$ /19/ the superconducting transition temperature has been reported to decrease from 2.64 to 2.43 K at a hydrogen concentration H/M = 0.12. The Curie temperature in $Fe_{91}Zr_9$ glasses /20/ has been found to increase from T_c = 230 K by more than 50 K at a hydrogen content H/M = 0.04.

For any application of these materials in a hydrogen technology mechanical properties are of utmost importance. Whereas usually metallic glasses are known to expand during hydrogen charging with the hydrogen content, very recently a surprising volume contraction at very low hydrogen concentrations (< 50 ppm H) /21/ has been reported. In Pd-Si glasses Young's modulus /5/ has been observed

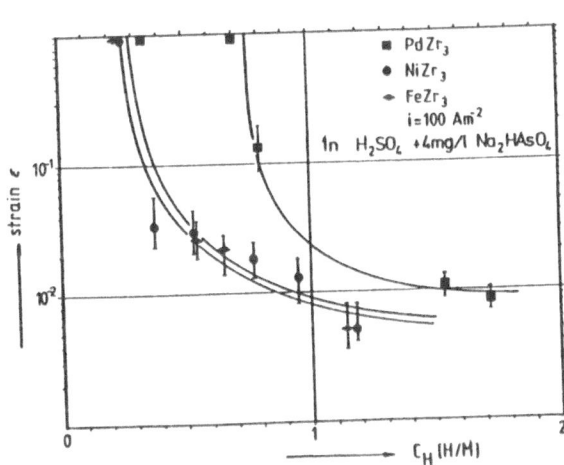

Fig. 3: Fracture strain vs. hydrogen content in Zr-based glasses

to be constant at low hydrogen contents up to a critical H/M which decreases with increasing Si-content of the glass; above this H/M, Young's modulus decreases linearly with increasing H/M with a rate of $7.1 \cdot 10^8$ MPa/at.%.

Hydrogen embrittlement occurs in all metallic glasses investigated so far; whereas even small quantities of hydrogen are able to embrittle iron-based metal-metalloid glasses severely, only moderate embrittlement has been observed in a number of intertransition metal glasses even at hydrogen contents higher than H/M = 0.8. In fig. 3 the fracture strain measured by a bending test /21/ has been plotted versus hydrogen content for a number of intertransition metal glasses. It is quite interesting, that the glass with the highest hydrogen solubility exhibits the smallest sensitivity against embrittlement. Fig. 4 shows the fracture surface of hydrogen embrittled metallic glasses; the vein pattern typically for the ductile state of glass has been observed to be replaced by a shell-like pattern or after further increasing of the hydrogen content by a very smooth fracture surface.

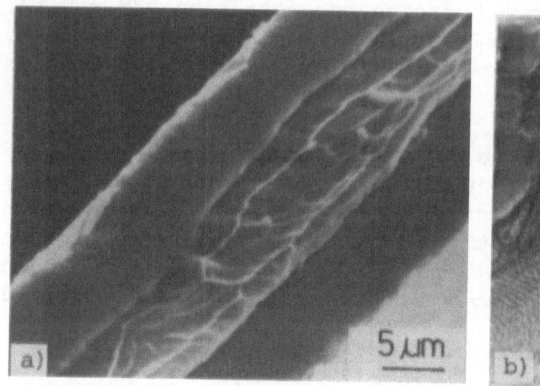 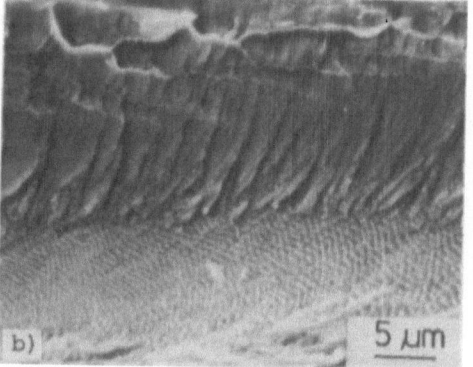

Fig. 4: Fracture morphology of hydrogen embrittled metallic glasses: (a) $Fe_{39}Ni_{39}B_{22}$; (b) $Pd_{25}Zr_{75}$

REFERENCES:

1. A.J.Maeland, "Hydrogen Absorption in Metallic Glasses", in: Metal Hydrides, ed. G.Bambakidis (Plenum Press, New York 1981), p.177

2. F.H.M.Spit, J.W.Drijver, S.Radelaar, Script. Met. 14 (1980), 1071

3. A.J.Maeland, L.E.Tanner, G.G.Libowitz, J.Less-Common Met. 74 (1980), 279

4. J.McBreen, L.Nanis, W.Beck, J.Electrochem.Soc. 113 (1966), 1218

5. R.S.Finochiaro, C.L.Tsai, B.C.Giessen, J.Non-Cryst.Solids 61&62 (1984), 661

6. K.Samwer, X.L.Yeh, W.L.Johnson, J.Non-Cryst.Solids 61&62 (1984), 631

7. A.C.Switendick, Z.Phys.Chem. N.F. 117 (1979), 89

8. B.S.Berry, priv. communication (1983)

9. U.Köster, H.-W.Schroeder, M.Blank, J.Non-Cryst.Solids 61&62 (1984), 673

10. H.Kaneko, T.Kajitani, M.Hirabayashi, U.Mitsuaki, K.Suzuki, Proc. Rapidly Quenched Metals 4, (Sendai 1981), p.1605

11. K.Aoki, A.Horata, T.Masumoto, Proc. Rapidly Quenched Metals 4, (Sendai 1981), p.1649

12. F.Spit, K.Blok, E.Hendricks, Proc. Rapidly Quenched Metals 4, (Sendai 1981), p.1635

13. R.C.Bowman, A.J.Maeland, Phys.Rev. B24 (1981), 2328

14. B.S.Berry, W.C.Prichet, Phys.Rev. B24 (1981), 2299

15. R.Kirchheim, F.Sommer, G.Schluckebier, Acta Met. 30 (1982), 1059

16. R.Kirchheim, Acta Met. 30 (1982), 1069

17. Y.Sakamoto, K.Baba, W.Kurahashi, K.Takao, S.Takayama, J.Non-Cryst.Solids 61&62 (1984), 691

18. S.Hatta, J.Nishioka, T.Mizoguchi, Proc. Rapidly Quenched Metals 4, (Sendai 1981), p.1613

19. E.Babić, B.Leontić, J.Lukatela, M.Miljak, M.G.Scott, Proc. Rapidly Quenched Metals 4, (Sendai 1981), p.1617

20. H.Fujimori, K.Nakanishi, K.Shirakawa, T.Masumoto, T.Kaneko, N.S.Kazama, Proc. Rapidly Quenched Metals 4, (Sendai 1981), p.1629

21. R.Kirchheim, priv. communication (1984)

22. H.-W.Schroeder, U.Köster, J.Non-Cryst.Solids 56 (1983), 213

MECHANICAL RESPONSES OF METALLIC GLASSES

Lance A. Davis

Materials Laboratory
Allied Corporation
P. O. Box 1021R
Morristown, NJ 07960

INTRODUCTION

Metallic glasses have been the subject of curiosity and fascination for slightly over 20 years now. Early research efforts focussed on the glass formation process (1,2), on investigations of structure (3) and on thermal transformation behavior (4). With the availability of ribbon-like "samples", studies of mechanical properties were facilitated (5,6). It then became evident that metallic glasses are exceedingly strong, as well as hard, materials and this realization encouraged speculations and investigations concerning possible structural applications (7). Unfortunately, achievement of such applications has proven elusive and continuing commercial development efforts have focussed on the magnetic responses of metallic glasses, most prominently on the development of a material to replace Si-Fe in the cores of power distribution transformers (8-10). Nonetheless, the mechanical properties of glassy alloys remain of interest on a number of counts. For example, by contrast with the properties of inorganic glasses and crystalline metals, they provide insight into the mechanics of materials, in general. Some of their properties are unique; in particular, they behave as elastic-perfectly plastic materials and, hence, provide a "test-bed" for model mechanics studies. Mechanical behavior dictates the need for fine control of the ribbon making process; the occurrence of nonhomogeneous deformation renders post-quench cold forming processes, e.g., cold rolling to improve surface finish, ineffective. Finally, the performances of metallic glasses in magnetic applications depend on the control of magnetoelastic properties; the magnetoelastic interaction must be minimized to reduce losses in transformer cores or maximized to enhance conversion of magnetic to mechanical energy in a transducer. In the following, we will touch on each of these subjects, as we review the fundamental mechanics of glassy alloys. For other reviews on mechanical behavior, the reader is referred to references 11 to 20. Unless otherwise indicated, data given below are for behavior at room temperature, which usually lies at least several hundred °C below the glass to supercooled liquid or glass to crystal transition.

ELASTICITY

A theoretical treatment of the elastic properties of an elemental metallic glass by Weaire et al. (21) predicts that the glass should be more compliant than the corresponding crystalline material. As the glass structure is random <u>close-packed</u> (22) {the density difference between glass and crystalline material (23) is only 1 to 2%}, the bulk stiffness (K) is expected to be, perhaps, only 5% less than that for the crystal. The shear stiffness (μ) of the glass is expected to be reduced by as much as 25 to 50%. This follows because the atoms of the glass are not constrained by a lattice to exhibit self-similar motion, i.e., they will displace in a somewhat random fashion under the stimulus of an applied shear stress. For the case of $Pd_{77.5}Cu_6Si_{16.5}$* glass, which may be converted to a single-phase crystalline solid, Golding et al. (24) found, in fact, that K increases by ~ 6% and μ increases by ~ 35% on crystallization.

A comparison of the stiffnesses of amorphous and crystalline solids of the same composition is usually, however, not so straightforward, as decomposition of the amorphous alloy to a multi-phase crystalline material typically occurs. For binary Fe-B glasses, K decreases from ~163 GPa to ~111 GPa as the B content decreases from 23 to 15 at % (25). (Fig. 1) By comparison with the bulk stiffnesses of the terminal crystalline phases, i.e., α-Fe (K=169 GPa) and metastable Fe_3B (K>163 GPa), it is apparent that the bulk stiffnesses of the glasses do not follow a simple rule of mixtures (suitably discounted ~5%) between those of the crystalline phases. If these results are indicative, any attempt to estimate the elastic constants of the glass, given some knowledge of the elastic properties of the constituent crystalline phases, must be held suspect. This is somewhat unfortunate as convenient determinations of the elastic constants of a ribbon sample by sonic techniques, other than of Young's Modulus, E, are precluded.

Metallic glasses are commonly formed in transition metal (e.g. Fe,Ni,Co) - metalloid (B,P,C) systems in the vicinity of a relatively deep metal rich eutectic (26) {see Fig. 2 for Fe-B (27)}. By definition, the eutectic liquid (17 at % B in Fe-B) is the lowest melting liquid between the terminal crystalline phases and, hence,

*Subscripts in atomic percentages.

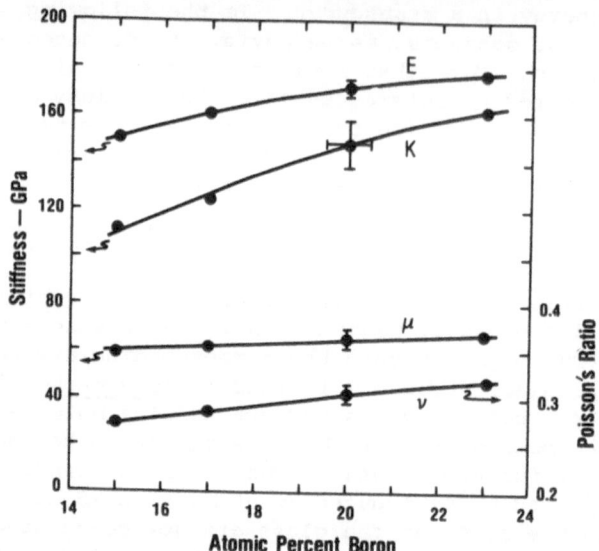

Fig. 1. Elastic properties of Fe-B glasses (25).

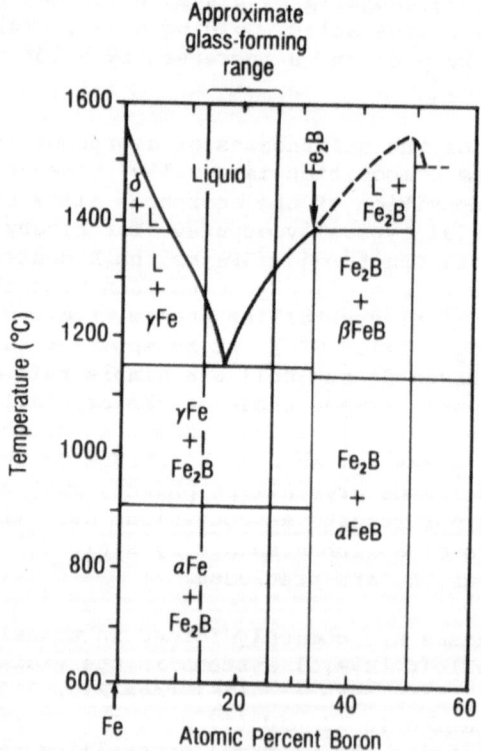

Fig. 2. Part of the Fe-B phase diagram. The glass-forming region
spans the deep eutectic composition at 17 at. % B (27).

corresponds to the alloy most resistant to crystallization. Gilman (28) has suggested that the eutectic liquid is characterized by a unique structure, comprised of "tightly" bonded molecular clusters, which interact only weakly with one another. This model follows from the very existence of the deep eutectic, which is thought to reflect a strong chemical interaction between metal and metalloid atoms in the melt, and from viscosity (η) data for metal-metalloid liquids, which indicate a decrease of η as the eutectic composition is approached. As the transition from liquid to glass is a continuous one, it must be expected, if one accepts the Gilman hypothesis, that clustering remains in the glass. In that regard, the low bulk stiffnesses of near eutectic Fe-B glasses may reflect such clustering (as opposed to a more continuous structure for near compound glasses). It may be noted that the variation of K in the Fe-B series is much stronger than for μ, so that the ratio K/μ decreases with decreasing B content, reaching 1.86 for $Fe_{85}B_{15}$ (25). This value is considerably lower than those commonly observed for glassy alloys (~3 to 5) (25) {though not as low as that observed for silica glass (~1.15) (29)} and approaches that expected (1.67) for a Cauchy solid, i.e., a solid bonded by centrally directed interatomic forces (30). This is consistent with the notion of clustering in near-eutectic glasses, such that the bulk compliances of these glasses are dominated by the relatively weak bonding between, rather than the strong bonding within, molecular clusters.

The variations of K and μ, as a function of varying metal content in fixed metalloid content glasses, have been examined by Chen et al. (31) for $(PdNi)_{80}P_{20}$ and $(PdFe)_{80}P_{20}$ and by Chou et al. (32) for $(FeNi)_{80}B_{20}$ and $(FeCo)_{80}B_{20}$. For the Pd based glasses, variations of μ and K track each other, roughly; for the FeNi and FeCo glasses, K shows a strong dependence on composition, while μ does not. (Fig. 3) The importance of electronic structure in the latter ferromagnetic glasses is indicated by the variation of K with electron/atom (e/a) ratio (Fig. 4); maximum stiffness is observed at the same e/a value. These data also indicate that K is sensitive to near neighbor interactions, while μ reflects longer range atomic interactions.

Variations of Young's modulus (E) with composition for Fe-P-C, Fe-B-Si and Co-B-Si wires have been examined by Inoue et al. (33). For $Fe_{75}P_{15}C_{10}$ glass, E is $\simeq$ 144 GPa and varies little with compositional variations of about 5 at % for each element of the glass. For Co-B-Si, for either fixed Si content (10-12.5 at %) or fixed B content (10-15 at %), E increases significantly (~170 to 190 GPa) as Co replaces B or Si, respectively. The reverse is true for Fe-B-Si glasses; E increases (~160 to 190 GPa) as the B and/or Si content is increased. (The values of E for Fe-B-Si glasses with low Si content appear consistent with those for Fe-B glasses reported in ref. 25.) In all these cases, however, it is uncertain whether changes of E reflect changes in bulk stiffness, shear stiffness or both, as

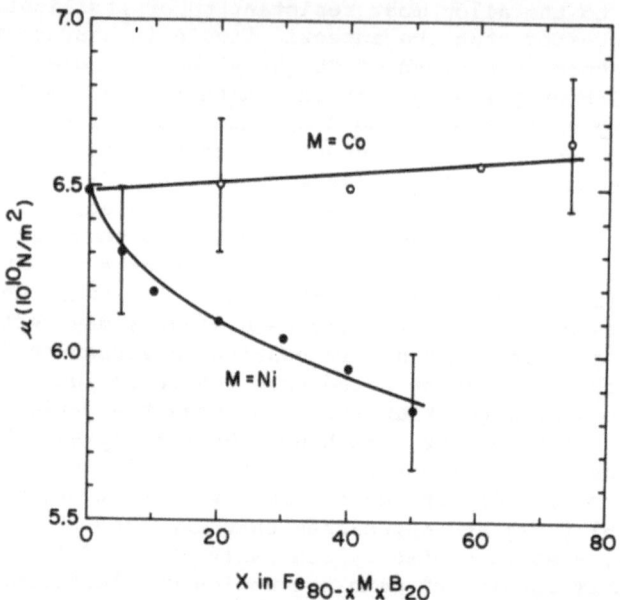

Fig. 3. Shear moduli for (FeNi)$_{80}$B$_{20}$ and (FeCo)$_{80}$B$_{20}$ glasses (32).

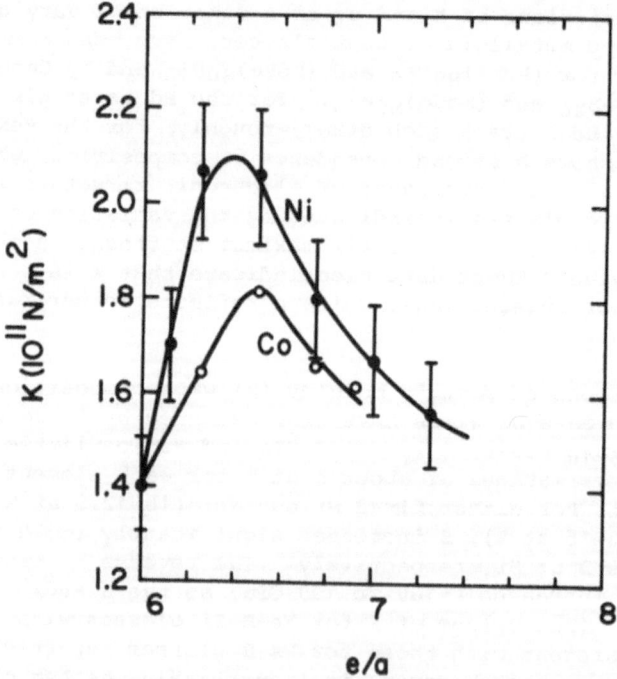

Fig. 4. Bulk moduli for above glasses vs. d-electron/atom ratio (32).

$E = 9K\mu/(3K+\mu)$. In Fe-B glasses (Fig. 1), E increases with increasing B content mainly because K increases.

Finally, as a matter of reference, it is worth noting that E for the stiffest Fe based glasses, e.g., 166 GPa for $Fe_{80}B_{20}$ glass, remains somewhat below that for iron or steel wire (~210 GPa) (34) and well above that for silica glass (~73 GPa) (35).

PLASTICITY

Deformation Mode

The plastic responses of metallic glasses may be _very_ roughly separated into homogeneous/high temperature/low strain rate behavior and inhomogeneous/low temperature/high strain rate behavior (14). In this instance, "homogeneous" denotes the situation where, on a microscale, each volume element experiences the same average deformation. Only in the limiting case of flow at or above Tg, the glass transition temperature, is plastic flow homogeneous on a near atomic scale. At temperatures above 0.7 to 0.8 Tg, tensile specimens deformed at intermediate strain rates (~ 10^{-4}/sec) exhibit diffuse necking (36). According to the definitions of mechanics, this is a classical example of inhomogeneous deformation, although it is commonly referred to as "homogeneous" in the metallic glass literature (14).

Deformation at low temperature (T<<Tg) occurs in the form of localized shear deformation bands, which can be astonishingly intense. For example, on examining surface replicas in a transmission electron microscope, Masumoto and Maddin (5) found 20 nm wide shear bands which could sustain displacements of 200 nm, or a shear strain of 10; the strain at which fracture (shear rupture) occurs through a band may be many times larger. An example of the shear deformation of a bent $Pd_{77.5}Cu_6Si_{16.5}$ wire is shown in Fig. 5.

Shear bands form under all manner of low temperature loading conditions, i.e., in tension (6,36,37) (Fig. 6), compression (38), torsion (19,39), during wire drawing (40,41), during indentation by a hard object (6,42) or during cold rolling (43,44). In the last case, it thus proves impossible to effect improved ribbon smoothness and flatness by cold rolling; hence, the need to achieve good process control while casting ribbons from the melt. It is, nonetheless, the ability to flow plastically at low temperatures under a variety of loading conditions, which sharply distinguishes metallic glasses from oxide glasses; the latter are capable of local plastic flow only under very specialized conditions, as, for example, when scratched by a hard stylus (45). Metallic glasses are intrinsically ductile, because of their nonstereospecific metallic bonding characteristics.

100

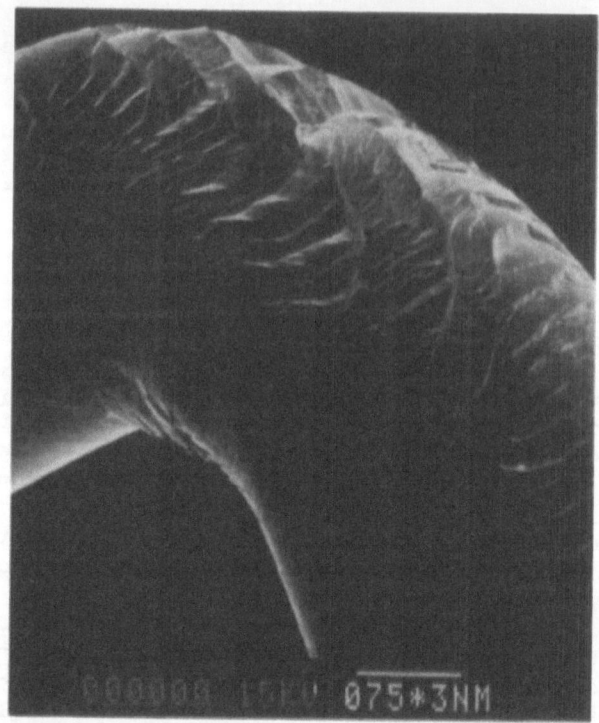

Fig. 5. Intense shear bands which appear on a bent $Pd_{77.5}Cu_6Si_{16.5}$ wire; scanning electron micrograph (SEM).

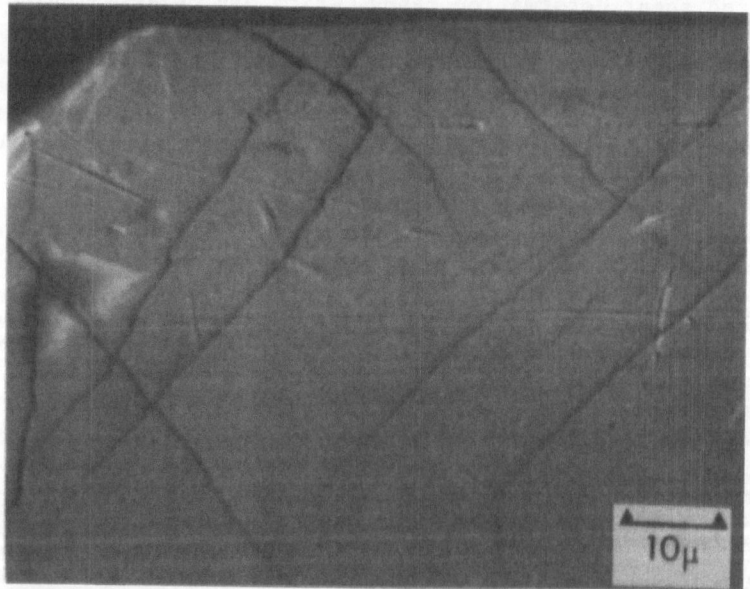

Fig. 6. Shear steps isolated near the tensile fracture surface (edge at upper left) of amorphous $Pd_{77.5}Cu_6Si_{16.5}$. Some bands terminate in the material; SEM (37).

The occurrence of highly intense shear deformation bands is indicative of a lack of work hardening, even in the absence of measured stress-strain behavior. As pointed out by Cottrell (46), this is simply deduced from the fact that the bands would continually expand into softer material if deformation in the bands produced hardening. By the same token, the appearance of <u>multiple</u> shear bands in compression or torsion reflects the absence of work softening; unstable deformation would proceed in the first band to form, if the plastic resistance of deforming material was less than that of undeformed material. Kimura and Masumoto (19,39) specifically noted that the density of shear bands steadily increases with increasing strain during the torsion deformation of $Pd_{77.5}Cu_6Si_{16.5}$, i.e., the deformation is stable. In the absense of work hardening and work softening, the stress for plastic flow remains constant at the yield stress. A constant flow stress is readily observed in compression (38) or torsion tests (19,39); it may be observed in a tension test only if the testing machine is exceedingly stiff (47). Otherwise, failure occurs coincident with yielding by a catastrophic shear instability (36,48).

The above comments not withstanding, references to the occurrence of work hardening and work softening do appear in the literature. Takayama and Maddin (43) reported a 3% increase in microhardness after cold rolling as possible evidence of work hardening; in fact, microhardness measurements lack the precision to establish such a small difference. Indeed, Murr et al. (49) have shown that $Fe_{80}B_{20}$ and $Fe_{38}Ni_{40}Mo_4B_{18}$ glasses show no increase of hardness when worked under much more severe conditions, i.e., when shock deformed to a peak pressure of 35 GPa (~10 times their yield stresses). A sample of 304 stainless steel, loaded in the same experimental sandwich, experienced about a 50% increase in hardness, as is typically seen, due to the generation of dislocations and other crystal defects. Hagiwara et al. (41) conducted tensile experiments on cold drawn $Fe_{75}Si_{10}B_{15}$ wires and observed about a 5 to 8% increase in tensile strength, independent of the degree of prior drawing deformation; this increase was suggested to result from drawing induced work hardening. As discussed previously (50), however, this is not a work hardening phenomenon, because it does not depend on strain; it is also not clear that it is not simply an artifact of testing. It has also been suggested (51) that one should calculate the fracture strength of a glass using the maximum load and the apparent load bearing area at failure (the "vein" region - see below). Since very large shear displacements can occur, this would imply massive hardening. However, this approach is incorrect, since the load on the sample at the instant of final separation is only a fraction (usually unknown) of the peak load.

Kimura et al. (52) examined the deformation of $Ni_{40}Fe_{40}P_{14}B_6$ glass in a diamond anvil cell. Their measurements, which are rather indirect, suggest that the heavily deformed (~20%) material exhibits a flow stress about one-half that of undeformed material. As noted above, other measurements under simple uniaxial compression (38) and torsion (19,39) conditions, albeit at more modest strains, show no work softening. Hence, one is led to suspect the validity of the assumptions underlying the calibration of the diamond cell measurements.

Microscopic Mechanism

Observation of a deformed specimen, for example, as in Fig. 6 (37), indicates the presence of shear bands which terminate within the material. The elastic discontinuity which exists at the boundary between deformed and undeformed material corresponds to a Somigliana dislocation (53). It seems most logical to conclude that this macroscopic dislocation is comprised of and propagates due to the presence of microscopic dislocations with Burger's vectors of the order of an atomic spacing. Gilman (54) was the first to suggest the existence of such atomic scale dislocations in a glassy solid. Since then, Li (16-18,53,55) has commented extensively on their likely existence and properties in glassy metals. For example, the shear displacement profile along a band was measured and it was demonstrated that these displacements are stable on annealing near Tg; hence, the mobile units responsible are intrinsically stable, as would be expected for a pile-up of line defects of the same sign (17). The creep recovery of $Pd_{80}Si_{20}$ was also analyzed, indicating that second order kinetics appear to be applicable (56).

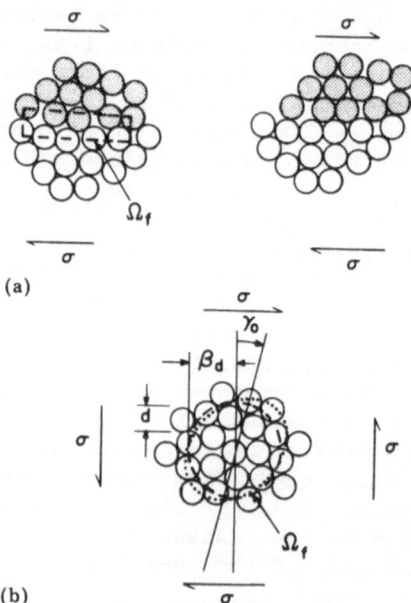

Fig. 7. Idealized shear transformations observed in sheared bubble rafts: (a) dislocation pair formation; (b) diffuse shear (60).

This would be expected if mobile "deformation units", such as dislocations, of opposite signs annihilate one another during recovery. The effects of prior annealing on creep kinetics may be similarly analyzed (17); in this case, annihilation of mobile defects can lead to a five-fold decrease of creep rate, while the activation energy for creep remains constant (57).

Speculation as to the existence of microscopic dislocations in metallic glasses would be unnecessary, of course, if they were amenable to direct observation. Unfortunately, as a result of the structural nonperiodicity of glasses, this does not appear to be possible; the detection of dislocations in crystalline materials by transmission electron microscopy depends on diffraction contrast associated with a periodic lattice (58). Dislocation-like structures do appear in bubble raft models (59-61) and in three dimensional computer models sheared to simulate deformation of a metallic glass (62). These are most readily visualized in the former (Fig. 7), where both sharp and somewhat diffuse shear transformation regions about 5 atom diameters across are observed, corresponding to the formation of isolated and slightly more extended dislocation loops, respectively. Within the context of the dislocation model, ideal plasticity, that is, the lack of work hardening and work softening, may be considered to result because the glassy structure corresponds to that of a highly dislocated and disclinated solid, insensitive to the presence of locally mobile dislocations. Deformation is localized in shear bands initiated at regions of minor stress concentration and sustained there because mobile dislocations are then present (53).

It has also been suggested that plastic flow in glassy alloys occurs due to a stress induced increase of free volume, which is imagined to make the material behave in a fluid like manner (14,63). As applied to homogeneous deformation (14), the free volume formalism appears, at least, physically sensible; as applied to inhomogeneous deformation, as in the model of Steif et al. (63), it does not. The latter authors postulate that the free volume in a shear band depends exponentially on stress. Taking the total strain rate as constant, they show that a sharp yield drop (decrease of flow stress) and rapid increase of plastic strain rate in the band will occur coincident with a rapid increase in free volume. The treatment is analogous to that used to model a yield drop in steel, for example, which occurs due to locally rapid multiplication of dislocations against a relatively low dislocation density background. A critical test of the Steif model is afforded by the torsion deformation experiments of Kimura and Masumoto (19,39), noted above; no yield drop is observed, whence one can conclude that the Steif model is not physically realistic.

This is not to say, however, that free volume could not play a somewhat more subtle role in nonhomogeneous deformation. In

attempts to support such a role, efforts have been made to establish
the existence of residual free volume in shear bands, including,
among others, transmission microscope observations of shear bands
and determinations of dilatency of deformed samples. The results in
the first case are ambiguous; Donovan and Stobbs (64) report that
shear bands exhibit enhanced low-angle electron scattering, suggesting the presence of small regions of reduced density. Sethi et al.
(65), in an earlier report, indicate, to the contrary, that one
can't even find shear bands in a deformed film, unless the film is
deformed <u>after thinning</u>, in which case they are observed only due to
thickness contrast. Determinations of volume changes in shear
bands, as deduced from measurements of density of cold drawn wires
(66) and by measurements of relative axial shrinkage on crystallization of virgin specimens and specimens deformed by compression
(67,68), indicate dilatency due to deformation. Such measurements
must be viewed with caution, however, as macrovoids can form in
intense shear bands, even in compression specimens (38). In any
event, even if some residual dilatency truly exists as a result of
deformation, it does not necessarily follow that the critical step in
deformation is the creation of such dilatency. If it were, one
would expect the flow stresses of glassy alloys to scale with their
bulk moduli and this is not observed (see below). Similarly, in

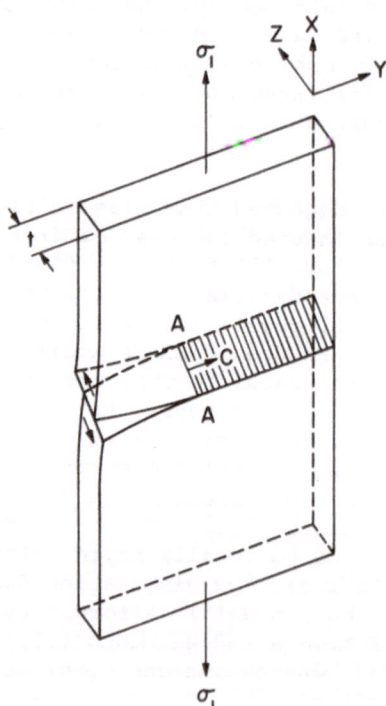

Fig. 8a. Schematic representation of tearing failure (antiplane
strain shear) of a metallic glass strip (13).

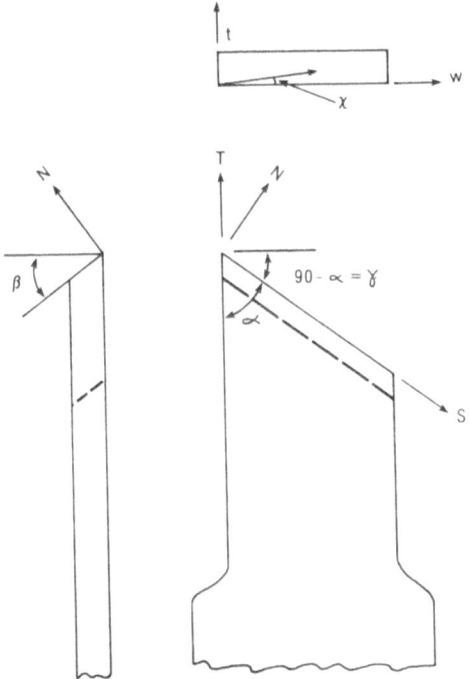

Fig. 8b. Samples which fail by tensile yielding in the low temperature regime (T/Tg≤0.8) exhibit a fracture surface orientation defined by the angles α and β, shown in the front and
side views of the figure. Vector T defines the tensile
direction and N the normal to the fracture surface; t and
w represent the thickness and width vectors of the sample. β and χ are typically ~0°, while α is ~53° (36).

computer models, deformation events appear to be associated with
regions of local shear stress, not with regions of local hydrostatic
stress (69).

YIELD STRENGTH

One may conveniently determine the yield strengths of glassy
alloys under conditions of compression, torsion or tension deformation. In the first two cases one requires a specimen on the order
of 1 mm in diameter and this is practical only for easily quenched
glasses, such as $Pd_{77.5}Cu_6Si_{16.5}$. In each case the yield stress is
taken as the plateau flow stress. Thin ribbon tensile specimens of
high strength materials, whether glassy or crystalline, tend to
fail by tearing (Fig. 8a) at a stress lower than the yield stress
(48). If one prepares a metallic glass strip with a reduced area
gage section, having an aspect ratio < 8:1, it is possible to avoid
tearing and reach the yield stress, σ_y (48). Yielding is evidenced

106

by the distinctive failure mode of the specimen, as schematically in Fig. 8b. Imperceptible necking occurs through the thickness along a zone of "zero extension" at ~53° to the tensile axis,
followed instantaneously (at T<<Tg and when using a standard tensile
testing machine) by a catastrophic shear instability through the
zone (36,47,48). Necking is clearly evident at higher temperature
(0.7 to 0.8 Tg) and is again followed by shear instability (36). In
either case, the yield stress is equal to the maximum tensile stress.

For axisymmetric glassy alloy specimens, such as wires or bars,
yielding still occurs through an oblique shear zone (36,47). This
is an instance where the properties of glassy alloys shed light on
the properties of crystalline materials. While oblique necking
occurs in polycrystalline sheets (70-73), it does not occur in bars,
where necking normal to the tensile axis appears. It has been considered that such "normal" necking occurs in the absence of the
lateral constraint provided by the sheet geometry. From the results
on glassy alloys, it is apparent, however, that absence of constraint is a necessary, but not sufficient, condition; normal necking
also requires the presence of work hardening to disperse slip.

A representative collection of yield strength values is shown
in Table I, whence it is clear that glassy alloys are exceedingly
strong materials. The yield strength of $Fe_{80}B_{20}$ approaches the tensile strength of hard drawn piano wire (~380 GPa). It is also ap-

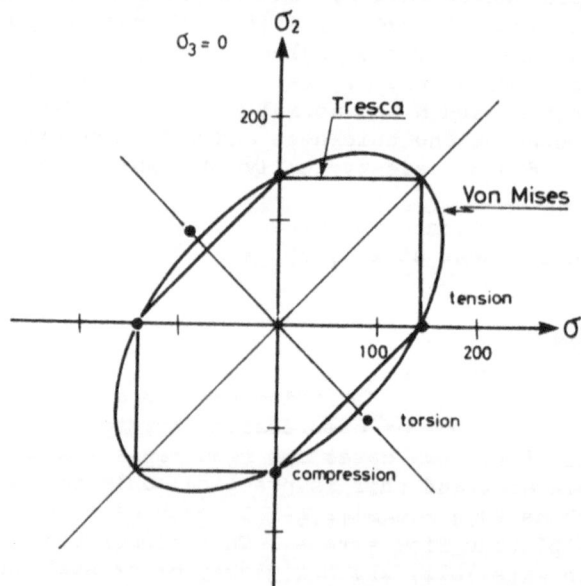

Fig. 9. The yield loci (σ_1-σ_2 yield surface) for samples of
amorphous $Pd_{78}Cu_6Si_{16}$, using the compressive, tensile and
torsional shear stresses (19).

TABLE I

MECHANICAL PROPERTIES OF METALLIC GLASSES

Alloy	H_v GPa	σ_y GPa	H_v/σ_y	E GPa	Ref.
$Ni_{36}Fe_{32}Cr_{14}P_{12}B_6$ (METGLAS® 2826A)	8.63	2.73 tension	3.16	141	
$Ni_{49}Fe_{29}P_{14}B_6Si_2$	7.77	2.38 tension	3.26	129	
$Fe_{80}P_{16}C_3B_1$	8.19	2.44 tension	3.35		
$Fe_{80}P_{12.5}C_{7.5}$	7.94	2.73 tension	2.91	140	(33)
$Fe_{80}Si_{10}B_{10}$	8.13	2.91 tension	2.79	158	(33)
$Fe_{80}B_{20}$ (METGLAS® 2605)	10.79	3.63 tension	2.97	166	
$Co_{77.5}Si_{12.5}B_{10}$	11.2	3.58 tension	3.13	190	(33)
$Pd_{77.5}Cu_6Si_{16.5}$	4.88	1.54 compression	3.17	88.0	
		1.47 tension	3.39		
		0.91 (τ_y) torsion			(19)
$Pd_{64}Ni_{16}P_{20}$	4.43	1.44 compression	2.86	91.9	

H_v and σ_y have uncertainties of the order of ±5%. METGLAS® is a registered trademark of Allied Corporation. Data summarized in Ref. 13,
except as noted.

parent that, whereas deformation characteristics are independent of composition, strength is highly composition dependent. Roughly speaking, strength tends to increase as the average electron/atom ratio of the metal component decreases, i.e., Pd=Ni→Co→Fe→Cr. The same general trend was observed for $(FeM)_{80}P_{13}C_7$ alloys (with M = Cu,Ni,Co,Mn,Cr,V,Ti) by Naka et al. (74). The non-metal elements also play a key role; glasses containing only B are stronger than those containing B and Si and far stronger than those containing P and C.

Table I and Fig. 9 show yield strength values for $Pd_{77.5}Cu_6Si_{16.5}$ in compression, tension and torsion. From these data it is apparent that the yield stress in torsion or shear, τ_y, is approximately equal to the yield strength in tension or compression (the latter appearing to be slightly larger) divided by $\sqrt{3}$. This indicates that metallic glasses behave as von Mises solids, i.e., yielding occurs when the root mean-square shear stress reaches a critical value (75). This is also evidenced by the observed 53° shear zone on yielding. For a Tresca solid, $\tau_y = \sigma_y/2$ and tensile yielding must occur on a plane of maximum shear stress, i.e., at 45° to the tensile axis (75).

HARDNESS

An extremely convenient means to sample intrinsic strength is through measurement of Vickers (136° diamond pyramid) microhardness (H_V), which provides a measure of flow resistance under conditions of plastic constraint. A material may accommodate indentation by slipping up around the sides of the indenter (6,42,76) or, as analyzed by Marsh (45,47), by compressing radially under the indenter. In the first case, $H_V \approx 3\sigma_y$; in the second case, the ratio of H_V to σ_y is less than 3, decreasing as the ratio σ_y/E increases. For metallic glasses (and polycrystalline metals) deformation by slip is favored, as discussed above, and on the average $H_V/\sigma_y \approx 3.2$ (Table I). Oxide and polymeric glasses possess directional bonding and are not close packed, so radial compression is favored and $H_V/\sigma_y < 3$ (45,77).

Given their high strengths, it follows that the hardnesses of metallic glasses are extremely high, reaching 10.8 GPa for $Fe_{80}B_{20}$ and 15 GPa for a glass such as $Ni_{40}Mo_{30}Cr_{20}B_{10}$ (78), which has a large refractory metal content. Ordinary steels exhibit hardnesses equivalent to ~2 to 7 GPa on the Vickers scale; for SiO_2 and mixed oxide glasses H_V falls in the range of 3 to 6 GPa (79).

Sargent and Donovan (80) have noted that σ_y/E values for metallic glasses are large (~0.02), falling in the same range as for oxide and polymeric glasses, and have concluded that metallic glasses are somehow anomalous because H_V/σ_y is not less than 3. This is a curious conclusion, which ignores the physics of these materials, i.e., they deform by shear. Marsh's analysis does not require that large σ_y/E materials accommodate indentation by radial compression;

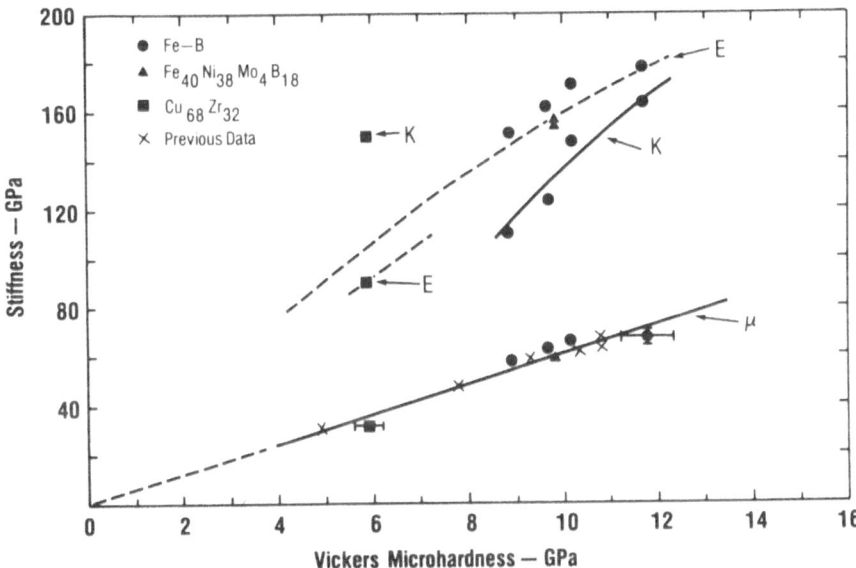

Fig. 10. Comparison of the elastic stiffnesses and microhardnesses
of metallic glasses. The data for µ shown by crosses are
from Refs. 32 and 83. For clarity, corresponding values
for K, where known, are not shown; they scatter in the vicinity of 160-170 GPa, independent of the hardness (ranging from 4.9 to 10.8 GPa) of the glass. The upper dashed
curve is intended to indicate the trend of a variety of
data for E for transition metal-metalloid glasses (Ref.
81). The short dashed curve indicates a slightly different trend observed for metal-metal glasses, such as
Cu-Zr, Cu-Ti, and Be-Ti-Zr (25).

it simply models the pressure under the indenter if radial compression <u>does</u> occur.

STRENGTH VS. STIFFNESS

It has been noted that hardness (81,82) and yield strength (33)
scale, at least approximately, with Young's Modulus, E. As E
reflects a combination of distortional and dilational deformation
and metallic glasses exhibit shear deformation, it follows that a
more direct correlation might exist between H_v (or σ_y) and µ.
Inspection of Fig. 10 indicates that this is, indeed, the case
(25,83). On the average, $\mu \approx 6.1 H_v$ and, therefore, $\sigma_y \approx \mu/19.5$.

Li (56) has estimated that the theoretical shear strength of a
glassy alloy, i.e., the stress for uniform shear flow, should be of
the order of µ/4 which is well above the observed value of $\tau_y \approx \mu/34$

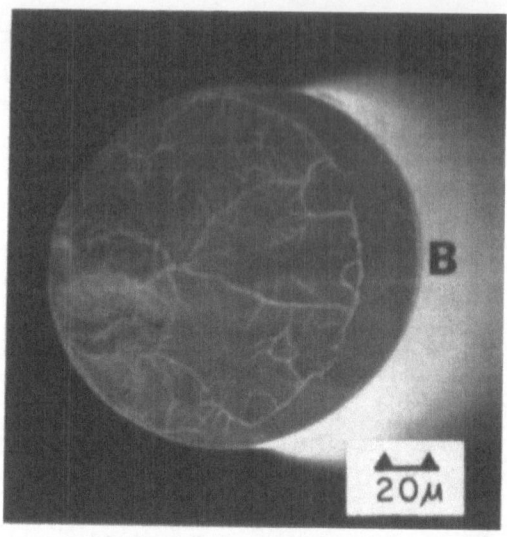

Fig. 11. Fatigue fracture surface of a reduced gauge section
$Pd_{77.5}Cu_6Si_{16.5}$ filament. The fatigue crack initiated to
the left of the surface. The remainder of the section
is marked by a smooth shear offset (near B) and vein
pattern typical of shear rupture. This indicates that
final failure occurred by yielding of the remaining section. On a macroscopic scale the fracture surface slopes
down from left to right; the normal to this surface lies
at ~ 35-40° to the wire axis (106).

(with $\tau=\sigma/\sqrt{3}$). If one follows Morris (84) and models the glass as
a highly dislocated and disclinated solid, and recognizes that struc
tural studies indicate the absence of atomic correlations beyond a
spacing of 5 atom diameters (a), then the disclination spacing must
be ~5a. Glide dislocations are considered to be pinned at disclinations and, therefore, the glide dislocation length is also ~5a and
one can calculate τ_y (or σ_y) from the stress for irreversible loop
expansion, i.e., $\tau_y=[\mu b/4\pi\sqrt{(Ns)}]\ln[\sqrt{(Ns)}/b]$, where b = slip vector
$\simeq a$ and $\sqrt{(Ns)}$ is the density of wedge disclinations = $\ell \simeq$ the dislocation loop length $\simeq 5a$. This calculation then yields $\tau_y \simeq (\mu/20\pi)\ln 5$
= $\mu/39$ or, with $\tau_y= \sigma_y/\sqrt{3}$, $\sigma_y=\mu/22.5$, in reasonable agreement with
the observed value.

FRACTURE MODE, FRACTURE TOUGHNESS AND EMBRITTLEMENT

Tensile specimens, which fail by shear, can exhibit featureless
fracture surfaces if they simply shear apart. More commonly, internal rupture occurs, through the formation of shear disk cracks, and
a characteristic "vein" pattern is left on the fracture surface
where these cracks impinge (Fig. 11). Such a pattern is essentially

identical to that which appears when two glass microscope slides, containing between them viscous grease, are separated. This has been repeatedly (14,51,85,86) cited as evidence for fluid flow in shear bands, such flow supposedly being induced by an increase in free volume. In addition to the comments above concerning such a model, it should be noted that similar shear fracture surface features are noted in steels (87). Accordingly, vein formation is a macroscopic phenomenon not limited to liquid films.

For the particular sample shown in Fig. 11, failure occurred by general yielding of the net section during sinusoidal tension-tension cyclic loading in the presence of a fatigue crack. This particular failure event reflects the influence of both sample shape (circular section) and the intrinsic ductility of $Pd_{77.5}Cu_6Si_{16.5}$ glass. This glass will flow plastically in tension under conditions of severe plastic constraint, i.e., a circular cross section tensile specimen, severely notched around its periphery, will yield through

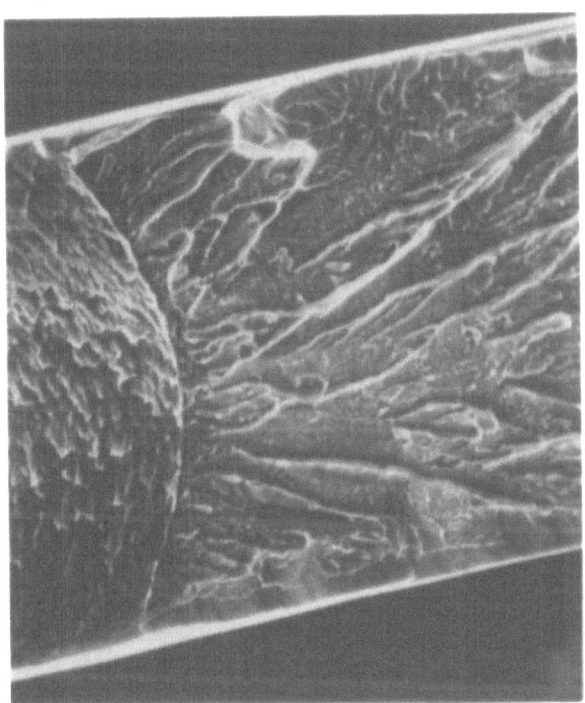

Fig. 12. A portion of the fracture surface of a sample of $Ni_{69}Cr_6Fe_3B_{14}Si_8$ glass ($K_m \simeq 34.3$ MN/m$^{3/2}$). The fatigue crack lies to the left and the rapid square fracture surface to the right. The scale of the micrograph is fixed by the ribbon width (top to bottom) which is 140 μm. The slight bowing of the final position of the fatigue crack front is negligible in comparison with the total length of the crack; SEM (88).

the net section at a net stress equal to ~2.4 times the uniaxial tensile yield stress (19).

Sheet specimens, subjected to cyclic tensile loading, will normally fail at low stress ($\sigma < \sigma_y$)in the antiplane strain tearing mode (Fig. 8a) or by Mode I (fracture surface nominally perpendicular to the tensile axis) in the presence of a sharp fatigue crack (88-90). The observed fracture behavior is most conveniently quantified in the terms of linear elastic fracture mechanics (91), where the nominal failure stress, σ_f (referred to the original gross specimen area), is given in the simplest case by $\sigma_f = K_m/\sqrt{(\pi a)}$, where K_m is the fracture toughness for a given mode, m. For a flat panel with a central hole and a fatigue crack emanating from either side of the hole normal to the tensile axis {center cracked panel (CCP) specimen}, 2a equals the total crack length. Under cyclic loading the specimen will fail when the propagating crack reaches a critical length dictated by the peak cyclic stress and the characteristic toughness of the specimen; alternatively, one may ramp-load a specimen with an existing fatigue crack, in which case failure will occur at a stress dictated by K_m and a. The sharp demarcation between the stable fatigue crack propagation region and an unstable Mode I crack in a metallic glass is shown in Fig. 12.

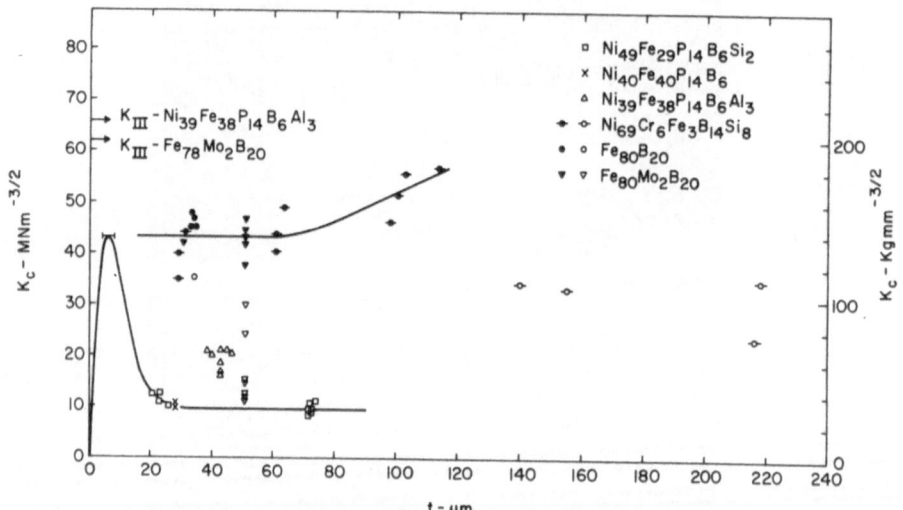

Fig. 13. Fracture toughnesses for various thickness metallic glass specimens. Open symbols (and X) indicate square fracture; shading indicates shear rupture failure. Data for the $Ni_{69}Cr_6Fe_3B_{14}Si_8$ (-o- ; -●-) and $Ni_{40}Fe_{40}P_{14}B_6Si_2$ (X) glasses and for the 20 to 26 μm thick samples of $Ni_{49}Fe_{29}P_{14}B_6Si_2$ (□) glass are from Ref. 88. Other data shown for the Ni_{49} glass and for $Ni_{39}Fe_{38}P_{14}B_6Al_3$ (Δ) glass are from Ref. 89. Data for $Fe_{80}B_{20}$ (O, ●) and $Fe_{78}Mo_2B_{20}$ (∇, ▼) glasses are from Ref. 13.

A collection of fracture toughness data, as a function of specimen thickness, for center cracked panel specimens of as-quenched Ni-B, Fe-B and Ni-Fe-P-B dominated glasses is shown in Fig. 13 (88). On the toughness axis the data are bounded by values of ~10 $MN/m^{3/2}$ and ~45 $MN/m^{3/2}$; the former is characteristic of a fully plane strain mode I failure (K_{IC}) and the latter is characteristic of antiplane strain shear (K_{45}) failure. Since τ_y/μ is ~ constant for metallic glasses, it would appear that both K_{45} and K_{IC} should be approximately independent of composition {although K_{45} may increase with thickness due to the suppression of buckling (88)}. The composition variations noted in Fig. 13 apparently reflect subtle structural variations in the as-quenched glasses. One expects that the well-quenched "ideal" glass, a true elastic continuum, would always fail by shear, regardless of thickness.

To put the toughnesses of glassy alloys in perspective, one may compare with those for oxide glasses and for steels. In the former, K_{IC} is of the order of 0.1 $MN/m^{3/2}$ and critical crack sizes are at least 10^4 times smaller than for metallic glasses (92). For example, according to the simple formula above, for $K_{IC}=10$ $MN/m^{3/2}$ and $\sigma \approx 1$ GPa, $a \approx 32$ μm for a metallic glass, while at the same stress level a is ~3.2 nm for an oxide glass; hence, the common identification of oxide glasses as brittle. For steels, K_{IC} ranges from ~30 $MN/m^{3/2}$ for high strength alloys (σ_y~2 GPa) to ~130 $MN/m^{3/2}$ for low to medium strength alloys (σ_y~1.2 GPa) (93). Metallic glasses are tougher than oxide glasses because their bonding characteristics permit them to readily sustain plastic flow; relative to steels, their lower toughnesses are consistent with their exceptionally high yield stresses and result because their plastic flow is so intensely localized.

The loss of easy handleability, as most conveniently quantified by loss of bend ductility (94) {which corresponds to a loss of toughness (95)}, is a matter of practical moment for metallic glasses. Ferromagnetic glasses must be annealed to induce uniaxial magnetic anisotropy, so as to minimize losses and maximize permeability (96), and this annealing can embrittle the material without crystallization. A core cannot then, for example, be readily unwound after annealing and rewound around a set of energizing and secondary coils without suffering chipping and cracking. The typical course of such embrittlement is indicated by Fig. 14 for $Fe_{80}B_{20}$ glass (97). After annealing for 2 hours at temperatures in excess of 225°C, ductility (as measured by bending at room temperature) is decreased considerably. Just prior to substantial crystallization (@ ~ 390°C), the surface strain matches the yield strain, i.e., the surface of the material reaches the yield stress, but the material can no longer sustain appreciable plastic flow. The crystallized material exhibits much less ductility still. The annealing temperature (T_b) at which such embrittlement first occurs is highly composition dependent. As noted above, T_b is ~ 225°C for $Fe_{80}B_{20}$,

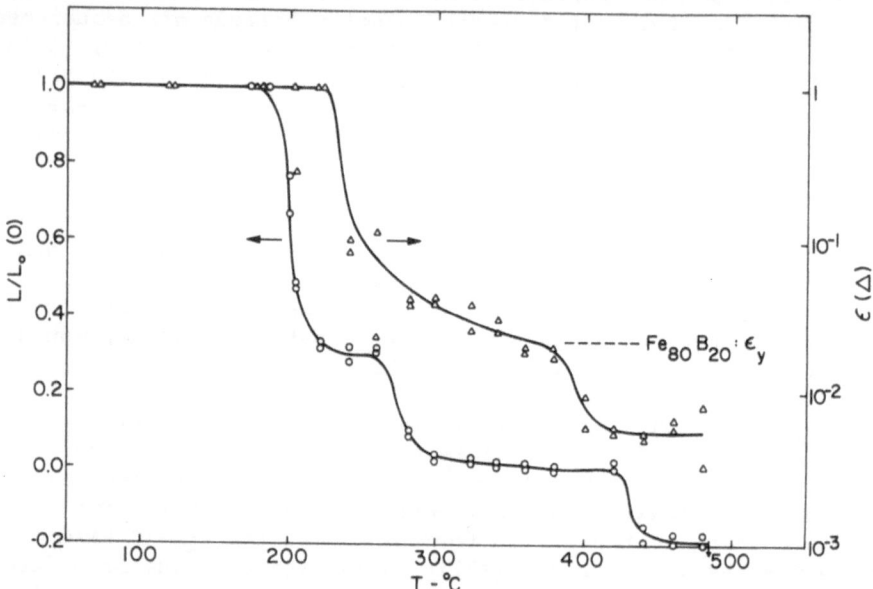

Fig. 14. Influence of thermal annealing (2 hrs. at T in vacuum) on
stress relief (L/L_0; circles) and ductility (ϵ; triangles)
of $Fe_{80}B_{20}$ glass (97).

while for Ni-Fe-P-B glasses it is only about 140°C (97); for Co-B
(13) and Ni-B dominated glasses, embrittlement first appears at
about 400°C or more when substantial crystallization occurs. Chen
(98) has shown that Fe-P and Fe-B are relatively less susceptible to
embrittlement than Fe-P-B glasses. Latuszkiewicz et al. (99) have
shown that T_b is much less than the crystallization temperature, T_c,
for Fe-rich $(FeNi)_{80}B_{10}Si_{10}$ glasses; for Ni content > 48 at % in
these glasses, $T_b \simeq T_c$, so ductility is lost only when crystallization occurs. The latter is also true for Pd-based glasses (100).

The atomic mechanism of low temperature embrittlement, i.e.,
$T_b<T_c$, remains a puzzle. As just indicated, it appears to occur
readily in mixed metal and/or mixed non-metal glasses, such as
Ni-Fe-P-B. This behavior appears analogous to the well known tendency for impurity metal-metalloid clustering in steels (101-103),
e.g., between Mn or Ni and P or S, which leads to their temper
embrittlement (104). Embrittlement can still occur, however, when
the only metal is Fe and when there is only one metalloid, B or P.
In this case, incipient phase separation, e.g., towards Fe and Fe_3P
regions, may generate internal stresses and subsequent embrittlement. It is noted that Fe atoms prefer to be eight fold coordinated, whereas ~ 12-fold coordination exists in the glass (22).
Crystalline Co and Ni are twelve fold coordinated; the driving force
for phase separation in their glasses should be less. In any event,

at the present time, there appears to be no means by which one can avoid thermal embrittlement of Fe-based glasses of practical importance. As suggested by the toughness data above, the very beginning of the structural changes responsible for embrittlement can actually occur during the initial quenching process (105). As a practical matter, then, one usually puts the coils around an annealed glassy magnetic core, rather than vice-versa.

FATIGUE

The fatigue lifetimes of a number of Ni-Fe-P-B base metallic glasses subjected to tension-tension loading are shown in Fig. 15 (106). The behaviors of Pd-Si (107), Co-Si-B (108), and Fe-B (108) dominated glasses appear to be essentially the same (although, these

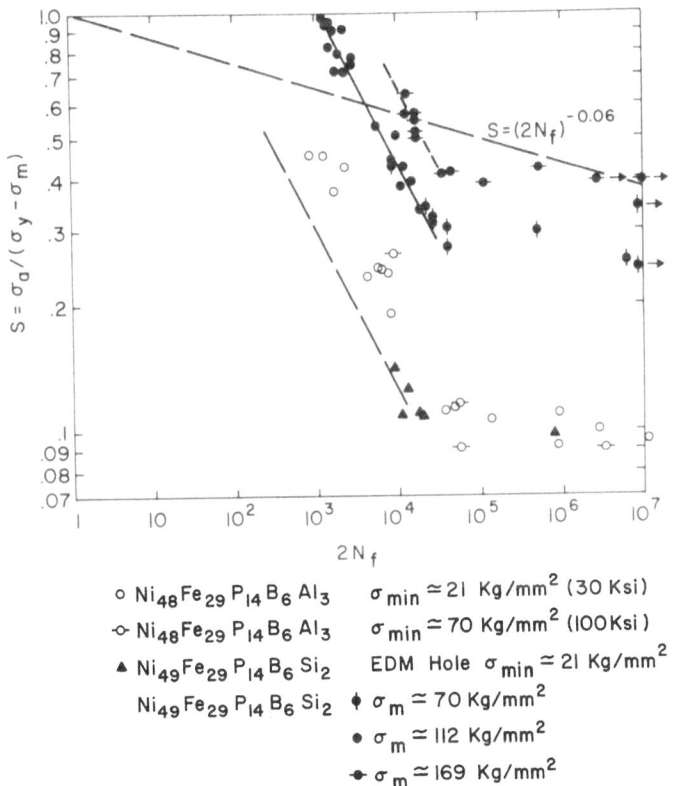

Fig. 15. Stress ratio vs. reversals to failure for Ni-Fe-P-B metallic glasses (106). The dotted line given by $S = (2N_f)^{-0.06}$ is the upper limit of fatigue lifetimes observed for high-strength steels. $S = \sigma_a/(\sigma_y - \sigma_m)$, where the stress amplitude $\sigma_a = (\sigma_{max} - \sigma_{min})/2$, σ_{max} and σ_{min} are the maximum and minimum cyclic stresses and the mean stress $\sigma_m = (\sigma_{max} + \sigma_{min})/2$.

materials should have modestly longer lifetimes if they fail in
shear, since the Ni-Fe-P-B glasses usually fail by mode I). It is
evident from the figure that fatigue lifetimes depend on the degree
of mechanical constraint. Samples with a central hole (solid
triangles) exhibit the shortest lifetimes; straight ribbons (open
symbols) suffer fretting with the grips, but last somewhat longer;
samples with a reduced area gage section avoid this constraint and
exhibit the longest lifetimes. The latter samples can sustain hundreds of cycles ($2N_f$, where N_f equals the number of stress reversals)
of stress to within a few percent of their yield (and fracture)
stresses, because, at least nominally, they behave elastically.
However, as the cyclic stress is reduced, sample lifetime does not
increase very rapidly. Eventually, at sufficiently low S, (defined
in the figure caption) it is no longer possible to effect crack initiation and infinite lifetime ($2N_f > 10^6$) is achieved. The case of
most practical interest, here, is for straight strips, where S is
of the order of 0.1 for $2N_f > 10^6$. For zero-tension loading, the peak
allowable stress for infinite lifetime will then be about 18% of σ_y,
which represents a relatively poor dynamic loading performance.

A comparison with the tensile fatigue behaviors of steels
(109) is shown by the upper dotted line in Fig. 15, marked S
$= (2N_f)^{-0.06}$. At high S steels experience shorter lifetimes, because they are loaded in their plastic regimes; as S is reduced,
sample lifetime increases rapidly as work hardening is more effective at thwarting persistent slip. As the fatigue crack propagation
rates in Ni-Fe-P-B glasses (89) are comparable to those in steels,
and, as it is known that ~ 90% of the fatigue lifetime of a steel
specimen is consumed in the crack initiation step, it is clear that
the short lifetimes of metallic glasses at low stresses reflect a
relative ease of crack initiation. This, once again, results
because of an absence of work hardening to disperse slip initiated
at surface or bulk inhomogeneities, which are, apparently, commonly
present.

The ability to resist crack initiation and raise the value of S
for "infinite" lifetime will depend, to some degree, on the toughness of a given metallic glass, but is likely to remain low in practical tensile loading situations. Improved performance appears to
be possible for the case of bending fatigue, where S can reach 0.35
to 0.75 (108); in experiments employing bending with some superposed
tension (110), the strain-lifetime behaviors of Fe-B-Si, Co-B-Si and
Ni-B-Si glasses appear to be about the same as for 304 stainless
steel. For both these sets of experiments, the improved fatigue
resistances of the glassy alloys reflect the greater difficulty of
crack initiation in the absence of large triaxial stresses.

The fatigue behaviors exhibited in Fig. 15 require cyclic
plastic deformation. Static fatigue failures, which commonly occur
in the ordinary laboratory environment in the case of oxide glasses

(111,112), only occur in extremely agressive environments, such as in acid solutions or liquid metals, in the case of metallic glasses (113,114).

Fatigue resistance is also a key issue in the case of metallic glass ribbon reinforced resin composites. One may produce such composites with high static strengths, especially under transverse loading conditions, consistent with the high strengths of the glassy alloy reinforcing elements (115-117). However, the fatigue characteristics of the ribbons also dictate the fatigue performances of the composites. Hence, under longitudinal cyclic loading, such composites can sustain peak stresses (σ max) of only ~ 11% of the composite strength (σ comp) for >10^6 cycles to failure (118). This compares rather unfavorably with (σ max/σ comp) ~0.7 for graphite epoxy composites (119).

MAGNETOELASTICITY

Magnetostrictive metallic glasses may experience elastic deformation when exposed to a magnetic field; when stressed they may experience a change of magnetization (120,121). The magnitudes of these effects are influenced by the saturation magnetostriction, λ_s, the magnetic domain configurations and the ease with which the domains may move; the apparent elastic stiffness will vary accordingly. If a ribbon containing magnetic domains oblique to its long axis is stressed in tension (positive magnetostriction material) parallel to its long axis, its apparent modulus will be reduced to $\sigma/(\epsilon+\epsilon_1)$, where ϵ is the intrinsic elastic strain and ϵ_1 is an additional magnetostrictive strain, which can achieve the maximum value of $3\lambda_s/2$. This effect will appear until either the stress is high enough or a high enough field is applied so that the material becomes magnetized parallel to the tensile axis. Hence, intrinsic elastic properties may be measured using acoustic techniques when a magnetic field is applied to saturate the material. This approach was used to measure the values of E cited above and reported in references 25,32 and 33. Poisson's ratio may also be measured by precision strain gauge techniques at sufficiently high stress to be free of magnetostrain effects (32).

Large differences between E in the saturated and demagnetized states, as determined by sonic techniques, have been observed for a variety of metallic glasses (120); values of E_s/E_d approaching 1.6 to 1.8 were observed for Fe-B glasses by Kikuchi et al. (122) {although the absolute values of E_s appear to be 10 to 20% higher than those reported in ref. 25 and inferred from ref. 33}. A strong interaction between elastic and magnetic properties also leads to Invar (low thermal expansion coefficient, α) and Elinvar (low temperature coefficient of Young's modulus, dE/dT) anomalies in the demagnetized state and near-zero values of α and dE/dT are also observed for Fe-B glasses (122-124).

118

Magnetoelastic effects are important in the applications of magnetostrictive metallic glasses as transformer cores. As-quenched Fe-based glassy metal strips contain local residual compressive stresses, inducing transverse domains which are difficult to magnetize parallel to the ribbon axis (125). To eliminate such domains, a magnetic core is annealed close to its Curie temperature in the presence of a circumferential magnetic field. This produces a material with broad domains lying along the ribbon length, each magnetized approximately parallel to the ribbon axis. In this condition, magnetization reversal occurs by domain wall translation, rather than by domain rotation (the net magnetostrain approaches zero), core saturation is readily achieved (the d.c. B-H loop is nearly square) and core loss and exciting power are minimized on low frequency (50-60Hz) excitation, where hysteresis losses account for about half the losses. At higher frequencies, on the other hand, eddy current losses become more important and must be minimized by domain size refinement. This is achieved in magnetostrictive metallic glasses by slight crystallization of the glass (126-128), as the crystals act as nucleation sites for domain walls. Such domains typically lie, however, transverse to the ribbon axis. Nonetheless, the fine domain size mitigates magnetostrain effects to minimize high frequency core losses and maximize permeability, except when critical combinations of frequency and core size lead to mechanical resonance of the core (129).

A transverse domain configuration is desirable when one wishes to take advantage of the magnetoelastic properties of metallic glasses to transduce magnetic to mechanical energy or vice-versa. Rare-earth iron materials can exhibit orders of magnetude higher magnetostrains and ceramic materials can exhibit large piezostrains, but, in each case, only at the expense of very large drive fields. In contrast, metallic glasses can exhibit quite low magnetic anisotropies (K_u), of the order of 100 J/m^3, or less, and can be magnetized relatively easily. Hence, since the magnetomechanical coupling factor, k, is proportional to $\lambda^2_s E_s/K_u$ (121), it can be quite large. Values ranging from 0.8 to 0.96 (k=1 corresponds to perfect coupling) have been reported for optimally annealed Fe-base metallic glasses (130-134); these compare with k=0.3 to 0.7 for crystalline magnetostrictive and piezoelectric materials.

CONCLUSION

Metallic glasses are a study in contrasts: stronger and harder than ordinary metals, stiffer and tougher than ordinary glasses, capable of enormous plasticity and yet brittle in the classical engineering sense, nominally isotropic and yet prone to local inhomogeneities which promote mode I failures and reduce fatigue lifetime. All these features reflect, to some degree, their unique combination of metallic bonding and dense, random-close-packed, non-crystalline structure. Deformation characteristics are sensitive to bonding and structure in the qualitative sense and, hence, are the same for all metallic glasses. Strength, stiffness and hardness depend on the specifics of bonding interactions between metals and nonmetals and on subtle local variations of structure; hence, they are strongly composition dependent. The resistances of metallic glasses to loss of ductility on themal annealing are also sensitive to bonding and structure specifics. Such specifics "cancel" out, however, when comparing the properties of a given glass; hence τ_y/μ, H_v/σ_y and fatigue lifetime, on cycling to a given fraction of the yield stress, are approximately constant for metallic glasses.

At this juncture, no straightforward load bearing application of a metallic glass appears to be in the offing. Nonetheless, the mechanical properties of glassy alloys are of practical importance in the development of metallic glass technology. The ability to catch and wind ribbons emanating from a casting wheel at ~ 1000 m/min depends on their strengths and toughnesses. An alloy which requires an excessive cooling rate to maintain ductility cannot be readily handled; nor would it be readily adapted for further use. Magneto-elastic properties play a key role in magnetic applications. To reduce losses in transformer cores, one must anneal the core materials to induce uniaxial magnetic anisotropy parallel to the direction of the applied field. No domain rotation will then occur; that is, the magnetostrictive strain will approach zero. In the opposite case, i.e., for a transducer, one anneals to produce transverse domains to maximize the magnetostrictive strain and the conversion of magnetic to mechanical energy, or vice-versa. Beyond these practical considerations, it is clear that the mechanical responses of metallic glasses will continue to be of interest, as part of a broader attempt to advance the basic understanding of structure-property relationships.

REFERENCES

(1) W. Klement, R. H. Willens and P. Duwez, Nature 187, 869 (1960)
(2) P. Duwez and R. H. Willens, Trans. AIME 227, 362 (1963)
(3) G. S. Cargill, III, J. Appl. Phys. 41, 12 (1970)
(4) H. S. Chen and D. Turnbull, J. Chem. Phys. 48, 2560 (1968)
(5) T. Masumoto and R. Maddin, Acta Met. 19, 725 (1971)
(6) H. J. Leamy, H. S. Chen and T. T. Wang, Met. Trans. 3, 699 (1972)
(7) D. E. Polk, B. C. Giessen and F. S. Gardner, Mat. Sci. Eng. 23, 309 (1976)
(8) R. Hasegawa, J. Mag. Mag. Mat. 41, (1984) in press
(9) D. Raskin and L. A. Davis, IEEE Spectrum, Nov. 1981, p. 28
(10) D. Raskin and C. H. Smith, in "Amorphous Metallic Alloys", F. E. Luborsky, Ed., Butterworths, London (1983) p. 381
(11) C. A. Pampillo, J. Mat. Sci. 10, 1194 (1975)
(12) L. A. Davis, in "Rapidly Quenched Metals", N. J. Grant and B. C. Giessen, Eds., MIT Press, Cambridge, MA (1976) p. 369

(13) L. A. Davis, in "Metallic Glasses", J. J. Gilman and H. J. Leamy, Eds., ASM, Metals Park, OH (1978) p. 190

(14) F. Spaepen, in "Rapidly Quenched Metals III", B. Cantor, Ed., The Metals Society, London (1978) V.2, p. 253

(15) H. S. Chen, Rep. Prog. Phys. 43, 353 (1980)

(16) J. C. M. Li, in "Treatise on Materials Science and Technology", H. Herman, Ed., Academic Press, NY (1981) V.20, p. 325

(17) J. C. M. Li, in "Rapidly Quenched Metals", T. Masumoto and K. Suzuki, Eds., Japan Inst. Metals, Sendai, Japan (1982) V.II, p. 1335

(18) J. C. M. Li, in "Chemistry and Physics of Rapidly Solidified Materials", B. J. Berkowitz and R. O. Scattergood, Eds., The Metallurgical Soc. of AIME, Warren, PA (1983) p. 173

(19) H. Kimura and T. Masumoto, in "Amorphous Metallic Alloys", F. Luborsky, Ed., Butterworths, London (1983) p. 187

(20) F. Spaepen and A. Taub, ibid, p. 231

(21) D. Weaire, M. F. Ashby, J. Logan and M. J. Weins, Acta Met. 19, 779 (1971)

(22) G. S. Cargill, III, in "Solid State Physics", F. Seitz, D. Turnbull and H. Ehrenreich, Eds., Academic Press, NY (1975) V.30, p.227

(23) R. Ray, R. Hasegawa, C. P. Chou and L. A. Davis, Scripta Met. 11, 973 (1977)

(24) B. Golding, B. G. Bagley and F.S.L. Hsu, Phys. Rev. Lett. 29, 68 (1972)

(25) L. A. Davis, Y. T. Yeow and P. M. Anderson, J. Appl. Phys. 53, 4834 (1982)

(26) D. Turnbull, J. Physique, Colloque 4, 35, 1 (1974)

(27) J. J. Gilman, Science 208, 856 (1980)

(28) J. J. Gilman, Phil. Mag. B37, 577 (1978)

(29) A. Makishima and J. D. Mackenzie, J. Non-Cryst. Solids 17, 147 (1975)

(30) C. Kittel, "Introduction to Solid State Physics", John Wiley, NY (1956) p.5

(31) H. S. Chen, J. T. Krause and E. Coleman, J. Non-Cryst. Solids 18, 157 (1975)

(32) C. P. Chou, L. A. Davis and R. Hasegawa, J. Appl. Phys. 50, 3334 (1979)

(33) A. Inoue, H. S. Chen, J. T. Krause, T. Masumoto and M. Hagiwara, J. Mat. Sci. 18, 2743 (1983)

(34) W. J. M. Tegart, "Elements of Mechanical Metallurgy", MacMillan Co., NY (1966) p. 94

(35) A. Makishima and J. D. Mackenzie, J. Non-Cryst. Solids 12, 35 (1973)

(36) L. A. Davis and Y. T. Yeow, J. Mat. Sci. 15, 230 (1980)

(37) L. A. Davis and S. Kavesh, J. Mat. Sci. 10, 453 (1975)

(38) C. A. Pampillo and H. S. Chen, Mat. Sci. Eng. 13, 181 (1973)

(39) H. Kimura and T. Masumoto, Mat. Sci. Eng., in press

(40) S. Takayama, Mat. Sci. Eng. 38, 41 (1979)

(41) M. Hagiwara, A. Inoue and T. Masumoto, Met. Trans. A 13A, 373

(1982)

(42) L. A. Davis, Scripta Met. $\underline{9}$, 431 (1975)

(43) S. Takayama and R. Maddin, Acta Met. $\underline{23}$, 943 (1975)

(44) S. Takayama, J. Mat. Sci. $\underline{16}$, 2411 (1981)

(45) D. M. Marsh, Proc. Roy. Soc. $\underline{A279}$, 420 (1964)

(46) A. H. Cottrell, "The Mechanical Properties of Matter", John Wiley, NY (1964) p. 322

(47) T. Murata, T. Masumoto and M. Sakai, in "Rapidly Quenched Metals III", B. Cantor, Ed., The Metals Society, London (1978) V.2, p. 401

(48) L. A. Davis, Scripta Met. $\underline{9}$, 339 (1975)

(49) L. E. Murr, O. T. Inal and S. H. Wang, Mat. Sci. Eng. $\underline{49}$, 57 (1981)

(50) L. A. Davis, Scripta Met. $\underline{16}$, 993 (1982)

(51) F. Spaepen, Acta Met. $\underline{23}$, 615 (1975)

(52) H. Kimura, D. G. Ast and W. A. Bassett, J. Appl. Phys. $\underline{53}$, 3523 (1982)

(53) J. C. M. Li, in "Metallic Glasses", J. J. Gilman and H. J. Leamy, Eds., ASM, Metals Park, OH (1978) p. 224

(54) J. J. Gilman, J. Appl. Phys. $\underline{44}$, 675 (1973)

(55) J. C. M. Li, in "Frontiers in Materials Science", L. E. Murr and C. Stein, Eds., Marcel Dekker, NY (1976) p. 527

(56) T. M. Ahn and J. C. M. Li, Scripta Met. $\underline{14}$, 1057 (1980)

(57) A. I. Taub and F. Spaepen, Scripta Met. $\underline{14}$, 1197 (1980)

(58) P. B. Hirsch, A. Howie, R. B. Nicholson, D. W. Pashley and M. J. Whelan, "Electron Microscopy of Thin Crystals", Butterworths, London (1965) p. 169

(59) A. Argon and H. Y. Kuo, in "Rapidly Quenched Metals III", B. Cantor, Ed., The Metals Society, London (1978) V.2, p. 269

(60) A. Argon and H. Y. Kuo, Mat. Sci. Eng. $\underline{39}$, 101 (1979)

(61) A. Argon, Phys. Chem. Solids $\underline{43}$, 945 (1982)

(62) M. Doyama, R. Yamamoto and H. Matsuoka, in "Rapidly Quenched Metals", T. Masumoto and K. Suzuki, Eds., Japan Inst. Metals, Sendai, Japan (1982) V.I, p. 233

(63) P. S. Steif, F. Spaepen and J. W. Hutchinson, Acta Met. $\underline{30}$, 447 (1982)

(64) P. E. Donovan and W. M. Stobbs, Acta Met. $\underline{29}$, 1419 (1981)

(65) V. K. Sethi, R. Gibala and A. H. Heuer, Scripta Met. $\underline{12}$, 207 (1978)

(66) Deguo Deng and Banghong Lu, Scripta Met. $\underline{17}$, 515 (1983)

(67) J. Megusar, A. S. Argon and N. J. Grant, in "Rapidly Quenched Metals", T. Masumoto and K. Suzuki, Eds., Japan Inst. Metals, Sendai, Japan (1982) V.II, p. 1411

(68) J. Megusar, A. S. Argon and N. J. Grant, in "Rapidly Solidified Amorphous and Crystalline Alloys", B. H. Kear, B. C. Giessen and M. Cohen, Eds., North Holland, NY (1982) p. 283

(69) D. Srolovitz, V. Vitek and T. Egami, Acta Met. $\underline{31}$, 335 (1983)

(70) R. Hill, "The Mathematical Theory of Plasticity", Oxford Univ. Press, London (1967) p. 300

(71) A. K. Chakrabarti and J. W. Spretnak, Met. Trans. A $\underline{6A}$, 733

122

(1975)

(72) Idem, ibid. 6A, 737 (1975)

(73) A. S. Argon, in "The Inhomogeneity of Plastic Deformation", ASM, Metals Park, OH (1973) p. 161

(74) M. Naka, S. Tomizawa and T. Masumoto, in "Rapidly Quenched Metals", N. J. Grant and B. C. Geissen, Eds., MIT Press, Cambridge (1976) p. 273

(75) A. H. Cottrell, "The Mechanical Properties of Matter", John Wiley, NY (1964) p. 313

(76) R. Hill, "The Mathematical Theory of Plasticity", Oxford Univ. Press, London (1967) p. 213

(77) D. M. Marsh, Proc. Roy. Soc. A282, 33 (1964)

(78) S. K. Das, E. M. Norin and R. L. Bye, in "Rapidly Solidified Metastable Materials", Materials Res. Soc. 1983 Annual Mtg., Boston, MA, Nov. 1983, to be published

(79) M. Yamane and J. D. Mackenzie, J. Non-Cryst Solids 15, 153 (1974)

(80) P. M. Sargent and P. E. Donovan, Scripta Met. 16, 1207 (1982)

(81) L. A. Davis, C. P. Chou, L. E. Tanner and R. Ray, Scripta Met. 10, 937 (1976)

(82) S. H. Whang, D. E. Polk and B. C. Giessen, in "Rapidly Quenched Metals", T. Masumoto and K. Suzuki, Eds., Japan Inst. Metals, Sendai, Japan (1982) V.II, p. 1365

(83) C. P. Chou, L. A. Davis and M. C. Narasimhan, Scripta Met. 11, 417 (1977)

(84) R. C. Morris, J. Appl. Phys. 50, 3250 (1979)

(85) F. Spaepen and D. Turnbull, Scripta Met. 8, 563 (1974)

(86) F. Speapen, Acta Met. 25, 407 (1977)

(87) Metals Handbook, J. A. Fellows, Ed., ASM, Metals Park, OH (1974) V.9, p.206, micrograph #3787

(88) L. A. Davis, Met. Trans. 10A, 235 (1979)

(89) L. A. Davis, J. Mat. Sci. 10, 1557, (1975)

(90) S. Henderson, J. V. Wood and G. W. Weidmann, J. Mat. Sci. Letts. 2, 375 (1983)

(91) J. F. Knott, "Fundamentals of Fracture Mechanics", Butterworths, London (1973)

(92) G. T. Hahn, M. F. Kanninen and A. R. Rosenfield, Ann. Rev. Mat. Sci. 2, 381 (1972)

(93) V. F. Zackay, E. R. Parker, J. W. Morris, Jr. and G. Thomas, Mat. Sci. Eng. 16, 201 (1974)

(94) F. E. Luborsky, J. J. Becker and R. O. McCary, IEEE Trans. Mag. MAG-11, 1644 (1975)

(95) D. G. Ast and D. Krenitsky, Proc. Second Int. Conf. on Rapidly Quenched Metals, Sec. II, Mat. Sci. Eng. 23, 241 (1976)

(96) R. Hasegawa in "Glassy Metals: Magnetic, Chemical and Structural Properties", R. Hasegawa, Ed., CRC Press, Boca Raton, FL (1983) p. 163

(97) L. A. Davis, R. Ray, C.-P. Chou and R. C. O'Handley, Scripta Met. 10, 541 (1976)

(98) H. S. Chen, Scripta Met. 11, 175 (1978)

(99) J. Latuszkiewicz, P. G. Zielinski and H. Matyja, in "Rapidly Quenched Metals", T. Masumoto and K. Suzuki, Eds., Japan Inst. Metals, Sendai, Japan (1982) V.II, p. 1381

(100) H. Kimura and T. Masumoto, Acta. Met. 28, 1677 (1980)

(101) M. Guttman, Surface Sci. 53, 213 (1975)

(102) Idem, Mat. Sci. Eng. 42, 227 (1980)

(103) Ph. Dumoulin and M. Guttman, Mat. Sci. Eng. 42, 249 (1980)

(104) A. Tetelman and A. J. McEvily, Jr., "Fracture of Structural Materials", John Wiley, NY (1967) p. 529

(105) U. Koster, U. Herold and H.-G. Hillenbrand, Scripta Met. 17, 867 (1983)

(106) L. A. Davis, J. Mat. Sci. 11, 711 (1976)

(107) T. Ogura, T. Masumoto and K. Fukushima, Scripta Met. 9, 109 (1975)

(108) A. L. Mulder, J. W. Drijver and S. Radelaar, in "Rapidly Quenched Metals", T. Masumoto and K. Suzuki, Eds., Japan Inst. Metals, Sendai, Japan (1982) V.II, p. 1361

(109) R. W. Landgraf, in "Achievement of High Fatigue Resistance in Metals and Alloys", STP467, ASTM, Philadelphia, PA (1970) p. 3

(110) M. Doi, K. Sugiyama, T. Tono and T. Imura, Jpn. J. Appl. Phys. 20, 1593 (1981)

(111) C. Gurney and S. Pearson, Proc. Roy. Soc. A192, 537 (1947-48)

(112) R. Adams and P. W. McMillan, J. Mat. Sci. 12, 643 (1977)

(113) M. D. Archer and R. J. McKim, J. Mat. Sci. 18, 1125 (1983)

(114) S. Ashok, N. S. Stoloff, M. E. Glicksman and T. Slavin, Scripta Met. 15, 331 (1981)

(115) J. R. Strife and K. M. Prewo, Nat. SAMPE Tech. Conf. Series 11, 719 (1979)

(116) Y. T. Yeow, Composites Technology Rev. 2, 17 (1980)

(117) Y. T. Yeow, in "Composite Materials: Testing and Design" STP 787, I. M. Daniel, Ed., ASTM, Philadelphia, PA (1983) p. 101

(118) J. R. Strife and K. M. Prewo, Rep. #AFWAL-TR-80-4060, Materials Laboratory, Air Force Wright Aeronautical Laboratories, Wright-Patterson AFB, OH, April 30, 1980

(119) B. G. Agarwal and L. J. Brontman, "Analysis and Performance of Fiber Composites", John Wiley, NY (1980) p. 241

(120) B. S. Berry, in "Metallic Glasses", J. J. Gilman and H. J. Leamy, Eds., ASM, Metals Park, OH (1978) p. 161

(121) J. D. Livingston, Phys. Stat. Sol.(a) 70, 591 (1982)

(122) M. Kikuchi, K. Fukamichi, T. Masumoto, T. Jagielinski, K. I. Arai and N. Tsuya, Phys. Stat. Sol.(a) 48, 175 (1978)

(123) K. Fukamichi, M. Kikuchi, S. Arakawa and T. Masumoto, Sol. State Comm. 23, 955 (1977)

(124) K. Fukamichi, in "Amorphous Metallic Alloys", F. E. Luborsky, Ed., Butterworths, London (1983) p. 317

(125) H. Kronmuller and B. Groger, J. Physique 42, 1285 (1981)

(126) L. A. Davis, N. J. DeCristofaro and C. H. Smith, in "Metallic Glasses: Science and Technology", C. Hargitai, I. Bakonyi and T. Kemeny, Eds., Central Research Institute for Physics, Budapest, Hungary (1981) p. 1

(127) R. Hasegawa, G. E. Fish and V. R. V. Ramanan, in "Rapidly Quenched Metals", T. Masumoto and K. Suzuki, Eds., Japan Inst. Metals, Sendai, Japan (1982) V.II, p. 929

(128) R. Hasegawa, V. R. V. Ramanan and G. E. Fish, J. Appl. Phys. 53, 2276 (1982)

(129) V. J. Thottuvelil, T. G. Wilson, I. Miyazaki and H. A. Owen, Jr., IEEE PESC RECORD - 1983, in press

(130) M. A. Mitchell, J. R. Cullen, R. Abbundi, A. Clark and H. Savage, J. Appl. Phys. 50, 1627 (1979)

(131) C. Modzelewski, H. T. Savage, L. T. Kabacoff and A. E. Clark, IEEE Trans. Mag. MAG-17, 2837 (1981)

(132) M. L. Spano, K. B. Hathaway and H. T. Savage, J. Appl. Phys. 53, 2667 (1982)

(133) H. T. Savage and M. L. Spano, J. Appl. Phys. 53, 8092 (1982)

(134) S. W. Meeks and J. C. Hill, J. Appl. Phys. 54, 6584 (1983)

HOMOGENEOUS FLOW AND ANELASTIC/PLASTIC DEFORMATION
OF METALLIC GLASSES

J. PEREZ

Groupe d'Etudes de Métallurgie Physique et de Physique
des Matériaux - L.A. au C.N.R.S. n° 341 - I.N.S.A.
Bât. 502 - 69621 - VILLEURBANNE CEDEX - FRANCE

ABSTRACT

Non elastic deformation of glassy metallic alloys is analysed.
Both the anelastic and viscoplastic aspect of deformation, the relation between strain rate and stress, and the effect of structural
relaxation are considered. A model of homogeneous deformation of
metallic glasses near Tg which takes into account all the experimental features is proposed. According to this model, plastic deformation is principally dependent on the recovery processes (implying
atomic diffusion) which occur after the shear microdomains are formed. By introducing a distribution of times, which is characteristic of the thermo-mechanical nucleation of the shear microdomains,
expressions, which quantitatively describe the experimental data,
are obtained.

INTRODUCTION

There have been numerous investigations on homogeneous flow in
metallic glasses by creep and stress relaxation measurements. Among
the theories which have been proposed, we shall mention that of
Spaepen (1) and that of Argon (2) who have applied transition state
theory to the flow process. Although some discrepancies in the measurements reported by different investigators have been clarified
by taking into account (i) the so called equilibrium or isoconfigurational properties of an amorphous system (3) and (ii) the
effect of preannealing (4), some questions on the following points
still remain :

- (i) the stress-strain rate relationship is generally linear at
low stress levels and non linear when stress is increased but large.

discrepancies. are found in the values of the strain rate sensiti-
vity $m = (\partial \ln \dot{\gamma})/\partial \ln \sigma$ which can vary between 1 and 10 (5).

- (ii) structural relaxation is accepted to be the origin of the
decrease of strain rate following annealing treatment (4), but the
details of the physical processes are not very clear.

-(iii) the models quoted above (1, 2) describe the plastic flow of
metallic glasses but actually, the situation is more complicated
as both anelastic and viscoplastic behaviours are often observed
simultaneously (6).

In previous papers (6, 7) some results of mechanical tests
obtained on iron based amorphous alloys illustrating the above mentionned points, were presented ; some new concepts were proposed in
order to take into account the so far unexplained experimental features ; such an analysis emphasized the role of atomic diffusion
resulting in recovery process and a balance between hardening and
recovery was shown to result in plastic flow. We will discuss a
quantitative improvement of this interpretation by first giving the
physical assumptions which allow us to develop numerical calculations for obtaining creep, recoverable deformation, stress-strain or
stress relaxation curves.

MODEL OF NON ELASTIC DEFORMATION IN METALLIC GLASSES

PHYSICAL BASIS

The experimental study of the non elastic deformation of metallic glasses at high temperatures is generally done either by mechanical tests (tensile or compression test, creep, stress relaxation)
or by internal friction measurements. In the former, attention is
given to the steady state regime of plastic deformation ; in the
latter anelasticity is generally considered. It is clear that a description of the entire non elastic deformation on the basis of a
unique set of physical assumptions is desirable.

In order to explain the plastic flow of glasses, the general
hypothesis according to which cooperative rearrangements of atoms
become possible as the temperature approaches Tg is often used. This
has lead to different models and relations which give the strain
rate $\dot{\gamma}$. Indeed, the idea about an elementary shear associated with
localized rearrangement of groups of atoms which results in a macroscopic deformation, has often been considered. This has been developed by Eyring (8) who, along with the rate kinetics, has proposed
an analysis of the flow of liquidds. The plastic flow can be described by an equation such as :

$$\dot{\gamma} = \frac{2N_o v_1 \Delta \gamma}{\tau_m} \sinh \frac{f v_a \sigma}{2kT} \tag{1}$$

where N_0 is the number of elementary domains of volume v_1 in which the local shear is $\Delta\gamma$ per unit volume of matter ; τ_m the average delay for one event ; v_a an activation volume defined by considering that the applied stress σ does the work of magnitude $fv_a\sigma$ (f is a factor relating σ to the effective shear component of the stress).

The more recent developments of such a description allow us to define N_0 : Spaepen (1) defines N_0 as the number of sites with high free volume per unit volume ; Chen (9) defines it as the density of sites with high configurational entropy. These satisfactorily explain the temperature dependence corresponding to either isoconfigurationnal or equilibrium condition as well as the structural relaxation effects.

As with most materials, the mechanism of non elastic deformation of metallic glasses is closely linked to their microstructure and we shall assume that microscopic events leading to shear strain imply "defects". Several types of defects may be present in amorphous metals and Spaepen (10) has distinguished different types of possible atomic rearrangement which depend on the details of the configuration of sites.

(i) a first set of sites produce a long range elastic strain field so that they can annihilate free volume and are therefore implied in structural relaxations ;

(ii) another set can exhibit local shear and contributes to plastic flow and

(iii) a third set of sites are those which upon rearrangements lead to a change in the local nearest neighbor configuration and hence contribute to the diffusion. Another definition of possible different structural defects has been proposed by Srolovitz et al (11), who have calculated the atomic-level stresses from a statistical analysis of a computer model. The defects are then defined as regions in which the different components of the atomic-level stress deviates significantly from their average value. Thus three classes of defects are found : (i) positive (p-type defects) and (ii) negative (n type defects) local density fluctuations (i.e. fluctuations in the atomic-level hydrostatic pressure) and (iii) regions of large shear stresses (τ-defects). p -and n- type defect could contribute to diffusion and structural relaxation could be the consequence of annihilation of p - and n -type defects. Plastic deformation could result from atomic rearrangements linked to properties of τ - defects.

QUALITATIVE ASPECT OF THE MODEL

We propose that the basic deformation mechanism is nucleation of shear microdomains (as we proposed several years ago (6)), which we regard as τ- defects. The nucleation occurs under the effect of the applied stress and is thermally assisted. But where can such a nucleation occur ? A general case of shear microdomain is illustrated in figure 1 : the shear is along the surface S and the cooperative atomic rearrangement occurs inside the volume of matter limited by the surface Σ. The curve C_n, defined by the intersection of Σ and S, separates the area S_1 where shear has occured from the non sheared part of S. In the mechanics of continuous media, the line C_n is a dislocation loop : in crystals such a dislocation is of Volterra type or, in a more realistic way, of Peieirls type ; in amorphous solids, dislocations, as far as this concept is valid, have to be of the Somigliana type. The nucleation rate has been calculated by several authors who also determined the height ΔG^* of the free energy barrier and the loop radius R. The activation free enthalphy is given by :

$$\Delta G = \Delta E_{el} + \Delta E_b - \Delta E_w \qquad (2)$$

with ΔE_{el} is the elastic energy of the line defect, ΔE_b is the difference between the binding energy, and ΔE_w is the work done by the applied shear stress τ. Bowden and Raha (12) solved equation (2) by neglecting the term ΔE_b to obtain,

$$\Delta G^* \simeq \frac{1}{4} \mu <b>^2 R^* \quad \text{and} \quad R^* \simeq 0,1 \frac{\mu<b>}{\tau}$$

with μ as shear modulus and $<b>$ the mean value of the shear vector. For metallic glasses with $\mu = 4 \times 10^{10}$ Pa, $\tau \simeq 10^8$ Pa and $<b> \approx 3.10^{-10}$ m we obtained $R^* = 1.2 \times 10^{-8}$ m and $G^* = 70$ eV. These values are unrealistic. Argon (2) has also taken into consideration equation (2) but the loop radius was considered constant, but determined by the structure, and the shear vector was used as the activation path parameter. This also seems questionnable. Furthermore the results obtained by Argon (as those of Bowden and Raha) are limited to high stress, since the reversibility of the reaction rate is ignored. In conclusion, it appears improbable that shear domains are nucleated everywhere in the amorphous solid. We therefore, propose that nucleation can occur only in those regions where resistance to shear is appreciably weaker than in the rest of the material. Such soft sites may be regarded as another type of defect here after called "shear-source", and the thermomechanical activation of a "shear-source" - defect may lead to the formation of a shear microdomain or τ - defect. When the stress is suppressed, the solid recovers its previous configuration and the τ- defect disappears : such a view would then explain the anelastic behaviour of metallic glasses. In order to obtain plastic deformation, the expansion of a τ-defect must be necessarily invoked, but line C_n which is boardening this defect and which could be considered as a somigliana-type dislocation as previously noticed, is a sessile defect. Nevertheless, such an expansion can be obtained either at a high stress or at high temperature through diffusionnal mechanism. This expansion

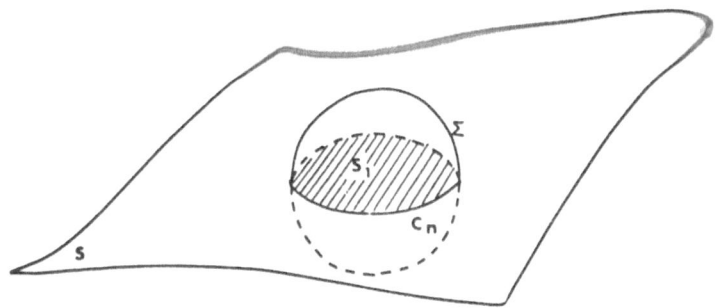

Figure 1 : Schematic representation of a shear microdomain : S_1 is the sheared area.

covers a distance of mean value, l_c, at which the line C_n losses its identity by combination with other similar lines formed by the neighbouring τ- defects and becomes ineffective. The "shear-source" can then be activated again as far as the stress is applied.

QUANTITATIVE DESCRIPTION

A set of equations based on the qualitative vew mentionned above was proposed elsewhere (13) and the main points will be recalled. If there are N_0 "shear-source" - defects per unit volume, n being non activated and N_0 - n activated (i.e. in the state of a τ defect), it is calculated through classical arguments that the application of the stress σ induces activation of ΔN defects (figure 2) with

$$\Delta N = n(0) - n(\infty)$$

and
$$\Delta N \simeq \alpha N_0 \frac{f\sigma v_a}{kT} \text{ with } \begin{cases} \alpha = 1/4 \text{ if } \Delta u \simeq 0 \\ \alpha = 1/2 \exp\left(-\frac{\Delta u}{kT}\right) \text{ if } \Delta u > kT \end{cases} \quad (3)$$

Once activated, the defect can expand as previously discussed and we have shown in figure (2b) variation of the free energy G(R) as a function of the radius R of the τ - defect where
dotted line corresponds to the variation of the elastic energy, given by,

$$\Delta E_{el} \simeq \mu <b>^2 R$$

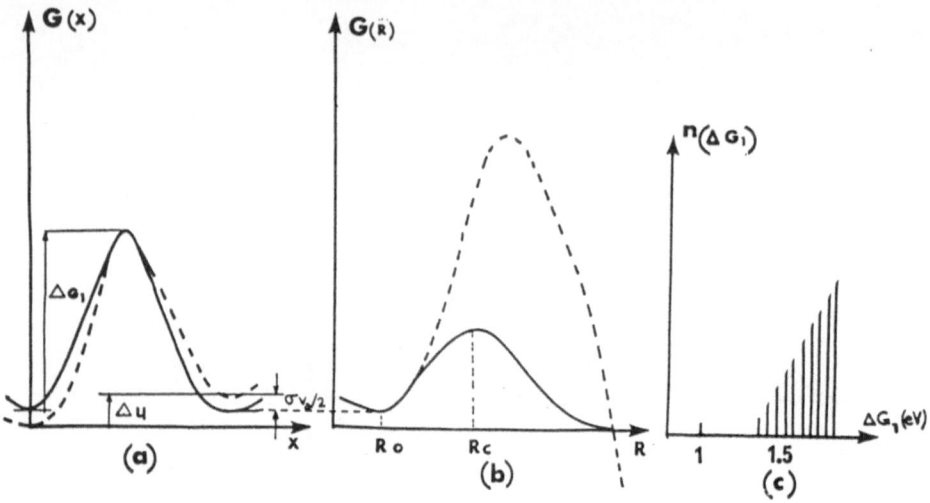

Figure 2 : a) Free enthalpy diagram corresponding to the whole set of atoms constituting the "shear source" - defect ; b) Variation of the free enthalpy with the radius of the sheared area assuming a constant shear vector (dotted line) or a shear vector decreasing toward zero (full line) ; c) Triangular distribution of ΔG_1

and that of the work of the stress, given by,

$$\Delta E_W \simeq - f\sigma <b> \Pi R^2$$

But, as said before, line C_n is annihilated after the expansion and the previous energy level of the non activated defect recovered, as shown by the continuous curve of figure (2b).

If the stress is maintained, the "shear-source" - defect can be activated again (activation 1) and the new τ - defect can expand (mechanism 2) and so on. Now the kinetic of such a mechanism can be developed. Let us write τ_1 as the mean time of activation 1, τ_2 the mean time of mechanism 2 and $N(t)$ the number of non activated "shear-source" defects per unit volume. The rate of variation of $N(t)$ is given by :

$$\frac{dN(t)}{dt} = - \frac{N(t) - n(\infty)}{\tau_1} + \frac{n(0) - N(t)}{\tau_2} \tag{4}$$

and the strain rate is given by

$$\gamma = \gamma_1 \frac{N(t) - n(\infty)}{\tau_1} + \gamma_2 \frac{n(0) - N(t)}{\tau_2} \tag{5}$$

$\gamma_1 = \Delta\gamma$. v_1 is the deformation induced by one elementar process 1 and $\gamma_2 = A <b>$ that due the mechanism 2 (A is the area swept by the loop C_n and $<b>$ is the mean shear vector of this loop).

From equation (4) and (5), the strain can be calculated (13)

$$\gamma = A\sigma \left[\frac{\tau_1(\tau_2 - \tau_1)}{(\tau_1 + \tau_1)^2} \{1 - \exp(-\frac{t}{\tau})\} + \frac{2t}{\tau_1 + \tau_2} \right] \qquad (6)$$

with $1/\tau = 1/\tau_1 + 1/\tau_2$ and $A = \alpha\gamma_0 f v_a No/kT$ $(\gamma_0 \simeq \gamma_1 \simeq \gamma_2)$

Equation (6) shows two terms : the first corresponds to the transient stage of creep and is anelastic in nature ; the second to the steady state of flow. On the removal of stress, the anelastic deformation is expected to be recovered. Then, when, $\sigma = 0$ at $t = t_1$ one has :

$$\gamma = \gamma_{plast}(t_1) + \gamma_{anel}(t_1) \exp(-\frac{t - t_1}{\tau}) \qquad (7)$$

The physical meaning of τ_1 and τ_2 must now be precised. As explained in ref. (6, 7), τ_1 can be written :

$$\tau_1 = \frac{n}{\nu_D} \exp\frac{\Delta G_1}{kT}$$

with n as the number of atoms moving cooperatively during activation 1 ; ν_D is the Debye frequency and ΔG_1 is defined in figure 2a. Due to the disorder, it is obvious that n and ΔG_1 cannot have a unique value and as is generally done, a distribution of the values of τ_1 must be taken into account mainly because of the distribution of ΔG_1. τ_2 is more difficult to define as the precise profile $G(R)$ (figure 2b)) is not known ; nevertheless some approximate values of τ_2 can be obtained by, $\tau_2 \simeq l_c/v$ with v given by Einstein's law, $v = DF/kT$, with D and F respectively, the diffusion coefficient of and force acting on the moving species. Two cases have to be considered : (1) the neighbouring loops act as sources and sinks of diffusionnal defects helping the expansion of the loops until they anihilate. These diffusionnal defects move by a mean distance l_c with the drift velocity v in the stress field of the loop C_n and $F \simeq \delta E_{el}/l_c$ (δ is a numerical constant between 0,1 and unity ; E_{el} : elastic energy per atomic length of the line C_n) (ii). The line C_n moves through some climbing mechanism until it becomes ineffective and $F \simeq f'. \sigma.<b>^2$. In the former case, we have

$$\tau_2 \simeq \frac{kT.l_c^2}{\delta E_{el}.D}, \qquad (8a)$$

and in the latter,

$$\tau_2 \simeq \frac{kT.l_c}{f'\sigma<b>^2 D} \qquad (8b)$$

Since the rate is determined by the easiest mechanism, at a low stress equation (8a) must be considered, and for a stress higher than a threshold value, equation (8b) applies leading to a value of τ_2 depending on the stress. Furthermore, at a high stress, the τ - defect expansion might not be irreversible when $R = 1_c$ but rather when $R = R_c$, as shown in figure 2b, with R_c depending on the value of σ (from the dotted line of figure 2b, one could classically obtain, $R_c - R_o \simeq \dfrac{\mu b}{2\Pi f \sigma}$).

Hence, there is a threshold stress σ_c obtained from equations (8a) and (8b)

$$\sigma_c \simeq \frac{\delta E_{el}}{f'<b>^2 \, l_c}$$

For $\sigma < \sigma_c$, τ_2 is independant of the stress and equation (6) shows that the strain is linearly related to the stress ; for $\sigma > \sigma_c$; τ_2 varies as σ^{-q} with $1 < q < 2$.

We note that activation 1 corresponds to a unique event (such as thermal vibration, or phonon-mode while the correlated movements of atoms in a "shear-source" defect corresponds to the jump of figurative point over the barrier ΔG_1, figure 2(a)), thus justifying a distribution of values of τ_1 ; on the contrary, mechanism 2 results from several elementary events of diffusion and τ_2 has a mean value. So we need not introduce a distribution for τ_2. In conclusion, by introducing the distribution of τ_1 in equation (6), we obtain an expression for the non elastic deformation in metallic glasses :

$$\gamma = A \sigma \sum_{i = 1}^{N} g_i \left\{ \frac{\tau_2^{(\tau_2 - \tau_1)}}{(\tau_{1i} + \tau_2)^2} \left[1 - \exp \left(- \frac{t}{\tau_i} \right) \right] + \frac{2t}{\tau_{1i} + \tau_2} \right\} \quad (9)$$

with $\sum g_i = 1$ (g_i : statistic weight)

APPLICATION : COMPARISON WITH EXPERIMENT

Theoretical creep curves and recovery curves have been calculated from equations (9) under isoconfigurational conditions with a triangular distribution of ΔG_1 as indicated in the figure 2-C , i.e. $\Delta G_1 = 1.8 - 0.05 (N - i)$ with $N = 10$ and $1 < i < N$; $g_i = 2i/N(N + 1)$; $\tau_2 = 2.10^{-13} \exp (1,8 \text{ eV}/kT)$; $A = 2.1.10^{-12}$ Pa-1; $\sigma = 2.9.10^8$ Pa ($\sigma < \sigma_c$). Figure 3 shows the curves for T between 500 and 625 K. These curves are comparable to experimental curves given in the literature (6, 7, 13). The slope of the curves $\gamma(t)$ has been determined and used to draw an Arrhénius diagram. Curve a (figure 4) shows that the apparent activation energy is nearly determined by the variation of τ_2 with temperature at high temperature

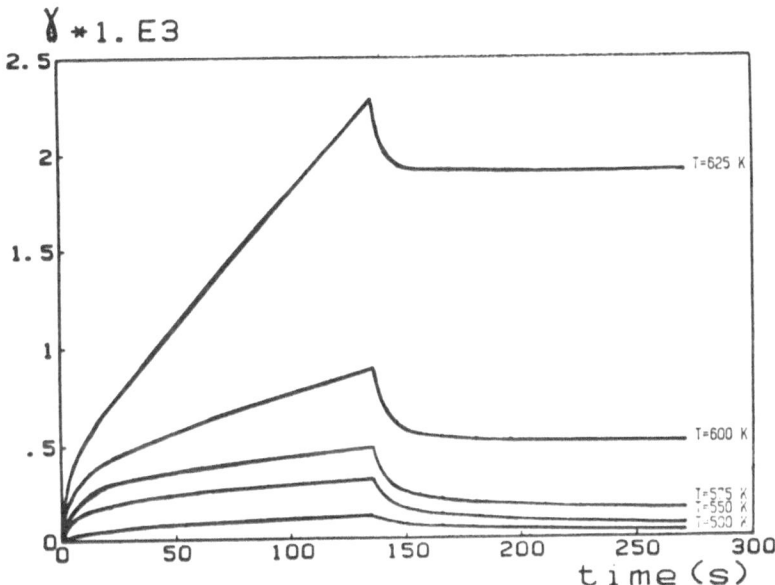

Figure 3 : Creep curves and recovery of deformation between 500 and 625 K (values of other parameters are indicated in the text).

at which plastic flow is predominant. The apparent activation energy decreases with temperature : this is well explained by the model as, at lower temperature the anelastic fraction of the deformation is predominant and, due to the distribution of τ_1, the apparent activation energy has no physical meaning. Thus, the discrepancy observed in the literature could be explained : for instance, in the case of Pd Si alloys, Maddin and Masumoto (14) mentioned an activation energy equal to 0.5 eV at 370 < T < 470 K (predominant anelastic effect) and with the same type of alloy, Taub and Spaepen (15) obtained about 2.1 eV at 500 K (predominant viscous flow) : in agreement with the preceding explanations. Similarly, curve (a) of figure 4 could explain the deviation of the strain rate from the Arrhénius plot observed by Patterson and Jones (16). Arrhenius plots from data in the transient region (curve b, figure 4) alone give a low activation energy. Although no precise physical meaning can be given to such a value it is note worthy that it can be roughly compared to the value obtained by measuring the variation of strain rate with temperature at the early stage of the transient in the case of an iron basis amorphous alloy (17). Two other features can be pointed out from figure 3 (i) the total anelastic strain obtained from the recovery curves increases with temperature, in agreement with the experiment of Chen and Goldstein (18), and (ii) the ratio $\gamma_{plast}/\gamma_{anel}$ increases with temperature, as experimentally observed. In fact equation (9) indicates that only viscoplasticity occurs when τ_2 is lower than τ_1 : this corresponds to the behaviour of liquids.

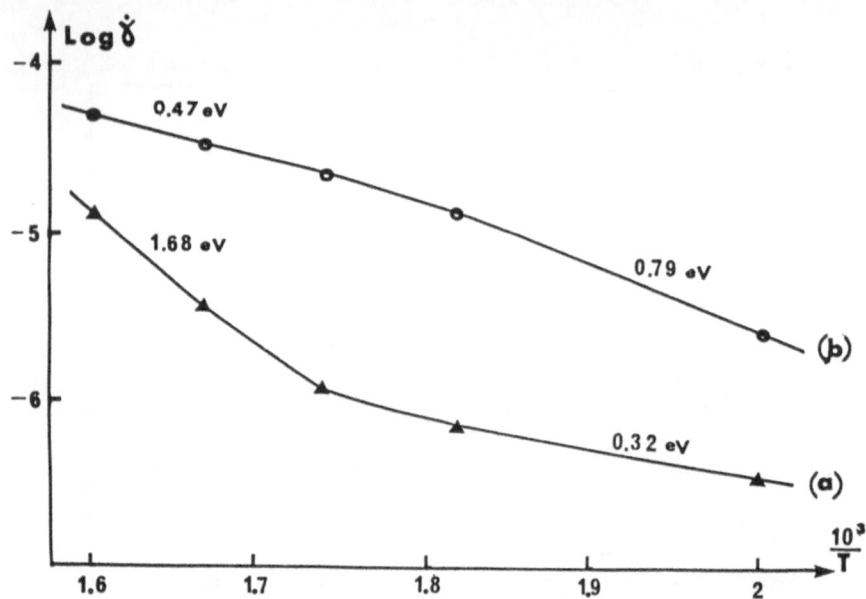

Figure 4 : Log Ẏ versus 1/T in the steady state (curve a) or at the begining of the transient (curve b). Figures indicate the apparent activation energy corresponding to the local slope.

The present analysis is limited to only iso-structural deformation in a stabilized structure. Nevertheless it is well known that creep behaviour is noticeably modified when structural relaxation occurs. We shall assume that, (i) structural relaxation in glassy metals results from the annealing of p and n-defects as proposed by Srolovitz et al (11) ; (ii) the diffusion coefficient is proportionnal to the concentration of p and n defects which in turn is (iii) proportionnal to the concentration of "shear-source" defects as the concentration of all these defects is dependant on disorder. Hence, in a first approximation the kinetics of defects annealing could be described by

$$\frac{dN_0(t)}{dt} = \frac{N_0(t) - N_0(\infty)}{\tau_{sr}}$$ (10)

with $\frac{1}{\tau_{sr}} = \lambda N_0(t)$ as τ_{sr} is inversely proportionnal to D. Equation (10) is typical of self delaying effects in glasses. By integrating equation (10) one obtains :

$$N_0(t) = \frac{N_0(0)\,\exp\,\lambda t}{1 + \frac{N_0(0)}{N_0(\infty)}\left(\exp(\lambda t) - 1\right)}$$ (11)

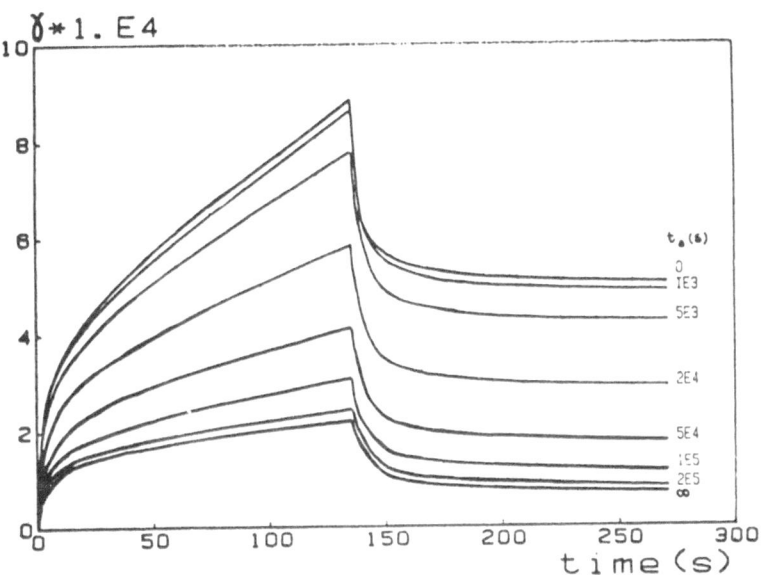

Figure 5 : Creep curves and recovery of deformation at 600 K after aging (figures indicate the aging time t_a (S)).

Taking into account of equations (6) and (8), equation (11) leads to

$$A(t) = \frac{A(0)\exp \lambda t}{1 + \frac{A(0)}{A(\infty)} (\exp(\lambda t) - 1)} \qquad (12\text{-}a)$$

and

$$\tau_2(t) = \tau_2(0) \frac{1 + \frac{\tau_2(\infty)}{\tau_2(0)}(\exp(\lambda t) - 1)}{\exp (\lambda t)} \qquad (12\text{-}b)$$

Equations (12a) and (12b) imply that the product $A(t).\tau_2(t)$ remains constant during structural relaxation. Experimental data for metallic and oxide glasses (19) show that it is rougly the case.

Thus a calculation of creep and recovery curves with equations (9) and (7) with A and τ_2 given by expressions (12a) and (12b) respectively, leads to the set of curves shown in figure 5. These curves are comparable to experimental curves (19). Several remarks can be made : (i) the ratio $\lambda_{plast}/\lambda_{anel}$ is decreased by 3 after aging experimentally this ratio is observed to be decreased by 5 in the case of an iron based metallic alloy and by 2.6 in the case of an oxide glass (19). (ii) the ration $\sigma/\dot{\gamma}$ (i.e. the viscosity) at the time when the stress is released, increases with aging time t_a. One can verify that for shorter aging times, this viscosity increases linearly with t_a in agreement with the observations of Taub and Spaepen (5) ; for longer values of t_a, there is a curvature and

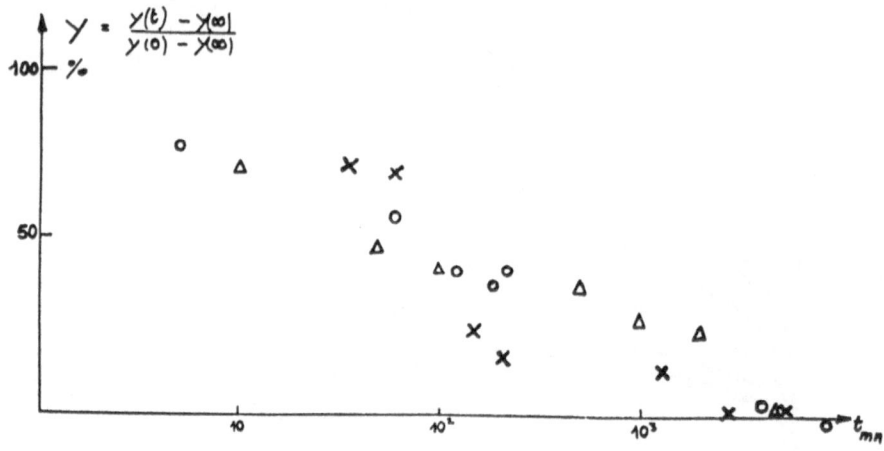

Figure 6 : normalized evolution of electric resistivity (Δ) brittelness (0) and concentration $N_0(X)$ of "shear-source" defects during
aging at 300°C. Experiments are make on the some iron based metallic glass.

this could explain the $t_a^{-(0,54 \pm 0,05)}$ law mentioned by
Patterson and Jones (16).

As a conclusion of this paper, the brittleness of metallic
glasses can be considered : it is well known that the rupture mechanism of a metallic glass specimen involves a competition between ductile processes and cracks propagation. In the latter case,
the velocity of this propagation is determined by the amount of
energy which is wasted in the bottom of the crack during its propagation. Thus, it is important to consider the non elastic deformation as this phenomenon results in wasted energy. As aging of the
material implies a decrease of anelasticity and viscoplasticity
(figure 5), an increase of brittleness can be expected, as observed
experimentally. For illustration, experimental results obtained on
iron based metallic alloy are shown in figure 6 : a good correlation is observed between (i) the decrease of the concentration
of "shear-source" defects (as deduced from creep and recovery
deformation curves), (ii) the variation of brittleness (as
observed through bending tests) and (iii) the variation of the
electric resistivity during the same thermal treatment.
In the present model, aging corresponds to an evolution of the
material towards a more stable (i.e. less disordered- state
thus implying a drecreasing concentration of "defects" (e.g.the
"shear-source" defects concentration N_0). Due to this evolution a
decrease of both the electric resistivity and the non elastic

deformation ability occurs, the latter causing an increase of brittelness.

Other consequences of the model e.g. effect of stress, stress relaxation, tension or compression $\sigma(\gamma)$ laws etc... are discussed elsewhere (13).

REFERENCES

1. Spaepen, F. A microscopic mechanism for steady state inhomogeneous flow in metallic glasses. Acta metal. 25 (1977) 407-415
2. Argon, A.S. Plastic deformation in metallic glasses. Acta metal. 27 (1979) 47-58
3. Spaepen, F. and. D. Turnbull. Atomic transport and transformation behavior, "Metallic glasses" (American Society for-metals, 1978) 114-127
4. Taub. A.I. and F. Spaepen. The kinetics of structural relaxation of a metallic glass. Acta Metal. 28 (1980)1781-1786
5. Taub. A.I., Stress-strain rate dependence of homogeneous flow in metallic glasses. Acta Metal. 28 (1980) 633-637
6. Perez, J. Cavaille J.Y., Etienne S. and F. Fouquet. Viscoelastic and plastic behaviour of metallic and other glasses near the glass transition. J. de Phys. C8 (1980) 850-855
7. Perez, J., Fouquet F. and Y. Ye. Homogeneous flow in metallic glasses. Phys. St. Sol.(a) 72 (1982) 289-299
8. Eyring, H. Viscosity, plasticity and diffusion as examples of absolute reation rates. J. Chem. Phys. 4 (1936) 283-291
9. Chen, H.S. Entropy model for flow behavior in metallic glasses. J. Non Cryst. Sol. 22 (1976) 135-143
10. Spaepen, F. Structural imperfections in amorphous metals. J. Non Cryst. Sol. 31 (1978) 207-221
11. Srolovitz, D. Maeda K., Vitek K. and T. Egami. Structural defects in amorphous solids. Statistical analysis of a computer model. Phil. Mag. 44 (1980) 847-866
12. Bowden, P.B. and S. Raha. A molecular model for yield and flow in amorphous glassy polymers making use of a dislocation analogue. Phil. Mag. 22 (1974) 149-166
13. Perez, J. Homogeneous flow and anelastic plastic deformation of metallic glasses. Submitted for publication to acta Metal.
14. Maddin, R. and T. Masumoto. The deformation of amorphous palladium. 20 at. % silicon. Mat. Sci. Eng. 9 (1972)153-162
15. Taub, A.I. and F. Spaepen, Ideal elastic, anelastic and viscoelastic flow in a metallic glass. J. Mat. Sci. 16 (1981) 3087-3092
16. Patterson, J.P. and D.R.H. Jones, Creep of amorphous $Fe_{40}-$

138

$Ni_{40} P_{14}B_6$ - Acta Metal. 28 (1980) 675-681

17. Perez J., F. Fouquet and G. Lormand. Propriétés mécaniques des verres métalliques. To be published in "Amorphous metals (Ed. Phys., 1984).

18. Chen H.S. and M. Goldstein. Anomalous viscoelastic behavior of metallic glasses of Pd-Si-based alloys. J. Appl. Phys. 43 (1972) 1642-1648

19. Borde C. Mai C. and J. Perez. Physical interpretation of creep and recovery tests made in glassy materials near Tg. J. Non Cryst. Sol. 56 (1983) 399-404

REAL AND POTENTIAL APPLICATIONS OF AMORPHOUS METAL RIBBONS

F. E. Luborsky

General Electric Corporate Research and Development,
Schenectady, NY 12301 USA

ABSTRACT

Both the United States and Japan have programs underway to exploit the large scale use of amorphous metals in distribution transformers. The background, current status and potential of these programs are described. A variety of smaller magnetic devices made from amorphous metal ribbons are now for sale in commercial equipments. These devices and their impact are briefly described. A final group of potential applications of amorphous alloys, which have been reported on in the technical literature, are mentioned.

INTRODUCTION

Amorphous metals have been called "the material of the century" in the United States and Europe (1,2) and "the dream material" in Japan (2). Both catch phrases are useful in the mass media for accenting the great potential of this new class of metallic alloys. Their great potential comes from their many truly unique characteristics — namely the variety of alloys which can be made to have one or more of the following characteristics; extremely low magnetic losses, zero magnetostriction, high mechanical strength and high hardness, radiation resistance, ease of manufacture in large volume, high chemical corrosion resistance or excellent catalytic properties.

The soft magnetic properties are the result of the lack of any magneto-crystalline anisotropy. Thus losses about twenty times smaller than obtained in the best quality Fe-3 1/4 Si have

140

been achieved in the laboratory (3,4) while losses three times smaller than Fe-3 1/4 Si are typically achieved in prototype transformers (5). These lower losses represent another large step in the decreasing losses achieved in transformer steels as a function of time. This progression is shown in Fig. 1 for the conventional steels and for the amorphous alloys. Coupled with these much lower losses in the amorphous alloys are much lower exciting power due to the extremely low coercivities and loop squareness. These properties, resulting in great savings in power and thus operating cost, are the driving force for applying these materials in the large volume distribution transformer application in the United States and in Japan. In Europe, where the distribution

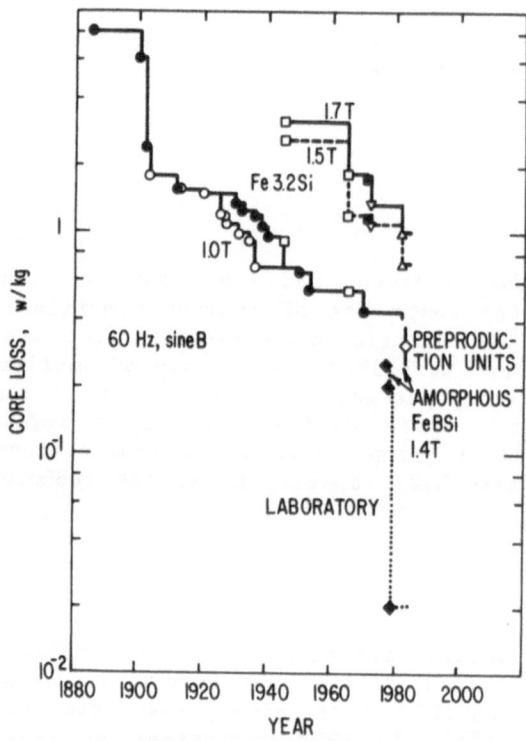

Fig. 1. Decrease in core loss with year of reported decrease. ●,■ M. Littman, IEEE Trans. Magn. MAG-7 48 (1971); ○ R. Becker and W. Doring, Ferromagnetismus (Springer, 1939) p. 409; □ R.D. Olsen, J. Appl. Phys. 37 1197 (1966); ▽ S. Taguchi, T. Yamamoto and A. Sakakura, IEEE Trans. Magn. MAG-10 123 (1974); △ S. Nakashima, K. Takashima, K. Kurodi and M. Harada, IEEE Trans. Magnetics MAG-18 1511 (1982); ◆ F.E. Luborsky, Amorphous Magnetism II, ed. R.A. Levy and R. Hasegawa (Plenum Press, N.Y., 1977) p. 345 and F.E. Luborsky and J.J. Becker, IEEE Trans. Magnetics MAG-15 1939 (1979); ◊ results from 25 kVA transformer made by General Electric Co.

system does not use the large number of transformers close to the customer, this application is not available for exploitation. However, there are many other magnetic devices which make use of the unique properties of these amorphous alloys.

There have been a number of reviews which describe potential commercial applications of amorphous metals based on engineering studies reported in the technical literature (6-12). The most recent and complete review has been given by Raskin and Smith (11). In this brief account we will concentrate our attention on the commercial or near commercial applications of amorphous metals in magnetic devices. Pulse power applications and thin film applications will be covered by other authors. For those interested in the preparation and properties of amorphous metal alloys, there have recently been several book chapters and books written which review these subjects (13-17).

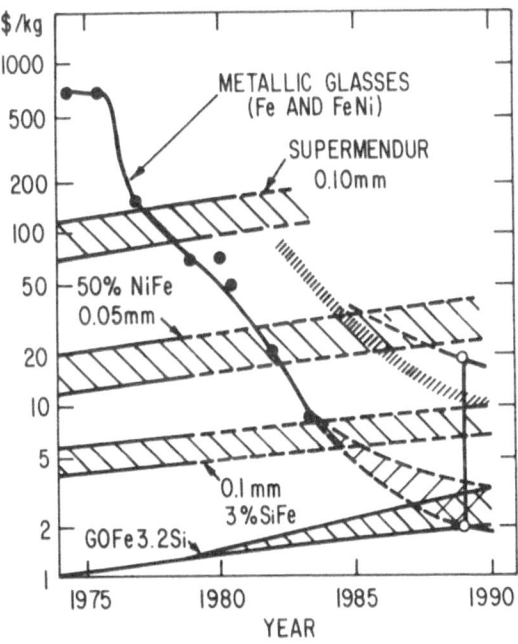

Fig. 2. Past prices and future predictions of prices for metallic glasses and some competitive crystalline alloys. Solid dots represent the minimum price quoted by Allied Corp. for large quantity orders. The open circles are minimum and maximum estimates made in 1981 from USA experts. The fine cross-hatched line is a prognosis from Vacuumschmelze. (Updated from D. Raskin and L.A. Davis, IEEE Spectrum 18 28 (1982).

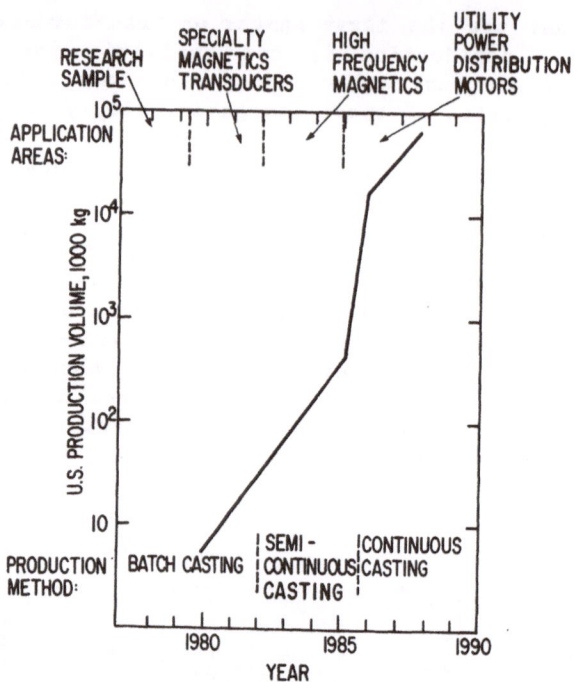

Fig. 3. Forecast of possible production volume. (Updated from D. Raskin and L.A. Davis).

BACKGROUND OF APPLICATIONS

With the application of the melt spinning techniques (18-20) to amorphous alloy in the beginning of the 1970's it became possible to envisage continuous production in large quantities. This set the stage for the decreasing price and increasing volume of production of metallic glasses shown in Figs. 2 and 3. These figures have been updated from previously published work (13,21). The projections depend strongly on a number of factors; primarily on the large scale use of amorphous ribbons in distribution transformers which in turn depends on the availability of ribbons at a low enough price and from more than one source, all with satisfactory quality. The past prices are indicated by the dots in Fig. 2 and represent the lowest price quoted by the Allied Corporation for large orders of iron based amorphous alloys.

The number of suppliers has increased substantially from the early days when only the Allied Corporation was in business selling amorphous metal alloy tapes. Since then tapes can be purchased from Vacuumschmelze, Hanau, Germany; from Hitachi Metals, Tokyo, Japan; from Nippon Amorphous Metals Co., Japan, a joint venture with Allied Corp; and from Nippon Steel Corp., Kawasaki

City, Japan. In addition smaller quantities and finished cores may be purchased from Arnold Engineering Co., Marengo, IL; Magnetic Metals, Camden, NJ; Magnetics Inc., Butler, PA and from SGL Electronics, NJ. Thus the stage is now set for the growth of this business as predicted by Fig. 3

DISTRIBUTION TRANSFORMER APPLICATION

As mentioned earlier in this paper, the driving force behind the attempt to use amorphous alloys in distribution transformers is the enormous savings in energy possible. For example, it can be shown that if all of the distribution transformers now installed in the United States were replaced with transformers made from amorphous alloys, a savings of electric energy costing

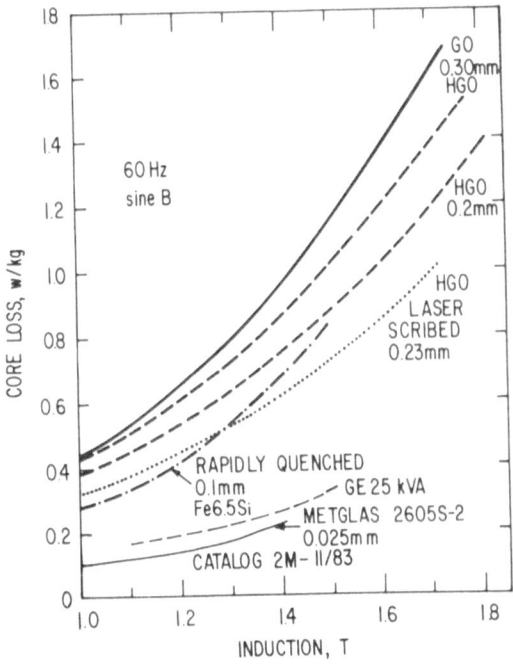

Fig. 4. Core loss as a function of induction for some typical oriented high quality Fe-Si alloys compared to typical amorphous alloys for transformers. The GO (grain oriented) and HGO (high permeability grain oriented) both are for 3.2 Si. The Fe6.5Si is from N. Tsuya et al, IEEE Trans. Magnetics MAG-16 728 (1980). The laser scribed curve is from S. Nakashima et al, IEEE Trans. Magnetics MAG-18 1511 (1982). The solid curve for METGLAS amorphous alloy is from the Allied Corp. catalog 2M-11/83; the dashed curve are results from a 25 kVA transformer made by General Electric Co.

144

about a quarter of a billion dollars a year would result due to the lower losses of the amorphous alloys. These core losses are compared to several varieties of Fe-Si alloys as a function of operating flux level in Fig. 4. The grain oriented (GO) and the high permeability grain oriented (HGO) are now available and are used in transformers. But these have losses of more than three times that of the amorphous alloys. The high permeability laser scribed material is just now becoming available commercially and shows a significant improvement over the HGO alloys but still is no competition for the amorphous alloy. The rapidly quenched Fe-6.5% Si steel, although better at lower flux levels, still does not compete with the amorphous alloys and it is not yet available commercially. Two curves are shown in Fig. 4 for the amorphous alloys; the lower one is for losses reported in the Allied Corporation catalog; the higher curve represents a curve from one of

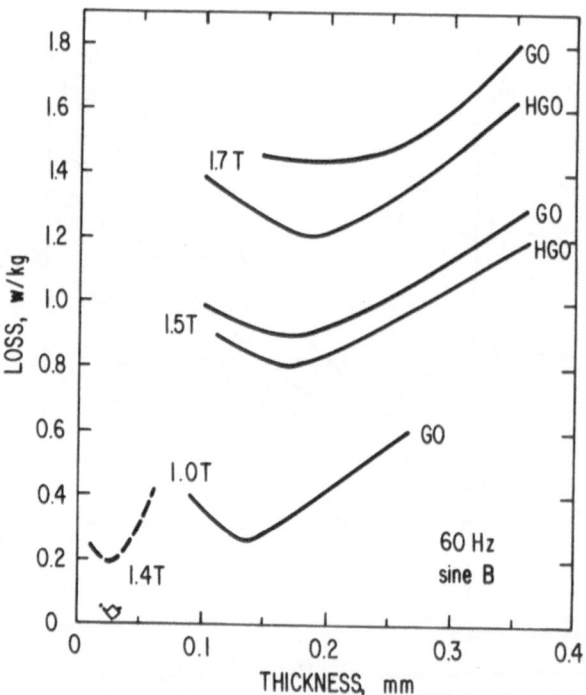

Fig. 5. Loss as a function of thickness measured at various flux levels for Fe-3.2Si and for amorphous alloys. Solid lines for Fe-3.2Si, GO = grain oriented, HGO = high permeability grain oriented; dashed line for amorphous $Fe_{81.5}B_{14.5}Si_3C_1$ ("Rapidly Solidified Amorphous and Crystalline Alloys", Edited by B.H. Kear, B.C. Giessen and M. Cohen, Elsevier, NY 1982 p. 211); dotted line for amorphous $Fe_{83}B_{15}Si_2$ (F.E. Luborsky, P.G. Frischmann and L.A. Johnson, J. Magnetism Magnetic Materials 19 130 (1980).

the first 25 kVA distribution transformers made by G.E. The
difference between the two curves represents the manufacturing
loss for this particular transformer at this stage of development.

It is apparent in Fig. 4 that the decrease in sheet thickness
decreases the loss. We can see this effect in more detail in Fig.
5. Clearly, the losses for Fe-3.2 Si decrease to a minimum with
decreasing thickness. This is due to the inability to develop the
highly oriented crystal texture during the secondary recrystallization in the processing of the steel. The amorphous alloys also
show losses which depend on thickness as shown in Fig. 5. The
reason for this minimum in the amorphous alloys, however, is different. The losses first decrease with increasing thickness
because of the reduction in surface pinning of domain walls. At
some critical thickness the cooling rate during the solidification
on the casting wheel on the upper surface of the ribbon decreases
to the point where the alloy is no longer amorphous and thus the
high coercivity of the crystalline phase starts to play an
increasing role in the measured coercivity.

Clearly, with all of the potential advantages noted above,
there will be considerable activity in this area. This is borne
out by the number of different companies who have actually built
full size prototype distribution transformers. These are listed
in Table 1.

In the latter part of 1982 the General Electric Distribution
Transformer Department was awarded a contract from the Electric

Table 1. Prototype or Engineering Model Distribution
Transformers Built

Company	Rating kVA	Approximate Year built
General Electric	1/2, 1, 25	1979–1981
Allied Corp.	15, 25	1980–1981
Osaka Transformer	10, 30	1981–1982
Westinghouse	10	1981
McGraw Edison	10	1981
Takaoka Transformer	20	1982
Hitachi	10	1983

Power Research Institute to design and build 1000 distribution transformers rated at 25 kVA. We believe that this contract may be said to represent the beginning of actual production. Before the awarding of this contract the G.E. Distribution Transformer Department had experimented with various design concepts by building many small 0.5 to 1 kVA transformers. Some characteristics of these designs are shown in Table 2. They ended this work by building three 25 kVA transformers all from 2.5 cm wide ribbons; two were built from ribbons purchased from Allied and one from material produced in the G.E Corporate Research and Development Center. A shell-type cruciform design (13) was used for all three and all of them performed satisfactorily. They contained cores that weighed about 90 kg each. A summary of some of the characteristics measured is shown in Table 3. Also listed for comparison are the characteristics obtained for Fe-3.2Si steel transformers. The transformer made with G.E. material was then shipped to the Duke Power and Electric Co. and installed in their distribution system in early 1982. It has been operating satisfactorily for the past 2 years.

Table 2. Characteristics of Small Model Amorphous
Core Transformers[a]

Rating	0.5 kVA	1.0 kVA
Primary/Secondary Volts	120/240	360/240
Unit Weight (kg)	7.3	10.7
Core Weight (kg)	4.3	4.8
Core Space Factor (%)	78	67
Flux Density (T)	1.275	1.430
Core Loss (W)	1.22	1.27
(W/kg)	0.28	0.265
Exciting Current (%)	0.71	0.52
Winding Loss at 85°C (W)	26.3	67.6
Audio Noise (dB)	34.2	--
Telephone Influence Factor (IT/kVA)	5.5	8.5

Note: Industry specifications for audio noise are less than 48dB; and for TIF, less than 22 IT/kVA.

(a) : Reference 5

Table 3. Performance Comparison of the 25 kVA (7200 V–120/240 V) Transformers Made From Amorphous Metals With a Conventional Silicon Steel Design

Test	Amorphous Metal	M-4 Silicon Steel
Flux density	1.4 T	1.6 T
Core loss	28 W	86 W
	(0.32 W/kg)	(0.96 W/kg)
Exciting current	0.3%	0.9%
Audio noise	38 dB	48 dB (limit)
TIF 100%/110%	3/14	22/66 (limit)
Short circuit	passed (full 33 times)	passed (40 times)
In-rush magnetizing current: 0.01 s/0.1 s	17.8	25/12
Impulse LV and HV	passed	passed
Core weight	88 kg	90 kg

The first phase of the contract of EPRI with G.E. required that we build 25 distribution transformers using the same design as we used to build the three precontract units. These were then to be shipped to various utility companies throughout the United States for installation in their distribution systems. This first phase has all been completed satisfactorily and the units are now in service and being tested periodically using standard test equipment. Initial field tests after installation has been completed on 18 of the units and tests after about three months of installed operation have been made on 9 units. None of the results showed any significant changes.

The second phase of work under this contract calls for an evaluation of many different design concepts for the transformers. At the end of this evaluation two or three designs with the best trade-offs for optimum performance at minimum total manufacturing cost are to be chosen for construction of three engineering models together with a detailed economic sensitivity analysis and technical analysis. The result of this final analysis will be the selection of one model for construction of all 1000 transformers. These transformers will then be shipped to various utility companies for installation in their electrical distribution systems.

Japan has a very similar program aimed at eventual production of distribution transformers partially sponsored by their JRDC called the Shin-Nitetsu Project. Both the U.S. and Japanese programs are scheduled to be completed in 1985. Makino stated (2) "When both projects are finished successfully, we shall find ourselves in the dawn of the industrial age of amorphous metals". However, I believe that the dawn of the industrial age of amorphous metals has already arrived as evidenced by the many applications of amorphous metals in small devices to be described.

APPLICATIONS IN SMALL DEVICES

The first amorphous metal device sold in a commercial system was by Sony in the moving magnet phonograph cartridge where an amorphous metal core was used. This was first introduced in 1980. The amorphous core transformed the mechanical motion of a needle on a record into electrical signals through the motion of a magnet in the core, on the end of a cantilever, attached to the needle. A zero magnetostrictive cobalt based amorphous alloy was used. Permalloy had been used but the amorphous core has several advantages. The output of the amorphous core is higher throughout its frequency range due to its higher permeability. The amorphous core has almost no phase delay over the entire frequency range as compared to Permalloy, thus resulting in much reduced signal distortion. Finally the noise from the amorphous core is about 5 db lower than from the Permalloy core.

Amorphous heads for audio and computer tape recording are also on the market by Matsushita Electric (Panasonic), TDK and Sony. These cores are made from a zero magnetostrictive amorphous alloy. This material has a $4\pi M_s$ of more than 9 kG, thus ensuring excellent recording properties. Laminations of 30 to 50 μm thickness are used to obtain a small recording current and high frequency response. The alloy is extremely hard which results in a high resistance to wear or gap dropping due to the result of the motion of the tape against the head. There is very little magnetic distortion, so there is little effect on the properties of the alloy due to stresses and finally this alloy is highly corrosion resistant.

There are also a number of other devices on the market for use in audio recording and playing equipments. For example, in Japan, a dynamic microphone with an amorphous transformer is being made. The CBS-Sony Corporation is now using an amorphous head for duplicating music tapes. An amorphous core is being used as a transformer for the amplifier of a moving coil cartridge. These applications are driven by the exceptional combination of properties available from amorphous metal tapes, namely high magnetization, high permeability, high hardness, low losses, etc.

Toshiba, Hitachi and several European firms have on the market a switched mode power supply using an amorphous magnetic amplifier (22,23). In such an application the most important material characteristics are the high saturation induction, together with low losses and high Curie temperatures. These characteristics are all met by amorphous alloys; they have values of $4\pi M_s$ up to 16 kG with very low high frequency losses because of the high resistivity. Thus with the same thickness ribbons the high frequency losses are lower because these losses are dominated by the eddy current losses.

In Table 4 we have listed all of these, and other commercially available applications of devices using amorphous metals, together with the supplier.

Table 4. A Partial List of Magnetic Devices Sold Commercially Using Amorphous Alloys[a]

Approx date of introduction	Device	Manufacturer	Country
1980	Core in moving magnet recording cartridge	Sony	Japan
1981	Audio, computer, cassette heads	Matsushita, TDK, Sony	Japan
	Transformer in dynamic microphone	Sony	Japan
	Recording head for tape duplication	CBS-Sony	Japan
	Transformer for moving coil recording head	Sony	Japan
	Magnetic amplifier in a switched mode power supply	Toshiba	Japan
	Delay lines in a data tablet	Sony	Japan
	Multigap head for height gage	Sony	Japan
1982	Sensor in electronic "GO" machine	National	Japan
	Step-up transformer for moving coil cartridge	Matsushita	Japan
	Recording head in an electronic still camera	Sony	Japan

	Cores for transductor chokes, high frequency small transformers, magnetic switches		Europe
	Strip for magnetic recording heads, theft protection devices, magnetic shielding		Europe
1983	Magnetic amplifier output regulator in switched mode power supply	Coutant, Hitachi	England Japan
	Turns ratio transformer	Biddle	
	"C" cup cores on converter output	Northrop	USA
	Current spike suppressors using saturable cores	Toshiba	Japan
	Core in power supply modulator for laser	Lambda Physik	Germany
	Digitizing pad	Wacom	Japan

(a) Other than distribution transformers

APPLICATIONS CLOSE TO COMMERCIALIZATION

In addition to the devices now in various stages of commercial production described above there have been a wide variety of devices evaluated and reported in the technical literature which have not yet reached commercial exploitation. For example, force, pressure and displacement gages have been reported in a variety of configurations, all making use of the combination of high magnetostriction and high yield strength (24). In fact, the two measures of performance in such applications, that is their coupling coefficients and efficiency are both higher than any other known material. In security systems (25), the high magnetic permeability makes them good generators of harmonics in an a.c. or d.c. field. This characteristic makes them useful as anti-theft tags. Conventional Permalloy can also be used but is much more prone to degradation due to mechanical deformation in handling. The amorphous alloys are also offered as magnetic shielding materials (26). Allied Corporation sells large sheets made by weaving narrow ribbons together. All kinds of transformers and cores have been described (22,23,25,27) for example for lamp ballasts, ferroresonant transformers, magnetic modulators, active power filters, high frequency power transformers, and electronic transformers. Finally an induction motor has been fabricated at

G.E. using an amorphous metal stator (28). This was a 1/3 hor-
sepower, two pole, 230 V, three phase stator using stacked laminations. The fundamental 60 Hz loss for the amorphous iron stator
was 1 watt compared to 5 watts for a companion stator made from
silicon-iron and 10 watts for common iron. This clearly demonstrated that the low losses reported for amorphous alloys can be
achieved in a motor, however the costs appear prohibitive.

CONCLUSIONS

In the past few years we have seen the birth of a new soft
magnetic materials technology — the amorphous metals — as it
emerged from the laboratory to commercially available devices.
This represents the first new class of soft magnetic material to
be found since the discovery of the soft ferrites in the 1940's.
It is expected that the commercial applications will grow at an
increasing rate in the next few years.

REFERENCES

1. U. Gonser, Fifth International Conference on Liquids and
   Amorphous Metals, Aug. 1983, Univ. California, Los Angeles,
   CA.
2. Y. Makino, Fifth International Conference on Liquids and
   Amorphous Metals, Aug. 1983, Univ. California, Los Angeles,
   CA.
3. F.E. Luborsky and J.J. Becker, IEEE Trans. Magnetics MAG-15
   1939 (1979).
4. F.E. Luborsky, P.G. Frischmann and L.A. Johnson, J. Magnetism
   Magn. Mat'ls. 19 130 (1980).
5. L.A. Johnson, E.P. Cornell, D.J. Bailey and S.M. Heggi, IEEE
   Trans. Power Appar. Systems PAS-101 2109 (1982).
6. F.E. Luborsky in "Amorphous Magnetism II", edited by R.A.
   Levy and R. Hasegawa, Plenum Press, NY, 1977, pp. 345-368.
7. F.E. Luborsky, J.J. Becker, P.G. Frischmann and L.A. Johnson,
   J. Appl. Phys. 49 1769 (1978).
8. F.E. Luborsky, IEEE Trans. Magnetics MAG-14 1008 (1978).
9. F.E. Luborsky, P.G. Frischmann and L.A. Johnson, J. Magnetism
   Magn. Matls. 8 318 (1978).
10. F.E. Luborsky, P.G. Frischmann and L.A. Johnson, J. Magnetism
    Magn. Matls. 19 130 (1980).
11. F.E. Luborsky and L.A. Johnson, J. de Physique 41 C8-820
    (1980).
12. D. Raskin and C.H. Smith in "Amorphous Metallic Alloys",
    edited by F.E. Luborsky, Butterworths, London, 1983, Chapter
    20.
13. F.E. Luborsky, editor, "Amorphous Metallic Alloys", Butter-
    worth Publishers, London, 1983.

14. R. Hasegawa, editor, "Glassy Metals: Magnetic, Chemical and Structural Properties". CRC Press, Boca Raton, Florida, 1983.

15. T.R. Amantharaman, editor, "Metallic Glasses: Production, Properties and Applications". Trans Tech Publications, Switzerland, 1983.

16. T. Kaneyoshi, "Amorphous Magnetism", CRC Press, Boca Raton, Florida, 1983.

17. H.-J. Güntherodt and H. Beck, editors, "Glassy Metals 1", and "Glassy Metals II", Springer-Verlag, NY 1981 and 1983.

18. R. Pond and R. Maddin, Trans. Metall. Soc. AIME $\underline{245}$ 2475 (1969).

19. S. Takayama and T. Oi, J. Appl. Phys. $\underline{50}$ 4962 (1969).

20. M.C. Narasimhan, U.S. Patent 4 142 571 (1979).

21. D. Raskin and L.A. Davis, IEEE Spectrum $\underline{18}$ 28 (1982).

22. R. Boll and H. Warlimont, IEEE Trans. Magnetics $\underline{MAG-17}$ 3053 (1981).

23. W. Kunz and D. Gratzer, J. Magnetism Magn. Mat. $\underline{19}$ 183 (1980).

24. K. Mohri and S. Korekoda, IEEE Trans. Magnetics $\underline{MAG-14}$ 1071 (1978).

25. C.H. Smith, IEEE Trans. Magnetics $\underline{MAG-18}$ 1376 (1982).

26. L.I. Mendelsohn, E.A. Nesbitt and G.R. Bretts, IEEE Trans. Magnetics $\underline{MAG-12}$ 924 (1976).

27. M. Milkovic, F.E. Luborsky, D. Chen and R.E. Tompkins, IEEE Trans. Magnetics $\underline{MAG-13}$ 1224 (1977).

28. W.R. Mischler, G.M. Rosenberry, P.G. Frischmann and R.E. Tompkins, IEEE Trans. Power Apparatus Systems $\underline{PAS-100}$ 2907 (1981).

APPLICATIONS OF THIN FILM AMORPHOUS MAGNETIC MATERIALS

Richard J. Gambino

IBM T. J. Watson Research Center, Yorktown Heights, NY 10598

INTRODUCTION

At about the same time that Duwez and coworkers first reported (1) methods for preparing amorphous alloys by rapid liquid quenching, methods were also discovered for preparing amorphous alloys by vapor quenching as shown by Mader et al.. (2) A few years later, in the mid 1960's both liquid (3) and vapor quenched amorphous alloys (4) were found to order magnetically. The applications of both classes of materials first received intense scrutiny in the early 1970's with the discovery by Chaudhari et al. that vapor deposited, amorphous rare earth-transition metal alloys will support magnetic bubble domains (5). It was subsequently found that similar films with slightly different composition can be used as a medium for thermomagnetic storage with magneto-optical readout. (6) Also, in the early 70's it was reported that liquid quenched ribbons of metal-metalloid amorphous magnetic alloys can be fabricated by high speed production methods (7) and that these materials have properties of interest for low loss inductor core applications. (8)

At the present time, many types of amorphous alloys have been prepared by both methods (9-23) as shown in Table I. In some cases, the same magnetic alloy has been prepared by both liquid and vapor quenching so that their magnetic properties can be compared (24). In general, the intrinsic properties of the alloy are independent of the method of fabrication. Vapor quenching usually allows a wider range of compositions to be prepared in the amorphous state probably because it provides a faster, effective quench rate ($10^{14} K/$ sec). The compositions that are readily liquid quenchable are frequently close to a eutectic composition but there is a reasonable latitude in compositional variations with both methods. The choice of a thin film amorphous magnetic material is dictated by other considerations. The need for a thin film, or broad area coverage for a storage medium is one consideration. Compatibility with microelectronic processing methods as for thin film recording head fabrication is another possible factor. In some cases, however, the overriding factor is that the deposition process itself gives a means of controlling an essential magnetic property such as magnetic anisotropy for the application (6). In that sense, vapor deposition offers the widest range of magnetic properties in amorphous alloys.

Magnetic materials can be broadly separated into classes according to the magnitude of three important properties: saturation magnetization ($4\pi M_s$), anisotropy field (H_k) and coercivity (H_c). The outline of this paper is to first define the materials requirements in terms of these three properties. Second, it will be discussed how properties are controlled by compositional variations. Third, the effects of deposition control parameters on these properties will be explained in the context of simple models of short range atomic ordering anisotropy (25) and a simple inhomogeneity model of coercivity. (26) Fourth, the secondary magnetic properties of amorphous films will be described and lastly, the effects of post deposition annealing will be described.

PROPERTY REQUIREMENTS

In Figure 1, the ranges of properties for each class of materials are shown graphically. The orthogonal axes represent $H_k, 4\pi Ms$ and H_c. For simplicity, only the magnitude of H_k is shown although positive (meaning perpendicular easy axis) and negative (perpendicular hard axis) could be included. The $4\pi M_s$ axis goes to 20 kG which is slightly less than the saturation value of pure bcc Fe. In fact, it would be difficult to obtain a stable amorphous alloy with a saturation as high as 20 kG because the alloying elements which are used to stabilize the amorphous alloy also lower $4\pi M_s$. The H_k and H_c axes are limited in magnitude to the range available in amorphous alloys and partially crystallized films. Very high magnetocrystalline anisotropy fields (over 100 kOe) are found in some single crystals but these materials are not included in this discussion.

Low Coercivity Materials - $H_c \cong 0$ kOe

Magnetic bubble domains (27,28) are an example of magnetic materials requiring low coercivity. For most bubble devices, a wall motion coercivity over 2 or 3 Oe is unacceptable and coercivities of less than an oersted are preferable. In addition, the material must have a perpendicular uniaxial anisotropy sufficient to hold the magnetization normal to the film plane. The latter requirement has been embodied in a quality factor called Q which is defined as $Q = H_k/4\pi M_s$. If Q is greater than 1, the anisotropy field H_k is just sufficient to hold the magnetization normal to the film plane in opposition to the demagnetizing field of the thin film which has the magnitude $4\pi M_s$. After the bubble quality factor was defined in this way, it was discovered that when Q = 1, bubble domains can exist but they are very unstable with respect to strip out during device operation. For practical material, Q=2 is recommended. (28)

In Figure 1, the range of properties of materials suitable for practical bubble device applications are shown. These materials are mapped on the $H_c = 0$ plane (a good approximation on a scale of 10 kOe) and on the low magnetization side of the Q = 2 line. Of course, there are many other factors in the design of a practical bubble material which may limit the range of properties of interest. For example, the domain size increases with decreasing saturation magnetization and increasing H_k so only the region close to the Q = 2 line is of much interest for high density bubble memories which require small domains. Also, there are a number of important dynamic properties which do not depend on $4\pi M_s$, H_k or H_c but which will make it necessary to exclude some materials which meet all the requirements in terms of these three properties. The regions in Figure 1 delineate the range of properties necessary but not sufficient for the indicated application.

Table I

System	Example	Liq.	Vapor	Other	Ref.
TM-M	Ni-P			electro.	9,10
	Pd-Si	x			11
	Fe-B	x	x		12,13
TM-TM	Zr-Cu	x	x		14
	Nb-Ni	x			15,16
	Mo-Co		x		17
	Fe-Zr	x	x		18,19
RE-TM	Gd-Co	x	x		5,20
	RE-Fe	x			21
Act-TM	U-V	x			22
	U-Cr	x			22
AE-B	Mg-Zn	x			22
	Ca-Al	x			22
AE-AE	Ca-Mg	x			23
	Sr-Mg	x			23

TM = transition metal
M = metalloid
RE = rare earth metal
Act = actinide metal
AE = akaline earth metal
B = B group metal

156

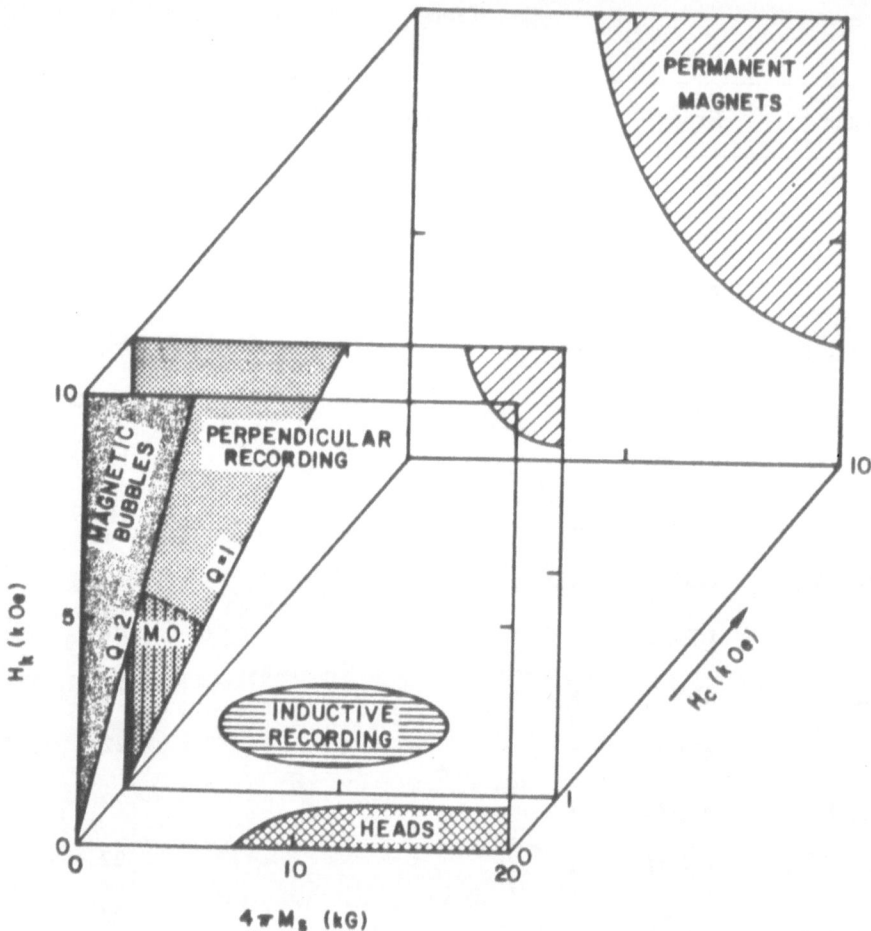

Fig. 1.   Applications of magnetic materials mapped according to saturation magnetization, $4\pi M_s$, anisotropy field, H_k and coercivity, H_c. The three vertical planes represent $H_c \cong 0$ kOe (magnetic bubbles and recording heads), $H_c \cong 1 kOe$ (recording media) and $H_c \cong 10 kOe$ (permanent magnets).

Another class of application requiring very low H_c are thin film inductive heads. (29) For this application, a large saturation and a low anisotropy field is required. In addition, the material should have a number of secondary properties such as low magnetostriction so that strains caused by processing will not cause large variations in properties. Also, the gap region of the head must be exposed so that it can be brought into close proximity to the storage medium. This also exposes the head to abrasion and chemical attack so that corrosion and wear resistance are desirable properties. Also, eddy current losses become increasingly significant with frequency so that for high data rate applications it is advantageous to have a high resistivity head material.

Materials for magnetoresistive (30) heads have similar requirements in terms of H_k, M_s and H_c. In this case, the signal does not depend on the flux change through the magnetic materials but rather on the change in resistance when the magnetization is switched from perpendicular to parallel to the direction of current flow. For this reason, the magnitude of the change ($\Delta\rho/\rho$) is an important materials property. A possible source of noise in these devices is thermal spike noise which depends on $d\rho/dT$. It has been found that in amorphous metals, $d\rho/dT$ can be made very small but for unrelated reasons, $\Delta\rho/\rho$ is also always very small.

Medium coercivity - $H_c \cong 1 kOe$

In magnetic recording, a bit is written by creating a region of reversed magnetization. The bit must be stable - it must not grow, shrink or disappear until it is intentionally erased. This stability is obtained by using a material with a coercivity in the intermediate range. If a field is applied which is larger than H_c, the magnetization can be switched. If H_c is too large, it is difficult to make a head that will provide enough field to switch the storage layer. If H_c is too low, data will be lost because stray fields will be large enough to create unintentional regions of reversed magnetization. For most applications, H_c values of 0.3 to 0.8 kOe have been used. Future applications may use somewhat higher values as will be discussed below, so all recording media applications have been plotted on the $H_c = 1 kOe$ plane.

There has recently been a great deal of interest in storage layers suitable for perpendicular recording (also known as vertical recording). The anisotropy field required, as in the case of magnetic bubble films, can be defined in terms of Q. In vertical recording, however, the coercivity provides much of the stability to the bit so that Q = 1 is sufficient. This is indicated in Figure 1. The bit size is also to a great extent determined by the size of the writing head so the range of H_k and $4\pi M_s$ is less restricted than in the bubble case. The signal from the recorded bit depends on the saturation or more accurately, the remanent magnetization of the storage layer, however, so high $4\pi M_s$ compositions are favored. In order to get a large signal from a small bit, it is essential that the pickup head come in close proximity to the medium. This makes it impossible to effectively coat the medium for corrosion protection. Furthermore, corrosion products which cause asperities on the surface of the medium can cause catastrophic failures. For these reasons, corrosion resistant alloys are sought.

In thermomagnetic recording with magneto-optical (MO) readout, the bit is written by heating a small spot on the medium with a focused laser beam. (6) The coercivity decreases with increasing temperature especially as the Curie temperature is approached so the magnetization in the heated spot can be switched. The bit is detected by means of the magneto-optical rotation of the plane of polarization of a laser beam of lower intensity. The medium still must satisfy the $Q \geq = 1$ condition, but in this case, the signal depends on the Kerr, θ_k, or Faraday , θ_f, rotation, not on the magnetization. This is shown in Fig. 1 by the region marked M.O. to the left of the $Q = 1$ line at low magnetization. Also, the optical readout allows the storage layer to be encapsulated so that corrosion resistance is not as essential. A relatively low T_c or at least a strongly temperature dependent H_c is essential.

In conventional inductive recording (also known as longitudinal recording) the bits are written as regions of reversed inplane magnetization along a track. A small inplane anisotropy helps keep the magnetization from becoming dispersed. A large saturation magnetization and a large remanent magnetization is important for a strong signal. Corrosion resistance is also desirable because of the low head flying heights needed for high density recording.

High Coercivity - $H_c = 10$ kOe

Thin film permanent magnets have relatively few applications but are included in this discussion for completeness. Also, amorphous materials in general have been considered as precursors for the formation of hard magnetic materials by crystallizing them. Recently, this approach has met with some success in the formation of $Nd_{15}Fe_{77}B_8$ intermetallics from melt spun ribbons.(31)

The main figure of merit of a permanent magnet material is the energy product - the product of the saturation induction and the coercive field. As will be shown below, a leading term in the coercivity is $2K_u/M_s$ so it is essential to have a large H_k in order to have a large H_c.

Other factors in choosing a permanent magnet material are low cost and high T_c. Low cost usually involves trying to make a material without Co or expensive rare earths like Sm. This effort, in part, explains the recent interest in $Nd_{15}Fe_{77}B_8$.

COMPOSITION CONTROL OF PROPERTIES

The saturation magnetization is directly controlled by film composition. The coercivity, on the other hand is only indirectly composition dependent. That is, the same composition can have very different coercivities depending on its microstructure. The anisotropy field is also largely dependent on preparation conditions but certain constituents, such as the non-S-state rare earths, have a major effect on H_k.

Control of $4\pi M_s$:

The magnetic 3d transition metals Fe, Co and Ni must be alloyed in order to obtain amorphous phases that resist crystallization at room temperature and above. These alloy constituents lower the saturation magnetization. The systematics of this composition dependence has recently been elucidated by Williams et al (32). The magnetic moment per atom in Bohr magnetons is given by the sum of the average magnetic valence (V_m) and a small non-integer contribution of the sp band (N_{sp}).

$$M = N_{sp} + V_m \qquad (1)$$

where N_{sp} is typically 0.6 and V_m is the negative of the chemical valence (all the electrons outside of a rare gas core) except for Fe, Co and Ni which have magnetic valencies of 2, 1 and 0 respectively. The average Vm is defined by:

$$V_m = (x_1)V_{m1} + (x_2)V_{m2}\cdots \tag{2}$$

where x_i and V_{mi} are the atomic fraction and valence of the ith constituent of the alloy. The magnetic valence method is based on the Friedel-Terakura-Kanamori model which assumes that the alloy displays strong band magnetism (i.e. the down spin band is completely full). In many Fe rich amorphous alloys there is a tendency for weak ferromagnetism which leads to lower moments than predicted by the magnetic valence method. In fact, pure Fe is predicted to have a moment of 2.6 as compared to the observed value of 2.2 μ_B. The value of the method is that it gives a simple way of predicting approximate values of $4\pi M_s$ of a wide range of compositions.

Alloying the transition metals with the rare earths leads to ferrimagnetic materials in most cases. The net magnetization is the difference between the RE and TM subnetworks. (26) In addition, the two subnetworks have different temperature dependencies so that in some compositions the two subnetworks become equal and opposite at a particular temperature - summing to zero. The temperature at which this occurs is called a compensation point T_{comp}. Some compositions happen to be compensated at room temperature - these are called compensated compositions. In these ferrimagnetic systems, the saturation magnetization is a sensitive function of composition. The light rare earths (e.g. Nd) have a net moment which is dominated by the orbital moment. The spin is coupled antiparallel to the transition metal net moment but the RE spin is also antiparallel to its own orbital moment. Thus, the net moment of Nd is parallel to the TM moment. This mode of coupling among the light RE's has been used to advantage in tailoring the temperature dependence of RE-TM alloys (32). In a binary alloy such as GdCo, the strong temperature dependence of M arises from the big difference in the individual Gd and Co subnetworks. By using a ternary alloy, of the type NdGdCo, it is possible to partially cancel the Gd subnetwork with the antiparallel Nd subnetwork which has a similar temperature dependence.

For applications, we are generally concerned with the room temperature properties; so we must also consider the effects of alloying on T_c. In general, the T_c's of amorphous Fe alloys are considerably lower than Co alloys. Among the rare earths, Gd alloys have the highest T_c because the T_c follows the de Gennes factor which has a maximum at Gd. An example of the composition dependence of the Curie temperatures of a series of ternary GdCoMo alloys is shown in Fig. 2.

With regard to coercivity, we can consider two types: domain wall motion and domain nucleation. Wall motion coercivity is largely controlled by process determined factors and will be discussed below under control of properties by deposition conditions. Nucleation coercivity is a measure of the barrier to nucleation of a reverse domain in an easy axis saturated (single domain) sample. Magnetization reversal can occur by coherent rotation, buckling or surface nucleation. In each case, nucleation involves aligning some spins out of the easy axis. The barrier to this hard axis rotation is the anisotropy energy, K_u. The anisotropy energy can be equated to 1/2 the anisotropy field H_k, saturation magnetization M_s, product . That is:

$$(H_k M_s/2) \; = \; K_u \text{ or } H_k \; = \; 2K_u/M_s \tag{3}$$

which suggests that the nucleation field, H_n, is equal to the nucleation coercivity and is equal to the anisotropy field H_k. At the compensation point, M_s goes to zero, K_u remains finite and thus H_c undergoes a discontinuity. An example of this behavior is shown in Fig. 3.

The anisotropy field H_k and thus the nucleation coercivity is also composition dependent because of the important role of the single ion anisotropy of the non-S-state rare earths. In general, the f orbitals have non spherical spacial electron density probability distributions. The exceptions are the S-state like ions with empty (La+3,Ce+4), filled (Lu+3, Yb+2) or half filled (Eu+2, Gd+3) f orbitals which are spherical to a good approximation. With the non-S-state ions, the orbital moment interacts through the crystal field with its coordination surroundings. In the amorphous alloys, the crystal field is in generally assumed to be axial and random. (35) Each non-S-state rare earth atom has an easy axis direction which is to a first approximation randomly oriented in the sample. The exchange interaction (J) on the other hand tends to make all the spins parallel and unidirectional as does the applied field. Depending on the relative magnitudes of the random anisotropy (D) and the exchange the spin system can be fully aligned (J>>D) or distributed over a hemisphere (D>>J). In the latter case, the rare earth moment one measures in typical laboratory fields (20 kOe) is just 1/2 the fully aligned moment. The magnitude of the single ion anisotropy is very large and is strongly temperature dependent, generally going as the square of the RE subnetwork magnetization. Furthermore, the easy axes are not completely random in a deposited thin film because the growth direction is in general unique. This slight deviation from randomness can give rise to a rather large macroscopic anisotropy.

The coercivity associated with domain wall motion can be treated with simple models such as the one presented below. (26) Consider a perpendicular domain that is square in shape with edge length D. The square shape is somewhat unphysical but it simplifies the model to emphasize the physics of the wall motion coercivity. The height of the domain is h, the film thickness. If the domain moves a distance dx, the volume swept out is 2Dhdx and the work done is:

$$\Delta E_i \; = \; 2MH2Dhdx \tag{4}$$

where M and H are the saturation magnetization and external field respectively. If the domain wall is in a local potential well the energy to move it is:

$$\Delta E \; = \; \frac{\partial E}{\partial x} \, dx \tag{5}$$

To obtain H_c, the field at which the domain just moves, we set Eq. (4) equal to Eq. (5) with $H = H_c$.

$$4MH_c Dhdx \; = \; \frac{\partial E}{\partial x} \, dx \tag{6}$$

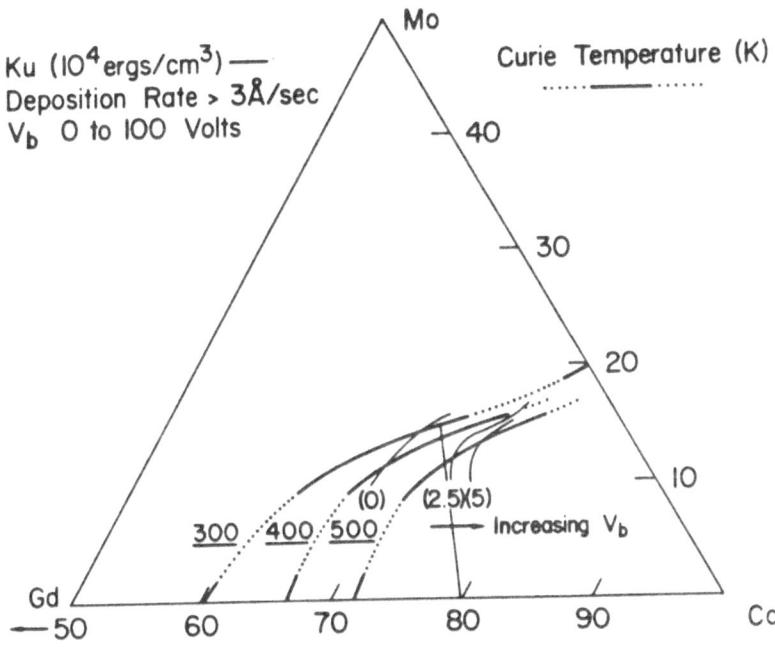

Fig. 2.   Composition dependence of Curie temperatures in sputtered amorphous films in the ternary system Gd Co Mo is shown. The direction
of composition change with increasing bias voltage is shown by the
arrow. The lighter contour lines indicate compositions of equal anisotropy energy, K_u, achieved by bias sputtering.

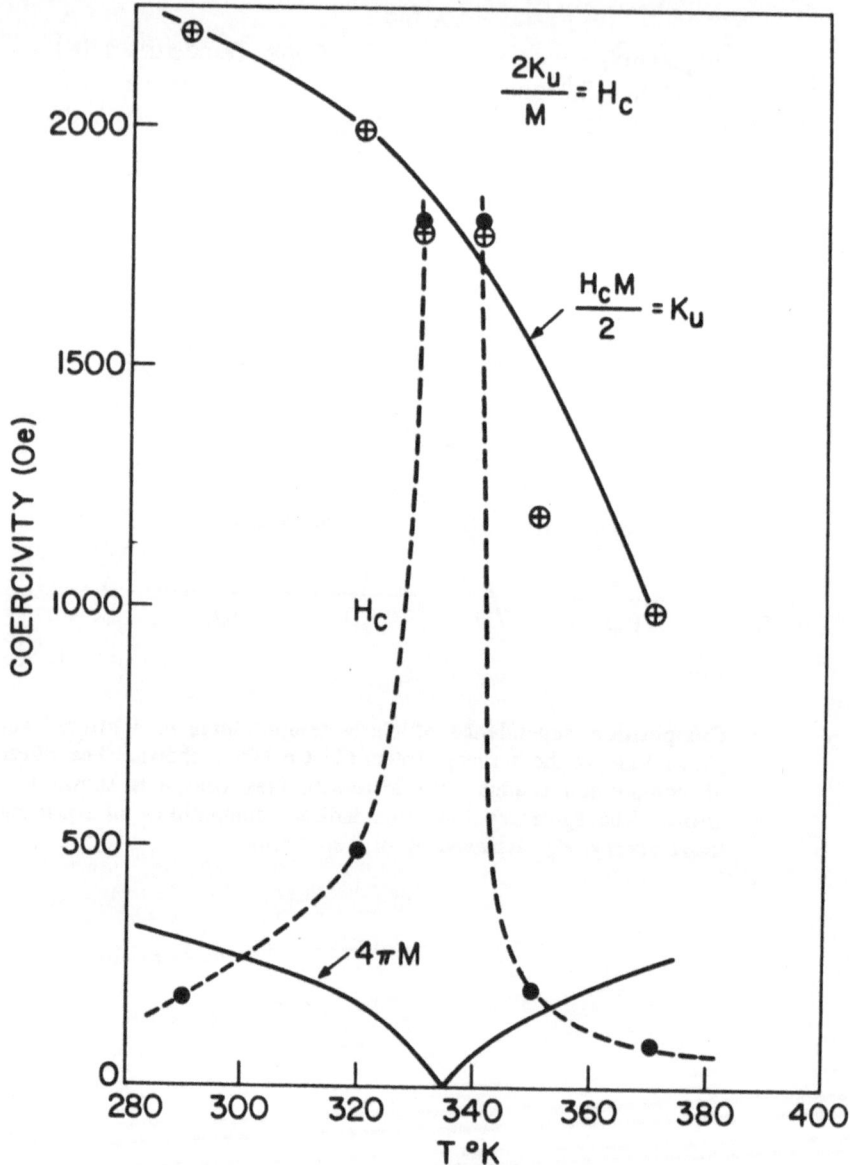

Fig. 3. Coercivity as a function of temperature in the vicinity of a compensation point in amorphous $Gd_{21}Co_{79}$. The temperature dependence of the saturation magnetization and the anisotropy energy calculated assuming $K_u = H_c M/2$ are also shown.

The energy of the wall is:

$$E_x = 4\sigma_w hD = 16(AK_u)^{1/2} hD \tag{7}$$

where σ_w is the wall energy per unit area which can be expressed in terms of the exchange stiffness, A, and the uniaxial anisotropy energy, K_u. Taking the appropriate partial derivatives leads to the following expression for H_c:

$$H_c = \frac{\sigma_w}{M}\left[1/2\left(\frac{\partial K_u}{K_u \partial x} + \frac{\partial A}{A \partial x}\right) + \frac{\partial D}{D \partial x} + \frac{\partial h}{h \partial x}\right] \tag{8}$$

To get a better physical insight into the meaning of this expression, we consider a simple saw tooth shaped potential well of half width δ, where δ is the domain wall width. The justification for considering this particular case is that only perturbations on a scale comparable in size to the wall width will interact strongly with the wall. If the saw tooth has a very large wavelength, $\partial E/\partial x$ over the wall width will be small. If the wavelength is very much less than δ then $\partial E/\partial x$ will average to zero over the wall width. We thus rewrite Eq. (8) to obtain:

$$H_c = \frac{\sigma_w}{M}\left[\frac{1}{2}\left(\frac{\Delta K_u}{\delta K_u} + \frac{\Delta A}{\delta A}\right) + \frac{\Delta D}{\delta D} + \frac{\Delta h}{\delta h}\right] \tag{9}$$

$$= \frac{\sigma_w}{M\delta}\left[\frac{1}{2}\left(\frac{\Delta K_u}{K_u} + \ldots\right)\right]$$

The wall width can also be written in terms of A and K_u as $\delta = \sqrt{A/K_u}$ so that:

$$\frac{\sigma_w}{\delta} = \frac{4\sqrt{AK_u}}{\sqrt{A/K_u}} = 4K_u \tag{10}$$

which leads to the final expression for H_c:

$$H_c = \frac{2K_u}{M} \sum_i \frac{\Delta x_i}{x_i} \quad \text{where } x_i = K_u, A, D, h \tag{11}$$

Hasegawa (36) used this expression for the coercivity of liquid quenched ribbons with inplane domains. In spite of the very different domain geometry, he obtained an excellent fit of H_c versus temperature using equation (11). Paul (37) has done a more detailed calculation leading to a general expression of the same form as Eq. (11).

Note that as with nucleation there is a $2K_u/M$ term in the wall motion coercivity. Also, we can consider the effects of compositional fluctuations on H_c. In a system in which M is a strong function of composition and M becomes very small, as near a composition point, the $\Delta M/M$ term will make a large contribution to H_c. We can estimate the effects of purely statistical fluctuations for example on the anisotropy.

Assuming that the anisotropy arises from pair ordering and that there are n pairs within the wall width region, the fluctuation in n from statistical mechanics gives:

$$\frac{\Delta K_u}{K_u} = \frac{2\sqrt{n}}{n} \tag{12}$$

Even though the concentration of pairs is relatively small, the contribution of statistical fluctuations only gives about 0.1 Oe in a 2 μm bubble material. A recent calculation has considered random fluctuations in single ion anisotropy systems. Even though these fluctuations are on an atomic scale, a scale much smaller than the wall width, the tails of the distributions can give significant contributions to wall motion coercivity. (38) When intermediate coercivities are needed as in vertical recording or magneto-optical recording it is often necessary to use a non-S-state rare earth and these single ion anisotropy effects. The resulting large K_u acts as a barrier to reverse domain nucleation and fluctuations provide a large $\Delta K_u / K_u$ term to make the written bit stable with respect to wall creep.

Other effects that can give increased coercivity can be understood in terms of the inhomogeneity model. For example, if a small concentration of a reactive gas (e.g. N or O) is present during the sputter deposition of GdCo higher H_c films are obtained but the films are still amorphous. The reason is that the alloy phase separates into an amorphous oxide and a Co rich amorphous metal. The scale of the phase separation is generally smaller than the wall width but $\Delta M/M$ can have a significant impact on H_c.

PROCESS CONTROL OF PROPERTIES

As pointed out above, $4\pi M_s$ is essentially only a function of composition; however, the composition is to some extent determined by the processing conditions. For example, in bias sputtering, the bias voltage controls the fraction resputtered. (39) The alloy constituents have different sputtering yields so the composition is a function of the bias voltage. (40,41) In GdCo amorphous alloys, 100 volts bias corresponds to about a 10 atomic percent decrease in Gd concentration which can cause a significant change in $4\pi M_s$. Bombardment of the surface of the growing film also helps to reject impurities such as oxygen so the impurity content is also a function of the bias voltage. (39) If the bias voltage is above about 25 volts, however, the impurity effects are no longer significant. Another consequence of the ion bombardment caused by the application of a bias voltage is implantation of atoms of the sputtering gas into the surface of the growing film. The concentration of gas incorporated goes as the square of the bias voltage up to a maximum at about 150 volts. (42) The sputter gas incorporation depends on the atomic radius of the inert gas used and on the pressure. For argon at 30 mTorr the concentration can be kept below 1 atomic percent using a bias voltage of about 50 volts. For control of magnetization, therefore, the most important effect is the resputtering of one alloy constituent, particularly in the ferrimagnetic systems. The impurity and inert gas effects can be reduced to a minimum by operating in the 25 to 50 volts bias range.

In very thin films, another factor in composition control is preferential oxidation of one constituent at surfaces and interfaces. (43) Typical examples of systems that show selective oxidation are RE-Co systems, Zr-Co or Ti-Co. In these systems the RE, Zr or Ti are oxidized leaving behind a Co rich layer. If the film is thick, 0.1 μm or greater, the self limiting thickness of the Co layer is a relatively small perturbation. In films of only a few hundred angstroms, the effect of surface oxidation can increase the average $4\pi M_s$ by a factor of 2 or more. Overcoating of the film with a dielectric does not always help because there is sometimes an interfacial reaction between the dielectric and the film which also causes perferential oxidation of the more reactive constituent. It has been found that plasma oxidation is sufficiently aggressive that both constituents are oxidized and the troublesome cobalt layer left behind by selective oxidation is avoided. (44) The plasma oxidized surface can be overcoated without degradation. This treatment is particularly useful in magneto-optical Kerr effect applications which use only the near surface region within the penetration depth of the light.

The anisotropy of an amorphous film can be controlled to a large extent through the deposition process. Bias sputtering provides a large measure of control particularly in the RE-Co system. The models of growth induced anisotropy in bias sputtering are based on short range ordering from the site selective resputtering of one species. (46, 47, 48) For example, in Gd-Co it appears that Gd when bonded to Co has a lower sputter yield than when it is bonded to other Gd atoms. This leads to a preferential distribution of GdCo pairs oriented perpendicular to the plane of the film and a corresponding tendency for CoCo pairs to lie in the plane of the film. Cargill and Mizoguchi (49) have shown that even a simple dipole interaction considering all the possible types of pairs and their spacial distribution is sufficient to explain the magnitude of the observed anisotropy even if the deviation from a random distribution is very small (less than 1%). It has been observed that the anisotropy increases with bias up to a maximum then decreases rapidly at higher bias. The reasons for this fall-off are not known with certainty. It may be that the sputter yield threshold of the other species has been exceeded so that the selectivity is lost (45) or that resputtering from subsurface layers becomes important so that the induced anisotropy in atom distribution is lost. (48) It has also been shown that in Gd-Fe alloys there is a strong dependence of anisotropy on bias voltage but that it has essentially the opposite sign - i.e. the anisotropy goes through a peak in negative K_u at about 150 V bias. This indicates that there is a large contribution to the anisotropy that does not arise from the dipole term, which would be of the same sign for Gd-Co and Gd-Fe, but there must also be a pseudodipole term which has a different sign in each system. Of course, if a non-S-state RE is used, there is also a single ion anisotropy contribution.

In evaporated films, there is less control over the anisotropy. There are atomic self shadowing effects that can give rise to anisotropic atomic distributions as well as columnar microstructures. Films sputtered without bias bombardment are similar to evaporated films and also tend to be columnar. When these films are exposed to air, the preferential oxidation discussed above can also occur on the sides of the columns. This distribution of high saturation TM rich phase on the walls of the columns can also give a dipolar contribution to the anisotropy. (49) This mechanism is not applicable to the bias sputtered films, however, which have been shown to have very little small angle x-ray scattering. (49) Also, the fact that some Gd-Co evaporated films have a hard perpendicular axis anisotropy suggests that short range atomic ordering can occur just as a consequence of the deposition process.

Stress induced anisotropy can be an important factor in some compositions such as Gd-Fe or Fe-B alloys. The Gd-Co alloys have very small magnetostriction so the stress anisotropy contribution is negligibly small in this system. In systems like Fe-B, alloys with large magnetostriction, the films tend to have perpendicular anisotropy in the as deposited state. (13) When the films are annealed below the crystallization temperature but at a high enough temperature to allow stress relief, the magnitude of the anisotropy can be markedly reduced. Even in zero magnetostriction compositions, e.g. CoFeNiBSi there is a small growth induced anisotropy, probably caused by short range atomic ordering. In all these films, with short range ordering anisotropy, it is possible to anneal out the anisotropy or even to anneal in a new easy axis in the plane by annealing in an applied field. Alloys with single ion anisotropy using the non-S-state RE's tend to have very large magnetostriction coefficients which can cause large stress induced anisotropy effects.

The effects of anneal can be very large if partial crystallization or full crystallization occurs. In fact, it is possible to make an amorphous alloy composition useful as a permanent magnet by crystallizing it into a highly anisotropic crystal structure. This seems to be the case in the recently reported $Nd_{15}Fe_{77}B_8$ compositions (31, 50). In a relatively narrow range of Nd and B compositions, annealing of liquid quenched ribbons leads to high anisotropy crystallites of a particular tetragonal phase. In this phase, it appears there are strong crystal field effects acting on the Nd giving a very large anisotropy. In the crystalline case, all the RE atoms can be in the same crystal field with the same orientation in the unit cell. In the $CaCu_6$ structure, for example, this can lead to anisotropy energies of 10^{+8}ergs/cm^3. This is in contrast to the amorphous film deposited under bias bombardment which has at most 1% deviation from a random distribution. In the amorphous case, the anisotropy energy is limited to about 10^{+6}ergs/cm^3 or about 1% of the single crystal value. When the alloy is crystallized, e.g. $Nd_{15}Fe_{77}B_8$, each crystallite may have a very high anisotropy but if their easy axes are randomly distributed only those favorably aligned to the applied field give a large contribution of the magnetization. For this reason, the processing of an amorphous ribbon into a permanent magnet involves in addition to crystallization of the ribbon, the steps of powdering the crystallized material, aligning the powder in an applied field and sintering the particles or otherwise fixing them in the aligned condition. It has, in fact, been shown that the liquid quenching step is not required in the case of $Nd_{15}Fe_{77}B_8$, the appropriate composition can be simply induction melted. (50) The liquid quenching method, however, may be a convenient way to insure good mixing and small particle size.

The process for making $Nd_{15}Fe_{77}B_8$ permanent magnets does not appear to be suitable for making thin film permanent magnets. Applications are rather limited for thin film magnets but there are a few. Bias magnet layers for bubble films requiring a perpendicular easy axis is one possibility. Another is a bias magnet for a thin film MR head where an inplane magnetization is needed (30). The alignments are difficult if not impossible to control during crystallization so the approach of depositing the film in the amorphous state and then crystallizing it is not very satisfactory in this case. It is better to deposit the film in the crystalline state in the first place with the desired preferred orientation. This can be done by deposition of thermalized atoms onto hot substrates (51). Good results have been obtained with $SmCo_5$ and a number of other compositions

CONTROL OF SECONDARY PROPERTIES

A number of secondary properties are very important for applications so it is important to know how to adjust them and what effect adjusting the secondary properties has on the primary magnetic properties $4\pi M_s$, H_k and H_c. For almost all applications thermal stability is important. This is especially the case if short range order anisotropy is being used because the lower the crystallization temperature the lower will be the annealing temperature at which the anisotropy will be lost. This is illustrated in the case of GdCoMo, GdCoAu, and GdCoCu. The addition of Mo increases both the crystallization temperature and the annealing temperature and makes it possible to thermally cy cle the material to 150C repeatedly without loss of growth induced anisotropy. This is generally true of multivalent additives like the refractory metals. According to the magnetic valence model for determining the moment of the transition metal, these high valence metals will have a large influence on the TM moment.

In the case of iron-TM alloys, it has been shown that the concentration of alloy addition needed to obtain an amorphous film is directly proportional to the size ratio, r_{Fe}/r_{TM}. When the amount of TM needed is small, it is because the TM is large. (19) This means that the magnetic dilution effect of the TM is about the same at the limiting concentration in all cases. Of course, the magnetic valence effects must also be taken into account.

In order to use these materials for thermomagnetic writing, it is necessary to have a large magneto-optic Kerr effect. To this end, it is important to have well aligned subnetworks. In this application, however, it is also essential to have a high H_c which is obtained by using a non-S-state RE which tends to fan out the magnetic moments. Also, it is important to have a low Curie temperature for this application, but with a low Curie temperature the subnetworks are far from saturated at room temperature. Fortunately, there is enough flexibility in these amorphous systems to allow adjustment of these many variables. For example, a commonly used alloy for thermomagnetic writing is TbFe. (52) It has a low T_c, high H_c and about $15'$ of Kerr rotation which is too little for a good signal to noise ratio. Sakurai et al. (53) have shown that the rotation increases when part of the Fe is replaced by Co. This is for two reasons: the T_c of the alloy has gone up so that room temperature is a smaller fraction of T_c and the subnetworks are more nearly saturated. Also, the stronger Co-Co exchange tends to improve the alignment of the Fe subnetwork. The stronger Co-Co exchange also increases the T_c, however, so that higher laser power is needed to write a bit.

Other approaches to increasing the Kerr rotation without increasing the laser power involve multilayered films. (54) One layer, TbFe, provides the high H_c and low T_c and serves as a storage layer. The other layer, GdCo, is exchange coupled to the storage layer and serves as a readout transducer because it has a higher Kerr rotation. One possible disadvantage of this approach is that the layers may inter-diffuse after repeated thermal cycling. The GdCo can serve as a storage layer by itself if used in the compensation point writing mode rather than the Curie point writing mode. The H_c is not as high, however, and the bits are only stable in a narrow range near T_{comp}. The essential feature of both Curie point and compensation point writing is the temperature dependence of H_c. In the temperature interval, ΔT, provided by the laser heating, the coercivity must decrease from a value 2 to 3 times the switching field, H_s, to below the switching field. The switching field is provided by an external electromagnet or by the closure field of the surrounding material. This can be achieved with a combination of single ion anisotropy and compensation point coercivity. The coercivity associated with the single ion anisotropy increases monotonically with decreasing temperature. The compensation point system shows a peak in H_c at T_{comp} because M_s goes to zero. Since $H_c = 2K_u/M_s$ a high K_u at T_{comp} increases the magnitude of H_c and keeps it high above the compensation point. The necessary change in H_c over a small temperature range can be achieved without a low T_c and without degrading the Kerr rotation.

The magneto-optic effects and the Hall effect in these amorphous metals are closely related phenomena (55) . The Hall effect is large in these materials (56) because of the short mean free path and the large number of scattering events of carriers which also gives rise to the high resistivity of these materials. The Hall effect has been proposed for sensor applications but the necessity for a device with at least three terminals makes Hall sensors less attractive then the simpler two terminal magnetoresistive, MR, sensors. The anisotropic magnetoresistance effect is used in heads with polycrystalline permalloy as the sensing element. (30) The MR effect in permalloy films is typically 3%, $\Delta\rho/\rho$. In the amorphous alloys, $\Delta\rho$ is about the same as in crystalline films but ρ is about an order of magnitude larger making $\Delta\rho/\rho = 0.3\%$ too small for device applications. (57) If a high MR amorphous composition could be found, it would have the advantage that these high resistivity alloys have temperature coefficients of resistivity close to zero. A major source of noise in MR heads is thermal spike noise. When the head comes into contact with the moving recording medium, it heats up and the voltage spike created is difficult to discriminate from the MR signal. The Hall effect has been used to move magnetic domains in amorphous alloys via the domain drag effect. (58)

CONCLUSIONS

Amorphous materials provide a wide range of magnetic properties that give them potential applications in many devices. In thin film form, amorphous materials are useful for thermomagnetic recording media. The bit stability, recording density and long term stability of these media seem to be suitable for high density data storage applications. The main limitation is the signal to noise ratio (SNR) of these materials which limits the data rate. (59) Progress is being made toward improving the SNR without degrading the other desirable properties such as low power threshold.

Another potential application for soft magnetic materials is thin film heads. At present, the advantages of these amorphous materials have not been sufficient to displace permalloy even though higher saturation magnetization and lower coercivity materials are available. (29)
In magnetic bubble applications, a similar situation prevails. Amorphous materials are available suitable for small bubble applications but garnets will probably continue to be used until higher densities are needed and the lithographic techniques for submicron bubbles are well under control as production methods.

In the near term, by far the most important application of thin film amorphous magnetic materials will be as a thermomagnetic recording medium. Present materials are suitable for audio and video digital recording applications. (60) A relatively small improvement in SNR will make these materials important for digital data recording with its more stringent requirements of high data rate and low error rate.

References

1. Klement, W., R. H. Willens and P. Duwez, Nature, 187 (1960) 869.

2. Mader, S., H. Widmer, F. M. d'Heurle, and A. S. Nowick, Appl. Phys. Lett. 3 (1963) 201.

3. Mader, S., and A. S. Nowick, Appl. Phys. Lett. 7 (1965) 57.

4. Tsuei, C. C., and P. J. Duwez, J. Appl. Phys. 37 (1966) 435; P. Duwez, and S. C. H. Lin, J. Appl. Phys. 38 (1967).

5. Chaudhari, P., J. J. Cuomo, and R. J. Gambino, IBM J. Res. and Dev., 17 (1973) 66.

6. Chaudhari, P., J. J. Cuomo and R. J. Gambino, J. Appl. Phys. Lett. 22 (1973) 337; P. Chaudhari, J. J. Cuomo, R. J. Gambino and T. R. McGuire, U.S. Patent 3,949,387.

7. Pond, R. Jr. and R. Maddin, Trans. Met. Soc. AIME, 245 (1969) 2475.

8. Egami, T., P. J. Flanders and C. D. Graham, Jr., AIP Conf. Proc., 24 (1975) 697; F. E. Luborsky, AIP Conf. Proc. 29 (1976) 209.

9. Brenner, A., D. E. Couch, and C. K. Williams, Res. Nat. Bur. Stds., 44 (1950) 109.

10. Cargill, G. S. III, J. Appl. Phys. 41, (1970) 12

11. Duwez, P., R. H. Willens and R. C. Crewdson, J. Appl. Phys. 36 (1965) 2267.

12. Hasegawa, R., R. C. O'Handley, L. E. Tanner, R. Ray and S. Kavesh, Appl. Phys. Lett. 29 (1976) 219.

13. Kobliska, R. J., J. A. Aboaf, A. Gangulee, J. Cuomo and E. Klokholm, Appl. Phys. Lett. 33 (1978) 473.

14. Ali, A., W. A. Grant and P. J. Grundy, Phil. Mag. B, 37 (1978) 353.

15. Ray, R., B. C. Giessen and J. J. Grant, Scripta Met. 2 (1968) 357.

16. Ruhl, R. C., B. C. Giessen, M. Cohen and N. J. Grant, Acta Met. 15 (1967) 1693.

17. Wang, R., M. D. Merz and J. L. Brimhall, Scirpta Met. 12 (1978) 1037.

18. Ohnuma, S., K. Shirakawa, M. Nose and T. Masumoto, IEEE Trans. Mag. MAG16 (1980) 1129.

19. Fukamichi, K. and R. J. Gambino, IEEE Trans. Mag. MAG 17 (1981) 3059.

20. Fukamichi, K., M. Kikuchi, T. Masumoto and M. Matsuura, Phys. Lett. **A73** (1979) 436.

21. Buschow, K. H. J., and A. M. van der Kraan, J. Magn. and Mag. Mat. **22** (1981) 220.

22. Polk, D. E. and B. C. Giessen, Metallic Glasses, ASM, Metals Park, Ohio (1978).

23. R. St. Amand, and B. C. Giessen, Scripta Met. **12** (1978) 1021.

24. Bayreuther, G., G. Enders, H. Hoffmann, U. Korndorfer, W. Oestreicher, K. Roll and M. J. Takahashi, J. Magn. and Mag. Mat., **31-34** (1983) 1535.

25. Gambino, R. J., and J. J. Cuomo, J. Vac. Sci. Technol. **15** (1978) 296.

26. Gambino, R. J., P. Chaudhari, and J. J. Cuomo, AIP Conf. Proc. **18** (1974) 578.

27. Chang, H. (ed.), Magnetic Bubble Technology, IEEE Press, New York 1975.

28. Eschenfelder, A. H., Magnetic Bubble Technology, Springer-Verlag, New York 1980.

29. Yamada, K., T. Maruyama, H. Tanaka, H. Kaneko, I. Kagaya, and S. Ito, J. Appl. Phys. **55** (1984) 2235.

30. Tsang, C., J. Appl. Phys. **55** (1984) 2226.

31. Croat, J. J., J. F. Herbst, R. W. Lee and F. E. Pinkerton, Appl. Phys. Let. **44** (1984) 148.

32. Williams, A. R., V. L. Moruzzi, A. P. Malozemoff and K. Terakura, IEEE Trans. Mag. **MAG19** (1983) 1983.

33. Heiman, J., J. Kazama and K. Lee, J. Appl. Phys. **50** (1979) 4891.

34. Craik, D. J. and R. S. Tebble, Ferromagnetism and Ferromagnetic Domains, North - Holland, Amsterdam (1965).

35. Harris, R., M. Plischke and M. J. Zuckermann, Phys. Rev. Lett. **31** (1973) 160.

36. Hasegawa, R., AIP Conf. Proc. **29** (1976) 216.

37. Paul, D. I., AIP Conf. Proc. **29** (1976) 545.

38. Thiele, A. A. and P. Asselin, J. Appl. Phys. **55** (1984) 2584.

39. Cuomo, J. J., R. J. Gambino and R. Rosenberg, J. Vac. Sci. Technol. **11** (1974) 34.

40. Cuomo, J. J. and R. J. Gambino, J. Vac. Sci. Technol. **12** (1975) 79.

41. Harper, J. M. E. and R. J. Gambino, J. Vac. Sci. Technol. **16** (1979) 1901.

42. Cuomo, J. J. and R. J. Gambino, J. Vac. Sci. Technol. **14** (1977) 152.

43. Argyle, B. E., R. J. Gambino and K. Y. Ahn, AIP Conf. Proc. **24** (1975) 564.

44. Gambino, R. J. and R. R. Ruf, unpublished.

45. Gambino, R. J. and J. J. Cuomo, J. Vac. Sci. Technol. **15** (1978) 296.

46. Mizoguchi, T., R. J. Gambino, W. N. Hammer and J. J. Cuomo, IEEE Trans. Mag. **MAG13** (1977) 1618.

47. Nishihara, Y., T. Katayama, Y. Yamaguchi, S. Ogawa and T. Tsushima, Jap. J. Appl. Phys. **17** (1978) 1083; 18 (1979) 1281.

48. Muller, H. R. and R. Perthel, Phys. Stat. Sol. **87** (1978) 203.

49. Cargill, III, G. S., and T. Mizoguchi, J. Appl. Phys. **49** (1978) 1753.

50. Sagawa, M., S. Fujimura, N. Togawa, H. Yamamoto and Y. Matsuura, J. Appl. Phys. **55** (1984) 2083.

51. Cadieu, F. J., T. D. Cheung, S. H. Aly, L. Wickramasekara and Pirich, R. G., IEEE Trans. Mag. **MAG19** (1983) 2038.

52. Tanaka, S. and N. Imamura, IEEE Trans. Mag. **MAG19** (1983) 1751.

53. Tsujimoto, H., M. Shouji and Y. Sakurai, IEEE Trans. Mag. **MAG19** (1983) 1757.

54. Tsunashima, S., H. Tsuji, K. Kobayashi and S. Uchiyama, IEEE Trans. Mag. **MAG17** (1981) 2840.

55. Hartmann, M. and T. R. McGuire, Phys. Rev. Lett. **51** (1984) 1194.

56. McGuire, T. R., R. J. Gambino and R. C. O'Handley, Hall Effect and Its Applications, C. L. Chien and C. R. Westgate (ed.), Plenum Publishing, New York (1980).

57. Fukamichi, K., R. J. Gambino and T. R. McGuire, J. Appl. Phys. **53** (1982) 8254.

58. Deluca, J. C., R. J. Gambino and A. P. Malozemoff, IEEE Trans. Mag. **MAG14** (1978) 500; J. C. Deluca, R. J. Gambino, A. P. Malozemoff and L. Berger, J. Appl. Phys. **52** (1981) 6168.

59. Bell, A. E., Computer Design, Jan. (1983) 133.

60. Togami, Yuji, IEEE Trans. Mag. **MAG18** (1982) 1233.

GLASS FORMATION AND GROWTH IN SOLIDS

W. L. Johnson, B. Dolgin and M. Van Rossum

California Institute of Technology
Pasadena, California 91125   USA

I. FORMATION OF METALLIC GLASSES

Traditionally, a glass is defined as an undercooled liquid brought below a characteristic freezing temperature referred to as the glass transition temperature, T_g. The formation of glass requires that the process of crystallization be bypassed during cooling through the thermodynamic melting point T_m, and the undercooled temperature range to below T_g. The nucleation and growth of crystals in an undercooled melt involves atomic rearrangements which occur on a time scale $\tau_x(T,P,C,Y)$ which depends on temperature T, pressure P, melt composition C, and possible other less conventional parameters collectively labeled Y. Very roughly, cooling through the undercooled region must take place on a time scale τ_c such that $\tau_c \lesssim \tau_x$ in order to achieve the glassy state. Detailed theories which lead to specific criteria have been put forth by Turnbull and others for the case of metallic liquids.[1]

Modern techniques for rapid quenching applied to metallic alloys, lead to cooling rates ranging from 10^4 Ks^{-1} in splat cooling to 10^{10} Ks^{-1} in pulsed laser melting experiments.[2,3] For undercooling through temperature intervals of 10^2–10^3 K, this gives $\tau_c \sim 10^{-1}$–10^{-7} (sec.). For a pure liquid metal, such values of τ_c are insufficient to bypass crystallization. For some alloys, typically those with composition lying near a deep eutectic (or low lying liquidus feature of the phase diagram), such τ_c values are sufficient to produce metallic glass. The "deep eutectic" criteria for glass formation is associated naturally with the fact that the undercooling interval T_m–T_g or reduced undercooling interval $\Delta = (T_m$–$T_g)/T_m$ is typically smaller for alloy compositions near deep eutectics.

The above discussion emphasizes the historical view that the liquid state is the parent phase of the glass. Glass formation theories based on this view have generally focussed on kinetic processes in liquids. On the other hand, it has been known since the pioneering studies of Buckel and Hilsch[5] that amorphous alloys can form when, for example, a metallic vapor or alloy vapor is quenched onto a relatively cold substrate. Here, the parent phase is the vapor and the kinetics of interest involve vapor transport and the rearrangements of atoms in and on the surface of the growing condensed film. The direct connection with the liquid phase is lost, and materials so produced are historically referred to as amorphous rather than glassy. Systematic structure and property studies comparing compositionally identical amorphous alloys (vapor-quenched) to their glassy counterpart (liquid-quenched)[6,7] in fact show that these materials are often very similar if not practically indistinguishable. In this sense, use of the two terms metallic glass and amorphous alloy reflects the two methods of preparation as opposed to two fundamentally different materials. To emphasize this similarity, the term glass will herein be used to refer to an amorphous material irrespective of its origin.

The above mentioned two methods for glass synthesis proceed via the liquid and vapor states. It is only natural to consider the third generic possibility of glass synthesis via the solid state. The approach would follow a similar line of reasoning as used above. It necessarily involves a process in which the nucleation and growth of more stable competing crystalline phases must be bypassed. The following "thought experiment" illustrates one possible case. Consider a binary alloy system $A_{1-x}B_x$ at temperature T_1 lying everywhere below the liquidus curve of the equilibrium A-B phase diagram shown in Fig. 1. One assumes that pure solids A and B have crystal structures α and β respectively, and can exothermically react to form several crystalline intermetallic compounds (e.g. γ-AB, ε-AB_2 and δ-A_2B). That is, the solid elements A and B have a large negative heat of mixing with respect to the crystalline compounds of structure (γ, δ and ε). In contrast, assume that the heat of mixing for the terminal solutions α and β are characterized by a large and positive value. This situation is, as will be seen, characteristic of a large number of binary alloy systems. We add to this still rather general case an additional assumption that the binary liquid $A_{1-x}B_x$ is characterized by a large negative heat of mixing with respect to the pure liquids. This is again a rather common situation (see for example, Miedema[8]). At temperature T_1, a free energy diagram corresponding to Fig. 1. is shown in Fig. 2. The free energy vs. composition curves for the α, β, γ, δ, ε phases are all shown. In addition, the free energy of the liquid state is shown and labeled by λ. Curve λ represents a <u>metastable</u> liquid state since T_1 lies below the temperature range where a thermo-

174

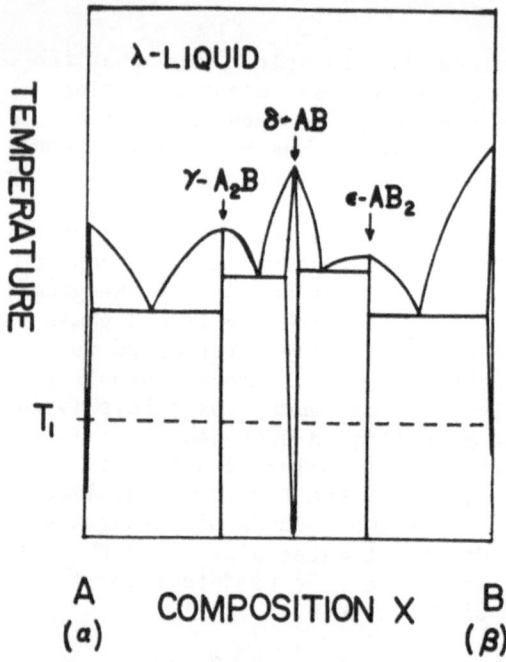

Fig. 1. An example of a binary phase diagram with intermetallic compounds γ-A_2B, δ-AB and ε-AB_2.

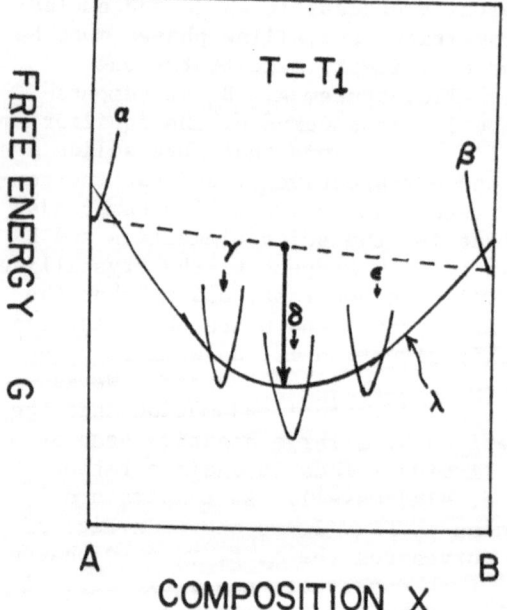

Fig. 2. Free energy diagram at temperature T_1 corresponding to phase diagram in Fig. 1.

dynamically stable liquid can exist. For $T_1 < T_g$, the λ-curve will be used to represent a glass. We do not here address the question of justifying the existence of this metastable curve. For a discussion of the justification, the reader is referred for example, to Penrose and Liebowitz.[9] Notice that the terminal points of λ-curve represent undercooled liquids of pure A and B at temperature T_1. The difference between these terminal points and those of the α- and β-curves give the free energy of crystallization of the pure undercooled metals A and B. Following Turnbull[1], this difference can be simply approximated by

$$\Delta H_{\alpha-\lambda}(T_1) = \left[\frac{T_m^\alpha - T_1}{T_m^\alpha}\right] \Delta H_m^\alpha \qquad (1)$$

where T_m^α and ΔH_m^α are the melting point and heat of fusion for pure α-A (or β-B). More sophisticated approximations could be used as well.[10] In any case, this allows us to fix the endpoints of the λ-curve.

We now consider a problem depicted in Fig. 3. A layer of liquid _or_ glass $A_{1-x_o} B_{x_o}$ is brought into simultaneous contact with layers of pure solids A and B at temperature T_1. How will such a system evolve in time? For convenience consider A and B to be in the ratio $(1-x_o)/x_o$. The equilibrium state of this system is given by Fig. 1 and depends only on x_o and T (we ignore pressure). But the achievement of equilibrium requires in general the nucleation and growth of new crystalline phases $(\gamma, \delta, \varepsilon)$. Suppose that a time τ_x is required for one of these phases to nucleate (or grow). Now consider a second time constant of this problem. This is the time τ_D required to transport A or B atoms by diffusion across the glass layer. It depends on an interdiffusion coefficient[11] which is defined by

$$\tilde{D}_G = [x_o D_A^o + (1-x_o)D_B^o] \, [1+H_{mix}(x_o)/K_B T] \qquad (2)$$

for A and B in the glass layer. Here D_A^o and D_B^o are intrinsic diffusivities of A and B in the glass and $H_{mix}(x_o)$ is the heat of mixing of A and B in the glass $A_{1-x_o} B_{x_o}$. Eqn.(2) assumes that the glass can be treated as a regular solution. One defines

$$\tau_D = (d_G^2/\tilde{D}_G) \qquad (3)$$

as the second characteristic time constant. Both τ_x and τ_D depend on temperature. To keep things simple, one can ignore interfacial diffusion barriers which might introduce a third time constant.

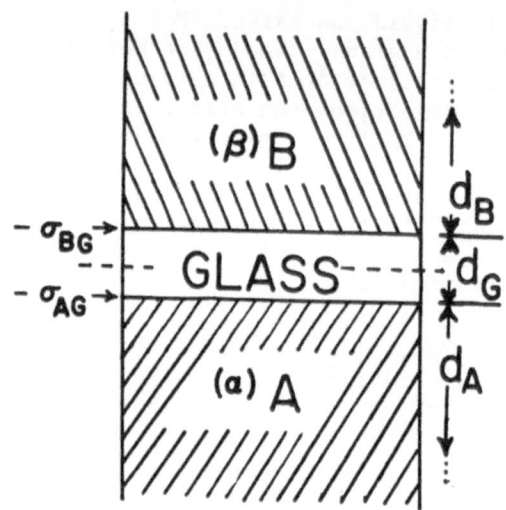

Fig. 3. Binary diffusion couple containing a glassy interlayer of composition $A_{1-x_o}B_{x_o}$. Interface energies σ_{AG} and σ_{BG} characterize the structure.

Returning to the original question, how does the system evolve in time? The answer depends on the relative magnitudes of τ_x and τ_D. In particular for the case $\tau_D \ll \tau_x$, the glass layer can grow (providing that interfacial diffusion barriers are ignored). Dolgin and Johnson[12] have solved a more general version of this problem for diffusion controlled growth in the case where interfacial effects are included. The general problem involves a coupled set of non-linear differential equations which are solved numerically. However, in limiting cases, the solution is simple. For example, when the layer is sufficiently thick so that τ_D is long compared with interfacial transit times, one obtains the simple asymptotic form:

$$d_G = -d_o + \sqrt{Ct}\qquad(4)$$

where d_o is a characteristic thickness up to which interface effects are significant and C a constant related to $\tilde{D}_G$. This is the usual $t^{\frac{1}{2}}$ law for diffusion controlled layer growth with a shifted thickness origin. For details of this argument, see ref. 12.

Several additional facts are worth explicit mention. First, the net thermodynamic driving force for amorphous layer growth is provided by the decrease in free energy when A and B react to form the glass. The total drop in free energy following total consump-

tion of A and B by the growing glass layer and subsequent homogen-
zation of the amorphous phase is given by the downward arrow in
Fig. 2. Second, this total process requires a time which is several
units of the time constant τ_D (eqn. (3)) with d_G taken as the final
thickness of the glass layer after complete reaction. Provided that
this time is still much smaller that τ_x, the entire couple will
react to form glass prior to the appearance of crystalline compounds.
Third, the possible complications which arise when the glass forming reaction involves a change in total volume of the system have
been ignored. This will result in strain energy effects etc.
Finally, an obvious question was not addressed. In a virgin binary
A-B diffusion couple (i.e. $d_G = 0$ initially), how would the assumed
glass layer form or "nucleate"? Is there a characteristic time τ_G
required to form the initial glass layer? How would this time τ_G
compare with τ_x? One can follow a classical nucleation argument to
investigate these questions. Formation of a uniform glass layer of
composition x_o (one can ignore concentration gradients in the glass
layer for simplicity) involves a free energy change per unit area
of layer given by

$$\Delta G_{tot} = \Delta g d_G + (\sigma_{AG} + \sigma_{BG}) - \sigma_{AB} \tag{5}$$

where Δg is the negative Gibbs free energy change per unit volume
on formation of glass from elemental A and B at composition x_o
(downward arrow of Fig. 2). The surface energies σ_{AG}, σ_{BG} and σ_{AB}
refer to the A-glass, B-glass and A-B interfaces. From eqn. (6),
we obtain a critical thickness $d_G^c = -[\sigma_{AB} + \sigma_{BG} - \sigma_{AB}]^{-1} \Delta g$ which
measures the thickness of a glass layer required to overcome the
surface energy associated with the creation of two interfaces at
the expense of one. One can attempt to estimate d_G^c. For a pure
metal[13], the crystal melt interface energy is roughly given by

$$\sigma_{AL} = \alpha_o H_f (N\bar{V})^{-1/3} \tag{6}$$

where H_f is the heat of fusion, $\bar{V}$ the molar volume, and α_o is a
dimensionless parameter which is found to be $\alpha_o \approx 0.5-0.6$ for a
pure metal in contact with its own melt. Ignoring the fact that
one has a pure metal in contact with an amorphous alloy $A_{1-x_o} B_{x_o}$,
rather than its own melt, one can estimate $H_f \sim 10$ kJ/mole,
$\bar{V} \sim 10 cm^3$, and $\sigma_{AG} \sim \sigma_{BG} \sim 1.5 \times 10^{-5}$ J/cm^2. Using
$\Delta g \sim 50$ kJ/mole $= 5 \times 10^3$ J/cm^3, a value which will later be seen
to be appropriate, and for the moment ignoring σ_{AB}, one gets
$d_G^c \lesssim 10^{-8}$ cm. Including a positive σ_{AB} would reduce the estimate of
d_G^c. In any case, one can see that d_G^c is less than one atomic monolayer thickness. It could even be the case that $\sigma_{AB} > \sigma_{AG} + \sigma_{BG}$.

This would imply that the original interface is intrinsically
unstable. Summarizing, it is unlikely that the formation of a glass
interlayer will be nucleation limited. The large negative value of
Δg precludes this. One can conclude that only atomic rearrangement
rates are likely to limit τ_G. In particular, atomic mobility in the
vicinity of the interface will control the kinetics.

What about the relative magnitude of τ_G and τ_x? The above
argument could equally well be applied to initial formation of
crystalline compound layers. Will glass layers form more or less
rapidly than crystalline compound layers? One can gain some insight
by considering the degree of correlated atomic rearrangement required to form an amorphous vs. crystalline compound layer. To
form an amorphous interlayer requires the development of short
range order extending at most a few atomic distances in space. To
form a crystalline compound layer requires the development of long
range order along the interface. This should tend to kinetically
favor amorphous interlayer formation. When kinetics are as restrictive as in the case of solid state diffusion at low temperature,
this may lead naturally to the appearance of the amorphous interlayer in the limit of short times. Unfortunately, space restricts
a more extensive discussion of this fascinating question. Instead,
we now turn to an examination of the experimental feasibility of
glass growth in real diffusion couples.

II. FEASABILITY OF GLASS GROWTH IN SOLID DIFFUSION COUPLES

From the discussion in section I, one can draw the following
conclusion concerning favorable conditions for solid state glass
formation and growth:

(I) Glass formation and growth must be a thermodynamically
downhill process. A (preferably large) negative value of Δg
should characterize the process.

(II) Atomic mobility must be adequate to allow alloy formation
and growth in practical times τ_p (which we take to be
10^4-10^6 sec.).

$$\tau_G \sim \tau_D \ll \tau_p \sim 10^4\text{-}10^6 \text{ sec.} \tag{7}$$

(III) Furthermore, we know that metallic glasses, once formed,
crystallize in practical times for temperature $T \approx T_g$. This
leads to a third constraint which for isothermal reactions at
temperature T can be stated.

$$\tau_G \sim \tau_D \; (T \lesssim T_g) \ll \tau_x(T_g) \tag{8}$$

One now turns to several cases of actual binary diffusion couples. As examples, the Au-La[14,15], Zr-Ni[16] and Hf-Ni[17] binary alloy systems are chosen.

A semiquantitative free energy diagram for the Au-La system constructed in ref. 14 is shown in Fig. 4. The diagram was constructed using published thermodynamic data for the metals together with Miedema's[8] regular solution theory for the liquid (glass) phase. The diagram shows a large negative value of $\Delta g \sim 90$ kJ/mole for reaction of the pure metals to form glass near x = 0.5. Note that a further drop in free energy $\Delta g' \approx 10$ kJ/mole can occur when a glass with x = 0.5 is crystallized to form the compound AuLa. Note that $\Delta g' \ll \Delta g$. In other words, most of the "excess chemical energy" present initially in a Au-La diffusion couple is released when glass forms. Crystallization releases only a small additional "heat of crystallization". Criteria (I) above is seen to be easily satisfied in any case.

Evidence that criteria (II) and (III) can be satisfied comes indirectly from diffusion data for interdiffusion of the parent metals and from crystallization and T_g data on melt-quenched Au-La metallic glasses.[18] Au is known[19] to be an "anomalous fast diffuser" (AFD) in crystalline La metal. This is illustrated in Fig. 5 which shown the self diffusion constant for La in crystalline La together with that for Au in crystalline La. We shall denote these by D_{La}^{La} and D_{Au}^{La} respectively. One should note that

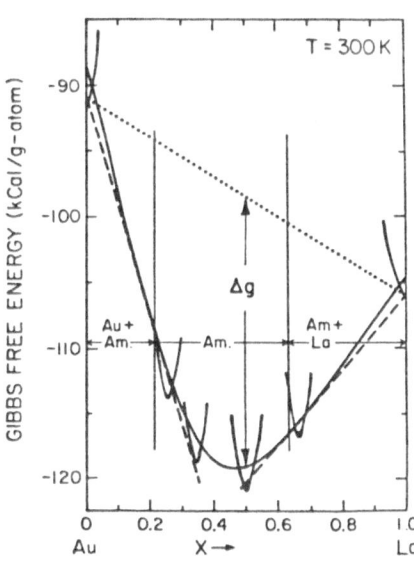

Fig. 4. An empirically derived free energy diagram for the Au-La binary system.

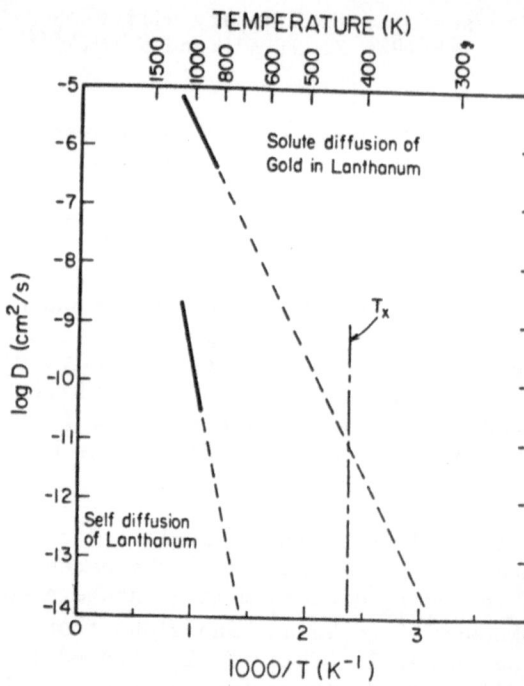

Fig. 5. Diffusion data for diffusion of Au in La metal and self diffusion in La metal. Also shown is the approximate T_g of Au-La glass.

the solubility of Au in La is small [20] (< 1 at.%), but that Au atoms are highly mobile at relatively low temperature. In particular, the T_g of Au-La melt-quenched metallic glass is known to be $T_g \approx 180°C$. [18] Fig. 5 shows the diffusivity of Au in La at this temperature to be $D_{Au}^{La} \sim 10^{-12} cm^2/sec$. In $\tau_p \sim 10^4$ sec., such a value gives a characteristic length $d_D = \sqrt{4 D_{Au}^{La} \tau_p} \sim 10^{-5} cm$. But note that from the earlier discussion $\tau_D = d_G^2 / \tilde{D}_G$ where $\tilde{D}_G$ is not the diffusivity of dilute Au in crystalline La metal, but rather that for interdiffusivity of Au(La) in a $Au_{1-x} La_x$ glass layer. Fig. 5 is not of direct applicability unless $\tilde{D}_G$ and $^oD_{Au}^{La}$ can be related. This point was unfortunately not emphasized in ref. (14) To go further we must examine interdiffusion data in the glassy state. Such data have been recently been reviewed by Cahn and Cantor. [21,22] In particular a diagram from ref. (21) is reproduced here in Fig. 6 showing measured diffusion data for a number of "early transition metal-late transition metal" (ETM-LTM) glasses. In the diagram, most of the data are for "Zr-base" metallic glasses. The data are conveniently plotted in reduced temperature units (T_g/T) where T_g is measured for each of the several glassy alloys used in the respective tracer experiments. For diffusion of LTM atoms

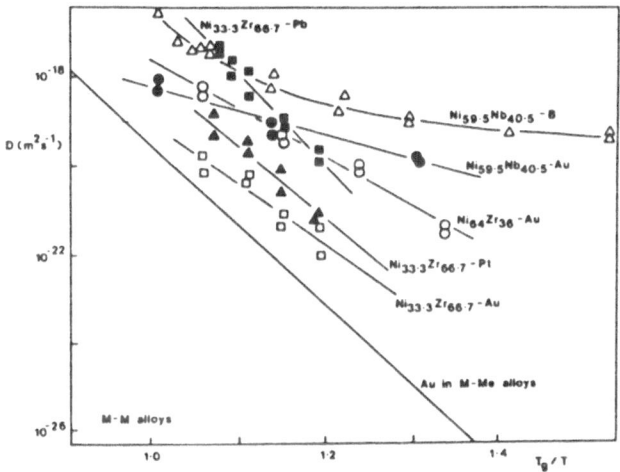

Fig. 6. Diffusion data for ETM-LTM glasses taken from ref. 21.

(e.g. Ni, Pt) with relatively small Gouldschmit radii in ETM-LTM glasses with the ETM (e.g. Zr, Y, La) having a relatively large Gouldschmit radius, one tends to find a rather large value of $\tilde{D}_G \approx D_{LTM}^0 \sim 10^{-13}$ cm^2/s for T near T_g. Further, one finds rather low activation energies ($\lesssim$ 1eV) for such diffusion. In fact, a comparison of diffusion data for AFD[19] metal pairs (e.g. Ni in Zr) with that for interdiffusion in the corresponding ETM-LTM glass (e.g. $Zr_{1-x}Ni_x$) shows that near T = T_g there is a striking similarity. It is further rather curious that many AFD binary systems[19] also exhibit ready glass formation in the corresponding concentrated liquid alloys by melt quenching. Unfortunately space does not permit further discussion of this latter point. Rather one can simply state an important conclusion. In ETM-LTM glasses near T_g, it is frequently possibly to have $\tilde{D}_G$ values ranging up to 10^{-12}-10^{-13} cm^2/s. It is further clear that the "small" LTM atom is the mobile species in these cases. In quantitative terms, one obtains interdiffusion coefficients given by

$$\tilde{D}_G = [(1-x)\ D_{LTM}^0 + x(D_{ETM}^0)]\ [1 + \frac{H_{mix}}{k_B T}] \approx (1-x)\ D_{LTM}^0\ [1 + \frac{H_{mix}}{k_B T}].$$

$$(9)$$

where D^o_{LTM} (D^o_{ETM}) is the intrinsic diffusivity of the LTM (ETM) in the glass. For AFD of the same LTM in dilute solution in the same crystalline ETM one finds values of D^{ETM}_{LTM} which are of roughly the same order of magnitude at $T = T_g$ as the $\tilde{D}_G$ values obtained by tracer diffusion in the ETM-LTM glasses.

These facts provide one with the required information to determine whether criteria (II) and (III) can be satisfied. We find for $d_G \sim 10^{-6}-10^{-3}$ cm, we have

$$\tau_D = d_G^2/\tilde{D}_G = [\frac{10^{-12}-10^{-6}}{(10^{-12}-10^{-14})}] \ (sec.) \tag{10}$$

or $\tau_D = 10^{-1}-10^7$ sec.

It should be possible to grow ETM-LTM glasses by solid state reaction of the metals to thicknesses of several microns in practical times! This will be true provided that a glass interlayer forms prior to crystalline compound interlayers.

III. EXPERIMENTAL RESULTS

Several experiments on different ETM-LTM binary systems have now been carried out which confirm what we have labored to establish above. The first results were reported for the case of Au-La in ref. (14). Fig.7A, reproduced from that reference shows the results of reacted thin-film multilayers (6-16 layers of 300-500 Å per layer). Curves (a), (b) and (c) show x-ray diffraction patterns of samples having total average compositions which are Au-rich, La-rich, and roughly equiatomic respectively. The samples were reacted at a temperature of $T = 80$ C (at least 100 C below T_g) for 4 hours. Curve (d) is an unreacted sample held at $T = 0$ C during the same period. The remaining unreacted metals (Au and La) in curves (a) and (c) respectively are explained by the fact that the overall average sample composition lies outside the single phase amorphous region. This region can be defined as shown in ref. (14) by the common tangents of the amorphous and pure metal free energy curves shown in Fig. 4. Similar reactions have more recently been carried out on Au-La bilayers and multilayers with layer thickness ranging up to 2000-4000 Å. Complete reactions are obtained for such thick layers after reaction at $T = 100$ C for times of order 10 hours.(12) TEM studies can be conveniently made on (Au-La-Au) films and show the reacted samples to be entirely amorphous. A typical electron diffraction pattern is shown in Fig. 7B. Resistivity measurements have also been used to study the reaction with in-situ films. An example from ref. (12) is shown in Fig. 8.

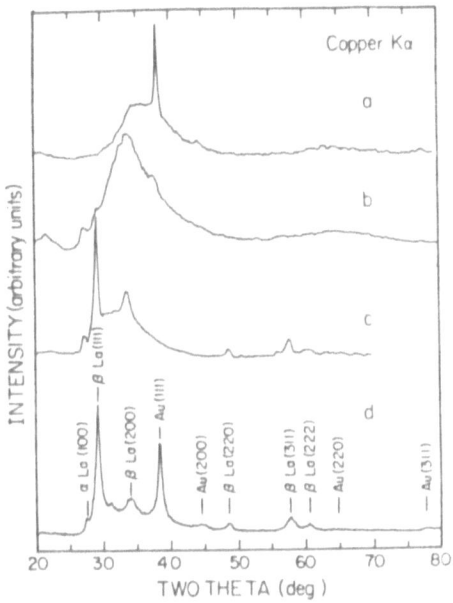

Fig. 7(A) X-ray diffraction patterns of unreacted alternating Au-La multilayers (curve (d)) and reacted alternating multilayers for Au-rich[a], equiatomic[b] and La-rich[c] overall compositions. Individual layers were 300-500 Å thick.

Fig. 7(B) Transmission electron diffraction pattern of fully reacted Au-La-Au trilayer sandwich lying near the equiatomic composition.

Using measurements of resistivity vs. time at various temperatures, it was possible to monitor the progress of the reaction. It was shown that an amorphous layer can grow until the supply of pure metals is consumed. The glass layer then undergoes a homogenization. The layer growth and homogenization are both found to be characterized by a thermally activated time constant τ_D which has an activation energy of $\Delta = 0.9 \pm 0.1$ eV. This value is remarkably close to that observed for anomalous diffusion of Au in La metal suggesting that the atomic jump barriers are similar for Au diffusing in La metal or in amorphous Au-La.

In separate studies, the Ni-Zr and Ni-Hf binary systems have been investigated.[16,17] Clemens et al.[16] used Auger depth profiling to first show that amorphous interlayers in Ni-Zr grow uniformly along the interface. This important result establishes that the model of uniform layer growth used in ref. (12) is appropriate. Van Rossum et al.[17] have studied the Ni-Hf system using the Rutherford α-backscattering technique to profile composition in a reacting bilayer or multilayer. A typical set of spectra for a Hf-Ni-Hf trilayer are shown in Fig. 9. The spectra show a well defined uniformly thick amorphous interlayer which grows with time at T = 320 C (for this alloy $T_g \approx 450$ C). Note that there is a graded Ni-concentration profile in the growing amorphous interlayer. This was shown to correspond to composition gradient within the growing layer.[17] Such large gradients follow naturally from the diffusion controlled growth model of ref. (12). From more recent α-backscattering data on thick bilayers (~ 2500 Å Ni and ~ 2500 Å Hf) it has been possible to estimate $D_G \approx 3 \times 10^{-13}$ cm^2/s in the growing Ni-Hf glassy layer at T = 350 C. Direct comparison of the growth model (ref. 12) with such data can also be carried out and used to obtain model growth parameters for the glassy interlayer. A full description of these results will be given in an upcoming publication.[23]

IV. CONCLUSIONS AND FUTURE PROSPECTS

It has been shown that metallic glasses can form and grow when pure metals characterized by a large negative heat of mixing are brought into contact under suitable kinetic conditions. A set of thermodynamic and kinetic criteria for the occurrence of such glass forming reactions were given which appear to be well satisfied in the case studied. The observed kinetics of growth of glass interlayers can be understood in terms of a diffusion controlled layer growth picture modified to include interface effects. It was pointed out that binary systems exhibiting this behavior also exhibit "anomalous fast diffusion" behavior of one of the metals in the crystalline dilute parent solution of the other. It remains to

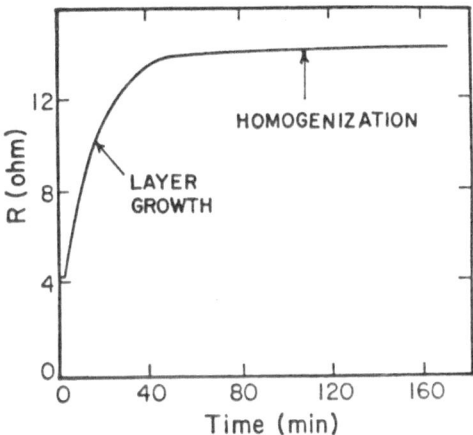

Fig. 8. Resistance vs. time for a reacting AuLa bilayer sandwich. Note the large increase in resistance during glass growth.

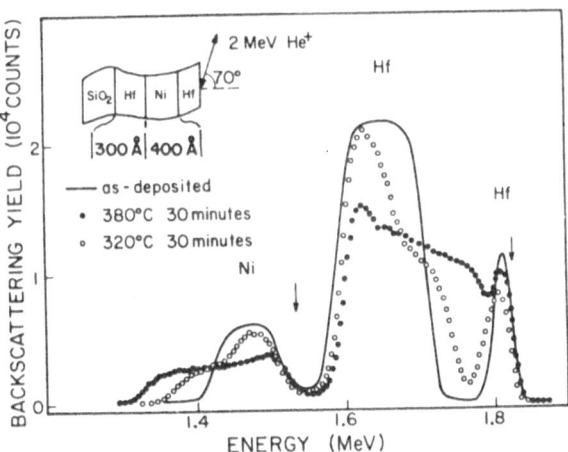

Fig. 9. Rutherford α-backscattering spectra for a Hf-Ni-Hf trilayer sandwich as reported in ref. 17.

demonstrate a fundamental connection for these two diffusion phenomena.

This novel method of glass synthesis raises a number of un-answered questions. Foremost is the question of why the glassy phase forms initially as opposed to more stable crystalline inter-metallic compounds. No satisfying answer has yet been provided. Once formed, the glass layer appears to be subject to similar (perhaps not identical) crystallization kinetics as bulk melt-quenched glasses. In particular, the glassy layers are stable up to $T \approx T_g$. With such a temperature constraint, one might inquire as to the maximum thicknesses to which glassy layers can grow in practical time. One can estimate (see eqn. (10)) that thicknesses in the micron range are possible. This leads naturally to the question of three dimensional structures. A three dimensional analogue of the layer experiments can be realized in powder metal-lurgical methods. It should in fact be possible to synthesize bulk structures via powder metallurgical route involving some type of mechanical consolidation of pure metal powders followed by a solid state reaction. Work aimed toward this objective is currently in progress.

ACKNOWLEDGEMENTS - The authors would like to thank Dr. Konrad Samwer and Prof. Peter Haasen for several valuable discussions which contributed to ideas presented here. The continued support of the U. S. Department of Energy under Contract No. DE-AM03-76SF00767 is greatfully acknowledged.

REFERENCES

1. Turnbull, D., Comtemp. Phys., 10, 473 (1969).
2. Lin, C. J. and Spaepen, F., J. Non-Cryst. Sol., 61&62, 767 (1984).
3. Bloembergen, N. in: Laser-Solid Interactions and Laser Process-ing, ed. S. D. Ferris, H. J. Leamy, and J. M. Poate, (Academic Press, New York, 1979) p. 1.
4. From a review see H. S. Chen, Rep. Prog. Phys., 43, 353 (1980)
5. Buckel, W. and Hilsch, R., Z. Phys., 138, 109 (1954).
6. Mehra, M., Johnson, W. L., Thakoor, A. P. and Khanna S. K., Sol. State Comm., 47, 859 (1983).
7. Johnson, W. L. in: Glassy Metals I, ed. by P. Beck and H. J. Guntherodt, (Springer-Verlag, Heidelberg, 1981) Ch. 9.
8. Miedema, A. R. and De Châtel, P. F. in: Theory of Alloy Phase Formation, ed. L. H. Bennett, (American Metals Society, Warrendale, PA., 1979) p. 344.

9. see for example, O. Penrose and J. L. Liebowitz in: <u>Fluctuation Phenomena</u>, ed. by E. W. Montroll and J. L. Liebowitz (North Holland, New York, 1979) Ch. 5.

10. Thompson, C. V. and Spaepen, F., <u>Acta Metall.</u>, <u>27</u>, 1855 (1979).

11. see P. G. Shewmon, <u>Diffusion in Solids</u>, (McGraw Hill Book Co., New York, 1963) p. 26.

12. Dolgin, B. and Johnson, W. L., submitted to Phys. Rev. B, (1984).

13. Spaepen, F. and Turnbull, D. in Proc. 2nd Int. Conf. on Rapidly Quenched Metals, ed. by N. J. Grant and B. C. Giessen (M.I.T. Press, Boston, 1976) p. 205.

14. Schwarz, R. and Johnson, W. L., <u>Phys. Rev. Lett.</u>, <u>51</u>, 415 (1983).

15. Schwarz, R., Wong, K. and Johnson, W. L., <u>J. Non-Cryst. Sol.</u>, 61&62, 129 (1984).

16. Clemens, B. M., Johnson, W. L. and Schwarz, R., <u>J. Non-Cryst. Sol.</u>, 61&62, 817 (1984).

17. Van Rossum, M., Johnson, W. L. and Nicolet M-A., <u>Phys. Rev. B</u>, <u>29</u> (in press) (1984).

18. Johnson, W. L., Poon, S. J. and Duwez, P., <u>Phys. Rev. B</u>, <u>11</u>, 150 (1975); also A. R. Williams Ph.D. Thesis, Calif. Inst. of Tech. (1981).

19. Le Clair, A. D. in <u>Properties of Atomic Defects in Metals</u>, ed. N. L. Peterson and R. W. Siegel, (North Holland, New York, 1978) p. 70

20. Shunk, F. A., <u>Constitution of Binary Alloys</u>, (McGraw-Hill, New York, 1969) p. 73.

21. Cahn, R. W. in <u>Physical Metallurgy</u>, ed. by R. Cahn and P. Haasen (North Holland, Amsterdam, 1983) p. 1823.

22. Cantor, B. and Cahn, R. in <u>Amorphous Metallic Alloys</u>, ed. by F. Lubosky (Butterworth, London, 1983) p. 487.

23. Dolgin, B., Van Rossum, M. and Johnson, W. L. (results to be published).

METALLIC GLASSES IN HIGH-ENERGY PULSED-POWER SYSTEMS

Carl H. Smith

Materials Laboratory
Allied Corporation
P.O. Box 1021R
Morristown, NJ 07960

INTRODUCTION

The recent need to advance the state-of-the-art in high-reliability, high-power pulse sources for accelerators and lasers has led several laboratories to refine and scale up in power pulse compression techniques first developed for radar in the 1950's (1). These efforts occured at the same time as the commercial availability of large quantities of ferromagnetic metallic glass ribbons. These new magnetic materials were found to be excellently suited to this resurrected technique.

Magnetic amplifiers and other devices which use the saturation of a ferromagnetic core material to control power have long been known for their extremely high reliability and for their high voltage and current capabilities. However, the use of non-linear magnetic materials in saturable inductors for the control of power was largely supplanted by the rapid development of inexpensive and simple to use SCR's and other semiconductor devices in the 1960's. The control action in a saturable inductor takes place in the conductor itself, due to an induced reverse voltage, rather than at the junction as in a semiconductor device. Since switches are often the most critical component in determining the reliability and lifetime of a pulse-power system, magnetic switching may play a significant role in making high-power laser systems and linear induction accelerators into practical industrial tools.

Utilization of metallic glasses in pulse-power systems has been the subject of considerable research and developmental activity (2-4). High-power pulse sources for linear induction particle accelerators have been developed (5-7), as have induction modules for coupling energy from the pulse source to the beam of such accelerators (8-10). Superpower generators using magnetic switching have been developed as drivers for inertial confinement fusion research (11-13). Magnetic modulators utilizing metallic glasses have also found a place as drivers for high-power excimer lasers (14-16).

Ferromagnetic metallic glasses possess several properties which excellently suit them to pulse-power applications. The method of production results in a thin, strong ribbon which can be wound into cores. Alloys with high saturation inductions have been developed which allow minimum core volume. Resistivities of amorphous metals are 2 to 4 times those of crystalline magnetic metals. High resistivities, together with ribbon thinness, allow low losses under the fast magnetization reversals. Finally, the development of large scale manufacturing capability for these materials based on their application in distribution transformer cores promises relatively low prices, which can be an important consideration in large scale projects.

This paper will briefly review the principles behind magnetic switches and magnetic modulators as well as fast magnetic reversals in magnetic ribbons. The magnetic properties of metallic glasses relevant to this application will be discussed and compared to conventional crystalline materials. Finally, results achieved in some specific projects will be described.

THEORY

Large changes in permeability of square-loop ferromagnetic materials upon saturation allow saturable inductors to be used as "closing" switches as shown in Figure 1. The applied voltage initially appears across the inductor due to its high unsaturated reactance. After the magnetic material saturates, the reactance of the inductor decreases to a low value due to the low saturated permeability of its core, and the voltage appears across the load. The time for the inductor to saturate, or the "hold off" time, in MKS units is just:

$$t = \Delta BNA/V \qquad (1)$$

where ΔB is the change in magnetic induction, N the number of turns of conductor around the core, A the cross section area of the core and V the average voltage. The change in magnetic induction depends upon the previous state of magnetization of the core material. If the core has been demagnetized, ΔB is equal to the saturation magnetization. If the core has been reset in the opposite sense, ΔB is the sum of the remanent and saturation inductions.

MAGNETIC SWITCH

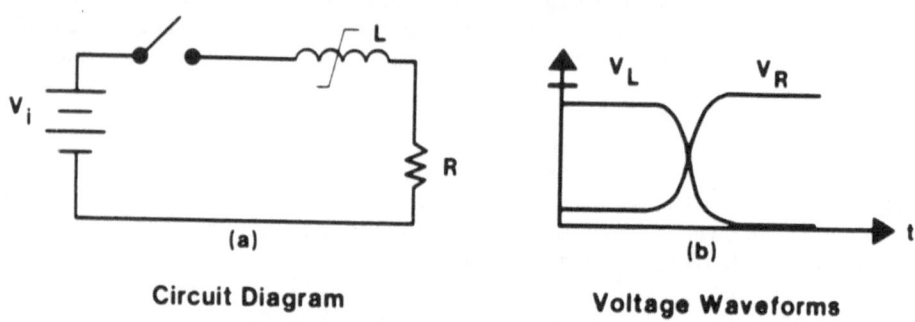

Circuit Diagram   **Voltage Waveforms**

Figure 1. A saturable inductor used as a magnetic switch. Circuit diagram a) and Voltage waveforms b).

In the simple magnetic switch described above, little energy is stored in the switch itself, because the current and the inductance do not simultaneously have large values. Such magnetic switches are used to protect other switching devices such as SCR's from large values of inrush current (dI/dt) (17) or to reduce switching losses by delaying large currents until switching devices such as thyratrons are fully turned on (18).

In order to use magnetic switches to compress a voltage pulse in time and, thereby, increase the peak power, devices for energy storage must be added between successive saturable magnetic switches. The series magnetic pulse compressor shown in Figure 2a utilizes identical value capacitors in each stage and decreasing value inductors to successively compress the voltage pulse in time. Typical voltage waveforms for successive stages are shown in Figure 2b. In a well designed system, efficiency is close to unity, and the output voltage is close to the input voltage. The increase in peak current can be substantial. Pulse compression in time by a factor of 3 to 5 per stage is typical with an increase in current and peak power by a similar factor.

Pulse compression per stage can be calculated from the ratio of the times to charge the capacitors in successive stages through the saturated inductors. As mentioned earlier, all capacitors are equal to store equal energy at each stage. The time to charge a given capacitor through the preceeding saturated inductor is easily calculated as:

$$t_n = (L_{n-1}^{sat} C/2)^{1/2}. \tag{2}$$

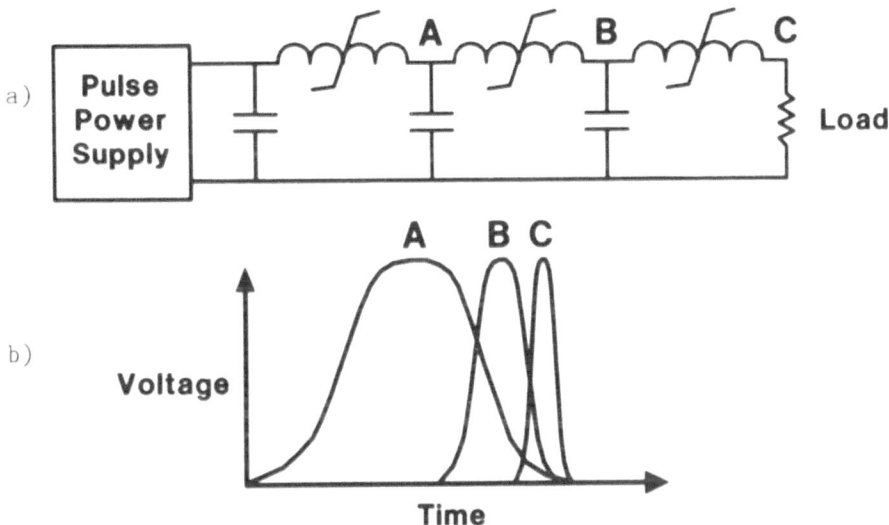

Figure 2.   A series magnetic pulse compressor. Circuit diagram a)
and Voltage waveforms across successive capacitors b).

The gain is just the ratio of the times to charge two successive
stages:

$$G_n = t_{n-1}/t_n = (L_{n-1}{}^{sat}/L_n{}^{sat})^{1/2}. \qquad (3)$$

To achieve high efficiency, not much current can be allowed to leak
through to the n+1th capacitor while the nth capacitor is charging.
Therefore the n+1th inductor's unsaturated inductance must be much
greater than the nth inductor's saturated inductance.   Using a
criterion of 20 for "much larger than" (1,4) we find that:

$$G_n = (L_n{}^{unsat}/20\, L_n{}^{sat})^{1/2} = (\mu^{unsat}/20\, \mu^{sat})^{1/2}. \qquad (4)$$

We have also substituted the values for saturated and   unsaturated
incremental permeabilities (change in magnetic induction divided by
change in field) in the above equation by assuming that only μ in
the expression for L changes upon saturation and not the geometric
factors.   In fast pulse magnetization, unsaturated permeability
depends on losses, or the area in the dynamic B-H loop.   In the
simple model of saturation wave behavior described below, with
half-cycle losses E_{sw}, the permeability is just:

$$\mu_{sw} = \Delta B^2/2E_{sw}. \qquad (5)$$

For high gain per stage, therefore, we need a material with low
losses and also a low saturated permeability.

Loss mechanisms in conducting ferromagnetic ribbons at high magnetization rates are often described in terms of the saturation wave or sandwich domain theory (19). The higher mobility of the portion of each domain wall near the surface of the ribbon soon results in a single domain wall which encircles the interior of the ribbon (20). Macroscopic eddy currents around the exterior of the encircling domain shield the interior from the applied field. The domain wall velocity is determined by the applied voltage per turn, ribbon thickness, core area, and saturation induction. In this model the H-field required to magnetize the material is:

$$H_a(\Delta B) = H_c + (d^2/4\rho)(\Delta B/2B_s)(dB/dt) \qquad (A/m) \qquad (6)$$

where $H_a(\Delta B)$ is the magnetizing H-field as a function of ΔB, H_c the dc coercive field, d and ρ the ribbon thickness and resistivity respectively, B_s the saturation induction and dB/dt the rate of magnetization. The half-cycle energy lost in magnetizing the core to ΔB with constant applied voltage (constant dB/dt) is (20):

$$E_{sw} = H_c \Delta B + (d^2/8\rho)(\Delta B^2/2B_s)(dB/dt). \qquad (J/m^3) \qquad (7)$$

Notice that the only material related quantities are the thickness, resistivity, dc coercive field, and saturation induction.

The saturation wave model predicts that the magnetizing H-field will increase linearly with ΔB until saturation. This behavior represents a constant permeability for cases in which the coercive field can be ignored. However, observations of dynamic magnetization curves reveal a much more complicated behavior as shown in a schematic representation in Figure 3. In region I, after a rapid increase usually limited by stray inductances, H reaches a maximum and then actually decreases in many cases. This peak, most noticeable in materials with extremely square B-H loops, is associated with the establishment of bar domains spanning the ribbon thickness. Magnetization progresses by the motion of these bar domain walls in region II. A minimum in H near the beginning of region II represents the establishment of an efficient wall spacing (21). The greater mobility of domain walls near the surface of the ribbon results in severe domain wall bowing which, at sufficiently large magnetization rates or great enough ribbon thickness, results in an encircling domain wall and saturation wave type behavior in region III. Finally, the H-field increases rapidly in region IV as saturation is approached and the permeability rapidly decreases. The straight line represents the saturation wave model. Note that since losses are equal to the area between the curve and the B axis, the losses in the case shown exceed those for saturation wave theory. The magnetization behavior of amorphous ribbons approaches saturation wave behavior only for relatively thick ribbons or for saturation times of less than one microsecond (22).

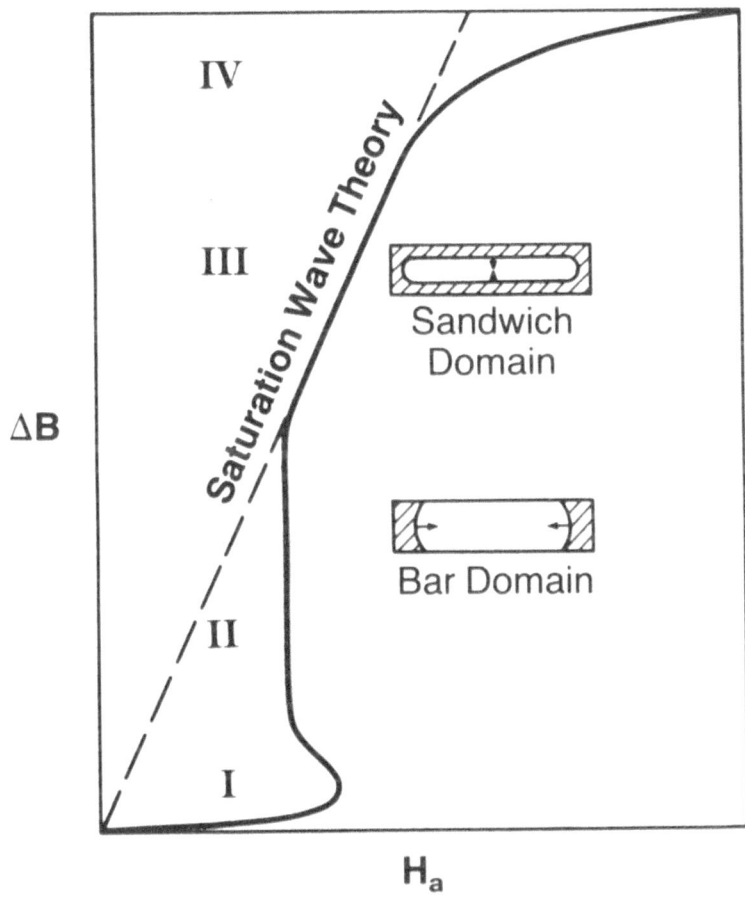

Figure 3.   Schematic dynamic magnetization curve with regions of I-
    domain nucleation, II-bar domain magnetization, III-sandwich
    domain magnetization, and IV-approach to saturation shown.

    To compare different materials for pulse applications it is
useful to have a figure-of-merit which can be easily calculated
from basic material properties.   The most important properties
discussed above are the resistivity, saturation induction, and the
ribbon thickness. Unsaturated permeability, which has a large role
in determining gain, depends upon the magnetization losses which in
turn depend upon the resistivity times the saturation induction
divided by the thickness squared.   Total volume of core material
can be shown to depend upon the square of ΔB (3), which can
approach twice the saturation induction.   Considering all of these
factors, an often used figure-of-merit is:

$$\text{FOM} = \Delta B^2 \rho / d^2. \tag{8}$$

Table I. Properties of Magnetic Materials for Pulse Power Systems.

Material	Elements	B_s (T)	B_r (T)	H_c (A/m)	ρ ($\mu\Omega$m)	T_c (°C)	FOM (25 μm)
METGLAS[R]2605CO 1)	FeCoBSi	1.8	1.6	4.0	1.30	415	15
METGLAS 2605SC	FeBSiC	1.6	1.4	3.6	1.25	370	11
METGLAS 2826MB	NiFeMoB	.9	.6	1.2	1.60	350	3.6
Silicon Steel	FeSi	2.0	1.6	24	.50	730	6.5
50% Ni-Fe	NiFe	1.6	1.5	8.0	.45	480	4.3
80% Ni-Fe	NiFe	.8	.7	.4	.60	400	1.2
Ni-Zn Ferrite	NiZnFeO	.48	.14	16	10^{12}	210	–

MAGNETIC PROPERTIES OF METALLIC GLASSES

Ferromagnetic amorphous alloys in ribbon form have attractive properties for pulse power applications due both to their magnetic and mechanical properties. Alloys with high saturation inductions can be cast into ribbons as thin as 13 μm. Resistivities are typically 2 to 4 times those of polycrystalline soft magnetic materials. Tensile strengths of typically 1/2 to 1 GPa allow even thin ribbons to be handled and wound into large cores. Magnetic properties of several amorphous alloys, together with conventional magnetic materials, are given in Table I. The figures-of-merit show the superiority of nominal 25 μm amorphous materials. The figures-of-merit for 18 μm materials would be twice and for 13 μm materials four times the values shown. Conventional magnetic materials thinner than 25 μm are not readily available in large quantities and are very difficult to handle without degrading their magnetic properties. The properties given in the table for amorphous materials represent values for field annealed materials in which the cast-in stresses and winding stresses have been annealed out, and a preferred magnetic direction has been established by applying a field during annealing.

Although annealing reduces the coercive field and increases the remanence and, therefore, ΔB, not all pulse applications use annealed ribbon. At rapid magnetization rates considerable induced voltages develop between layers and large interlaminar eddy current losses would exist without some form of interlayer insulation. The simplest methods of dip or spray coatings suffice to a few volts per lamination which corresponds to about 1 μs time to saturation for typical ribbon dimensions (23). More rapid magnetization usually requires a dielectric insulating film wound between layers

---

1. METGLAS[R] is Allied Corporation's registered trademark for amorphous alloys of metals.

of the core. Few of the available thin plastic films can withstand the required 350 to 450 °C annealing temperatures. Therefore, many applications utilize unannealed amorphous alloys with somewhat lower remanence and, therefore, lower ΔB in order to use mylar or other plastic film insulation.

Losses have been measured for amorphous alloys under fast pulse conditions (16,20-22,24-26). Figure 4 shows losses as a function of time to saturation. At fast magnetization rates, losses increase as the reciprocal of the time to saturation as predicted by equation (7). The thickness dependence of losses for saturation wave behavior go as the thickness squared. However, saturation wave domains are not achieved for thin ribbons except at very fast magnetization rates. Figure 5 shows the thickness dependence of normalized losses. The hysteresis losses, $H_c \Delta B$, have been subtracted due to different values of coercive field, and the losses have been normalized by ΔB squared to compensate for differing values of remanence and hence ΔB in these samples (22).

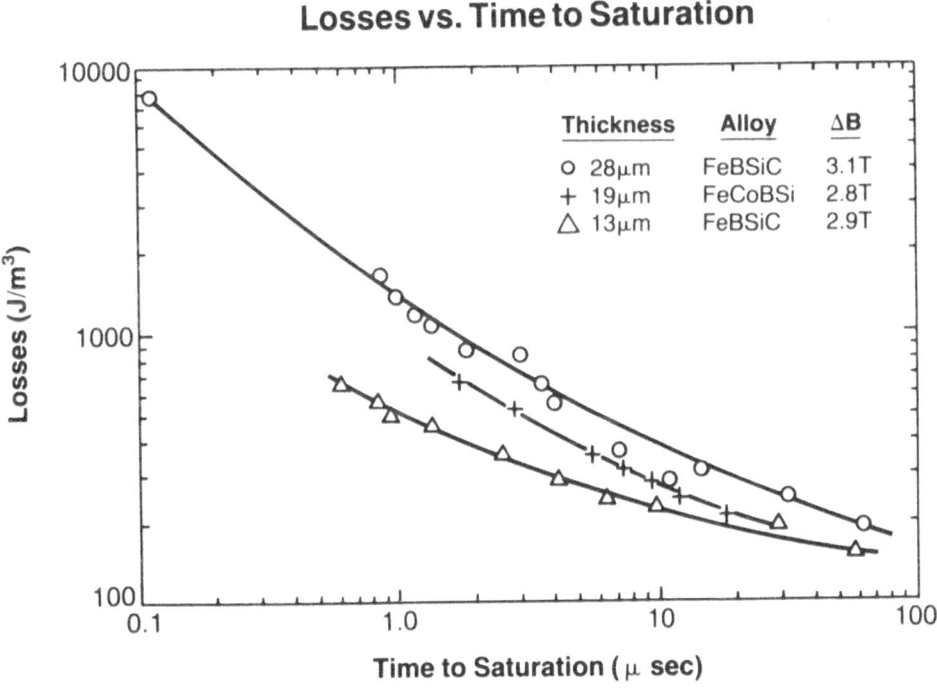

Figure 4. Half-cycle losses as a function of magnetization time. Two different alloys and three different thicknesses are shown. Constant dB/dt excitation was used.

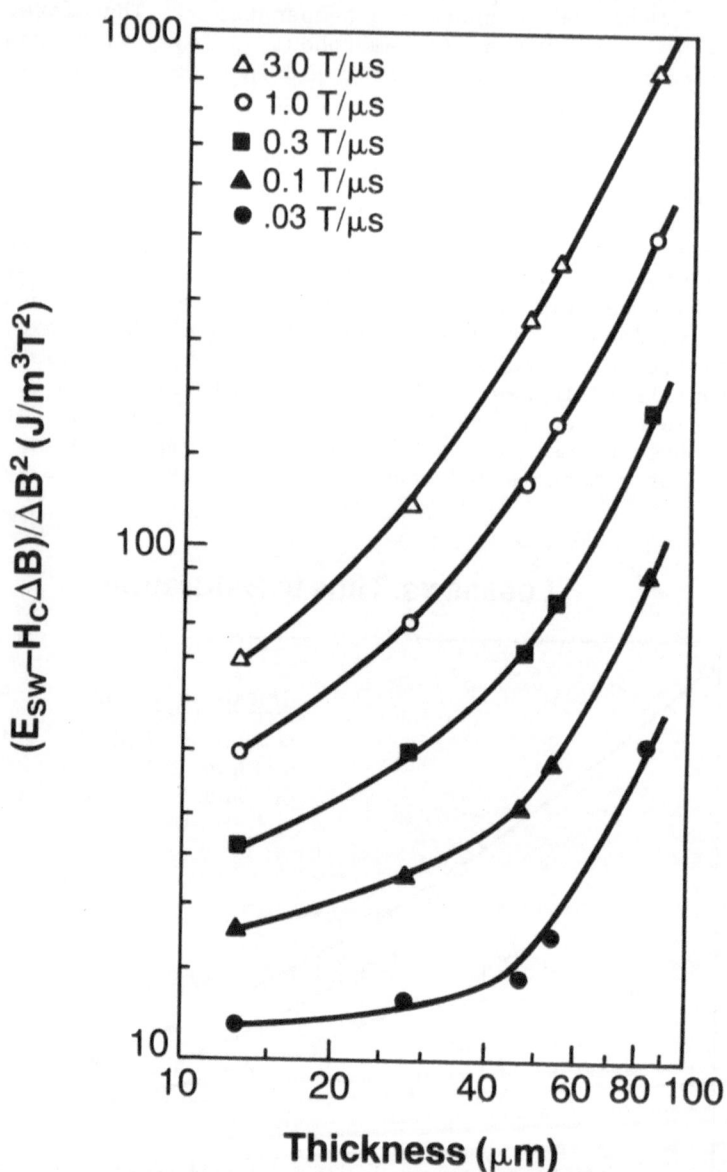

Figure 5. Normalized half-cycle losses as a function of ribbon thickness for constant dB/dt excitation. Hysteresis losses have been subtracted and the result normalized by ΔB^2 to compensate for differing values of ΔB.

PULSED POWER SYSTEMS

Several pulsed power systems have been constructed using amorphous alloy ribbons. In this section a few of these systems will be described to illustrate the results which have been achieved.

One of the earliest laboratories to experiment with metallic glasses in magnetic switches was the Particle Beam Fusion Accelerator Group at Sandia National Laboratories in Albuquerque, NM (11-13). This light-ion fusion test facility accelerates a very high current light-ion beam across a very high voltage diode to achieve multi-terawatt beam energies, ultimately to be focused on a fusion target. To eventually scale up such a concept to a commercial power plant would require numerous parallel superpower switches operating at 1 to 10 Hz for years with extremely high reliability. To investigate magnetic switching as an alternative to the existing water spark gaps, a small magnetic modulator was built. Two stages of amorphous alloy cores in a coaxial water transmission line delivered a 100 kV, 60 ns full width at half maximum (FWHM) pulse into a 1 ohm load resulting in a peak power of 10 GW. More recently a large one-stage magnetic switch was built as a possible low-loss substitute switch with repetition rate capability in place of the self-closing water switches designed for the new 100 TW PBFA-II accelerator. The prototype COMET module utilized an almost 2 meter diameter toroidal core containing approximately 2000 kg of METGLAS 2605SC ribbon wound with mylar interlayer insulation. The module produced a 2 MV, 150 ns FWHM output pulse into a 2.2 ohm load. The 0.9 MA peak current produced a peak power of 1.8 TW. A second stage is expected to result in a 3.25 MV 40 ns pulse. Extremely low jitter and low mechanical noise are additional advantages of this switch over the conventional water switches used in PBFA-I (13).

Lawrence Livermore National Laboratory in Livermore, CA is a second laboratory which has developed magnetic switching for a particle accelerator -- in this case, a linear induction electron accelerator (4-6). The advanced test accelerator (ATA) is a 10 kA, 50 MeV accelerator capable of operating at 1 kHz in bursts. Pulsed power is provided to the induction modules by blown-gas spark gaps. The master trigger for these spark gaps is metallic glass cored magnetic pulse generator as are the grid driver and the cold-cathode drivers (5). A requirement to upgrade the ATA to a 10 kHz rep-rate in bursts resulted in the design of magnetic switches with the desired repetition rate. Each metallic glass magnetic switch is also capable of replacing two of the previous blown-gas spark-gap switches. It was thought that this repetition rate would be unattainable with conventional switching techniques, even with extremely large blowers. The MAG-1 prototype delivered 10 GW peak power in 80 ns FWHM, 300 kV, 800 J pulses. The peak burst rate was 15 kHz and the efficiency was 80% (6).

Another possible driver for inertial confinement fusion uses heavy ions accelerated by a linear induction accelerator. The Accelerator and Fusion Research Division of Lawrence Berkeley Laboratory in Berkeley, CA is investigating such an accelerator. Large quantities of ferromagnetic material are required for the induction modules which act as one-to-one transformers coupling the pulse energy to the changed particles in the beam. The high resistivity and thin ribbon form of amorphous alloy ribbons result in considerably lower losses and therefore fewer capacitors required in the drive pulsers than do silicon-steel (8-9). Furthermore large quantities of material are required to provide approximately 1/2 volt-sec of magnetic material per meter of an accelerator which could be several hundred meters long in its final configuration. Prices of $5/kg for large volume orders, stimulated by the needs of the distribution transformer industry, provide a superior material at an attractive price for this project.

High-energy excimer lasers require reliable high-power pulse sources with repetition rates of 100's of Hz in continuous operation. Magnetic modulators offer the only posibility of the required reliability for industrial or space based applications. A prototype magnetic modulator with amorphous metal cores was built by Maxwell Labortories in San Diego, CA for a mercury bromide discharge laser for space-based communications. The unit, utilizing meter-long racetrack-shaped cores deliverd 50 to 70 kV, 40 ns rise-time pulses into a 0.5 ohm load at 70 to 77% efficiency (16). High energy excimer lasers are becoming commercially available, and magnetic switching will provide them with the reliability and lifetime required in industrial use.

SUMMARY

Requirements for pulse power which exceeded the state-of-the-art resulted in combining the old technique of magnetic pulse compression with new amorphous metals to provide pulse-power systems with repetition rates and reliabilities which were not previously attainable. These magnetic pulse-power techniques may help bring these laboratory accelerators and lasers more rapidly into commercial use.

REFERENCES

1. W.S. Melville, Proc. Inst. of Electrical Engineers (London) 98, Part 3 (Radio and Communication), No. 53, 185 (1951)
2. S.E. Ball and T.R. Burkes, Proc. 3rd IEEE Int'l Pulsed Power Conf., Albuquerque, NM, June 1981, p. 269

3. E.Y. Chu, <u>Proc. 4th IEEE Int'l Pulsed Power Conf.</u>, Albuquerque, NM, June 1983, p. 242

4. D.L. Birx, E.J. Lauer, L.L.Reginato, D. Rogers Jr., M.W. Smith and T. Zimmerman, <u>Proc. 3rd IEEE Int'l Pulsed Power Conf.</u>, Albuquerque, NM, June 1981, p. 262

5. D.L. Birx, E. Cook, S. Hawkins, S. Poor, L.L. Reginato, J. Schmidt and M. Smith, IEEE Trans. Nuc. Sci. <u>NS-30</u>, 2763 (1983)

6. D.L. Birx, E. Cook, S. Hawkins, S. Poor, L.L. Reginato, J. Schmidt and M. Smith, <u>Proc. 4th IEEE Int'l Pulsed Power Conf.</u>, Albuquerque, NM, June 1983, p. 231

7. <u>Energy and Technology Review</u>, Lawrence Livermore National Laboratory, August 1983, p.11

8. A. Faltens, M. Firth, D. Keefe and S. Rosemblum, IEEE Trans. Nuc. Sci. <u>NS-30</u>, 3669 (1983)

9. Heavy Ion Fusion Staff, Univ. of Calif. Lawrence Berkeley Lab., PUB 5031, Sept. 1979

10. A. Faltens, presented at INS Int'l Symp. on Heavy Ion Accelerators and Their Applications to Inertial Fusion, Tokyo, Japan, January 1984

11. J.P. VanDevender and R.A. Reber, <u>Prod. 3rd IEEE Int'l Pulsed Power Conf.</u>, Albuquerque, NM, June 1981, p. 256

12. M. Stockton, E.L. Neau, and J.P. VanDevender, J. Appl. Phys. <u>53</u>, 2765 (1982)

13. E.L. Neau, <u>Proc. 4th IEEE Int'l Pulsed Power Conf.</u>, Albuquerque, NM, June 1983, p.246

14. W.C. Nunnally, <u>Proc. 3rd IEEE Int'l Pulsed Power Conf.</u>, Albuquerque, NM, June 1981, p. 210

15. W.C. Nunnally, J. Power, T.E. Springer, A. Litton, P.N. Mace and K.W. Hanks, <u>IEEE Conf. Record 15th Power Modulator Symp.</u>, Baltimore, MD, June 1982, p. 28

16. E.Y. Chu, B. Hofmann, H. Kent and T. Bernhardt, <u>IEEE Conf. Record 15th Power Modulator Symp.</u>, Baltimore, MD, June 1982, p. 32

17. G. Hinz, <u>Proc. 7th Int'l PCI Conf.</u>, Geneva, Switzerland, Sept. 1983, p. 372

18. E.J. Lauer and D.L. Birx, <u>IEEE Conf. Record 15th Power Modulator Symp.</u>, Baltimore, MD, June 1982, p. 47

19. A.G. Ganz, AIEE Trans. <u>65</u>, 177 (1946)

20. C.H. Smith, IEEE Trans. Nuc. Sci. <u>NS-30</u>, 2918 (1983)

21. R.M. Jones, IEEE Trans. Mag. <u>MAG-19</u>, 2024 (1983)

22. C.H. Smith, D. Nathasingh and H.H. Liebermann, Intermag Conf., Hamburg, W. Germany, April 1984, to be published in IEEE Trans. Mag.

23. C.H. Smith, <u>IEEE Conf. Record 15th Power Modulator Symp.</u>, Baltimore, MD, June 1982, p. 22

24. R.M. Jones, A.J. Collins and N.G. Cleaver, IEEE Trans. Mag. <u>MAG-17</u>, 2707 (1981)

25. R.M. Jones, IEEE Trans. Mag. <u>MAG-18</u>, 1559 (1982)

26. P. Williams and J.E.L. Bishop, J. Magn. and Magn. Mat. <u>20</u>, 245 (1980)

SPHERE MODELS FOR GLASSES

J. BLETRY

SAINT-GOBAIN RECHERCHE
39, Quai Lucien Lefranc
93304 AUBERVILLIERS (FRANCE)

ABSTRACT

The possibilities and limitations of adhesive sphere models for the structural description of metallic, covalent and ionic glasses are discussed.

1 - CLOSE -PACKED METALLIC GLASSES

Chemically disordered binary alloys made of randomly mixed A and B spheres are first studied. Using computer simulation, it is shown that their maximum packing fraction : $\gamma m = 0,63_7$ does not depend on either the concentration x in A spheres or the ratio of atomic diameters : $\delta = d_B/d_A$
(if δ lies in the interval $1,5 > \delta > 0,65$).
Chemically ordered alloys are then studied and the first neighbour order parameter which measures departures from perfect disorder is defined by : $\xi = (N_{AB}^d - N_{AB})/N_{AB}^d$
(where the numbers of first neighbour A − B pairs in disordered and ordered alloys are respectively represented by N_{AB}^d and N_{AB}).

At a constant packing fraction γ_m and for given values of δ and x, ξ varies between + 1 (total demixing) and a minimum value ξ_m which corresponds to the "maximum order" situation.
Three concentration regimes, corresponding to three kinds of maximum order, are found :
i) In the interval $0 \leqslant x \leqslant x_c^A$ maximum order is a "type A total order" where A − A contacts are forbidden. When x increases from 0 to the "critical concentration" : $x_c^A = 0.29 + 0.41 \ (\delta -1), \xi_m$ decreases from 0 to a critical value : $\xi_c = -0.40$ which is independent from δ $(1.5 > \delta > 0.65)$.

ii) In the interval $x_c^A \leqslant x \leqslant x_c^B$ maximum order is only partial i.e.
A - B contacts are favoured but A - A (or B - B) contacts cannot
be forbidden.
ξ_m remains equal to ξ_c within this concentration interval.
iii) When x increases from x_c^B to 1, maximum order is a type B total order, and ξ_m increases from ξ_c to 0.
It is obviously possible to keep total order in the interval $x_c^A < x < x_c^B$
at the expense of a packing fraction decrease. Such a density decrease starting at x_c^A has been reported in the case of iron - boron
alloys.
Finally, analytical expressions of the partial coordination numbers
are derived which involve four parameters :
    γ (< 0.64) , δ (1.5 > δ > 0.65) , x (1 > x > 0) and ξ (< 1).

2 - DIRECTIONALLY BONDED COVALENT GLASSES

    The combination of size effects ($\delta \neq 1$) with chemical order
($\xi < 0$) is shown to produce polyhedral or directionnal atomic arrangements with covalent character.
Partial pair distribution functions of the maximum (partial) order
alloy ($\xi = -0.40$) made of equal size spheres ($\delta = 1$) in equal proportion (x = 0.5) can be identified with the pair distribution
functions of atoms and holes in amorphous tetravalent semiconductors. Agreement with experimental structure factors of silicon and
germanium is good, and it should be interesting to measure the hole
structure factor by "decorating" them with foreign atoms. Hole
study by positon annihilation experiments should provide a new
approach to the problems of hydrogenated amorphous silicon.

3 - IONIC "NON ADDITIVE" GLASSES

    The structure of ionic liquids or glasses can be represented
by generalized geometrical models which only depend on first neighbour distances : d_{AA}, d_{AB}, and d_{BB}.
These models correspond to non additive adhesive sphere potentials.
Using solid phase physical properties and a non additivity parameter :
$$\lambda = \frac{2\,d_{AB}}{d_{AA} + d_{BB}} - 1$$

equal to -0.26, we can describe the main features of the partial
structure factors of molten sodium chloride.
    We conclude that liquid or glassy structures can be represen-
ted by unified and simple geometrical models regardless of the nature of the chemical bonds between the constituents.

Abstract only

MAGNETIC PROPERTIES AND MEDIUM-RANGE ORDER IN METALLIC GLASSES

J. DURAND

Physique des Solides (LA 155)
Université de NANCY I   (France)

Magnetic properties of metallic glasses are primarily determined by both the chemical composition and the atomic arrangement. Alloying effects were found in many cases to be predominant on basic magnetic properties, so that a proper evaluation in the structural effects can be achieved only by a direct comparison between an amorphous alloy and a crystalline compound of same composition. Such a comparison has been made for a rather large number of compounds. The structural information that can be thus extracted is most often analyzed in two parts namely the "short-range" order for the first atomic shells and the "long range" disorder for the more remote environment. However, some indications about atomic arrangements over a medium scale ranging from about 15 Å up to few thousands Å can be found in basic magnetic properties such as distribution of hyperfine fields, approach to saturation, temperature dependence of bulk magnetization, critical and pseudocritical phenomena. These structural information concern various phenomena ranging from concentration or density fluctuations up to possible phase separation. They are reviewed here under the loose term of "medium range order". Their crucial importance for applications oriented properties such as macroscopic anisotropy, coercivity and permeability is briefly stressed.

CHAPTER III: SOL-GEL PREPARATIVE METHODS

THE GEL-GLASS PROCESS

J. ZARZYCKI

Laboratory of Materials Science and CNRS Glass Laboratory
University of MONTPELLIER - FRANCE

The gel route to glass formation is based on the possibility of forming a glass network by chemical polymerization of suitable compounds in the liquid state at low temperatures. In this way a "precursor" material (gel) is formed from which glass may be obtained by subsequent elimination of the unwanted residues (water, organic compounds, etc.) and collapse of the structure at temperatures much lower than those required for the direct melt formation by fusion of the constituent oxides.

This is of interest in the case of many systems which are difficult to prepare directly because of the too high temperature involved and/or high viscosity of the resulting melts which prevent satisfactory homogeneization.

By this method a high degree of homogeneity is directly obtained on a molecular scale and the absence of contamination during fusion makes it possible for high-purity glasses to be obtained, without problem.

From a theoretical standpoint, approaching the glassy state from lower temperatures instead of from higher ones may imply that some phase transitions (unmixing, crystallisation) may be circumvented and new glasses impossible to prepare by quench may become accessible.

In contrast to the "thermal" polymerization during melt formation "chemical" polymerization at low temperatures offers more possibilities of building the glass network. This route to glass formation should give inorganic glasses some of the flexibility of the organic polymer synthesis which was lacking in the direct way.

Technically, the method is readily applicable to the formation of thin coatings, fibers and hollow spheres ; special care is however, required if monolithic pieces of glass are to be obtained.

Intensive investigation has been carried out in this field. For

general reviews which contain numerous references see refs(1) to (7).
     The main steps in the gel-glass technique are essentially :
     . gel formation
     . drying of the wet gel
     . densification of the dry gel
     In nearly all cases silica is the essential component of the
gel.

1 GEL FORMATION

     There are two ways of obtaining silica based gels :
     a) Destabilisation of silica sol, (e.g. Ludox^R) pure or contai-
ning other metal ions added in the form of aqueous solutions of salts.
(Method I).
     b) Hydrolysis and polycondensation of organo-metallic compounds
dissolved in alcohols in the presence of a limited amount of water.
(Method II).
     Both methods lead to non-crystalline materials (precursors) con-
taining substantial amounts of water and/or organic residues which
can be eliminated by suitable curing treatments. The dried and purified gels are essentially porous materials and a densification treatment is necessary to convert them into solid glasses devoid of residual porosity. The overall scheme fo the process in presented schematically in Fig. 1.

1.1 Gel Formation by Destabilization of Sols

1.1.1 Silica sols. Silica sols may be prepared either by mechanical
or electrical dispersion of the material or by chemical condensation methods, starting from solutions of Na-silicates, K-silicates,
NH_4-silicates or hydrolysable products such as $SiCl_4$ or $Si(OR)_4$ where R is an alkyl group. It is known (8) that the formation of silicic acid in aqueous solutions is followed by polymerization of monomers $Si(OH)_4$ when its concentration exceeds 100 ppm, the limiting
solubility in water at 25°C.
     The polymerization reaction is based on the condensation of si-
lanol groups :

$$- Si-OH + HO-Si - \longrightarrow - Si-O-Si - + H_2O \qquad (1)$$

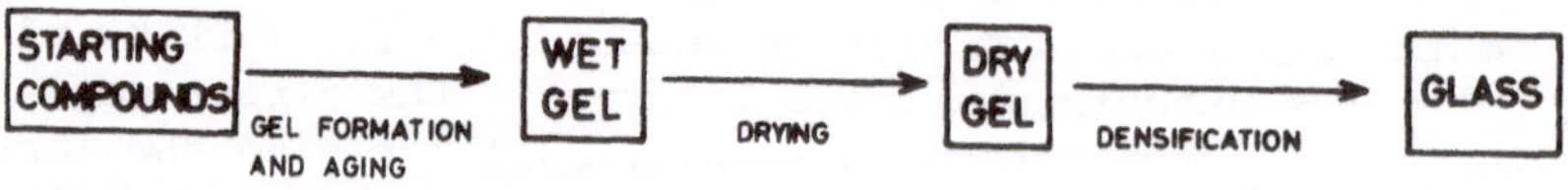

Fig.1 : The gel-glass process.

The following steps may be distinguished:
. formation of dimers and higher molecular species
. condensation of these to form primary particles
. growth of the particles
. linking of the particles together in chains and then into
  three-dimensional networks.

Reaction (1) intervenes in the primary particle formation, in their growth and in their subsequent linking during the gel formation. Above pH = 2 the rate of polymerization is proportional to the concentration of OH, below pH = 2 to that of the H^+ ions. The tendency is to produce a maximum of Si-O-Si bonds and a minimum of uncondensed Si-OH groups. The condensation leads to a most compact state, the Si-OH groups being placed at the outside of the condensate.

These amorphous spheroidal groupings of about 1 to 2 nm are formed by a <u>nucleation</u> process similar to that which occurs in the formation of crystalline precipitates.

Because of the size differences, <u>Ostwald ripening</u> then sets in, the smaller particles which have a higher solubility dissolve and the silica is redeposited on the larger ones, the total number of particles decreasing.

At low pH values, particles growth stops once the size of 2-4 nm is reached. Above pH = 7 particle growth continues at room temperature until particles of about 5-10 nm in diameter are formed, then it slows down. At higher temperatures particle growth continues, especially for pH > 7.

For the pH range between 6 an 10.5 the silica particles are negatively charged and they repel each other - growth continues without aggregation, resulting in the formation of stable <u>sols</u>.

At low pH the particles have little ionic charge ; they can collide and form by aggregation continuous networks leading to <u>gels</u> (Fig.2). This process may involve primary particles of different sizes according to the pH condition and the presence of salts.

Condensation may be controlled and even stopped when the particles reach the required size. The addition of stabilizing ions prevents their further condensation.

Commercial silica hydrosols (e.g. Ludox[R]) are stable sols with 20-50 wt % SiO_2. They are made up of dense silica particles with an average diameter of between 7 and 21 nm. The pH is between 9 and 11.

1.1.2 Destabilisation. To obtain a <u>gel</u> from a stable sol this latter must be destabilized either by temperature increase or by the addition of an electrolyte.

Increase of temperature reduces the amount of intermicellar liquid by evaporation and increases thermal agitation which induces collisions between particles and their linking by condensation of surface hydroxyls.

By electrolyte addition the pH of the sol may be modified in order to reduce the electric repulsion between the particles, (depending on the zeta potential). This is accomplished by adding an acid

206

to diminish pH to 5-6 and to induce gel formation by aggregation. This conversion of sol into gel is progressive, the growing aggregates (microgel) invading progressively the whole volume of the sol. When about half of the silica has entered the gel phase a rapid increase in viscosity is noted.

The mechanism of interparticle bonding leading to microgels and gels involves the attachment of two neighbouring silica particles *via* the formation of Si-O-Si bonds (reaction 1). The presence of soluble silica or monomer near the points of contact may contribute to the cementing together of the particles.

1.1.3 Aging. A further step is the <u>strengthening</u> of the network of the particles by a mechanism involving the partial coalescence of the particles. The negative radius of curvature at the <u>neck</u> joining the two particles implies that the local solubility is less there than near the surface of the particle. A transport and deposition of silica occur there preferentially, leading to a thickening of the neck (Fig.3).

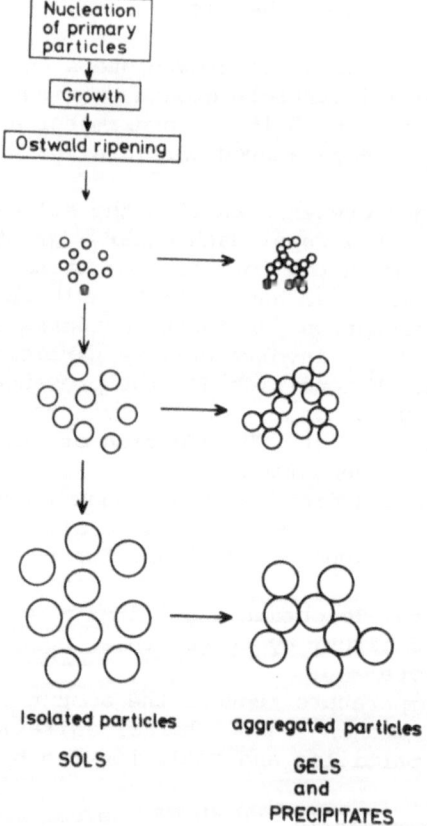

Fig.2 : Polymerization steps leading to the formation of sols and gels. From ref. (9).

NECK FORMATION

Fig.3 : Strengthening of particulate chains by the deposition of silica at the necks. From ref. (9).

This occurs in aging treatments where chains of particles may
be converted into more or less "fibrillar" structures by this rearrangement mechanism. Particle size in sols may be increased by adding
"active" silica in the form of particles less than 2 nm or even smaller polymer species ; they redissolve in the presence of larger particles and redeposit on them. This "nourishment" of particles is at
the base of the so-called "build-up" process (Fig.4).

The sol gel transition should be distinguished from a precipitation (or floculation) mechanism in which separate aggregates are
formed in contrast to the continuous three-dimensional particle networks (Fig.5).

Colloïdal particles will form gels only if there are no·active
forces which would promote coagulation into aggregates with a higher
silica concentration than the original sol. Metal cations, especially the polyvalent ones, may lead to precipitation rather than gelling;
this is the case of some multicomponent gels.

1.1.4 Multicomponent gels. At least one gellifying constituent (generally silica sol) is required. Other constituents may be added in the
form of soluble salts (nitrates, sulfates, etc.) or organometallic
compounds.

By adjusting the temperature, concentration and especially the
pH of the resulting sol, a homogeneous solution is obtained which
may then be gelled in a controlled way in order to avoid precipitation.

SECONDARY   DEPOSITION

Fig.4 : Strengthening of chains during aging. From ref. (9).

208

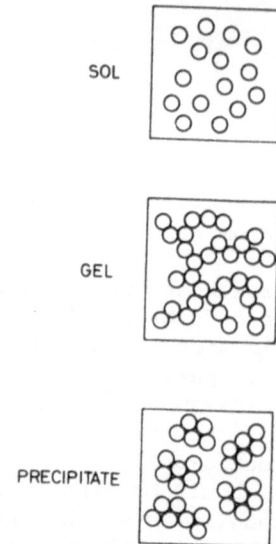

SOL

GEL

PRECIPITATE

Fig.5 : Difference between gelling and precipitation. From ref. (9).

1.2 Gel Formation from Organometallic Compounds

Metal alcoholates, also called metal alkoxides $M(OR)_n$ where M is a metal and R an alkyl group, react with water and undergo hydrolysis and polycondensation reaction which lead to the formation of metal oxide. The overall reaction scheme consists, at least formally, of two steps :

$$M(OR)_n + n\ H_2O \longrightarrow M(OH)_n + nR(OH) \qquad (2)$$

$$p\ M(OH)_n \longrightarrow p\ MO_{n/2} + pn/2\ H_2O \qquad (3)$$

The resulting oxide is produced in the form of extremely small particles ($\simeq 2$ nm) which may form a gel.

In reality the situation is more complex ; reactions (2) and (3) proceed simultaneously and are generally incomplete. Hydrolysis may be achieved using a quantity of water less than the stoechiometric one and a number of radicals R remain unreacted. The polycondensation is arrested and the final product rather corresponds to the formula :

$$(MO)_x\ (OH)_y\ (OR)_z \qquad (4)$$

In the case where several different compounds, e.g. M(OR)
M'(OR)$_{n'}$, are reacted, a complexation step may precede reactions (2)
and (3).

In this way complex networks involving one or several different
cations M, M', e.g. :

$$-M-O-M'-O-M-$$ (5)

may be produced. The use of alkoxides of Si, B, Ti, Zr, etc. leads
to the formation of complex gels formed of small particles and which
prefigurate the network of corresponding oxide glasses.

Table I gives the list of organometallic compounds most frequently used in the synthesis of glasses by this method. The reagents are
dissolved in alcohol, generally methyl or ethylalcohol, and the water necessary for the hydrolysis is either taken from the atmosphere or added to the starting solution to accelerate the process. Some
of the cations may also be introduced in the form of alcoholic or
aqueous solutions of salts (nitrates, acetates, etc.). A carefully
controlled amount of acid acting as a catalyst (e.g.HCl, glacial ace-
tic acid, etc.) is added. The gelling time depends on the pH, temperature and the amounts of H$_2$O and catalyst added.

Little is known about the detailed reactions which take place
during gel formation, especially in the case of multicomponent systems.

2 DRYING OF GELS

2.1 Phenomenological Approach

The freshly prepared gel is formed of a network of particles
holding an interstitial liquid - the solvent trapped during the gelling step, water in the case of hydrogels prepared by Method I, mixtures of alcohols and water for the alcogels prepared by Method II.

Elimination of these liquid phases leads to dry gels, namely
xerogels.

When a "wet" gel is dried the following sequence of events is
generally observed on a macroscopic scale :
. progressive shrinkage and hardening
. stress-development
. fragmentation.

The chief difficulty is encountered when massive pieces of gel
are required. This is the central problem of monolithic gels and
much research has recently been devoted to this problem. Monolithic
pieces are required if the subsequent consolidation into glass is to
be accomplished without the use of hot-pressing, in which technique
a particulate material may still be compressed into a solid body.

Cracking during in the drying stage is the result of non-uniform

M	$M(OR)_n$
Si	$Si(OCH_3)_4$ $Si(OC_2H_5)_4$
Al	$Al(O\text{-}iso\ C_3H_7)_3$ $Al(O\text{-}sec\ C_4H_9)_3$
Ti	$Ti(O\text{-}C_2H_5)_4$ $Ti(O\text{-}iso\ C_3H_7)_4$ $Ti(O\text{-}C_4H_9)_4$ $Ti(O\text{-}C_5H_7)_4$
B	$B(OCH_3)_3$
Ge	$Ge(O\text{-}C_2H_5)_4$
Zr	$Zr(O\text{-}iso\ C_3H_7)_4$ $Zr\ (O\text{-}C_4H_9)_4$
Y	$Y(O\text{-}C_2H_5)_3$
Ca	$Ca(O\text{-}C_2H_5)_2$

TABLE 1 : Alkoxides used in gel synthesis

shrinkage of the drying body. This is a well-known problem in ceramic technology for which, however quantitative treatments are still scarce (10).

In any case the admissible rate of drying is inversely proportional to the linear dimension of the object.

In practice the necessary times for drying gels in order to preserve monolithicity are considerable. Fig.6 shows, for example, the drying curves of a SiO_2 gel produced by Method II. Curve A corresponds to an evaporation of a specimen covered by a plastic sheath, and curve B represents condition of accelerated drying which permit the obtention of monolithic gels in about 200 hours. The rate drops when the weight loss attains 60 % of the initial weight of the gel.

2.2 Structural Approach

In the phenomenological approach the body is considered as an isotropic continuum ; no attempt is made to correlate the stresses with the texture or the structure of the material.

In fact the stresses arise not only from the differences in expansion coefficient due to variable water content but, in the first place, from the <u>action of capillary forces</u> which become operative when the pores start to empty and a liquid-air interface is present in the form of menisci distributed in the pores of the drying gel.

2.2.1 Capillary forces and capillary potential. When a liquid evaporates from a porous material the solid phase is subjected to forces due to capillary phenomena at the liquid-gas-solid interfaces.

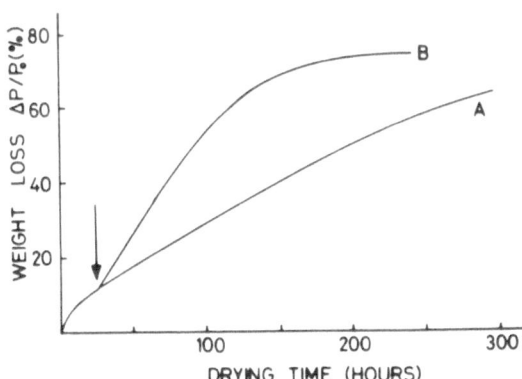

Fig.6 : Drying curves of a SiO_2 gel. A : evaporation of a specimen covered with a plastic sheath ; B : accelerated drying. The arrow marks the gelling point. After ref. (11).

The capillary pressure Δp developed across a curved interface with principal radii $R_1 R_2$ is given by the well-known Laplace's formula :

$$\Delta p = \gamma \ (\frac{1}{R_1} + \frac{1}{R_2}) \tag{6}$$

where γ is the surface tension of the liquid.

For a cylindrical capillary of radius r and a liquid having a wetting angle θ this pressure is :

$$\Delta p = \frac{2 \gamma \cos \theta}{r} \tag{7}$$

The behavior of a liquid in different capillaries can be characterized by the capillary potential defined as the potential energy of the field of capillary forces per unit mass of liquid.

A wetting liquid will tend to occupy a position with the highest capillary potential and a non-wetting liquid a position with the lowest potential. A wetting liquid moves spontaneously from a wide into a narrow capillary.

In a drying gel there are capillary pores and gaps of different sizes and shapes ; the value of the capillary potential differs at different points of the system. The liquid will tend to occupy positions ensuring the minimum energy of the system as a whole.

The magnitude of capillary forces depends on the size of the capillaries in the system and generates stresses that may reach considerable values. Stresses developed by the capillary forces depend, in the first place, on the capillary pressure Δp. Fig.7 shows variations of Δp as a function of the radius r for (a) water ($\gamma = 0.073$ N/m) and (b) methyl alcohol ($\gamma = 0.022$ N/m). The wetting angle taken was $\theta = 0$.

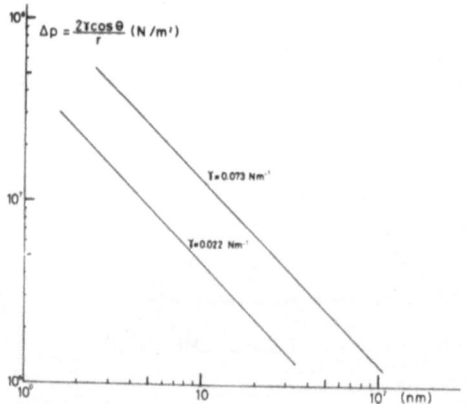

Fig.7 : Capillary pressure, Δp, as a function of pore radius, r. From ref. (9).

The local variation of Δp across the structure and stress concentration effects due to porosity make a detailed calculation impracticable but an estimate can be obtained from these curves of the admissible pore size for a given rupture stress.

2.2.2 Rheological behavior of the gel. During the gelling and subsequent drying process the mechanical characteristics of the material undergo very significant changes.

The initial solution is a Newtonian liquid which changes during gelling to a Bingham material with a steady increase of viscosity. The resulting gel is first a visco-elastic liquid and then a visco-elastic solid which is progressively transformed into a purely elastic solid during the drying process.

The rigidifying process is irreversible in SiO_2-containing gels, the dehydration establishing new Si-O-Si bonds between particles and thus cementing them in new positions, resulting in a collapse and overall shrinkage of the skeleton of the gel.

The porosity of the system diminishes but at the same time the pore-size distribution is changed, the dry gel being thus a modified version of the initial hydrogel.

The capillary stress Δp will therefore affect smaller and smaller pores and, if the surrounding material is deformed viscoelastically, the radius of a pore will simultaneously decrease, delaying its emptying until the partial water pressure has lessened. At the same time, however, as the water content is decreased, the system hardens, the visco-elastic behavior being replaced by a purely elastic deformation. In this competition between closure of pores by capillary forces and the increase of rigidity, the smaller pores may undergo a complete collapse provided the rigidity has not reached an excessive value. This depends on the nature of the given gel.

2.3 Preparation of Monolithic Gels

If massive pieces of glass are to be obtained without the use of hot-pressing techniques, which have the disadvantage of limiting the specimen size, massive pieces of gels are required. The obtention of monolithic dry gels is thus an essential prerequisite.

The preceding analysis of the phenomena accompanying drying has shown the importance of capillary forces and of differential stresses which operate during shrinking.

All actions which tend to minimize these stresses and increase the mechanical resistance of the network should enhance the probability of monolithic gel formation. The following are possible :
. Strengthening the gel by reinforcement.
. Enlarging the pores.
. Reducing the surface tension of the liquid.
. Making the surface hydrophobic.
. Operating in hypercritical conditions where the liquid-vapor interface vanishes.
. Evacuating the solvent by freeze-drying.

214

Tensioactive substances which lower the surface tension of the
liquid may be used as well as substances which diminish the wetting
of the solid phase. Ammonia and its organic derivatives have been
proposed to decrease the wetting of silica particles.

The increase in the radius of the pores may be obtained by varying the conditions of hydrolysis or by the addition of foreign
substances.

Enhancement of the mechanical resistance of the gel may be obtained before drying by an aging process or by the addition of "active silica" during the gelling process. The most efficient way,
however, of eliminating the destructive action of the surface tension of the liquid is to suppress the liquid-vapor interface, operating in hypercritical conditions. (9), (11).

The method consists of treating the gel in an autoclave in hypercritical conditions for the solvent.

The process is schematised in Fig. 8 which shows the equilibrium
curve between the liquid and the gas phase of the solvent. In order
to ensure the continuity of the transition liquid-gas, the path of
the thermal treatment must not cross the equilibrium curve. To cir-
cumvent the critical point C a path such as a b d e may theoretically be used.

In practice this path is modified in the following way : the
open container which contains the gel is placed inside an autoclave
and, to obtain hypercritical conditions, a given quantity of solvent
(methanol) is added to the autoclave. This is closed and electrically heated. When the critical temperature of methanol is exceeded,
successive flushings with dry argon eliminate the last trace of alcohol. The autoclave is then cooled down and the gel removed at ambient temperature (path a d e). The gel the pores of which are filled with air is termed aerogel.

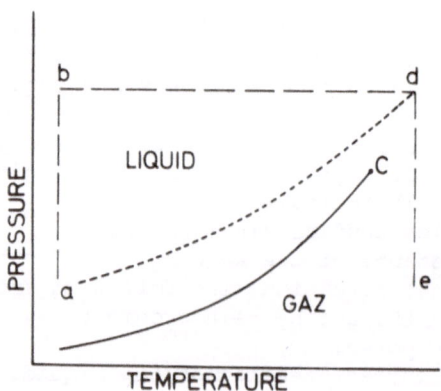

Fig.8 : Paths for hypercritical solvent evacuation (Schematic). After ref. (11).

A number of experiments has shown that monolithicity depends
on many variables, namely :
- the speed of heating
- the proportion of the additional solvent
- the concentrations of organometallic compound and water of
  hydrolysis
- the geometry of the sample and its size
- the previous aging of the gel.

Optimizing these variables, monolithic samples were obtained
with 100 % certainty.(11).Fig.9 shows as an example cylindrical samples of monolithic silica aerogels up to 40 mm in diameter and 250 mm
in length.
The gels are hydrophobic and contain an appreciable percentage
of organic adsorbed radicals. Their mechanical resistance is, however, sufficient for these to be eliminated by subsequent thermal
treatment without loss of monolithicity and finally converted into
a clear glass of excellent optical quality.

2.4 Structure of Dried Gels (Xerogels and Aerogels)

The final structure of the dry gel will depend on the structure
of the wet gel originally formed in solution ; it will be a contracted or a distorted version of the latter.

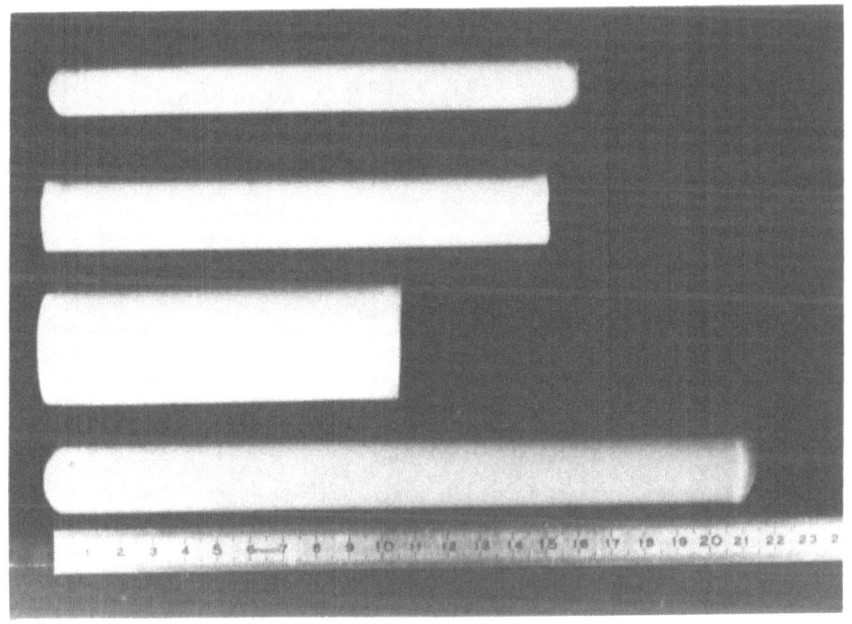

Fig.9 : Examples of monolithic aerogels obtained by hypercritical
solvent evacuation.

The models proposed comprise (Fig.10) : (a) aggregates of particles of approximately the same size, essentially massive in nature ; (b) aggregates of particles formed by primary particles (with ultrapores) ; (c) aggregates of a more complex nature in which three levels of particles can be distinguished as well as micro-and macropores.

The dried amorphous gel differs from a glass by its texture. The gel is essentially an agglomerate of elementary particles, the size of which may be of the order of 10 nm arranged more-or-less compactly. The porosity may vary considerably according to the method of preparation.

The residual space represents the pores which may be closed for dense arrangement of particles (Fig.11b) or open when the texture consists of more-or-less regular "lattices" of particles leaving large interstices (Fig.11b).

The constituent particles are coated with residual OH and OR groups which have to be eliminated during the transition from a particulate texture towards a continuous solid ; these groupings may be detected and analysed by conventional infra-red spectroscopic techniques.

The systems present different porosity and specific surface, according to the packing geometry.

The final characteristics of a dried gel are thus determined by the physico-chemical conditions at every step of the preparation :
. the size of primary particles at the moment of aggregation
. the concentration of particles in solution.
. the pH, salt concentration, temperature and time of aging or other treatment in the wet state.
. mechanical forces during drying.
. temperature, time and atmosphere during drying.

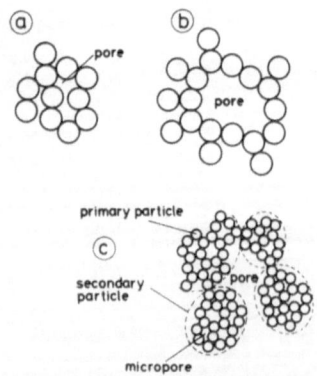

Fig.10 : Different structural models of dried gels, (a) Massive aggregates ; (b) aggregates ; with macropores ; (c) secondary aggregates of primary particles. From ref. (9).

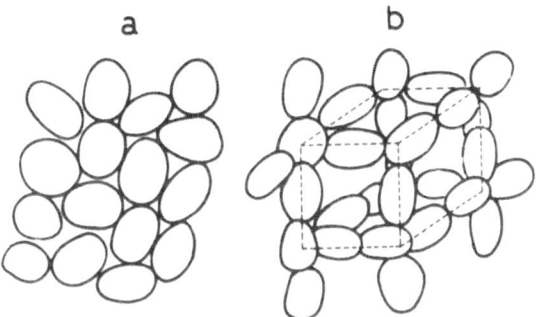

Fig.11 : Texture of gels (schematic) : (a) Dense agglomerate of particles with closed pores, (b) Lattices of particles with
open pores. From ref. (12).

3. DENSIFICATION PROCESS

3.1 Elimination of Residues

To transform the particulate structure of a dried gel into continuous glass the elementary particles must weld together which results in progressive pore elimination. This is achieved by heating
the gel in order to promote diffusion phenomena and viscous flow. During this heat-treatment the residual OH and OR groups will tend to
be eliminated in the form of H_2O and ROH which is accompanied by an
additional polymerization of the system :

$$- \overset{|}{\underset{|}{Si}} - OR + HO - \overset{|}{\underset{|}{Si}} - \longrightarrow \overset{|}{\underset{|}{Si}} - O - \overset{|}{\underset{|}{Si}} - + ROH \qquad (8)$$

The escape of residual products from <u>closed</u> pores may be a problem ; the organic residues are finally carbonized at a sufficient
temperature which brings about a coloration of the gel and leaves
carbonaceous particles in the glass. It is therefore important to favour the escape of residues before the closure of the pores and oxydation treatments are often necessary to eliminate certain organic
groups.

In almost all the studies the progressive elimination of residues
is followed by DTA and TGA analyses. Losses of H_2O and organic volatiles during heat-treatment and influence of oxydising atmosphere were extensively investigated in order to determine the optimal heat-
treatment schedule which leads to a dense and clear glass. Escaping
products may be monitored by gas-chromatography.

Oxydation treatments were found necessary in some cases to eliminate carbon precipitates.

Infra-red transmission spectroscopy, particularly in the 3400-
3600 cm^{-1} and in the 1300 cm^{-1}, 400 cm^{-1} regions, are commonly used
to identify the remaining groups and follow the gel into glass conversion.

In general during the thermal treatment the following temperature zones are found :

50 - 265°C	departure of OH_2 and of ROH
265 - 300°C	oxydation of -OR groups
300 -1000°C	departure of -OH

It is important to define in each particular case the heating
schedule in order to eliminate the unwanted residues without impai-
ring monolithicity before the onset of the viscous flow phenomena.

Occluded-OH and H_2O may provoke bloating upon heating of the
gel. This will often be the case with gels dried slowly at low temperatures - even if they remained monolithic so far. Residual -OH
groups may be eliminated by chlorination treatments if very low OH
levels are required in the final glass.

On the other hand, occluded water may be used in foaming processes, e.g. in blowing gel particles into microballoons.

3.2 Sintering

Densification is essentially a sintering process by which the
pores of a dry gel are eliminated and the material converted into
clear, massive glass. After the elimination of residues the driving
force in this process is essentially supplied by the surface energy
of the porous gel. It tends to decrease the interface, thus eliminating the pores, the collapse being governed in the case of glasses
by Newtonian viscous flow. Extra pressure, as in hot-pressing techniques, may be applied to speed up the process. To study this transformation a model must be adopted to represent the texture of the porous solid. Two models were used in the case of gels :
a) the model of closed pores proposed by Mackenzie and Shuttleworth
in their theory of sintering (13).
b) the model of open pores devised by Scherer (14) to study "lattice-like", less dense textures encountered in the sintering of "soots"
in optical fibres technology.

These two models which idealise the situations of Figs. 11a and
11b are shown in Figs. 12a and 12b, respectively.

3.1.1 Closed pores model. Mackenzie and Shuttleworth assume that the
pores are identical spheres of initial radius r_i ; their number n
per unit volume of solid phase is supposed to remain constant during
densification.

If the relative density $D = \rho/\rho_s$ is defined as the ratio of the
apparent density ρ of the porous solid (gel) to the density ρ_s of the
solid phase (glass), the relation between n and r_i is :

$$n \frac{4\pi}{3} r_i^3 = \frac{1-D}{D} \qquad (9)$$

To evaluate n it is most convenient to consider the total surface $S = 4\pi r_i^2 n$ of the pores per <u>unit volume</u> of <u>solid</u> phase :

$$S = 4\pi \ (\frac{3}{4\pi})^{2/3} n^{1/3} (\frac{1-D}{D})^{2/3} \qquad (10)$$

Experimentally, using various techniques (BET), (SAXS), etc.. a specific surface $S_{sp} = S/\rho_s$ is measured (expressed generally in m^2/g).
   In the case of sintering of gels the solid phase is a glass which has a Newtonian viscosity η independent of the rate of strain and a surface energy γ_g. The kinetic equation of sintering is then found :

$$\frac{dD}{dt} = \frac{3}{2} \ (\frac{4\pi}{3})^{1/3} \ \frac{\gamma_g n^{1/3}}{\eta} \ (1-D)^{2/3} D^{1/3} \qquad (11)$$

which may be integrated to

$$\frac{\gamma_g n^{1/3} \Delta t}{\eta} = \frac{2}{3} \ (\frac{3}{4\pi})^{1/3} \int_0^D \frac{dD}{(1-D)^{2/3} D^{1/3}} \qquad (12)$$

where Δt is the time necessary to reach the reduced density D.
   A <u>reduced time</u> t_r is defined by the relation :

$$t_r = \frac{\Delta t \ \gamma_g \ n^{1/3}}{\eta} \qquad (13)$$

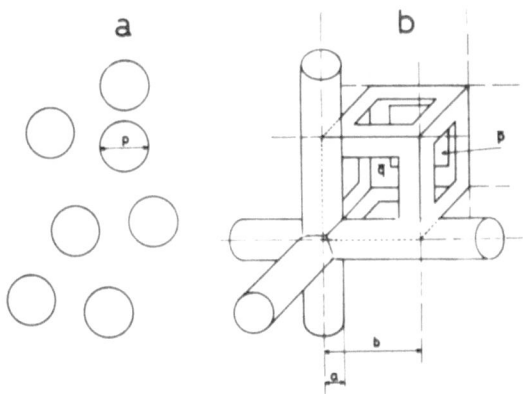

a   b

Fig.12 : Models of texture of gels (schematic) ; (a) closed pores,
         (b) open pores. From ref. (12).

If an external pressure P is applied during sintering, γ_g has to be replaced by the expression :

$$\gamma_g \left[1 + b \left(\frac{1-D}{D} \right)^{1/3} \right] \qquad (14)$$

where

$$b = \left(\frac{3}{4\pi} \right)^{1/3} \frac{P}{2 \gamma_g n^{1/3}} \qquad (15)$$

The parameter b is proportional to the pressure difference between the pores and the outside of the compact. In both cases it can be shown that sintering to D=1 occurs in a <u>finite</u> time. Fig. 13 shows the results of these calculations. (Sintering without applied pressure corresponds to b = 0). It can be seen in particular that the <u>reduced</u> sintering time from D = 0,5 to D = 1 is nearly equal to unity.

3.2.2 Open-pores model. For gels with an open "lattice-like" texture which implies an open porosity, the Mackenzie-Shuttleworth model is no longer applicable. Scherer proposed a model which consists of a regular cubic lattice of intersecting cylinders of radius a, the edge of the lattice being l. (Fig. 12b).

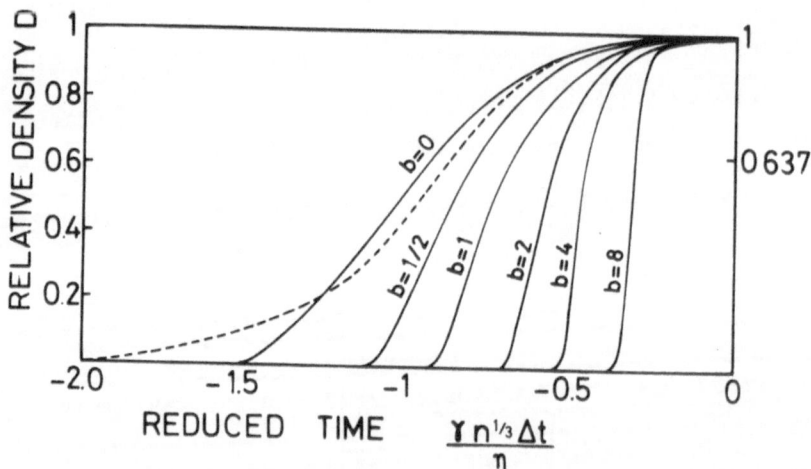

Fig.13 : Kinetics of sintering for different values of parameter b (see text). The full curves represent results for the Mackenzie-Shuttleworth closed pore model. The broken curve represents the Scherer's open model for b = 0. From ref. (12).

3.2.3 Hot pressing ; the Murray-Rodgers-Williams (M.R.W.) approximation. Murray et al. (15) have simplified the M-S model in the case
where P >> $2\gamma/r$ i.e. for b >> 1, the equation becoming in this case :

$$\frac{dD}{dt} = \frac{3P}{4\eta} (1-D) \qquad (16)$$

which may be integrated into :

$$\ln(1-D) = -\frac{3P}{4\eta} t + \ln(1-D_i) \qquad (17)$$

The consequence, however, is that the time t for complete densification is no longer finite, it is necessary to specify the final density D_f.
    This treatment is valuable for hot-pressing techniques. It ena-
bles a quick determination of the $\eta(T)$ relationship in the pressing
temperature interval by a method of linear temperature incrementation
(16) (17).
    In the case of "flash" pressing, if it is assumed that the tem-
perature varies linearly with time in the temperature interval where
densification occurs, the basic M-R-W equation (16) may be written

$$\frac{d \log (1-D)}{dT} = \frac{3 P}{4 \eta x 2,3 v_o} \qquad (18)$$

where v_o is the rate of the increase in temperature. From the slope
of the log (1-D) = f(T) curves it is then possible to calculate the
viscosity η of the gel within the $10^{9.5}$ -$10^{11.5}$ viscosity interval.
The results of such determination for SiO_2 and silica gels are given
in Fig. 14. They show the marked difference between gels prepared
by Method I and II and vitreous silica.

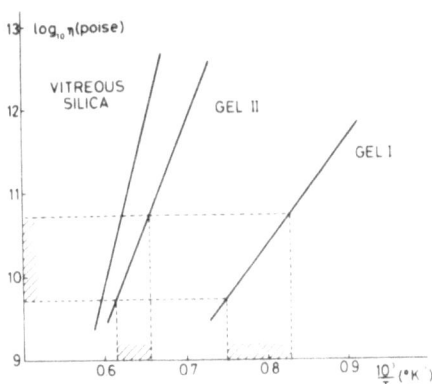

Fig.14 : Viscosity-temperature relationships for vitreous SiO_2 and
         silica gels prepared by methods I and II. Hot-pressing in-
         tervals are indicated. From ref. (18).

3.3 Devitrification Kinetics

During densification the gel will, at the same time, tend to devitrify. The successful conversion of gel into glass therefore depends on a competition between phenomena which lead to densification and those which promote crystallization.

These are best presented using the TTT (Time-Temperature-Transformation) diagrams which are a convenient way of studying the devitrification vs. compaction problem (18).

The TTT diagrams show the time t_y to reach a determined crystallized fraction y as a function of the temperature T. Treating y as a parameter, a set of C_y curves is obtained which represent the kinetic behavior of the system. In particular if y_o corresponds to the smallest crystallised fraction detectable by analytical techniques, the curve C_{yo} represents a frontier not to be crossed during a thermal treatment schedule if crystallization is to be avoided. (The value generally adopted for y_o is 10^{-6}, but other criteria are possible).

Using the method previously exposed, it is possible to evaluate the time for sintering of a given gel when the characteristics of the material and the thermal treatment program are specified.

The relative position of this thermal treatment path during densification and of the C_{yo} curve of the gel determines the possibility of obtaining either glassy or crystallised materials at the end of the compaction program (Fig. 15). If, for example, for a given gel corresponding to C_1 there is no danger of devitrification using path (a), this is no longer true for a gel corresponding to the curve C_2 and the same path would lead to crystallized material. The solution would then be either to shorten the time of sintering. e.g. by applying a suitable pressure P (path b) or to increase the temperature for a short time using the technique of "flash-pressing" (path c).

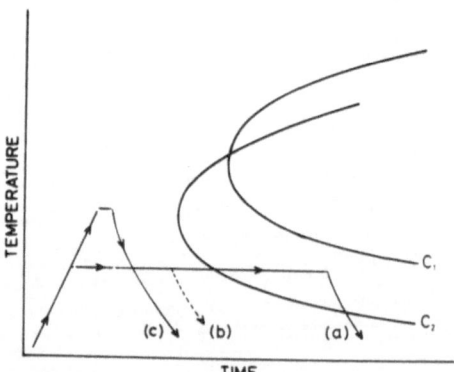

Fig.15 : Time-temperature-transformation (TTT) diagrams and thermal paths for compaction. From ref. (12).

In the case of gels, the process corresponds to <u>heterogeneous</u> nucleation, the position of the curves C strongly depends on the impurities of the material and in the first instance on water content which influences the viscosity as well as the surface tension γ_g of the material.

References

1. DISLICH H. Angew. Chem. Int. Ed. <u>10</u>, (1971), 363,
2. KAMIYA K., and SAKKA S. Res. Rep. Fac. Eng. Mie Univ. <u>2</u>, <u>87</u>, (1977).
3. MUKHERJEE S.P. J. Non Cryst. Sol. <u>42</u>, (1980), 477,
4. SAKKA S. and KAMIYA K. J. Non Cryst. Sol. <u>42</u>, (1980) 403,
5. PHALIPPOU J., PRASSAS. M, and ZARZYCKI J., Verres et Refr. <u>35</u>, (1981), 975.
6. SAKKA S., "Gel method for making glass" in Treatise on Materials Science and Technology vol.22, Glass III. pp.129-167 Tomozawa and Doremus eds., Acad. Press, N.Y. (1982).
7. ZARZYCKI J., "Processing of gel-glasses" in Glass Science and Technology vol.2, Uhlmann and Kreidl Eds. Acad. Press. N.Y. in press.
8. ILER R.K., The chemistry of silica - Wiley (N.Y.) (1979)
9. ZARZYCKI.J., PRASSAS M. and PHALIPPOU J., J. Mater. Sc. <u>17</u>, (1982) 3371.
10. ZARZYCKI J., Monolithic xero and aerogels for gel-glass processes in"Ultrastructure processing of ceramics glasses and composites Hench and Ulrich Eds., Wiley, (1984)
11. PRASSAS M. Thesis, Montpellier, (1981)
12. ZARZYCKI J., J. Non Cryst. Sol. <u>48</u>, (1982) 105,
13. MACKENZIE J.K., and SHUTTLEWORTH R., Proc. Phys. Soc. <u>62</u> (1949) 833.
14. SCHERER G. J. Amer. Cer. Soc. <u>60</u>, (1977) 236
15. MURRAY P., RODGERS E.P. and WILLIAMS A.E., Trans. Brit. Cer. Soc. <u>53</u>, (1954), 474,
16. DECOTTIGNIES M., PHALIPPOU J. and ZARZYCKI J., J. Mater. Sc. <u>13</u>, J. Mater. Sc. <u>13</u> (1978), 2605,
17. JABRA R., PHALIPPOU J., and ZARZYCKI J., J. Non Cryst. Sol. <u>42</u> (1980), 489,
18. ZARZYCKI J., "Nucleation in glasses from gels" in : Advances in ceramics IV p.204-217, Simmons, Uhlmann, Beall Eds. Amer. Cer. Soc. (Columbus)(1981).

SOL-GEL TECHNOLOGY IN THE GLASS INDUSTRY

JACK WENZEL

Direction de la Recherche
Compagnie de Saint-Gobain
Cedex 27, 92096 Paris-La Défense, France

Sol-gel technology has been used with success commercially for over thirty years by one glass company to produce thin film coatings with specialized optical or protective properties (1). However, the last ten years have seen a resurgence of academic interest in this process, which has led other companies to consider its economic and technical viability.

This paper will examine the applications of this technology in the glass industry, and will consider advantages and disadvantages of the process.

Three themes are developed :

- to make existing products using this new technology, high-priced products should be considered for replacement.

- for new products, the field is wide open, subject only to researchers' imaginations. Here one must exploit the unique properties of gels and the unique advantages of the sol-gel route.

- mastery of the process will be necessary for product quality and yield, and will be critical to economic success.

Table 1 - Retail Prices of Glass Products

Float glass	5 F/kg
Fiber glass insulation	10 F/kg
Alkali-resistant glass fiber	20-40 F/kg
Glass frit - TV & vacuum tubes	25-50 F/kg
Ceramic fiber (gel process)	700 F/kg
High purity glass frit	500-1 500 F/kg
Silica fiber	1 000 F/kg
Optical quality silica	1 000-10 000 F/kg
Low-density aerogel	75 000 F/kg
Graded-index optical fiber	400 000 F/kg

(For this comparison : 1 USD = 3 DM = 8 FF)

MARKET FOR GEL-GLASS PRODUCTS

Table 1 presents the approximate retail prices of some glass products, chosen to fall into two price ranges. Traditional glass products, such as flat glass and fiber glass insulation, are produced in megaton/year quantities and tend to retail for less than 50F/kg. On the other hand, sophisticated glass products, often characterized by low-volume production and high purity, may retail for over 400,000 F/kg. For comparison, a ceramic fiber made by the gel process lies in the middle, at about 700 F/kg.

Without going into the detailed economics of the sol-gel process (which would be premature--and premature economic analyses have sometimes killed good ideas), it seems that sol-gel technology should be aimed at manufacturing high-priced sophisticated goods. This is in part due to the high cost of raw materials used in the process, and in part to process yield.

Alternative raw materials sources should lower the costs of sol-gel technology, making more products accessible to manufacture by this process, while simultaneously improving product quality. For example, high-purity tetra-ethyl orthosilicate (TEOS) bought in lab-size quantities is expensive (over 1800 F per kg-equivalent of silica), but this is reduced to 75 F/kg for commercial grade TEOS, followed by in-house purification. The colloid route shows promise for lowering costs as well; fumed silica costing less than 20 F/kg is used as a raw material.

Lastly, in-house or custom synthesis of precursor material, whether metal-organic or fumed, is likely to lead to a cost decrease, and more importantly, to an increase in product range and quality. Germanium alkoxides have been synthesized for fiber optic preforms, indium alkoxides for ITO coatings, and titania-silica soots have been co-deposited for the manufacture of ULE glass.

For thin glassy or crystalline films, raw materials price is a small fraction of the price of the finished product. Float glass retails for about 70 F/m2, whereas the same glass with a sun-shielding thin film retails for 130 F/m2. The price difference of 60 F/m2 is much greater than the raw materials costs of less than 1 F/m2. Thus, as with high-priced bulk glass products, the raw materials price in thin films is not significant, and process reliability and control (and, derivatively, yield and product quality) are paramount.

ADVANTAGES OF SOL-GEL TECHNOLOGY

To produce existing products more cheaply and of higher quality, or to conceive new products, the unique properties of the sol-gel process must be exploited. These include :

1. High purity raw materials are available.

2. Glasses of many compositions may be made, including those not accessible through the melt route.

3. Highly homogeneous multi-component gels and glasses may be synthesized. The scale of this homogeniety (molecular level or greater) may be controlled (2).

4. The microstructure of gels may be controlled. Dry gels may be made with a wide range of densities, surface areas, and pore sizes.

5. Glasses may be prepared from gels by sintering at relatively low temperatures. Because of the high homogeniety and microstructure control of the gel, a properly prepared gel sinters to a glass at about the glass transition temperature Tg, whereas in normal glass practice the liquidus temperature Tl must be surpassed, often by hundreds of

degrees. Because of the empirical relation Tg=(2/3)Tl (3) it is possible to make a glass by the gel route at a temperature one-third lower than its liquidus temperature. This opens the possibility of the making of refractory glasses, and of bypassing either phase separation or crystallization. This should prove useful in preparing fluoride glass fiber optic preforms, which seem (to date) impossible to prepare crystal-free from the melt.

6. The rheological properties of sols allow the formation of fibers, films, and composites, by such techniques as spinning, dip-coating, injection, impregnation, or simple mixing and casting.

UNREALIZED ADVANTAGES OF SOL-GEL TECHNOLOGY

Among the oft-cited advantages of the sol-gel process are three which closer scrutiny shows to be without merit. Supposedly the sol-gel process, a low-temperature process, saves energy over traditional glass-making practice. It is true that lower temperatures are involved in the sol-gel method, but the temperatures are still relatively high. However, this is only half the truth, because the precursors used in the sol-gel process are energy-intensive. On the balance this process has not been shown to be energy-saving.

It is said that the sol-gel process minimizes evaporative losses on processing, thus reducing compositional changes and possibly atmospheric pollution. It is true that evaporative losses, for example of boron, occur in traditional glass practice, but similar evaporative losses occur in the sol-gel process, due to the high vapor pressure of trimethyl borate relative to other alkoxides. In either case compositional changes can be corrected by compensation.

It is also said that environmental pollution may be minimized through use of the sol-gel process, but this process is one for specialty glasses, with low volumes and little pollution, so the advantages of its adoption are likely to be small.

OBSTACLES TO THE ADOPTION OF SOL-GEL TECHNOLOGY

Three problems are encountered in utilizing the sol-gel process commercially : raw materials scarcity and expense ; health hazards ; and process control. The first two are soluble, given the economic incentive, while the third, mastery of the process, remains as the outstanding stumbling block to the wide-spread adoption of this technology by industry.

Many metal organic precursors are unavailable commercially, which has hindered sol-gel research. A wide range of precursors is necessary for two reasons. First, a precursor containing a target element is usually necessary simply to permit its incorporation in the gel -- although it is sometimes possible to use available organic salts instead of the more versatile alkoxides. More importantly, to master the sol-gel process it will be necessary to study systematically the influence of raw material and reaction sequence (among other parameters) on gel and glass properties. For example, the rates of hydrolysis of alkoxides vary widely with cation and with the R-group. A large selection of precursors would allow the hydrolysis rates of the various cations to be balanced, leading to a more refined control of molecular level structure.

For the colloid route, the raw materials situation is critical : only fumed silica and fumed titania are available commercially.

The metal-organic precursors of the polymer route and the organic solvents used in the colloid route are often toxic. This, however, is well known to those working in the area, and current industrial hygiene practice is well equiped to deal with toxicity problems, even upon large scale adoption of this process.

Process Disadvantages

The disadvantages in using the sol-gel technique to produce monolithic gels or glasses are serious, and have prevented the wide-spread adoption of this technology by the glass industry. Major difficulties include :

- maintaining a sol at constant viscosity due to ever-progressing gelation. A constant viscosity is necessary for thin film and fiberization applications.

- converting a wet gel to a monolithic dry gel, due to differential stresses created by the departure of the solvent. For most applications a dry, monolithic gel is a necessary step to the glass.

- transforming a dry monolithic gel to a dense, monolithic glass by heating, due to large shrinkage which causes cracking ;   due to residual pores, which makes final densification impossible ;   due to trapped gases (including water), which cause bloating or reboil;   due to residual water with unwanted infrared absorption, and due to residual organic material, which discolors the glass.

In addition the process as a whole has three drawbacks :   it is sometimes irreproducible , it is a batch process, and long processing times are common.

TOWARDS PROCESS MASTERY

Four recent developments offer hope that the disadvantages of the sol-gel process may be overcome :   stabilization of sol viscosity ;   and controlled drying of gels, including controlled solvent evaporation, hypercritical solvent extraction, and the use of chemical additives. In addition studies of the process which take into account the interrelations of the various steps may lead to reproducibility, shortened process times, and automation.

Control of sol viscosity during fiberization now seems possible. In the first fiberization studies on silica sols (4), sol viscosity increased rapidly with time and gelation soon followed, leading to unacceptably short times over which fiberization was possible - 2 hours or less. However, by modifying reaction conditions and by controlling the humidity to inhibit further gelation once a desired viscosity has been attained, sols have been produced with stable viscosities which remain fiberizable for over 300 hours (5).   Furthermore, by appropriate dilution and temperature changes, the sol viscosity may be made to assume any value between 100 - 1000 poise. Fiber quality after firing (particularly strength) must be improved next.

Gel monoliths now may be fabricated reproducibly by several methods, and sintering to a dense glass is usually possible. Controlled evaporation of solvents at temperatures up to 110° from an aged gel yields a monolithic dry gel if the gel contains pores of a uniform size. This method has been used to some extent with polymeric gels, but it is more appropriate for colloidal gels (6,7).

Evacuation of solvent from an aged gel under hypercritical conditions yields monolithic dry gels, even if there is a variation of pore sizes in the gel (8, 9, 10). The wet gel is heated in an autoclave to above the critical temperature and pressure of the solvent mixture (typically 100 bars and 300°C), and the solvent is then removed. This method is applicable to polymeric gels and is versatile : conditions may be chosen to produce monoliths suitable for sintering, or low-density monoliths for gel applications. Process times are less than two days in general.

The use of chemical additives together with an optimization of the stages of the sol-gel process is a new and promising method to produce monolithic gels (11). The additives, such as formamide and glycerol, give flexibility in initial hydrolysis reaction conditions, leading to decreased gelation time and a controlled, narrow pore-size distribution. The green gel is aged minimally to strengthen it, and drying by controlled solvent evaporation is just slow enough to avoid cracking. Processing times to gel monoliths of one day have been achieved, and are likely to be reduced. Densification to monolithic glass is possible, but it is too early to predict what range of gel properties is accessible.

APPLICATIONS

Potential industrial applications of sol-gel technology are numerous (12). For convenience they may be classified in six areas :

1. gels per se : catalyst supports, thermal insulation, Cherenkov radiators, and biological supports.

2. gels which are impregnated, then sintered : gradient index optics, nuclear waste storage, nitrogen glasses, and diphasic materials.

3. refractory glasses : optical silica, fiber optic waveguides, ultra-low expansion glass, and other refractory silicates.

4. a) optical thin films : sun-shielding, anti-emissive, conductive, and anti-reflective.

   b) protective thin films : oxidation resistant, anti-catalytic, scratch resistant, and weather resistant.

5. fibers : high-temperature insulation and reinforcement.

6. miscellaneous : microspheres, unsupported thin films, graded seals, dental materials, abrasives (soft or hard), organically modified silicates, optical fiber coupling, and raw material for flame-spraying.

CONCLUSIONS

The glass industry is interested in sol-gel technology for two reasons: to make existing products better and more cheaply ; and to imagine new products accessible only through this technology.

Although the sol-gel process has unique advantages over traditional glass-making practice, it also has major drawbacks. If these can be overcome, the technology should find acceptance by industry, since potential applications are numerous.

REFERENCES

1. H Dislich and P Hinz, J Non-Cryst Solids 48,11 (1982)

2. B Yoldas, J Non-Cryst Solids 63, 145 (1984)

3. S Sakka and J Mackenzie, J Non-Cryst Solids 6, 145 (1971)

4. S Sakka et al, Yogyo-Kyokai-Shi 90, 555 (1982)

5. W Lacourse and S Dahar, presented at the Am Cer Soc Annual Meeting, Chicago, IL, 1983

6. E Rabinovich et al, J Am Cer Soc 66, 683 (1983)

7. G Scherer and J Luong, J Non-Cryst Solids 63, 163 (1984)

8. T Woignier et al, J Non-Cryst Solids 63, 117 (1984)

9. W Schmitt, Master's Thesis, U of Wisconsin, 1982

10. S Henning and L Svenson, Phys Scripta 23, 697 (1981)

11. L Hench, private communication

12. See : J Non-Cryst Solids 48 and 63 ; Proc Int Conf on Ultrastructure Processing, Gainsville, FL, 1983 ; and Proc Mat Research Soc Spring Meeting, Albuquerque, NM, 1984

GLASSY THIN FILMS AND FIBERIZATION BY THE GEL ROUTE

Shyama P. Mukherjee
Jet Propulsion Laboratory
California Institute of Technology
Pasadena, California 91109
and
Jean Phalippou
Materials Science Laboratory and CNRS Glass Laboratory
University of Montpellier 2
France

ABSTRACT

A discussion of the physico-chemical principles involved in deposition of gel-films and formation of gel-fibers is followed by an evaluation of the advantages and limitations of various film deposition and fiberization techniques. Examination is made of an expression developed by Landau and Levich for estimating the thickness of a fluid layer, using solutions of varying viscosities, for the deposition of silica gel-films on a vertical moving wall. How the solution chemistry and thermal treatments influence the molecular structure, pore structure, density, and refractive index of films is discussed, as are the roles of viscosity, molecular configuration, and gelling rate in fiberization and in maintaining cross-sectional uniformity of fibers. Finally, applications of gel-derived films and fibers are summarized.

INTRODUCTION

Earlier as well as recent research activities in the field of sol-gel films(1-11) and fibers(12-20) indicate that these two areas of sol-gel technology are of present and future commercial importance. However, the present scientific understanding of the sol-gel processes for the thin film deposition and fiberization is empirical and limited. Hence, the technological growth of these fields will depend on the basic understanding of (a) the physico-chemical nature of sols that are suitable for the deposition films and formation of fibers of desired or predictable

physico-chemical properties, (b) the influence of coating or
fiberization technique parameters on the physical nature (such as
thickness uniformity, cross-sectional uniformity and physical
properties) of films and fibers. Hence, the objective of the
present work is to critically review and discuss the existing
knowledge in these areas, particularly in the context of recent
developments.

1. GLASSY THIN FILMS

1.1 Sols or Solution Preparation

There are two methods for the preparation of stable sols or
solutions for film deposition:

Method I: The first method is based on the preparation of stable,
colloidal sols in an aqueous or nonaqueous liquid medium using
noncrystalline particles of colloidal sizes. These particles can
be prepared by any one of the following techniques: (i) sol-gel
technique using metal halide as starting material(9), (ii) sol-gel
technique using metal alkoxides as starting materials(21), (iii)
vapor phase oxidation of metal halides and organometallic
compounds(22).

The physico-chemical nature of the constituent colloidal
units has a strong influence on the microstructure and
densification behavior of the films. The constituent colloidal
units can be either aggregated in the liquid phase or remain
essentially nonaggregated spherical particles. In a dilute
nonaggregated aqueous sol the individual particles are isolated
and surrounded by an anionic double layer. These sols give gels
of low porosity which have a large number of interparticle
contacts and which consequently densify readily on firing, whereas
the sols having aggregated primary particles produce gels of
considerably higher porosity than that of nonaggregated sols(9).

The preparation of bulk glass from gels obtained from the
nonaqueous sols produced by the dispersion of noncrystalline
oxide particles prepared by the vapor phase oxidation of metal
halides is recently reported by Scherer and Loung(22).
It seems that the sols prepared by similar methods having
appropriate wetability can be used for the deposition of
ultrapure glassy films of a wide variety of compositions.

Method II: This second method is based on the hydrolytic
polycondensation of metal alkoxides and alkoxysilanes in
nonaqueous solvents. The chemistry of hydrolytic polycondensation
reactions and gelation for producing bulk gels and glasses is
investigated and reported by many researchers(2,17,23). The

basic chemistry of the preparation of the sols for making bulk gels are similar to that of making gel-films. The physico-chemical characteristics of the metal alkoxide derived solutions that are suitable for the deposition of glassy films are reported by Schroeder(1) and are as follows:

(a) stability or shelf-life and constancy of rheological behavior during the period of film deposition in a particular deposition environment; (b) sufficiently small contact angle between the solution and the substrate to be coated; (c) starting compounds and the polymerization process such that the gel-film remains noncrystalline during the drying and densification process; (d) transformability of the sol-films to uniform gel-films and subsequently to defect free glassy films; (e) the influence of the pH of the solution on the developing different pore volumes and pore morphologies which influence the optical and mechanical properties of the gel-films.

One typical coating solution composition(24) which was used to determine the influence of withdrawal rates on the film thickness is given here: 15 mols of anhydrous ethanol and 0.8 mol of tetraethoxysilane followed by 5.8 mols H_2O 0.02 mol HCl and 0.02 mol HF per mol $Si(OC_2H_5)_4$ were mixed, in the same order, at room temperature and stirred for about 2 hours. The pH of the solution was close to 2. The viscosity of the solution was determined until gelation took place after about 310 hours. The viscosity changed from 1.8 cp to about 4 cp during the first 250 hours and then increased rapidly until gelation took place. The SiO_2 concentration was 0.05 gms/lit.

2. DEPOSITION TECHNIQUES

Three techniques: dipping, spinning, and spraying, are normally considered for depositing gel-films.

2.1 Dipping Technique

Schroeder(1) investigated the influence of the dip-coating parameters on film thickness. According to him, the thickness of a solid layer depends on the following parameters: μ, withdrawal speed; α angle of inclination of the coated surface relative to the horizontal line; c, the solution concentration; and K, a constant characteristic of each solution that recognizes properties such as viscosity, surface tension, and vapor pressure of the solution. Recently, Mukherjee(25) tested the applicability of a quantitative expression developed by Landau and Levich(26). According to Landau and Levich(26) the thickness of a layer of a liquid film obtained by withdrawal from liquids of low capillary numbers can be expressed as follows:

Thickness; $t=0.944 (N_{ca})^{1/6} (\eta\upsilon/\rho g)^{1/2} = K(\eta\upsilon/\rho g)^{1/2}$ (1)

where ρ is the density of the solution, g_c the gravitational acceleration, U, the withdrawal speed, N_{ca} the capillary number = $\eta\upsilon/\gamma$, γ being surface tension.

The influence of withdrawal speed on thickness of gel-films derived from polymeric tetraethoxy solutions (coating solution preparation procedure described earlier) was investigated[24]. Two high performance coating equipments were designed and constructed[24]. The first one, a flat plate mechanism, provides linear motion and is suitable for coating flat plates with different angles of inclination to the solution surface. The second one, a lens mechanism, provides rotational motion and is useful for coating curved surface. In order to maintain a constant angle between a constant radius lens and the solution interface, the lens must be rotated out of the solution with the center of rotation at the center of curvature of the lens. The mechanisms proved a reliable and repeatable means for withdrawing flat plates and lenses from solution at constant angle and velocity.

2.1.1 Film thickness as a function of withdrawal speed[24,25]. Two approaches were taken to investigate the effect of withdrawal speed on film thickness. The first approach was to use the lens dipping mechanism rotating a flat plate (10 cm wide) from the solution at constant angular velocity. Film thickness as a function of withdrawal speed was obtained for a 5% SiO_2 sol on a flat fused quartz plate for an angle of 90° with respect to solution surface. For a 10 cm wide plate, the range of speed is from 0.090 cm/sec closest to the axis to 0.270 cm/sec farthest from the axis. Two sets of data were recorded, one at 15° to the middle line of the plate and the other at 5° to the midline. Film thickness after drying at $70^\circ C$ for 12 hours was measured by the Talysurf technique. Film thickness is shown plotted as a function of withdrawal speed in Figure 1. The variation in thickness is parabolic with respect to speed which suggests the applicability of Equation(1). The data was, therefore, replotted in Figure 2 as t vs $u^{1/2}$.

The variation of film thickness in relation to the withdrawal speed was also investigated by the second approach of using the flat plate mechanism. Several substrates were withdrawn at different speeds at a constant sol viscosity. The withdrawal speed vs the film thickness data appears to follow the equation t = K $u^{1/2}$. However, at lower viscosities(1.825 cp) the data does not appear to fit a parabolic relationship. However, note that the Landau-Levich equation was developed for ideal viscous liquid. The transformation of partially polymerized metal alkoxides solution to solid gel-films during withdrawal is a complex dynamic

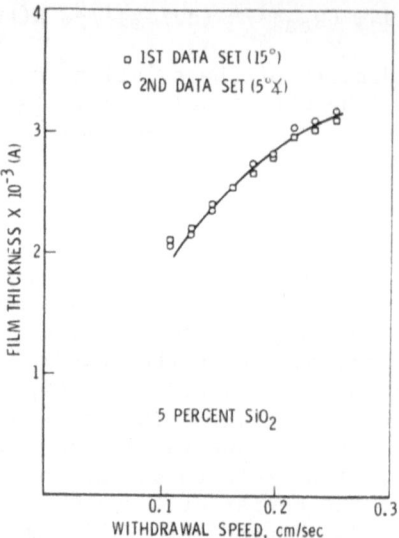

Figure 1. Film Thickness as a Function of Withdrawal Speed
Solution Viscosity = 2.835 cp

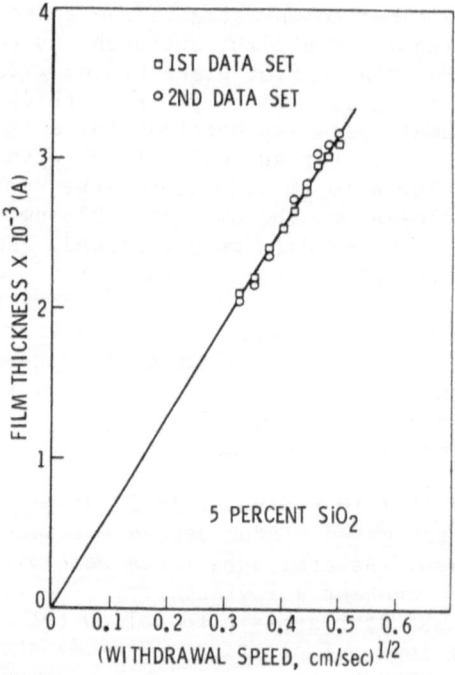

Figure 2. Film Thickness as a Function of (Withdrawal Speed)$^{1/2}$
Solution Viscosity = 2.835 cp

process controlled by several other factors such as solvent vapor pressure, substrate temperature, humidity, etc. The role of these factors on the change of viscosity, surface tension, and density should be considered, and their contributions are to be incorporated in the equation. The uniformity of film thickness on a 2" X 4" vycor substrate was also measured (see Table 1). It is evident that for single dipping, the thickness variation is negligible.

2.2 Spinning Technique

The spinning technique consists of uniformly spreading out the liquid film by pouring the solution on a rotating horizontal substrate. The film thickness of the heat treated TiO_2 films as a function of spin speed on 3 cm dia silicon wafers was determined by Yoldas and O`Keefe[4] and is shown in Figure 3. Increased spin speeds decrease the film thickness in a nonlinear manner. Similar results are also reported by Brinker and Harrington[6]. The influences of viscosity and gelling rates on the film thickness and film uniformity are not reported in the literature. However, it is claimed[4] that the technique gives uniform film thickness on small circular substrates.

2.3 Spraying Technique

The solution to be sprayed is squirted out of one or several stationary spray guns onto the preheated substrates which can be

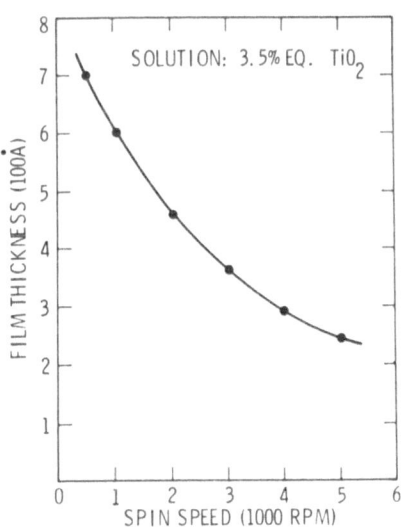

Figure 3. Film Thickness as a Function of Spin Speed (Taken From Reference 4)

Table 1. Film Thicknesses for SiO_2 Solution on a 2" x 4" Pyrex Substrate

Sample No.	Solution Viscosity (cp)	Withdrawal Rate (cm/min)	No. of Dippings	Film Thicknesses in Angstroms						Average Thkness
				Right Half			Left Half			
				Top	Center	Bottom	Top	Center	Bottom	
1	1.825	9.0	1	1020	1020	1020	1020	--	1020	1020
2	1.825	9.0	2	--	1910	1910	2030	1910	1910	1930
3	1.825	13.5	1	1270	1270	1270	1330	1400	1330	1310
4	1.825	13.5	2	--	2640	2730	2670	2540	2790	2670
5	1.825	22.5	1	2060	2100	2170	--	1910	2030	2050
6	1.825	22.5	2	4060	4380	4450	3810	4060	4320	418C
7	1.825	33.0	1	2670	2730	2920	2540	2730	2790	2730
8	2.835	9.0	1	1140	1140	1140	1140	1140	1170	1150
9	2.835	13.5	1	1910	1960	1910	--	1970	1970	1940
10	2.835	22.5	1	3050	3110	3050	3050	3050	3050	3060
11	2.835	33.0	1	--	3990	4000	4000	3850	3910	3950

moved across the jets at a predetermined speed. The spray gun should have a high atomizing capacity. This technique is only suitable for nonoptical coatings where precise thickness uni- formity is not needed(1). However, the deposition of antireflec- tion coatings by the spraying of the titanium isopropoxide on hot semiconductor wafers is reported by Hovel(3). Hovel's technique was to pass a nonoxidizing carrier gas such as N_2+He through a flask containing the liquid alkoxide held at 80 C. A second stream of carrier gas was bubbled through a flask containing water at room temperature. The two gas streams were brought together at the output of a nozzle which was held from 1 to 3 cm away from the substrate resting on a hot plate. It is reported that uniform AR coatings can be obtained by moving the nozzle or sample in a circular or figure eight pattern. However, the optical properties of the TiO_2 films depend very little on the deposition rate but depend strongly on the deposition temperature.

3. THERMAL TREATMENT OF FILMS

The Thermal treatment schedule and the atmosphere during thermal treatment are dictated by the nature of the films and desired physico-chemical properties. Usually gel-films after deposition are dried under an infrared lamp or in an air oven at $\leq$ 100 C. Finally they are heated in an air or oxygen atmosphere at the rate of about 10 - 20 C/min to the desired thermal treatment temperatures.

4. CHARACTERISTICS OF FILMS

The following features of the gel-derived films should be critically analyzed and understood in order to obtain desirable and predictable physico-chemical properties: (i) interface, (ii) change of refractive index, density and thickness with thermal treatment, (iii) pore volume, pore morphology and microstructures, (iv) structural evolution and kinetics of ordering during thermal treatment, and (v), stresses and microcracking. The resulting structures and properties of the films depend on the chemistry of solutions, the nature of substrate and ambient conditions during the film formation, and thermal treatment.

4.1 Interface

The following aspects of the interfacial phenomena should be examined: (a) the bond formation mechanism and nature of bonds between films and substrates, (b) the diffusion of ions (e.g., Na+ ions from glass) from the substrate to films during thermal treatment, (c) oxidation of metallic or semiconductor substrates during thermal treatment, (d) any preferred orientation of the first layer of macromolecules onto substrates, particularly on single crystal substrates.

At present, the information available about these aspects is limited. Schroeder(1) suggested that the strong adhesion of gel-films with substrates at an early stage of solidification is due to the formation of chemical bonds by the reaction of orthoesters or metallic acid esters with hydroxyl groups that are always on substrates. However, no experimental work is available about the extent and nature of surface hydroxylation needed for getting strong bonding with different types of substrates. Bertrand and Fleischauer(27) investigated the interface compositions of TiO_2 gel-films deposited under different experimental conditions on GaAs substrates. Results indicate that films deposited in high humidity exhibit much intermixing of semiconductor and oxide components and oxidation of the semiconductor; films deposited in low humidity exhibit neither mixing nor oxidation of the semi-conductor unless they are stored in the atmosphere for a prolonged period. Thus, it is evident that the chemistry of the coating solution as well as the ambient atmosphere during thermal treatment influence the interface reactions.

4.2 Refractive Index and Density

The change of refractive index and/or density of gel-films during baking is a complex function of the initial compounds, the chemistry and concentration of the solution, and the ambient conditions during film formation. The change of the refractive index as a function of baking temperatures in various oxide systems is investigated by several researchers(1,3,4,6,28). According to an empirical expression reported by Schroeder(1), n_f is the refractive index of a film $= F\ (\theta_\beta(t_\beta))$, where $\theta\beta$ is baking temperature, $t\beta$ is baking time. The process becomes more complex in the gel systems which tend to crystallize during baking. Similarly if the hydrolytic reactivities of alkoxides are different, the rate of formation of oxide layer is different.

Film thickness of silicate gel-films as a function of heat treatment temperatures was investigated by several researchers(1, 4,6,28). Results show that the film thickness decreases considerably ($\geq 30\%$) on heat treatment up to 400C. This high shrinkage plays a major role in limiting the thickness of gel-films by single dipping. Recently, Brinker and Harrington(6) investigated the spectral reflection of TiO_2-SiO_2 gel film as a function of heat treatment temperature. Their results indicate that spectral reflectance can be tailored by adjusting any of the three parameters: composition, film thickness, and heat treatment temperatures. Hovel(3) determined the refractive index change of TiO_2 films as a function of substrate and found that the index increases linearly between 120 and 240°C and saturates at 2.4

for temperatures above 400°C. A structural transformation from
an amorphous phase to a polycrystalline phase occured at a
substrate temperature between 200 and 300°C.

4.3 Porosity and Pore Morphology:

The microporosity, a characteristic feature of gels, develops
after the removal of solvents and volatiles. The pore volumes,
pore sizes, pore morphologies (controlled by the chemistry of the
sol), and heat treatment temperature influence the reflectance
of as-formed baked films and also of chemically leached films(25,
29). The formation of interconnected pores of uniform size
($\approx$100A) is normally associated with the silicate gel-films
deposited from sols prepared in alkaline medium, whereas the pores
of films deposited from sols prepared in acidic medium (pH $\leq$5) are
extremely fine (30-40A) and only a fraction of the pores appear to
be interconnected(25). Experimental work(25) indicates that the
development of single layer gradient index films will by leaching
be more effective with the interconnected pore morphologies that
are associated with the gel-films deposited from the higher pH
sols.

4.4 Structural Evolution on Thermal Treatment

The structural evolution of bulk gels which involves the formation
of more bridging oxygen and the structural ordering, i.e.,
crystallization, are reported by several workers(30,31). Similar
phenomena are anticipated to occur with films. However, the rates
of the processes might be different with films due to factors such
as (a) the influence of substrate surface from which cations might
diffuse into the films and change the crystallization rates(1),
(b) the high surface-to-volume ratio of films which can influence
the rate of removal of volatiles and hydroxyl groups, in turn
influencing densification and crystallization. The influence of
Na^+ ions diffusing from the glass substrate in enhancing the
crystallization of TiO_2 gel-films was demonstrated by
Schroeder(1). The IR reflection spectroscopic studies(32) of
aluminosilicate gel-films as a function of heat treatment
indicate that the reflection spectrum progressed from that of a
typical alkoxide starting material toward the spectrum expected
of a conventional glass. The peak due to the nonbridging species
$\equiv SiO^-$ around 950 cm^{-1} was removed on thermal treatment, but
an absorption due to $\equiv$S-O-Al$\equiv$ where Al^{+3} is in 4-fold
coordination, was seen to develop at 1030 cm^{-1}. Thus, the
structural transformation such as the crystallization or change
of coordination state can strongly influence the optical and
electrical properties of the films during thermal treatments.

4.5 Film Thickness and Stresses

It is suggested(1) that film molecules adjoining the substrate are strongly bound on certain sites of a ceramic or metallic surface. Therefore, the inner boundary zone of the film can hardly take part in the subsequent condensation process of the outer zone. If the entire film thickness is small (some hundreds of molecular layers) the upper particles will only have to undergo minor displacements in order to occupy subjacent vacancies. With increasing film thickness, the outer zones tend to condense irrespective of bonds to the substrate, and consequently reticular cracks may occur. However, the restriction with respect to obtainable film thickness is not effective to the same extent if the film consists of several layers, each remaining below a certain thickness limit (0.3μ). Hence, the deposition of thicker layers should be considered in the light of the following approaches: (a) the deposition of multiple layers from metaloxide solutions (b) the formation of concentrated sols using densified noncrystalline particles (as discussed in Section 1.1, Method I).

5. ADVANTAGES AND APPLICATIONS

Unique advantages of the sol-gel coating techniques in depositing thin glassy films are as follows: high purity and homogeneity, easy coating both sides of large or complex shape substrates, suitable for depositing high temperature materials at low temperatures, high surface smoothness, suitable for depositing composite films, suitable for depositing extremely thin ($\simeq$ 50A) films and porous films.

The applications of gel-derived films are based on the development of one or more of the following physico-chemical properties: (a) optical, (b) electrical, (c) protective, (d) surface active.

a. Optical Films Various types of gel-derived optical films and their applications are extensively reviewed by Schroeder(1), Dislich and Hussman(2). Recently, new applications have been reported, such as single layer AR coatings for silicon solar cells(3,4,16), laser damage resistant broad band AR coatings obtained after leaching multicomponent gel-films for use on optical components in high power laser systems(7), coatings for optical fiber couplers(11), and transparent heat reflecting coatings for polyester sheets(33).

b. Electrical Films The published work in this area is limited. However, the advantages and the potential applications of the gel technique in depositing semiconducting films are mentioned(1). Recent reports on gel-derived transparent IR-reflecting indium-tin oxide (ITO) semiconductor coatings(8), and transparent conductive cadmium stanate(34) coatings indicate that the gel technique could be modified to deposit semiconducting films of other metal oxides. Of another potertial technological importance in this area could be the gel-derived dielectric (electrically insulating) glass films. The higher purity and probably less intrinsic structural defects due to low temperature synthesis(29) may improve the dielectric properties of the glass films. The higher laser damage resistance of gel-derived films also indicates this possibility which has yet to be examined. The possibility of using gel-derived films for production of stable GaAs metal/oxide/semiconducing (MOS) devices is also suggested(27). The formation of fast ion conducting coatings is also of potential importance.

c. Protective Films The following protective gel-derived films are of technological importance: abrasion resistant, chemical corrosion resistant(34-36), high temperature oxidation resistant(37), and thermally insulating coatings.

d. Surface Active Films Because of the microporosity and high specific surface areas of gel-films, the porous films can be utilized for various applications that can be obtained either by the porous surface itself or by physically/chemically modifying the surfaces. Work on catalyst supports(9) and thin layer chromatographic plates(38) are some examples of this type.

6. GLASS FIBERS

Fiberization by the gel route can be achieved by the following techniques: (1) spinning, (2) freezing, (3) drawing from monolithic gel-preforms. Because the physico-chemical principles of the techniques are quite different, each technique and its principles will be reviewed separately.

6.1 Spinning

In polymer technology, the conversion of bulk polymer to fiber is done by three types of spinning(39): melt spinning, dry spinning, and wet spinning.

In melt spinning or melt extrusion a supply of molten polymer is pumped through a spinneret into air where the fibers cool and solidify. The success of this technique in organic polymer technology is based on the synthesis of linear polymers which soften at temperatures below their decomposition points. The formation of noncrystaline or crystalline silicon carbide(40-41) fibers by the melt spinning of polycarbosilane precursors shows the success of this approach with organometallics such as carbosilanes. With metal alkoxides, the instability because of gelation or network formation and decomposition during heating introduces serious limitations to this approach that are yet to be overcome. In dry spinning, dry polymer is dissolved in the selected organic solvent(s) and the solvent is removed by evaporation as the solvent emerges from the spinneret. The solidification is achieved by evaporation of the solvent involving the transfer of the latent heat of evaporation from the ambient atmosphere. The fiberization by the spinning of polymeric solution of metal alkoxides in nonaqueous solvent is somewhat similar to the dry spinning process of organic polymers. In this case, the polymer is generated in situ by polycondensations of monomeric reactants. Moreover, other phenomena that are characteristic of the gelling process such as network formation or microporosity make this technique more complex when compared with the spinning of simple soluble linear organic polymers. In wet spinning of organic polymers, the solvent from the wet fibers is removed by leaching out into another liquid which is miscible with the spinning solvent but is not itself a solvent for the polymer. This technique has not yet been applied for the fiberization of organometallic solutions.

6.2 Glass Fiber Formation from Metalalkoxide Solutions

The spinning of gel fibers from nonaqueous polymeric metal-alkoxide solutions and their subsequent densification to glassy state are controlled by the following inter-related factors: (a) the chemistry of polymeric solutions that determine the solution rheology, the kinetics of sol-gel transition and pore morphologies and residual organics; (b) the environmental conditions of spinning influences on the gelling rates, the solvent evaporation rates and the gel fiber surface; (c) the removal of residual organics from the micropores and surfaces of wet fibers, and (d) the gel to glass transformation without crystallization and bloating.

The production of silica glass fiber by the gel technique was reported by Mansmann and Winter(12). In their process, polyethylene oxide was used as a viscosity modifying agent for improving spinning of acid hydrolyzed tetraethoxy silane in alcohols. The formation of carbon-free as well as carbon-containing silica glass fibers produced the densification of gel fibers at around 1000 C. Kamiya, Sakka, and Ito(42) reported the fabricaation of glass fibers in the TiO_2-SiO_2 system by the spinning of metal alkoxide solutions without any viscosity modifying additions. Sakka and Kamiya(43,44) subsequently reported systematic investigations on the compositions and viscosities of polymeric

tetraethoxysilane solutions in relation to their fiber spinnability and fiber cross-sectional uniformity. Results show that in
acid catalyzed solutions, when the molar ratio of H_2O to
$Si(OC_2H_5)_4$ is less than 4, the viscosity of the solution
increases gradually as a function of time and the rheology before
gelling becomes suitable for fiber spinning. When the viscosity
reaches about 10 poises the solution becomes spinnable. The ease of
spinning increases with decreasing water content up to a molar ratio
($H_2O/Si(OC_2H_5)_4$) of 1.5. When the molar ratio of water is
higher than 4 or when a base catalyzed solution is used, the viscosity after a certain period increases abruptly and the solutions
transform into elastic gels without exhibiting spinnability. The
curves showing viscosity as a function of the reduced time T/T_{gel}
appear to be different in these two types of solutions.

According to Sakka and Kamiya(43-45), the spinnability of
acid-catalyzed, low water content solutions is due to the formation
of linear polymers, whereas the solutions that are not spinnable are
composed of three dimensional network polymers of spherical shapes.
To prove this concept, they determined(40) the intrinsic viscosity
and number average molecular weights of the acid catalyzed
solutions, having different water content and used the empirical,
Mark-Houwink equation $[\eta] = KM^a$, showing the relationship
between intrinsic viscosity $[\eta]$ and number average molecular
weight Mn. The magnitude of "a", for a specific polymer-
solvent system, is a measure of solvent quality on the hydrodynamic volume of the polymer molecule, i.e., the main chain
rigidity and branching. The "a" values of the solutions having
molar ratio of water 20, 5, 2, and 1 were measured(45) and found to
be 0.34, 0.2, 0.64 and 0.75 respectively. According to Sakka and
Kamiya, the higher values of "a" (0.5 to 1.0) in low water content
solution indicates the formation of linear polymers; whereas values
smaller than 0.5 and close to zero indicate discs or sphere
particles.

It may be noted, however, that both K and "a" are functions of the
solvent as well as of the polymer type. This relation is valid
only for linear polymers. The values of "a" vary from 0.5 to a
maximum of about 1.0 for randomly coiled polymers(46). Moreover,
little fundamental information is available concerning the structure, polymer-solvent interactions, and rheology of functionalized
(hydroxyl groups) condensed polysiloxanes in C_2H_5OH or C_2H_5OH -
H_2O solvent systems. The reported formation of cyclic
structures(47) and double chain structures(48) for condensed
polysiloxanes makes the picture more complex.

Recently, Mizuno, Phalippou, and Zarzycki(43) investigated the variation of viscosities of acid catalyzed ethanolic
solutions of $Si(OCH_3)_4$ having increasing water content as a
function of time. Figure 4 shows the variations of viscosities as

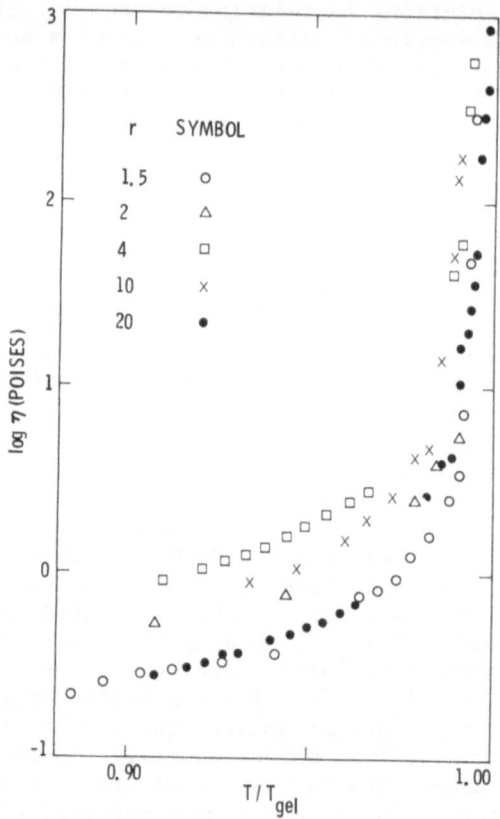

Figure 4. Variation of the Viscosity of Si(OCH₃)4 Solution with Increasing Molar Ratios (r) of Water, where $r = \dfrac{H_2O}{Si\,(OCH_3)4}$

a function of the reduced time T/T_{gel} where T_{gel} is gelling time. It is evident that the viscosities increase with increase of molar ratios of water up to 4; on the further increase of water, the effect is reversed and is not the same as that reported with Si(OC₂H₅)₄(43). The difference might be attributed to the higher rate of hydrolysis of Si(OCH₃)₄ as compared to Si(OC₂H₅)₄. The rate of hydrolytic polycon- densation reaches maximum at a water content of 4; further addition of water changes the rheology of the system, probably because of the change of solvent character due to the presence of excess water. Figure 5 shows the strong catalytic influence of acids on hydrolytic polycondensation reactions.

The cross-sectional uniformity of glass fibers is an important requirement for optical as well as mechanical applications. Sakka and Kamiya(44) reported that the cross-sectional

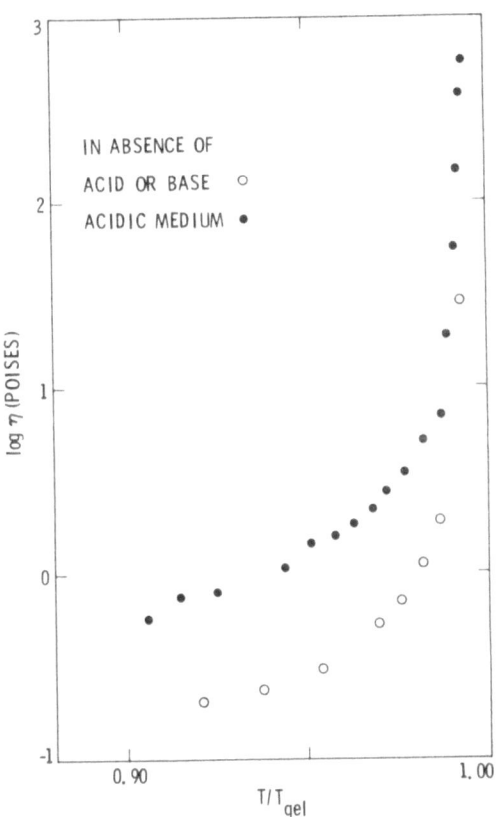

Figure 5. Variation of the Viscosity of Si (OCH3)4 Solution With
Molar Ration of Water Equal to 4 in Acid and Base Media

uniformity is influenced by the extent of shrinkage or volume
change that occurs on solidification at a particular holding
temperature. When the volume change is small, a circular section
develops, whereas a noncircular cross-section develops when the
shrinkage is large. They also observed that cross-sectional
uniformity improved with increase of molar ratio H_2O to
$Si(OC_2H_5)_4$ and that the volume shrinkage decreased with the
increase of molar ratio of water from 1 to 4.

The fiber formation on spinning occurs from skin onwards to
core during rapid evaporation of solvents. If sufficient water
is not internally available for gelation, the outer surface
becomes more rigid due to the adsorption of moisture, whereas the
inner core rigidifies at a later stage when the diffusion of
water or evaporation of solvents occurs. Consequently,
nonuniformity in cross-section develops due to nonuniform network
formation. When water content is high, the polymeric species,

having higher degrees of branching, solidify uniformly throughout the cross-section. Hence, the uniform gelation or network formation on rapid evaporation of solvents during spinning might be the required criterion for higher cross-sectional uniformity.

6.3 Unidirectional Freezing

When a liquid system close to a eutectic composition is frozen directionally, the shape, volume, and orientation of crystals depend on the heat transfer, diffusion kinetics and interfacial surface energy. In unidirectional freezing of eutectic superalloys the formation of well defined parallel fibers occurs(50). A similar formation of silica gel fibers by the unidirectional freezing of aqueous polysilicic acid is reported by Mahler and Bechtold(19). When a silicic acid gel prepared under certain conditions was unidirectionally frozen by lowering a cylinder containing the gel into a $-70^{\circ}C$ cold bath at $4 cmh^{-1}$ after thawing, all the silica was in the form of insoluble parallel fibers with polygonal cross-section. After drying at $150^{\circ}C$, the composition of fiber was $Si_3O_5(OH)_2$. The porous (surface areas 550 to 1200 m^2/g) amorphous fibers were transformed into amorphous silica fibers by heat treatment at $925^{\circ}C$ for 8 hrs. The density and tensile strength were 2.2 g cm^{-3} and 510± 216 MPa, respectively(51).

The formation of solids in a cellular substructure by directional freezing of melts is a well known phenomenon(50). According to Mahler(51), the growth of ice in a cellular substructure is accomplished by advancing the solid-liquid interface through the body of aqueous polysilicic acid at a rate which produces a region of constitutional supercooling ahead of the advancing interface. Constitutional supercooling occurs when the polysilicic solution which is ahead of the interface is below its freezing point but above the temperature of the interface; and under these conditions ice will form in cellular substructure. As freezing proceeds, water separates and polysilicic acid is packed between the growing ice crystals. Hence, the rates at which polysilicic acid separates from the ice and flows into the channels between ice crystals must be considered in selecting the directional freezing rates. With the change of polymeric structures of the gels, the above rates change. This is evident from the fact that highly polymerized polysilicic acid gels give fibers only when frozen at a slow rate (≈2 cm/hr).

This approach of forming gel fibers by unidirectional freezing is also applicable to other gel systems(52).

6.4 Drawing from Monolithic Gel-Preforms

A rod shaped monolithic gel-preform is produced by casting aqueous or nonaqueous colloidal sols or polymeric metal organic solutions.

Subsequently, the gel-preform is transformed into a glass-preform by appropriate heat treatment. Finally, continuous glass fibers are drawn from the gel-derived rod preforms by the fiber pulling techniques that are used for pulling fibers from the vapor deposited glass rod preforms.

Gel-preforms are produced by two methods. The first method is based on the gelation of aqueous or nonaqueous colloidal sols(22,53). The second method is based on the hydrolytic poly-condensation of metal alkoxides in nonaqueous solvents(17,18,20). The monolithicity of gel-preforms during air drying is conserved by the slow evaporation of solvents. The evaporation using supercritical drying conditions(54,55,56,57) is used to increase the drying rates and to maintain desired pore structures and pore morphology that are needed for easy removal of hydroxyl groups and complete densification without bloating. The removal of hydroxyl groups is achieved by the treatment of porous gel with reactive gases(58). The transition metal impurities in gel-derived glass is controlled by the purity of starting materials and by controlling the purity of the gel processing environments and containers(59). Purity of starting metal alkoxides can be improved by successive distillation of liquid metal-organic compounds(60). The high purity oxide powders for colloidal dispersion can be prepared by the vapor phase oxidation of metal halides or organometallic compounds(22). The recent result of attenuation measurement of gel-derived silica glass fibers show that the optical loss at 0.85 μm is 6dB/km(20).

7.   ADVANTAGES AND APPLICATIONS

Various high temperature oxide or nonoxide glass or glass-ceramic fibers can hardly be fabricated by the conventional technique of melt spinning. The sol-gel technique of fiberization can overcome such limitations and a wide variety of new high temperature glass/glass-ceramic fibers can be produced. The formation of porous gel-fibers directly from sols offers a unique approach to making microporous fibers which might find new applications. The low temperature synthesis from polymeric precursors might produce optical fibers having fewer intrinsic structural defects than those obtained with melt spun fibers.

The applications of gel-derived glass fibers are based on the following physico-chemical properties: (a) refractoriness and mechanical strength(12-16), (b) optical transparency(18,20,22), (c) chemical resistance(61), (d) low thermal expansion(42), and (e) surface activity(51).

250

REFERENCES

1. Schroeder, H. in Physics of Thin Films 5 Eds. G. Hass and
    R. E. Thun. Academic, N. York (1969), p. 87-141.
2. Dislich, H. and E. Hussmann. Amorphous or Crystalline Dip
    Coatings, Obtained from Organometallic Solutions, Procedure,
    Chemical Process, and Products; Thin Solid Films 77 (1/2/3)
    (1981) 129-139.
3. Hovel, H. J. TiO$_2$ Antireflection Coatings by a Low
    Temperature Spray Process, J. Electrochem. Soc. 125 (6)
    (1978) 983-985.
4. Yoldas, B. E. and T. W. O'Keefe. Antireflective Coatings
    Applied from Metalorganic Derived Liquid Precursors, Appl.
    Opt. 18 (18) (1979) 3133-3138.
5. Mukherjee, S. P. Gel-Derived Single Layer Antireflection
    Films with a Refractive Index Gradient, Thin Solid Films 81
    (1981) L89-L90.
6. Brinker, C. J. and M. S. Harrington. Sol-Gel Derived
    Antireflection Coatings for Silicon Solar Energy Mat. 5
    (1981) 159-172.
7. Mukherjee, S. P. and W. H. Lowdermilk. Gradient Index
    Antireflection Films by the Sol-Gel Process, Appl. Optics
    21, (1982) 293-296.
8. Arfsten, N. J. Sol-Gel Derived Transparent IR-Reflecting
    ITO Semiconductor Coatings - Properties and Technical
    Possibilities, J. Non-Crystalline Solids 63 (1984) 243-249.
9. Nelson, R. L. and J.D.F. Ramsay, J. J. Woodhead, J. A. Cairns
    and J. A. A. Crossley. The Coating of Metals with Ceramic
    Oxides via Colloidal Intermediates, Thin Solid Films 81
    (1981) 329-337.
10. Rothan, R. N., R. J. Ashley, B. G. Carter, and E. P. Short.
    Solution Deposited Thin Inorganic Glass Coatings for
    Environmental Protection of Anodized Aluminum, Trans. Inst.
    Met. Finishing 56 (1978) 37-40.
11. Tran, D. C. and K. P. Koo. Stabilizing Single Mode Fiber
    Couplers by Using Gel-Glass, Electronics Lett. 17 (5)(1981)
    187-188.
12. Mansmann, M. and G. Winter. Producing Novel Silicon Dioxide
    Fibers, Brit. Patent. 133, 5754 (1973).
13. Sowman, H. G. Refractory Fibers of Zirconia and Silica
    Mixtures, U.S. Patent 3,795,524 (1974).
14. Winter, G. Production of Inorganic Fibers, U.S. Patent
    3,846,527 (1974).
15. Sowman, H. G. Alumina-Chromia Metal (IV) Oxide Refractory
    Fibers Having a Micro Crystalline Phase, U.S. Patent (1978).
16. Kamiya, K., S. Sakka, N. Tashiro. Preparation of Refractory
    Oxide Fibers from Metal Alcohotes, Yogyo Kyokai Shi 84
    (1976) 614-618.
17. Mukherjee, S. P. Sol-Gel Processes in Glass Science and
    Technology, J. Non-Crystalline Solids 42 (1980) 477-488.

18. Puyane, R., A. L. Harmer, and C.T.R. Gonzalez-Oliver. Optical Fiber Fabrication by the Sol-Gel Method, Proc. 8th European Conf. on Opti. Comm. (Cannes) (1982) 623.

19. Mahler, W. and M. F. Bechtold. Freeze-Formed Silica Fibers, Nature 285 (27) (1980) 5759-61.

20. Susa, K., J. Matsuyama, S. Satoh, T. Sugnuma. Electronics Lett 18 (12) (1982) 499.

21. Stober, W. A., Fink, and E. Bohn, Controlled Growth of Monodisperse Silica Spheres in the Micron Size Range, J. Colloid and Interface. Sci. 26 (1968) 62-29.

22. Scherer, G. W. and J. C. Loung. Glasses from Colloids, J. Non-Crystalline Solids 63 (1984) 163-172.

23. Dislich, H. New Routes to Multicomponent Oxide Glasses, Angew. Chem. Int. Ed (Eng) 10 (6) (1971) 363-70.

24. Mukherjee, S. P., J. C. Debsikdar, J. F. Cordaro, K. J. Wurm, and G. T. Ruck. Final Report on "Single Layer Gradient Index Antireflection Films Deposited by the Sol-Gel Process" to Lawrence Livermore National Laboratory from Battelle-Columbus Laboratories, January (1983).

25. Mukherjee, S. P. Deposition of Noncrystalline Metal Oxide Coatings by the Sol-Gel Process. Int. Conf. on "Ultrastructure Processing of Ceramics, Glasses and Composites", Gainesville, Florida, (1983).

26. Landau, L. D. and V. G. Levich. Dragging of a Liquid by Moving Plate, Aeta. Phys-Chem. URSS 17 (1942) 42.

27. Bertrand, P. A. and P. D. Fleischuer. Chemical Deposition of TiO_2 Layers on GaAs, Thin Solid Films, 103 (1983) 167-175.

28. Brinker, C. J. and S. P. Mukherjee. Comparison of Sol-Gel Derived Thin Films with Monoliths in a Multicomponent Silicate Glass System, Thin Solid Films 77 (1981) 141-148.

29. Mukherjee, S. P. and W. H. Lowdermilk. Gel-Derived Single Layer Antireflection Films, J. Non-Crystalline Solids 48 (1982) 177-184.

30. Gottardi, V, M. Guglielmi, A. Bertoluzza, S. C. Fagnano, and M. Morelli. Further Investigations on Raman Spectra of Silica Gel Evolving Towards Glass, J. Non-Crystalline Solids, 63 (1984) 71-80.

31. Zarzycki, J. Gel-Glass Transformaton, J. Non-Crystalline Solids, 48 (1982) 105-116.

32. Strawbridge, I., J. Phalippou, and P. F. James. Characterization of Alkali Alumino Borosilicate Glass Films Prepared by the Sol-Gel Process on Window Glass Substrates, (to be published in Phys. Chem. Glasses).

33. Chiba, K. S. Sobajima and T. Yatabe. Transparent Heat Insulating Coatings on Polyester Film using Chemically Prepared Dielectrics, Solar Energy Materials 8 (1983) 371-385.

34. Dislich, H. and P. Hinz. History and Principles of the Sol-Gel Process and Some New Multicomponent Oxide Coatings, J. Non-Crystalline Solids 48 (1982) 11-16.

35. Rees, J. M. and R. N. Rothan. Fluorine Resistant Coatings for Use on the Silica Envelopes of Tungsten/Fluorine Lamps, Chem. and Industry, 1 July (1978) 479-480.

36. Ginsberg, T. Film Formation Mechanism of Alkylsilicate Zinc-Rich Coatings, J. Coatings Tech. 53 (679) (1981) 23-32.

37. Schlichting, J. and S. Neuman. GeO /SiO Glasses from Gels to Increase the Oxidation Resistance of Porous Silicon Containing Ceramics, J. Non-Crystalline Solids 48 (1982) 185-194.

38. Mendel, A. and R. W. Lange. Thin Layer Chromatographic Plates U.S. Patent 4,138,336 (1979).

39. Hill, R., Fibers from Synthetic Polymers, Elsevier Publ. Co. N. York (1953), p. 363, 379, 395.

40. Verbeek, W. Materials Derived from Homogeneous Mixtures of Silicon Carbide and Silicon Nitrides and Methods for Their Production, Ger. Offen 2,218 960 1973 (U.S. Pat. 3,853,567).

41. Yajima, S., J. Hayashi, and M. Imori. Continuous Silicon Carbide Fiber of High Tensile Strength, Chem. Lett. 9 (1975) 931-934.

42. Kamiya, K., S. Sakka, and I. Ito. Preparation Oxide Fibers from Metal Alcoholates, Yogyo - Kyokai - Shi 85 (1977) 599-605.

43. Sakka, S. and K. Kamiya. The Sol-Gel Transition in the Hydrolysis of Metal Alkoxides in Relation to the Formation of Glass Fibers and Films. J. Non-Crystalline Solids 48 (1982) 31-46.

44. Sakka, S. and K. Kamiya. Preparation of Shaped Glasses through the Sol-Gel Method, presented at the 19th Univ. Conf. on Ceramic Science "Emergent Process Methods for High Technology Ceramics", Univ. of North Carolina, U.S.A. Nov. 8-10, (1982).

45. Sakka, S., K. Kamiya, K. Makita, and Y. Yamamoto. Formation of Sheets and Coating Films from Alkoxide Solutions", J. Non-Crystalline Solids, 63 (1984). 223-236.

46. Billmeyer, F. W. Textbook of Polymer Science, Interscience, Wiley, N. York (1962) p. 82-83.

47. Peace, B. W., K. G. Mayhan, and J. M. Montle. Polymers From the Hydrolysis of Tetraethoxysilane Polymer, 14 (1973) 420-422.

48. Abe, Y. and T. Misono, Preparation of Polysiloxanes from Silicic Acid, J. Polymer Sci., Chem. 21 (1983) 41-53.

49. Mizuno, T., J. Phalippou, and J. Zarzycki, Evolution of the Viscosity of Solutions Containing Metal Alkoxides (to be published in Glass Technology).

50. Tiller, W. A., in the Art and Science of Growing Crystals, Ed. J. J. Gilman. p. 276-312, Wiley, New York (1963).

51. Mahler, W. Silican Fibers and Method of Preparing Them, U.S. Patent 4,122,041 (1978).

52. Kokubo, T., Y. Teranishi, and T. Maki. Preparation of Amorphous ZrO_2 Fibers by Unidirectional Freezing of Gel, J.

Non-Crystalline Solids 56 (1983) 411-416.

53. Rabinovich, E. M., D. W. Johnson, J. B. MacChesney, and E. M. Vogel. " Preparation of Transparent High Silica Glass Articles from Colloidal Gels", J. Non-Crystalline Solids 47 (1982) 435-439.

54. Prassas, M., J. Phalippou, J. Zarzycki, and J. Mater. Sci. (to be published).

55. Mukherjee, S. P. and J. C. Debsikdar. Influence of Gel Preparation Procedures and Drying Techniques on the Pore Structures of Dried Silica Gel-Monoliths, Ceram. Bull., 62 (1983) 413.

56. Woignier, T., J. Phalippou, and J. Zarzycki. Monolithic Aerogels in the System $SiO_2 - B_2O_3$ $SiO_2 - P_2O_5$, $SiO_2 - B_2O_3 - P_2O_5$, J. Non-Crystalline Solids 63 (1984) 117-130.

57. Pancrazi, F., J. Phalippou, F. Sorrentino, and J. Zarzycki. Preparation of gels in the CaO Al_2O_5 SiO_2 System from Metal Alkoxide. J. Non-Crystalline Solids 63 (1984) 81-93.

58. Phalippou, J., T. Woignier, and J. Zarzycki. Monolithic Silica Aerogels at Temperatures above 1000°C; Int. Conf. on "Ultrastructure Processing of Ceramics, Glasses and Composites" held in Gainsville, Florida, (1983).

59. Beam, T. L. and S. P. Mukherjee, Ultrapure Gel Processing in a Clean Room facility, Ceram. Bull 62 (3) (1983) 413.

60. Gossink, R. G., H.A.M. Coenen, A.R.C. Engelfriet, M. L. Verheijke and J. C. Verplanke. Mat. Res. Bull. 10 (1975) 35.

*   *   *   *   *   *

Publication support for this paper was provided by the Jet Propulsion Laboratory, California Institute of Technology under a contract with the National Aeronautics and Space Administration.

FINE CERAMICS FROM THE SOL-GEL PROCESS

VITTORIO GOTTARDI

Faculty of Engineering - University of Padova - Italy

It is well known and has also been exaustively demonstrated
during this meeting that, starting from solutions of the appropriate
metal/organic compounds, certain gels may be obtained which, upon
sufficient heating, are able to transform themselves into glass.
The conditions for this transformation are the following:
1) that the concentration-ratios of the component oxides fall into
   those limits currently ascertained for glass;
2) that chemical species do not have time or possibility to interact
   or in any way organize themselves in that crystalline arrangement
   foreseen as the maximum stability and equilibrium species.
This second condition is linked to the course of the densification process; that is, to the progressive reduction of porosity
acquired through the heating of the gels and in relation to
hydroxide elimination.
In fact, as greater are the specific surfaces, as easier
crystallization is achieved,ultimately being inversely proportional
to the densification process.
However, both phenomena are influenced by the presence of
hydroxides which promote reduction in viscosity and affect the
energy levels of the surfaces.
If the situation proceeds favorably towards the formation of
a glass, the product may be more or less stable analogously to
the material with equivalent composition but obtained through
fusion; the glass upon further heating may result in the occurrence
of crystals and ultimately devitrify.

With the sol-gel method, it is therefore possible to obtain a crystalline structure, in other words, a ceramic product, either by formation of a parent-glass which successively devitrifies, or by crystallization of the amorphous gel before real occurrence of the glassy state.

It is evident that the terms crystallization and devitrification are used here according to current terminology, even if it is clear, and will be made more clear further on, that the boundaries between the two terms are rather indistinct: actually, definitions should require further insight. This is also due to the fact that the chemical and physical properties of the gel evolve without any characteristic discontinuity in the transition from gel to glass; frequently, the devetrification may only be ascertained by the existence of a precise exothermic peak during the course of a Differential Thermal Analysis (DTA).

Today we have a considerable amount of available documentation in order to compare the behaviour of glasses obtained through melting with glasses of equivalent composition obtained through the sol-gel method

Main differences concern the kinetics of devitrification; the different concentration of hydroxides, which act as mineralizers, and the greater homogeneity of products, obtained from the parent compounds in solution, are important factors which substantiate different crystallization behaviours.

However, the stil insufficient availability of data and the multiplicity of variables that influence the vitrification kinetics taken into examination, prevents the arrangement of all findings in a single general picture.

Indeed, the experimental procedures used in the preparation of gels may effect not only the devitrification kinetics, but even the nature and the morphology of the phases obtained in the crystalline state. One of the most interesting aspects is the observed structural influence that different parent systems have upon the final product, so that frequently, the first product that devitrifies is not the most stable, but is the one with the lowest energy of nucleation. This renders the resort to sol-gel type techniques rather uncertain and problematic in the aim to study the equilibrium situations foreseen by the phase diagrams, although this new approach, according to R.Roy suggestions, is much quicker than the traditional systems and techniques.

The whole of the observations which have appeared up to now in research studies suggest a wide series of arguments for further research ranging from:

1. The possibility to get new and interesting information concerning the mechanisms of devitrification;
2. The use of the gel route in order to obtain glass-ceramics with chemical compositions requiring particularly high melting temperatures;
3. The study of the behaviour of those systems which, by starting from melt, are able to reach the glassy state only by means of particularly high cooling rates.

It would also be promising to study those systems unable to meet previously mentioned 1 and 2 conditions and, consequently,unable to transform themselves into a glass, since in these cases ceramic products are directly obtained from the amorphous-gel state by means of crystallization.

An example of this type of behaviour was discovered in the author's laboratory for the SiO_2-Fe_2O_3 system; at Fe_2O_3 concentrations 20% in weight, crystallization of hematite was observed to occur at 300°C, that is, immediately after the complete disappearance of organic products. Even for this quite peculiar composition, the accuracy of the mixture and the extreme purity resulting from the lack of contact with contaminating substances, may ultimately bear the achievement of ceramic products.

To mention some examples of this topic recalling direct experiences of the author and coworkers, it will be described the results obtained by using the sol-gel process in the case of AZS refractories usually achieved by electrocasting. With the classical composition Al_2O_3-SiO_2-ZrO_2-Na_2O, the formulation of glass was at first observed by heating the gel; this is followed by an abrupt devitrification well indicated by DTA, with appearance of the same type of crystals found in the products resulting from electrocasting processes. In reference to the above, the following points should be stressed:

1. the dimensions of the crystals are considerably reduced so that, as a consequence, the zirconia may be found in the tetragonal form at room temperature;
2. the amount of the phase that remain in vitreous form even after prolonged heatings at high temperature is much less important;
3. the quantity of mullite is considerably high; it is present in the needle-shaped form typical of product generated from components dispersed at molecular level; at variance, during the cooling of the melt mullite results from a peritectic transformation.

As for another classical composition, holding a noticeable percentage of Cr_2O_3, the glass state cannot be obtained even with pronounced and forced cooling.

Even the thermal evolution of the gel in these cases does not result in the formation of a glass; an amorphous material is obtained which bears crystalline phases with not very-well localized thermal effects.

The crystalline phases are the same as those observed in the electro-melted product;particularly interesting is the possibility to follow the occurrence of the solid solution Al_2O_3-Cr_2O_3. Indeed, the Al_2O_3-Cr_2O_3 crystals are highly homogeneous, at variance with those of the electrocasting product in which a gradual increase in Cr_2O_3 occurs going from the periphery to the crystal center.

The satisfactory correspondence observed between structural features of products obtained both by devitrification or by crystallization of gels and similar products derived from melt, somehow substantiate the reliability of the laboratory-scale experiments on the effects of those components added beyond traditional compositions.

These laboratory results would be much more difficult if performed directly on electric furnaces in a systematic method requiring a great number of trials. Thus , in the laboratory scale, it has been found that addition of magnesium oxide progressively leads to the formation of $MgAl_2O_4$ spinel; on the other hand, addition of cerium oxide results in the increase of the vitreous phase with consequent mullite reduction. However, the CeO_2 was not observed to have any stabilizing effect on the tetragonal ZrO_2.

This brief review on the possibilities to obtain totally or partially crystallized ceramic products by using the sol-gel techniques would not be complete without mentioning the work of Mazdiyasni and coworkers concerning the synthesis of ceramics from suitable alkoxyde solutions which, after hydrolysis, generate crystals with the desired composition.

Following this procedure, titanates and zirconates of barium and strontium in the form of extremely small particles with dimensions on the order of hundreths of Å have been prepared. These particles can be obtained directly in the crystalline form, or in amorphous form suitable of crystallization following appropriate thermal treatments. In this case a clear-cut transition through the gel state is not observed, but the similarity of the technological procedures suggests that this case may also be considered part of a unified panorama. This out-look may be summed-up by the following scheme giving an idea of the actual state of our knowledge as far as new possibilities of obtaining great purity ceramics at desired composition, with often features different from those of products obtained by traditional processes.

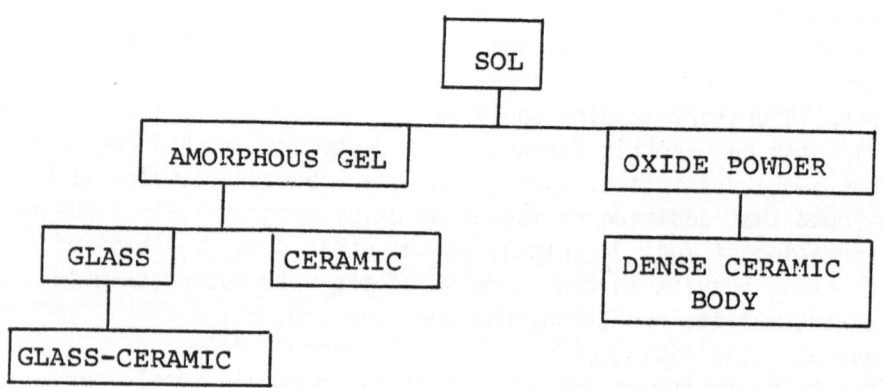

USE OF DRYING CONTROL CHEMICAL ADDITIVES (DCCAs) IN PRODUCING GEL MONOLITHS

L. L. Hench

Department of Materials Science and Engineering
University of Florida, Gainesville, Florida 32611

INTRODUCTION

Rapid, reliable production of gel monoliths is necessary to realize many of the potential advantages of this type of glass processing (1). Zarzycki has reviewed the many factors that lead to drying stresses and cracking of gels (2). He shows that the drying stress is a function of pore size and rate of evaporation of the pore liquor, which depends on the liquor vapor pressure. In a recent series of papers we have demonstrated the use of organic additions to alkoxide sols termed drying control chemical additives (DCCA), to control the pore liquor vapor pressure, pore size distribution, and drying stresses (3-8). By use of the DCCA's, which includes formamide (NH_2CHO) and glycerol ($C_3H_8O_3$), it is possible to produce a wide range of sizes and shapes of dried gel monliths of SiO_2, Li_2O-SiO_2, Na_2O-SiO_2, and Na_2O-B_2O_3-SiO_2 within a 1-2 day processing schedule with 100% reliability (Fig 1).

SiO_2 GELS

Figure 2 shows an optimized flow diagram for producing pure SiO_2 gels using formamide in the metal organic system. The DCCA reduces gelation, aging, and drying times, the drying stress and increases the size of gel monoliths that can be made up to 100 cm^3 after drying for 2 days. This is facilitated by acid catalysis, use of an optimum solvent volume, and optimum DCCA/solvent ratio. The acid catalysis causes rapid hydrolysis and a stronger gel to form. FTIR liquid cell spectroscopy confirms the extremely rapid formation of siloxane bonds in the acid catalyzed-formamide DCCA-TMS system. Physical properties of the DCCA controlled silica gels dried at 60°C are listed in Table I.

Table I
Physical Properties of DCCA Modified Silica Gel Dried at 60°C.

Density	1.42 g/cm^3
Hardness	4 DPH (100g load)
Tensile Strength	200 psi
Index of Refraction	1.442
Surface Area	750 m^2/g
Pore Volume	1.0 cm^3/g
Average Pore Size	40 A

Viscous sintering of the gels starts at about 800°C resulting in dense silica with a hardness of 400 DPH at 1000°C. X-ray diffraction showed no evidence of devitrification; however, FTIR analysis showed a 926 cm^{-1} SiOH peak still present. The major problem in this process is a tendency for residual formamide in the pores to react with water vapor. When the absorption occurs preferentially on the surface of the dried gel uneven stresses develop between the surface and the bulk and cracks develop. Therefore, it is essential to eliminate residual formamide without exposure to water vapor if full densification of monoliths is to be achieved.

Na_2O-SiO_2 GELS

A similar procedure to that outlined above was optimized for 20 Mol % Na_2O-80 Mol % SiO_2 (20 N) and 33 Mol % Na_2O-67 Mol % SiO_2 (33 N) gels. It was found that by using formamide and glycerol as DCCA's the critical shrinkage rate of the 20N gels was substantially increased. Consequently, the evaporation rate of the pore liquor was increased and the drying time shortened severalfold. In addition the DCCA's accelerate the aging process such that the gel strength is nearly doubled in a 12 hour period. A substantially higher temperature is required to remove the formamide compared with methanol which enables interparticle necks in the gel to develop sufficiently to resist appreciable drying stresses. The combination of aging times and drying temperatures which lead to large monolithic dried 20N gels are above the curve drawn in Fig. 3. The net effect is to reduce total processing time of 20N monoliths to 1-2 days with 100% reliability. Similar results have been obtained for the 33N and the 42S-30B-28N systems.

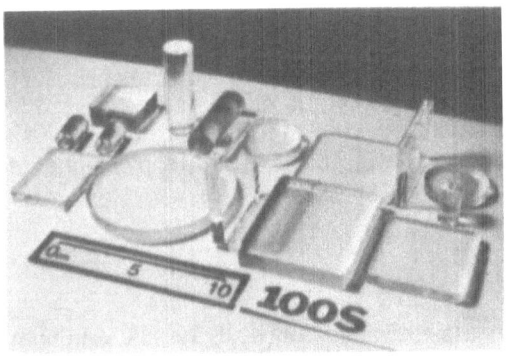

Fig. 1. Prototype optical components made with DCCA controlled SiO$_2$ gel process.

FLOW DIAGRAM OF OPTIMIZED 100S GEL MANUFACTURE

15 cm^3 TMS (0.10008 Moles)

12.5 cm^3 CH$_3$OH

12.5 cm^3 NH$_2$CHO

2 cm^3 Conc. HNO$_3$ (0.0477 Moles)

Stir Magnetically at 25°C

Add 18.17 cm^3 (1.008M) of H$_2$O (R=10)

Stir for 10 Minutes

Cast Into Sealed Polystyrene Container

Gel at 60°C in 1 Hr.

Age at 60°C for 12 Hrs.

Dry Unidirectionally

Fig. 2. Example of optimized flow diagram for gel processing.

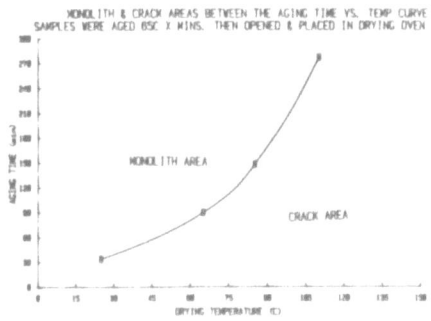

MONOLITH & CRACK AREAS BETWEEN THE AGING TIME VS. TEMP CURVE
SAMPLES WERE AGED 65C X MINS. THEN OPENED & PLACED IN DRYING OVEN

MONOLITH AREA

CRACK AREA

AGING TIME (mins)

DRYING TEMPERATURE (C)

Fig. 3. Processing diagram for drying and aging for a 20N gel.

262

Li_2O-SiO_2 GELS

With control of processing variables such as R, pH, solvent concentration, and lithia and metal organic precursors, it is possible to produce 20L gel monoliths in 2 days or less. $LiOCH_3$ is the preferred alkoxide precursor for lithia, $LiNO_3$ is the preferred inorganic lithia precursor, and TMS is the preferred silica precursor for 20L gel monoliths.

ACKNOWLEDGMENT

The author gratefully acknowledges the U.S. AFOSR Contract # F49620-83-C-0072 for support of this work.

REFERENCES

1. Mackenzie, J. D. "Applications of Sol-Gel Methods for Glass and Ceramics Processing" in Ultrastructure Processing of Ceramics, Glasses and Composites, L. L. Hench and D. R. Ulrich, eds., J. Wiley & Sons, Inc., N.Y. 1984.
2. Zarzycki, J., "Monolithic Xero and Aerogels for Gel-Glass Process" in Ultrastructure Processing of Ceramics, Glasses and Composites, L. L. Hench and D. R. Ulrich, eds., J. Wiley & Sons, Inc., N.Y. 1984.
3. (a) Wang, S.H. and L. L. Hench, "Processing Variables of Sol-Gel Derived (20N) Soda Silicates"
    (b) Orcel, G. and L. L. Hench, "Use of a Drying Control Chemical Additive (DCCA) in the Sol-Gel Processing of Soda Silicate and Soda Boro Silicates"
    (c) Wallace, S. and L. L. Hench, "Metal Organic Derived 20L Gel Monoliths" all in Proceedings of American Ceramic Society Symposium on Composites and Advanced Materials, Cocoa Beach, FL, Jan. 15-18, 1984.
4. (a) Wang, S. H. and L. L. Hench, "Processing and Properties of Sol-Gel Derived 20 Mol %-80 Mol % SiO_2 (20N) Materials,
    (b) Wallace, S. and L. L. Hench, "The Processing and Characterization of DCCA Modified Gel-Derived Silica"
    (c) Orcel, G. and L. L. Hench, "Physical-Chemical Variables in Processing $Na_2O-B_2O_3-SiO_2$ Gel Monoliths" all in Proceedings of MRS 1984 Spring Meeting, Albuquerque, N.M., Feb. 27-29, 1984.

MECHANISMS AND KINETICS OF THE HYDROLYSIS AND CONDENSATION OF ALKOXIDES

H.SCHMIDT and H.SCHOLZE

Fraunhofer-Institut für Silicatforschung, Neunerplatz 2,
D-8700 Würzburg, F.R.G.

ABSTRACT

Hydrolysis and condensation of alkoxides involve different reaction steps. Generally the first step is the dissolution of monomers in organic solvents like alcohols. The second step is hydrolysis, where in most cases condensation may not be separated. Dissolution may incorporate solvatation, coordination, complexation or polymerization. The addition of water leads to hydrolysis of Si-O-C bonds and subsequently condensation of silanoles takes place. Another possibility of reaction is the hydrolysis-free condensation. As a function of reaction conditions and starting material different possibilities and mechanisms may take place. General conclusions are hardly to be drawn, since the comparison of data is difficult. In the paper theoretical models are reviewed and compared with experimental data. Conclusions with respect to the formation of solid materials are drawn.

1 INTRODUCTION

The traditional way to prepare inorganic polymers is the use of solid compounds which generally are polymeric (crystalline or amorphous). In order to perform these reactions, high temperatures have to be applied, since thermal cleavage of the polymeric bonds is necessary in order to form new compounds. In the field of organic polymers in opposition to this, the general way is to use monomeric compounds and to apply one of the common crosslinking mechanisms (polycondensation, polyaddition, polymerization). The possibility to produce inorganic polymers by one of these mechanisms exists, too. Generally it is the polycondensation me-

264

chanism, since polymerization or polyaddition at least in the
case of non-metallic inorganic materials based on the metal to
oxide bond is not possible, if one excludes methods like chemical
vapour deposition. A simple mean to form metal oxide bonds by a
polycondensation mechanism is the condensation of metal hydroxides (eq.1):

$$Me-OH + HO-Me \longrightarrow Me-O-Me + H_2O \tag{1}$$

The preparation of metal hydroxides may be performed very simple
for example by use of aqueous solutions of metal salts and changing of the pH to a value, where hydroxides may be precipitated.
This reaction is well-known and often used for the preparation of
synthetic powders for ceramics. The disadvantage of this reaction
normally is that the reaction control is very difficult. Another
reaction which allows a better control is the hydrolysis of alkoxides (eq.2):

$$Me-OR + H_2O \longrightarrow Me-OH + HOR \tag{2}$$

The main advantage of this route is the possibility to run the
reaction in a way that precipitation can be avoided and gels of
high homogeneity can be formed. The possibilities to form materials for practical use first was systematically shown by the
work of [1] and [2]. In the past many authors discovered the
interesting possibilities of this route known as the sol-gel process [3-10]. The process leads (starting with monomers, in general alkoxides) through hydrolysis, condensation, densifying and
mostly firing to dense materials. If one considers the hydrolysis
and condensation reaction of alkoxides, it does not make any basic difference, if the chosen system leads to glasses, glass ceramics or ceramics. Moreover, the basic considerations are valuable, too, for alkoxides containing, non-hydrolzying Si-C bond
ligands, a possibility to introduce organic radicals into inorganic structures.

2 FUNDAMENTAL CONSIDERATIONS

The reaction path roughly described above indicates that
very different reaction steps are included and that the total
reaction is a very complex one. The simple reaction route (eqs.
3 and 4):

$$Si(OR)_4 + H_2O \longrightarrow SiO_2 + 4 HOR \tag{3}$$

$$2 Al(OR)_3 + 3 H_2O \longrightarrow Al_2O_3 + 6 HOR \tag{4}$$

does not represent in any way this complexicity. The pure ther-
modynamical standpoint leads to the conclusion that crystalline

SiO_2 and Al_2O_3 should be the final products. In reality in both cases at first amorphous gels are formed as a consequence of the crosslinking reaction of the reactive intermediates.

It is well known that, as a function of reaction conditions, the properties of these gels, especially with respect to their ability to be dried and densified, differ strongly [11-14]. The difference of these properties must be due to the different structures formed during the first step of polymer formation, that means, the first step of condensation. This fact leads to the question at which stage of the reaction stable structures may be formed or at which stage of the reaction parameters occur which influence the formation of these structures. Another question is to which extent these structures are able to survive the whole processing and determine properties of the formed materials.

A scheme which points out the main processing steps which not necessarily have to be carried out altogether, starting from hydrolysis and ending up with a molded material is given in table 1.

Table 1. Scheme of possible reaction steps of the material formation during the sol-gel process (The dots indicate to which extend the single reaction steps may expand.)

1. Hydrolysis (in solution) ...
2. .....Condensation ...
3. Drying.............
4. Densification.....   material
5. Sintering......
6. .... Crystallization...
7. ....Molding

Table 1 indicates that the different reactions are not necessarily chronological, but partly may take place at the same time, e.g. sintering and crystallization or condensation and molding (fiber drawing). In principle all these steps may include reactions with parameters influencing the material properties.

If one looks into the details these reaction steps become even more complex and the state of art with respect to the connection between reaction parameters and material properties shows a lack of data. One of the main reasons for this is the sensitivity of these reactions to reaction conditions, like solvents, catalysis, the role of water, and so on. This sensitivity makes it very difficult to compare data of different authors. In this paper the attempt is made to point out the basic problems of reaction mechanisms and reaction kinetic in the two first main steps of the sol-gel process, hydrolysis and condensation.

3 HYDROLYSIS AND CONDENSATION

3.1 Solutions of One-Component Systems

A one-component system, like $Si(OR)_4$ or $Al(OR)_3$, has the advantage that an interaction of two or more different components cannot take place. The first step generally performed to make materials by the sol-gel process from alkoxides, is to put the components into a solvent like alcohol. The question arises, if there are structure-forming reactions in solution. There are different possibilities: solvatation, coordination, coordination polymerization, ligand-exchange, and polycondensation.

$$Me(OR)_4 + HOR' \rightarrow \quad \text{solvatation} \quad (5)$$

$$Me(OR)_4 + HOR' \longrightarrow \quad \text{coordination} \quad (6)$$

$$Me(OR)_4 \longrightarrow \quad \text{coordination polymerization} \quad (7)$$

$$Si(OR)_4 + HOR' \longrightarrow R'O-Si(OR)_3 + HOR \quad \text{ligand exchange} \quad (8)$$

$$Si(OR)_4 + Si(OR)_4 \longrightarrow (RO)_3Si-O-Si(OR)_3 + ROR \quad \text{condensation} \quad (9)$$

Equation (6) and (7) only may take place, if the coordination number is increased. This is unlikely with Si under these conditions, but well-known with Al. Higher coordination numbers are not required with equation (5), (8), and (9), at least not, if one considers the over-all reaction only. The question of the formation of complexes in solution is an important one, since the complex formation may influence the hydrolysis rate strongly, and, if coordination polymerization structures are formed, they may be partially kept up in the polymers formed by condensation. Coordination polymerization is unlikely with tetraalkoxysilanes. At least viscosity measurements done by our own with mixtures of tetraethoxysilane and ethanol show an additive behaviour. But with aluminum alkoxide as shown by Bradley [15], a strong tendency of coordination polymerization is observed. These reactions strongly should be influenced by the type of central atom (e.g. Si, Al, Ti, B), the ligand (length of the chain "R"), and the solvent. Unfortunately there are not much structural details known about these reactions. This is understandable, since the evaluation of structures of liquids is rather difficult. Equation (5) (solvatation) should take place with most solvents, if alkoxides are dissolved. The reaction number (8) (ligand exchange) is described by different authors [16-18]. But almost no kinetic data are known. The reaction is important, since the rate of hydrolysis depends strongly on the type of ligand, that means, if an alkoxide is dissolved in an alcohol, where the alcoholic OR is not identical with alkoxide OR, ligand exchange will take place and the hydrolysis rate will change with time. The mechanisms of this reaction may be described by equations (10) and (11).

$$
\text{a) } Si(OR)_4 \longrightarrow \underset{OR \quad OR}{\overset{OR}{Si^+}} + RO^-
$$

(10)

$$
Si^+(OR)_3 + HOR' \longrightarrow R'OSi(OR)_3 + H^+
$$

$$
H^+ + RO^- \longrightarrow HOR
$$

$$
\text{b) } \underset{RO}{\overset{OR}{RO-Si-RO}} + R'OH \longrightarrow \underset{RO \quad OR}{\overset{R' \quad OR}{HO \rightarrow Si \rightarrow OR}} \longrightarrow H^+ + R'OSi(OR)_3 + RO^-
$$

(11)

$$
H^+ + RO^- \longrightarrow HOR
$$

Reaction (10) requires a trivalent transition state, reaction
(11) requires a pentavalent one. Different authors [19-21] postulate a pentavalent transition state, but there is no experimental proof for orthoesters. The stability of a trivalent sili-
conium ion depends very strongly on the type of ligand, e.g. the
+I-effect of these ligands. The pentavalent state should lead to
a change of conformation. This type of reaction was investigated
by Sommer et al. [21], but with optically active organoalkoxysilanes. But is not quite clear, as far as the reaction of organoalkoxysilanes may be transferred to alkoxysilanes. The reaction
(9) does not take place at room temperature [22], but at higher
temperatures it may add remarkably to the condensation.

3.2 Solution of Multicomponent Systems

In this case interaction between different types of alkoxides
has to be considered. Especially as a function of difference in
lewis acidity, complexation of two different components may occur
as it takes place, if one adds a Me(I) alkoxide to a Me(III) alkoxide as indicated in equation (12).

$$Me^I OR + Me^{III} OR \longrightarrow Me^{I+} \left[Me^{III}(OR)_4 \right]^- \qquad (12)$$

This very easily takes place e.g. with sodium alkoxide and aluminum or boron alkoxide. The reactivity, in case of hydrolysis of
these complexes, should be different to that of single components
so that a non-additive behaviour should occur. There are rather
good imaginations of structures of different (especially transition metals) alkoxides in solutions developed by Bradley [15].
But there are almost no data on interactions of mixtures of glass
and ceramics forming alkoxides and the influence on processing
and properties of these interactions.

3.3 Reaction with Water

The most common step for the performance of the sol-gel process is to add water to the solution of alkoxides. In the case of
a Me(I) alkoxide like sodium alkoxide, hydrolysis reaction is a
very simple one. In the case of multifunctional elements like
silicon a 4-step reaction has to take place for total hydrolysis
as it is shown in equation (13).

$$(13)$$

In this reaction mixture five different types of monomers are possible as they are indicated with the numbers I-V. For each reaction step different rate constants (K_I to K_{IV}) are possible. Since there is no necessity for identical reaction rates, and if these reaction rates are different there must be a population profile which is changing the time and which defines the number of reactive monomers. If the condensation rate is proportional to the number of reactive polymers, the formation of different structures occurs with time.

In order to get an imagination on mechanisms and since no data are available, one has to consider closer the basic steps of these reactions. The first question which arises is, what can happen, if the OR is changed to OH. The first, who investigated this question in detail, was Khaskin [23]. He used ^{18}O in order to find out, if a dissociative or a non-dissociative substitution of an OR group by an OH group takes place. His results show that with alkoxysilanes a dissociative reaction occurs.

$$\equiv Si^{18}OR + H_2O \longrightarrow \equiv Si-OH + H^{18}OR \qquad (14)$$

This dissociative reaction takes place with acid and basic catalysis and with organosilanes. But in the case of acyl ligands the reaction mechanism is a non-dissociative one.

$$\equiv Si-^{18}O-\underset{O}{\overset{\parallel}{C}}-R + H_2O \longrightarrow \equiv Si-^{18}OH + HO-\underset{O}{\overset{\parallel}{C}}-R \qquad (15)$$

From these facts one can follow that in the case of alkoxides, where an OH group exchanged by an OR group, a nucleophilic attack of OH$^-$ ion (basic catalysis) or an H_2O molecule (acid catalysis) takes place. In the case of alkoxides this attack aims to the silicon atom, since the silicon atom carries the highest positive charge (eqs. 16a and b).

$$\equiv \underset{\underset{H_2O}{\uparrow}}{Si}-^{18}O\overset{\overset{H}{\diagup}}{\diagdown}_R \longrightarrow \equiv Si-OH + H^{18}OR + H^+ \qquad (16a)$$

$$\equiv \underset{\underset{OH^-}{\uparrow}}{Si}-^{18}OR \xrightarrow{+H_2O} \equiv Si-OH + H^{18}OR + OH^- \qquad (16b)$$

In the case of the acyl compound the nucleophilic attack aims to the carbonyl carbon atom as the carrier of the highest positive charge.

As indicated in equation (16a) in the case of proton catalyzed reactions, an oxonium complex will be formed by the protonation of the oxygen atom of the OR group. This increases the
positive charge of the Si-O bond of the silicon atom and enables
the H_2O molecule despite of its low nucleophilic power to attack
the silicon atom. The dissociation of this transition complex
leads to the cleavage of the Si-O bond, while a new Si-O bond
from the H_2O molecule will be formed. In the case of the basic
catalysis the OH⁻ ion is able to attack the silicon atom directly
as a consequence of its higher nucleophilic power. These mecha-
nistics can be deducted from the results of Khaskin, but this
does not allow to say anything about the conformation of the
transition state. But it is possible to deduct prognosises with
respect to parameters which may influence the reaction kinetics.
That means that all facts which increase concentration of transition complexes from equation (16a) or (16b) should increase the
reaction rate. These parameters should be the concentration of
H_2O, if there are no superposing effects, the concentration of
catalyst and parameters who increase the positive charge of the
silicon atom. Since the concentrations of water and the catalyst
might be chosen, the electronic state of the silicon atom is dependent on the type of the ligands. It is well known that the
length of an alkyl chain determines the +I (electron pumping)
effect. Longer chains show stronger +I effects. That means that
the reaction rate should decrease as a function of the chain
length, if the other parameters are kept constant. Measurements
of Aelion et al., Akerman, Schmidt et al. [24-26] show that the
hydrolysis and condensation rate decrease with increasing chain
length. Another explanation for this might be the steric hindrance of longer chains, which makes it more difficult for a
water molecule to attack the silicon atom. The results of Aelion
and Akerman are given in table 2 and 3.

Table 2. Dependance of hydrolysis rate on the chain length of
R ($\equiv$SiOR)

R	$k \cdot 10^2$, liters$\cdot$mole$^{-1} \cdot$sec$^{-1} \cdot H^{+-1}$
C_2H_5	5.1
C_4H_9	1.9
C_6H_{13}	0.83
$\begin{matrix} CH_3 \\ \searrow \\ CH_3 \end{matrix} CH(CH_2)_3CH(CH_3)CH_2$	0.30

Table 3. Dependance of gelling time on the chain length of
R (≡SiOR)

Formula	Hydrolysis time up to gel formation, hr
$(C_2H_5O)_4Si$	2
$(n-C_4H_9O)_4Si$	32
$(CH_3(CH_2)_6O)_4Si$	25
$(CH_3CH_2CH(CH_3)CH_2O)_4Si$	75
$(n-C_4H_9O)(t-C_4H_9O)_3Si$	236
$(sec-C_4H_9O)_4Si$	500

Figure 1 shows the hydrolysis rate of tetraethoxysilane and te-
tramethoxysilane with different proton and water concentrations.
The surprising effect of this experiment is that the hydrolysis
rate decreases with increasing amounts of water. This might be
due to the special experimental conditions and the water acting
as a proton acceptor which decreases the proton activity.

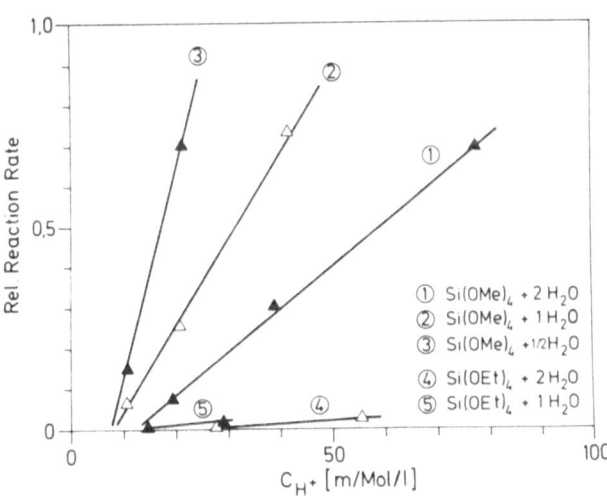

Fig.1 Dependence of hydrolysis rate on type of silane, H_2O
and proton concentration; solvent: ethanol; volume ratio
solvent: silane = 1:1; T = 25°C; catalyst: HCl

Extensive investigations on the hydrolysis rates of different silanes the first time were carried out by Aelion and coworkers. They found a second-order reaction for the hydrolysis of tetra-alkoxysilanes in dioxane as solvent, if excess of water is used (eq.17).

$$\frac{dx}{dt} = k.c_{H_2O}.c_{silane} \qquad\qquad \begin{array}{l} x = \text{number of reacting} \\ \quad\ \ \text{silane molecules} \end{array} \qquad (17)$$

In the basic catalyzed case a first-order reaction was observed (eq.18).

$$\frac{dx}{dt} = k.c_{silane} + C \qquad (C = \text{constant}) \qquad (18)$$

These rate laws cannot be generalized, since they are collected under very special experimental conditions and the general validity is not proved yet.

The rate dependence of k follows equation (19).

$$k = 0.051 \cdot c_{H^+} \qquad (19)$$

Aelion also points out that the extent of hydrolysis depends on the H^+ concentration, that means that there is a pH dependent equilibrium which indicates that the role of the proton is not only a catalytic one. Measurements of Schmidt et al. [26,27] do not support these data, since there the hydrolysis equilibrium was found to be independent of the proton concentration. But these results are obtained with understochiometric amounts of water and in ethanol as solvent. So these results are not quite comparable. Aelion postulates a pentavalent transition state without giving experimental proof for this.

The question, if a pentavalent or trivalent transition state is possible, is very difficult to be answered, since direct experimental data are lacking. From the work of Sommer [21] with trialkylalkoxysilanes with optical activity it is shown that hydrolysis in the most cases does not lead to racemization, but in some cases it does (eqs.20,21).

$$(-)R_3Si^*OCH_3 \xrightarrow[\text{NaOH}]{H_2O \text{ in acetone}} (+)R_3Si^*OH \ \begin{array}{l}(\text{change of} \\ \text{conformation})\end{array} \qquad (20)$$

$$(-)R_3Si^*OCH_3 \xrightarrow[\text{CH}_3\text{OH}]{CH_3ONa (10^{-3}\text{moles/l})} \begin{array}{l}\text{racemization;} \\ \text{too fast to measure}\end{array} \qquad (21)$$

That means, in the second case the racemization is only explain-
ble, if a trivalent transition state will be formed. But the
question cannot be answered, if these results obtained from
alkylalkoxysilanes are valid for tetraalkoxysilanes.

The influence of an alkyl ligand on the silicon atom should
be of a +I type. But since in the case of the oxygen silicon bond
a partially π-bond is possible, in the case of a Si-C bond it is
not since carbon does not have any π-orbitals. So it is very difficult to estimate the differences between the electronic effects
of OR or R bonds on the silicon atom.

Since there is an experimental proof that with increasing
chain length of the OR groups the hydrolysis rates slows down,
one can think about the influence of the substitution of an OR by
an OH group in the tetraalkoxides of silicon. The OH group exhibits a lower +I effect as an OCH_3 or OC_2O_5 group. So a partially
hydrolyzed silicon tetraalkoxide should have a higher positive
charge on the silicon atom and should show an increased hydrolysis rate, compared with the silane containing one OR group more.
On the other hand, the probability of the hydrolyzation of an OR
group in a silicon group is proportional with the number of OR
groups. So that even if the rate constants are the same, one OR
group of the tetraalkoxysilane would be hydrolyzed as much as one
OR group of a silane containing two OR groups. The question is
which distribution of monomers, as a function of reaction condition and time, will occur and, if this distribution may be influenced in order to receive useful properties. The hydrolysis
reaction of tetraalkoxysilane which leads to monomers with only
two hydrolyzed OR groups are able to form linear polymers (in
analogy to dimethyldiethoxysilane). This type of polymers should
be very useful for fiber drawing, and investigations of Sakka [9]
show that under certain reaction conditions, especially with acid
catalysis and understochiometric amount of water, spinable polymers are formed, indicating a certain linearity. Rheological measurements support this hypothesis. From Brinker et al. recently
[28] experimental data were given, pointing out that hydrolysis
and condensation rates depend very strongly on pH over a wide
field, and in addition to that two types of structures could be
evaluated as a function of pH (Fig.3 and 4). These experiments
were carried out with tetraethoxysilane. From Schmidt et al. similar results were obtained (Fig.5) with alkylalkoxysilanes and
alkoxysilanes [29]. Brinker points out that in the acid case a
more linear chain growth takes place then in the ammonia catalyzed, indicating that in the acid case monomers with two con-
densable (OH) groups are formed preferably. But up to now no mechanistical hypothesis exists in order to explain these facts in
a satisfying way. This theory of Brinker may be supported by work
of Bechtold et al. [30], who concludes from light scattering

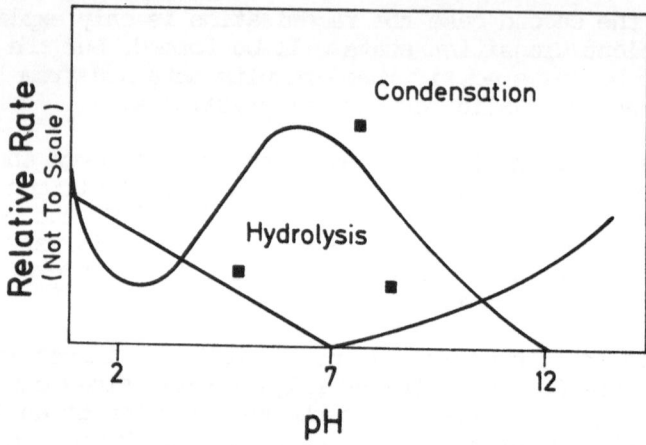

Fig.3 Rate dependence of hydrolysis and condensation of tetra-alkoxysilane on pH

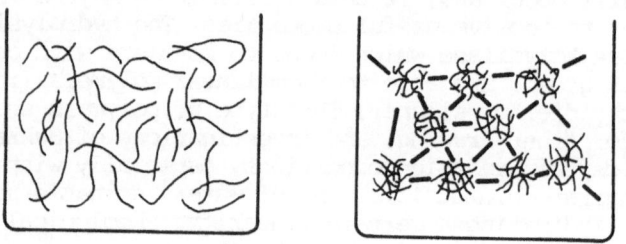

Fig.4 Structure evaluation of gels from (a) acid and (b) basic catalysis of hydrolysis and condensation of tetraalkoxy-silane

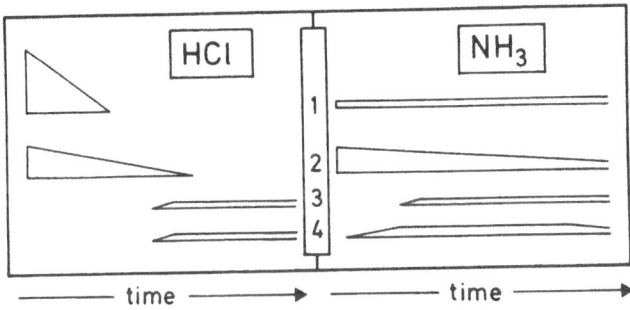

Fig.5 Scheme of chronological order of the single reaction steps
of gel formation with HCl and NH$_3$ catalysis. The ordinate
expansion of the bars refers to the reaction rate.
Hydrolysis of (CH$_3$)$_2$Si(OC$_2$H$_5$)$_2$ (1) and Si(OC$_2$H$_5$)$_4$ (2)
Condensation of (CH$_3$)$_2$Si(OC$_2$H$_5$)$_2$ (3) and Si(OC$_2$H$_5$)$_4$ (4)

experiments and statistical calculations that units of the type

$$
\begin{array}{ccc}
OH & OH & OH \\
| & | & | \\
Z-Si-O-Si-O-Si-Z & & \\
| & | & | \\
OH & OH & OH \\
\end{array}
\qquad Z = OH, \; OR
$$

exist.

One possibility to follow the hydrolysis reaction of alkoxysilanes in one-component systems would be to monitor the methyl-
ene protons by NMR spectroscopy. Hopeful attempts were made by
Assink et al. [31], who are able to distinguish between the me-
thylene protons of an alkoxysilane and of alcohols. Up to now the
resolution of the NMR experiments is not high enough to distinguish quantitatively between the different hydrolyzed steps.

A behaviour similar to tetraalkoxysilane is reported from tetraalkoxytitanium [6,15,32]. The hydrolysis rates of titanium alkoxide are far higher than those of the analogous silicon compounds. Unfortunately only few data on the hydrolysis and condensation rates of other alkoxy compounds are available. The knowledge of the relative rates of the different components is very important in multicomponent systems, since the concentration of the polymerizable monomers is a function of the hydrolysis rate of their starting compounds. By this, the structure of a polymer can be influenced very strongly. If one compares e.g. the system of tetraethylsilicate and tetraethyltitanate in ethanol and one adds water, titanium oxide will precipitate and the ethyl silicate will hydrolyze only very slowly. By variation of reaction conditions it is possible to avoid precipitation of titanium and to get homogeneous looking gels, but there is no proof, how homogeneous they really are. There might be structure elements which mainly consist of titanium oxide chains, as a consequence of the higher hydrolysis and condensation rate of Ti-alkoxides compared to the Si.

Experiments of Schmidt and Scholze [27] showed that in a borosilicate system sodium alkoxide and boron alkoxide show the far quickest hydrolysis rate. If the three-component system is hydrolyzed, non-additive behaviour occurs indicating some inter action between the single components, where a type of complexation (formation of a sodium borate complex) is probable. This shows again that structures may be formed in a very early stage.

3.4 Reaction without Water

As indicated in equation (9) in principle it is possible that alkoxysilanes react to a polymer network without the addition of water. Two basic types of reaction may be distinguished: There is the direct reaction of two Si-OR groups which under special conditions may form a Si-O-Si group under the formation of ether. This reaction takes place at temperatures of about 200°C and is described by Noll.

Another type of reaction is described by Voronkov et al 16 where the solvent is involved (eq.22). In this reaction, no water is added to the reaction mixture, but water is formed as an intermediate:

$$\equiv Si-OR + HX \longrightarrow \equiv Si-X + HOR$$

$$ROH + HX \longrightarrow RX + H_2O$$

$$\equiv Si-OR + H_2O \longrightarrow \equiv Si-OH + ROH \qquad (22)$$

$$\equiv Si-X + H_2O \longrightarrow \equiv Si-OH + HX$$

$$2\equiv Si-OH \longrightarrow \equiv Si-O-Si \equiv + H_2O$$

The condensation reaction then is the reaction of the Si–OH group with one Si–OR group under the formation of alcohol. Schwarz et al. [33] described a reaction where by the use of alkoxides with longer chains Si–O groups under the formation of olefines are described. These Si–O groups then lead to condensation with other OR groups. Mazdiyasni et al. [3] used this reaction for the formation of reactive powders from different alkoxides. This reaction takes place at temperatures around 200°C and may be used for the formation of polymers based on the metal oxide bond starting from alkoxides and without using water.

Another possibility to form a inorganic polymer structure is the silanolysis reaction as described by Schott et al. [34]. They found out that especially phenylalkoxysilanes may be used for this reaction where the Si–O group attacks the silicon carbon bond which is cleaved by the formation of an silicon oxygen bond.

$$\equiv Si\text{-}OH + C_6H_5\text{-}Si \equiv \longrightarrow \equiv Si\text{-}O\text{-}Si \equiv + C_6H_6 \qquad (23)$$

This reaction in special cases may be used, too, for the formation of polymers.

Reaction conditions without the use of water can be successfully applied in all cases, where the additon of water leads to complications, like precipitations, inhomogenities based on extremely differnt reaction rates or the demand of low water contents in gels.

4 CONCLUSIONS

If one starts to think about the formation of the materials by the sol-gel process and wants to use alkoxysilanes as starting materials, a couple of mechanistical questions arises. The first is the correlation of reaction rates and material properties. Therefore detailed data of the reaction rates, hydrolysis and condensation rates of different alkoxysilanes should be available. Unfortunately only a few data on alkoxysilanes are known and since they depend on very special reaction conditions, they are not able to be generalized. So the present state is that each researcher has to work out his own set of reaction data.

In order to be able to use the nice tool of a sol-gel processing for the material preparation better than it is possible in the moment, these data should be evaluated as soon as possible. Table 4 gives a survey on the present state over this field.

Table 4. Reaction Parameters Influencing Hydrolysis and Condensation

Parameter	Effect	Data available	Consequences
Type of Metal	Rate Structure	Only on Si alkoxides	Kinetics should be determined, especially on other alkoxides
Hydrolyzable Ligand	Rate	Detailed investigations often on organoalkoxy-silanes	Reaction rates can be influences
Solvent	Equilibrium Rate	—	Determines number of unhydrolyzed residues, structure of gels
Water	Equilibrium	Only on Si alkoxides	Degree of hydrolyzation structure of glass, data on others required
Catalyst	Rate Branching Structure of gels	Only on Si alkoxides (kinetics) Structure models evaluated as a function of H + and OH -	Data on others required
Temperature	Rate Structure?	—	Data to be evaluated
Vapor Pressure (H$_2$O)	Rate (at low pressures), Structure (at high pressures)	On Si alkoxides and on aerogels	More data could provide good means for tayloring

5 REFERENCES

1 Roy R., J.Amer.Ceram.Soc. 52 (1969) 344.

2 Dislich H., Angew.Chem. 83 (1971) 428.

3 Mazdiyasni,K.S., R.T.Dolloff and J.S.Smith, II.
   J.Amer.Ceram.Soc. 52 (1969) 523.

4 Mukherjee,S.P. and J.Zarzycki. J.Mater.Sci. 11 (1976) 341.

5 Yoldas,B.E. J.Mater.Sci. 12 (1977) 1203.

6 Kamiya,K. and S.Sakka. Res.Rep.Fac.Eng.Mie Univ. 2 (1977)
   87.

7 M.Nogami and Y.Moriya. J.Non-Cryst.Solids 37 (1980) 191.

8 Carturan,G., V.Gottardi and M.Graziani. J.Non-Cryst.Solids
   29 (1978) 41.

9 Sakka,S. and K.Kamiya. J.Non-Cryst.Solids 43 (1980) 403.

10 Mackenzie,J.D. J.Non-Cryst.Solids 48 (1981) 1.

11 Yoldas,B.E. J.Non-Cryst.Solids 51 (1982) 105.

12 Brinker.C.J., K.D.Keefer, D.W.Schaefer and C.S.Ashley.
   J.Non-Cryst.Solids 48 (1982) 47.

13 Partlow,D.P. and B.E.Yoldas. J.Non-Cryst.Solids 46 (1981)
   153.

14 Zarzycki,J. J.Non-Cryst.Solids 48 (1982) 105.

15 Bradley,D.C. et al. Inorganic Polymers. (Academic Press, New
   York 1962)

16 Voronkov,M.G., V.P.Mileshkevich and Y.A.Yuzhelevskii. The
   Siloxane Bond (Plenum, New York, London 1978).

17 Andrianov,K.A. Organic silicon compounds (State
   Sci.Publ.House for Chemical Literature, Moscow, 1955)

18 Sheefer,D.W. Proc.Mater.Res.Soc.Spring Meeting 1984,
   Albuquerque, USA

19 Swain,C.G., R.M.Esteve Jr and R.H.Jones. J.Amer.Chem.Soc. 71
   (1949) 965.

20   Keefer,M.D. Proc.Mater.Res.Soc.Spring Meeting 1984,
     Albuquerque, USA

21   Sommer,L.H. and C.F.Frye. J.Amer.Chem.Soc. 82 (1960) 3796.

22   Noll,W. Chemie und Technologie der Silicone. 2.Auflage
     (Verlag Chemie, Weinheim, 1968).

23   Khaskin,I.G. Dokl.Akad.Nauk SSSR 85 (1952) 129.

24   Aelion,R., A.Loebel and F.Eirich. J.Amer.Chem.Soc. 72 (1950)
     5705.

25   Akerman,E. Acta Chem.Scand. 10 (1956) 298; 11 (1957) 298.

26   Schmidt,H. and A.Kaiser. Glastechn.Ber. 54 (1981) 338.

27   Schmidt,H.,H.Scholze and A.Kaiser. J.Non-Cryst.Solids 48
     (1982) 65.

28   Brinker,C.J. et al. Proc.Mater.Res.Soc.Spring Meeting 1984,
     Albuquerque, USA.

29   Schmidt,H., A.Kaiser and H.Scholze. J.Physique 43 (1982)
     275.

30   Bechtold,M.F., R.D.Vest and L.P.Plambeck Jr.
     J.Amer.Chem.Soc. 90 (1968) 4590.

31   Assink,R.A. Proc.Mater.Res.Soc.Spring Meeting 1984,
     Albuquerque, USA.

32   Kamiya,K. et al. J.Mater.Sci. 15 (1980) 1765.

33   Schwarz,R. and K.G.Knauff. Z.anorg.allg.Chem. 275 (1954)
     176.

34   Schott,G. and H.Berge. Z.anorg.allg.Chem. 297 (1958) 32.

CHAPTER IV: OPTICAL APPLICATIONS OF GLASS

CHALCOGENIDE GLASSES FOR OPTICAL APPLICATIONS

J A Savage

Royal Signals and Radar Establishment, Malvern, UK

ABSTRACT

In the past 20 years chalcogenide glasses have been actively researched in order to determine their suitability as passive bulk optical component materials for 3-5 μm and 8-12 μm infrared applications and as active electronic device components in photocopying and electronic switching applications. Much theoretical and experimental work has been done which has led to a greater understanding of the glass forming and general physical properties of these materials. This review concentrates upon the optical and general properties of these bulk chalcogenide glasses.

1. INTRODUCTION

The chalcogenide glasses are so named because they contain one or more of the chalcogenide elements S, Se or Te together with one or more of the elements Ge, Si, As, Sb and a number of others. They are mainly covalently bonded materials with room temperature resistivities between 10^3 and 10^{13} ohm cm. For instance As_2S_3 has a resistivity of $\sim 2 \times 10^{12}$ ohm cm (130°C), As_2Se_3 a resistivity of $\sim 1.5 \times 10^8$ ohm cm (130°C), Se a resistivity of $\sim 2 \times 10^4$ ohm cm (120°C) and As_2SeTe_2 a resistivity of $\sim 3.5 \times 10^3$ ohm cm (130°C). The conductivity activation energy of these glasses varies from 0.3 eV to 1.25 eV while the optical energy gap approximates to that of the crystalline analogues where these exist (Edmond 1968).

For optical applications the chalcogenide glasses possessing the higher resistivities mainly sulphides, selenides, and mixed selenide tellurides are utilised and during the 1950s major work

first centred on arsenic trisulphide as an optical material for the
near and middle infrared. This work led to commercial exploitation
and arsenic trisulphide is now well known as an optical component
material. During the early 1960s to the early 1970s it was shown
that selenide glasses and mixed selenide-telluride glasses were
suitable for optical component applications in the far infrared and
the commercial exploitation of these glasses is now established.
Generally the sulphides offer some limited visible transmittance,
while the selenides and tellurides are opaque in the visible part
of the spectrum. For useful thicknesses of a few millimetres,
sulphides offer transmittance from 0.7 to 12 μm, selenides from 1.0
to 15 μm and tellurides 2.0 to 20 μm. The infrared refractive
indices are in the range 2 to 3 and the reciprocal dispersive power
V_{8-12} ranges from around 100 to 200 depending upon the glass composition. Reported infrared absorption coefficients range from
4×10^{-1} to 7×10^{-3} cm^{-1} depending upon the wavelength, purity and
chemical composition. Extrinsic absorption can be a problem with
these optical glasses and in particular oxygen impurity must be
kept $\leqslant$ 1 ppm in the final product to avoid excessive absorption
particularly between 8 and 12 μm.

While the optical and electrical properties of the chalcogenide glasses are reasonably well known, other properties such as
thermal conductivity, hardness, the elastic moduli and the mechanical properties are less well known. There is some information on
how these latter properties vary with chemical composition amongst
the sulphide (Tsuchihashi et al 1968) and selenide glasses (Tille
et al 1977)(Michels and Frischat 1982). For instance Hilton et al
(1975) have shown that amongst Ge-As-S glasses knoop hardness can
vary from 200 to 280 Kg/mm^2 while Young's Modulus varies from
20×10^9 Nm^{-2} to 41×10^9 Nm^{-2} and amongst the selenide glasses the
knoop hardness can vary from 100 to 200 Kg/mm^2, while Young's
Modulus varies from 14×10^9 Nm^{-2} to 27×10^9 Nm^{-2}. The fracture
toughness of Ge-As-Se glasses ranges from 5.5 to 9.4 Nmm$^{-3/2}$
(Michels and Frischat 1981a) while the four point bending strength
varies from 1 to 2.5×10^7 Nm^{-2} (Michels and Frischat 1981b). The
thermal conductivity can vary from 14×10^{-4} W cm^{-1} $^{\circ}$C^{-1} for pure
selenium glass to 38×10^{-4} W cm^{-1} $^{\circ}$C^{-1} for a glass of composition
Ge35As40S25 (Hilton et al 1975). Clearly these glasses are much
less physically and thermally robust than oxide glasses but nevertheless still retain sufficiently acceptable thermal and mechanical
properties to be used as optical components with the exception of
window components interfacing with rugged environments.

2. THE NEED FOR INFRARED OPTICAL MATERIALS

It is fortunate that the two most common gases in the atmosphere, nitrogen and oxygen, are homopolar and possess neither a
permanent nor an induced dipole moment, and hence do not exhibit

infrared molecular vibrations which would result in the absorption of infrared radiation. In the initial region of the infrared spectrum from 0.75 μm to 14.0 μm the absorptions of the minor atmospheric constituents, water vapour and carbon dioxide result in three main 'windows' in the atmosphere (Kruse et al 1962), one from 0.75 μm to 2.5 μm (near infrared), another from 3.0 μm to 5.0 μm (middle infrared), and a third from 7.5 μm to 14.0 μm (far infrared). From the black body spectral emittance curves shown in Fig 1 it is clear that to detect relatively hot objects the 3.0 μm to 5.0 μm window is most suitable and to detect objects at room temperature the 7.5 μm to 14.0 μm (8 μm to 12 μm hereafter) window is most suitable. There is a major interest in thermal surveillance systems at present and these practical uses of infrared radiation are concerned with wavelengths up to about 12.0 μm. In order to focus this infrared radiation onto detection systems, windows, lenses and telescopes are required made from materials exhibiting adequate infrared transmission in this wavelength region. Generally the requirements for infrared transmitting materials are set primarily by the atmospheric transmission and secondarily by the operational wavelength range of the sources and detectors and by the power handling requirements of particular systems. For low power surveillance or thermal imaging applications where chalcogenide glasses are useful, materials are required in sizes up to 150 mm diameter and up to 20 mm thick. At present the requirements for optical component materials tend to be for single thermal band operation ie 3-5 μm or 8-12 μm but it is possible that future requirements may call for multi thermal band operation eg 3-5 μm and 8-12 μm. Thus the reader so far will have gained some insight as to how requirements for infrared optical materials arise and the likely component sizes necessary. The next points to consider are those governing the transparency of insulators and semiconductors to understand why chalcogenide glasses are suitable as infrared optical materials.

3. LOSS MECHANISMS IN INFRARED MATERIALS

Metallic materials not exhibiting an energy gap possess a high density of free electrons in their structure. These electrons are of such low inertial mass that they can freely respond to electromagnetic radiation over a wide frequency range and thus metals are opaque in significant thicknesses. Insulators and semiconductors with energy gaps > 0.1 eV will exhibit some near, middle and far infrared transmission within their transparency range. The short wavelength cut off frequency in these solids is determined by electronic transitions across the band gap while the long wavelength cut off results from lattice absorptions due to the vibrational modes of the atoms or ions of the solids. The chalcogenide glasses can be classified amongst this latter group of substances.

When electromagnetic radiation is incident upon and passes through an insulator various loss mechanisms operate. Some of the radiation is reflected at the interfaces between a solid and its environment. The amount reflected is determined by the refractive index of the solid and that of the medium in which it is immersed. This is a basic property of the material but may be partially overcome by means of anti-reflection coatings applied to the surfaces of the solid. Some of the radiation may be scattered at the surface of the solid and/or in the bulk. The surface scattering is likely to be due to lack of adequate care in surface preparation, but bulk scatter can arise from defects, inclusions or perturbations in the refractive index particularly in a complex solid consisting of several atoms of different mass as is the case with chalcogenide glasses. Some of the radiation may be absorbed at the surface of the solid or in the solid. The mechanisms which give rise to bulk absorption may be classified as intrinsic or extrinsic ones. The intrinsic absorption mechanisms are those which result in electronic and vibrational structural absorptions in vitreous material of specific chemical composition. Extrinsic mechanisms are those associated with impurity atoms and molecules and deviations from stoichiometry. The intrinsic mechanisms define the region of transparency to electromagnetic radiation in a solid and the ultimate transmission achievable within this region, while the extrinsic mechanisms generally determine the percentage of the theoretical level of transparency achievable in practice within this region.

3.1 Intrinsic Absorption Mechanisms

Intrinsic absorption mechanisms in a solid define its region of transparency to electromagnetic radiation. In order to transmit infrared radiation a solid must be an insulator or a semiconductor exhibiting an energy gap Eg (ie the nominal energy to excite bound electrons to a conduction band). Only materials possessing a band gap larger than the infrared wavelengths of interest (0.75 to 14.0 μm) need be considered since it is the band gap that sets the transmission limit at the short wavelength end of the spectrum of a solid as illustrated in Fig 2. This short wavelength cut off is given by the relationship shown in equation 1.

$$\lambda_c = \frac{hc}{Eg} \tag{1}$$

where h = Planck's constant and c = velocity of light.

The low frequency tail of this short wavelength cut off extends slightly into the transparent region of the solid. This is known as the Urbach tail (Urbach 1953)(Hopfield 1968) and is of the form shown in equation 2.

$$\beta \propto e^{CW/KT} \tag{2}$$

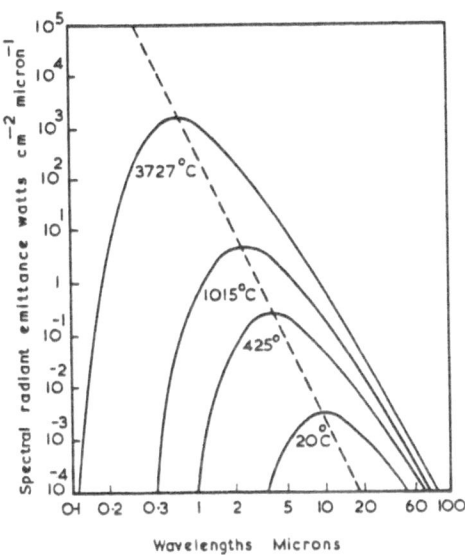

Figure 1   Black body spectral emittance curves.

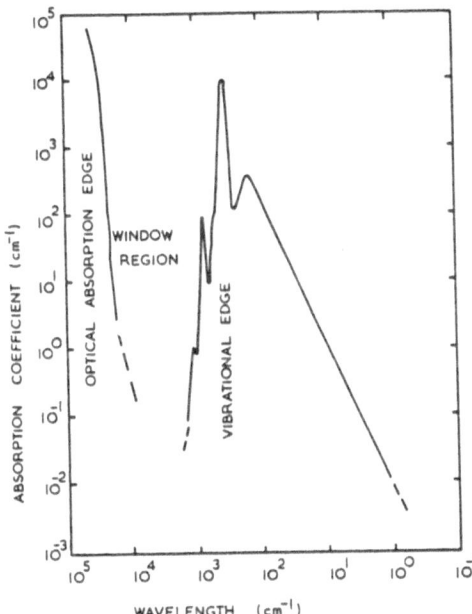

Figure 2   A plot of absorption coefficient against wavelength showing intrinsic absorption in As_2S_3 glass.

where W = frequency, K = Boltzmann's constant, T = absolute temperature and β = absorption coefficient.

This exponential tail would only be of major significance for most infrared applications if it was very close to one of the infrared atmospheric 'windows' discussed in para 2. An explanation of this exponential tail in the UV interband edge has been recently put forward. It is known that the random structure in glasses gives rise to varying local electric fields on a microscopic scale. Theoretical work by Dow and Redfield (1971)(1972) and by Tauc (1975) provide evidence that such local microfields cause intrinsic absorption in chemically pure materials in what is normally the transparent region below the fundamental interband absorption edge. The mechanism is due to local field induced broadening of exciton levels which are created in optical absorption energies close to but below the interband edge.

The long wavelength cut off in a solid is set by lattice absorptions. These lattice absorptions result from the vibrational modes of the atoms in the solid. In ionic crystals vibrations of large amplitude will occur when incident radiation is of the same frequency as the resonant frequency of the atomic units and this is termed the reststrahl frequency. These atomic units must possess a permanent dipole moment which can be activated by the oscillating electric field of the incident radiation. Non polar solids would be expected to be infrared inactive but usually exhibit induced dipole effects. For example a non ionic solid can have an effective charge and thus a dipole moment if the atoms are not identical. The fundamental absorption frequency can be calculated for a linear polar diatomic molecule consisting of two point masses m_1 and m_2. The frequency of vibration V, of the simple harmonic motion of the two masses along a line joining them is given by equation 3 (Kittel 1956).

$$V = \frac{1}{2\pi} \left(\frac{K}{M}\right)^{\frac{1}{2}} \tag{3}$$

where K is the force constant and M is the reduced mass $\left(\frac{1}{m_1} + \frac{1}{m_2}\right)$.

If an anharmonic oscillator is considered then in addition to the fundamental frequency a series of overtone vibrational bands arise. The long wavelength cut off in a solid is usually set by the first overtone of the fundamental lattice absorption. In the case of multi element chalcogenide glasses of somewhat indeterminate structure equation 3 can be used to provide general rules as to the likely long wavelength transmission limit. For instance it is clear that the smaller the force constants or the weaker the bonding in the glass, and the larger the atomic masses in the glass and the higher the coordination number then the lower will be the frequency of the fundamental absorption. Thus other things being equal selenide glasses will transmit farther into the IR than sulphide glasses and telluride glasses will transmit farther than sulphide or selenide glasses.

Recently several investigations have shown how the infrared absorption decreases as the frequency becomes much greater than the fundamental lattice absorptions and overtone frequencies in alkali halides (Mitra and Bendow 1975) and semiconductors (Deutch 1975). Highly purified samples of these materials exhibit an absorption coefficient, β, which reduces exponentially (Bendow 1975). This exponential tail can be represented by the expression shown in equation 4.

$$\beta (W) \approx Ae^{-\gamma W} \tag{4}$$

where β = absorption coefficient, A and γ are material dependent parameters and W = frequency.

In such materials multiphonon absorption takes place when a high energy phonon couples weakly with a transverse optical mode of the material. This transverse optical mode decays into two or more lower energy phonons of frequencies corresponding to fundamental vibrational modes. The most probable multiphonon absorption process involves production of the minimum number of final state phonons (Pohl and Meier 1974). For the alkali halides the one phonon density of vibrational states is sizeable over a broad range of frequencies and this has been shown (Pohl and Meier 1974) to lead to a predicted exponential absorption spectrum in the multiphonon region at ambient temperature in agreement with the experimental results. In the case of chalcogenide glasses such as As_2S_3, As_2Se_3 and GeS_2, a molecular model for the vibrational properties has been proposed (Lucovsky and Martin 1972). In this model it is presumed that the AsY_3 pyramidal groups or the GeY_4 tetrahedral groups are only loosely coupled to one another by the bridging chalcogenide atoms. As a consequence of this the infrared spectra are expected to correspond to those of isolated AsY_3 or GeY_4 molecules superimposed on less intense spectra due to the bridging As-Y-As or Ge-Y-Ge groups. The one phonon density of states is predicted by this molecular model to consist of a collection of discrete vibrational modes broadened slightly due to the amorphous state. This means that the multiphonon absorption processes in chalcogenide glasses should be analogous to combination and overtone bands in isolated molecules and experimental results have been shown to be in agreement with this model (Mitra and Bendow 1975).

Thus the fundamental absorption processes that limit the transparency range of chalcogenide glasses are due to electronic transitions across the band gap at short wavelengths and lattice vibrations at longer wavelengths. A sufficiently wide spectral window exists between these two limits, where these materials are usefully transparent as shown in Fig 2 where the absorption coefficient of As_2S_3 is plotted as a function of frequency. The first region of Fig 2 shows a sharply rising absorption at the electronic band edge followed by a relatively transparent region ($\sim$ 2000 to 10,000 cm^{-1}). Multiphonon absorption occurs between 500 cm^{-1} and 2000 cm^{-1} and one phonon peaks occur between 100 cm^{-1} and 500 cm^{-1}. This is followed by a region extending up to microwave frequencies where the absorption varies approximately as the frequency squared (Strom et al 1974). Within the window illustrated by Fig 2 the transparency is limited by the tails of the optical and vibrational absorption edges, possibly free electron absorption in some chalcogenide glasses and by intrinsic scatter.

3.2 Intrinsic Scatter Mechanisms

From thermodynamic considerations some degree of intrinsic scatter is likely in all homogeneous infrared optical glasses due to the natural perturbations in the refractive index caused by compositional fluctuations frozen in during the glass synthesis. Scattering theory is complex (Stacey 1956) but three cases of wavelength dependence can be distinguished. If the scattering centres are $\lesssim \lambda$ then Rayleigh scattering theory can be used and the backward scatter is $\alpha \; \lambda^{-4}$; if the scattering centres are $\approx \lambda$ Mie forward scattering theory which is a complex function of λ can be used; and if the scattering centres are $\gg \lambda$ then the scattering can be described as non selective and is independent of λ. It can often be difficult to identify and distinguish between scatter inducing defects in materials but a number of informative measurements can be made. The wavelength dependence at a fixed angle can provide data on the size of the inhomogeneity responsible for the scatter. The angular dependence of the scatter may also aid in identifying the relative size and shape of the inhomogeneity. Measurements of polarised scattering can provide data on strain induced inhomogeneities. Much useful data on scatter problems is published and referenced in SPIE (1982).

3.3 Extrinsic Loss Mechanisms

Extrinsic loss mechanisms determine the percentage of the theoretical level of transparency achievable in a glass in practice at certain wavelengths within its available window region. These mechanisms are scatter and absorption arising from the fabrication process employed and the chemistry of the glass in relation to certain specific impurities. For instance OH, hydrocarbons and oxygen in chalcogenide glasses can give rise to unacceptable absorptions at specific wavelengths and inadequately homogenised glass can give rise to unacceptable scatter in the high frequency regions of the near and middle infrared. Chalcogenide glasses can be synthesised from the melt or by a distillation process and components can be manufactured by a pressure slumping or cutting and grinding process utilising a variety of containment vessels such as silica, carbon or metal. Since the extrinsic loss mechanisms are specific to a particular glass or family of glasses and the method of synthesis they will be discussed for individual materials. In this way an appreciation of the problem of extrinsic loss in relation to particular applications will be obtained. An indication of the likely absorbing species is given in Fig 3 and their effects on the transmittance of chalcogenide glasses is shown in Fig 4. Useful works of reference to find the absorbing frequencies of these impurities are Miller and Wilkins (1952), Nyquist and Kagel (1971) and Nakamoto (1963).

4. SULPHIDE GLASSES

Most commercial optical glasses produced for use in the UV and visible region of the spectrum exhibit effective transmission in the 0.75 μm to 2.5 μm wavelength region hence there is no need to develop special materials for this region. However these optical glasses are not effectively transparent beyond 2.5 μm due to the Si-O bond vibration (Adams 1960), and special materials have been developed for use in the 3-5 μm spectral band which are also transparent in the near infrared and sometimes completely or partially transparent in the visible. The need for these special materials for optical applications first stimulated research on sulphide glasses and resulted in arsenic trisulphide being developed as a bulk optical material.

Arsenic trisulphide is the most widely known and used sulphide glass. It was manufactured in the USA (USP 1957) and in the UK during the mid 1950s and has been found useful ever since that time as an optical component in many 3-5 μm optical systems. The manufacture of this glass to optical standards set new technical problems since arsenic and sulphur are toxic and volatile and their oxides act as major extrinsic impurities thus reducing the infrared transmission of the material. In a bold departure from the then current glass technology practice the volatility of these elements was used to advantage in that the raw arsenic metal and sulphur were first prereacted in a closed steel or silica vessel to form a crude solid which was broken to a small particle size and then heated in silica apparatus under an inert atmosphere to such a temperature ($\geqslant$ 700°C) that final reaction, melting and distillation took place. The vapours were condensed at a low enough temperature (300-500°C) to maintain the As_2S_3 in a liquid state but high enough ($\geqslant$ 193°C) to keep As_2O_3 in its vapour state. Thus the bulk of the oxide contamination present in the starting materials was swept out of the still as SO_2, SO_3 and As_2O_3. This synthesis was operated as a batch process with batches of 3-5 Kg being collected, mixed by stirring at 625°C, then finally annealed. Components were either

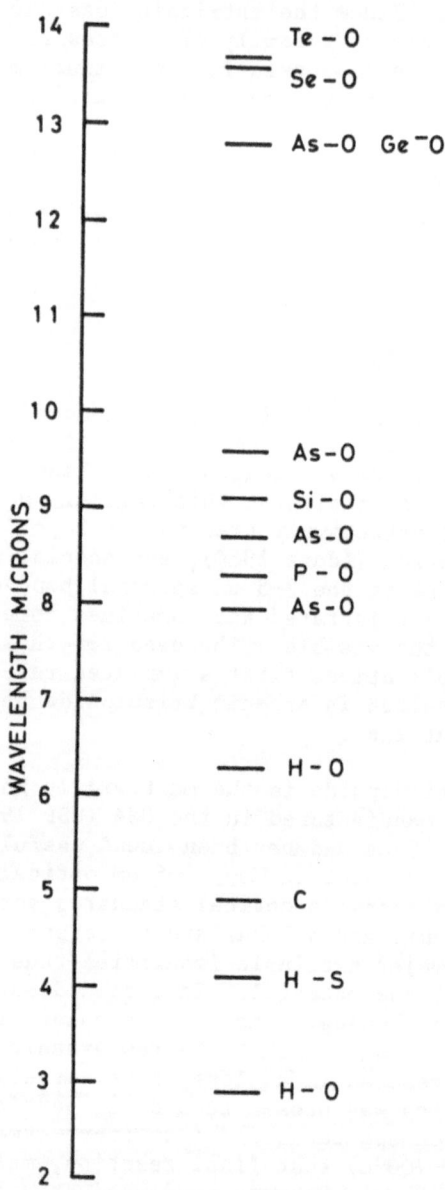

<u>Figure 3</u> An indication of the extrinsic absorptions seen in chalcogenide glasses.

cut or heat slumped from the glass boules. This material being required for low power applications was not produced to a very high standard of purity. Commercial grade arsenic and sulphur were used so that cation impurities were present in the material as well as traces of H_2S and oxides. Nevertheless the quality was adequate for 3-5 μm low power applications and a stock is still available (Billard and Cornillault 1962). A particular production problem in such a distillation process was maintaining constancy of chemical composition and therefore of optical properties in the finished product. The material produced by the ICI company in the UK was designated A or B (Billard and Cornillault 1962) depending upon whether the distillate was collected at the beginning or end of a batch distillation. Fibre optic components such as fused faceplates and light pipes have been made from As_2S_3 with a cladding glass containing a different sulphur concentration to provide the required refractive index difference. These have demonstrated effective transmission (⩾ 40%) in lengths up to 8 cm between 1 and 5 μm (Kapany 1967). The full potential of the material was not realised since the glasses contained extrinsic impurities such as Fe, H_2O and H_2S. Some of the physical properties of type B As_2S_3 are given in Table 1 and 2 and refractive index data on USA material is given in the literature (Rodney et al 1958). Details of the vibrational absorption edge in As_2S_3 are reported by Onomichi et al (1971) and these are broadly in agreement with the data shown in Fig 2. V_{TO} is given as 301 cm^{-1} and V_{LO} as 339 cm^{-1}. A number of other uses have been proposed for As_2S_3 glass such as the moisture passivation of NaCℓ components (Young 1970), and as a transparent optical film on $LiNbO_3$ active substrates for integrated optics applications (Klein 1974).

For those interested in glass formation data the following indicates some of the information available for sulphide glasses; As-Tℓ-S (Flaschen et al 1960a), As-I-S (Flaschen et al 1960b), Ge-As-S (Savage and Nielsen 1965a), As-Te-S (Kolomiets 1964), Ge-P-S, Si-Sb-S (Hilton et al 1964) and Ge-Sb-I-S (Turjanitsa et al 1972). The optical properties of sulphide glasses depend markedly upon chemical composition. Hence for the present purposes it has been decided to take the example of the Ge-As-S system which is one of the most useful from the point of view of bulk optical glass manufacture. It has been shown that Ge-As-S glasses (Savage and Nielsen 1965a) demonstrate excellent infrared transmission from 1.0 to 11.5 μm as indicated in Fig 4 and that the wide composition range allows glasses with differing optical, thermal and mechanical properties to be prepared. Binary sulphide glasses can be made containing 10% to 40% As or 15% to > 30% Ge and ternary sulphide glasses can be made containing as little as 30% S. Compositions containing more than 60% of sulphur are transparent in the visible, the position of the cut on wavelength depending upon the arsenic germanium/sulphur ratio. Glasses containing less than 60% sulphur show little infrared fine structure between 7 and 13 μm, but still show the strong Ge-S or As-S absorption between 11 and 15 μm as is seen in Fig 4. Materials containing more than 60% sulphur show a complex infrared spectrum. Between 7 and 15 μm the number of absorption bands increases as the sulphur content of the glasses increases. There is good agreement between the positions of these absorption bands when extrinsic oxygen absorption has been removed and the absorption bands observed in crystalline sulphur due to combination tones of the S_8 fundamental absorptions (Bernstein and Powling 1950). Glasses from which to manufacture optical components would be best chosen from those containing ⩽ 60% of sulphur. These materials are likely to yield a transmission curve similar to that shown in Fig 4. The dashed curve in this figure shows clearly the problem of extrinsic absorption due to water, oxide and carbon impurities. If a typical component is say 10 mm thick then an examination of Fig 4 shows that the level of impurity indicated by the dashed curve would destroy the useful infrared transmission of the component. Impurities such as these need to be controlled at ⩽ a few ppm wt in order to achieve an adequate transmission. Fuxi et al (1983) have reported devitrification and property studies in the Ge-As-S glass system. They found that because of the stable glass formation in this system devitrification was difficult mainly occurring at low ~ 10% As content. The frequency of the main IR absorption peak (cm^{-1}) was given for several ternary glasses, Ge30 As10 S60 378(s) 330(m), Ge30 As15 S55 375(s) 325(m), Ge25 As15 S60 378(s) and Ge20 As20 S60 375(s) 330(s). Some optical property data for four ternary sulphide glass compositions is given in Table 1 and some other physical property data is given in Table 2. Structural and physical property data for other glasses in the Ge-As-S system is reported by Andreichin et al (1976). Three useful general reviews of the optical properties of chalcogenide glasses including sulphide glasses appear in the literature (Savage and Nielsen 1965b) (Hilton 1966) (Hilton 1970).

Silica fibres for optical communications have been developed to their practical limit of ~ 0.2 dB/Km in the wavelength range 1.3 to 1.55 μm. Residual intrinsic losses are due to Rayleigh scatter proportional to λ^{-4} and particularly due to the increasing phonon loss in SiO_2 based materials. The use of an alternative material with a phonon loss edge situated farther into the infrared would allow fibre operation at longer wavelength and also result in lower Rayleigh scatter losses thus offering the promise of losses perhaps as low as 0.001 dB/Km (Tebo 1983). Thus there is a major drive to develop a mid IR (~ 4 μm) optical communications glass fibre from the recently discovered fluoride glasses reviewed by Drexhage et al (1982). However a number of workers are considering sulphide glasses for this or shorter length applications in the mid infrared. The first glasses to be considered (Kapany 1967) were arsenic sulphides particularly As_2S_3 glass but as previously mentioned the purity of these was insufficient for them to be considered for anything other than very short length applications such as fused

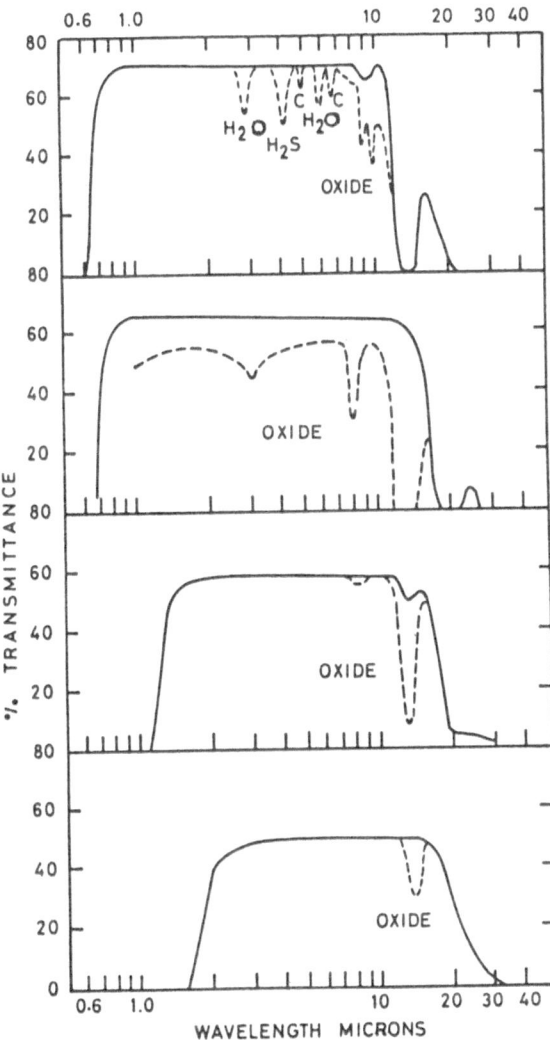

Figure 4 The infrared transmittance of chalcogenide glasses

a) sulphide glass Ge30 As20 S50 1.86 mm thick.
b) selenide glass Ge34 As8 Se58 1.80 mm thick.
c) selenide-telluride glass Ge30 As13 Se27 Te30 2.3 mm thick.
d) telluride glass Ge10 As50 Te40 1.62 mm thick.

The effect of extrinsic impurity is indicated by the dashed curves.

faceplates. Vasiliev (1983) has reported that the minimum optical loss wavelength of As_2S_3 is near 5.15 μm and that the material dispersion is zero at 4.89 μm and does not exceed 20 πs/nm Km in a wide spectral region around this zero dispersion wavelength. Hence this and other sulphide glasses are promising candidates for mid infrared optical communications applications. Dianov (1982) has considered the Rayleigh scattering and the infrared absorption loss of As_2S_3 glass and has predicted a minimum loss of ~ 0.05 dB/Km in the wavelength region of 4 to 5 μm. CO laser calorimetry of high purity bulk glass indicated the absorption losses in glass to be ~ 70 dB/Km. Miyashita and Terunuma (1982) have drawn unclad fibre from high purity rods of As_2S_3 glass and have found that the optical loss was limited by impurity absorption (eg SH at 4.1 μm) but nevertheless a loss of 170 dB/Km was measured at 5.25 μm. Kanamori et al (1983) have reported fibre losses of 64 dB/Km at 2.4 μm for As_2S_3 glass and Katzir and Arieli (1982) indicate that losses of 1 dB/m at 5 μm have been measured in a Ge-S glass. Shibata et al (1980) have shown that optical losses in Ge-P-S glass may be as low as 10^{-1} to 10^{-2} dB/Km at 5.5 μm. Thus it remains to be seen whether there is sufficient further interest in these chalcogenide glasses for mid infrared applications to allow their theoretical loss levels to be realised.

5.   SELENIDE GLASSES

As can be seen from Fig 4 bulk sulphide glasses do not fully cover the far infrared spectral band. Hence the attention of researchers looking for glasses for use in this band was directed towards selenides. Elemental selenium was known to be a glass former and to transmit over the required spectral range. Further work on the absorption and reflection spectra (Vasko 1965) of pure selenium glass confirmed this. Selenium and its derivatives in thin film form have been used in a multi-billion dollar photocopying industry and this glass must be regarded as the single most important chalcogenide in terms of commercial exploitation. It has also been used in rectifying applications and as a photovoltaic detector for visible radiation. However, because of its poor general physical properties it was found wanting for bulk optical applications. This stimulated research on selenide glass formation and an indication of some of the information available is as follows: As-Tℓ-Se, As-S-Se (Flaschen et al 1960a), As-Sb-Se, As-Tℓ-Se (Kolomiets 1964), Ge-As-Se (Kolomiets 1964)(Savage and Nielsen 1964), Ge-P-Se, Si-Sb-Se (Hilton et al 1964) and Ge-Sb-Se (Hilton et al 1966a). The physical properties of selenide glasses like those of sulphide glasses depend upon the chemical composition (Hilton and Hayes 1975) but the most work has gone into glasses in the Ge-Sb-Se and Ge-As-Se systems from which commercial bulk optical materials have been manufactured. Hence this review will mainly concentrate on the bulk optical properties of these glasses.

First of all it must be borne in mind that when compared to oxide
optical glasses, the chalcogenide glasses are classified as weak
materials with low softening temperatures. Therefore it is particularly important that the thermal and mechanical properties are
optimised. During the mid 1960s a useful review was published
(Savage and Nielsen 1965b) which indicated that selenide glasses
with acceptably high glass transition temperatures > 150°C could be
made. It was also established that the majority of the absorptions
exhibited by chalcogenide glasses between 1 and 6 μm and 8 and
13 μm resulted from traces of H_2O, H_2Se and other oxide impurities
all of which could be eliminated from the glasses if sufficient
care was taken during the preparation and processing of these
materials. The transmission curve of a Ge-As-Se glass showing the
positions of some of the oxide impurity bands is given in Fig 4.
Further reviews were published yielding much more detailed information on physical properties (Hilton et al 1966a), additional
information on absorption by oxide impurities (Hilton and Jones
1966) and investigations of the atomic structure of selenide and
other chalcogenide glasses (Hilton et al 1966b). At the time these
four reviews neatly summarised the general physical property data
of most known chalcogenide glasses and indicated that in principle,
glass compositions with physical properties suitable for 8-12 μm
requirements were possible. In addition the requirement was for an
optical glass to correct chromatic aberration in 8-12 μm germanium
lens systems. The refractive index of the glass was required to be
~ 2.5, the reciprocal dispersive power, $V_{8-12} = (n_{10} - 1)/(n_8 - n_{12})$,
was required to be ~ 100. Other requirements were that the refractive index be maintained within the range $\pm$ 0.005 to $\pm$ 0.0001 in a
batch of material and from batch to batch of material, the T_g
$\geqslant$ 150°C, the mechanical strength and hardness were required to be
high and the thermal expansion was required to be low. Hence work
was done to establish the detailed physical properties of Ge-Sb-Se
(Hilton and Hayes 1975)(Savage et al 1978) and Ge-As-Se (Webber and
Savage 1976)(Savage et al 1977) glasses to enable industrially
makeable materials to be identified. From this work it became
clear that all of the requirements could only be met by a glass
containing of the order of 30% Ge and 10 to 20% As or Sb. The
refractive indices of several Ge-As-Se glasses are given in Table 1
so that the effect of As/Ge content and ratio on the refractive
index and reciprocal dispersive power can be seen. Other physical
properties are given in Table 2. However much work has been done
particularly in relation to extrinsic absorption problems on the
glass Ge28 Sb12 Se60 in the USA and glass Ge30 As15 Se55 in the UK
both of which have been produced commercially. Selenide glasses
are synthesised from the elements inside sealed evacuated silica
tubes at temperatures of the order of 950°C (Ford and Savage 1976)
in quantities of 25-100 gm in research and in quantities up to
several kilogrammes in production. The glasses are then annealed
and cooled to room temperature for use, or heat slumped into
particular component shapes followed by further annealing and

cooling. This sealed tube process has the merit of retaining compositional integrity, but requires extrinsic impurity free starting materials. An alternative technique of distillation of lower grade elemental material in hydrogen gas was investigated and proved to be practical for a Ge-As-Se ternary glass (Kettlewell et al 1977) for batches of 1.5 Kg. However the batch to batch compositional variation and hence refractive index variation was too great for the present requirement so that emphasis was placed on solving the extrinsic impurity problems associated with the sealed tube process.

Inspection of Fig 4 reveals the importance of oxide removal from the raw materials and reaction vessel surfaces in the sealed tube process. A synthesis technique was evolved which reduced the oxide impurity level in Ge-As-Se glasses to the order of 1 ppm wt. Essentially, this was to remove surface oxide from the silica reaction tube and the Se and As raw materials by baking them in a vacuum. The reactants and tube together with the germanium were then subsequently handled in an argon glove box until the reaction tube was finally evacuated and sealed prior to glass melting (Savage et al 1977). The effect of oxygen impurity on the absorption coefficient of two selenide glasses is shown in Fig 5. The oxygen levels were measured by a gamma photon activation analysis technique (Savage et al 1977).

Similar purification techniques were adopted for Ge28 Sb12 Se60 glass but with this material the purification process was taken further to establish the absorption limit for the glass and to lower the absorption at 10.6 μm (Hilton et al 1975). After removal of oxygen and carbonaceous matter it was shown that the transmission was limited by "silica" in the glass originating from the ampoule sealing process. In further studies (Rechtin et al 1975) after eliminating extrinsic absorption effects it was shown that by reducing the fraction of Ge-Se bonds in the glass composition to those in Ge23.5 Sb18 Se58.5 glass a value of absorption coefficient at 10.6 μm of 8×10^{-3} cm^{-1} could be achieved.

Russian workers have provided refractive index data for Ge-As-S-X glasses where Z = S, Sn, Te, Tℓ, Pb and Sb and came to the conclusion that the experimental results support the view of chalcogenide glasses as covalently bonded polymers with saturated valence bonds (Aio et al 1978). In an interesting quantitative study of infrared absorption in the 250-4000 cm^{-1} region of As_2Se_3 glass (Moynihan et al 1975) it was demonstrated that three distinct oxide impurity species existed in the glass from the relative intensities of the extrinsic absorption bands. Oxide bands at 1125 cm^{-1} and 650 cm^{-1} were assigned to oxide incorporated in the As_2Se_3 network, bands at 1050, 1265, 1340 and 785 cm^{-1} were assigned to As_4O_6 molecules dissolved in the glass and a band at 965 cm^{-1} was considered to be separately but not unambiguously

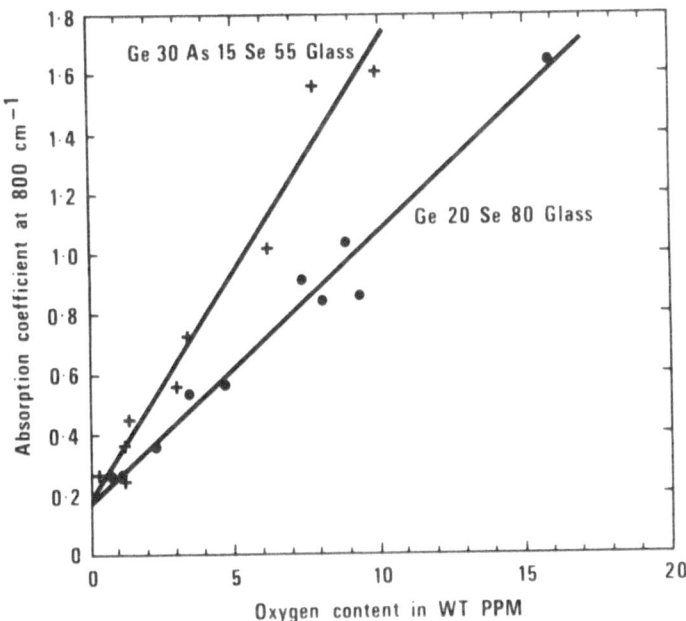

Figure 5 A plot of absorption coefficient at 800 cm^{-1} against oxygen extrinsic impurity content for two selenide glasses.

assignable. It was also concluded that the absorption at 10.6 μm was limited by intrinsic multiphonon processes to a value of the order of 10^{-2} cm^{-1}. Selenide glasses in the Ge-As-Se and Ge-Sb-Se systems were considered to be sufficiently structurally similar to possess similar multiphonon absorptions and hence similar absorption coefficients at 10.6 μm. This is in broad agreement with the data given above for Ge-Sb-Se glasses and that below for Ge-As-Se-Te glasses.

Glasses Ge20 Sb12 Se60, Ge33 As12 Se55 (Hilton 1978)(Hilton et al 1975) and Ge30 As15 Se55 have been produced in batch quantities of several kilogrammes by the sealed tube method from semiconductor grade raw materials. After synthesis the glass boules are either annealed and cooled or sometimes subjected to a further homogenisation process (Hilton 1970) before annealing and cooling to room temperature. Components are then either cut from the glass or heat slumped to shape from slices of the glass. The main problem in achieving optical homogeneity requirements has been found to be in maintaining chemical compositional homogeneity after synthesis and mixing during the subsequent cooling process. This is due to the fact that the vapour composition is not necessarily the same as that of the liquid glass. While this is not a problem for experimental melts of the order of 100 gm it becomes more serious for melts of several kilogrammes since the available vapour space above such melts is greater and the temperature gradients tend to be larger. If any condensate of different chemical composition to the bulk is allowed to contaminate the liquid glass after homogenisation during the cooling process before annealing, then the viscosity of the glass is still sufficiently low for limited intermixing and hence local compositional variations to occur. Such variations in composition can lead to refractive index variations of the order of 3×10^{-3} or greater. However, these problems have now been largely overcome (Worrall 1979). Large 200 mm dia and 50 mm thick cast glass plates of high quality for submarine window applications have recently been reported by Stachiw and Bertic (1982) made by the Hilton process (Hilton et al 1975). This indicates that there is no fundamental problem in attaining good optical homogeneity if good working practices are maintained.

Kapany (1967) reported that chalcogenide glasses were potential materials for fibre applications where transmittance up to 14 or 15 μm was required. Since that time much data on glass formation and physical property measurement has become available and a number of workers are now investigating selenide glasses as candidate fibre materials. Bornstein et al (1982) synthesised As_2Se_3 glass by the sealed tube technique and unclad fibres 100-500 μm in diameter and up to 20 m long were drawn from the melt at speeds of 0.5 to 5 m/min inside a glove box containing an inert atmosphere. Using a CO_2 laser and pyroelectric detector the loss in the fibre was found to be 0.1 dB/cm. These authors considered that improved losses could be achieved by further reducing extrinsic absorption and by modifying glass composition to minimise intrinsic phonon absorption. Brehm et al (1982) have synthesised plastic clad fibres from Ge30 As15 Se55 glass. Rods 10 mm in diameter and 80 mm long were made from glass synthesised by the sealed tube technique and these were used for fibre drawing at 400° in an argon atmosphere to avoid oxidation. Fibres 200 μm in diameter and 100 m in length were drawn at a speed of 10 m/min. The fibres were coated in polyolefin plastic and placed in a heat shrinkable polyethylene tube to improve the handling characteristics. The packaged fibre either 1.8 or 3 mm OD had a minimum bending radius of 30 mm, a breaking strength $>$ 1 daN and an optical loss in the 4–11 μm band of $\sim$ 10 dB/m. Takahashi et al (1983) have reported a loss of 4.5 dB/m at 10.6 μm for a selenide glass teflon FEP clad fibre of composition As38 Ge5 Se57 made from oxide impurity reduced raw materials.

Katsuyama et al (1982) have disclosed the synthesis of solid or hollow core fibre of high stability and low optical loss from selenide glass deposited by the MCVD inside tube deposition process. An object of the work was to avoid contamination from impurities in the raw materials and from the containing vessel by avoiding prolonged melting. Argon gas carrying $GeCl_4$, $SbCl_5$ and Se_2Cl_2 was passed through a lead glass substrate tube of composition mole % SiO_2 57, Na_2O + K_2O 12, PbO 30 and of dimensions 12–13 mm ID, 14 mm OD. Heating to 600°C during deposition was achieved using a traversing oxyhydrogen burner. After a glass of composition mole % Ge 28, Sb 12, Se 60 was deposited in the tube it was collapsed and drawn into a conventional fibre or hollow core fibre at 800°C. The hollow core fibre had a measured loss of 0.7 dB/m at 10.6 μm and the conventional fibre had a loss of 0.1 dB/m. When a core glass of composition mole % Ge 28, Sb 12, Se 60 and a cladding glass of composition mole % Ge 23.5, Sb 12, Se 64.5 were deposited in a lead glass tube and drawn into a solid fibre as before, the measured loss at 10.6 μm was 0.01 dB/m. This patent claim represents a breakthrough in reducing the loss in chalcogenide glass. If this achievement can be repeated for a commercial cabled product, selenide glass fibres are very promising for far infrared applications.

6.   MIXED SELENIDE TELLURIDE GLASSES

Having established basic industrial products in the Ge-Sb-Se and Ge-As-Se glass systems researchers turned their interest to examining the possibilities of extending the range of optical properties in terms of refractive index but particularly of increasing the reciprocal dispersive power, V_{8-12}. Clearly from Table 1 the composition As40 Se60 offers a useful increase in V_{8-12} but at the expense of rather poor thermal and mechanical properties.

Edmond (1968) suggested that selenide and telluride glasses were compatible and that it would be possible to make high quality stable mixed selenide-telluride melts. Muir and Cashman (1967) also indicated that a glass of composition (Ge Se Te)92 As8 was thermally stable, and the refractive index at 10 μm was found to be approximately 2.71. Savage et al (1980) considered that work on mixed selenide-telluride glasses offered the most useful approach to extending the range of optical properties of the basic selenide glasses for 3-5 μm and 8-12 μm applications. Additions of tellurium to the Ge-Sb-Se system rapidly led to devitrification problems therefore it was decided that the Ge-As-Se system would offer the most stable glasses in which to substitute tellurium for selenium in order to retain sufficiently robust thermal and mechanical properties for the intended applications (Savage et al 1980). The glasses in the Ge-As-Te (Savage 1971) system possess lower values of T_g and are much less thermally stable than those in the Ge-As-Se system (Savage and Nielsen 1966). Hence tellurium additions to the latter glasses would be expected to decrease their glass transition temperatures and reduce their thermal stabilities so that very stable base glasses containing 20 to 30% germanium and 10 to 30% arsenic were initially chosen for tellurium substitution. About 40 glass melts were made and analysed by differential thermal analysis and on the basis of the thermal properties one base glass Ge30 As13 Se57 was chosen for investigation of optical properties. It was found that $\leqslant 30\%$ Te could be substituted for Se in this glass but amounts of Te $> 30\%$ caused devitrification. A further extrinsic impurity problem was encountered with these quaternary glasses in that Te cannot be purified from TeO_2 by thermal baking since the vapour pressure of the metal is greater than the oxide. In this case an acid etching technique (Savage et al 1980) was developed which allowed glass of adequate purity to be made by the sealed tube technique. Several melts of glass Ge30 As13 Se27 Te30 were analysed for oxygen by a gamma photon activation technique and within the measured range of 0 to 10 ppm wt the oxygen content of this glass was found to be given by expression 5.

$$Y = 0.078 + 0.182 \ x \qquad\qquad (5)$$

where x = oxygen content in ppm wt
      Y = absorption coefficient at 780 cm^{-1}

An absorption coefficient of 7×10^{-3} cm^{-1} was obtained for this glass at 10.6 μm by laser calorimetry which correlates very well with the values given for Ge23.5 Sb18 Se58.5 and As40 Se60 glasses in 5.0. Refractive index and reciprocal dispersive power data are given in Table 1 where it can be seen that the range of V_{8-12} is 110 to 185 when the base glass is substituted with up to 30% Te for Se. All of these glasses possess very acceptable general physical properties for infrared optical applications as can be seen in Table 2. The effect of Te substitution on the short wavelength end of the spectral window is to move the absorption edge from 0.6 - 0.7 μm for a glass containing no tellurium to 1.1 - 1.2 μm for a glass containing 30% Te

A good general review of mixed $As_2(Se\ Te)_3$ glasses containing some data on optical properties is given by Thornburg (1973), and this complements data already given in this review on As_2S_3 and As_2Se_3 glasses. McLauchlan and Gibbs (1977) have reported thin film data for mixed Se-Te glasses.

7.   TELLURIDE GLASSES

In the early 1960s before selenide glasses had become established preliminary research was also conducted on telluride glasses as alternative materials for 8-12 μm applications. As is seen in Fig 4 telluride glasses were shown to transmit farther into the infrared than sulphide or selenide glasses and in the 8-12 μm spectral region appeared to be less prone to multiphonon absorptions and in the case of Ge-P-Te and Ge-As-Te glasses also less prone to oxide absorptions. This work and the later work for switching glass applications in the early 1970s has resulted in much information on glass formation, an indication of which is as follows: Si-As-Te (Hilton and Brau 1963), As-I-Te (Peck and Dewald 1964), Ge-As-Te, Ge-P-Te (Savage and Nielsen 1966)(Hilton et al 1966a), Ge-Te, As-Te (Savage 1972a), Si-Ge-As-Te (Savage 1972b) and quaternaries based on Si-As-Te (Anthonis et al 1973/74). From this work some conclusions can be drawn concerning telluride glasses. The major glass forming region was found in the Si-As-Te system while the Ge-P-Te and Ge-As-Te systems only offered minor glass glass forming regions. The effect of oxygen extrinsic impurity on the transmission of Si-As-Te glasses (Hilton et al 1966c) was similar to that in the case of sulphide and selenide glasses. Trace oxide impurity has a much less deleterious effect on the transmission of Ge-As-Te and Ge-P-Te (Savage and Nielsen 1966) glasses as is noticeable in Fig 4. This was perhaps due to a restricted solubility in the case of Ge-As-Te glasses and a negligible solubility in the case of Ge-P-Te glasses. These latter facts are very advantageous but the limited glass forming regions means that commercial production is likely to be difficult and the achievement of a family of glasses with a range of optical properties unlikely. Germanium raw material can be incorporated in chalcogenide glasses readily at 950°C with no reaction tube corrosion problems. On the other hand silicon is difficult to incorporate into a glass melt in the sealed tube process since the glass melting temperature of around 1000°C is considerably lower than the melting point of silicon and in addition attack on the silica melt tube is observed. Therefore in spite of the fact that the Si-As-Te glass forming region is large these glasses are difficult to melt and homogenise and very difficult to make with low intrinsic oxygen impurity due to corrosion of the silica melt tubes. Due to these problems and the fact that selenide glass manufacture became established no further work was done towards telluride glass production for optical applications. Comprehensive data on the properties and structure of As-Te glasses are given in the literature (Cornet and Rossier 1973a,b,c) and this complements the data already referred to on As_2S_3 and $As_2(SeTe)_3$ glasses.

REFERENCES

Adams, R.V. 1960 The absorption of infrared radiation and the structure and constitution of various oxide glasses. PhD Thesis, Dept Glass Technology, Sheffield University.

Aio, L.G., Efimov, A.M. and Korkorina, V.F. 1978 J Non Cryst Solids 27 299.

Andreichin, R., Nikiforova, M., Skordeva, E., Yurakova, L., Grigorovici, R., Manaila, R., Papescu, M. and Vancu, A. 1976 J Non Cryst Solids 20 101-122.

Anthonis, H.E., Kreidl, N.J. and Ratzenback, W.H. 1973/74 J Non Cryst Solids 13 13.

Bendow, B. 1975 Multiphonon Infrared Absorption in the Highly Transparent Regime of Solids. A review AFCRL LQ Tech Memo 29.

Bernstein, H.J. and Powling, J.J. 1950 Chem Phys 18 1018.

Billard, P. and Cornillault, J. 1962 Acta Electronica 6 Special IR Cahier No 3.

Bornstein, A., Croitova, N. and Marom, E. 1982 SPIE 320 Advances in IR Fibres.

Brehm, C., Cornebois, M., Le Sargent, C. and Parant, J.P. 1982 J Non Cryst Solids 47 251-254.

Cornet, J. and Rossier, D. 1973a J Non Cryst Solids 12 61.

Cornet, J. and Rossier, D. 1973b J Non Cryst Solids 12 85.

Cornet, J. and Rossier, D. 1973c Mat Res Bull 8 9.

Deutch, T.F. 1975 J Electronic Materials 4 663.

Dianov, E.M. 1982 SPIE 320.

Dow, J.D. and Redfield, D. 1971 Phys Rev Lett 26 762.

Dow, J.D. and Redfield, D. 1972 Phys Rev B5 594.

Drexhage, M.G., El-Bayoumi, O.H. and Moynihan, C.T. 1982 SPIE Proceedings 320 27.

Edmond, J.T. 1968 J Non Cryst Solids 1 39.

Flaschen, S.S., Pearson, A.D. and Northover, W.R. 1960a J Amer Ceram Soc 43 274.

Flaschen, S.S., Pearson, A.D. and Northover, W.R. 1960b J Appl Phys 31 219.

Ford, E.B. and Savage, J.A. 1976 J Phys E - Scientific Instruments 9 622.

Fuxi, G., Xilai, M. and Peihang, W.H.Y. 1983 J Non Cryst Solids 56 309-314.

Hilton, A.R. and Brau, M. 1963 Infrared Physics 3 69-76.

Hilton, A.R., Jones, C.E. and Brau, M. 1964 Infrared Physics 4 213.

Hilton, A.R. and Jones, C.E. 1966 Phys Chem Glasses 4 112.

Hilton, A.R., Jones, C.E. and Brau, M. 1966a Phys Chem Glasses 7 105.

Hilton, A.R., Jones, C.E., Dabrott, R.D., Klein, H.M., Bryant, A.M. and George, T.D. 1966b Phys Chem Glasses 4 116.

Hilton, A.R., Jones, C.E. and Brau, M. 1966c Infrared Physics 6 183.

Hilton, A.R. 1966 Applied Optics 5 1877.

Hilton, A.R. 1970 J Non Cryst Solids 2 28.

Hilton, A.R., Hayes, D.J. and Rechtin, M.D. 1975 J Non Cryst Solids 17 319.

Hilton, A.R. and Hayes, D.J. 1975 J Non Cryst Solids 17 339.

Hilton, A.R. 1978 SPIE 131.

Hopfield, J.J. 1968 Comments on Solid State Physics 1 16.

Kanamori, T., Terunuma, Y. and Miyashita, T. 1983 Integrated Optics and Optical Fibre Communications, Tokyo.

Kapany, N.S. 1967 Fiber Optics Principles and Applications. Academic Press NY.

Katsuyama, T., Matsumura, H. and Suganuma T. 1982 European Patent Application 82301088.9 Publication No 0 060 085.

Katzir, A. and Arieli, R. 1982 J Non Cryst Solids 47 149-158.

Kettlewell, B.R., Kinsman, B.E., Wilson, A.R., Pitt, A.M., Savage, J.A. and Webber, P.J. 1977 J Mat Sci 12 451.

Kittel, C. 1956 Solid State Physics. J Wiley and Son Inc NY.

Klein, R.M. 1974 J Electronic Materials 3 79-99.

Kolomiets, B.T. 1964 Phys Stat Sol 7 359.

Kruse, P.W., McGlauchlin, L.D. and McQuistan, R.B. 1962 Elements of Infrared Technology. J Wiley and Son Inc NY.

Lucovsky, G. and Martin, R.M. 1972 J Non Cryst Solids 8-10 185.

McLauchlan, A.D. and Gibbs, W.E.K. 1977 NBS Publ 509 222-228.

Michels, B.D. and Frischat, G.H. 1981a J Amer Ceram Soc 64 C150-C151.

Michels, B.D. and Frischat, G.H. 1981b Glastechnische Berichte 54 302-306.

Michels, B.D. and Frischat, G.H. 1982 J Mat Sci 17 329-334.

Miller, F.A. and Wilkins, C.H. 1952 Anal Chem 24 1253-1294.

Mitra, S.S. and Bendow, B. (eds) 1975 Optical Properties of Highly Transparent Solids. Plenum Press NY.

Miyashita, T. and Terunuma, Y. 1982 J Appl Phys 21 L75-L76.

Moynihan, T.C., Macedo, P.B., Maklad, M.S., Mohr, R.K. and Howard, R.E. 1975 J Non Cryst Solids 17 369.

Muir, J.A. and Cashman, R.J. 1967 Opt Soc of America 57 1.

Nakamoto, K. 1963 Infrared Spectra of Inorganic and Coordination Compounds. J Wiley and Son Inc NY.

Nyquist, R.A. and Kagel, R.O. 1971 Infrared Spectra of Inorganic
 Compounds. Academic Press NY and London.
Omonichi, M., Arai, T. and Kudo, K. 1971 J Non Cryst Solids 6 362.
Peck, W.F. and Dewald, J.F. 1964 J Electrochem Soc 1 561.
Pohl, D.W. and Meier, P.F. 1974 Phys Rev Lett 32 58.
Rechtin, M.D., Hilton, A.R. and Hayes, D.J. 1975 J Electronic
 Materials 4 374.
Rodney, W.S., Malitson, I.H. and King, T.A. 1958 JOSA 48 633.
Savage, J.A. and Nielsen, S. 1964 Phys Chem Glasses 5 82.
Savage, J.A. and Nielsen, S. 1965a VII International Congress on
 Glass, Brussels.
Savage, J.A. and Nielsen, S. 1965b Infrared Physics 5 195.
Savage, J.A. and Nielsen, S. 1966 Phys Chem Glasses 7 56.
Savage, J.A. 1971 J Non Cryst Solids 6 964.
Savage, J.A. 1972a J Mat Sci 11 121.
Savage, J.A. 1972b J Mat Sci 7 64.
Savage, J.A., Webber, P.J. and Pitt, A.M. 1977 Applied Optics 16
 2938.
Savage, J.A., Webber, P.J. and Pitt, A.M. 1978 J Mat Sci 13 859.
Savage, J.A., Webber, P.J. and Pitt, A.M. 1980 Infrared Physics
 20 313.
Shibata, S., Terunuma, Y. and Manabe, T. 1980 Japan J. Appl Phys
 19 L603-L605.
Stacey, K.A. 1956 Light Scattering in Physical Chemistry,
 Butterfields.
Stachiw, J.D. and Bertic, S.L. 1982 NOSC Tech Report 634.
 ADA 119495.
SPIE 1982 362 Scattering in Optical Materials.
Strom, U., Hendrickson, J.R., Wagner, R.J. and Taylor, P.C. 1974
 Solid State Commun 15 1871.
Takahashi, S., Kanamori, T., Terunuma, Y. and Miyashita, T. 1983
 4th International Conference on Integrated Optics and Optical
 Fibre Communication, Tokyo, Japan.
Tauc, J. 1975 in Optical Properties of Highly Transparent Solids
 Eds Mitra, S.S. and Bendow, B. Plenum Press NY. 245-260.
Tebo, A.R. 1983 Electro Optics June 41-46.
Thornburg, D.D. 1973 J Electronic Materials 2 495.
Tille, U., Frischat, G.H. and Leers, K.J. 1977 J Non Cryst Solids
 Ed Frischat, G.H. (Trans Tech Publications Aedermannsdorf,
 Switzerland) 631-638.
Tsuchihashi, S., Kawanato, Y. and Adachi, K. 1968 J Ceram Soc
 Japan 76 103-106.
Turjanitsa, I.D., Mihalinets, I.M., Kaperljas, B.M. and
 Kopinets, I.F. 1972 J Non Cryst Solids 11 173.
Urbach, F. 1953 Phys Rev 92 1324.
USP 2,804,378 27/8/57.
Vasco, A. 1965 Czechoslovak J Phys 15 170.
Vasiliev, A.V., Dianov, E.M., Plotnichenko, V.G., Sysoer, U.K.,
 Bagrov, A.M., Baikalov, P.I., Devyatykh, G.G., Scripacher, I.V.
 and Churbanov, M.F. 1983 Electron Lett 19 589-590.

Webber, P.J. and Savage, J.A. 1976   J Non Cryst Solids 20 271.
Worrall, A.J. 1979   SPIE 163 8-12.
Young, P.A. 1970   Thin Solid Films 6 423.

TABLE 1   OPTICAL PROPERTIES OF CHALCOGENIDE GLASSES

Glass atomic %	n_2	n_3	n_4	n_5	V_{3-5} †	n_8	n_{10}	n_{12}	V_{8-12} *	dn/dT °C $\times 10^{-5}$
As40 S60 type B	–	2.395	2.390	2.386	154	–	–	–	–	–1
Ge15 As25 S60	2.30	–	–	–	–	–	–	–	–	–
Ge25 As15 S60	2.22	–	–	–	–	–	–	–	–	–
Ge30 As15 S55	2.25	–	–	–	–	–	–	–	–	–
Ge40 As15 S45	2.30	–	–	–	–	–	–	–	–	–
As40 Se60	–	–	–	–	–	2.7789	2.7789	2.7728	159	–
Ge20 Se80	–	–	–	–	–	2.4071	2.4027	2.3973	143	–
Ge10 As20 Se70	–	–	–	–	–	2.4649	2.4594	2.4526	119	–
Ge10 As30 Se60	–	–	–	–	–	2.6254	2.6201	2.6135	135	–
Ge10 As40 Se50	–	–	–	–	–	2.6108	2.6067	2.6016	176	–
Ge20 As10 Se70	–	–	–	–	–	2.5628	2.5583	2.5528	156	–
Ge30 As10 Se60	–	–	–	–	–	2.4408	2.4347	2.4271	104	–
Ge30 As15 Se55	–	–	–	–	–	2.4972	2.4914	2.4840	113	–
Ge30 As20 Se50	–	–	–	–	–	2.5690	2.5633	2.5560	120	–
Ge33 As12 Se55	–	–	–	–	–	2.5002	2.4942	2.4867 (11.0 µm 2.5962)	111	–
Ge28 Sb12 Se60	–	2.6263	2.6200	2.6165	165	2.6083	2.6002	–	99	+8
Ge30 As13 Se57	–	2.4936	2.4887	2.4859	193	2.4784	2.4724	2.4650	110	+7
Ge30 As13 Se47 Te10	–	2.6118	2.6057	2.6024	171	2.5952	2.5897	2.5829	129	+7
Ge30 As13 Se37 Te20	–	2.7412	2.7342	2.7305	162	2.7229	2.7178	2.7117	154	+11
Ge30 As13 Se27 Te30	–	2.8818	2.8732	2.8688	144	2.8610	2.8563	2.8509	185	+15
Si25 As25 Te50	–	–	–	2.93	–	–	–	–	–	+1
Ge10 As20 Te70	–	–	–	3.55	–	–	–	–	–	–
Si15 Ge10 As25 Te50	–	–	–	3.06	–	–	–	–	–	+17

† $V_{3-5} = \dfrac{n_4 - 1}{n_3 - n_5}$ * $V_{8-12} = \dfrac{n_{10} - 1}{n_8 - n_{12}}$

TABLE 2   THERMAL AND MECHANICAL PROPERTIES OF CHALCOGENIDE GLASSES

Glass Composition Atomic %	Tg °C	Thermal Expansion Coefficient $\times 10^{-6}$/°C	Density Kg/m^{-3} $\times 10^3$	Hardness K-Knoop V-Vickers Kg/mm^2	Thermal Conductivity Cal/cm Sec °K	Rupture Modulus MPa	Young's Modulus GPa	K_{Ic} N mm$^{-3/2}$	Viscosity Fulcher Equation $10^5 - 10^{13}$ P
As40 S60 type B	-	26.1	3.15	109 (K)	-	-	-	-	-
Ge15 As25 S60	-	19.4	3.05	159 (K)	-	-	-	-	-
Ge25 As15 S60	425	12.8	3.00	200 (K)	-	-	-	-	-
Ge30 As15 S55	400	9.6	3.17	216 (K)	-	-	-	-	-
Ge40 As15 S45	-	7.7	3.53	276 (K)	-	-	-	-	-
As40 Se60	178	21.0	4.62	-	-	-	-	-	$\log_{10} \eta = -4.44 + 2764/(T°C - 22.25)$
Ge20 Se80	154	24.8	4.37	147 (V)	-	-	-	-	-
Ge10 As20 Se70	159	24.8	4.47	154 (V)	-	-	16.5	6.7 ± 0.4	-
Ge10 As30 Se60	210	19.0	4.51	176 (V)	-	-	18.0	7.1 ± 0.6	-
Ge10 As40 Se50	222	20.9	4.49	173 (V)	-	-	15.9	7.4 ± 0.8	-
Ge20 As10 Se70	209	20.5	4.41	186 (V)	-	-	16.1	-	-
Ge30 As10 Se60	345	13.7	4.36	236 (V)	-	-	18.6	7.7 ± 0.4	-
Ge30 As15 Se55	351	12.8	4.42	245 (V)	-	-	-	-	-
Ge30 As20 Se50	361	11.7	4.47	266 (V)	-	-	21.3	-	-
Ge33 As12 Se55	-	13.0	4.40	170 (K)	0.60	17.2	22.1	-	$\log_{10} \eta = -4.97 + 2824/(T°C - 122.41)$
Ge28 Sb12 Se60	277	15.8	4.67	150 (K)	0.72	17.3	21.8	-	$\log_{10} \eta = -4.71 + 4070/(T°C - 116.13)$
Ge30 As13 S57	342	13.0	4.40	237 (V)	-	-	-	-	$\log_{10} \eta = -5.91 + 4627/(T°C - 67.49)$
Ge30 As13 Se47 Te10	308	13.2	4.56	234 (V)	-	-	-	-	$\log_{10} \eta = -9.74 + 6466/(T°C - 5.06)$
Ge30 As13 Se37 Te20	285	12.9	4.77	228 (V)	-	-	-	-	$\log_{10} \eta = -8.19 + 4868/(T°C - 35.52)$
Ge30 As13 Se27 Te30	262	12.8	4.91	226 (V)	-	-	-	-	-
Si25 As25 Te50	-	13.0	4.76	167 (K)	-	-	-	-	-
Ge10 As20 Te70	-	18.0	-	111 (K)	-	-	-	-	-
Si15 Ge10 As25 Te50	-	10.0	-	179 (K)	-	-	-	-	-

HALIDE GLASSES

Jacques LUCAS

Université de Rennes, Campus de Beaulieu, Laboratoire de Chimie
Minérale D, Avenue du Général Leclerc, 35042 RENNES CEDEX (France)

1. INTRODUCTION

In this tutorial presentation of a very fast growing field,
only the most general references on halide glasses science will be
mentionned (1, 2, 3, 4, 5). Glasses which have been proved to
exist, but which are very unstable or difficult to handle, will be
mentionned but not discussed in depth because the potential of
their industrial applications is very weak. For the same reasons,
informations on structure and physical properties of these marginal glasses are fragmentary and they will be commented only when
they provide important informations for the fundamental understanding of the mode of formation of the glassy state.

The history and actual situation of the glass chemistry are
largely dominated by oxides systems. The glass forming ability
which requires the creation of a strong tridimensional 3D covalent
bond is specific of small highly charged cations as found in oxides such as B_2O_3, SiO_2, P_2O_5... The lowering of energy due to the
intrinsic disorder coming from the aperiodicity of the glassy lattice is balanced by the very strong metal-oxygen bond originating
from the overlapping of the atomic orbitals of M and O in the 3D
directions.

The situation is different with halides X = F, Cl, Br, I,
which are the most electronegative elements. When associated with
a metal, they generally have a great tendency to monopolize the
bonding electrons and to give a pure ionic bond and consequently a
crystalline state whose stability is governed by coulombic forces.
The difficulty in stabilizing halide glasses is demonstrated by
their very recent discovery and their absence as a natural material.

By analogy with oxides, one can easily imagine that, if a

small polarizing cation is combined with fluorine, the formation of a covalent bond is possible. The challenge to obtain a glassy state is twofold, firstly to avoid the formation of individual volatil molecules like UF_6 or NbF_5, SiF_4, $FeCl_3$ for example, and secondly to make the competition glass versus crystal to go in the glass direction. This requires the formation of a polymeric melt and some quenching precautions to avoid the cinetic of the halide diffusions leading to an organized situation during the cooling of the melt. This challenge is very difficult and is only won by a few halides MX_y which could give a 3D aperiodic framework stabilized by a strong covalent bond coming from an important electrons transfert from X to M as indicated here.

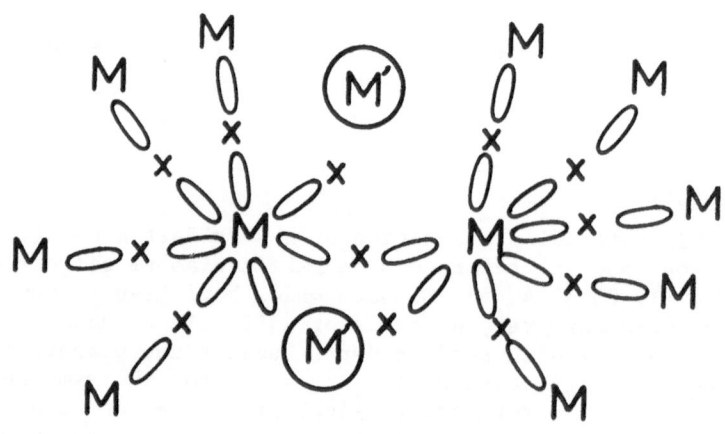

M = small cations ; network former

M' = large cations ; network modifier

2. GLASS FORMING ABILITY OF METALLIC HALIDES

2.1. Heavy halide systems ; vitreous chlorides, bromides, iodides

2.1.1. Divalent chlorides

Pure $ZnCl_2$ has long been known as a glass former chloride. Potential applications of this glass is due to the interesting transmitting properties in the 10.6 μm region (CO_2 emitting laser) and the predicted ultra-low loss in the 4 μm region. Therefore, this interest is highly restricted by high hygroscopicity of the glasses even if the composition is modified by additional chloride. In this low melting glass, T_g = 120° C, T_c = 190° C, T_f = 310° C. The basic structural unit is the $ZnCl_4$ tetrahedra making an aperiodic framework by corner-sharing.

Practical application of $ZnCl_2$ glass is a difficult challenge due to the very poor chemical durability in moisture ; small contamination by O^{2-} or OH^- severely degrades the optical transparency. The chloride $CdCl_2$ has also be claimed to be a glass former material for instance in the ternary composition $CdCl_2-BaCl_2-NaCl$, but the fast crystallization rate requires severe quenching and only thin hygroscopic samples have been obtained.

2.1.2. Trivalent, quadrivalent chlorides

The glass forming ability of $BiCl_3$ and $ThCl_4$ has been proved in the $BiCl_3-KCl$ system and in the $ThCl_4-NaCl-KCl$ ternary diagram. In opposition with $ZnCl_2$, these chlorides are not glass former alone, and must be combined with modifier. As $ZnCl_2$, they are low melting materials and hyghly hygroscopic.

For a $BiCl_3$ based glass, the typical temperatures are :
T_g = 30-50° C and fusion T_f = 160-200° C. For a typical $ThCl_4$ glass T_g = 130° C and T_f = 290° C.

2.1.3. Bromides, iodides

Far I.R. transmitting glasses based on the glass forming properties of $ZnBr_2$ and CdI_2 have been described. The addition of bromides or iodides like PbI_2, TlI, CsI increases their resistance to humidity, however their low glass transition temperature, 10-35° C, leads to relatively poor stability.

2.2. Fluoride glasses

Evidence of glass forming ability has been proved for MF_2, MF_3, MF_4 fluorides. In the case of fluorine, which is the most electronegative element, the competition with the crystalline state governed by coulombic forces is very great and the covalent bond necessary to polymerized the melt could only be achieved in certain situations. M must be highly polarizing this means small and charged to delocalize a part of the outside electrons of the F^- anion. This is achieved when $M = Be^{2+}$ for the divalent fluorides, $M = Cr^{3+}$ Fe^{3+}, Al^{3+}, Ga^{3+}... for the MF_3, $M = Zr^{4+}$, Hf^{4+}, Th^{4+}, U^{4+} for the quadrivalent fluorides. In certain cases, two of them must be combined for instance MF_2 (M = Zn, Mn) and ThF_4 to obtain stable glasses.

Monovalent fluorides like LiF must not be excluded from this systematic. Although LiF is not by itself a glass former fluoride, it gives stable glasses when combined with a small amount of P_2O_5 or other fluorides ; note that Li^+ is isoelectric with Be^{2+}.

In the case of MF_5 and MF_6 fluorides, the covalency of the M-F bond is so strong that individual volatil molecules are made preferentially. This hinders the formation of an associated viscous melt, first and necessary step for the glass formation.

2.2.1. MF_2 based glasses

It is well known that crystalline BeF_2 is isotypic with SiO_2 and that his crystalchemistry is based on the tetrahedral coordination of Be^{2+}. Like SiO_2, molten BeF_2 is highly viscous near the liquidus temperature ; at 550° C, the viscosity is around 10^6 poises, indicating a large degree of polymerization of the melt due to the corner sharing of elementary BeF_4 tetrahedra. Consequently, the ionic mobility and crystallization kinetic are extremely slow and the glass formation very easy.

The only specific potential application of BeF_2 glasses is related to their very low refractive index $n_D = 1,27$ and high Abbe number $\nu = 106$. Combined with a low non-linear refractive index, these characteristics make Nd^{3+} doped BeF_2 glasses very attractive as optical components in high power laser fusion. Large disks of such glasses having good optical qualities have been produced by Corning Glass Work. Therefore because of toxicity and moderate hygroscopicity, the future of such materials for industrial applications appears to be more or less problematic.

2.2.2. MF_3 based glasses

Although no MF_3 fluoride could by itself be a glass former, when combined with appropriate fluorides they could lead to interesting glassy materials. Fluorides such as AlF_3, FeF_3, GaF_3, CrF_3, VF_3, InF_3... have been proved to give glasses when they are associated with a divalent transition metal fluorides and another fluoride playing the role of network modifier. For example (as indicated in figure 1), glasses are found in the following ternary systems : CrF_3-NaF-SrF_2, FeF_3-MnF_2-PbF_2, AlF_3-CaF_2-BaF_2, GaF_3-ZnF_2-PbF_2.

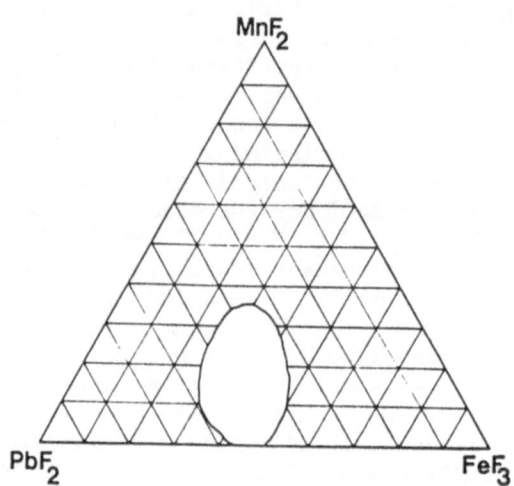

Figure 1

Example of glass formation in the ternary system
MnF_2-FeF_3-PbF_2.

The T_g of such glasses are in the range 250-280° C, T_c around 300-400° C and T_f about 500-550° C. As discussed later, the aperiodic framework of such glasses originated from the corner sharing of MF_6 octahedra.

In addition to their good I.R. transparency, up to 6-7 μm, this type of material could contain an extremely rich amount of magnetic cations such as Fe^{3+}, Mn^{2+} giving rise to spin glass interactions at low temperatures. Their preparation requires severe dry box conditions due to the hygroscopicity of starting trivalent fluorides. Small hydrolysis of the melt generates oxides which increase rapidly the devitrification phenomena.

2.2.3. MF_4 based glasses

Zirconium tetrafluoride ZrF_4 is one of the most interesting glass formers although by itself it gives amorphous microcrystalline material, but not glass. If a modifier fluoride such as BaF_2 is combined with ZrF_4, glasses can be obtained as indicated in figure 2. The composition 2 ZrF_4-1 BaF_2 is the most favorable and only moderate quenching is necessary to stabilize vitreous material. This simple composition is of particular interest for structural investigations as discussed in the next chapter.

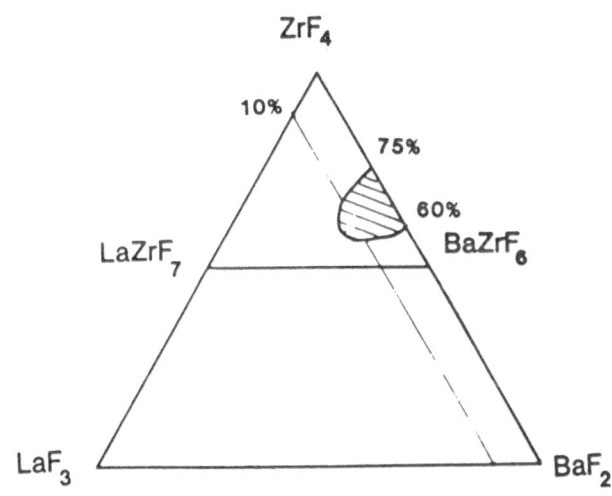

Figure 2 : Vitreous area in a ZrF_4-based system

To decrease the devitrification rate, a third fluoride must be used in a proportion not superior to 10 percent : the most efficient fluorides belong to the uranides or lanthanides series : ThF_4 or LaF_3 being the most appropriate.

In order to stabilize ZrF_4 based glasses, a great number of multicomponent materials has been investigated. The aim was to decrease the crystallization rate and to obtain a glass in which the viscosity slope was suitable for fiber drawing after T_g. In fact, in the competition glass-crystal, the F diffusion responsible for the nucleation and crystallisation is very critical and depends on the viscosity evolution.

Among the different compositions suitable for fiber drawing or for the preparation of optical quality bulk samples, we will select one of the most simple ones directly derived from the vitreous area mentionned in figure 2. This "good" multicomponent glass has the composition : 54 ZrF_4, 36 BaF_2, 6 LaF_3, 4 AlF_3. The addition of a small amount of AlF_3 has a significant effect on the polymerization process in the melt in lowering the liquidus temperature. At the same time, the difference $T_c-T_g \simeq 100$, which is one of the highest ever obtained, is favorable to avoid crystals formation during the annealing operation. The following table shows the typical temperatures of the best glass in a binary, ternary and quaternary system.

Table

Glass composition	T_g	T_c	T_f	$T_c - T_g$
66,6 ZrF_4, 33,3 BaF_2	298	345	509	47
62 ZrF_4, 30 BaF_2, 8 LaF_3	312	385	540	73
57 ZrF_4, 34 BaF_2, 5 LaF_3, 4 AlF_3	315	395	513	80

Thorium fluoride ThF_4 by itself, or even combined with BaF_2, cannot be vitrified. Therefore, if ThF_4 is associated with a low melting fluoride such as ZnF_2 or MnF_2, vitrification conditions are obtained. For instance, the vitreous domain is indicated in the diagram $ThF_4-ZnF_2-BaF_2$ in the figure 3.

Investigations to optimize such ThF_4 based glasses indicate that multicomponent composition involving a large amount of lanthanides Ln = Y, Yb, Lu or indium fluorides is necessary to avoid devitrification and obtain pieces of glass having a maximum thickness of 1 cm. A typical composition is :

The interest of such glasses compared to zirconium based material is their strong resistance to moisture and a best I.R. transparency in the 8 μm region.

These two families of heavy metal fluoride glasses could be easily prepared by the direct melting of the appropriate fluorides in a vitreous carbon or Pt crucible. Therefore, a very convenient

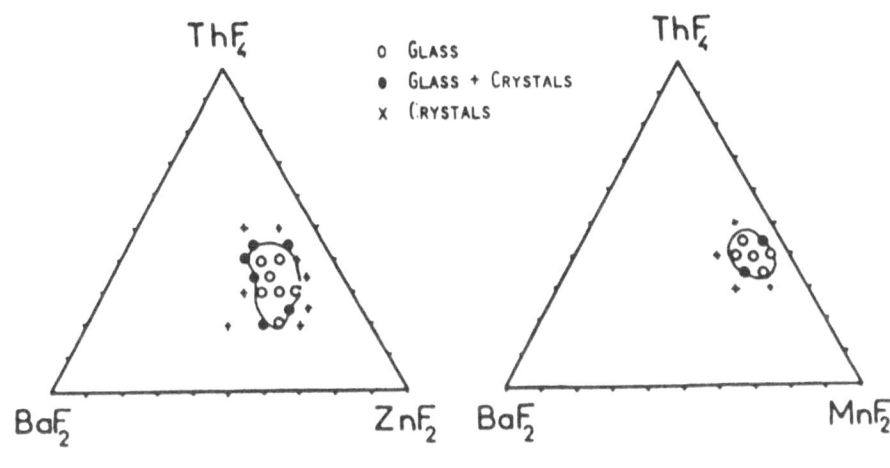

Figure 3

Glassés in two ThF$_4$ based system

way to prepare these glasses is to use the NH_4F HF process which converts oxides used as starting materials in fluorides at 300° C, the operation being followed by the melting operation between 700 and 900° C depending on the composition.

3. HALIDE GLASS STRUCTURES

X Ray or neutron diffraction studies are among the most power-ful techniques to describe the short range order in a glassy mate-rial. In case of halide glass, even if containing three or four elements, the pair interactions are sufficiently different in dis-tances and intensities to be interpreting with a rather good relia-bility. The informations obtained by this method are :

- the anions-cations, cations-cations and anions-anions distances
- the coordination number around cation and anion.

Halides glasses because of the presumption of the partial ionicity of the bond are also good materials for dynamic molecular calcula-tion using computer simulation.

In addition to this, all kinds of spectroscopical methods such as U.V., visible, I.R., absorption of fluorescence, N.M.R... using or not local probes are very interesting to obtain local informations. One of the most powerfull tool is the Fluorescence Line Narrowing (FLN) spectroscopy allowing the site to site des-cription of a glass by means of selective laser excitation of a local probe such as Eu^{3+}

3.1. MF$_2$ based glasses

3.1.1. BeF$_2$ based glass structure

Vitreous BeF$_2$ and alkali fluoroberyllates have been largely investigated and the results interpreted by analogy with the SiO$_2$ model used as a reference. It has been proved that the Be-F distance is very short ($\simeq$ 1.54 Å) indicating a very strong covalent interaction, not surprising in view of the very large ionization energy of Be atom. As expected, the coordination number around Be is 4 in agreement with the strongly directional bonding predicted by the sp$_3$ hybridization of Be. The aperiodic network built up by BeF$_4$ tetrahedra sharing corners is isomorphous with SiO$_2$ glass and could be described by reference with the cristoballite or quartz structure which has been proved to exist for crystalline BeF$_2$.

3.1.2. ZnCl$_2$ based glasses

Despite their great tendency to crystallize and their high hygroscopicity, ZnCl$_2$ glasses have been investigated and compared to the three crystalline forms. In both case Zn^{2+} is tetrahedrally coordinated as expected in view of the electronic structure of Zn^{2+}. It is now well established that the glass structure consists of a random network of corner sharing ZnCl$_4$ tetrahedra. The short Zn-Cl bond distance observed (2.29 Å), compared to the sum of ionic radii r Zn^{2+} = 0.70 Å and r Cl$^-$ = 1.80 Å, indicates a strong overlapping of atomic orbitals.

3.2. MF$_3$ based glasses

Only fragmentary informations have been published on AlF$_3$ and others MF$_3$ such as FeF$_3$ based glasses. Therefore, a lot of structural works have been published on the corresponding crystalline materials. All the preliminary results converge on describing these glasses as deriving from the corner sharing of elementary MF$_6$ octahedra as found in the ReO$_3$ structural type. These observations are in good agreement with the usual coordination number found in crystalline materials derived from MF$_3$.

3.3. ZrF$_4$ based glasses

It is well known that Zr^{4+} has an ambiguous coordination number in crystalline fluoride materials ranging between 7, 8 and sometimes 6. X.Ray diffraction as well as computer simulation calculation have been used to investigate glasses of the binary system ZrF$_4$-BaF$_2$. The situation is favorable here because the pair interactions Zr-F, Ba-F, Zr-Zr, Ba-Ba, Ba-Zr are well separated, and because of the high scattering factor of the heavy metal. The following results have been well established : Zr-F bond is 2.1 Å and coordination number = 7.5 indicating that statistically an identical number of ZrF$_7$ and ZrF$_8$ polyhedra sharing corners but also edges are responsible of the aperiodic framework. As expected, the Ba^{2+} playing the role of modifier interacts with two kinds of fluorine bridging and non bridging. The proposed model is in good agreement with the situation found in the two reference crystalline compounds ZrF_4 and $BaZrF_6$.

4. CHEMICAL DURABILITY

The strongest and most usual corroding factor for halide materials, crystalline or vitreous, is the water molecule H_2O, either in vapour phase or in liquid state.

4.1. Chlorides, bromides, iodides glasses

The most obvious feature concerning the few glasses previously mentionned and deriving from heavy halides is their extremely poor resistance to corrosion by water.

This situation, which is not specific of the vitreous state, is also found in all crystalline halides MX_y, specially when M is small and highly charged. The local site available for M in a tetrahedral and more in an octahedral coordination of large halides anion like Cl^-, Br^-, I^- is very broad. It gives rise to a poor M-X chemical bond which competes severely with the very well known stable situation of the hydrated cation $M(H_2O)_n$.

The only way to stabilize non hygroscopic chloride glasses would be to use large non polarizing cation, but in this case the material obtained is purely ionic and crystalline.

4.2. Fluoride glasses

The situation is here quite different because F^-, H_2O, OH^- have almost the same size ; therefore, OH^- and H_2O are strong dipoles and are better ligand than F^- which is one of the poorest. The OH^- groups are natural poisons for fluoride materials because of the identity of charge and size with F^- and also because they are strongly fixed to the lattice by a hydrogen bond.

4.2.1. Corrosion by H_2O liquid

The leach rates of different families of fluoride glasses have been investigated and the main results could be summarized as follows :

a) Zirconium fluoride based or thorium fluoride based glasses are corroded by water but the second family exhibits a hundredfold improvement in leach resistance. The zirconium glasses could be compared to the Na_2O-2 SiO_2 glasses and the thorium glasses to the Li_2O-2 SiO_2 glasses.

b) During the corrosion process, alkali (Na, Li) or alkaline earth accelerate the dissolution

c) the highest leaching rate of Zr based glass could be attributed to the well known large hydratation enthalpy of Zr^{4+}, which is a rather small and charged cation.

316

4.2.2. Corrosion by atmospheric vapour

During the melting process, the pyrohydrolysis of the melt is significant : $M\text{-}F + H_2O \rightarrow M\text{-}OH + HF$. At these temperatures, the OH^- diffusion is important and the corrosion is a bulk phenomenon.

During the annealing operation of OH free glasses, some corrosion by atmospheric vapour could occur before T_g but the OH^- groups remain at the surface and could be eliminated by polishing.

REFERENCES

1. Poulain, M. and Lučas, J. Une nouvelle classe de matériaux : les verres fluorés au tétrafluorure de zirconium, Verres et Réfractaires, 32, 4 (1978) 505-513
2. Poulain, M. Halide glasses, Journal of Non-Crystalline Solids, 56 (1983) 1-14
3. Baldwin, C.M., Almeida, R.M. and Mackenzie, J.C. Halide Glasses, Journal of Non-Crystalline Solids, 43 (1981) 309
4. Proceedings of the first and second international symposium on halide glasses. I : Cambridge, march 1982. II : Troy (N.Y.) august 1983
5. Proceedings of the international and VII University conference on Glass Science, Clausthal, july 1983, Journal of Non-Crystalline Solids, volume 56 (1983).

Abstract only

RECENT DEVELOPMENTS IN OPTICAL WAVEGUIDES

John R. GANNON

Materials Research Department - Corning Glass Works
Corning, NY 14831 (U.S.A.)

It is now more than 13 years since the first optical waveguide having a transmission loss below 20 dB/km was produced. Since that time, in what has become one of the world's fastest growing technologies losses have been further reduced by over two orders of magnitude through attention to materials purity and processing conditions.

In this paper, an introduction to the principles of optical waveguide transmission will be given and each key fabrication technique will be described. The major optical, physical and chemical properties of materials used in the construction of waveguides will then be reviewed and correlated with actual performance data. Environmental effects, such as hydrogen gas permeation and ionizing radiation have been shown to adversely effect transmission behaviour, under certain conditions. The materials interaction of these effects will be discussed.

Finally, the status of future fiber materials (eg. halides) and novel structures will be presented, together with a review of alternative processing techniques such as sol-gel and soot casting.

OPTICAL AND RELATED PROPERTIES OF BULK HEAVY METAL FLUORIDE GLASSES

MARTIN G. DREXHAGE

Solid State Sciences Division
Rome Air Development Center
Hanscom AFB, Massachusetts, 01731 USA

ABSTRACT

The recent emergence of new families of non-oxide glasses has expanded the realm of materials available to the optical designer. This tutorial review endeavors to survey the fundamental optical properties of one such group, which can be readily prepared in bulk and fiber form, namely multicomponent glasses derived from the fluorides of heavy metals. Since much of the applications interest in fluoride glasses stems from their ability to transmit light in the 1-5 micron wavelength region, emphasis is given to mid-infrared optical phenomena.

INTRODUCTION

The term "heavy metal fluoride glasses" or "HMFG" defines a large family of glasses which may incorporate (to a greater or lesser extent) virtually any metallic fluoride in the periodic table. In practice, however, only a limited number of compositional fields have been discovered which allow the preparation of bulk castings, e.g., pieces several cm in diameter by 1-2 cm thick. Illustrative of these are the multicomponent fluorozirconate or fluorohafnate glasses, composed primarily of ZrF_4 or HfF_4, along with lesser amounts of, e.g., BaF_2, LaF_3, AlF_3, PbF_2, and alkali metal fluorides. Also of interest are the so-called "barium-thorium fluoride" glasses, derived from mixtures of, e.g., BaF_2, ZnF_2, LuF_3 and ThF_4. The HMFG share a number of physical and optical attributes which make them potentially suitable for use as multispectral optical components (e.g., lenses, windows, prisms, fibers), particularly in the long wavelength region not accessible with silicate or other oxide-based glasses. They exhibit a broad range of high transparency which spans the ultraviolet to the mid-infrared. In the vicinity of 2-4 μm, the fluoride glasses show a minimum in absorption coefficient; losses of 8.5 dB/km (~ 2 x 10^{-5} cm^{-1}) have been demonstrated and are expected to decrease further. To supplement this examination of the optical behavior of HMFG, the reader is referred to several reviews which describe the preparation and other properties of the materials (1-4).

OPTICAL PROPERTIES: BASIC CONCEPTS

The optical properties of HMFG have served as a catalyst for much of the research activity directed at these materials since their discovery in 1975. A disparity presently exists between the theoretically predicted optical loss behavior of the fluoride glasses and that observed in the laboratory, analagous to that encountered in during the development of high purity silicate-based glasses in the late 1960's. Since the concept of very low optical absorption serves as the driving force for many applications envisioned for the HMFG, it is instructive to briefly examine the basic mechanisms responsible. The total intrinsic optical absorption (α) as a function of wavelength (λ) in an ideal vitreous solid may be described by (3):

$$\alpha = A/\lambda^4 + B_1 \exp(B_2/\lambda) + C_1 \exp(-C_2/\lambda) \qquad (1)$$

where A, B_1, B_2, C_1, and C_2 are constants. The first term indicates losses due to light scattering arising from microscopic density and composition fluctuations in the material; these effects are seen to decrease rapidly with increasing wavelength. The second and third terms in Eqn. 1 describe, respectively, losses due to ultraviolet absorption from the electronic band edge (Urbach tail) and infrared (IR) edge losses arising from multiphonon absorption. The multiphonon edge in transparent solids results from convolutions and overtones of the far-IR fundamental vibrational frequencies of the cations and anions in the material. When taken together, the terms in Eqn. 1 describe a "vee"-shaped transparency curve for the hypothetical vitreous solid, as shown in Fig. 1. The wavelength of the minimum in the vee-curve and the magnitude of the absorption coefficient at that wavelength depends on the slope and separation of the three loss-inducing factors. In the case of HMFG, it is generally agreed that the intrinsic absorption minimum will occur near 2.5-3.5 μm, with an anticipated absorption coefficient of 10^{-2} - 10^{-3} dB/km (2,5,6). By contrast, the minimum in the vee-curve for fused silica occurs near 1.5 μm, and losses corresponding to the theoretical limit of about 0.2 dB/km have been observed in fibers. In reality, the idealized

320

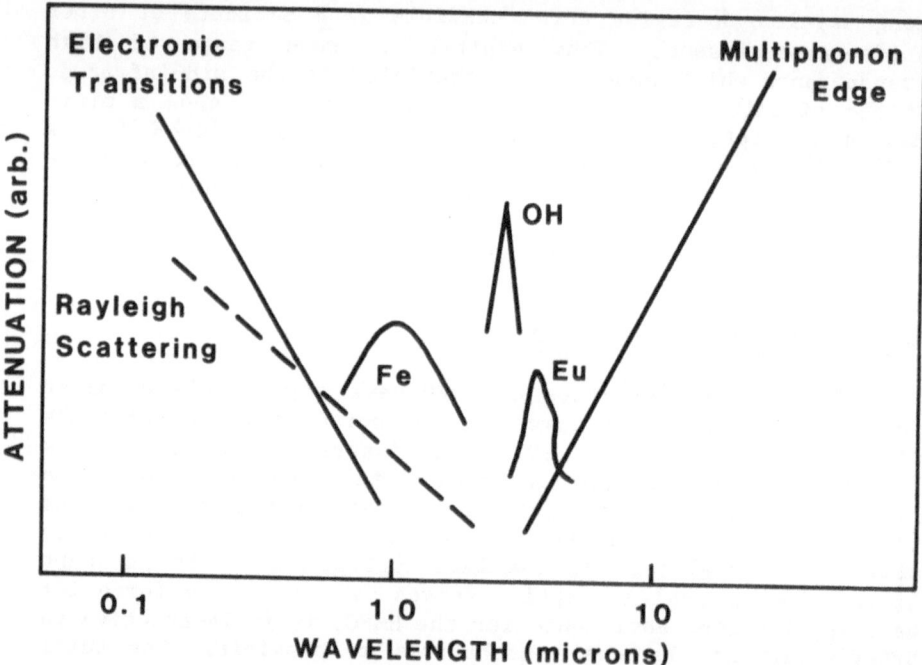

Fig. 1   Schematic of loss mechanisms in a hypothetical vitreous solid.

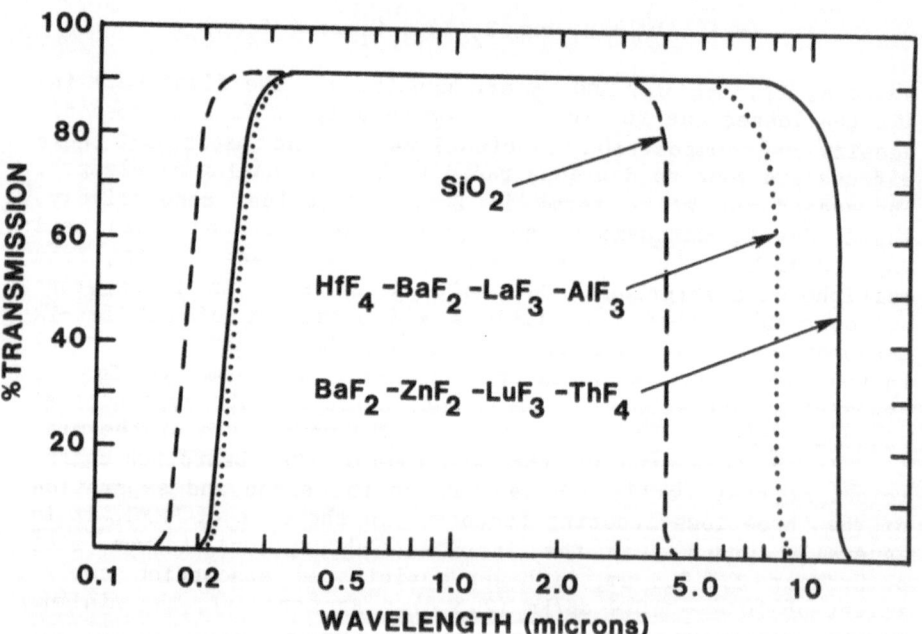

Fig. 2   Percent transmission vs. wavelength for several HMFG and fused silica.

absorption behavior discussed above is distorted by a variety of extrinisc factors, some of which are shown schematically in Fig. 1. A significant portion of the studies devoted to optical properties of HMFG have been concerned with the identification and elimination of extrinsic impurities such as crystallites and inclusions (which give rise to non- λ^{-4} scattering behavior), hydroxyl species, oxide impurities, transition metals, and rare earths. These efforts have been made to replicate in practice the theoretical vee-curve of Fig. 1 and Eqn. 1.

TRANSMISSION CURVES AND ULTRAVIOLET BEHAVIOR

The multispectral transparency which characterizes many HMFG is indicated by the percent transmission versus wavelength curves in Fig. 2. The data, taken on polished specimens about 5 mm thick, were chosen to illustrate the approximate limits attainable with composition changes. For both the fluorohafnate and barium/thorium glasses in Fig. 2, the onset of high transmission occurs at around 0.3 μm in the UV. Throughout the visible and near-IR, the transmission is nearly constant at 92% due primarily to reflection losses (~ 4%) at each surface of the sample. In the hafnate glass, the absorption edge begins at around 6 μm. If ZrF_4 is substituted for HfF_4, the IR edge onset would occur at slightly shorter wavelengths (5.5 μm) in a sample of identical thickness. The high transmission plateau for the barium/thorium glass extends to the vicinity of 8 μm. By contrast, Fig. 2 shows that the overtones of vibrational modes associated with oxygen limit the transmission of fused silica (and other oxide-based glasses) to wavelengths of less than about 3.5 μm, although the UV transparency is superior to that of the HMFG.

The unavailability of ultra-high purity HMFG has limited investigations of their intrinsic UV properties, although some first order observations have been reported. Fig. 2 shows that the cut-off wavelength in both types of fluoride glasses occurs near 0.21 μm (5.9 eV). Examination of a variety of fluorozirconate compositions has confirmed this approximate value, with cut-offs reported in the range of 0.20 to 0.25 μm (4,7,8). For comparison, the UV cut-off for vitreous SiO_2 lies near 0.16 μm (8.0 eV) whereas data for multicomponent fluoroberyllate glasses indicates values between 0.15 and 0.16 μm. Raw materials, crucibles, and furnace atmosphere can also affect the UV behavior of the HMFG; it has been shown that the CCl_4 atmospheres often used to rid fluoride glasses of hydroxyl species can give rise to absorption bands on the UV edge due to chlorine incorporation (8).

LIGHT SCATTERING

Fig. 1 and Eqn. 1 suggest that, in the absence of other sources of extrinsic absorption, the fraction of light scattered by a glass in the mid-IR will be the key factor limiting the attainable absorption, i.e., the point of intersection of the scattering curve and the multiphonon edge. In HMFG, two types of scattering have been observed, one due to "extrinsic" factors, the other being considered "intrinsic" to the vitreous matrix. An example of the former is scattering from inhomogeneities of size comparable to the wavelength of light, characterized by a λ^{-n} wavelength dependence, where n is equal or close to 2. In bulk glasses, microcrystals formed during cooling, crystal nuclei, and inclusions due to incomplete melting can give rise to this type of behavior. In fibers, wavelength independent scattering can arise due to fiber diameter variations and core-clad defects. In principle, it should be possible to rid glasses and fibers of these sources of λ^{-2} type scattering; in practice it has proven very difficult. The vitreous fluorides should intrinsically exhibit the Rayleigh-Brillouin light scattering which characterizes other glassy solids. Theoretical models suggest the intrinsic light scattering limit in HMFG should be lower than that of silicate glasses, while retaining the classical λ^{-4} wavelength dependence (5,6). This is due to the low glass transition temperature of the materials, their anion valence, and their moderate refractive index.

Experimental studies in fluoride glasses have yielded evidence for both Rayleigh-type scattering and scattering due to larger particles in the glass. Several reports of fluorozirconate fibers have noted or presume a λ^{-2} dependence of the scattering intensity which the authors attributed to the drawing-induced growth of the microcrystals (9,10). In the 2 μm region, the magnitude of this scattering loss has ranged from $10-10^4$ dB/km. Rayleigh-type scattering approximating the theoretical limit has also been observed in a limited number of glasses and fibers. Bulk fluorozirconate glasses recently prepared in our laboratories have exhibited scattering intensities 1/3 to 2/3 that of fused silica with a λ^{-4} dependence. Tran et. al. (11, 12) have reported similar results in both glass rods and fibers, some of which are reproduced in Fig. 3. The bulk glass data of Fig. 3 exhibit a Rayleigh scattering character with a magnitude approaching that of high quality fused silica. Extrapolation of the data to 4 μm yielded a scattering loss of 2.7×10^{-3} dB/km, only slightly higher than that estimated by Poignant (6), whose theoretical curve is also shown in the figure. The fiber data in Fig. 3 also shows a λ^{-4} dependence, though with a greater magnitude than observed in the bulk glass, presumably because of fiber drawing-

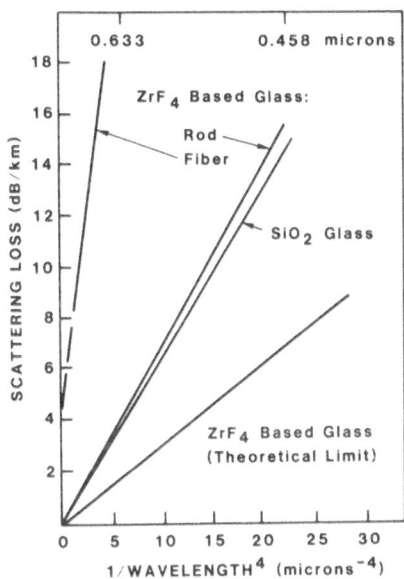

Fig. 3   Scattering loss vs. 1/wavelength^4 for a fluorozirconate glass rod and fiber. Data adapted from Refs. 6, 11, 12.

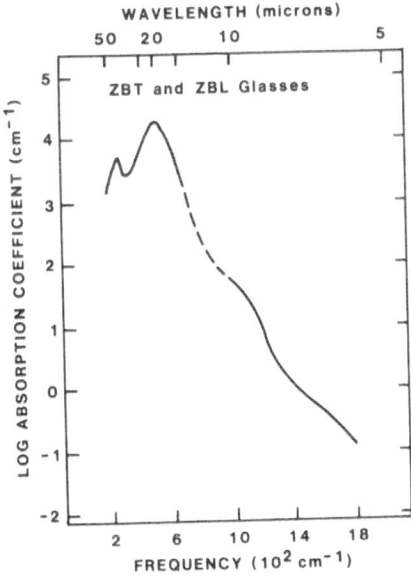

Fig. 4   6-50 micron absorption coefficient for a $58ZrF_4 - 33BaF_2 - 9ThF_4$ (ZBT) or 9 LaF_3 (ZBL) glass. Adapted from Ref. 13.

induced defects.

INFRARED ABSORPTION BEHAVIOR

The position, magnitude and wavelength dependence of the IR or multiphonon absorption edge is the second factor which governs the long wavelength transparency of the HMFG. The fundamental vibrational modes of the "lattice" in fluoride glasses occur in the far-IR at wavelengths of about 14 to 50 μm (700-200 cm^{-1}), which is often termed the "1-phonon" region. Fig. 4 shows the absorption coefficient versus frequency (or wavelength) for some typical 3-component fluorozirconate glasses. It has been suggested that the 500 cm^{-1} peak in Fig. 4 is due to stretching vibrations of Zr complexes with fluorine, while the smaller peak may originate from Ba-F stretching vibrations and/or bend modes of Zr with F (13,14). In hafnium containing glasses, a similar fundamental spectrum is observed, but with the peaks shifted to slightly lower frequencies (18). To a first approximation, the IR edge observed at shorter wavelengths (e.g., the 5-10 μm region in Fig. 4) is caused by overtones and harmonics ("multiphonon absorption processes") of the fundamental vibrational modes. As higher order overtones and harmonics become active, the intrinsic IR edge becomes progressively more featureless with increasing frequency, resulting in an overall exponential dependence of the absorption coefficint on wavelength (or frequency), as suggested by the third term in Eqn. 1. Experimental studies of the temperature dependence of the absorption coefficient, coupled with the "straight line" behavior of the log absorption-frequency data above about 1200 cm^{-1} (Fig. 4) observed in many HMFG compositions, lend credence to the belief that the IR edges of carefully prepared samples are intrinsic, and that projections to shorter wavelengths can be made on the basis of an exponential decrease in the absorption coefficient. Data along the IR edge must, however, be viewed with caution as absorption caused by impurities can easily lead to deviations from intrinsic behavior.

The compositional dependence of the IR edge in HMFG has been extensively examined (13-15). The general trends exhibited by the transmission curves of Fig. 2 manifest themselves in both the fundamental and multiphonon regions. Barium/thorium glasses show a two-peaked fundamental spectra shifted to lower frequencies with respect to fluorozirconate or fluorohafnate glasses. Addition of aluminum and other relatively light metallic fluoride species has been shown to have a detrimental effect on the IR absorption coefficient, although they are often helpful from a glass formation viewpoint. Fig. 5 illustrates some of the basic compositional effects in the 6-10 μm region. The fluorohafnate glass shows a somewhat lower absorption coefficient than its fluorozirconate analog, due (in

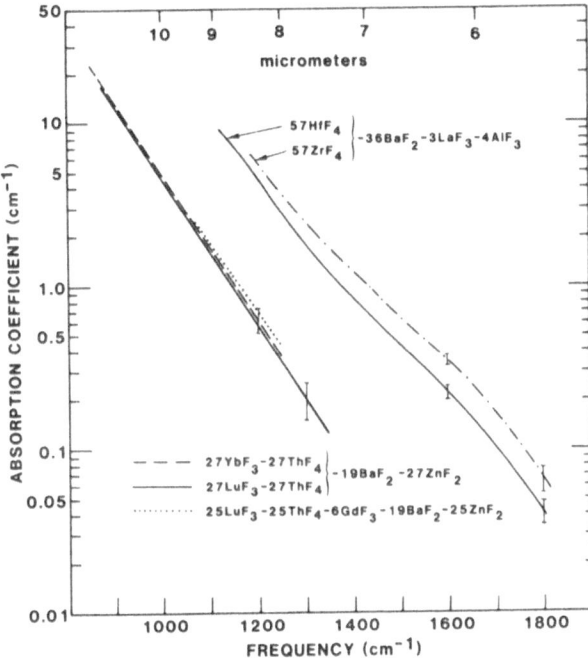

Fig. 5   Absorption coefficient for various HMFG.

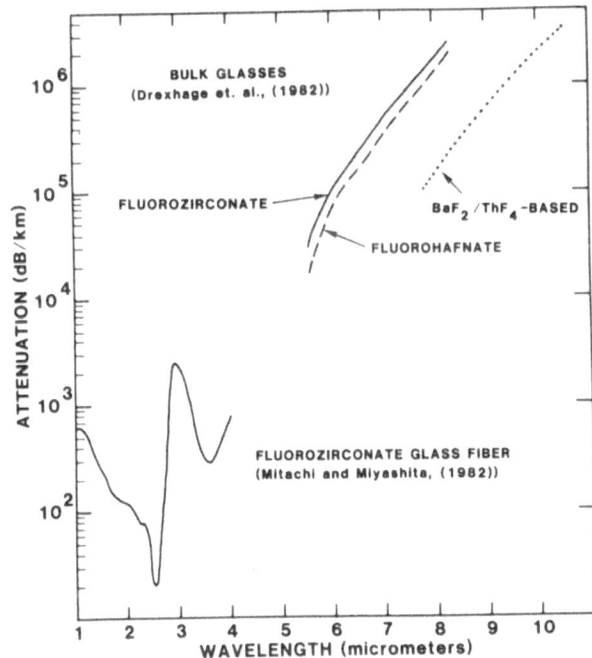

Fig. 6   Attenuation in bulk fluorozirconate glasses and fibers.

part) to the higher atomic weight of hafnium. These glasses contain 4 mol% AlF_3; had data for aluminum-free glasses been included, a small but perceptible decrease in absorption coefficient would be noted. At the other extreme, in Fig. 5 are the barium/thorium glasses. The absorption coefficient of these materials is observed to be an order of magnitude or more lower than the Zr/HfF_4 glasses. This reflects the trends alluded to above; a shift to lower frequencies in the fundamental or "1-phonon" region manifests itself as the 200-300 cm^{-1} displacement of the absorption curves in Fig. 5, which span the "2-3 phonon" regime.

Determination of the IR absorption edge at frequencies greater than about 1700 cm^{-1} (i.e., < 6 μm) requires fiber samples with a long optical path length. Fig. 6 endeavors to show the relationship between such fiber data and that derived from bulk glasses (i.e., Fig. 5). When viewed as a whole, the data of Fig. 6 serves as a first order illustration of the "vee-curve" transparency concept for the HMFG expressed in Eqn. 1 and Fig. 1. At wavelengths below 7 μm, the absorption coefficient decreases rapidly until the strong band due to OH species near 2.8 μm is encountered. In this particular fluoro-zirconate fiber sample, a minimum in attenuation of about 20 dB/km is observed on the short wavelength side of the OH peak near 2.6 μm (10); fibers with losses of 8.5 dB/km have recently been reported (9). The loss increases towards shorter wavelengths as a result of the combined effects of light scattering and impurity absorptions resulting from transition metals and rare earths. These sources of extrinsic absorption distort both the magnitude (10^{-2} - 10^{-3} dB/km) and the wavelength position (2.5 - 3.5 μm) of the theoretically estimated absorption minimium. Fig. 6 also serves to illustrate that mid-IR fibers and optical components may benefit from the use of other compositions. Waveguides prepared, e.g., from the barium/thorium glasses would potentially have their absorption minima displaced to somewhat longer wavelengths, reducing the effects of the 2.8 μm hydroxyl peak.

CONCLUSIONS AND FUTURE PROSPECTS

The deviation from intrinsic optical behavior caused by the presence of impurities presents a major challenge to current applied research on HMFG. Some progress in this and related areas has been made. Alternative methods of glass fabrication (e.g., chemical vapor deposition processes) are under examination at several laboratories. Extensive studies of the refractive index and dispersion behavior of the materials have been carried out; glasses with indices between about 1.48 and 1.53 are now available (16). The wavelength position and absorptivities of various transition metals and rare earths in fluoride glasses have been determined; it is now recognized that such contaminants will have to be reduced to sub-ppb levels (17). Successful efforts to reduce hydroxyl and oxide impurities have focused on the use of reactive atmospheres such as carbon tetrachloride. For certain applications (e.g., short length fibers for image or power transmission, simple optical components), current mid-IR loss levels in the HMFG may be adequate, but until the glass chemistry-related problems are solved, the goal of ultra-low optical absorption may remain an elusive one.

REFERENCES

1. C.M. Baldwin, R.M. Almeida and J.D. Mackenzie, J. Non-Cryst. Solids 43, 309 (1981).
2. M.G. Drexhage, O. El-Bayoumi, and C.T. Moynihan, Proc. SPIE 320, 27 (1982).
3. T. Miyashita and T. Manabe, IEEE J. Quan. Elec. QE-18, 1432 (1982).
4. M. Poulain and J. Lucas, Verres Refract. 32, 505 (1978).
5. S. Shibata, M. Horiguchi, K. Jinguji, S. Mitachi, T. Kanamori, and T. Manabe, Electron. Lett. 17, 775 (1981).
6. H. Poignant, Electron. Lett 17, 973 (1981).
7. J. Lucas, M. Chanthanasinh, M. Poulain, M. Poulain, P. Brun, and M. Weber, J. Non-Cryst. Solids 27, 273 (1978).
8. R.N. Brown, B. Bendow, M.G. Drexhage, C.T. Moynihan, Appl. Opt. 21, 361 (1982).
9. S. Mitachi, Y. Terunuma, Y. Ohishi, and S. Takahashi, Jpn. J. Appl. Phys. 21, L537 (1983).
10. S. Mitachi and T. Miyashita, Electron. Lett. 18, 170 (1982).
11. D.C. Tran, K.H. Levin, C.G. Fisher, M.J. Burke, an G.H. Sigel, Electron. Lett. 19, 15 (1983).
12. D.C. Tran, G.H. Sigel, K.H. Levin, and R.J. Ginther, Electron. Lett. 18, 1046 (1982).
13. B. Bendow, M.G. Drexhage, H. Lipson, P. Banerjee, J. Goltman, S. Mitra, and C.T. Moynihan, Appl. Opt. 20, 2875 (1981).
14. B. Bendow, P. Banerjee, M.G. Drexhage, J. Goltman, S. Mitra, and C.T. Moynihan, J. Am. Ceram. Soc. 65, C8 (1982).
15. M.G. Drexhage, O. El-Bayoumi, C.T. Moynihan, A.J. Bruce, K.H. Chung, D.L. Gavin, and T.J. Loretz, J. Am. Ceram. Soc. 65, C168 (1982).
16. S. Mitachi and T. Miyashita, Appl. Opt. 22, 2419 (1983).
17. Y. Ohishi, S. Mitachi, T. Kanamori, and T. Manabe, Phys. Chem. Glasses 24, 135 (1983).

INFRARED GLASS OPTICAL FIBER TECHNOLOGY: A BRIEF REVIEW

D. C. Tran and G. H. Sigel, Jr.

Optical Sciences Division, Naval Research Laboratory
Washington, DC 20375, USA

ABSTRACT

A brief review of the present state of research in infrared (IR) glass optical fibers is provided. This includes candidate fiber materials, fiber fabrication techniques, typical fiber properties, and IR fiber applications.

1. INTRODUCTION

For the past decade, much attention has been given to IR optical fibers which theoretically possess intrinsic transparencies several orders of magnitude better than the conventional silica and doped silica glass fiber waveguides. Such fibers are essential for many promising military and industrial applications such as long distance, repeaterless data links, long-length fiber optics sensor systems, remoting of IR focal planes, and IR optical power delivery systems. This paper will provide a brief review of the present state of IR fiber technology including material considerations, fiber fabrication techniques, typical fiber properties, and potential IR fiber applications.

2. IR FIBER MATERIALS

The fundamental optical attenuation in wide bandgap solids is dominated by Rayleigh scattering (which varies as $1/\lambda^4$) at shorter wavelengths, and by multiphonon absorption at longer wavelengths. Maximum transparency can thus be found in between the intrinsic scattering and infrared absorption edges. Halide materials possess both lower fundamental scattering levels than silica as well as greatly enhanced total transparency, and therefore are most suitable for the preparation of ultra-low loss IR fibers. Halide crystals potentially can offer the highest optical transparency, but crystalline fibers generally exhibit extremely high scattering losses associated with grain boundary defects or extrusion induced scattering centers. ZrF_4-based glasses (or fluorozirconate glasses), discovered by Lucas and Poulain several years ago [1], have emerged as leading candidates for high-transparency optical fibers in the 2 to 5 μm spectral region. They generally contain between 55 to 70 mol% ZrF_4 as glass former, 18 to 30 mol% BaF_2 as network modifier, and lesser amounts of rare-earth fluorides, $AℓF_3$, PbF_2, and alkali-metal fluorides which serve as glass stabilizers. The intrinsic minimum attenuation in fluorozirconate glasses has been predicted to be as low as 10^{-3} dB/km as compared to ~ 0.16 dB/km at 1.6 μm for silica [2]. In addition, ZrF_4-based glasses possess attractive properties which are relevant to fiber drawing, practical handling and use a relatively wide working range of about 100°C; a lesser tendency toward devitrification than the $AℓF_3$-based glasses which recently were also considered to be strong candidates for IR fiber materials [3,4]; a relatively low activation energy for viscous flow of ~ 87 kcal/mole within the fiber drawing temperature range [5]; an adequate stability toward environmental effects; and non-toxicity characteristics. Typical fluoride glass compositions that have been investigated for fiber drawing are listed in Table I.

Table 1

Typical Fluoride Glasses Used in the Fabrication of IR Fibers

Composition (mol%)	References
51 ZrF_4-16 BaF_2-5 LaF_3-3 $AℓF_3$-20 LiF-5 PbF_2	5
61 ZrF_4-32 BaF_2-3.9 GdF_3-3.1 $AℓF_3$	6
56 ZrF_4-30 BaF_2-5 LaF_3-4 ThF_4-5 $AℓF_3$	7
60 ZrF_4-19 BaF_2-6 LaF_3-15 NaF	8
27 ZrF_4-27 HfF_4-23 BaF_2-8 ThF_3-4 LaF- 2 $AℓF_3$-3 LiF-3 NaF-3 PbF_2	9

Heavy halide glasses are also potential candidates for IR fiber technology. Chloride glasses such as $ZnCℓ_2$, and $BiCℓ_3$-$KCℓ$, and iodide glasses based on CdI_2, are also attractive because of their higher IR transparency than the fluoride glasses, but the high degree of hygroscopicity limits their practical use. Chalcogenide glasses also exhibit longer cut-off wavelengths than the fluorides. They transmit at wavelengths between 1 and 18 μm and are therefore potentially useful for optical devices operating with CO_2 lasers. They are more stable against moisture attack and crystallization than the fluorozirconates, but exhibit much higher intrinsic scattering levels because of their high refractive indices of ~ 2.4; the minimum attenuation in these glasses was predicted to reach only ~ 10^{-2} dB/km at 4.54 μm [2]. Typical chalcogenide glasses investigated for IR fiber drawing have been derived mainly from the As_2S_3, As_2Se_3, $As_2S_3-As_2Se_3$, GeS_3, and $Ge_3PS_{7.5}$ compositions. Heavy-metal oxide glasses based on $GeO_2-Sb_2O_3$ are also of interest since they can easily be prepared from a vapor phase approach [10], but their theoretical minimum is as high as ~ 0.06 dB/km between 2.2 to 2.4 μm.

Up to now, investigations on IR glass optical fibers have been emphasized around fluorozirconate glasses because of their stability and relatively high transparency as well as their excellent fiber drawing characteristics.

3. FABRICATION OF IR GLASS OPTICAL FIBERS

Both crucible and preform approaches have been used to draw fluoride and chalcogenide fibers. The crucible approach is desirable from a point of view of being a continuous process; moreover, drawing induced crystallization can be substantially suppressed if the halide melt can be rapidly quenched into solidified fibers. In actual practice however, fiber distortion and non-uniformity and/or devitrification often arise during the drawing process, since the viscosity of halide glasses changes very rapidly with respect to temperature as the metal is quenched through the drawable temperature [5]. This sharp viscosity-temperature relationship is much more pronounced in fluorides than in chalcogenides. The crucible technique has been investigated by many workers [5,11], but low-loss fluoride glass fibers prepared by this approach have not yet been achieved.

In the preform approach, fluoride and chalcogenide core melts have been cast into a cladding-glass tube. The tube can be formed from casting the cladding-glass melt into a heated metal mold which is then inverted [12], or by rotating the mold containing the melt at high speed until solidification takes place [13]. Germanate glass fibers have been fabricated by the vapor axial deposition (VAD) method [10]. $GeCl_4$ has been used as the main starting material, and $SbCl_2$ has been incorporated as the dopant to control the refractive index profile. IR fiber preforms have been drawn using both induction and resistance heating.

4. PROPERTIES OF IR GLASS OPTICAL FIBERS

To date, IR fibers prepared from preforms have shown much lower loss than those obtained from crucible approach. Optical loss spectra for $ZrF_4-BaF_2-GdF_3-AlF_3$, As-S, and $GeO_2-Sb_2O_3$ glass fibers are illustrated in Figure 1 [14]. A loss minimum of 8.5 dB/km at 2.2 μm was observed for the fluorozirconate glass fiber, 64 dB/km at 2.4 μm for the chalcogenide, and 4 dB/km at 2 μm for the germanate. The absorption in the 1 to 2.5 μm region is attributed to

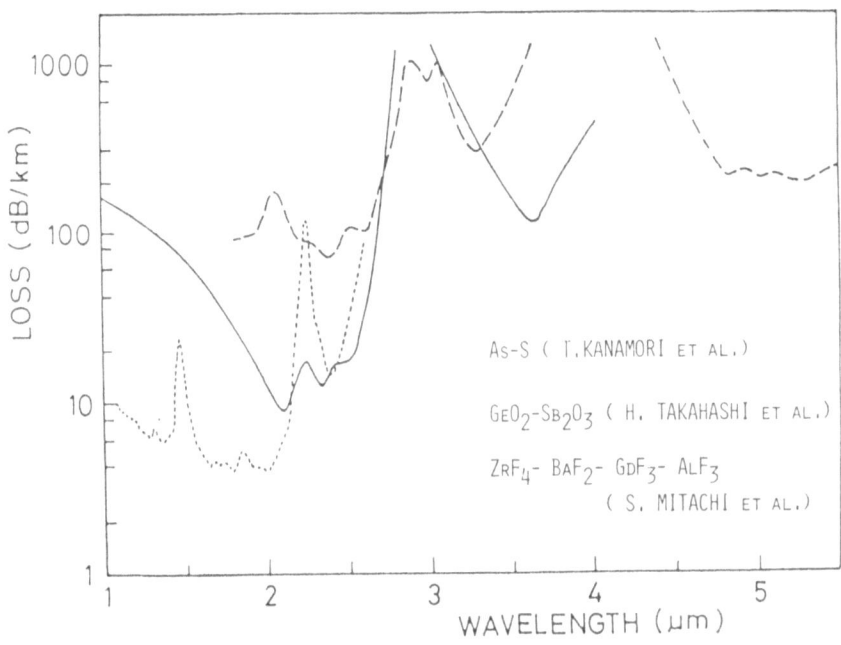

Figure 1. Spectral losses of IR glass optical fibers.

transition metal impurities, and the distinct peaks at 1.4 μm, 2.2 μm, and 2.9 μm to water contamination. The scattering loss component in ZrF_4-based glass fibers has been demonstrated elsewhere to vary as $1/\lambda^4$ and the extrapolated scattering loss data resulted in a low value of ~ 0.01 dB/km at 4 μm [15]. This Rayleigh scattering dependence is reflected by a sharp drop in optical attenuation between ~ 2 to 2.55 μm as shown in Figure 2 for a ZrF_4 BaF_2-LaF_3-$A\ell F_3$-LiF clad/ZrF_4-BaF_2-LaF_3-$A\ell F_3$-LiF-PbF_2 core fiber prepared at NRL.

In Figure 2, the data point representing a minimum loss of 6.8 dB/km at 2.55 μm was obtained from an average of three separate "cut-back" measurements using long integrating time and without scanning the monochromator to minimize sources of error; the error was ± 1 dB/km.

The strength of fluoride glass fibers has been recently investigated by several groups. British Telecom reports a breaking strength of ~ 70 kpsi (strains of ~ 1%), as shown in Figure 3 [16].

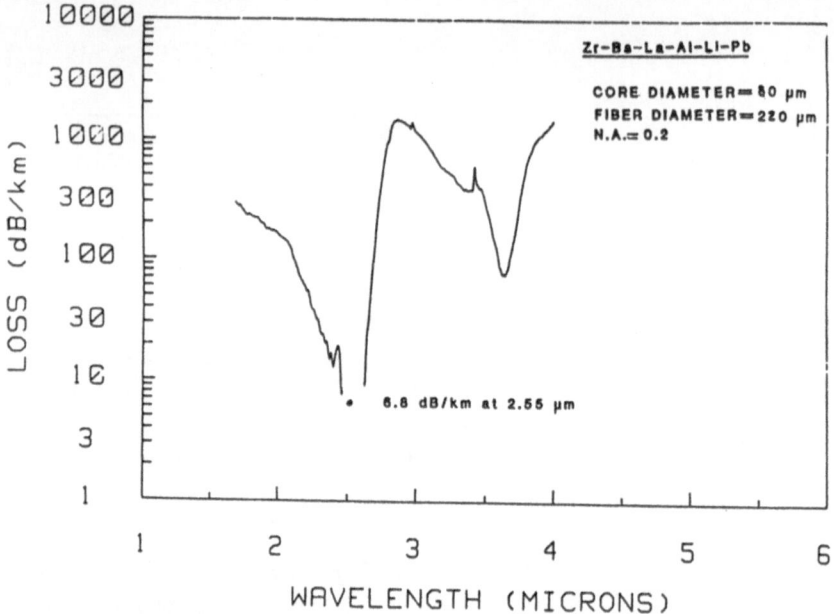

Figure 2. Loss spectrum of a ZrF_4-BaF_2-LaF_3-AlF_3-LiF-PbF_2 glass fiber.

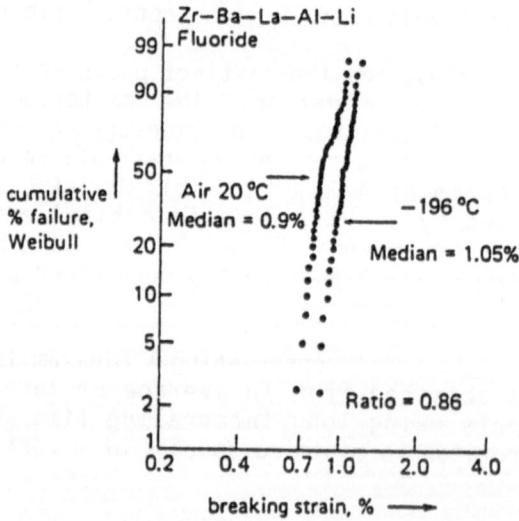

Figure 3. Weibull plots for fluorozirconate glass fibers.

The UCLA group reports strength of up to 100 kpsi for teflon FEP-coated fluorozirconate fibers [17]. Fractographic analysis recently conducted at Sandia Laboratories indicates relatively high values of fracture toughness for fluorozirconate fibers, namely K_{IC} ~ 0.5 MPam$^{1/2}$[17]. This value leads to projected strength in the 10^6 psi range for pristine fluoride fibers. Note that practical strength in silicate fibers is about 1/3 of the theoretical value; assuming the same results apply to fluoride glass fibers, then practical strength in carefully prepared fluoride fibers can approach a respectable value of 500 kpsi.

5. APPLICATIONS OF IR GLASS OPTICAL FIBERS

Various potential applications of IR fibers are summarized in Table 2.

Table 2

Potential Applications of IR Glass Optical Fibers

- Long distance repeaterless data links
- Optical power transmission
- Remoting of focal plane arrays
- Laser annealing and processing
- Low noise fiber sensors
- Remote IR spectroscopy
- Medical surgery and cauterization
- High-bandwidth color multiplexing

Fluoride glass can possess an intrinsic minimum loss several orders of magnitude lower than silica, and can be drawn into long-length optical fibers. These fibers are then essential for repeaterless long-haul communication links, and for long-length fiber sensor systems. Fluoride glass fibers can also operate over an extremely broad wavelength range -- from ~ 0.3 μm to ~ 7 μm -- and are useful for high-capacity wavelength multiplexed fiber systems. Both fluoride and chalcogenide glass fiber bundles can be used for remoting of IR focal planes. The fluorides can be used for 1-6 μm and multispectral sensor arrays, while the chalcogenides can be used for 1-6 μm and 8-12 μm sensor arrays. Other prospective applications include power delivery for medical applications and laser processing using CO_2 laser and chalcogenide glass fibers, and HF/DF lasers, for example, for fluoride glass fibers.

6. CONCLUSIONS

Fluoride glasses are considered prime candidates for ultra-low loss fiber applications in the 2-4 μm region because of their relatively high intrinsic transparencies and their adequate resistance toward devitrification and moisture attack. Chalcogenide glasses, on the other hand, are essential for numerous medium/short-length fiber applications in the 8 to 12 μm region. Both crucible and preform techniques have been used successfully to draw IR transmitting glass optical fibers. Scattering losses were observed to vary as $1/\lambda^4$ in fluoride glass fibers and extrapolated scattering loss data were reported to be as low as 0.01 dB/km; but total fluoride fiber losses are still higher than the theoretical values due to transition metal and OH absorptions. Chalcogenide glass fibers, to date, exhibit both higher scattering and absorption losses. Germanate glass fibers presently possess the lowest minimum loss of 4 dB/km -- as compared to 6.8 dB/km and 64 dB/km for fluoride and chalcogenide fibers, respectively -- but their prospective applications are undoubtedly quite limited since their IR absorption edge is only slightly shifted to a longer wavelength as compared to silica glasses. Preliminary investigations on the mechanical properties of fluoride glass fibers suggest that their practical strength levels are adequate for most applications.

1. Poulain, M., M. Poulain, J. Lucas, and P. Brun. Verres fluores au tetrafluorure de zirconium. Mat. Res. Bull. 10 (1975) 243-246.
2. Shibata, S., M. Moriguchi, K. Jinguji, S. Mitachi, T. Kanamori, and T. Manabe. Prediction of loss minima in infrared optical fibres. Electron. Lett. 17 (1981) 775-777.
3. Takahashi, S., et al. New fluoride glasses for IR transmission, Adv. in Ceramics, Physics of Fiber Optics (Edited by B. Bendow and S. Mitra, Eds. ACS, 1981).
4. Fontereau, G., F. Lahaie, and J. Lucas. Une nouvelle famille de verres fluores transmetterus dans l'infrarouge: fluorures vitreaux daus les systemes ThF_4-BaF_2-MF_2(M=Mn, Zn). Mat. Res. Bull. 15 (1980) 1143-1147.
5. Tran, D. C., R. J. Ginther, and G. H. Sigel, Jr. Fluorozirconate glasses with improved viscosity for fiber drawing. Mat. Res. Bull. 17 (1982) 1177-1184.
6. Miyashita, T., and T. Manabe. Progress in fluoride glass fiber research and development in Japan. paper 33, 2nd Int. Symposium on Halide Glasses, Troy, NY (1983).
7. Maze, G., V. Cardin, and M. Poulain. Fluoride glass infrared fibers for light transmission up to 5 μm. paper 484-16, SPIE's Technical Symposium East '84, Arlington, VA (1984).
8. Ohsawa, K., T. Shibata, K. Nakamura, and S. Yoshida. Fluorozirconate glasses for infrared transmitting optical fibers. in Tech. Digest, ECOC '81 (1981).
9. Poignant, H., J. F. LeMellot, and Y. Bosis. Infrared fluorozirconate and fluorohafnate glass optical fibers. in Tech. Digest, ECOC '82 (1982).

10. Takahashi, H., and I. Sugimoto. Preparation of Germanate glass by vapor-axial deposition. J. Am. Ceram. Soc. 66 (1983) C66-C67.
11. Mimura, Y., H. Tokiwa, and O. Shinbori. Fabrication of fluoride glass fibres by the improved crucible technique. Electron. Lett. 20 (1984) 100-101.
12. Mitachi, S., T. Miyashita, and T. Kanamori. Fluoride-glass cladded optical fibres for mid-infrared ray transmission. Electron. Lett. (1981) 591-592.
13. Tran, D. C., C. F. Fisher, and G. H. Sigel, Jr. Fluoride glass preforms prepared by a rotational casting process. Electron. Lett. (1982) 657-658.
14. Yoshida, S.. Progress in infrared fibers. in Tech. Digest, OFC '84, New Orleans (1984).
15. Tran, D. C., K. H. Levin, C. F. Fisher, M. J. Burk and G. H. Sigel, Jr. Rayleigh scattering in fluoride glass optical fibers. Electron. Lett. 19 (1983) 165-166.
16. France, P. W., J. Williams, J. F. Carter, and K. J. Beales. Mechanical properties of IR transmitting fibers. paper 11, 2nd Int. Symposium on Halide Glasses, Troy, NY (1983).
17. Mecholsky, J. J., et al. Fracture analysis of fluoride glass fibers. paper 32, ibid. (1983).

PREPARATION, CHARACTERIZATION AND USES
OF THIN FILMS IN OPTICS

A. CACHARD

Laboratoire Traitement du Signal et Instrumentation
ERA/CNRS/n° 996 - U.E.R. de SCIENCES
23, rue du Dr Paul Michelon, 42023 Saint-Etienne Cédex, France

1. INTRODUCTION

Remarkable advances have been made over the last decade in the scientific understanding of the nature of the process used to deposit thin films. These films can now be fabricated routinely and they play a vital role in nearly all electronic and optical devices.

An exhaustive review of this subject is obviously beyond the scope of this paper and we will not attempt to reproduce here that which already exists in general reviews and texts (1-3). This article will concentrate upon some aspects which are fundamental in respect with the subject : amorphous (glass – like) thin films in optics.

2. WHY GLASS-LIKE THIN FILMS ?

In the field concerned with optics (light emission, light manipulation, light detection) the use of thin films is often the only way to achieve the desired properties. As an example a single layer of materials whose refractive index is close to the square root of the substrate index gives a very low reflectance at the wavelength for which the film is about a quarter-wave thick (4). That means thickness will be of the order of 100 nm for visible light. In fact it is well known that one makes use of multilayers of quarter-wave thicknesses for improved performances.

Another example is given by guided waves in integrated optics. In a film of thickness t_f and refractive index n_f deposited on a substrate of index n_s, the number of guided modes at a wavelength λ_o is roughly given (5) by :

$$N = 2\ t_f(n_f^2 - n_s^2)^{1/2}\ \lambda_o^{-1}$$

This formula implies that monomode planar guides have a thickness of the order of $\lambda_o/2$.

In these two cases, the important parameters are the refractive index and the optical losses of the film. They are highly dependent on the preparation conditions which govern the structure. Amorphous films are often required in optics for a number of reasons.

First there are differences in optical properties of materials between crystalline and amorphous states. Figure 1 shows the evolution in absorption spectrum (actually imaginary part of dielectric constant) between c-Si and a-Si (6). Amorphous spectrum is reminiscent of a broadened crystalline spectrum, but there are modifications in the absorption edge leading to changes in the optical gap and modifications in the absorption tail at low energy which is highly sensitive to impurities and preparation conditions. One takes advantage of this feature to match the absorption properties to the radiation to be used (7).

Secondly amorphous materials may have new properties. An example is given by the photo-induced refractive index changes which are observed in amorphous $As_2 S_3$ and are related to topological rearrangements of atoms in the amorphous solid (8).

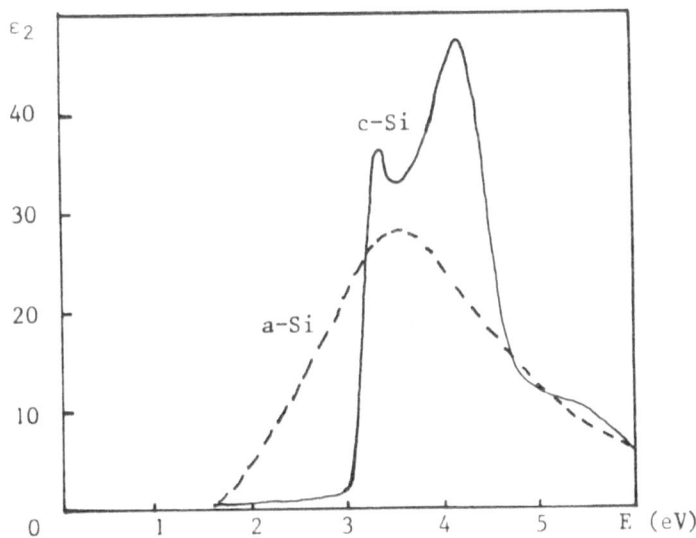

Fig. 1 - Imaginary part of the dielectric constant of c- and a-Si (adapted from (6))

A third reason why to use amorphous materials is given by other than optical properties which are characteristic of some amorphous solids : mechanical (hardness), chemical (corrosion), etc ...

Finally and perhaps the most important, as-deposited thin films are highly disordered (unless special preparation conditions as for epitaxy). In fact this is the main difficulty for their use in optics : How to control the microstructures of the films which to a large extend command the refractive index value and the losses ? The goal is to reach a perfect disorder (amorphous) which means no grain boundaries, no density fluctuations (homogeneity in composition and structure) and low roughness (highly correlated to substrate preparation).

The **table I** emphasizes the importance of these problems. In this table values of effective losses are given for current optical materials and devices. Equivalent absorption constant α, extinction coefficient k and absorption A (for a thickness of 1 μm) are also given. The important feature is the enormous gap between the performances achieved in glass bulk materials (fibers) and in amorphous thin films.

Optical materials	losses dB.cm^{-1}	α cm^{-1}	k λ=1.06μm	A for 1μm %
Optical fibers	$0.5.10^{-5}$	$1.2 \ 10^{-6}$	$< 10^{-11}$	$< 10^{-8}$
Planar waveguides	0.5	0.12	10^{-6}	0.0012
A.R. Coatings "laser grade"	1	0.24	2.10^{-6}	0.0024
A.R. Coatings "classical"	< 40	< 10	$< 10^{-4}$	< 0.1

Table I

Clearly good performances and improvements in optical quality of thin films depend on materials available and on the proper choice of the deposition technique and also on the methods of characterization. In the next two parts we shall briefly discuss these two points.

3. PREPARATION OF THIN FILMS

The challenge is to prepare a film with a priori optical properties. That means one has to control precisely the thickness and the value of complex refractive index $\tilde{n} = n(\lambda)-i \ k(\lambda)$, where $n(\lambda)$ is the real refractive index and $k(\lambda)$ the extinction coefficient.

In addition there exists constraints on mechanical, chemical or electrical properties.

Clearly, this technological problem appears as very complicated due to the number of processes by which can one deposit films and to the high number of parameters involved in each process. In fact there are few basic methods and one can distinguish two main techniques : Physical vapor deposition and chemical deposition.

3.1. Physical vapor deposition

Reduced to their essence these methods involve three steps : formation of a vapor from source materials, transport of the vapor from the source to the substrate and condensation of the vapor on the substrate.

The two methods used to form the vapor from the source are evaporation or sputtering. In principle the vapor is transferred, condensates on the substrate and the film is a reproduction of the source. In fact, it is possible to vary the composition of the film or to compensate for deviations by introducing a gas in the transfer zone (for example O_2 to avoid the reduction of an oxyde). One speaks of reactive evaporation or reactive sputtering. It is also possible to enhance the reactivity near or on the substrate by bombarding the substrate which is the case in assisted reactive evaporation, ion plating or bias sputtering.

	Evaporation	Sputtering
Vapor formation	*Joule heated boats *RF induction *e̅ beam *Laser beam	*Glow discharge .diode $\begin{cases} DC \\ RF \end{cases}$ *Triode *Magnetron *Ion beam
Vapor transfert	*Vacuum *Gas pressure .(reactive) *Plasma .(reactive or not)	*Gas pressure .inert .reactive *Vacuum (I.B.)
Vapor condensation	*Substrate .cold .heated .biased *Auxiliary I.B.	*Substrate .cold .heated .biased *Auxiliary I.B.

Table II - Physical vapor deposition

Table II gives a summary of the different variants that are currently encountered. Obviously all these variants are to be carefully designed in order to produce the desired film properties. In particular one has to take care of the quality of the starting materials, of the growing conditions (deposition rate, substrate temperature, residual atmosphere, ion bombardment), of the nature and the preparation of the substrate and of the effect of post-deposition treatment (1-3). Moreover even if general trends exist, each film is a new problem.

3.2. Chemical deposition

Methods of film formation by purely chemical processes are summerized in the Table III.

Gas phase chemical deposition

        - CVD
        - plasma polymerization
        - thermal oxidation

Liquid phase chemical deposition

■ Electrochemical
        - anodization
        - ion exchange
        - electroplating

■ Mechanical deposition of chemically reacting films
        - spray coating
        - sol-gel process

■ Chemical deposition
        - reduction plating
        - conversion coating
        - electroless plating

Table III - Chemical film formation

Gas phase chemical deposition : CVD is a materials synthesis process whereby constituents of the vapor phase react chemically to form a solid product (9). The chemical reactions are of several types : pyrolysis, reduction, oxidation, hydrolysis, synthesis, etc ... They are most often heterogeneous reactions and lead to good quality films. Thin film materials that can be prepared by CVD cover a tremendous range of elements and compounds.

Plasma polymerisation in a glow discharge (usually RF) can be regarded as a variant of low pressure CVD. The inlet gas (volatile organic materials) is decomposed by the glow discharge mainly at surfaces (electrodes, substrate) leaving the condensate species as a thin solid film.

Fondamental principles of these techniques involve reaction chemistry, thermodynamics, kinetics, transport mechanisms, film growth phenomena and reactor engineering. In particular, the understanding of nucleation phenomena is fundamental for the control of the structure of the growing film.

Liquid phase chemical deposition : Electrochemical processes involve anodisation, ion exchange and electroplating. Ion exchange in glass is probably the most convenient method for fabricating gradient index devices (10).

Chemical deposition processes concern reduction plating, electroless plating, conversion coating and displacement deposition. The oldest application of chemical reduction plating is the well known reaction for silvering glass and plastics to produce mirrors using silver-nitrate solutions together with one of various reducing agents such as hydrazine (11).

In mechanical deposition of chemically reacting films, reagent solutions are deposited on the surface by spraying, dipping and draining, spinning, brushing, etc ... Chemical reaction of the coating residue by thermal oxidation, hydrolysis or pyrolysis produces the desired thin solid film. Spray coating is probably the most widely used production method (12). Spinning or dipping of colloïdal solutions (sol-gel route) which will be discussed elsewhere in this school is a promising technique to produce amorphous films for optics.

3.3. What method for optics ?

In research laboratories almost all techniques have been employed The choice depends on your own problem and on financial supports.

On an industrial scale, electron beam evaporation is used extensively in instrumental optics. It ensures high evaporation rates and a great range of materials and refractive indexes are accessible. Sputtering has also industrial applications for surface treatment of windows. Low pressure CVD and spray coating are used mainly in photothermal solar energy conversion and solar cells.

Whatever the technique, the deposition of thin films is a highly multiparameter problem. Their control has to be rigorous in order to achieve the required properties. Such a control can be obtained only through efficient characterization techniques.

4. CHARACTERIZATION OF THIN FILMS

The high quality in present-day thin film technology is in part due to the development of performant characterization methods. The properties of interest are optical (refractive index and absorption coefficient) and analytical (composition, structure and bondings). Rather than consider in detail the capabilities of the available techniques, we will instead discuss briefly their performances according to the properties which are probed and then illustrate their usefulness in few examples.

4.1. Optical properties

One has to determine the thickness, the refractive index value $n(\lambda)$ and the absorption coefficient $\alpha(\lambda)$. Three methods are mainly used.

In spectrophotometry the reflectance $R(\lambda)$ or/and the transmittance $T(\lambda)$ of the film are measured. A comparison between experimental values and a model calculation allows us to deduce the thickness, n and α (13). The index accuracy is of the order of $\Delta n = 10^{-2}$. α values have to be higher than 50 cm^{-1} and thickness higher than 100 nm for good precision.

Ellipsometry makes use of the fact that in general an incident plane polarized beam is reflected in a state of elliptical polarization. Analysis of the ellipticity of the reflected beam can lead to an evaluation of the optical constants of the surface or film (14). Very thin films can be analysed (monolayer range) and the refractive index is determined with $\Delta n \simeq 10^{-3}$.

Guided wave spectrometry is the analysis of the propagating modes of light in a transparent film (15). A good precision implies a thickness higher than 300 nm. Index is measured with an incertitude lower than 10^{-3} and absorption coefficients in the 10^{3} cm^{-1} range are accessible. By varying the polarization state (TE or TM modes) birefringence of films can be determined.

In principle for these three methods, refractive index profiles can be extracted from the measurements, but the results are physical model dependent (13,16).

4.2. Non-optical properties

They concern the composition of the films, their crystalline structure and the chemical bondings.

One can determine the atomic composition of thin films by Rutherford backscaterring (RBS), Auger-electron spectroscopy (AES), secondary-ion mass spectroscopy (SIMS) or electron probe analysis (EPA) (17,18). In-depth analysis is possible using ion erosion for AES and SIMS ; RBS in-depth analysis is non destructive. Table IV summarizes the performances and limits of these techniques. Figure 2 is an illustration of typical RBS spectra using 2 MeV He^{+} beam : a) is relative to a film of aluminium evaporated through a 2.5 10^{-3} Torr N_2 partial pressure ;

343

	Stoechiometry	Impurity	In-depth analysis	Lateral resolution	Remarks
RBS	Yes – 1% Low mass resolution No chemical information	Yes $Z_i > Z_s$ 1% – 0.01%	Yes Non destructive 5 – 20 nm	0.1 mm	No sample degration Give mass/cm^2 NRA for $Z_i < Z_s$
AES	Yes – few % Chemical shift	Yes 0.1%	Yes Destructive 5 nm	50 nm	Sample modification possible by $\overline{e}$ bombardment Matrix effects (sputtering)
SIMS	Yes – few % High mass resolution Chemical effects	Yes ppm	Yes Destructive 5 nm	1 μm	Matrix effects possible in sputter etching
EPA	Yes – few % Chemical shift	Yes 0.1%	No 1 μm	few μm	In-depth analysis by LEEIXS

Table IV

344

b) concerns the same evaporation with similar N2 pressure but discharge assisted. These spectra show that nitridation of aluminium accurs only with ionized nitrogen (19).

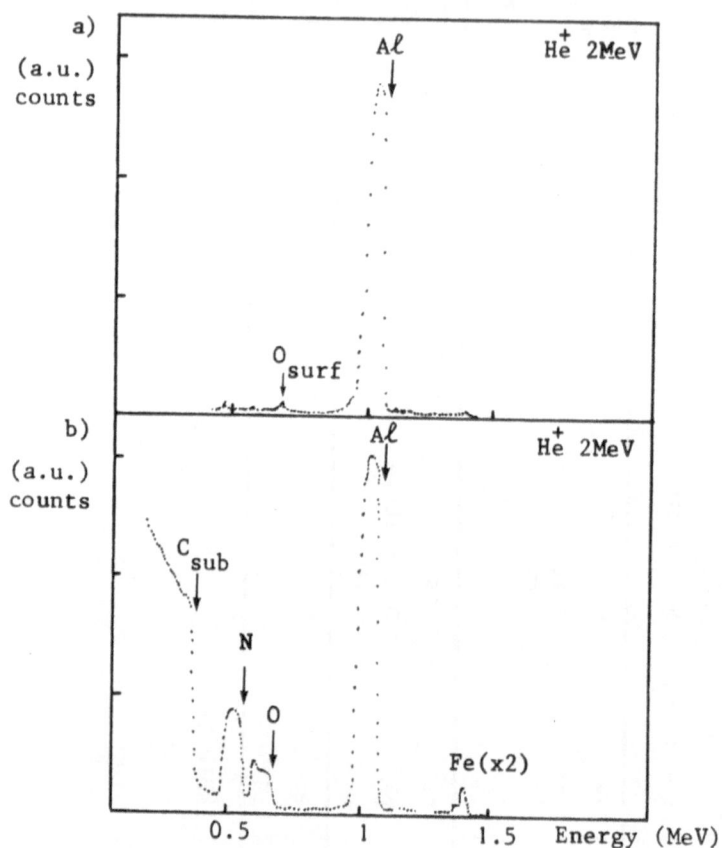

Fig. 2 - RBS spectra for aluminium compounds

The structural analysis of films is the domain of electron microscopy or diffraction and of X-ray diffraction. Transmission electron microscopy (TEM) gives structure details with a resolution of 0.5 to 2 nm using thinning or replica (20). This technique has revealed the columnar structure of numerous films (as TiO_2) which is the source of optical losses. Electron and X-ray diffraction can provide microcrystal identification or radial distribution functions of amorphous films (21). Neutrons can be used in principle but only for very thick films due to their low scattering cross-section.

To determine chemical boundings IR spectrophotometry is a classical technique even for solids where it gives rather broad peaks. The usefulness of this technique is illustrated by figure 3 where

IR absorption peaks of AℓN and Aℓ$_2$O$_3$ appear as clearly different
(19). Photoelectron spectroscopy (ESCA) is a sensitive method for
surface chemical analysis which is capable of in-depth information
when coupled with ion etching (18). Extended X-ray absorption fine
structure (EXAFS) which probes the next neighbours (first coordi-
nance sphere) gives a good chemical site characterization (22).
Low energy electron induced X-ray spectroscopy (LEEIX) provides
in-depth chemical analysis without erosion (23).

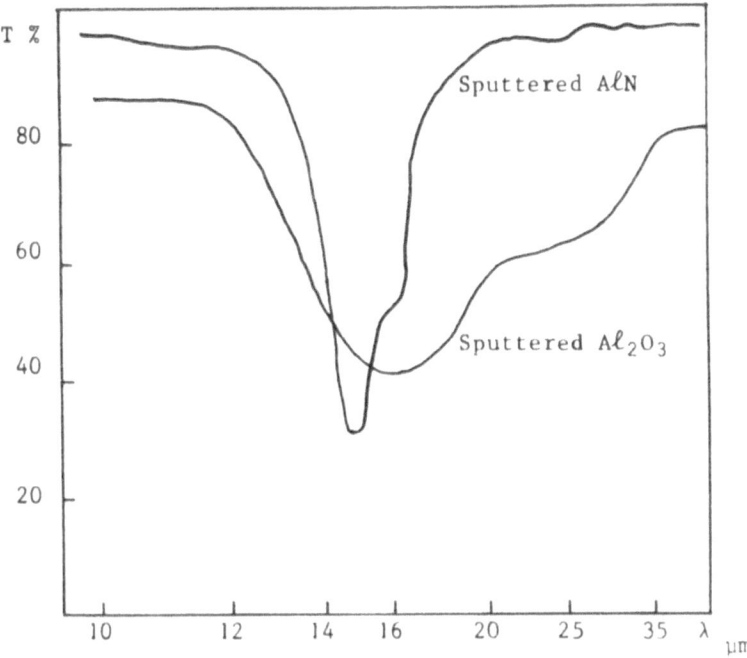

Fig. 3 - IR absorption peaks for aluminium compounds

5. THIN FILMS FOR OPTICS : USES AND PROBLEMS

Thin films play a vital role in nearly all optical devices. They
have long been familiar to modify the reflection or transmission of
optical systems : reflection coatings for laser mirrors or antireflection coatings for window glass, video screens and camera lenses.
Their use covers now a wide spectrum :
▪ antireflection coatings
▪ enhanced reflection mirrors (lasers)
▪ beam splitters : neutral or coloured
▪ filters : broad band, narrow band, polarizer
▪ absorbers : absorption or scattering
▪ waveguides : planar, channel.

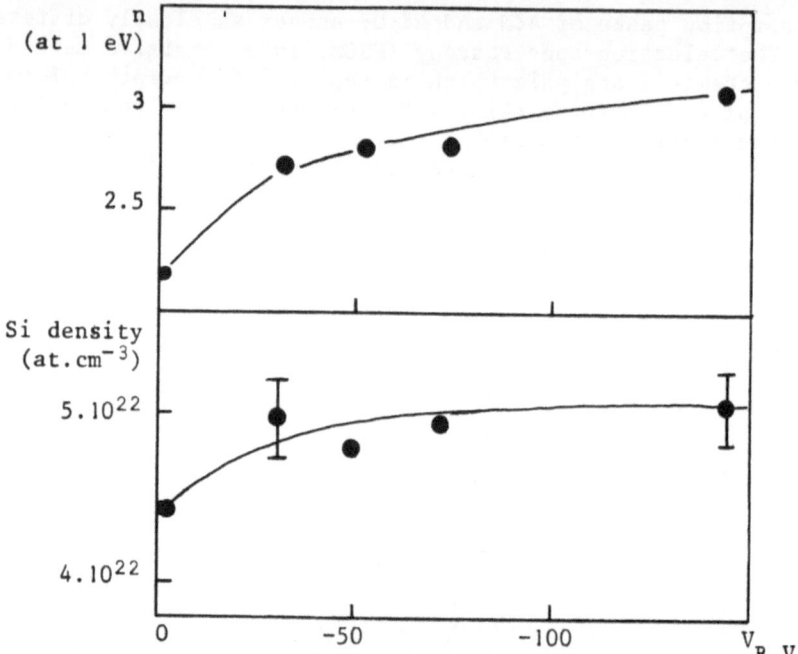

Fig. 4 - Influence of substrate bias on refractive
index and density of sputtered a-Si : H

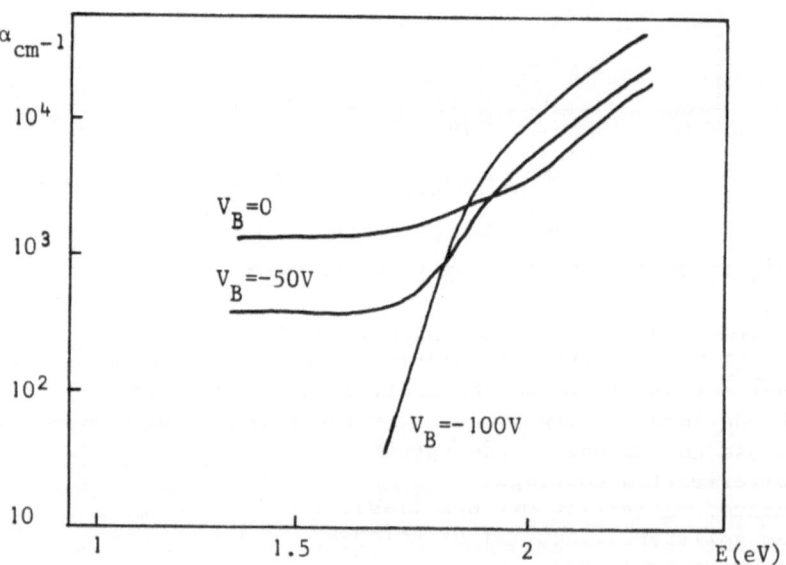

Fig. 5 - Absorption edge of sputtered a-Si : H

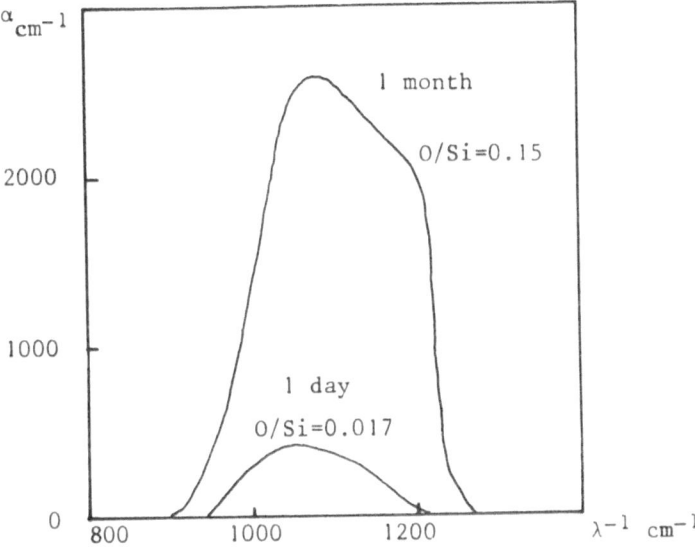

Fig. 6 - Post-oxidation of zero bias sputtered a-Si : H

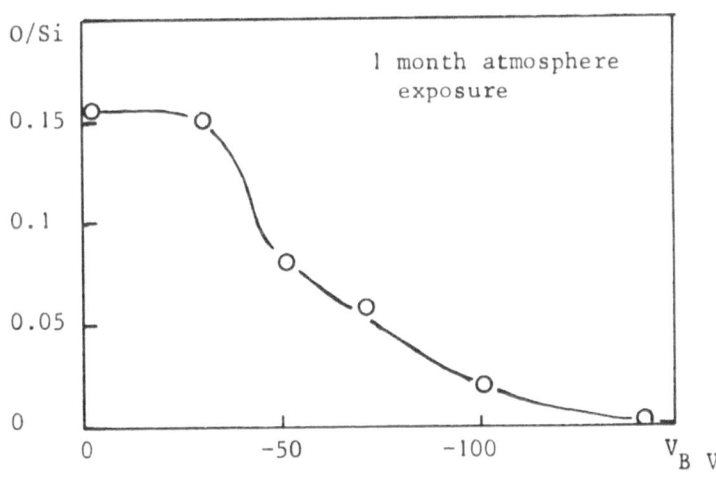

Fig. 7 - Influence of substrate bias on aging at room atmosphere

In spite of recent progresses in optical thin film technology, they are still largely imperfect and non well controled. Their microstructure is far from that of bulk materials. It consists in microcrystallites more or less interconnected which leads to a stacking coefficient lower than 1. This parameter depends on the materials to be deposited and also on methods and conditions of the deposition. The consequences concern the density, the refractive index and the losses (by scattering) of the film as well as its aging effects.

As an illustration we report here some results concerning a-Si : H thin films deposited by DC reactive sputtering in argon-hydrogen mixture (H_2/Ar 50 %, total pressure 50 mT, anode voltage 3.2 kV, substrate temperature 250 C). The variable parameter was the substrate negative bias which was warying between 0 and -140 volts in order to ensure a variable ion bombardment during film growth (24). Figure 4 shows the variation of silicon density and of refractive index as deduced from RBS and near IR spectrophotometry measurements. Figure 5 gives the evolution of absorption edge. The substrate bias has clearly a benefic influence on optical properties which appears correlated to reduction of microstructures. This is confirmed by the aging effect as measured by the evolution of oxygen content in the films. Figure 6 indicates that, in case of a zero bias film, oxygen concentration (as deduced from IR and RBS) increases by an order of magnitude after one month air exposure. This oxygen appears homogeneously distributed in-depth. Figure 7 shows the variations versus substrate bias of oxygen content after one month air exposure. The 140 volts biased films are very stable. These results are interpreted in terms of densification of the growing films due to ion bombardment. The consequences are an increase in refractive index, a decrease in scattering losses and a better resistance to oxidation induced aging effects.

Same kind of effects has been reported (25) on the properties of electron-beam evaporated thin films. Packing densities of SiO_2, TiO_2 and ZrO_2 have been increased by ion bombardment of the growing films, which has led to low absorption and scattering losses. In a ZrO_2 - SiO_2 multilayer interference filter the wavelength of the peak transmittance was stabilized when using ion-beam assisted deposition.

6. CONCLUSION

The application of thin films in optical technology is now well established. The goal of the deposition process is to produce thin films with accurately reproducible refractive index and thicknesses and with losses as low as possible. One has also to be aware of the importance of non-optical qualities such as mechanical (hardness, thermal expansion coefficient) and chemical (corrosion, materials compatibility).

While remarkable advances have been made, there are still a number of areas in which research is necessary. Improvements are needed in :
- minimizing aging effects in multilayer devices
- increasing laser damage threshold
- extending the wavelength range to X-UV region
- fabricating low-loss planar waveguides
- active guided wave devices for integrated optics
- highly performing filters for optical telecommunications

The problems encountered are closely related to the micro-structures and the microheterogeneities of the films which to a large extend are governed by the quality of the substrate prepa-ration and the proper choice and design of the deposition method.

REFERENCES

1. Maissel L.I. and R. Glang, eds, Handbook of thin film technology (Mc Graw-Hill, New-York, 1970).
2. Vossen J.L. and W. Kern, eds, Thin film processes (Academic, New-York, 1978).
3. Ritter E., in Physics of thin films, vol. 8 (G. Hass, M.H. Fran-combe and R.W. Hoffman, eds, Academic, New-York, 1975) 1.
4. Heavens O.S., in Physics of thin films, vol. 2 (G. Hass and R.E. Thun, eds, Academic, New-York, 1964) 193.
5. Kogelnik H., in Integrated Optics (T. Tamir, ed, Topics in applied physics, Springer-Verlag, Berlin, 1979) 13.
6. Aspnes D.E., Thin solid films, 89 (1982) 249.
7. Seraphin B.O. and A.B. Meinel, in Optical properties of solids : New developments (B.O. Seraphin, ed, North-Holland, 1976) 928.
8. De Neufville J.F., in Optical properties of solids : New develo-pments (B.O. Seraphin, ed, North-Holland, 1976) 438.
9. Sedgewick T.O. and H. Lydin, eds, Chemical vapor deposition (Proc. 7th Int. Conf. the Electrochem. Soc., Inc. Princeton, 1979).
10. Walker R.G., C.D.W. Wilkinson and J.A.H. Wilkinson, Appl. Opt., 22 (1983) 1923.
11. Vossen J.L. and W. Kern, Physics today, 33 (1980) 26.
12. Chapman B.N. and J.C. Anderson, eds, Science and technology of surface coating (Academic, London, 1974).
13. Borgogno J.P., B. Lazarides and E. Pelletier, Appl. Opt., 21 (1982) 4029.
14. Aspnes D.E., in Optical properties of solids : New developments (B.O. Seraphin, ed, North-Holland, 1976) 800.
15. Tien P.K., Rev. Modern Phys., 49 (1977) 361.
16. Cachard A. and A.M. Bouchoux, to be published.
17. Dupuy C.H.S. and A. Cachard, eds, Physics of non-metallic thin films (Plenum, New-York, 1976).

18. Thomas J.P. and A. Cachard, eds, Materials characterization using ion beams (Plenum, New-York, 1978).
19. Cachard A., unpublished results.
20. Guenther K.H., Appl. Opt., 20 (1981) 3487.
21. Temkin R.J., W. Paul and G.A.N. Connell, Advan. Phys., 22 (1973) 581.
22. Koch E.E., R. Haensel and C. Kunz, eds, Vacuum ultraviolet radiation physics (Pergamon, 1974).
23. Roche A., A. Cachard et al, J. Microsc. Spectrosc. Electron, 4 (1979) 351.
24. Tardy J., Thesis (Lyon, 1982).
25. Martin P.J., H.A. Macleod et al, Appl. Opt., 22 (1983) 178.

FUNCTIONAL GLASSES

J. BLETRY

SAINT-GOBAIN RECHERCHE 39, Quai Lucien Lefranc
93304 AUBERVILLIERS (FRANCE)

ABSTRACT

Basic properties and traditional functions of oxide glasses are first reviewed. Industrial research efforts to improve these properties or to achieve new functions are then presented. Emphasis is put on surface treatments, new glassy systems or compositions and composite products.

These general trends of research are finally illustrated with the example of coatings for energy saving glazings.

I - INDUSTRIAL RESEARCH FRAMEWORK

1. FUNCTION ALL

- Cheep 2.5 F/kg $\simeq$ potatoes
- Large scale production : 1 float unit $\simeq$ 15 000 000 m^2/year
  - $\rightarrow$ number of inhabitants x human size
- Reproducible and resistant product : 10 years warranty
       for French buildings

2. TRADITIONAL FUNCTIONS OF OXIDE GLASSES

2.1 DECORATION

- -5 000 BC Coloured glasses in Egypt

2.2 OPTICAL TRANSPARENCY OVER THE VISIBLE SPECTRUM

- -30 BC Glazings in Pompei

2.3 DURABILITY : CHEMICAL, MECHANICAL, UV IRRADIATION

- -1 500 BC : Bottles in Thebes
- Chemical resistance to : wine, beer, cider, oil, vinegar, perfumes, drugs ...
- Hardness
  - $\rightarrow$ Protection against climate variations : rain, wind, sun
- Easily transported by road and sea

3. TRADITIONAL WEAKNESS OF OXIDE GLASSES

3.1 FRAGILE :

. Breaks in tension
. Fracture starts from surface defects (GRIFFITH)

3.2 ORDER OF MAGNITUDE

. Theoretical resistance (bond strength) : 10^{10} Pa (Steel : 10^9 Pa !.)
. Usual annealed glass : compression : 10^9 Pa, tension : 10^8 Pa.

II – HOW TO IMPROVE TRADITIONAL FUNCTIONS OR TO OBTAIN NEW FUNCTIONS

1. SURFACE TREATMENTS

1.1 TO IMPROVE OPTICAL PROPERTIES

. Washing (or no washing !)
. Polishing : industrial decrease since the introduction of float glass process

1.2 TO IMPROVE MECHANICAL PROPERTIES

. Thermal treatments

- Annealing
- Toughening → prestressed glass with flexion strength x 5 (1929)

2. MODIFICATION OF THE SURFACE COMPOSITION

2.1 TO IMPROVE ESTHETIC OR OPTICAL PROPERTIES

. Ionic exchange under an electric field

2.2 TO IMPROVE MECHANICAL PROPERTIES

. Toughening by ionic exchange ($Na^+ \rightleftharpoons K^+$) with molten salts → prestressed glass with flexion strength x 10

3. COATINGS

. Cost of layer ≃ cost of glass → allows the use of expensive raw materials (Ag, Au) in very thin layers

3.1 TO IMPROVE ESTHETIC PROPERTIES

. Enamelling (15th century)
. Mirors (Pb 300 BC, Sn 16th, Ag 18th)
    → 3 F/m^2, 1 factory several millions m^2/year
    → sensibilisation of glass (nucleation centers for
      chemical deposition, 700 - 1000 Å silver, 2 varnish
      layers
    → Resist : 500^H, 50°C, 100 % H_2O or 50^H 120°C or

$$\text{cycling} \begin{cases} 100\ \%\ H_2O \\ 0.4\ \%\ SO_2 \\ 40°C \end{cases}$$

. Coloured layers

3.2 TO MODIFY OPTICAL OR INSULATING PROPERTIES

    → Static function → monolayer prefered
. Non reflective interference layers (MgF$_2$) :
    shops, paintings ...
. Reflection layer for sun control : high reflection in the
  visible spectrum, low transmitance in the solar infrared
  (involve Fe, Cr, Co, Ti oxides or "silicon") → buildings
. Far infrared reflection layers for thermal insulation :
  doped semiconductors (SnO$_2$-F...) or metals (Ag ...) → buildings
. Conductive layers (heating glazings : power between 5 and
  100 W/dm^2, voltage : between 12 and 300 V) - Antistatic layers (aircrafts)
. Magnetic ?

    → Regulation function → monolayer possible
. Photochromism → spectacles
. Thermochromic : metal-insulator transition (vanadium
  oxide) → buildings ?

    → Controlled function → multilayer necessary

$$\text{Displays} \rightarrow \begin{cases} . \text{ Electrochromism (solid) ITO - WO}_3 \text{ -MgF}_2 \text{ - ITO} \\ . \text{ Fluorescent ZnS - Mn (solid)} \\ . \text{ Liquid crystals} \end{cases}$$

3.3 DISPLAYS

    → All involve transparent electrode (ITO) on glass :
      actual price 200-250^F/m^2 !
    → Electronics market :
      Japan 1982: 3 100 000 m^2
    → Possible car market : technical aspects
      150 x 100 mm, red-green-yellow, contrast 7 : 1, re-
      sponse time 0.1 s, works during 4000^H and between
      -30 and + 85°C, cheep !

→ Decorative displays ? advertizing, indicators
(already in airports) : do not need such a good quality as electronic displays (→ 50 μm resolution)

3.4 TO IMPROVE MECHANICAL PROPERTIES :
SAFETY GLASSES

→ Plastic laminated glasses : keeps glass pieces together after breakage
→ Polyvinylbutyral film (PVB) : 15-20 F/m^2, 15 000 T/year
in Europ (1 500 T for buildings)
→ technical specifications : traction resistance 10-30
MPa, maximum elongation 150 to 300 %, optical transparency in the visible region > 85 %, "good" adhesion
to glass, works between - 20 and + 80°C
→ Results : can withdraw :
  · (sleeping) chicken 1,8 kg, 150 m/s
  · (running) mother in law 50 kg, ≈ 5 m/s
  · gangster attacks
→ Improvement with an inside layer of polyurethane: high
security windshield

4. COMPOSITION "DOPING" OF OXIDE GLASSES

Price of impurities inversely proportional to their concentration

4.1 TO MODIFY ESTHETIC PROPERTIES

Coloured impurities : Cr, Mn, Fe, Co, Ni, Cu, Ce, Nd,
Se ...

4.2 TO MODIFY OPTICAL PROPERTIES

Photochromic glasses : Ag Halogenides
response time may be slow (buildings), reasonable
(≈ 0.1/s for glasses or cars), fast (to protect against
laser or nuclear irradiation)

5. NEW COMPOSITIONS

5.1 FOR OPTICAL APPLICATIONS

→ Protection against X or γ irradiation : lead glasses,
against neutrons : B or Cd glasses ...

→ Transmission in the infrared : chalcogenides glasses...

5.2 FOR THERMOMECHANICAL APPLICATIONS

→ Low dilatation coefficient, resistance to high temperatures and thermal shocks :
boron oxide glasses, vitrocerams ...

→ Metallic glasses

→ Decreased fragility (and hardness) : organic glasses

5.3 TO CONTROL CHEMICAL DURABILITY

→ Fertilizing glasses (slowly leaching K^+, PO_4^{--}, oligo elements ...)

5.4 GLASSES FOR ELECTRONIC APPLICATIONS

→ Amorphous semi conductors (photovoltaïc cells with hydrogenated Si)

→ Optical fibers (doped silica)

5.5 MAGNETIC METALLIC GLASSES FOR TRANSFORMERS

6. GLASS CONTAINING COMPOSITE MATERIALS

6.1 TO IMPROVE MECHANICAL PROPERTIES

→ Armoured glasses

→ Glass fiber reinforced plastics

→ Glass covered plastics (hardness) ?

6.2 TO IMPROVE INSULATION PROPERTIES

→ Glass wool

III - WINDOW COATINGS FOR ENERGY SAVING : AN EXAMPLE OF BASIC INDUSTRIAL RESEARCH

1. INTRODUCTION

. 1973 : energy crisis, 25 % for heating buildings, 30 % window losses

2. CHALLENGES

. On float line (10-20 m/min) → coating rate 1000-10000 Å/s

. Corrosion resistance : 10 years
. Color uniformity

3. THE THREE PARTNERS

. Sun $\simeq$ Black body at 5800 K, 50 % at 0.72 μm, 700 W/m^2 + climate (number of sun days : may differ from 52)

. Housing spectrum $\simeq$ black body at 300 K,λ_{max} (μm) $= \dfrac{2898}{T}$ 50 % at 13.5 μm, 400 W/m^2

. Eye sensitivity curve

4. THERMAL BALANCE

+ Sun radiative contribution

− Radiative losses in the infrared

5. HOW TO IMPROVE THE RADIATIVE BALANCE

→ With infrared reflective coatings transparent in the visible spectrum

→ Use electronic plasma reflectivity (Incoming wave electric field → electronic plasma oscillations → reradiated electric field)

→ Two solutions :

Heavily doped semiconductors   ,   Metals

6. THEORETICAL SURVEY (DRUDE)

6.1 DIELECTRIC CONSTANT

. Incoming wave electric field → electronic plasma oscillation :

$$m^{*} \left(x'' + \frac{x'}{\tau} \right) = -eE$$

. Polarization : $P = -Nex$

. Dielectric constant : $\varepsilon = 1 + 4\,\Pi\,\dfrac{P}{E} = \varepsilon' - i\varepsilon''$

$$\begin{cases} \varepsilon' = \varepsilon_\infty \left[1 - \dfrac{\omega_p^2}{\omega^2 + \tau^{-2}} \right] \\[2em] \varepsilon'' = \varepsilon_\infty \dfrac{1}{\omega\tau} \dfrac{\omega_p^2}{\omega^2 + \tau^{-2}} \end{cases}$$

. Plasma frequency

$$\omega_p^2 = \frac{4 \Pi N e^2}{\varepsilon_\infty m^*}$$

6.2 WAVE PROPAGATION

. Maxwell

$$\nabla^2 E = \frac{1}{c^2} \varepsilon \frac{\partial^2 E}{\partial t^2}$$

. Alternating solutions

$$E_o \exp i\omega(t - \frac{x}{v})$$

with : $(\frac{c}{v})^2 = \varepsilon = (n - ik)^2$

$\rightarrow \quad \exp \left[i\omega(t - \frac{nx}{c}) \right] \exp(- \frac{k\omega x}{c})$

Phase shift $\qquad\qquad\qquad\qquad$ Attenuation

6.3 SKIN DEPTH AT LARGE WAVELENGTH

$\rightarrow$ if $\omega \tau \ll 1$ (and $\delta \gg v_F \tau = \ell$ mean free path)

$$\delta \approx \sqrt{\frac{\varepsilon_o c\lambda}{4 \Pi \sigma}} = 1.45 \ 10^{-2} \sqrt{\frac{\lambda(\mu m)}{\sigma}} \quad \text{(MKSA)}$$

$\rightarrow$ Examples : For $\lambda = 10 \ \mu m$

	F doped SnO_2	Silver
	$\sigma \approx 10^5 \ \Omega^{-1} m^{-1}$	$\sigma \approx 6.2 \ 10^7 \ \Omega^{-1} m^{-1}$
δ	1500 Å	60 Å

6.4 <u>THIN LAYER TRANSMISSION AND REFLECTIVITY</u>

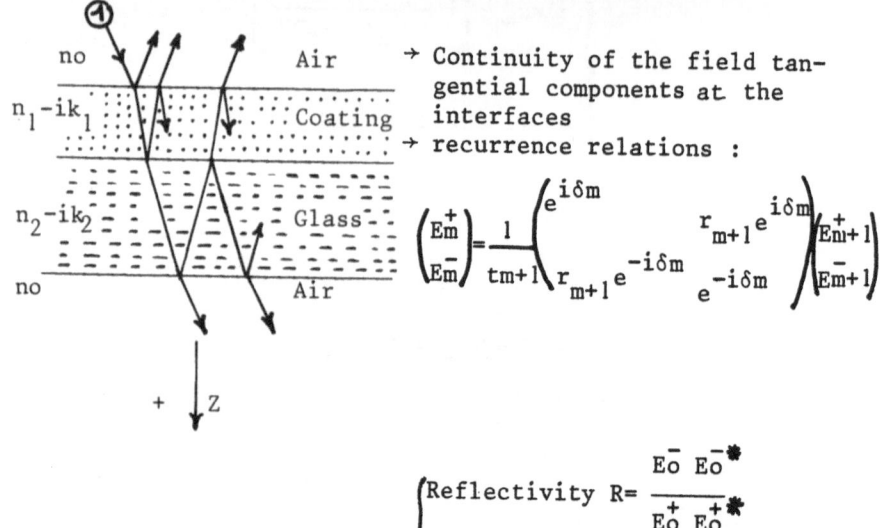

→ Continuity of the field tan-
  gential components at the
  interfaces

→ recurrence relations :

$$\begin{pmatrix} E_m^+ \\ E_m^- \end{pmatrix} = \frac{1}{t_{m+1}} \begin{pmatrix} e^{i\delta m} & r_{m+1} e^{i\delta m} \\ r_{m+1} e^{-i\delta m} & e^{-i\delta m} \end{pmatrix} \begin{pmatrix} E_{m+1}^+ \\ E_{m+1}^- \end{pmatrix}$$

→
$$\begin{cases} \text{Reflectivity } R = \dfrac{E_0^- \, E_0^{-*}}{E_0^+ \, E_0^{+*}} \\[3ex] \text{Transmitance } T = \dfrac{E_{m+1}^+ \, E_{m+1}^{+*}}{E_0^+ \, E_0^{+*}} \end{cases}$$

→ Involve factors like :

$$\exp(i\delta m) = \underbrace{\exp(i\tfrac{2\Pi}{\lambda}n_m d_m)}_{\text{Interference } e^{i\phi m}} \underbrace{\exp(-\tfrac{2\Pi}{\lambda}k_m d_m)}_{\text{Absorption } e^{-\alpha m}}$$

→ Thin coatings on thick glass

$$\begin{cases} \bar{R} = \dfrac{1}{2\Pi} \int R \, d\phi_2 \\[3ex] \bar{T} = \dfrac{1}{2\Pi} \int T \, d\phi_2 \end{cases}$$

6.5 <u>RESULTS</u>

→ Optical properties depend on 3 parameters :

. Layer Thickness → position and intensity of
  interference oscillations in the visible
  spectrum

. Carrier density : $N = \dfrac{\varepsilon_\infty m^* \omega_p^2}{4\Pi e^2}$

  → position of the reflectivity edge

. Carrier mobility : $\mu = \dfrac{e\tau}{m^*}$

→ slope of the reflectivity edge

→ Example : F doped SnO_2

. $\varepsilon_\infty = 4.3$, $m^* = 0.27m$, $\underset{\mu m}{\lambda} p = 3.7 \ 10^{10} \ \underset{cm^{-3}}{\sqrt{N}}$

7. COMPARAISON BETWEEN DIFFERENT MATERIALS

	OXIDE SEMICONDUCTORS	METALS
THICKNESS $\simeq$ SKIN DEPTH	10 times larger	10 times smaller
	Interference	$\simeq$ No interference
	Corrosion resistant ↓	Handle with care ↓
	Mono layer	Multilayer : nucleation layer + protective layer

8. COATING IMPROVEMENTS

. Thickness increase : not interesting (cost, transmission, reflectivity)
. Increase of carrier density : for semi-conductors
. Mobility increase : very interesting
. Since $\mu \propto \tau$ and :

$$\frac{1}{\tau} = \frac{1}{\tau phonons} + \frac{1}{\tau impurities} + \frac{1}{\tau gain \ boundaries}$$

→ Irreducible , Avoid parasitic , Increase grain size
               impurities

9. THERMAL CHARACTERISTICS

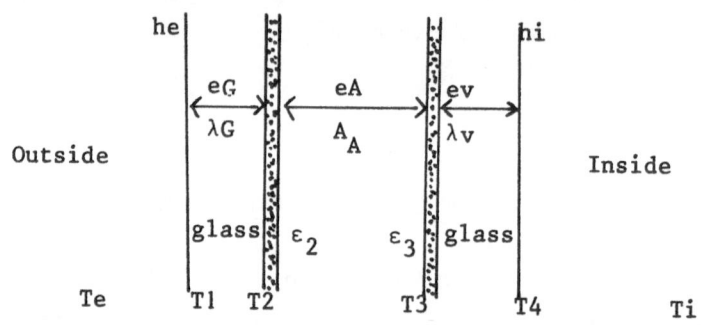

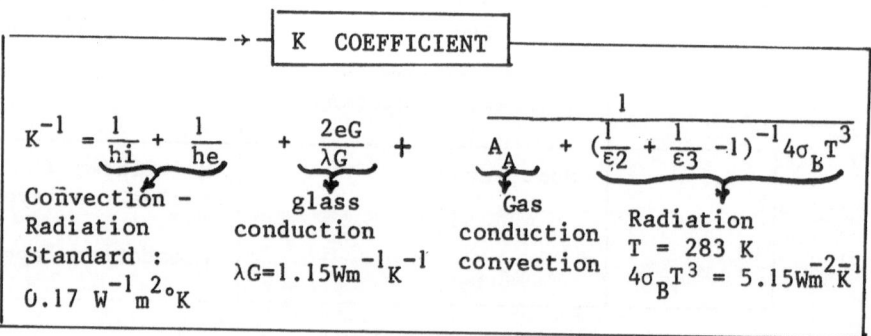

$$K^{-1} = \underbrace{\frac{1}{hi} + \frac{1}{he}}_{} + \underbrace{\frac{2eG}{\lambda G}}_{} + \underbrace{\frac{A}{A}}_{} + \cfrac{1}{\underbrace{(\frac{1}{\varepsilon 2} + \frac{1}{\varepsilon 3} - 1)^{-1} 4\sigma_B T^3}_{}}$$

Convection –
Radiation
Standard :
$0.17 \ W^{-1} m^2 {}^\circ K$

glass
conduction
$\lambda G = 1.15 Wm^{-1} K^{-1}$

Gas
conduction
convection

Radiation
$T = 283 \ K$
$4\sigma_B T^3 = 5.15 Wm^{-2} K^{-1}$

$\rightarrow$FS SOLAR FACTOR IN %

10. TEST HOUSES

Chantereine (near misty Compiègne – FRANCE)
$100m^2$, $20m^2$ south equivalent glazings, $250m^3$

$\frac{K}{(Wm^{-2}K^{-1}}$	FS %	External heat sources K W H	Losses K W H	Needs K W H
5.7 (4mm single glazing)	0.88	3851	14 193	10 342
3.1 (4-12-4mm double glazing)	0.75	3056	11 356	8 306
2.3 (4-12-4mm double glazing with EKO layer)	0.74			7 352
2.5 (4-6-4-6-4mm triple glazing)	0.70			7 796
1.8 (4-12-4mm double glazing +ε=0,1 coating)	0.50	1958	9 649	7 691

CONCLUSION

- New glass functions
- New solutions
- →Involve strong research in solid state physics and chemistry
- Transfers from other industries (Electronics)
- New application :Displays

SPECTROSCOPY OF OPTICAL GLASSES

David L. Griscom

Naval Research Laboratory
Washington, DC 20375, USA

1. INTRODUCTION

Spectroscopy is the study of the interaction of radiation with matter as a function of the energy or wavelength of the absorbed or emitted particles—most commonly, photons or electrons. The application of spectroscopy to glasses and other materials without long-range atomic order is a specialized subfield which has been treated in at least one monograph [1] and will be the subject of the present discussion. Structural studies based on the <u>diffraction</u> of the impinging radiation by the material system are sometimes considered to be outside of the field of spectroscopy and for brevity will not be considered here. It is sometimes convenient to order the various spectroscopic techniques according to the energy regimes probed. From the highest to the lowest energies, these include: Mössbauer (γ ray), x-ray, ultraviolet (UV), visible, infrared (IR) and Raman, electron spin resonance (ESR), nuclear magnetic resonance (NMR), and acoustic. However, numerous derivative spectroscopies have been developed involving various combinations of the preceding, e.g., radioluminescence, cathodoluminescence, photodetected ESR, electron-nuclear double resonance (ENDOR), and photoacoustic spectroscopy.

There are very few "routine" spectroscopies, particularly as applied to glasses. Useful information can be derived from spectroscopic data only when all instrumental effects are thoroughly understood, the impurity contents of the samples and their spectral influences are known, and a reliable theoretical framework exists to relate the intrinsic features of the spectra to the glass structure [1,2]. Different techniques probe different aspects of this structure. Mössbauer, ESR and NMR give information about local atomic arrangements and about the electric and magnetic fields and their gradients at the positions of the atoms, ions, or electrons being probed. X-ray, UV and visible spectroscopies--together with photoconductivity, electron energy loss (ELS), photoelectron spectroscopy (PES) and Auger electron spectroscopy (AES)--probe the electronic structure of the material, i.e., its band structure. At the low end of the probe beam energy range, IR, Raman, and acoustic spectroscopies provide data more strongly related to vibronic structure.

Clearly, comprehensive discussions of even a small fraction of these methods is prohibited by the present abreviated format, as is also any attempt to cover a wide range of glass types and compositions. The visible and IR spectroscopies of many complex fluoride and chalcogenide glasses of current interest are extensively treated elsewhere in this volume by J. A. Savage, J. Lucas, J. A. Gannon, M. G. Drexhage, and D. C. Tran. Therefore, as "core material" for the present round table discussion 1), it was decided to offer an overview of some spectroscopic investigations of pure SiO_2 (c.f., Ref. 3) and, to a lesser extent, two of its structural analogues BeF_2 and $ZnC\ell_2$. The primary aim of the written summary is to define and briefly discuss many of the generic spectroscopic approaches, using these simple one-component network glasses as examples. Several specific spectroscopic techniques, representing the particular research interests of the author and the other panel members will be highlighted.

2. HIGH-ENERGY SPECTROSCOPIES

For present purposes, high-energy spectroscopies will be defined as those for which the incident probe particles have kinetic energies exceeding the optical band gap E_G of the material (for silica, $E_G \approx 9$ eV). Mössbauer spectroscopy will be excluded from discussion because relatively pure silica and its halide analogues generally contain insufficient quantities of Mössbauer-active nuclides (e.g., ^{57}Fe, ^{119}Sn, ^{121}Sb, and ^{125}Te) for practical study. Useful discussions of the application of Mössbauer spectroscopy to glasses have been given elsewhere [1,4].

The x-ray spectroscopies are logically divided into absorption, emission, and photoelectron techniques, with a further subdivision according to whether valence bands or core levels are being probed. Figure 1a schematically illustrates these different types of spectra while defining the essential nomenclature. X-ray absorption spectroscopy generally involves electronic transitions from a core level

---

1. Participating in this round table discussion were C. A. Angell, P. H. Gaskell, J. Lucas, and J. R. Gannon. Their contributions are gratefully acknowledged.

364

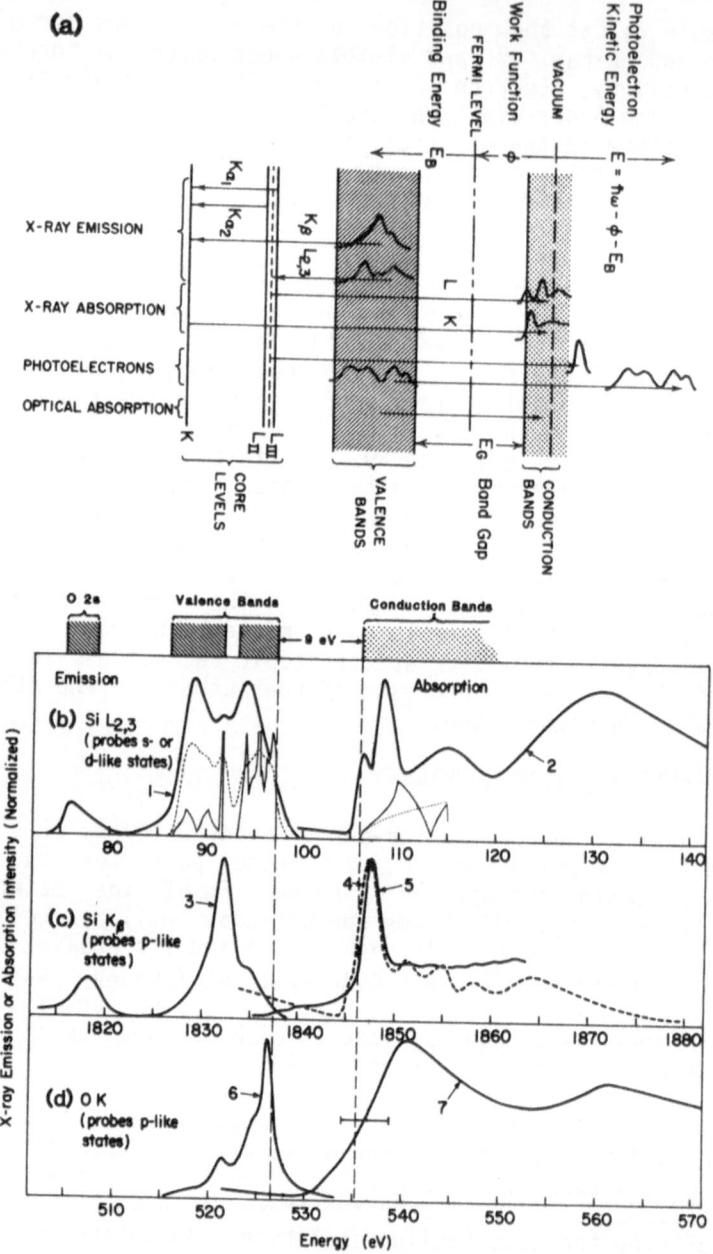

Fig. 1. (a) Schematic of various x-ray and photoelectron spectroscopies. (b), (c), and (d) x-ray absorption and emission spectra of silica and their relationships to energy band structure. (After Ref. 3.) Data due to the following sources: (1) Ref. 7; (2) and (7) Ref. 8; (3) and (6) Ref. 9; (4) Ref. 10; (5) Refs. 11 and 12.

to the conduction band, with the wavelength of the incident x-rays being scanned. However, detection may be accomplished by monitoring the x-ray fluorescence at another (fixed) wavelength. On the other hand, in x-ray emission spectroscopy the exciting wavelength is held fixed and the fluorescence emission is scanned.

In PES (sometimes referred to as electron spectroscopy for chemical analysis, or ESCA), the exciting quantum $\hbar\omega$ is again held fixed, but it is the emitted photoelectrons which are energy analyzed and counted. If the incident x-rays should be replaced by a mono-energetic electron beam (with a small modulation $\lesssim 1$ eV) and the transmitted (or reflected) electrons are similarly analyzed, the experiment becomes ELS. Strong analogies can be drawn between ELS spectra and the corresponding x-ray or optical absorption spectra.

If the incident x-rays or electrons are sufficiently energetic as to create holes in core states and the appropriate emitted electron energy regime is scanned, the sharply-peaked Auger electron spectrum will be recorded (not illustrated in Fig. 1). Additional details concerning the various high energy spectroscopies can be found in a number of sources: Ref. 3 offers an expanded overview for the nonspecialist with a particular orientation toward fused silica, while Refs. 5 and 6 provide more comprehensive treatments of x-ray and photoelectron spectroscopies, respectively.

Figure 1b, c, and d illustrate the results of some x-ray absorption and emission studies of vitreous silicon dioxide (v-SiO_2). These data from a variety of sources [7-12] have been registered with one another on a common energy scale by means of additional spectroscopic measurements as described in Ref. 3 (see also Ref. 13). The relative positions of the conduction and valence bands are indicated above Fig. 1b. The x-ray spectra show that the O 2s level is somewhat broadened and admixed with Si 3s and 3p states due to bonding, but that on the other hand there is no $\underline{sp}$ hybridization of the O 2s and 2p orbitals (see discussion in Ref. 3). Faintly drawn as a dashed curve in Fig. 2b is the valence band density of states as determined by ESCA. Note how the Si $L_{2,3}$, Si K, and O K x-ray emission spectra each sample separate characteristics of the valence band states. Note also that if the experimental resolutions of the various x-ray spectroscopies were believed to be intrinsic properties of the material, the absence of a band gap would be inferred. But such an inference is belied by the well-known transparency of silica. Indeed, theoretical calculations of the densities of states (faintly drawn unbroken curves in Fig. 1b) show the top of the valence band and bottom of the conduction band to be very sharply defined (Ref. 14 and literature cited in Ref. 3). Thus, as will be expanded upon in Sec. 3, the band gap of silica stands at ~ 9 eV. Various peaks in the x-ray absorption spectra (right hand side of Fig. 1b, c, and d) were initially interpreted as corresponding to critical points in the conduction band density of states [3], but in analogy with conclusions regarding the ultraviolet absorption spectrum [14], these x-ray peaks are probably best ascribed to exciton effects (binding of the excited electron to the core level hole).

Additional ESCA studies [15-17] have centered on the influences of alkali modifier additions to silicate glasses. Also, a wide range of high energy spectroscopies and theoretical calculations have recently been brought to bear on the question of the electronic structure of beryllium fluoride glass [18].

3. OPTICAL SPECTROSCOPIES

Figure 2 displays the optical absorption (and scattering) loss spectra of high purity fused silica, as gleaned from a number of sources [19-28]. As indicated by the large horizontal arrows in this figure, different spectral regions generally require differing experimental techniques and/or sample geometries. For example, in the far UV and IR spectral regions, extremely thin samples are required for practical transmission of light, but even then the absorptivity cannot be unambiguously separated from the effects of reflectivity. In these cases, it is customary to perform a Kramers-Kronig analysis of the reflectivity spectrum in order to extract the index of refraction $n(\omega)$ and extinction coefficient $k(\omega)$ (e.g., Refs. 19, 29 and 30). The absorption coefficient α plotted on the right hand ordinate of Figure 2 is given by $\alpha = 4\pi k/\lambda$, where λ is the wavelength of the probe light. In the visible and near IR regions of the spectrum, silica is so transparent that calorimetric techniques or fiber optic geometries are usually required for reliable measurements. Attenuation measured in optical fibers is customarily expressed in units of decibells/km (left hand ordinate in Fig. 2).

The ultimate transparencies of glasses comprise a question of critical interest from both the scientific and technological points of view [31]. The band gap of silica has been determined from photoconductivity to lie in the range 9.0 [20] –9.3 eV [21]. Thus, as is also supported by theoretical calculations [14], the intense absorption taking place in the range ~ 8-9 eV is inferred to be due to excitons. Absorption by excitons can be expected in general to give rise to an exponential tail, known as the Urbach edge [32]. The probable position of this edge is indicated in Fig. 2 by the short-dashed extrapolation of the loss spectrum in the range ~ 8.0 – 8.5 eV. Other "UV tail" absorptions of lower intensity and slope have been predicted as possible consequences of vitreous disorder [31], and the loss data of Ref. 24 were ascribed to the existence of such intrinsic "tail states" (long dashed curves in Fig. 2). However, the similarity of the variously measured "UV tails" to the radiation-induced absorption spectrum of high purity silica suggests a defect origin for such manifestations. Defects are also the most

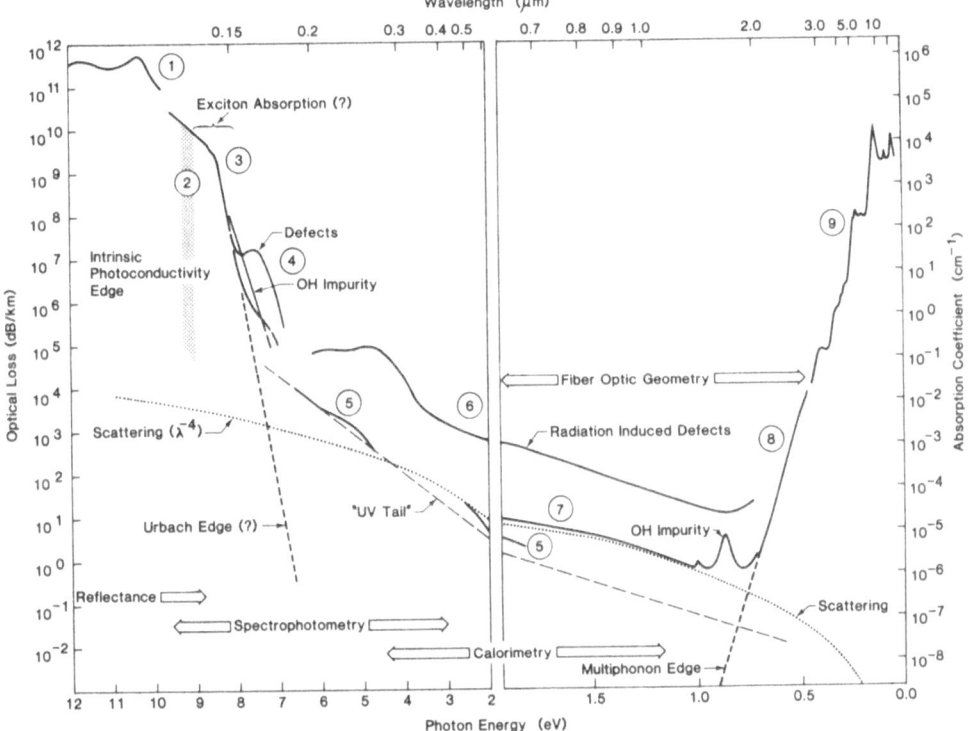

Fig. 2 Optical attenuation spectra of pure fused silica (1)-(6),
(9) and germanium-doped-silica optical fibers (7), (8).
Data due to the following sources: (1) Ref. 19; (2) Refs.
20 and 21; (3) Ref. 22; (4) Ref. 23; (5) Ref. 24; (6) Ref.
25; (7) Ref. 26; (8) and (9) Ref. 28.

probable cause of the absorption peak near 7.6 eV which varies in
intensity from sample to sample of high-purity low-OH silicas [23].
The data show clearly that the presence of OH in the material tends
to suppress this band, replacing it with a peak (or tail) at
somewhat higher energy.

It appears that the ultimate transparency of v-SiO$_2$ occurs at
the intersection of the Rayleigh scattering (dotted) and multiphonon
edge (short dash) curves near 2.0 μm (0.75 eV). Clearly, the last
vestiges of OH impurities must be removed in order to achieve this
ultimate low loss of ~ 0.15 dB/km (e.g., Refs. 26 and 27), but
OH-related losses may increase with time after fiber drawing due to
in-diffusion of hydrogen (e.g., Ref. 34). The Rayleigh scattering
losses arise from "frozen-in" density fluctuations which are linearly dependent on the fictive temperature of the glass (~ 1700K for
silica) [31].   The multiphonon edge invariably seems to take on an
exponential dependence on frequency in the high frequency limit
[33], justifying the dashed extrapolation in Fig. 2.

Careful analyses of the IR restrahlen bands in quartz and fused silica have been given in Ref. 29, where these results were interpreted in terms of various models for the glass structure. Raman spectra spanning the same frequency range are also important in forming unified structural interpretations, since IR and Raman spectroscopies are related to the same densities of vibrational states $g(\omega)$ via different frequency-dependent transition moments. Some recent Raman studies of v-SiO$_2$ can be found in Refs. [35-37]. $g(\omega)$, in turn, has been determined from neutron inelastic scattering [38]. Some IR, Raman, and cold neutron scattering results for BeF$_2$ and ZnCl$_2$ are summarized in Ref. 1.

4. MICROWAVE AND RADIO-FREQUENCY SPECTROSCOPIES

The nuclear and electron-spin-resonance experiments are most commonly carried out by bathing the sample in a fixed-frequency radiation field (microwave for ESR, R.F. for NMR) and examining the absorption of energy from this field as a function of the magnitude of an externally applied magnetic field (see, e.g., Refs. 1,2,4 or 39). Resonance absorption is observed between the Zeeman levels of nuclei with non-zero magnetic moments (NMR) or between those of unpaired electrons (ESR). For SiO$_2$ the only stable magnetic nuclei in nature are ^{29}Si (4.7% abundant) and ^{17}O (0.037% abundant). To date, the low abundance and other factors have weighed against ^{29}Si NMR in v-SiO$_2$, although some useful data have been obtained for alkali silicate glasses [39]. However, the ^{17}O NMR spectrum of an isotopically-enriched amorphous SiO$_2$ specimen has been recently obtained and interpreted in terms of the effects of statistical variations in the Si-O-Si bond angles [40].

ESR studies of v-SiO$_2$ have led to the characterization of numerous defects and impurities [41,42]. Although certain of these defects can result from the drawing of silica into fibers [43], most are induced by ionizing radiations. Intrinsic radiation-induced defects in v-SiO$_2$ include the E$^{\prime}$ center (a hole trapped at an oxygen vacancy) [44], the nonbridging oxygen hole center ($\equiv$ Si-O$\cdot$) [45], and the peroxy radical ($\equiv$Si-O-O$\cdot$)[46]. In addition, some extrinsic defects have been observed in nominally high-purity v-SiO$_2$ following irradiation; these include the formyl [47] and methyl [48] radicals. Figure 3a illustrates the ESR spectra of the formyl radical in silica [47]; spin Hamiltonian parameters extracted by means of the dotted computer line-shape simulations can be related to details of the structural model (Fig. 3b). Isochronal annealing experiments (Fig. 3c) performed following x-irradiation at 77K have resulted in important insights into the radiation chemistry of v-SiO$_2$ containing OH groups. In particular, the role of radiolytic molecular hydrogen in the (diffusion-limited) thermal bleaching processes occurring above 200K has been inferred [47, 49].

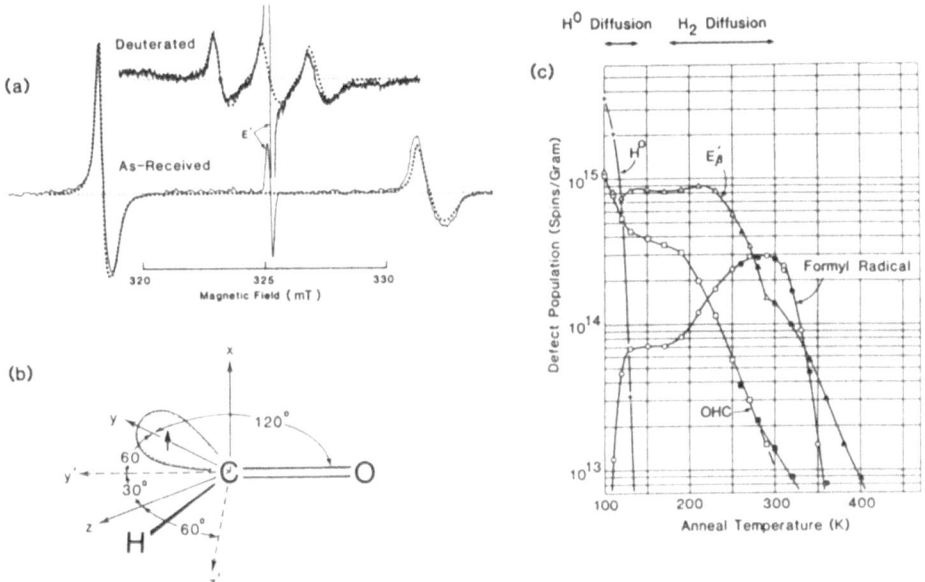

Fig. 3. ESR of the formyl radical in x-irradiated v-SiO₂, due to the reaction of radiolytic H° with $\lesssim$ 0.1 ppm CO impurity (after Ref. 47).

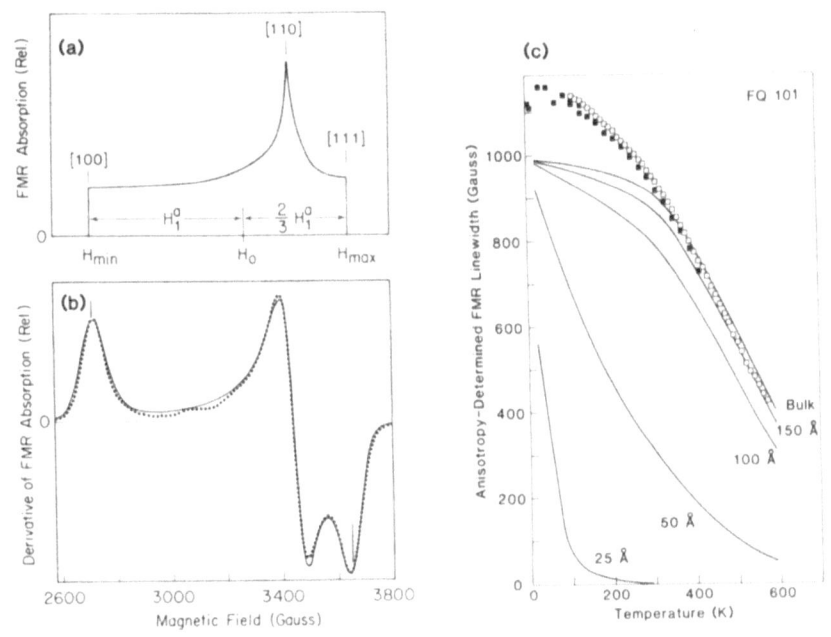

Fig. 4. FMR of single-domain metallic iron precipitated in v-SiO₂ containing 120 ppm Fe impurity (after Refs. 50-52).

Under certain processing conditions, fused silica containing $\sim 1 -100$ ppm iron (e.g., fused natural quartz) can be caused to precipitate fine-grained spherical particles of metallic iron [50]. When studied in a standard ESR spectrometer, such materials exhibit well-defined ferromagnetic resonance (FMR) spectra (Fig. 4b), which by means of powder pattern analysis (Fig. 4a) and computer line-shape simulation can be related to the chemical purity of the iron and other details [51]. The width of the resonance is given by $\Delta H_{pp} \approx (5/3)H_{a_1}$, where H_{a_1} is the anisotropy field of the ferromagnetic material. Figure 4c provides a comparison of the experimental anisotropy-determined linewidth for rather pure metallic iron preci-pitates (data points) with the theoretically predicted temperature dependences for various particle sizes (curves) [52]. Particle diameters were estimated from the high-temperature data to be $\sim 150A$; an interfacial rind of iron oxide is thought to account for the low-temperature discrepancies [52].

BeF_2 and $ZnC\ell_2$ glasses have also been studied by ESR, albeit to a lesser extent than SiO_2. A paramagnetic Mn^{2+} impurity and a ferrimagnetic precipitate (probably nickle ferrite) have been obser-ved in a distilled BeF_2 glass [53]. Irradiation of $ZnC\ell_2$ glass at low temperatures gives rise to the spectrum of $C\ell_2^-$ molecular ions [54]. The ESR spectrum of Mn^{2+} in $ZnC\ell_2$ glasses and compounds has been extensively characterized [55].

1. J. Wong and C. A. Angell, Glass Structure by Spectroscopy (Marcel Dekker, New York, 1976), 864 pp.
2. D. L. Griscom, in Borate Glasses, eds. L. D. Pye, V. D. Frechette, and N. J. Kreidl (Plenum, New York, 1978), p. 11.
3. D. L. Griscom, J. Non-Cryst. Solids 24 (1977) 155.
4. P. C. Taylor, in Treatise on Materials Science and Technology, Vol. 12, Glass I, M. Tomozawa and R. H. Doremus, eds. (Academic Press, New York, 1977) p. 223.
5. L. V. Azároff, ed., X-Ray Spectroscopy (McGraw-Hill, New York, 1974).
6. T. A. Carlson, Photoelectron and Auger Spectroscopy (Plenum, New York, 1975).
7. O. A. Ershov, D. A. Goganov, and A. P. Lukirskii, Sov. Phys.--Solid State 7 (1966) 1903.
8. O. A. Ershov and A. P. Lukirskii, Sov. Phys.--Solid State 8 (1967) 1699.
9. G. Klein and H.-U. Chun, Phys. Stat. Sol. (b) 49 (1972) 167.
10. Y. Cauchois and C. Bonnelle, C. R. Acad. Sci. (Paris) 242 (1956) 1596.
11. C. Senemaud, M. T. Costa Lima, J. A. Roger, and A. Cachard, Chem. Phys. Letters 26 (1974) 431.

12. M. T. Costa Lima and C. Senemaud, Chem. Phys. Letters 40 (1976) 157.
13. G. Wiech, E. Zöpf, H.-U. Chun, and R. Brückner, J. Non-Cryst. Solids 21 (1976) 251.
14. R. B. Laughlin, Phys. Rev. B 22 (1980) 3021.
15. R. Brückner, H.-U. Chun, and H. Goretzki, Glastechn. Ber. 51 (1978) 1.
16. D. J. Lam, A. P. Paulikas, and B. W. Veal, J. Non-Cryst. Solids 42 (1980) 41.
17. R. Brückner, H.-U. Chun, H. Goretzki, and M. Sammet, J. Non-Cryst. Solids 42 (1980) 49.
18. K. L. Bedford, R. T. Williams, W. R. Hunter, J. C. Rife, M. J. Weber, D. D. Kingman, and C. F. Cline, Phys. Rev. B 27 (1983) 2446.
19. H. R. Philipp, J. Phys. Chem. Solids 32 (1971) 1935.
20. T. H. DiStefano and D. E. Eastman, Solid State Commun. 9 (1971) 1560.
21. Z. A. Weinberg, G. W. Rubloff, and E. Bassous, Phys. Rev. B 19 (1979) 3107.
22. A. Appleton, T. Chiranjivi, and M. Jafaripour-Ghazvini, in The Physics of SiO$_2$ and Its Interfaces, ed. S. T. Pantelides (Pergamon, New York, 1978) p. 94.
23. I. P. Kaminow, B. G. Bagley, and C. G. Olson, Appl. Phys. Lett. 32 (1978) 98.
24. D. A. Pinnow, T. C. Rich, F. W. Ostermayer, Jr., and M. DiDomenico, Jr., Appl. Phys. Lett. 22 (1973) 527.
25. D. L. Griscom and E. J. Friebele, Rad. Effects 65 (1982) 63.
26. H. Osani, T. Shioda, T. Moriyama, S. Araki, M. Horiguchi, T. Izawa, and H. Takata, Elect. Lett. 12 (1976) 549.
27. T. Miya, Y. Terunuma, T. Hosaka, and T. Miyashita, Elect. Lett. 15 (1979) 106.
28. T. Izawa, N. Shibata, and A. Takeda, Appl. Phys. Lett. 31 (1977) 33.
29. P. H. Gaskell and D. W. Johnson, J. Non-Cryst. Solids 20 (1976) 153; ibid., 171.
30. H. R. Philipp, J. Appl. Phys. 50 (1979) 1053.
31. J. Tauc, in Optical Properties of Highly Transparent Solids, eds. S. S. Mitra and B. Bendow (Plenum, New York, 1975) p. 245.
32. J. D. Dow, Ref. 31, p. 131.
33. T. C. McGill, Ref. 31, p. 3.
34. N. Uesugi, T. Kuwabara, M. Ohashi, Y. Ishida, and N. Uchida, Electron. Lett. 19 (1983) 842.
35. F. L. Galeener, Solid State Commun. 44 (1982) 1037.
36. G. E. Walrafen and P. N. Krishnan, Appl. Optics 21 (1982) 359.
37. A. G. Revesz and G. E. Walrafen, J. Non-Cryst. Solids 54 (1983) 323.
38. A. J. Leadbetter and M. W. Stringfellow, in Neutron Inelastic Scattering (IAEA, Vienna, 1974) p. 501.

39. P. J. Bray, A. E. Geissberger, F. Bucholtz, and I. A. Harris, J. Non-Cryst. Solids 52 (1982) 45.
40. A. E. Geissberger and P. J. Bray, J. Non-Cryst. Solids 54 (1983) 121.
41. D. L. Griscom, Ref. 22, P. 232.
42. D. L. Griscom, J. Non-Cryst. Solids 40 (1980) 211.
43. E. J. Friebele, G. H. Sigel, Jr., and D. L. Griscom, Appl. Phys. Lett. 28 (1976) 516.
44. D. L. Griscom, Phys. Rev. B 20 (1979) 1823; ibid. 22 (1980) 4192.
45. M. Stapelbroek, D. L. Griscom, E. J. Friebele, and G. H. Sigel, Jr., J. Non-Cryst. Solids 32 (1979) 313.
46. E. J. Friebele, D. L. Griscom, M. Stapelbroek, and R. A. Weeks, Phys. Rev. Lett. 42 (1979) 1346.
47. D. L. Griscom, M. Stapelbroek, and E. J. Friebele, J. Chem. Phys. 78 (1983) 1638.
48. E. J. Friebele, D. L. Griscom, and K. Rau, J. Non-Cryst. Solids 57 (1983) 167.
49. D. L. Griscom, J. Non-Cryst. Solids, (in press).
50. D. L. Griscom, E. J. Friebele, and D. B. Shinn, J. Appl. Phys. 50 (1979) 2402.
51. D. L. Griscom, J. Magn. Res. 45 (1981) 81.
52. D. L. Griscom, IEEE Trans. on Magnetics, MAG-17 (1981) 2718.
53. D. L. Griscom, M. Stapelbroek, and M. J. Weber, J. Non-Cryst. Solids 41 (1980) 329.
54. D. L. Griscom, J. Non-Cryst. Solids 31 (1978) 241.
55. A. Chatelain and R. A. Weeks, J. Chem. Phys. 52 (1970) 3758.

ROUND TABLE DISCUSSION ON THE FUTURE OF OPTICAL WAVEGUIDES

A. Cachard

Université de Saint-Etienne, 23, rue Paul Michelon,
42023 Saint-Etienne Cedex, France.

This round table, devoted to recent developments in optical
waveguides, was centered on the lectures of J. Lucas, J.N. Gannon,
M.D. Drexhage and D.C. Tran. It was designed to give an informal
debate mainly in the field of fluoride glasses and I.R. transmitting optical fibers. It was led by J. Lucas and the intervenants
were J.N. Gannon, M.D. Drexhage, D.C. Tran, C.T. Moynihan,
J.A. Savage, H. Poignant, N.J. Kreidel, J. Perez and D. Griscom.

A large number òf questions were debated. They will be
presented in three parts :   The first concerns problems with
materials, the second the fiber drawing, and finally the industrial
interest was considered.

Materials.

Interest was focused on fluoride glasses (fluorozirconate)
with questions concerning the stability of glasses, purity, optical
quality and reproducibility.

The stabilization of the glasses is generally obtained by the
addition of fluorides like AlF_3, PbF_2, rare earth fluorides or
alkali fluorides. For AlF_3 there is a critical concentration of a
few percent above which crystallization occurs. Experiments are
cited where indium is used instead of Al. One observes an increase
of the viscosity plateau due to the weight of In, IR transmission
remains excellent. In the case of PbF_2, used as dopant for refractive index increase in preforms, concentrations up to 15 moles
percent are possible because of the covalency increase of the
bonding which tends to stabilize the glass.

Concerning purity, it turns out to be mostly a problem of raw materials. The impurities like iron, other transition metals and rare earths must be reduced to less than a few ppm. Considerable precautions are necessary during manipulation, because there is no easy way to purify the glass after fabrication, and because contaminations induce heterogeneous crystallization.

The main optical quality to be studied is the scattering loss around 1 μm. Losses of 3-6 dB.km^{-1} are attributed to intrinsic properties of the glass matrix. Rayleigh-Brillouin scattering has been measured only on bulk glass. Absorption lossses have not yet been determined.

The manufacturing reproducibility of fluoride glasses is now good in terms of the IR edge, the glass transition temperature, thermal properties, optical absorption and light scattering. As an example the IR edge absorption coefficient is the same within 1 % for 20 different samples. Obviously such a good reproducibility needs proper control of the atmosphere during the glass manufacture and of the purity of the starting materials.

Fiber drawing.

At the present time the ultimate theoretical performance in terms of optical losses are far from being achieved.

The main problem is to manufacture the pre-form microcrystals which may act as crystallisation centers (less than 5 microcrystals for 15g of glass). At present the maximum fiber length achievable is of the order of 100 m at a speed of about 5 ms^{-1}. This is amplified by the problem of fiber splicing which is not yet satisfactorily handled.

The mechanical strength in fluoride fibers is around 1 GPa less than silicate glass fibers, but this value appears to be sufficient for practical applications. For industrial use protective coatings will need to be applied to fluoride fibres. Polymers which are the usual coatings could be replaced by silicon nitride coatings which act as a good barrier for hydrogen even under a pressure of 65 atmospheres.

Industrial interest in optical waveguides.

The problem is a general one for optical fibers. It is now evident that fibers will be used more and more judging by the increasing number of experimental systems on an industrial scale (as in Biarritz in France) as well as those projected for the near future (the transatlantic cable TAT-8).

For the industrial applications of fluoride glass fibers, a problem is with their cost compared with silica based fibers. The cost difference between silica glasses and fluoride glasses in terms of the raw materials is only a factor of two at the moment. This means that the costs will not be a problem.

In the future the composition of useful floride glasses will probably be significantly different from those discussed in this Summer School.

CHAPTER V: ELECTRONIC AND IONIC DEVICES

ELECTRONIC TRANSPORT IN AMORPHOUS CHALCOGENIDE SEMICONDUCTORS

A.E. OWEN

Department of Electrical Engineering,
University of Edinburgh,
King's Buildings,
Edinburgh EH9 3JL, Scotland, U.K.

1. BACKGROUND

1.1 Introduction

The one-dimensional one-electron density-of-states distribution in an ideal amorphous semiconductor is as shown schematically in figure 1(a) [1] [2].   An ideal amorphous structure is a fully connected three-dimensional disordered network of atoms of finite size, i.e. a degree of short-range order is imposed but there are no unsaturated bonds.   The label ES in figure 1(a) means Bloch-type extended states above an energy E_C in the conduction band and below an energy E_V in the valence band.   The label T indicates localised tail states which are split-off from the Bloch-type states of the valence and conduction bands but which retain the characteristics of their parentage; G indicates gap states of indefinite origin which are assumed present in low but uniform density across the energy gap (e.g. $\sim 10^{14}$ cm^{-3}).   The tail and gap states are localised in the sense that electrons in these states have wave functions ψ_{loc} which decay exponentially in space, i.e. in one-dimension (x), $\psi_{loc} \sim$ exp 0 (αx) where α is a decay factor and is typically $\sim$ 0.1 Å^{-1}.

A real amorphous semiconductor is thought to have a one-dimensional density-of-states more like that depicted schematically in figure 1(b).   There are still extended, tail and gap states but in addition there is evidence for energetically rather well-defined maxima in the gap state distribution and these features are attributed to specific structurally-related defects such as dangling (broken) bonds, deviations from stoichometry and/or wrong

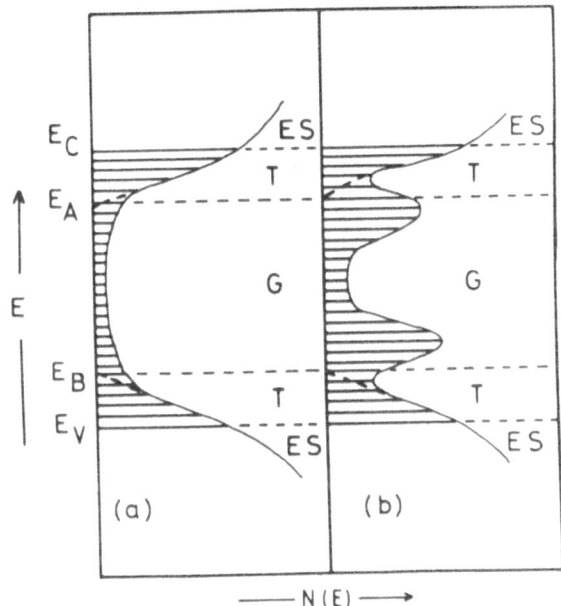

FIGURE 1.   Schematic one-electron one-dimensional density-of-
            states diagrams, -
            (a) for an ideal amorphous semiconductor, and
            (b) for a real amorphous semiconductor.

bonds in the case of compounds, and even perhaps to impurities.

1.2 The Mobility Gap Model

Figures 1(a) and 1(b) lead to the mobility-gap notion which
has been the accepted basis for interpreting electronic transport
in amorphous semiconductors since the early days of the subject
and which is shown diagrammatically in figure 2.   In extended
states just above E_c, or just below E_v, carrier transport is
essentially a diffusive Brownian-type motion and hence the carrier
mobility μ is approximately [2],

$$\mu \approx \frac{1}{6} \frac{ea^2}{kT} \nu_{el} \qquad (1)$$

where a is the average interatomic distance and ν_{el} is an electronic
frequency of the order of 10^{15} s^{-1}.   The estimated drift mobility
in extended states just above E_c or just below E_v is therefore
about 1 cm^2 v^{-1} s^{-1} at room temperature.   In localised states

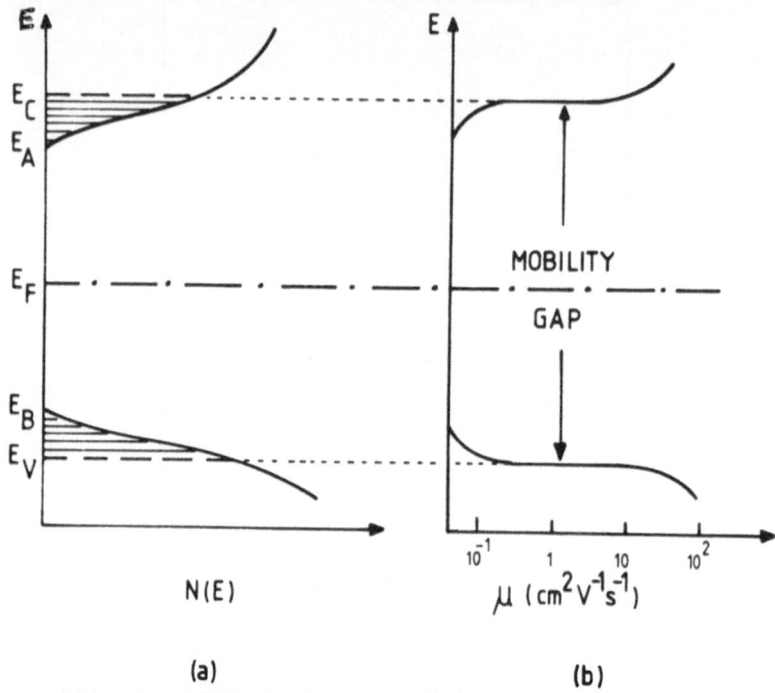

FIGURE 2.   Illustrating the mobility gap model for an amorphous
            semiconductor.
            (a) The one-electron density-of-states (the hatched
                 regions indicate the tails of localised states in
                 which the carrier mobility is low).
            (b) The mobility as a function of energy, corresponding
                 to (a).

(T or G) transport can only occur by phonon-assisted hopping and
the mobility is given by

$$\mu(E) \approx \frac{eR^2(E)}{kT} \; \nu_{ph} \; \exp\,(-2\alpha R)\,\exp(-W/kT) \qquad\qquad (2)$$

where R is the average hopping distance which depends on the
density-of-states distribution and is a function of energy (N.B.
R may be greater than a), ν_{ph} is a phonon frequency ($\sim 10^{13}$ s^{-1})
and W (the activation energy) is the energy difference between
the initial and final localised states involved in the hopping
motion.   Close to E_C and E_V, R → a and W → 0, hence $\mu \sim 10^{-2}$ cm^2
V^{-1} s^{-1} at room temperature.   Thus, near E_C and E_V the carrier
mobility changes by three orders of magnitude or more and this
defines the mobility gap, illustrated schematically in figure 2(b)

where μ is plotted as a function of energy.

1.3 General Conduction Mechanisms - Temperature Dependence

Accepting the mobility gap idea, four mechanisms of conduction may be expected and each will dominate the d.c. conductivity in an appropriate range of temperature [3].   Starting at high temperatures, the four processes are as follows:

(a) Conduction by carriers excited into extended states just above E_C or just below E_V.   In the case of hole transport, for instance, the conductivity will be given by

$$\sigma = C_0 \exp[-(E_F-E_V)/kT] \qquad (3)$$

Optical measurements usually show that the band gap of amorphous semiconductors decreases approximately linearly with temperature, i.e.

$$(E_F-E_V) = E(0) - \gamma T \qquad (4)$$

where $E(0)$ is the value of (E_F-E_V) at $T = 0$ K and γ is its temperature coefficient.   Thus,

$$C_0 = \sigma_0 \exp(\gamma/k) \qquad (5)$$

In chalcogenide glasses, the temperature coefficient of the fundamental optical absorption edge, which is approximately equal to the mobility gap (see section 3), is usually found experimentally to be in the region of 4×10^{-4} eV deg^{-1}. Moreover, the Fermi level E_F is invariably situated near the middle of the gap and hence values of γ roughly half that magnitude are to be expected; the magnitude of $\exp(\gamma/k)$ is therefore likely to be in the range 10-100.   The constant σ_0 is generally equated with σ_{min} the so-called minimum metallic conductivity.   Mott defined σ_{min} as the smallest non-zero value the conductivity can have at absolute zero. i.e. the lowest value of the conductivity contributed by carriers just at E_C (in the case of electrons) before the start of activated processes.   Mott derived the simple relationship

$$\sigma_{min} = \text{Constant} \ (e^2/ha) \qquad (6)$$

where a is the interatomic distance [1].   The constant depends somewhat on structure and is in the region 0.03 to 0.1 so that σ_{min} is 200-800 ohm^{-1} cm^{-1}, if a = 3 Å.   For conduction in extended states therefore the pre-exponential constant C_0 (equations (3) and (5)) should be roughly 10^3-10^4 ohm^{-1} cm^{-1}.

(b) Transport by carriers excited into the tail of localised states at energies close to E_A or E_B (see figure 2) and migrating by a hopping mechanism. Assuming conduction by electrons again,

$$\sigma = C_1 \exp [-(E_F-E_B+W_1)/kT)] \tag{7}$$

where W_1 is the activation energy for hopping. It is not easy to make an estimate of C_1 but the lower mobility and the lower density-of-states near E_A compared with E_C, will make it several decades smaller than C_0. The energy difference (E_F-E_B) is also expected to depend upon temperature but that is again difficult to determine.

(c) At low temperatures a significant number of carriers is not excited but if the density-of-states at the Fermi level is finite there will be a contribution from carriers with energies near E_F hopping between localised states. In this case,

$$\sigma = C_2 \exp (-W_2/kT) \tag{8}$$

where W_2 is the appropriate hopping energy and $C_2 < C_1$.

(d) At still lower temperatures it is probable that carriers will tend to hop beyond their spatially nearest neighbour states to states which are closer energetically. This is the so-called variable range hopping mechanism and Mott [3] showed that if the density-of-states at E_F is $N(E_F)$, -

$$\sigma = C_3 \exp [-(T_0/T)^{1/4}] \tag{9}$$

with $T_0 \cong [18 \, \alpha^3/kN(E_F)]$.

1.4 Polaron Transport

If a charge carrier remains in the vicinity of a particular site long enough its field will tend to displace or polarize the surrounding atoms and in its bound state the carrier cannot move unless the polarisation cloud also moves with it [4,5]. The trapped carrier and the surrounding polarized region can be treated as an entity known as a polaron; if the polarisation cloud extends over only a few interatomic distances the particle is called a small polaron. The polaron has a lower energy than a free electron but a larger effective mass since it must carry the induced deformation when it moves from site to site; the decrease in energy relative to that of the electron in an undistorted lattice is called the polaron binding energy W_p.

In a crystal, the small polaron states may overlap sufficiently to form a polaron band in an analogous way to electron energy band

formation in the undistorted lattice [6]. The small polaron band
is usually narrow and its width decreases exponentially with
temperature. In general therefore a small polaron can move by
two different mechanisms. At low temperatures band conduction
without phonon interaction is possible; at higher temperatures
the small polaron can migrate only by hopping between equivalent
sites. The deformation of the lattice to form equivalent
adjacent sites requires energy from phonons and the hopping
motion can therefore be regarded as phonon assisted tunnelling
between sites.

The possibility of small polaron formation in amorphous
semiconductors and insulators has been propounded particularly
by Emin [4][5][7] and an account of the theory is given in the
accompanying article in this volume by Livage [8].

An important parameter in polaron mechanisms is the Hall
mobility, μ_H. The calculation of the Hall mobility is difficult
and it depends on the local geometry of sites. For a triangular
lattice, Friedman and Holstein have derived for the non-adiabatic
case [9],

$$\mu_H = \frac{ea^2}{\pi} \left(\frac{\pi}{12kTW}\right)^{\frac{1}{2}} J . \exp\left(- \frac{W}{3kT}\right) \tag{10}$$

where a is the intersite spacing, J the electronic overlap integral
between sites, e is the electronic charge and W is the total hopping
energy. In an amorphous material

$$W = W_H + \tfrac{1}{2} W_D$$

where W_D is the site-to-site disorder energy and W_H, the polaron
hopping energy, is equal to $(W_p/2)$. The theory also predicts
a sign anomaly. If conduction is due to the hopping of small
polaron holes, for example, the sign of the Hall effect is
negative while the sign of the thermopower is positive (as
expected from the polarity of the carrier).

2. THE ELECTRONIC BAND STRUCTURE OF CHALCOGENIDE GLASSES

2.1 Introduction

In all chalcogenide compounds the chalcogen atoms (S, Se or
Te) are normally in two-fold coordination. The compounds of most
interest are those formed with pnictide elements, such as arsenic,
which normally bond in three-fold co-ordination but chalcogens
and pnictides can vary their valency and atoms in abnormal co-

ordination configurations are important, electronically, in the formation of defects.

The band structure of chalcogenide glasses is best approached in terms of normal chemical bonding however and this will be considered in the next section, taking Se and As_2Se_3 as examples.

2.2 Bonds and Bands

The outer electronic configuration of atomic As is $(4s)^2 (4p)^3$ and of Se $(4s)^2 (4p)^4$. The essential features of the band structure of solid Se and As_2Se_3 in crystalline or glassy forms can be seen from simple molecular orbital considerations with slightly different results depending on whether or not hybridization of the atomic s- and p-states is assumed.

(a) Se

If it is assumed that the s- and p-states are first hybridized to form an sp^3-state then on bonding the atomic sp^3-state is split into a lower σ bonding state (b), an upper σ* anti-bonding state (a) and two doubly occupied sp^3-states which form an intermediate lone-pair (LP) non-bonding (n) level in the molecular orbital scheme.

If there is no hybridization and the molecular orbitals are formed from pure s- and p-states the p-states again split into a σ lower bonding state (b), an upper σ* anti-bonding state (a) and an intermediate non-bonding (n) lone-pair (LP) level. The atomic s-state forms a molecular level well below the bonding p-states but in this case the LP level has half as many states (and electrons) as in the hybridized scheme.

Tutihasi and Chen [10] and Chen [11] have proposed detailed molecular orbital models for the band structure of trigonal and monoclinic Se, and for amorphous Se containing rings (monoclinic) and chains (trigonal). An important feature of their calculations is that in solid Se (crystalline or amorphous) there is considerable overlap and mixing of the σ (b) and LP (n) molecular levels and hence both bonding and non-bonding states contribute to the uppermost filled band, i.e. to the valence band. Kastner [12], following Mooser and Pearson [13], proposed a band model for chalcogenide semiconductors based on the absence of hybridization. The important feature of this picture is that in the solid state, the lone pair levels remain well separated from the σ bonding states, and hence the uppermost filled band, conventionally called the valence band in semiconductor terminology, is made up entirely of non-bonding electrons. Strictly speaking the term "valence band" is a misnomer.

(b) As_2Se_3

Calculations by Chen indicate strong mixing of bonding and non-bonding orbits [11], and hence also in the uppermost filled (valence) band of solid As_2Se_3. By contrast, the picture based on pure s- and p-states leaves the lone pair (non-bonding) band separated from and above the σ-bonding band. It is not clear which of the two molecular orbital schemes - sp^3-hybridization, or pure s- and p-state orbitals - is correct but evidence from optical spectra favours pure p-state bonding [14].

2.3 Defect States in the Band Gap of Chalcogenide Elements and Compounds

Chalcogenide glasses are diamagnetic and show no e.s.r. response, i.e. there appear to be no dangling bonds in the gap. On the other hand the Fermi-level appears to be pinned, suggesting a finite density of gap states. It was largely to resolve this contradiction that models of defect states were first proposed.

The earliest model was suggested by Mott Davis and Street (the so-called MDS model) [15][16][17] and it has proved successful in explaining a wide variety of phenomena in chalcogenides, such as the constant activation energy of the DC conductivity (i.e. the pinning of the Fermi-level), magnetism, luminescence, drift mobility and a.c. conductivity. In this model, bonding defects can have three charged states denoted D^+, D^- and D^0, associated with different local atomic configurations. The neutral dangling bond (D^0), say a singly-coordinated chain-end Se atom, would normally possess spin but it is unstable. Provided the temperature is high enough so that some atomic movement is possible, one of the end atoms moves closer to and cross-links with the LP p-electrons of a Se atom in a neighbouring chain. The three coordinated Se atom now gives up an electron to the remaining dangling bond. The former becomes D^+, the latter D^-, and both are now diamagnetic. The reaction

$$2D^0 \rightarrow D^+ + D^- \tag{11}$$

is exothermic because of a negative correlation (or Hubbard) energy, the potential energy decrease resulting from spin-sharing more than compensating the Coulomb-repulsive energy increase. The decrease in energy can also be regarded as the result of converting a non-bonding lone pair of electrons into a lower lying σ bonding state by dative bonding. The D^+ and D^- centres lie in the gap and determine the position of the Fermi-level, rather like donors and acceptors in a compensated semiconductor.

In the model of Kastner, Adler and Fritzsche (KAF) the chalcogen atom is denoted by C and its coodination by a subscript [18]-[23].

The dangling bond C_1^0 is considered to cross-link with the nearest neighbour on an adjoining chain, so that its bonds are satisfied. It therefore becomes a normal C_2^0, and the neighbour must use one of its LP electrons to become the three-fold coordinated neutral C_3^0. If atomic movement is possible while the glass is formed, this rearrangement is spontaneous. The reason is that C_1^0 and its neighbour C_2^0 are involved in only three bonds between them, but C_2^0 and C_3^0 form five bonds, a gain of two bond energies. In contrast with the MDS model, therefore, the common neutral defect in the glass is C_3^0. The unsatisfied fourth p-electron of C_3^0 renders it paramagnetic however and still relatively unstable. For the same reasons as in the MDS model the exothermic reaction

$$2C_3^0 \rightarrow C_3^+ + C_1^- \tag{12}$$

then leads to a "valence alternation pair" (VAP). Its members are identical to the D^+ and D^- centres of equation (11) and have the same properties.

Group V elements (pnictides) can also undergo valence alternation. The situation is more complicated than that of the chalcogens, because non-bonding s electrons become available for over-coordination only after hybridization. In a pnictide element, the lowest energy neutral defect, corresponding to C_3^0, is considered to be P_4^0 formed by hybridizing the s and p states to create three equivalent sp^3 orbitals containing the 5 valence electrons, two of which remain in a non-bonding lone pair. In close analogy with the chalcogens, the lowest energy defects are charged VAPs, not neutral P_4^0. The negative correlation energy gained by transferring charge is larger in pnictides than in chalcogens, so that although the initial creation energy for P_4^0 is greater than for C_3^0, the net energy gained in forming VAPs is comparable for the two groups of elements. As a first step, P_4^+ and P_4^- defect states are created by transferring an electron from one P_4^0 to another and this involves a positive correlation energy, but P_4^- is unstable and the following exothermic reaction takes place:

$$P_3^0 + P_4^- \rightarrow P_2^- + P_3^0 \tag{13}$$

This involves breaking a bond to one of the four neighbouring P_3^0s, placing the two anti-bonding electrons into lone-pair (p-like) orbitals, and dehybridizing the P_4^0 to make it a normal P_3^0. The complete exothermic reaction becomes

$$2P_4^0 \rightarrow P_4^+ + P_2^- \tag{14}$$

the VAP P_4^+, P_2^- being the equivalent of D^+ and D^- in this case.

When both chalcogen and pnictide elements are combined in a glass such as As_2Se_3, all the stable defects, viz. P_4^+, C_3^+, P_2^-, C_1^- will be present though in different concentrations. Kastner and Fritzsche [18] believe that C_1^- is always the dominant negative defect, but that either C_3^+ or P_4^+ could be the dominant positive partner depending on a number of considerations. They also argue that under preparation conditions where the defects are allowed to come into thermal equilibrium, P_4^+ and C_1^- will be the predominant VAPs although C_3^+ and P_2^- centres may also be present in lesser concentrations.

Mott and Street have suggested the possible clustering of D^+, D^- centres due to their Coulomb attraction [17]. Such an overlapping D^+D^- pair has been called an "intimate" valence alternation pair (IVAP) and an IVAP is a neutral (actually a dipole) centre. IVAPs can annihilate each other even at temperatures well below T_g. Since separated VAPs represent charged centres, transport properties are more sensitive to VAPs than IVAPs. Even in a glass in which the ratio of VAP to IVAP density is small, the density of VAPs may be sufficiently high to pin the Fermi-level. Unlike VAPs, IVAPs are not likely to act as traps and therefore are not expected to play a large part in trap controlled transport or conductivity. Approximate calculations by Adler and Yoffa [24] suggest that the density of VAPs is of the order of 10^{18} cm^{-3} and of IVAPs, 10^{19} cm^{-3}.

3. D.C. CONDUCTIVITY, THERMOPOWER AND HALL EFFECT

3.1 Over a wide temperature range, the d.c. conductivity of most chalcogenide glasses obeys an equation of the form,

$$\sigma = C \exp(-E_\sigma/kT) \qquad (15)$$

in which E_σ denotes the activation energy for conduction [25][26] [27]. Depending on composition, the activation energy E_σ is in the range of a few tenths of an eV to 1 eV or more, and almost invariably values of $2E_\sigma$ are close to the photon energy E_{opt} corresponding to the onset of strong optical absorption [25][26] [27]. With few exceptions C is 10^2 ohm^{-1} cm^{-1} or greater, and for many typical chalcogenide glasses such as Se and As_2Se_3 it is in the range $10^3 - 10^4$ ohm^{-1} cm^{-1}.

Also with few exceptions (some examples will be mentioned later), reasonably linear plots are normally obtained over the whole experimental temperature range. In particular variable-range hopping conduction, behaving even approximately according to equation (9), is not generally observed. Nor, usually, is there any evidence of a lower activation energy at low temperatures although it must be recognised that the relatively low conductivities and large activation energies of the majority of chalcogenide glasses makes it difficult to extend measurements to very low temperatures.

The thermoelectric power has been measured for many chalcogenide glasses and it is always found to be positive [25][26][27]. It has a magnitude typical of semiconductors (mV K^{-1}), and it decreases with temperature according to the simple equation established for semiconductors, i.e.

$$S = -\frac{k}{e}\left(\frac{E_S}{kT} + A\right) \qquad (16)$$

where E_S is the activation energy for thermopower and A is a constant (often $\sim$ 1).

It was noted in section 1.3 that values of the pre-exponential constant in the range 10^3 - 10^4 ohm^{-1} cm^{-1} are consistent with conduction in extended states at the mobility edge. Thus, C and E_σ in equation (15) are to be equated with C_0 & (E_F-E_V) in equation (3); the evidence is therefore that in most chalcogenide glasses, the main transport mechanism involves holes at or close to the mobility edge and that the Fermi-level is close to the centre of the mobility gap.

A critical question is whether the activation energy for conduction, E_σ, is the same as that for thermopower, E_S. For As$_2$Se$_3$ the answer is inconclusive. Thermopower measurements on As$_2$Se$_3$ have been reported by Hurst and Davis [28], Seager and Quinn [29] and Chiu [35], and although there is some difference in the magnitude of S, Hurst and Davis and Chiu agree that E_S = 0.90 eV, which is essentially the same as E_σ, while Seager and Quinn find a substantially lower value of E_S = 0.60 eV. On the other hand there is well documented evidence for differences $(E_\sigma-E_S)$ of about 0.15 eV for glassy "alloys" in the As$_2$Te$_3$Si$_x$ and As$_2$Te$_{(3-x)}$Se$_x$ systems [27][31][32].

The low conductivities and mobilities of chalcogenide glasses make Hall effect measurements extremely difficult but there are nevertheless several reports of the Hall-effect mobility μ_H. The Hall coefficient of these p-type materials is normally negative and this anomaly can be understood either in terms of the random-phase diffusive type transport of carriers in a 3-site motion through extended states at a mobility edge, or in terms of polaron hopping. Despite general agreement on the sign anomaly however, quantitatively the experimental situation for As$_2$Se$_3$ is again not clear. According to Mytilineou and Roilos [31], and Nagels et al [27][32], μ_H in vitreous As$_2$Se$_3$ is small ($\sim$ 10^{-1} cm^2 v^{-1} S^{-1}) and unactivated, while Klaffke and Wood [33] report a small activation energy.   For the $As_2Te_3Si_x$ and $As_2Te_{(3-x)}Se_x$ systems however there is general agreement that the Hall mobility is activated, i.e.

$$\mu_H = \mu_{Ho} \exp(-E_H/kT) \qquad (17)$$

with E_H in the range of 0.03 - 0.05 eV.   It is notable that these are the same materials for which consistent values of $(E_V - E_S)$ ~ 0.15 eV are also reported.

At the present time therefore the weight of the evidence seems to be that in the case of As_2Se_3 equation (3) is applicable, that conduction occurs by hole transport in extended states close to the valence band mobility edge E_V, and that

$$E_\sigma = (E_F - E_V) = E_S = 2E_{opt},$$

implying that the Fermi-level is close to the centre of the mobility gap.

However, in at least a number of chalcogenide glasses, of which the alloys $As_2Te_3Si_x$ and $As_2Te_{(2-x)}Si_x$ are typical, there is clear evidence that: (i) the thermopower has a smaller temperature dependence than the d.c. conductivity with $(E_\sigma - E_S)$ ~ 0.15 eV, and (ii) the Hall mobility is activated with an activation energy in the range 0.03 - 0.05 eV.   Two possible interpretations of these observations are currently considered to be likely explanations.

Nagels et al have proposed a two-path conduction process in which transport may occur almost simultaneously by holes in extended states just below E_V and by holes hopping in localised states just above E_V [27][32].   Using the subscript 1 to indicate extended state conduction and 2 for hopping conduction then

$$\sigma_1 = C_{01} \exp[-(E_F - E_V)/kT] \qquad (18)$$

and

$$\sigma_2 = C_{02} \exp[-(E_F - E_B + W)/kT] \qquad (19)$$

with the total conductivity σ,

$$\sigma = \sigma_1 + \sigma_2 \qquad (20)$$

Provided the rate constants of σ_1 and σ_2 do not differ greatly, a log σ vs. (1/T) plot will change its slope only gradually, and over at least a limited range of temperature it will not differ appreciably from a straight line.   The thermopower is the weighted sum, -

$$S = \frac{(S_1\sigma_1 + S_2\sigma_2)}{\sigma} \qquad (21)$$

with

$$S_1 = \frac{k}{e} \left[\frac{(E_F - E_V)}{kT} + A_1 \right] \qquad (22)$$

and

$$S_2 = \frac{k}{e} \left[\frac{(E_F - E_B)}{kT} + A_2 \right] \qquad (23)$$

Because the two thermopowers are similar for $(1/T) = 0$, the form of S in the transition region depends sensitively on the sharpness of the transition in σ and the slope of S vs. $(1/T)$ has little significance in the transition region. The Hall mobility is also given by the weighted sum, -

$$\mu_H = \frac{\mu_1\sigma_1 + \mu_2\sigma_2}{\sigma} \qquad (24)$$

Nagels et al assume that $\mu_2 = 0$, so that -

$$\mu_H = \frac{\mu_1\sigma_1}{\sigma} \qquad (25)$$

When conduction occurs <u>mainly</u> in extended states μ_H is equal to μ_1, but when hopping also contributes the Hall voltage is generated by the few carriers remaining at the mobility edge E_V and from equations (18), (19) and (25)

$$\mu_H = \mu_1 \left[1 + \frac{C_{02}}{C_{01}} \exp (E_B - E_V - W)/kT \right]^{-1} \qquad (26)$$

The alternative interpretation, first proposed by Emin and co-workers [5][34], and favoured by Seager and Quinn [29], is that transport is by small polarons (see section 1.3). In that case the difference $(E_\sigma - E_S)$ gives the hopping energy W directly, and the activation energy for the Hall mobility is $(W/3)$ (see equation (10)). From the available experimental data it seems as though the Hall mobility activation energy is approximately $[(E_\sigma - E_S)/3]$ and Emin [7] points out that C_0 values of $\sim 10^3$ ohm^{-1} cm^{-1} are also compatible with small-polaron hopping transport.

Thus trap-modulated hopping of holes at or near the mobility edge E_V, OR polaron hopping, are equally valid interpretations of transport measurements on a number of the more complex chalcogenide glasses and a definitive experiment is lacking at the present time. Mott [35] suggests that indirect evidence against the polaron model is provided by the ON-state of threshold switching devices based on complex chalcogenide glasses. The high ON-state conductance implies mobilities of ~ 10 cm^2 V^{-1} S^{-1} and although it is not known which species (electrons or holes) is the major current carrier, it is clear that either electrons or holes, at least, are moving in quasi-free Bloch states.

The conclusions reached from measurements of d.c. conductivity, thermopower, Hall mobility and the optical absorption edge are summarised diagrammatically in figure 3. These experiments reveal nothing, directly, about states well within the mobility gap, such as the defect centres discussed in section 2.3, and the influence of gap states is considered in the next section.

4. THE INFLUENCE OF STATES IN THE GAP

4.1 Introduction

Localised states in the mobility gap influence the electronic transport properties of chalcogenide glasses in a variety of ways [36], e.g.:

1. Pinning of the Fermi level E_F at or close to the middle of the mobility gap. Fritzsche [37] has pointed out that the characteristic straight line plots of log σ vs. (1/T) over a wide range of temperature is evidence that E_F is fixed (pinned) by a large density of localised states, rather than of intrinsic conductions. Marshall and Owen [38][39] have proposed specific models for As_2Se_3, As_2Te_3 in which E_F is pinned midway between donor- and acceptor-like states.

2. The creation of a high concentration of space-charge density at interfaces and metal contacts. As a result the field-effect conductance in chalcogenide glasses is very small, and again this can be interpreted in terms of E_F positioned midway between large densities of localised states at relatively discrete energies [39][40]. A concomitant effect associated with a large density of gap states is a very short screening length (e.g. ~ 100 Å) and this explains the apparent ohmic behaviour of most metal-chalcogenide contacts [41].

3. Localised gap states provide several possibilities for enhanced carrier hopping between neighbouring sites, i.e. for

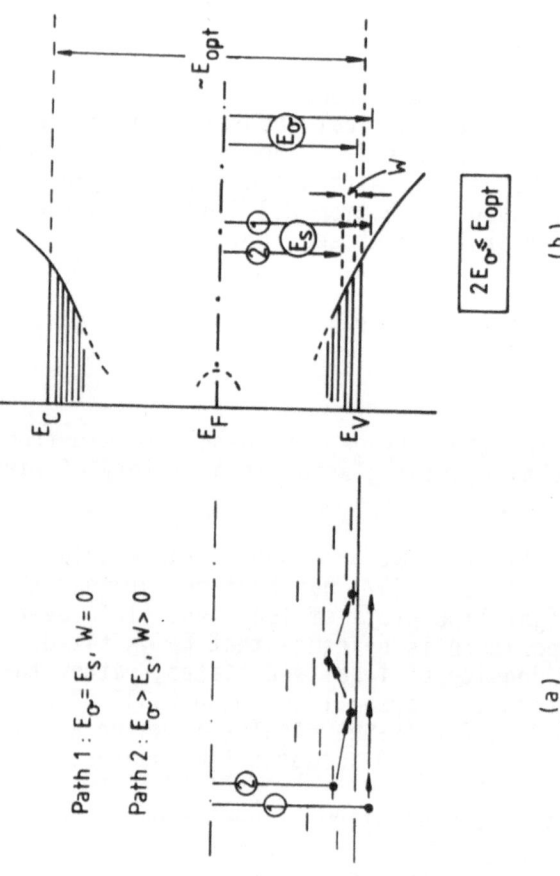

FIGURE 3. A diagrammatic summary of the information derived on transport (a) and band structure (b) from measurements of d.c. conductivity, thermopower, Hall mobility and optical absorption (see text, section 3).

E_σ is the activation energy for d.c. conduction (equation (15)), E_S is the activation energy for thermopower (equation (16)), W is a hopping energy (equation (7)), and the optical gap E_{opt} is obtained from the fundamental optical absorption edge.

dipolar activity which contributes to the a.c. conductivity and dielectric behaviour [42][43].

4. The carrier drift mobility in chalcogenide glasses is determined by interactions with localised states and in some circumstances mobility measurements can give information on the energies and distributions of gap states which can be correlated with the models of defect states described in section 2.3.

4.2 Amorphous Se and As$_2$Se$_3$: A "Defect-State" Approach

Of the many chalcogenide glasses, transit-time drift mobility measurements have been applied most thoroughly to amorphous Se and As$_2$Se$_3$. Comprehensive reviews of the experimental data and their interpretation have been published (see Enck and Pfister [44], and Owen and Spear [45]) and it is not necessary to repeat the details here but a brief account will be presented of a particular view which interprets the influence of gap states on transport phenomena in terms of specific electronic defect states of the kind discussed in section 2.3.

Amorphous Se is unusual amongst the chalcogenide glasses in that although holes are the more mobile carrier (i.e. Se is p-type, like most of the chalcogenides), the electron drift mobility is also easily measurable over a wide range of temperature. At least nine different groups have reported data on the electron or hole mobility and in general the agreement is surprisingly good [44][45]. Above about 200 K a well-defined transit time is observed, but below 200 K the pulse shape becomes progressively more dispersive. At about 300 K or a little higher, the hole mobility in amorphous Se tends to a constant (temperature independent) value, μ_0, in the region of 0.3 to 0.4 cm^2 V^{-1} S^{-1} while for electrons μ_0 is about 0.05 cm^2 V^{-1} S^{-1}. Below about 270 K the mobility for both electrons and holes is activated, i.e.

$$\mu = \mu_0 \exp(-E/kT) \tag{27}$$

with E in the region of 0.28 - 0.30 eV for holes and approximately 0.33 eV for electrons [44][45]. It is important to note that plots of log μ vs. (1/T), or log (inverse transit time) vs. (1/T) are continuous, with the same values of activation energy, from relatively high temperatures (> 270 K) where the transit pulses are Gaussian, to low temperatures (< 270 K) where dispersive pulses are observed.

The μ_0 value for holes is consistent with diffusive transport in extended states (see section 1.2), and while the value for electrons is lower it is still above the lower limit for diffusive mobility estimated by Cohen [2]. Thus, the evidence is that electron and hole motion in amorphous Se occurs by a process of trap-limited transport in extended states just beyond or very close to their respective mobility edges. The traps responsible for limiting the mobility could occur over relatively discrete range of energies (as in figure 1(b)) or they could be part of a broad distribution of traps, decreasing in density with increasing depth. Marshall and Owen [46], and Owen and Spear [45] have argued for the former possibility and this implies that at low temperatures ($<$ 270 K) a distribution of trap depth over a small range of energies is responsible for the dispersion of the transient current pulses. Combining the drift mobility experiments with d.c. conductivity, optical absorption and space-charge-limited current data, Owen and Spear have proposed the electronic density-of-states distribution shown in figure 4(a). The mobility gap (E_c-E_v) = 2.1 eV is equated with the fundamental optical absorption edge E_{opt}, and the position of the Fermi level is determined by the activation energy for d.c. conductivity.

Mott [47] has tentatively correlated the electron and hole trapping states in figure 4(a) with the defect states discussed in section 2.3 (using the Kastner, Adler and Fritzsche notation): If valence alternation pairs (VAPs) act as traps, the capture of an electron or a hole produces the same centre, C_0. The trap depth for electrons, ε_1, is determined by the reaction

$$C_3^0 \rightarrow C_3^+ + \text{electron}$$

and the trap depth for holes, ε_2, by -

$$C_3^0 \rightarrow C_1^- + \text{hole}$$

Thus an electron falling into a trap yields an energy ε_1 and a hole yields an energy ε_2, with the formation of two C_0 centres. To return the system to normal, an energy E is gained through the reactions

$$2C_3^0 \rightarrow C_1^- + C_2^+ \tag{28}$$

and

$$E + \varepsilon_1 + \varepsilon_2 = E_g \tag{29}$$

where E_g is the band gap. Mott equates the energy E with the chain scission energy of Se, less the energy to form the VAP. From NMR data, the Se chain scission energy is 1.4 eV at room temperature, or ~ 2 eV at 0 K; the atomic fraction of VAPs at 300 K is about 10^{-5} and hence $\exp(-W/2 kT) \cong 10^{-5}$, giving W = 0.5 eV for the energy of VAP formation. Thus E is 1.5 eV. Adding the trapping energies ε_1 and ε_2 (see figure 4(a)) gives, -

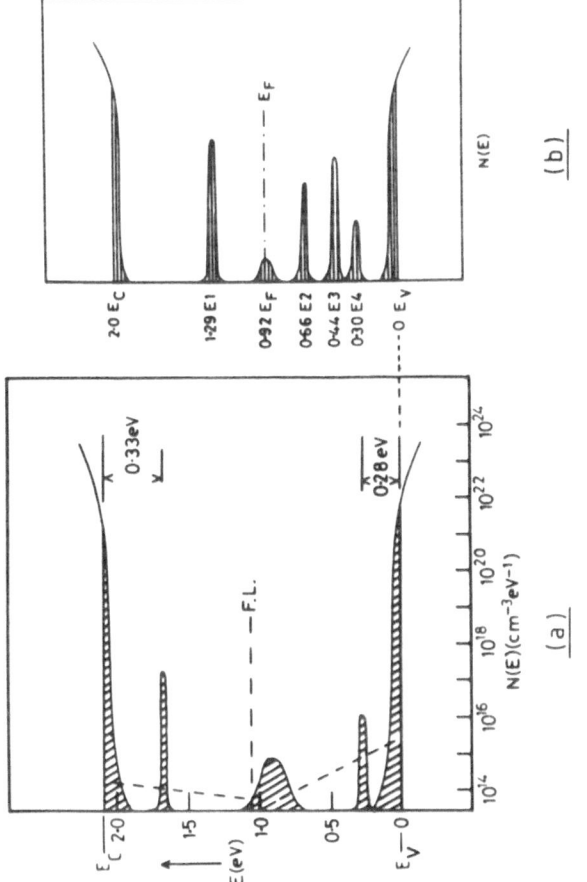

FIGURE 4. Electronic density-of-states diagrams constructed from a
variety of transport and optical measurements interpreted
in the spirit of defect states.
(a)   Amorphous Se.
(b)   Amorphous As₂Se3.

(See Ref. 45)

$$E_{opt} = 1.5 + 0.33 + 0.28 = 2.1 \text{ eV}$$

in good agreement with the experimental value (see also figure 4(a)).

Drift mobility data for As$_2$Se$_3$ are more difficult to interpret, on two counts. Firstly, only hole transits are observable, and electron mobilities are not therefore determinable, and secondly even at the highest temperature at which measurements have been made (~ 370 K) the hole transient times are two to four orders of magnitude longer than in Se. Nevertheless, hole transport in amorphous As$_2$Se$_3$ is clearly an activated process and Marshall et al [48][49] have observed that the activation energy depends on the time scale of the measurement. For transit times < 100 μs they report an activation energy of 0.43 eV, while for transit times > 200 μs it is 0.63 eV. These trapping energies are in good agreement with the more detailed studies of steady-state and transient photoconductivity which it is possible to make on amorphous As$_2$Se$_3$, which also provide evidence for electron traps at about 0.71 eV from the conduction band mobility edge, E_c [50].

The conclusion from section 3 was that amorphous As$_2$Se$_3$ is a p-type semiconductor and that hole transport occurs in states at or very close to the valence band mobility edge E_v. Thus, the hole transport phenomena observed in drift mobility and photo-conductivity experiments are trap-limited processes. Owen and Spear [45] have collected together the data from measurements of d.c. conductivity, optical absorption, drift mobility, photo-conductivity and space-charge-limited current, and interpreted it in terms of the electronic density-of-states model shown in figure 4(b). It is relevant to note here that transport experiments on amorphous Se with progressive additions of As indicate that some "Se-like" features are retained in the As-Se compositions, while very effective hole traps are also introduced by the addition of As [45][51]. There is, in addition, evidence that the electron traps at 0.33 eV in amorphous Se move farther away from the conduction band mobility edge E_c as As is added in concentrations of up to 8 atomic % [52]. It is clear from section 2.3 that there is a variety of likely electronic defect states in a chalcogenide compound such as As$_2$Se$_3$ and several of them are likely to be located within the band gap. At the present however it is not possible to attempt even the tentative correlation between the proposed band model and specific defect states as was described above for amorphous Se.

5.   CONCLUSIONS

Charge transport in the relatively simple amorphous chalco-genide semiconductors such as Se and As$_2$Se$_3$ most likely involves trap-limited processes. Carriers migrate via extended states or via states at or close to the mobility edges, E_c and E_v. The evidence from a variety of experiments is consistent with the view that the traps which limit carrier mobilities are located at relatively well-defined energies within the mobility gap, and that these traps originate from structurally-related electronic defects characteristic of various bonding abnormalities associated with the chemistry of chalcogenide elements and compounds.

The situation in more complex multi-component and/or non-stoichiometric chalcogenide glasses is more problematical. The additional compositional disorder associated with the chemical complexity will certainly affect transport mechanisms, perhaps encouraging multi-path conduction processes, polaron formation and probably also smoothing out the density-of-states distribution.

6. REFERENCES

1. Mott, N.F. and E.A. Davis. Electronic Processes in Non-Crystalline Materials, 2nd Edition, Chap. 1, pp.39-52 and Chap. 6, pp.209-215.

2. Cohen, M.H. J. Non-Cryst. Sol. $\underline{4}$, 391, (1970).

3. Ref. 1. Chap. 6, pp.219-222, (1979).

4. Emin, D. Adv. in Phys. $\underline{22}$, 57, (1973).

5. Emin, D. in Electrical and Structural Properties of Amorphous Semiconductors, Ed.: Le Comber, P.G. and J. Mort, pp.261-328 (Academic Press), (1973).

6. Holstein, T. Ann. Phys. (NY), $\underline{8}$, 343, (1959).

7. Emin, D. Comments in Solid-State Physics, $\underline{11}$, 35 and 72, (1983).

8. Livage, J. - in this volume.

9. Friedman,L. and T. Holstein. Ann. Phys. (NY), $\underline{21}$, 494, (1963).

10. Tutihasi, S. and I. Chen. Phys. Rev. $\underline{158}$, 623, (1967).

11. Chen, I. Phys. Rev. B, $\underline{8}$, 1440, (1973).

12. Kastner, M. Phys. Rev. Lett., $\underline{28}$, 355 (1972).

13. Mooser, E. and W.B. Pearson, in Progress in Semiconductors, Vol.5, p.104, (Heywood and Co., London), (1960).

14. Weiser, G. in The Physics of Selenium and Tellurium, Ed.: Gerlach, E. and Grosse, P., p.230, No.13 of Springer Series in Solid-State Science (Springer-Verlag), (1979).

15. Street, R.A. and N.F. Mott. Phys. Rev. Lett., $\underline{35}$, 1293, (1975).

16. Mott, N.F., E.A. Davis and R.A. Street. Phil. Mag. $\underline{32}$, 961, (1975).

17. Mott, N.F. and R.A. Street. Phil. Mag. $\underline{36}$, 33, (1977).

18. Kastner, M., D. Adler and H. Fritzsche. Phys. Rev. Lett., $\underline{37}$, 1504, (1976).

19. Kastner, M. and H. Fritzsche. Phil. Mag. $\underline{B37}$, 199 (1978).

20. Fritzsche, H. Proc. 7th Int. Conf. Amorphous and Liquid Semiconductors, p.3, Ed.: Spear, W.E. (CICL, Edinburgh, U.K.) (1977).

21. Fritzsche, H. J. Phys. Soc. Japan, 49, Suppl. A, 39, (1980).

22. Kastner, M. J. Non-Cryst. Sol. 31, 223, (1978).

23. Kastner, M. J. Non-Cryst. Sol. 35-36, 807, (1980).

24. Adler, D. and E.J. Yoffa. Can. J. Chem. 55, 1920, (1977).

25. Owen, A.E. Contemp Phys. 11, 227, 257, (1970).

26. See for example: Ref. 1 Chap.9, pp.452-460.

27. Nagels, P. in Amorphous Semiconductors, Ed.: Brodsky, M.H., pp.113-159, Vol.36 of Topics in Applied Physics, (Springer-Verlag), (1979).

28. Hurst, C.H. and E.A. Davis. J. Non-Cryst. Sol. 16, 343, (1974).

29. Seager, C.H. and R.K. Quinn. J. Non-Cryst. Sol. 17, 396, (1975).

30. Chui, D.M. Photo- and Thermal Effects in Arsenic Chalco-genides, M.Sc., Thesis (University of Edinburgh), (1976).

31. Mytilineou, E. and M. Roilos. Phil. Mag. B, 37, 387, (1978).

32. Nagels, P., R. Callaerts and M. Denayer, in Proc. 11th Int. Conf. Phys. of Semicond., Ed.: Miasek, M., p.549 (Polish Scientific Publishers, Warsaw), (1972).

33. Klaffke, G.R. and C. Wood in Proc. 4th Int. Conf. on Physics of Non-Crystalline Solids, Ed.: Frischat, G.H., p.236, (Trans. Tech. Publ.), (1977).

34. Emin, D., C.H. Seager and R.K. Quinn. Phys. Rev. Lett. 28, 813, (1972).

35. Mott, N.F. J. Phys. C., 13, 5433, (1980).

36. See for example: Ref. 1, pp.460-490.

37. Fritzsche, H., pp.55-125 of Ref. 5.

38. Marshall, J.M. and A.E. Owen. Phil. Mag. 24, 1281, (1971).

39. Marshall, J.M. and A.E. Owen. Phil. Mag. 33, 457, (1976).

40. Frye, R.C. and D. Adler. Phys. Rev. Lett. 46, 1027, (1981).

41. Wallace, A., A.E. Owen and J.M. Robertson. Phil. Mag. 38, 57, (1978).

42. Ref. 1, Chap.6, pp.223-235.

43. Owen, A.E. J. Non-Cryst. Sol. 25, 370, (1977).

44. Enck, R.C. and G. Pfister in Photoconductivity and Related Phenomena, Ed.: J. Mort and D.M. Pai, Chap.7, pp.297-302, (Elsevier), (1976).

45. Owen, A.E. and W.E. Spear. Phys. Chem. Glasses, 17, 174, (1976).

46. Marshall, J.M. and A.E. Owen. Phys. Stat. Sol. (a), 12, 181, (1972).

47. Mott, N.F. J. Phys. C. Solid St. Phys. 13, 5433, (1980).

48. Marshall, J.M. and A.E. Owen. Phil. Mag. 24, 1281, (1971).

49. Fisher, F.D., J.M. Marshall and A.E. Owen. Phil. Mag. 33, 261, (1976).

50. Main, C. and A.E. Owen. Ref. 5, pp.527 545, (1973).

51. Schmottmiller, J., M. Tabak, G. Lucovsky and A. Ward. J. Non-Cryst. Sol. 4, 80, (1979).

52. Marshall, J.M., F.D. Fisher and A.E. Owen. Phys. Stat. Sol. (a), 25, 419, (1974).

PHOTOSTRUCTURAL CHANGES IN AMORPHOUS CHALCOGENIDE MATERIALS.

S.R. Elliott

Department of Physical Chemistry, University of Cambridge,
Lensfield Road, Cambridge CB2 1EP, Great Britain.

ABSTRACT

Amorphous chalcogenide materials, formed from alloys of S or
Se with As or Ge (or similar elements) exhibit a wide variety of
changes upon irradiation by bandgap light. Such changes can be
structural, chemical or optical in nature, and depending on the mode
of preparation, the photo-induced effects can be irreversible or reversible (i.e. removed by heating to T_g). These photo-induced
changes will be discussed in relation to the structure and electronic
configuration of these materials, and mechanisms for the effects
will be proposed. Technological applications of these phenomena
are also discussed.

1. INTRODUCTION

Chalcogenide glasses, i.e. alloys of S, Se or Te with elements
such as Ge, Si, As or P, are semiconductors with bandgaps in the
range 1-3eV. Two features, flexibility of structure and high-lying,
lone-pair electron states, render them susceptible to photo-induced
changes. Firstly, they are invariably low-dimensional solids, at
least locally. In other words, their structure cannot be described
by means of a continuous random network which is *isotropic* in three
dimensions, as is the case for a-Si, for example. As_2S_3 and As_2Se_3
and GeS_2 and $GeSe_2$ (1) layer-like, and pure S and Se are chain-like.
Thus, for all these materials, there is considerable flexibility of
the structure as a result of the Van der Waal's bonding, the flexibility increasing with increasing chalcogen content, meaning that
changes in the structure can be relatively easily accommodated. The
other distinguishing feature of chalcogenide materials is that the
top of the valence band is composed predominantly of chalcogen lone
pair (p-π) states. These are the states which are preferentially

excited optically, and as a result can become chemically reactive and cause structural rearrangements.

2. TYPES OF PHOTO-INDUCED PHENOMENA

Changes may be induced by light in the structure, in the opto-electronic and chemical properties of amorphous chalcogenides. Furthermore, these changes can in general be either *irreversible* or *reversible* (by annealing to the glass-transition temperature, T_g), depending upon the preparation process and thermal history of the amorphous material. It should be noted, however, that chemical changes are always irreversible.) In discussing optically induced phenomena, it should be stressed that we are dealing only with *metastable* changes (i.e. which remain on cessation of optical exci-tation). Table 1 summarizes the types of photo-induced changes observed in amorphous chalcogenide solids, together with an indication of the type of material in which they are observed (film or glass), whether they are irreversible or reversible, and examples of the mag-nitude of changes for certain chosen materials.

3. MECHANISMS FOR PHOTO-INDUCED CHANGES

3.1 Structural Changes

From Table 1, it is clear that *irreversible* photostructural changes occur in virgin, vapour-deposited films, and in fact two types of such changes can be distinguished, depending on the mode of preparation of the film. For example, for the case of evaporation of a-As_2S_3, the stable vapour species are the molecules As_4S_4, As_4 S_3 and S_2 (not the molecule As_4S_6 which preserves the stoichiometry of the starting material). Thus, a film formed by vapour conden-sation will consist of an aggregate of such molecules held together by weak Van der Waal's bonds. The action of light (or equivalently heat) is to break the strained intramolecular covalent bonds and to form cross-linking intermolecular bonds, thereby polymerizing the film. An illustration of the effect of such photo-polymerization on the X-ray scattering intensity for the case of an a-As_2Se_3 film is shown in fig. 1(a). The sharp first peak at $2\theta \approx 17°$ (or $Q \simeq 1\text{Å}^{-1}$) in the as-deposited film has been ascribed to inter-molecular scat-tering (5) (as in molecular liquids), and the change in this peak upon photo-induced polymerization of the film is therefore large.

The other type of irreversible photostructural effect occurs in as-deposited films prepared by oblique evaporation. Such films in-variably exhibit *columnar growth*, viz. columns of dense material at an angle β to the film, interspersed with a low-density tissue. It is found that β is not equal to α, the angle of deposition, but rather the following empirical relationship is obeyed (6) : $\tan \alpha$ = $2 \tan \beta$. Such low-density, porous films densify (i.e. contract) upon irradiation, the effect being largest for films deposited at the largest values of α. Fig. 1(b) shows this behaviour for the case

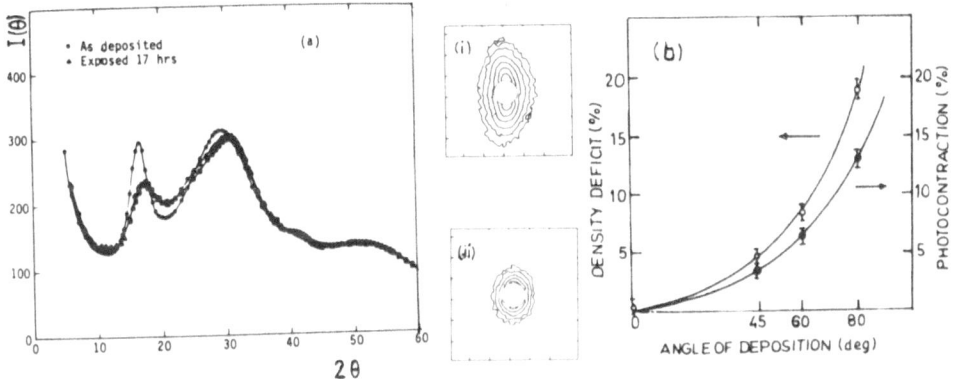

Fig.1(a)   X-ray scattering intensity for an evaporated a-As₂Se₃ film, both as-deposited (solid line) and after illumination with white light (dashed line).(4)

Fig.1(b)   Density of obliquely evaporated GeSe₂ films for as-deposited films (solid line), and after illumination by a Hg lamp (dashed line).(7)   Shown in the insets are the SANS contour plots for (i) an as-deposited (α=80°)a-GeSe film, and (ii) after illumination by a Xe arc lamp.(8)

of a-GeSe₂ films.(7) The insets to fig.1(b) show the small-angle neutron scattering from an as-deposited (α=80°) GeSe₃ film, together with the scattering after optical exposure.(8) After irradiation, the scattering intensity and the degree of anisotropy decrease, supporting the idea that the photo-contraction is caused by the photo-induced collapse of the elongated voids, leaving only a few, more spherical voids after exposure.

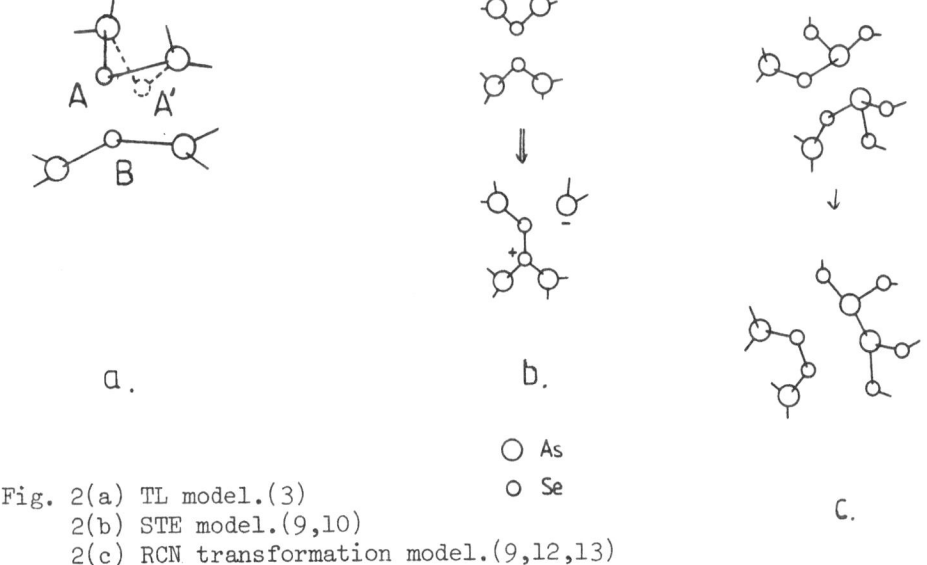

Fig. 2(a) TL model.(3)
      2(b) STE model.(9,10)
      2(c) RCN transformation model.(9,12,13)

We turn our attention now to the *reversible* photostructural effects observed in melt-quenched glasses or well-annealed thin films; these phenomena are much less well understood than are the irreversible changes discussed above, principally because the size of the effect is so small (see Table 1). Diffraction experiments are difficult to perform, although some progress in this area has been made(9). At least three distinct models have been proposed for the effect, namely the two-level model, the self-trapped exciton model and the random covalent network transformation model, respectively.

The two-level (TL) model (3) supposes that optical excitation of one $p\pi$ orbital (A) affects the interaction with the orbital on the neighbouring atom (B), and as a result the chalcogen atom moves to position A'. which is the lowest energy configuration for that electronic state. Thus the chalcogen atom can occupy one of two positions corresponding to two levels in the potential energy-configurational coordinate space (see fig. 2(a)); annealing the amorphous material to $T \simeq T_g$, can cause release of the trapped, optically excited electron and subsequent recombination of the $p\pi$ orbital, resulting in a return to state A.

The self-trapped exciton (STE) model assumes that an optically created electron-hole pair (exciton) can become trapped by means of the rupture of one bond and the formation of another, resulting in the creation of an over-coordinated, positively charged structural defect and an under-coordinated, negatively charged defect. This model was originally formulated for the case of a-Se (10), in which the defect centres are C_3^+ and C_1^- in the Kastner-Adler-Fritzsche nomenclature (11), the C_3^+ centre resulting from the formation of a dative bond employing the chalcogen lone-pair orbital. For the case of alloy chalcogenide glasses, the situation is more complicated, since defects involving any of the atoms in the alloy can in principle be formed. As an example, the formation of the pair of defects C_3^+, P_2^-, which can be produced in a-As_2S_3 or As_2Se_3, is shown schematically in Fig. 2(b), together with the configurational-coordinate diagram for this process. Note that for such binary alloys, *four* distinct sets of charged defects are in principle possible, namely (C_3^+, P_2^-), (C_3^+, C_1^-), (P_4^+, P_2^-) and (P_4^+, C_1^-). However, the formation of P_4^+ centres is perhaps less likely than that of C_3^+, since creation of the former requires sp^3-hybridisation of the deep-lying, non-bonding s-states in the pnictide atom, a circumstance not necessary in the employment of the high-lying $p\pi$ lone-pair states to form C_3^+. (Note that analogous overcoordination *cannot* occur at Ge or Si atoms in chalcogenide alloys containing these atoms if only s-p bonding is permitted and d-orbital mixing is excluded.)

Finally, there is the random covalent network (RCN) transformation model (9,12,13), in which the (stoichiometric) alloy is assumed to be chemically ordered before irradiation; it is then supposed that exposure to light breaks two of the hetero-polar bonds, and two *homo* polar bonds are formed instead, thereby forming locally a random

covalent network (see fig. 2(c)). The STE mechanism may be an intermediate step in such a process. (12) Note that such a mechanism is
inappropriate for the case of a-Se or S, where only the TL or STE
models can operate.

There is some experimental evidence that the reversible photostructural transformation involves the formation of "wrong" homopolar bonds in alloy glasses. Raman scattering measurements on a-
As_2S_3 have shown that two small peaks on either side of the main As-S
stretching peak appear after optical irradiation (13); these peaks
correspond in position to the As-As and S-S stretching bands observed in the pure elements, and this is therefore strong evidence for
the formation of homopolar bonds in the reversible photostructural
effect. Diffraction and EXAFS experiments (9) support this finding
and are consistent with the bonding rearrangements in either the STE
or the RCN models shown in figs. 2(b) and 2 (c), respectively; there
appears to be little experimental support at present for the TL
mechanism shown in fig. 2(a).

One final type of photo-structural change observed in amorphous
chalcogenides is photo-induced crystallization. This can occur in an
irreversible fashion, notably in a-Se where the low value of T_g
($\approx 30^{\circ}C$) undoubtedly facilitates the transformation if irradiation
takes place at room temperature, or even in a reversible manner.
Such laser-induced, reversible quasi-crystalliz tion has been observed in a-GeSe$_2$ using Raman scattering (14), where sharp features
characteristic of crystalline spectra are seen after irradiation;
these features disappear on annealing the sample, indicating that
the photo-induced crystallites are below the critical size for
nucleation and growth.

3.2 Optical Changes

Photo-induced optical changes are manifested most clearly in
the optical absorption edge; fig. 3 shows the *reversible* changes in
the edge observed in a-As_2S_3 (3). (Note that *no* photo-induced optical changes are observed in the corresponding crystal; this effect
appears to be unique to the amorphous phase.) In general, the absorption edge suffers a red-shift upon optical excitation (for both
reversible and irreversible changes), and the materials are therefore
observed to "photo-darken". An understanding of the processes responsible for the photo-induced optical changes can be gained from a
consideration of the relevent *structural* changes.

Irreversible optical changes can, like their structural counterparts, be divided into three types. Changes associated with the
photo-polymerization of molecular aggregates in as-deposited films
occur because of the difference in types of bond in virgin and
illuminated films; e.g. As-As bonds occur in As_4S_4 molecules, and
hence in as-deposited films, whereas predominantly As-S bonds are
expected in the cross-linked, irradiated material. The size of bandgap of a material is governed to a large extent by the strength of

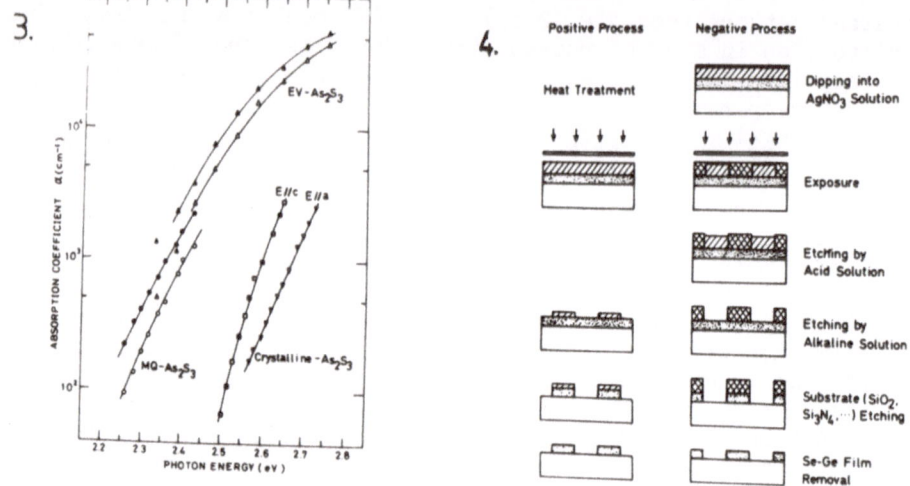

Fig. 3. Photo-darkening in an a-As$_2$S$_3$ film.(3) Also shown are the optical absorption edges for c-As$_2$S$_3$, which do not exhibit photo-darkening.

Fig. 4. Schematic illustration of the use of a-GeSe/Ag films as photo-(or electron-beam) resists.

the bonds which are present, (e.g. a-As has a gap of ≈1.2eV, whereas a-S has a gap of ≈3eV); hence changes in the bond-type will affect the gap and hence the optical absorption edge. The changes observed in as-deposited, obliquely evaporated chalcogenide films could also arise from the above mechanism. However, more important is the increase in *optical* density which accompanies the structural densification caused by the photo-induced collapse of voids in such materials. Photo-induced crystallization would also be expected to affect the optical absorption edge since crystalline and amorphous forms even of the same substance do not have the same optical properties (principally a result of the loss of k-conservation in the non-crystalline state).

The origin of reversible photo-darkening is uncertain, and this of course is related to the lack of understanding concerning the reversible photostructural effect. If the structural change is accompanied by the formation of "wrong" bonds, then the absorption edge would be expected to shift for the same reasons as given above for irreversible changes caused by similar bond rearrangements. However if the STE model is appropriate, then the influence on the absorption edge is by a different mechanism. It has been proposed (15) that the exponential part of the edge (the Urbach edge) results from an exciton line broadened by internal microfields. The formation of charged structural defects in the STE model would therefore increase the magnitude of such internal microfields, thereby increasing the optical absorption concomitantly.

3.3. Chemical Changes

As shown in table 1, optically induced chemical changes in amorphous chalcogenide materials can be of two very different types, both irreversible in nature. The first, changes in dissolution (or etching) rate of the chalcogenide is particularly pronounced in as-deposited, obliquely evaporated films, and is a direct consequence of the photo-densification characteristic of such films. The second type, namely photo-dissolution of metals, is unrelated. Noble metal (Ag or Au) overlayers dissolve into an underlying amorphous chalcogenide layer if irradiated by bandgap light; the effect is optical and not thermal (i.e. diffusive) in origin. Even more remarkable, perhaps, is the finding that such metals can photo-dissolve in chalcogenide films over distances of several microns with no lateral spreading. Irradiation by electrons or ions causes a similar dissolution of metal. The precise mechanism for this remarkable phenomenon is not known at present. Irradiation from either the metal or the chalcogenide side is effective in causing dissolution, and recent kinetic experiments in our laboratory indicate that the light causing the effect is absorbed near the metal-chalcogenide interface. Ionization of the metal atom (at least for the case of Ag) appears to take place, and the dissolution front (which is *sharp*, not having a graded concentration profile characteristic of diffusion) advances by means of rapid metal ion transport.

4. APPLICATIONS OF PHOTO-INDUCED PHENOMENA

A variety of technological applications can be envisaged for amorphous chalcogenide materials utilizing the various photo-induced phenomena discussed above. For convenience, these are listed in Table 2. Perhaps the most important potential application is the prospect of very high-resolution lithography using a-GeSe/Ag layers as a resist. The resolution limit of such a resist is potentially $\simeq 0.1\mu m$, and apart from this, such inorganic resist systems offer the advantage over conventional organic resists (such as PMMA) of the possibility of all processing (i.e. deposition, exposure and plasma development) taking place in vacuo. A schematic illustration of the steps involved in the photo-lithography process is shown in fig. 4.

5. CONCLUSIONS

Amorphous chalcogenide materials exhibit a wide variety of photo-induced phenomena. Reversible and irreversible changes in structure-related properties, optical behaviour and even chemical properties can be observed. The magnitude and type of such changes depend on the preparation method, as well as the chemical composition, of the amorphous chalcogenide material, and as such "engineering" of the photo-induced change for a specific application is possible.

TABLE 1.

Photo-induced phenomena in amorphous chalcogenide materials

Property		Type of Material	Reversibility	Magnitude and Material	Ref.
Structural	Density	Virgin, obliquely evaporated film	I	20%:GeS_2(80°)	(2)
	"	Glass or annealed film	R	0.6%:As_2S_3	(3)
	Structure factor	Virgin, evaporated film	I	40%(Q=1Å⁻¹):As_2Se_3	(4)
	"	Glass or annealed film	R	12%(Q=1Å⁻¹):As_2S_3	(3)
Optical	Absorption edge	Virgin, obliquely evaporated film	I	10%:GeS_2(80°)	(2)
	"	Glass or annealed film	R	2.4%:GeS_2	(3)
Chemical	Metal diffusion	Glass or film	I	Ag in $GeSe_2$	–
	Etching rate	Glass or film	I	–	–

(I = irreversible ; R = reversible)

TABLE 2.

Applications of photo-induced phenomena in amorphous chalcogenides

Application	Photo-induced Change Employed	Type of Material
High-density optical memory	Photo-darkening	Annealed film
Holographic grating	Photo-darkening and densification	Obliquely evaporated film.
Offset lithography	Photo-chemical (etching rate)	Thin film
Resist for photo - or electron-beam lithography.	Photo-dissolution	Thin film + Ag layer

6. REFERENCES

1. B.A. Weinstein, R.Zallen, M.L.Slade and A. de Lozanne, Phys. Rev. B24, 4652 (1981)
2. S.Rajagopalan, K.S.Horshavordhan, L.K.Malhotra, and K.L.Chopra, J. Non-Cryst. Sol. 50, 29 (1982)
3. K.Tanaka, J.Non-Cryst. Sol. 35-36, 1023 (1980)
4. J.P.De Neufville, S.C.Moss and S.R.Ovshinsky, J. Non-Cryst. Sol. 32, 271 (1973/4).
5. M.F.Daniel, A.J.Leadbetter, A.C.Wright and R.N.Sinclair, J.Non-Cryst. Sol. 32, 271 (1979)
6. H.J. Leamy, G.H.Gilmer, and A.G.Dirks in "Current Topics in Materials Science", vol. 6, ed. E.Kaldis (North-Holland:1980) p.309.
7. K.L.Chopra, L.K.Malhotra and K.S.Harshvardhan, Proc. Symp. Inorganic Resist Systems, eds. D.A.Doane and A.Heller (Electrochemical Soc : 1982), p.129.
8. T.Rayment and S.R.Elliott, Phys. Rev. B28, 1174 (1983)
9. S.R.Elliott, J. Non-Cryst.Sol. 59-60, 899 (1983)
10. R.A.Street, Sol. St. Comm. 24, 363 (1977).
11. M.Kastner, D.Adler and H.Fritzsche, Phys. Rev. Lett. 37, 1504 (1976)
12. R.Grigovovici and A.Vancu, J. de Physique, C4-391 (1981).
13. M.Frumar, A.P. Firth, and A.E. Owen, J. Non-Cryst. Sol. 59-60, 921 (1983)
14. J.E.Griffiths, G.P.Espinosa, J.P.Remeika and J.C.Phillips, Phys. Rev. B25, 1272 (1982)
15. J.D.Dow and D.Redfield, Phys. Rev. B1, 3358 (1970).

SMALL POLARONS IN

TRANSITION METAL OXIDE GLASSES

J. Livage

Spectrochimie du Solide - Université Pierre et Marie Curie
4, place Jussieu - 75005 Paris - France

ABSTRACT : Transition metal oxide glasses are mixed valence compounds. Their semiconducting properties arise from a small-polaron hopping between metal ions in different valence states. They can also be reduced by electrochemical insertion of alkaline ions, a property that could be used for electrochromic displays or reversible cathodes.

The electronic properties of Transition Metal Oxide (T.M.O.) glasses arise from the fact that transition metal ions may exhibit several valence states so that electron transfers from low to high valence states can take place. These transfers can be either thermally or optically activated and TMO glasses will exhibit electrical and optical properties. Strong interactions between the unpaired electrons and the polar network are usually observed in TMO and the charge carrier should rather be described as a "small polaron". In this paper, we shall first describe the small polaron model and then discuss some of the main properties of TMO glasses.

1. SMALL POLARON THEORY

1.1 Formation of a small polaron

According to Emin (1), a small polaron is an extra electron, or a hole, severely localized within a potential well that it creates by displacing the atoms that surround it. In order to illustrate this, let us consider an extra electron in an ionic lattice. At a large distance, the potential energy of another electron would be either $-e^2/k_s r$ or $-e^2/k_\infty r$ depending on wether the ions are allowed to move or not. k_s and k_∞ are respectively the static and high frequency dielectric constants. Thus, the potential energy $V_p(r)$ in the potential well arising from the displacement of the ions around the extra electron is given by (2)

$$V_p(r) = - \frac{e^2}{k_p r} \text{ with } \frac{1}{k_p} = \frac{1}{k_\infty} - \frac{1}{k_s}$$

This potential can be taken as the self-trapping potential of the electron itself. The electron digs its own potential well.

As a rough approximation, this relation can be considered as valid up to a distance r_p, called the polaron radius. The electron is then trapped into a potential well given by (fig. 1)

$$V_p(r) = - \frac{e^2}{k_p r} \quad \text{for } r > r_p$$

$$V_p(r) = - \frac{e^2}{k_p r_p} \quad \text{for } r < r_p$$

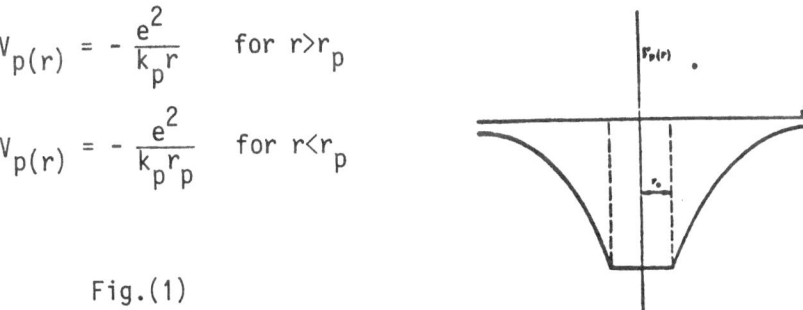

Fig.(1)

Beyond this point, there are two limiting approximations.

. If the effective mass m^* of the electron is small, its kinetic energy is important and r_p is quite large. We have a "large polaron".

. If m^* is very high, the kinetic energy due to the electron localization becomes negligible and r_p must be somewhat less than the inter-ionic distance. We have a "small polaron".

The formation of a small polaron involves two components :

. The energy required to polarize the medium $e^2/2k_p r_p$

. The lowering of the potential energy of the electron $-e^2/k_p r_p$.

The small polaron binding energy W_p is then : $W_p = e^2/2k_p r_p$.
It usually lies around 0.5 eV in most TMO.
The small polaron radius could be estimated if one knows the phonon spectrum (3)

$$r_p = \frac{1}{2}(\frac{\pi}{6N})^{\frac{1}{3}}$$

where N is the number of sites per unit volume.

An electron-phonon constant γ can be defined as $\gamma = W_p/h\nu_0$, where ν_0 is the mean optical phonon frequency. It corresponds to the number of optical phonons involved in the trapping process of the small polaron (4). We see that because of its interaction with the polar lattice (electron-phonon coupling), the extra electron will be trapped in its owen potential well. The electron-phonon coupling parameter γ will then lead to a localization of the charge carriers, the larger γ, the deeper the trap will be. Values of γ ranging between 5 and 10 are quite usual in most TMO glasses.

1.2 Localization or delocalization in mixed valence compounds

Electron delocalization between two metallic sites A and B can be easily described by using an Hush's diagramm (5). Such a diagramm (fig. 2a) gives the potential energy as a function of a configurational parameter $q = q_a - q_b$ corresponding to the anti-symmetric combination of the vibrational modes on both sites. Electron delocalization between A and B can actually occur only if the overlap between the two electronic wave functions ϕ_a and ϕ_b is different from zero (fig. 2b). This electronic interaction removes the degeneracy at the crossing point of the potential energy curves, leading to two energy levels separated by 2J, where J is a transfer integral.

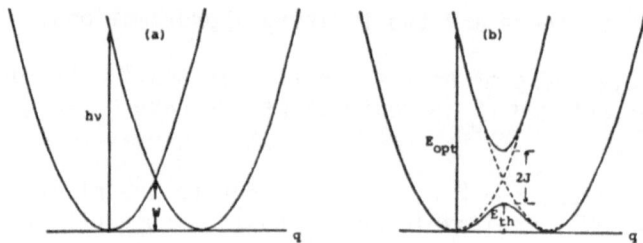

Fig.(2) : Potential energy diagram of an electron hopping between two transition metal ions :
a : J=0, b : J≠0

Two main parameters are involved in mixed valence compounds:
. The electron-phonon coupling γ leading to the formation of the small polaron and the localization of the charge carriers.

. The transfer integral J, lowering the potential energy between the two configurations and leading to the delocalization of the charge carriers.

Localization or delocalization then occur depending on the relative magnitude of γ and J, leading to the well known classification suggested by Robin and Day (6). No transfer is observed when J=0, valence states remain localized and the compound behaves as an insulator (class I). Electron exchange between localized states occurs when J<γ (class II). These compounds usually exhibit a deep blue color arising from intervalence transfer,(photochromism of amorphous WO_3) and semiconducting properties arising from small polaron hopping (vanadate glasses). Delocalization occurs when J>γ (class III) leading to a metallic behaviour as in the well known tungsten bronzes Na_xWO_3.

An additional parameter W_d has to be taken into account in TMO glasses. It corresponds to the energy difference between sites A and B, arising from the random disorder of the amorphous network. As for γ, it leads to a localization of the charge carriers. In TMO glasses, the d overlap is usually quite small so that the localization parameters prevail and these glasses belong to the class II.

1.3 Small polaron hopping in TMO glasses

The potential energy diagram (fig. 2a) shows that the thermal activation energy W_h for electron transfer should be $W_h = 1/2\ W_p$ (in the limit where J→0) (2). The chance per unit time that an electron jumps from one site to the other is then proportional to $\exp(-W_h/kT)$. This thermal activation energy becomes $W = W_h + 1/2\ W_d + W_d^2/16\ W_h$ when the disorder term is taken into account. In most cases, $W_h > W_d$ and the last term can be neglected.

The electron can also be optically transfered, from A to B. According to the Franck-Condon principle, i.e. without moving the ions, the optical activation energy is $E_{op} = 2\ W_p$ (2).

The small polaron diffusion through a lattice is somewhat more complicated to describe. The two centers model however points out that electron transfer only occurs when the configurational parameter q=0, i.e. when the two potential wells have the same energy. This can only be achieved by lattice distorsion and phonons must be involved in the small-polaron hopping process (fig.3).

The basic expression relating the drift mobility μ to the hopping rate P and the site separation R is (7)

$$\mu = \frac{eR^2}{kT} \cdot P$$

412

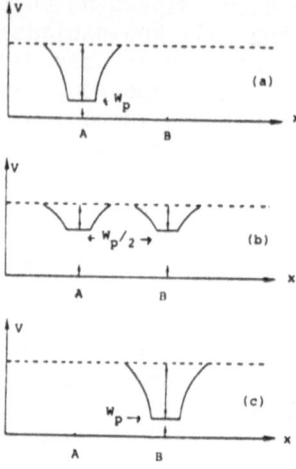

Fig.(3) : Small polaron hopping
between potential wells in TMO:
a : before hopping
b : thermally activated transi-
    tion state
c : after hopping.

The hopping rate P, depends on the number of vacant sites $(1-C)$ and on the probability P_1 and P_2 of two events :
$P = P_1 . P_2 (1-C)$.
. C is the concentration of reduced ions reported to the total concentration of transition metal ions.

. P_1 is the probability for the two potential wells to have the same energy. It can be expressed as $P_1 = \nu_0 \exp(-W/kT)$ where ν_0 is a mean phonon frequency usually related to the Debye temperature Θ by $h\nu_0 = k\Theta$.

. P_2 is the probability for the electron transfer to occur during the coincidence of the potential wells.

Two cases have then to be discussed :

The "adiabatic regime" where the electron can always fol-low the lattice motion. $P_2 = 1$ and μ is given by

$$\mu = \frac{eR^2 \, \nu_0 (1-C)}{kT} \, \exp(-\frac{W}{kT})$$

The conductivity $\sigma = n\mu e$ can be expressed as

$$\sigma = \frac{\nu_0 \, e^2 \, C(1-C)}{R.kT} \, \exp(\frac{-W}{kT}) \qquad (1)$$

where the number of charge carriers n per volume unit is given by $n = C/R^3$.

The non adiabatic regime when the time required for an electron transfer is large compared with the coincidence of the potential wells. The probability $P_2 \ll 1$ could be expressed as (4)

$$P_2 = \frac{1}{\hbar \nu_0} \left(\frac{\Pi}{4WkT}\right)^{\frac{1}{2}} J^2$$

This regime is observed when $J \ll h\nu_0$ and the mobility becomes :

$$\mu = \frac{eR^2}{kT} \cdot \frac{1}{\hbar} \left(\frac{\Pi}{4WkT}\right)^{\frac{1}{2}} (1-C)J^2 \exp\left(\frac{-W}{kT}\right)$$

Normally μ is dominated by the exponential term and the conductivity σ is given by

$$\sigma = \frac{\sigma_0}{T^{\frac{3}{2}}} \exp\left(\frac{-W}{kT}\right) \qquad (2)$$

where

$$\sigma_0 = \frac{e^2}{\hbar kR} \left(\frac{\Pi}{4kW}\right)^{\frac{1}{2}} J^2 C(1-C)$$

Electrical conductivity in TMO glasses is usually described by the model suggested by Austin and Mott (2). According to these authors, the transfer integral J depends on R as $J \sim I \exp(-\alpha R)$ where I is of the order of the ionization energy of the transition metal atom and α is the rate of the wave-function decay. We then have :

$$\mu = \nu \frac{eR^2}{kT} (1-C) \exp(-2\alpha R) \exp\left(-\frac{W}{kT}\right)$$

where

$$\nu = \frac{1}{\hbar} \left(\frac{\Pi}{4WkT}\right)^{\frac{1}{2}} \cdot I^2$$

The conductivity   is then given by the well known formula (2) :

$$\sigma = \nu \frac{e^2}{RkT} C(1-C) \exp(-2\alpha R) \exp\left(-\frac{W}{kT}\right) \qquad (3)$$

One of the most striking feature of the conductivity of TMO glasses, is that the activation energy W decreases with the temperature. A typical $\log(\sigma T)$ vs. (T^{-1}) plot is shown in fig.(4). It does not follow an Arhenius law as suggested by equation (3) and it is usually difficult to separate the measured activation energy W into a polaron term W_h and a disorder term W_d.

. At high temperature $(T > \Theta/2)$, the small polaron hopping is activated by an optical multiphonon process and the activation energy is given by $W = W_h + 1/2 W_d$ (2).

414

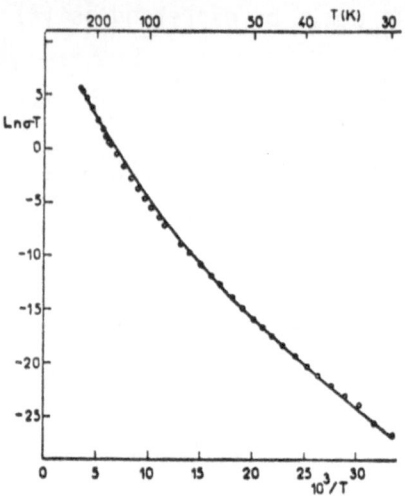

Fig.(4) : Temperature dependence of the dc conductivity in TMO glasses.

. As the temperature is lowered, the phonon spectrum freezes out and the polaron term drops continuously down to zero. According to Schnakenberg (8) a Boltzman distribution of the phonon spectrum has to be taken into account, leading to :

$$\sigma T = \sigma_0 \exp(-\frac{W_d}{2}) \exp -\frac{W_h}{kT} \cdot \frac{\tanh x}{x} \qquad (4)$$

with $x = \frac{h\nu_0}{4kT}$ and $h\nu_0 = k\Theta$ (Debye temperature).

. At very low temperature (T<Θ/4), an acoustical phonon assisted hopping takes place and the activation energy becomes $W = 1/2 W_d$. According to Mott, a variable range hopping process should even be observed when T $\longrightarrow$ 0 (9).
Due to the high resistivity of TMO glasses, the very low temperature regime can usually not be observed and conductivity measurements have rarely been performed below 77K. Some experiments have however been reported by Bullot et al down to 40K on amorphous V_2O_5 layers deposited from gels. They lead to $W_d \sim 0.1$ eV (10). W_d could also be estimated from low temperature measurements of the thermo-electric power as reported for V_2O_5-TeO_2 glasses where a value of about 0.02 eV was found (11).

2. PROPERTIES OF TMO GLASSES

2.1 Electrical properties

Two localization processes occur in TMO glasses. The first one arises from the formation of small polarons (W_h) and the second one from the random disorder (W_d). These glasses are therefore low-mobility n-type semiconductors. Their dc conductivity is usually expressed by eq.(3) at high temperature ($T > \Theta/2$) and eq(4) at lower temperature ($T < \Theta/2$). We shall not give here a detailed discussion of the electrical properties of TMO glasses. Good reviews will be found in the following papers (2)(7) (12-15). We would like only to point out some of the main features.

As shown by eq(3), the conductivity depends on the concentration C of reduced transition metal ions. It should go through a maximum for C = 0.5. This is however never observed and reported maxima for vanadate glasses usually lie between 0.1 and 0.2 (11) (16). Various explanations have been suggested, taking into account trapping effects (16) or polaron-polaron interactions (17). This discrepancy should however mainly arise from phase separation in the glass.

The importance of the tunnelling term $\exp(-2\alpha R)$ is not obvious. Sayer and Mansingh have shown that it was almost constant for $WO_3-P_2O_5$ and $V_2O_5-P_2O_5$ glasses (17). They noted that a plot of log σ versus the activation energy W allows one to distinguish between adiabatic and non-adiabatic hopping and judge the relative importance of the two exponential terms in eq(3). According to Murawski et al (7), $V_2O_5-P_2O_5$ and possibly $WO_3-P_2O_5$ glasses should be adiabatic while most other systems are not.

TMO glasses are usually obtained by quenching of the molten oxide with a glassformer such as P_2O_5, GeO_2 or TeO_2 (13). The nature of this glassformer appears to be very important. The electrical conductivity of $V_2O_5-TeO_2$ glasses for instance is 3 orders of magnitude larger than that of the corresponding $V_2O_5-P_2O_5$ glass (11)(18). The role of the glassformer appears to be predictable in first order in its effect on the effective dielectric constant k_p. However it seems clear that the structural match of both oxides should be of importance. In the case of vanadate glasses, for instance, Chung et al (19) have shown that the partial charge of the glass-forming oxide in the glass network, greatly affects the hopping process between transition metal ions. Adding a glassformer also increases the mean distance between transition metal ions. A strong relationship between activation energy W_h and site spacing R has been evidenced for vanadates glasses (14). According to Mott (9) the polaronic term can be modified to account for the overlap of polarisation wells at high carrier densities, leading to : $W_h = e^2/4.k_p(1/r_p - 1/R)$.

2.2 Optical properties

Small-polaron theory predicts an optically activated electron-transfer in mixed valence compounds. It corresponds to a broad Gaussian-like absorption at $h\nu = 2 W_p$. The oscillator strength being approximately (12)

$$f = z(M_0 R^2/\hbar^2) \, 4J^2/h\nu$$

where z is the number of nearest neighbours. The small-polaron binding energy being of about 0.5 eV, optical absorption occurs around 1 eV, i.e. in the red part of the spectrum. Mixed-valence compounds usually exhibit a typical deep blue color that could be used, for instance in electrochromic display devices based on amorphous WO_3 thin films (20).

One of the main problem, when dealing with the optical spectra of TMO glasses, is to make the difference between small-polaron absorption bands and d-d transitions. Electro-optical studies however could help to distinguish between an internal optical excitation at a single site, and an excitation between two different sites, since the latter is more sensitive to an external electric field (21).

In the case where W_d and J are small, optical absorption should occur at $h\nu \cong 4 W_h$. The theory also predicts that the conductivity at the maximum of the peak (σ_{max}) is roughly (2) :

$$\sigma_{max}/\sigma_{dc} = (kT/4W_h) \, \exp W_h/kT$$

Very few optical studies have however been performed in order to confirm the theory.

2.3 Electrochemical properties

Fast ion conduction has been reported in rapidly quenched TMO glasses (22), but the nature of the transition metal ion does not seem to have any noticeable effect on the conductivity. The most important parameter appear to be the alkali concentration. A maximum ionic conductivity of about $10^{-3} \, \Omega^{-1}cm^{-1}$ is observed at 500K, this value beeing reached for different Li concentrations in different glasses (23). No electronic conductivity is observed, all transition metal ions beeing in their higher valence state.

Mixed conducting glasses can however be obtained by chemical or electrochemical insertion of alkaline ions into TMO glasses (24)(25). In this case Li^+ insertion is accompagnied by the reduction of transition metal ions and both ionic conductivity arising from Li^+ diffusion and electronic conductivity arising from small polaron hopping are observed (25). Such glasses could be used as positive electrodes in solid state batteries, but, despite the significant advantages they could offer, the literature dealing with the electrochemical properties of TMO glasses is still remarkably sparse. The only published works deal with V_2O_5-P_2O_5 (25)(26), V_2O_5-B_2O_3 (27) and V_2O_5-Li_2O (28) glasses. Coulometric experiments performed with a $Li(POE)_8$ $LiClO_4$/ V_2O_5-P_2O_5 cell show that the open circuit voltage is significantly higher compared with crystallized vanadium oxides. It ranges between 3.6 V and 3.0 V for lithium insertion up to LiV_2O_5 (26).

Much more studies have however been performed in order to use the optical properties of amorphous WO_3 thin films in display devices (20). Amorphous WO_3 obtained by vapor deposition, sputtering or anodic oxidation, exhibits two stable states. One is transparent and highly resistive while the other one is deep blue and much less resistive. Coloration can be obtained by electrochemical reduction (29). This last process is highly reversible and appears to ve very promising for making electrochromic display devices. In such a device, the amorphous thin film, deposited onto a transparent conductive electrode (I.T.O), is placed into an electrochemical cell. Coloration and bleaching appear within 200 ms upon applying a negative or positive voltage of about 2 volts. The currently adopted model for electrolytic coloration is that of a double injection process of M^+ ions (M = H, Li) and electrons into the layer (30). The blue coloration arises from intervalence transfers between W(V) and W(VI) ions. The electrochromic properties of such films appear to be very dependant upon the water content into the electrolyte or the layer (31). Very good results have recently been obtained with amorphous WO_3 layers made by the sol-gel process (32).

REFERENCES

1. D. Emin, Physics today, 34 (1982).
2. I.G. Austin and N.F. Mott, Adv. in Physics, 18, 41 (1969).
3. V.N. Bogomolov, E.K. Kudinov and Y.A. Frisov, Soviet Physics, Solid State, 9, 2502 (1968).
4. T. Holstein, Ann. Phys. (N.Y.), 8, 325 (1959).
5. N.S. Hush, Prog. Inorg. Chem., 8, 391 (1967).
6. M.B. Robin and P. Day, Adv. Inorg. Chem. and Radiochem., 10, 247 (1967).
7. L. Murawski, C.H. Chung and J.D. Mackenzie, J. Non-Cryst. Solids, 32, 91 (1979).
8. J.Schnakenberg, Phys. Stat. Sol. b, 28, 623 (1968).
9. N.F. Mott, J. Non-Cryst. Solids, 1, 1 (1968).
10. J. Bullot, P. Cordier, O. Gallais, M. Gauthier and J. Livage Phys. Stat. Sol. a, 68, 357 (1981).
11. B.W. Flynn, A.E. Owen and J.M. Robertson in Amorphous and Liquid Semiconductors, Ed. W.E. Spear (C.I.C.L. University of Edinburgh) 678, (1977).

12. I.G. Austin and E.S. Garbett, Electronic and Structural Properties of Amorphous Semiconductors, ed. P.G. Lecomber and J. Mort, Academic Press, 393 (1973).
13. C.H. Chung, J.D. Mackenzie and L. Murawski, Rev. Chimie Minérale, 16, 308 (1979).
14. M. Sayer and A. Mansingh, J. Non-Cryst. Solids, 58, 91 (1983).
15. J. Livage, J. de Physique, 42, C4-981 (1981).
16. G.S. Linsley, A.E. Owen and F.M. Hayatee, J. Non-Cryst. Solids, 4, 208 (1970).
17. M. Sayer and A. Mansingh, Phys. Rev. B6, 462, (1972).
18. V.K. Dhawan, A. Mansingh and M. Sayer, J. Non-Cryst. Solids, 51, 87 (1982).
19. C.H. Chung and J.D. Mackenzie, J. Non-Cryst. Solids, 42, 357 (1980).
20. R.J. Colton, A.M. Guzman and J.W. Rabalais, Acc. Chem. Res., 11, 170 (1978).
21. I.G. Austin, J. Phys. C. Solid State, 5, 1687 (1972).
22. K. Nassan, A.M. Glass, M. Grasso and D.H. Olson, J. Am. Ceram. Soc., 127, 2743 (1980).
23. K. Nassan, R.J. Cava and A.M. Glass, Solid State Ionics, 2, 163 (1981).
24. P.D. Power, J. Applied Electrochem., 1, 91 (1971).
25. T. Pagnier, M. Fouletier and J.L. Souquet, Mat. Res. Bull., 18, 609 (1983).
26. T. Pagnier, M. Fouletier and J.L. Souquet, Solid State Ionics, 9-10, 649 (1983).
27. A.C. Leech, J.R. Owen and B.C.H. Steeh Solid State Ionics, 9-10, 645 (1983).
28. K. Nassan and D.W. Murphy, J. Non-Cryst. Solids, 44, 297 (1981).
29. S.K. Deb, Phil. Mag., 27, 801 (1973).
30. B.W. Faughnan, R.S. Crandall and P.M. Heyman, R.C.A. Review, 36, 177 (1975).
31. O. Bohnke and G. Robert, Solid State Ionics, 6, 115 (1982).
32. A. Chemseddine, R. Morineau and J. Livage, Solid State Ionics, 9-10, 357 (1983).

TRANSITION METAL OXIDE GELS

J. Livage

Spectrochimie du Solide - Université Pierre et Marie Curie
4, place Jussieu - 75005 Paris - France

ABSTRACT : Transition metal oxide gels can be obtained by acidification of inorganic aqueous solutions. They can be easily deposited onto a substrate by dip-coating. These layers exhibit electronic properties arising from the mixed valence oxide and ionic properties arising from the solvent water molecules. Both kind of properties may be modified by the interface between solid and liquid. Ion exchange may also occur leading to intercalation chemistry and electrochromic devices.

Transition metal oxide gels have been known for almost a century. The first V_2O_5 gel for instance was reported by A. Ditte in 1885. They have however almost never been used in the sol-gel technology. In this paper, we would like to describe some properties of transition metal oxide coatings derived from gels and show that they could lead to new developments of the sol-gel technology (1)(2).

Transition metal oxide are usually mixed valence compounds. Metal ions may exhibit several oxidation states so that electron transfer, from low to high valence states, can take place. This leads to specific electrical and optical properties : semiconducting V_2O_5 layers (3) electrochromic WO_3 display devices (4). Magnetic interactions may also occur when the two valence states have non zero magnetic moments as in ferrofluids (5).

Transition metal oxide gels can also be considered as hydrated oxides. Ionic properties arise from the ionization of water molecules, trapped in the gel, by the oxide surface. Some hydrous oxides show high proton conductivities at room temperature (6) and could behave as inorganic ion exchangers (7). This is the case for vanadium pentoxide gels that exhibit a layered structure and could be considered as a host lattice for intercalation of ionic species (8).

1. SYNTHESIS OF TRANSITION METAL OXIDE GELS

Transition metal oxide gels can be obtained by hydrolysis and polycondensation of metal alkoxides. This process has already been widely studied by Bradley (9) and applied to the synthesis of oxides such as TiO_2 or ZrO_2. In this paper, we shall rather focus our discussion on the synthesis of gels from aqueous solutions of inorganic salts. Such basic compounds are cheaper than the metal-organic ones and the resulting gels are free of carbon, leading to better glasses or ceramics.

Let us consider a water molecule bonded to a metallic ion M^{z+}. The formation of a $M-OH_2$ bond, by overlapping of the water σ orbital with the empty metallic d orbitals, draws electrons away from the O-H bonds, and weakens them. The coordinated water molecules then behave a stronger acids than the solvent water molecules (10). Depending on the σ transfer, we may have :

$$M-OH_2 \rightleftharpoons M-OH + H^+ \rightleftharpoons M-O + 2H^+$$

This equilibrium depends on the metal ion, particularly its charge, size, electronegativity and ionization potentials. It also depends on the pH. For a given ion, we then have pH intervals in which water, hydroxide or oxide are common ligands to the central ion (11). Usually $M-OH_2$ bonds are observed for low valent states in acidic medium, while $M-O$ bonds occur for high valence states in basic medium.
In an intermediate pH range, M-OH bonds are formed. One of the main properties of these hydroxo groups, is that they could lead to polycondensation reactions such as :

Olation 2M-OH $\longrightarrow$

$$M \overset{\displaystyle OH}{\underset{\displaystyle OH}{<>}} M$$

or oxolation 2 M-OH $\longrightarrow$ M-O-M + H_2O

These polymerization processes appear to be quite general. Polycations can be obtained by increasing the pH of an aqueous solution of a low valent ion. Polyanions can be obtained by decreasing the pH of an aqueous solution of a high valent ion such as VO_4^{3-}.

$$2\ VO_4^{3-} \xrightarrow{\ 2H^+\ } 2HVO_4^{2-} \longrightarrow V_2O_7^{4-} + H_2O$$

Decavanadate ions $(V_{10}O_{28})^{6-}$ are obtained below pH 6 (fig. 1). Such isopolyanions have already been studied in great details (10) (12). More or less condensed species are obtained depending on concentration and pH, as shown in figure 1.

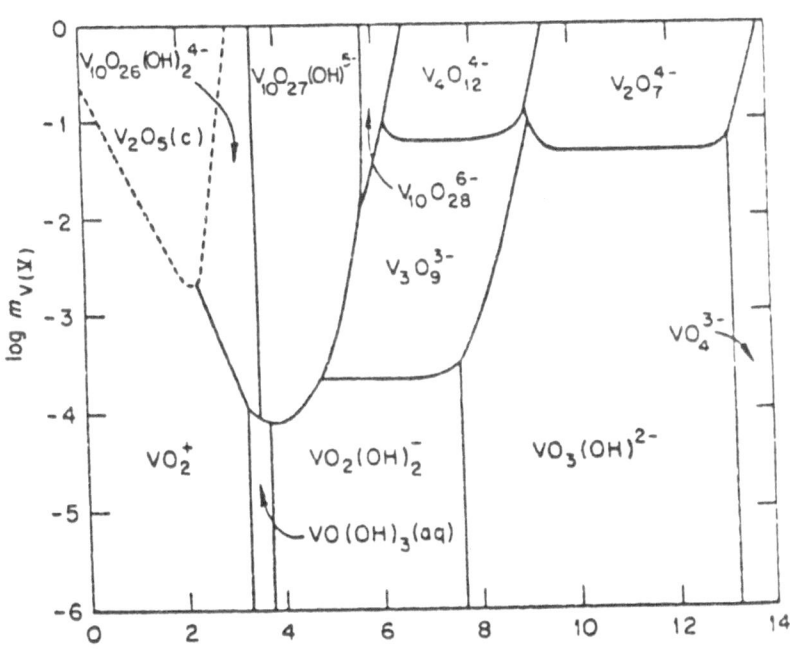

Fig. 1 : Stability diagram of vanadate polyanions

Such a diagram shows that for a given vanadium concentration, for instance 0.1 M.1^{-1}, condensation increases when the pH decreases. The total negative charge per vanadium decreases upon protonation, leading finally to the positive VO_2^+ ion. Around pH = 2, we go through a point of zero charge where vanadium pentoxide would precipitate. V_2O_5 could be considered as an infinite polymer in wich the total charge is zero. Stable colloidal species or gels could be obtained if the pH remains slightly higher than 2. These species are then negatively charged and electrostatic repulsions prevent further collisions and floculation.

Vanadium pentoxide gels are usually obtained by adding nitric acid to a vanadate salt. The gel then contains foreign ions such as NO_3^- and Na^+. Polyvanadic acid solutions can also be obtained by exchange in a sulfonic resin (Dowex 50 W-X2, 50-100 mesh) from sodium metavanadate solutions (13). The freshly prepared acid is yellow and decacondensed. Polymerization occurs spontaneously at room temperature, within a few hours, leading to a red viscous gel. Molecular weight determination, by light scattering and ultracentrifugation experiments, shows that the colloidal species are highly condensed ($M \approx 2.10^6$). They bear a negative charge of about -0.2 per vanadium. The viscosity of the gel depends on the concentration. A sol-gel transition occurs for a vanadium concentration of about 0.1 Mole.l^{-1} (14).

Vanadium pentoxide layers can easily be deposited onto a glass or polymeric substrate by dip-coating, spraying or screen-printing. After drying at low temperature, below 100°C, a xerogel is obtained that corresponds to the rough formula V_2O_5, nH_2O. The amount of water in the gel depends on the drying procedure : n = 1.8 in air and n = 0.5 under vacuum.

2. ELECTRONIC PROPERTIES OF T.M.O. GELS

TMO gels can be considered as biphasic systems containing a liquid phase (water) trapped into a solid network (oxide). We may then expect electronic properties arising from the mixed valence oxide and ionic properties arising from the ionization of water molecules.

The electronic properties of T.M.O. gels arise from the hopping process of unpaired electron between transition metal ions in different valence states : V(IV)-V(V) or W(V)-W(VI). Electron transfer may be either thermally or optically activated and TMO gels may exhibit both electrical and optical properties.

2.1 Semiconducting V_2O_5 gels

Semiconducting layers can be deposited from V_2O_5 gels (3) (15). Most of the solvent readily evaporates under vacuum and a rather hard coating is obtained corresponding to V_2O_5, 0.5 H_2O (16). The electrical behaviour of these layers appear to be quite ohmic. No transient regime is observed when a d.c. voltage is applied. The a.c. and d.c. conductivities are identical : $\sigma = 4.10^{-5}$ Ω^{-1}cm^{-1} at 300K for a xerogel containing 1% of V(IV) ions. This conductivity actually increases quite fast when the amount of reduced ions increases : $\sigma = 3.10^{-3}\Omega^{-1}$cm^{-1}, at 300K, for a xerogel containing 10% of V(IV) ions. These experiments suggest that the conductivity is mainly electronic. The charge carrier mobility, deduced from $\sigma = n\mu e$, leads to a very low value indeed : $\mu \cong 4.10^{-6}$ cm^2.V.s.

The temperature dependence of the d.c. conductivity, plotted as $\log(\sigma T)$ versus T^{-1} is shown in (fig. 2). The non linear variation, together with the very low mobility, are typical of small polaron hopping in transition metal oxides (17). The small polaron formation arises from a strong electron-phonon coupling and the conductivity can be expressed as (18) :

$$\sigma = \frac{\nu_0 e^2}{RkT} \ c(1-c) \ \exp(-2\alpha R) \ \exp(-W/kT)$$

where ν_0 is a phonon frequency related to the Debye temperature by $h\nu_0 \stackrel{\cong}{=} k\Theta$. R is the average hopping distance. C is the ratio $V(IV)^0 / \ V(IV)+V(V)$. The $\exp(-2\alpha R)$ term corresponds to a tunnelling transfer between two vanadium sites.

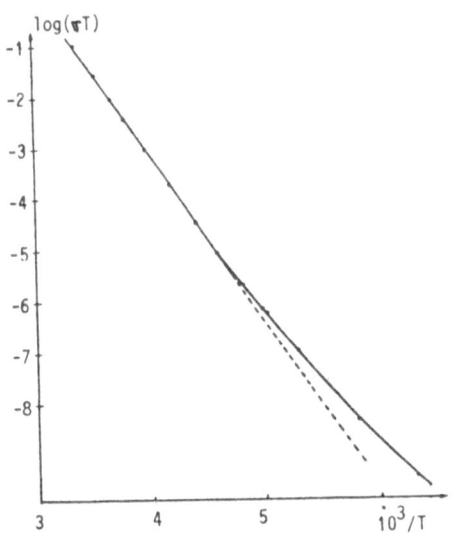

Fig. 2 : Temperature dependance of the dc conductivity of a $V_2O_5,0.5 \ H_2O$ xerogel.

According to the small polaron model, hopping at high temperature is thermally activated by an optical multiphonon process. The activation energy W is given by $W = W_h + 1/2 \ W_d$. W_h is a polaronic term and W_d corresponds to the potential energy distribution arising from the random disorder in the gel. A linear plot is actually observed above 220K, leading to W = 0.32 eV. As the temperature is lowered, the phonon spectrum freezes out and the polaronic term W_h continuously drops down leading to a decrease of the observed activation energy W below $\Theta/2$. An acoustical phonon assisted hopping takes place at lower temperatures $(T<\Theta/4)$ and the activation energy should become $W = 1/2 \ W_d$ (43).

Several patents have already been taken, suggesting new applications of the electrical properties of V_2O_5 layers deposited from gels. Their high conductivity, at room temperature, could make them usefull as antistatic coatings on the back of photographic films (19). Such coatings appear to be much less sensitive to ambient moisture than the polymeric antistatic coatings previously used in the photographic industry.

Switching devices, bases on V_2O_5 layers made by the sol-gel process, have recently been developped (20)(21). A coating, about 1μm thick is deposited onto a glass substrate and gold electrodes are evaporated at the surface of the xerogel. A typical Intensity-Voltage characteristic of such a device is shown in (fig. 3).

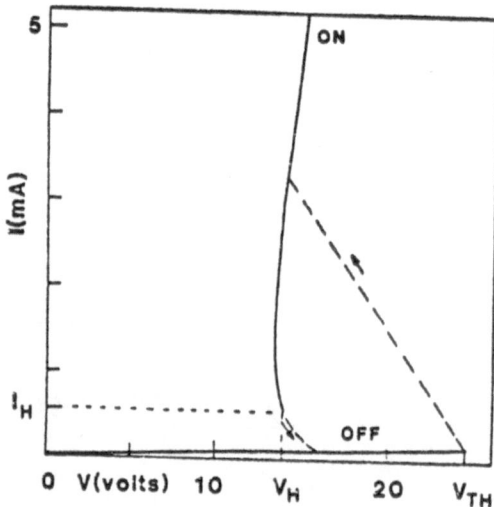

Fig. 3 : I = f(V) curve of a switching device based on a V_2O_5 layer deposited from gels.

Switching toward an "ON" state occurs for a threshold voltage V_t of about 20 volts. No memory effect is observed and the device switches back to the "OFF" state when the intensity falls below a minimum value (holding current) of about 2 mA. The ON/OFF ratio can reach two orders of magnitude and the switching effect appears to be highly reversible. It can operate more than 10^8 times without failure. The switching characteristics depend on the V(IV) content and the temperature. Switching disappears above 350K, or when the V(IV)/V(V) ratio becomes larger than 0.06.

2.2 Electrochromic WO_3 gels

Electrochromic display devices based on amorphous WO_3 thin film have been extensively studied (22). About ten years ago, Deb showed that such films could exhibit two stable states. One is transparent and highly resistive while the second one is deep blue and much less resistive. Fast reversible coloration can be obtained if the amorphous oxide is deposited onto a conductive electrode and placed into an electrochemical cell. Such devices exhibit a clear image, insensitive to the viewing angle and keep a long-term open circuit memory. They open new possibilities in the field of digital display devices.

Amorphous WO_3 thin films are usually obtained by vacuum evaporation or sputtering. They could also be made by the sol-gel
process. Colloidal tungsten oxide is often obtained as a transient
intermediate during the polymerization process of tungstic acids.
This colloidal phase is nevertheless not stable and precipitation
occurs leading to hydrated tungsten oxides. Such oxides do not
exhibit any electrochromic property. Stable colloidal solutions
can be obtained by ion exchange from sodium tungstate solutions
(23). A few drops of this colloidal solution were then deposited
onto a conductive transparent electrode (Indium Tin Oxide coated
glass substrate). The solvent readily evaporates and a coating is
obtained.

Electrochromic properties of colloidal tungsten oxide layers
were studied in a cell having the following configuration :
$ITO/WO_3/LiClO_4(M)-PC/Pt$. With such a device, a blue coloration is
obtained within 0.3s by applying a negative voltage of -4 volts.
The blue coloration remains for about 12 hours after the voltage
has been removed. The initial transparent state can be restored
within 0.4s upon application of a reverse voltage (+ 4 volts).
Under these conditions, the energy consumption for coloration and
bleachting is about the same and corresponds to 8 mC/cm^2 for an
optical density of 0.3. The device appears to be quite reversible,
coloring and bleaching cycles have been repeated over 10^5 times
without failure.

Compared with usual amorphous WO_3 thin films used in other
electrochromic display devices, colloidal tungsten oxide exhibits
two main advantages.
The first one is due to the special nature of the colloidal state.
Thin layers can be very easily made at low temperature. The sol-
gel process should then be much cheaper than vapor deposition or
sputtering.
The second one arises from the fact that colloidal tungsten oxide
exhibits a large surface area and contains water molecules.
According to the literature, the characteristics of WO_3 thin films
are very sensitive to the method of preparation. It has been
observed for instance that it is much more difficult to color and
bleach sputtered films than evaporated ones. This is probably due
to the smaller amount of water contained into sputtered films (24).
Knowles et al. (25) have shown that some water has to be incorporated into the WO_3 film in order to obtain any coloration. It thus
appears that the presence of water in the electrochromic cell may
greatly improve its characteristics.

3. IONIC PROPERTIES OF T.M.O. GELS

T.M.O. gels are made of charged particles held together by
water molecules and either H_3O^+ or OH^- ions. According to J.B.
Goodenough, they could be classified as particle hydrates (7).

They should then exhibit some typical properties of such compounds i(i) high proton conductivity, (ii) ion exchange properties, (iii) easy densification by cold pressing.

3.1 High proton conductivity

Conductivity measurements performed on a V_2O_5, 1.8 H_2O xerogel show a strong decrease of the intensity when a d.c. voltage is applied. This transient regime appears to be quite long and a steady state can be obtained after a few days only. Such a behaviour is typical of ionic conduction. A.c. conductivity measurements, using blocking electrodes, were performed in the temperature range 320-200K.

The a.c. impedance diagrams are typical of a simple fast-ion conductor (16). They show Debye behaviour and Warburg contributions, i.e. semicircles at high frequencies and straight lines with slopes of 43-45° relative to the real axis at low frequencies. The a.c. conductivity was deduced from extrapolation of this linear low-frequency portion. The room temperature conductivity of the layers is quite high : $\sigma = 10^{-2}\Omega^{-1}.cm^{-1}$. Vanadium pentoxide gels appear to be as good proton conductors, as the well known H.U.P $(HUO_2PO_4, 4H_2O)$. A plot of log (σT) versus T^{-1} shows two linear regimes (fig.4). A kink is observed around 260K. The activation energy above this temperature $E_1 = 0.35$ eV is smaller than the activation energy $E_2 = 0.44$ eV below 260K. This kink could correspond to the temperature at which the intercalated water layer is freezing.

A.c. conductivity of the V_2O_5, nH_2O layers strongly depends on the amount of water in the gel, and therefore on the partial water pressure above the layer. The conductivity decreases down to $10^{-6}\Omega^{-1}cm^{-1}$ when the sample is kept in a dry atmosphere with P_2O_5.

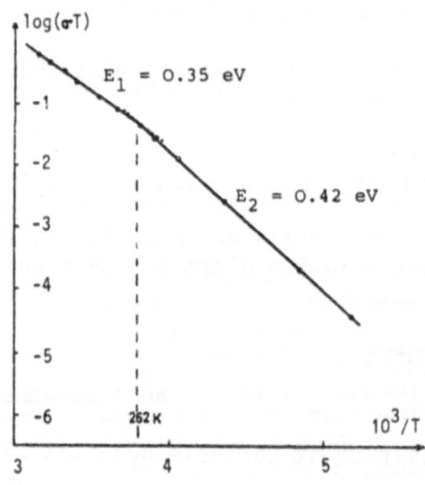

Fig. 4 : Temperature dependance of the ac conductivity of a V_2O_5, 1.8 H_2O xerogel.

A high proton conductivity necessitates a large concentration of mobile protons and a high proton mobility. The first factor is optimized in highly acidic or basic compounds. V_2O_5 gels appear to be good candidates. They exhibit a large oxide-water interface. They are negatively charged and produce acidic solutions (PH 2) when equilibrated with water.

It is more difficult to find a correlation between the proton mobility and the acid/base properties of the oxide. Two mechanisms are probably involved (26) :

. A vehicle mechanism corresponding to the drift of H_3O^+ ions between the fibres.

. A grotthus mechanism in wich a proton tunnelling occurs from a H_3O^+ ion to a H_2O molecule.

In both cases, the lamellar structure of V_2O_5 gels (27) provides a good conduction pathway for proton diffusion.

3.2 Cationic exchange properties

V_2O_5 gels appear to be good candidates for intercalation. They exhibit a layered structure with very weak interactions between adjacent layers (27). They are fast conductors and should therefore behave as good inorganic ion exchangers (7). They are mixed valence oxides and V(V) could easily be reduced into V(IV) (28). Ion exchange can be performed either chemically (8) or electrochemically (29).

Ion exchange, with metallic cations can readily be obtained, at room temperature, by dipping a layer of V_2O_5, 1.8 H_2O coated onto a glass substrate, into an aqueous solution of a metallic chloride (0.1 Mole.1^{-1}). The pH of the solution decreases and about 0.3 M^+ or 0.15 M^{++} cations per V_2O_5 can be intercalated. The basal spacing d, between the layers, depends on the nature of the metallic ion, mainly its positive charge and ionic radius. Large monovalent cations (Na^+, K^+, Rb^+, Cs^+, NH_4^+,...) lead to a d spacing of about 11 Å, while divalent cations (Ca^{++}, Mn^{++}, Mg^{++}, Fe^{++}, Co^{++},...) or Li^+ lead to a d spacing around 13.7 Å.

The difference d = 2.7 Å between these two basal spacings corresponds to the thickness of a water layer. This suggests that, because of the competition between the hydration energy of the cations and the energy required for lattice expansion, weakly polarizing cations can intercalate with one water layer only while more polarizing cations can be solvated with two layers.

Larger ions, such as alkylammonium cations $C_nH_{2n+1}N^+(CH_3)_3$ can also be intercalated into the layered structure of

V_2O_5 gels (30). Competition then arises between the electrostatic interactions of the oxide layers and the Van der Waals attractions of the alkyl chains. Depending on the number of carbon atoms, one interaction or the other prevails. Alkyl chain orientation in the interfoliar space can be deduced from the basal d-spacing variation with the chain length. The chains remain parallel to the layers up to C_7. They are perpendicular to then above C_{11}. In between they make an angle ranging from 42° to 53° with[11] the layer plane.

REFERENCES

1. J. Livage and J. Lemerle, Ann. Rev. Mater. Sci., 12, 103(1982).
2. J. Livage, Studies in Inorganic Chemistry, R. Metselaar, H.J.M. Heijligers and J. Schoonman eds. (Elsevier, Amsterdam 1983) 17.
3. C. Sanchez, F. Babonneau, R. Morineau and J. Livage, Phil. Mag. B, 47, 279 (1983).
4. A. Chemseddine, R. Morineau and J. Livage, Solid State Ionics, 9-10, 357 (1983).
5. R. Massart, I.E.E.E. Trans. on Magnetics, 17, 1241 (1981).
6. D.J. Dzimitrowicz, J.B. Goodenough and P.J. Wiseman, Mat. Res. Bull., 17, 971 (1982).
7. W.A. England, M.G. Cross, A. Hamnett, P.J. Wiseman and J.B. Goodenough, Solid State Ionics, 1, 231 (1980).
8. P. Aldebert, N. Baffier, J.J. Legendre and J. Livage, Rev. Chimie Minérale, 19, 485 (1982).
9. D.C. Bradley, R.C. Mehrotra and D.P. Gaur, Metal Alkoxides, Academic Press, London, (1978).
10. D.LE Kepert, The early transition metals, Academic Press, London (1972).
11. C.F. Baes and R.E. Mesmer, the hydrolysis of cations, John Wiley, New-York (1976).
12. P. Souchay, Ions minéraux condensés, Masson Ed. Paris, (1969).
13. J. Lemerle, L. Nejem and J. Lefebvre, J. Chem. Res. 301 (1978).
14. N. Gharbi, C. Sanchez, J. Livage, J. Lemerle, L. Nejem and J. Lefebvre, Inorg. Chem. 21, 2758 (1982).
15. J. Bullot, P. Cordier, O. Gallais, M. Gauthier and J. Livage, Phys. Stat. Sol.(a), 68, 357 (1981).
16. P. Barboux, N. Baffier, R. Morineau and J. Livage, Solid State Ionics, 9-10, 1973 (1983).
17. J. Livage, J. Phys., 42, C4, 981 (1981).
18. I.G. Austin and N.F. Mott, Adv. in Physics, 18, 41 (1969).
19. Kodak Pathé, French Patents BF 2318 442 (1977) and BF 2429 252 (1979).
20. J. Bullot and J. Livage, French Patent, 81, 13665 (1981).
21. J. Bullot, O. Gallais, M. Gauthier and J. Livage, Phys. Stat. Sol.(a), 71, K1-5 (1982).

22. R.J. Colton, A.M. Guzman and J.W. Rabalais, Accounts of Chem. Res. 11, 170-6 (1978).
23. A. Chemseddine, R. Morineau and J. Livage, Solid State Ionics, 9-10, 357 (1983).
24. O. Bohnke, and G. Robert, Solid State Ionics, 6, 115-20, (1982).
25. T.J. Knowles, H.N. Hersh and W. Kramer, in : 19th Electronic Materials Conf. Aime, Cornell, N.Y. (1977).
26. K.D. Krener, A. Rabenau and W. Weppner, Angew. Chem. Int. Engl. Ed., 21, 208-9 (1982).
27. P. Aldebert, N. Baffier, N. Gharbi and J. Livage, Mat. Res. Bull., 16, 669 (1981).
28. P. Aldebert and V. Paul-Boncour, Mat. Res. Bull., 18, 1263 (1983).
29. B. Araki, C. Mailhé, N. Baffier, J. Livage and J. Vedel, Solid State Ionics, 9-10, 439 (1983).
30. A. Bouhaouss and P. Aldebert, Mat. Res. Bull., 18, 1247 (1983).

APPLICATIONS OF AMORPHOUS SILICON

P.G. LeComber

Carnegie Laboratory of Physics,
University of Dundee, Dundee,
Scotland, U.K.

ABSTRACT

The preparation of amorphous silicon with a low density of defect states by the glow discharge decomposition of silane and the ability to control its electrical conductivity over many orders of magnitude by the addition of phosphine or diborane to the silane, stimulated a worldwide interest in this material and in its possible applications. This chapter begins with a description of the preparation technique and a brief review of some of the important properties of the material. The fabrication and characteristics of a-Si thin-film field effect transistors will be described and followed by a discussion of the applications of these devices in large area liquid crystal displays, in simple logic circuits and in addressable image sensors. The ability to prepare a-Si p-n junctions has led to its commercial development as a photovoltaic power source for many consumer products. This development and the possibility of the application of a-Si to large area power generation will be reviewed. Finally, the use of a-Si in memory devices will be briefly described.

1. INTRODUCTION

This article is concerned with the preparation, properties and possible applications of amorphous silicon (a-Si) prepared by the glow discharge (gd) decomposition of silane. There were a number of important fundamental developments during the 1970s which established the considerable applied potential of this material and led to the present rapid growth in its use in commercial products. The first of these was the discovery [1 - 3], as early as 1972, that a-Si prepared by the gd technique possesses a very low density of states in the mobility gap, probably the most important single factor in its applied potential. A direct result of this has been the development of the a-Si field effect transitor (FET), discussed in section 3, which could find applications in large area addressable displays, in addressable image sensing arrays, and in logic circuits.

The second crucial development was the discovery [4, 5] in 1975 that the electronic properties of gd material could be controlled over a very wide range by substitutional doping from the gas phase. This opened up a number of applications in junction devices, including its use in photovoltaic energy conversion. This discovery was followed by a worldwide interest in the material and this had led to its likely application in electrophotography, vidicons, CCDs, memories, etc. [6].

In the next section we shall give an introduction to the preparation of the gd a-Si.

2. GLOW DISCHARGE PREPARATION OF a-Si

Plasma techniques are not new: as long ago as 1857 Siemens [7] described a plasma method for the preparation of ozone. However, it was not until the 1960s that Sterling and his collaborators [8, 9] applied the technique to the production of thin films of a-Si and a-Ge from silane and germane gases respectively. A schematic diagram of the preparation unit developed in our laboratories [10] is shown in Fig.1. Doping of the material can be achieved by adding small but accurately determined amounts of phosphine or diborane to the silane. The silane is mixed with the doping gases in the cylinders C_1 and C_2 to produce mixtures for n- and p-type layers respectively. The gas flow rate is measured by the mass flow controller F and decomposition takes place in the rf glow discharge between plates A and B of the chamber. The substrates S are heated during deposition to about 300°C. The pressure in the chamber is a few tenths of a torr and the rf power usually very small, typically 5 to 10 Watts for a 30 cm diameter system. Films prepared in this way are extremely hard and possess a remarkably low density of localized gap states [1 - 3, 11, 12]. The films generally contain a few atomic percent of hydrogen, mainly in the form of Si-H bonds in samples deposited at 300°C. Their room temperature conductivity can be controlled over 10 orders of magnitude, from about $10^{-12}(\Omega cm)^{-1}$ to over $10^{-2}(\Omega cm)^{-1}$ with either n- or p-type doping [4, 5, 11]. One present disadvantage of the gd technique is the relatively slow deposition rate, of the order of $1\mu m/hr$. Considerable effort is being devoted to increasing this, particularly for those applications such as electrophotography where 15 - 20μm thick films are required. The preliminary results are encouraging, suggesting that rates as high as tens of microns per hour may be possible without serious degradation in the properties of the material [6].

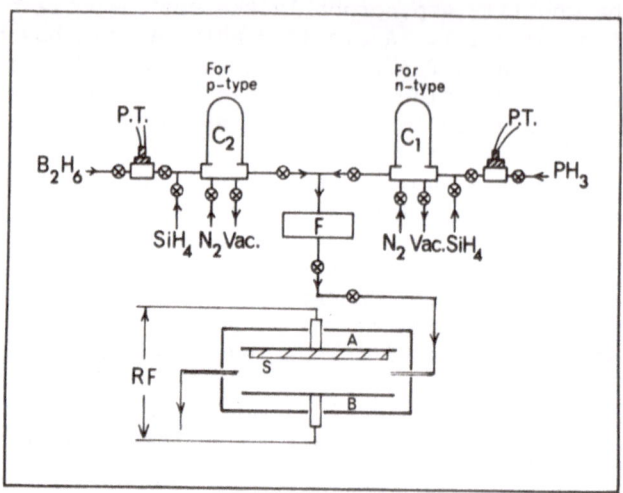

Fig.1. Schematic diagram of the preparation unit for n- and p-type
a-Si and a-Ge specimens developed by the Dundee group. (After Ref. 5)

3.   a-Si FETs

The field effect experiments carried out in our laboratory
since 1972 [1, 2, 3] demonstrated the remarkably low overall density
of states of glow discharge amorphous silicon (a-Si) and also formed
the basis for the subsequent development of the a-Si field effect
transistor (FET). The original aim of this work was to explore the
possibility of using an array of thin film FETs in addressable
liquid crystal displays [13, 14, 15] although attempts have also been
made to explore their use in other applications [15, 16]. Recently
a series of experiments has been carried out to optimise the performance of the FETs and to obtain further information on the electron
transport in the ON-state [17].

The basic design of the elemental a-Si FET is shown schematically
in the insert to fig. 2. A metal electrode, typically 30μm wide, is
evaporated onto a glass substrate to form the gate electrode. A
thin insulating layer, typically about 0.3μm of amorphous silicon
nitride (Si-N), is then deposited by the glow discharge technique
to form the gate dielectric. An a-Si layer, also about 0.3μm thick,
is next deposited onto the Si-N again by the glow discharge process.
The final step consists of depositing the required pattern of source
and drain contacts on to the a-Si surface.

The dc performance of an elementary device with a 4μm channel
length and a 500μm channel width, produced by photolithographic
techniques, is illustrated by the transfer characteristics shown in
fig. 2. The source-drain current I_D is plotted logarithmically

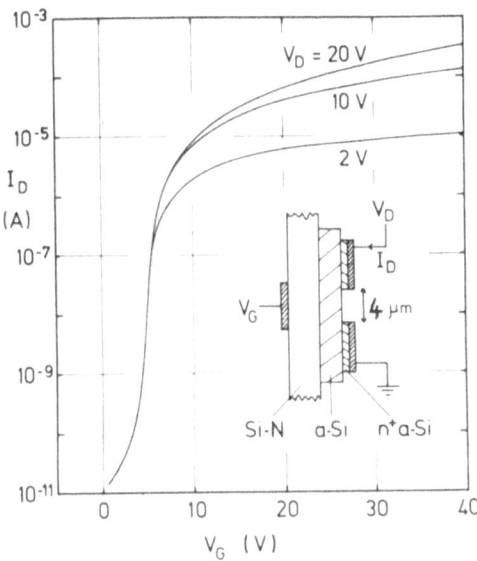

Fig.2. Transfer characteristics of a-Si FET element. (Ref. 17)

against the gate voltage V_G for drain potentials of 2V, 10V, and
20V. It can be seen that with +15V on the gate, drain currents in
excess of 10μA can be achieved for drain voltages as low as 10V.
In the off-condition, with V_G = 0, the current through the device
drops below 10^{-11}A. The remarkable rise in I_D is caused by an
electron accumulation layer formed at the Si/Si-N interface, which
produces an efficient current path between the source and drain
electrodes.

Fig. 3 shows a schematic diagram of a number of elements in a
liquid crystal display panel. A transistor is incorporated in a
corner of each element of the array. The FETs are interconnected
by means of X and Y buses, G_1, G_2 ... and S_1, S_2 ..., linking gate
and source contacts, respectively. The drain contact of each
transistor is connected to the ITO squares, D. From the section
through the panel in Fig. 3b it can be seen that the liquid crystal
material is sandwiched between the substrate carrying the FETs and
an ITO coated glass top plate which is normally returned to ground.
The liquid crystal element is therefore in series with the drain
circuit and behaves electrically as a capacitor C_{LC} with some leakage
resistance. Fig. 4a shows a section through an individual a-Si
device and fig. 4b illustrates the design of the FET in part of the
matrix array.

To simulate the operation of the FETs in an addressable array
[14, 15, 17], the response of single elements to typical pulse
voltages on the gate and source buses was investigated. In these

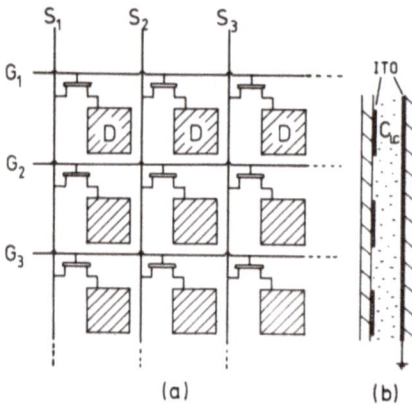

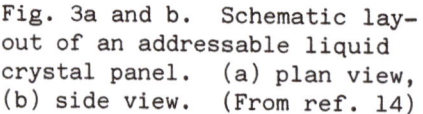

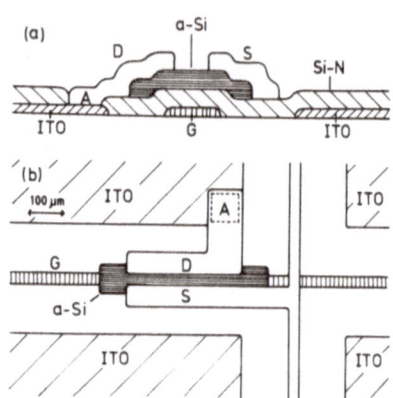

Fig. 3a and b. Schematic lay-out of an addressable liquid crystal panel. (a) plan view, (b) side view. (From ref. 14)

Fig. 4a and b. Design of a-Si field effect transistor element (a) section through device, (b) FET in part of the matrix array. (from ref. 14)

experiments the drain was connected to ground through a 10 pF capacitor to simulate the liquid crystal capacity C_{LC} of the 1 mm ITO elements (see fig. 3). An important parameter, as far as multiplexing is concerned, is the rise time of the potential V_{LC} across the liquid crystal element. Even with a 10μm channel length a LC drive potential is reached in about 10μs. With a frame time of 25ms it is therefore possible with the present devices to scan more than 1000 lines of display in this time.

An important factor in any development of a-Si FETs will be the reproducibility and uniformity of the characteristics of arrays of elements. The evidence available at present looks encouraging and would not indicate that this should be a limiting factor [14, 15, 17]. As a check on the stability of the devices, the output from an a-Si FET was monitored in the period from September 1980 to September 1983, corresponding to about 8×10^9 switching operations [17]. During three years of continuous operation the peak voltage level changed by only 10 – 15%. This is encouraging, especially as this device was totally unpassivated and unencapsulated.

Irradiation experiments on a-Si FETs with γ –ray doses of up to 5 Mrad (Si) have recently been reported [18]. Even at these levels, changes of less than 3V in threshold voltage and less than 10% in transconductance were observed. It was therefore concluded that the a-Si FETs are remarkably resistant to the effect of radiation, despite the relatively thick dielectric layer used at present in these devices.

Matsumura, Hayama and co-workers at the Tokyo Institute of Technology first showed that an integrated inverter circuit [19] and an image sensor [20] could be made from a-Si FETs. Their work demonstrated the feasibility of the application, but the comparatively poor characteristics of their early FETs reflected adversely on the device performance. We have published [15, 16] the design of an integrated a-Si inverter consisting of an a-Si FET in series with an a-Si load resistor. The output logic of this circuit was clearly demonstrated: the output potential changing from 14.5V to 2V as the input increased from about 5V to 15V. By extending the above circuit, we have produced logic circuits such as NAND and NOR gates, bistable multivibrators and also a shift register [15, 16]. The latter is of particular interest as an integrated drive circuit for the liquid crystal panels.

The high photoconductivity of a-Si also makes it possible to integrate light-sensitive elements into the circuitry. As an example, the performance of addressable image sensing elements has been invest-igated [15, 16, 21]. These experiments demonstrated that the output current of an integrated elementary device varied from about 10µA at a white light intensity of 10^{17} photons $s^{-1}cm^{-2}$ to about 10nA at 10^{13} photons $s^{-1}cm^{-2}$ [21]. This large dynamic range, providing an excell-ent grey scale, could be achieved with read-times of about 20µs.

4.   THE a-Si PHOTOVOLTAIC CELL

Early in 1976 our group at Dundee demonstrated that a-Si p-n junctions could be prepared by the glow discharge deposition method [22]. Preliminary measurements of the photoresponse of these devices showed the typical characteristics of photovoltaic cells. Shortly afterwards, Carlson and Wronski [23] reported more detailed photo-voltaic characteristics of a-Si p-i-n structures. Although these early devices were relatively inefficient, these papers aroused considerable interest in the possible applications of the material to cheap, large area photovoltaic devices and many industrial labora-tories entered the field. The advantages of a-Si alloys in this technology have been reviewed in a recent paper by Hamakawa [24].

Progress since 1976 has been rapid, with conversion efficiencies increasing to 9 to 10% for small area cells [24]. For larger area cells, 10cm x 10cm in size, more than 7% conversion efficiency is now commonly attained, even in industrial production. The Sanyo Electric Co. were the first company [25] to put a-Si solar cells into industrial production and they alone now fabricate nearly 1 million units per week. As well as Sanyo, Fuji, Sharp-ECD and Teijm are also in production, mainly producing cells for the consumer market in products such as calculators, watches and battery chargers. The use of a-Si in large scale power applications probably requires 8 to 10% conversion efficiencies over large areas and awaits the solution of a number of problems. However, several experimental solar houses and roof top test arrays are already in operation and their application for example in third-world irrigation systems appears a real possibility in the not too distant future.

5. APPLICATION TO A MEMORY DEVICE

In this section we should like to discuss briefly a new a-Si memory device [26, 27] developed jointly by the Edinburgh and Dundee groups. It is an electrically programmable, non-volatile memory element which in terms of speed, retention time, operating voltages and stability compares very favourably with crystalline MNOS (metal-nitride-oxide-semiconductor) or FAMOS (floating-gate-avalanche-metal-oxide semiconductor) devices currently used for non-volatile programmable storage.

The most promising configuration investigated so far is of the p-n-i type. After an initial forming step the structure operates as a non-volatile memory. Immediately after forming the device is in its on-state with a resistance of a few hundred ohms. On increasing the reverse potential (i.e. a negative voltage applied to the p-doped region) a reverse threshold voltage V_{ThR} is reached beyond which the device switches to an off-state with a resistance of the order of 1 MΩ. The reverse threshold $V_{ThR} \simeq 1V$ and the off-state is stable for voltage swings of ± 4V. If the forward potential is now increased beyond a value of V_{ThF}, the forward threshold voltage, the device switches back into its high conductivity state, in some cases through an intermediate state.

The above cycle has been repeated up to 10^5 times without observable changes in characteristics or threshold voltages. Devices set in on- and off-states have been monitored for several months without detectable change in either characteristic. Operation at temperatures of up to $180^\circ C$ shows little change in the threshold voltages.

The dynamic characteristics of the p-n-i devices has been investigated with 10V forward and reverse pulses of 100 ns duration. The experiments demonstrate the important result that both the on-off and off-on transitions are completed within 100 ns and that there appears to be no observable time delay in the switching response. It is estimated that the energy absorbed during either transition is extremely low, typically in the range 10^{-6} to 10^{-8} J.

Although the mechanisms underlying the switching phenomena described above are unknown at present, we believe on the basis of the present experimental evidence it is likely that the on-state of the a-Si memory devices involves the formation of a current filament [26, 27].

6. CONCLUDING REMARKS

The last decade has been an extremely exciting and rewarding experience for those of us involved in the fundamental and applied work on a-Si. In that time the field has developed from a purely academic one, to large scale commercial production of devices for consumer products. These applications have been made possible by the remarkable electronic properties of glow discharge Si, which had been clearly recognised in our fundamental work during the early 1970s. The glow discharge plasma has become a versatile and promising tool for the deposition of electronically viable a-Si and related materials. With the growing interest in this approach it is likely that important new possibilities in materials preparation will be opened up by the further development of this technique.

REFERENCES

1a. W.E. Spear and P.G. LeComber, J. Non-Crystal. Solids, 8-10 (1972) pp. 727-738.

1b. A. Madan, P.G. LeComber and W.E. Spear, J. Non-Crystal. Solids, 20 (1976) pp. 239-257.

2. A. Madan and P.G. LeComber, Proc. 7th International Conf. on Amorphous and Liquid Semiconductors, Edinburgh, ed. W.E. Spear (CICL, Univ. of Edinburgh, 1977) pp. 377-381.

3. P.G. LeComber, Fundamental Physics of Amorphous Semiconductors, ed. F. Yonezawa (Springer, 1981) pp. 46-55.

4. W.E. Spear and P.G. LeComber, Solid State Commun., 17 (1975) pp. 1193-1196.

5. W.E. Spear and P.G. LeComber, Phil. Mag., 33 (1976), pp. 935-949.

6. See, for example, the Proceedings of the 10th International Conference on Amorphous and Liquid Semiconductors, Tokyo, 1983, published in J. Non-Crystal. Solids, Vols. 59 and 60,(1983).

7. W. von Siemens, Pogg. Ann. 102 (1857) pp. 120.

8. H.F. Sterling, R.C.G. Swann, Solid State Electron. 8, (1965) pp. 653-654.

9. R.C. Chittick, J.H. Alexander and H.F. Sterling, J. Electrochem. Soc., 116 (1969) pp. 77-81.

10. See, for example, refs. 5 and 12.

11. P.G. LeComber and W.E. Spear, Topics in Appl. Physics 36, Chp.9, (1979) pp. 251-285.

12. W.E. Spear and P.G. LeComber, Topics in App. Physics, 55 (1984), at press.

13. P.G. LeComber, W.E. Spear, and A. Ghaith, Electron. Lett., Vol. 15, (1979) pp. 179-181.

14. A.J. Snell, K.D. Mackenzie, W.E. Spear, P.G. LeComber and A.J. Hughes, Appl. Phys. Vol.24, (1981) pp. 357-362.

15. P.G. LeComber, A.J. Snell., K.D. Mackenzie and W.E. Spear, J. de Physique, Vol.42, Supp.C4, (1981) pp. 423-432.

16. A.J. Snell, W.E. Spear, P.G. LeComber and K.D. Mackenzie, Appl. Phys., Vol.A26, (1981) pp. 83-86.

17. K.D. Mackenzie, A.J. Snell, I. French, P.G. LeComber and W.E. Spear, Appl. Phys., A31, (1983) pp. 87-92.

18. I.D. French, A.J. Snell, P.G. LeComber and J.H. Stephen, Appl. Phys., A31 (1983) pp. 19-22.

19. M. Matsumura, and H. Hayama, Proc. IEEE 68 (1980) pp. 1349-1350.

20. M. Matsumura, H. Hayama, Y. Nara and K. Ishibashi, IEEE, EDL-1 (1980) pp. 182-184.

21. A.J. Snell, A. Doghmane, P.G. LeComber and W.E. Spear, Appl. Phys. A (1984) at press.

22. W.E. Spear, P.G. LeComber, S. Kinmond and M.H. Brodsky, Appl. Phys. Lett. 28 (1976) pp. 105-107.

23. D.E. Carlson and C.R. Wronski, Appl. Phys. Lett. 28, (1976) pp. 671-673.

24. Y. Hamakawa, J. Non-Crystal. Solids, 59 and 60 (1983), pp. 1265-1272.

25. Y. Kuwano and M. Ohnishi, J. de Physique, 42, Supp. C4 (1981) pp. 1155-1164.

26. A.E. Owen, P.G. LeComber, G. Sarrabayrouse and W.E. Spear, IEE Proc. 129 (1982) pp. 51-54.

27. A.E. Owen, P.G. LeComber, W.E. Spear and J. Hajto, J. Non-Crystal. Solids 59 and 60, (1983) pp. 1273-1280.

IONIC CONDUCTIVE GLASSES

D. RAVAINE

Laboratoire d'Energétique Electrochimique
LA CNRS 265 - ENSEEG - BP 75
38402 Saint Martin d'Hères - FRANCE.

INTRODUCTION.

The ionic character of the conductivity in oxide glasses was established a century ago when Warburg experimentally demonstrated the transport of sodium between two amalgams separated by a glass membrane (1). In the meantime, many experimental results concerning traditional oxide-glasses have been collected (2 to 4). The conductivity of these glasses is cationic and highest for alkali and silver cations. Until about ten years ago, the best results in conductivity were around 10^{-8} $(\Omega cm)^{-1}$ at room temperature which made ionic conductive glasses suitable only for very specific applications, such as glass membranes in pH sensitive electrodes.

Recent progress and practical requirements have now focussed more attention on glassy electrolytes. New sulfide glasses and doped inorganic glasses have led to much higher conductivities than traditional oxide-based glasses, from 10^{-5} to 10^{-3} $(\Omega cm)^{-1}$ at ambient temperature for some lithium conductive glasses (5 to 7). In a short time, glasses have been synthesized that achieve performances comparable to those of the best solid electrolytes.

In the field of battery applications, glasses offer a large number of advantages over crystalline solids. First, the ionic conductivity is isotropic and does not involve any grain boundary effects like in polycrystalline materials. Also, the electronic contribution to the total conductivity is usually very weak which is a consequence of the aperiodic potential fluctuations imposed by the disordered structure. Moreover, metal impurities are unable to enhance the electronic conductivity since they can exist in their own distinct local environment in the glass structure. Electronic leakage is then unlikely to occur in electrochemical devices using a glass membrane as ionic separator.

From a practical point of view, glasses have the advantage of being easily obtainable in thin film configurations ; layered micro-batteries can then be expected in a near future by using thin film glassy electrolytes. Extremely good contacts can be obtained between electrode materials and ionic conductive glasses of low glass tran-sition temperatures Tg : this condition is expected to be especial-ly critical for intercalation-type electrodes whose volume varies during charge-discharge cycles. Finally, one of the interesting fea-tures of glasses as electrochemically active materials is the pos-sibility of continuously changing the composition through appropria-te techniques. Bulk glass samples can be obtained with ionic trans-port properties at one end and electronic transport properties at the other end of the sample. Promising results have been obtained by chemical intercalation of alkali ions in a phospho-vanadate glass (8). This opens the way to make a compact battery with improved per-formance through the delocalization of the electrode/electrolyte in-terfaces.

From a more fundamental point of view, varying the composition has been widely used as a means to investigate the ionic transport properties in glasses. This has led to an original approach to the interpretation of the ionic conduction mechanism in solid state con-ductors based on the existence of dissociation equilibria.

The applicability of the weak electrolyte theory to glassy electrolytes (9) has attracted a revival of interest for the inter-pretation of poorly understood results like the mixed alkali effect (10), the magnitude of conductivity variations with the concentra-tion of network modifiers or the electric field dependance (11).

I IONIC CONDUCTIVE GLASS COMPOSITIONS.

1) Oxide glasses.

In Table 1 are summarized the different oxide glass composi-tions which are known to exhibit ionic transport properties. These are by far the most studied among amorphous electrolyte materials (2-4, 12). For simple glass compositions, such as binary network former/network modifier oxides, three dimensional covalent macromo-lecules are formed by an assembly of elementary units (SiO_4, BO_4, PO_4... tetrahedra) in which at least one atom of oxygen is shared (BO's). Some oxygen atoms that are non bridging (NBO's) are negati-vely charged keeping in their vicinity the alkali or silver (M) ca-tions introduced by the network-modifier oxide (Fig. 1.a.).

Vitreous domains involving a wide concentration range in ca-tion (M) content are obtained by dissolving alkali or silver salts in an oxide-base glass in order to increase the number of charge carriers (13-17). In that case, only large size anions (halides, sulfate) can be incorporated without affecting the macromolecular structure of the glassy network (Fig. 1.b.). For this reason, the term "doped" has been used to describe these types of more complex glass compositions although, in some cases, very high levels (up to 80 m/o) of doping level can be attained (18). For some silver glass compositions, this expression is not related to any structural evidence : due to the higher glass forming ability of silver glass, rapid quenching techniques can be used to obtain glass formation with unusual glass formers (see Table 1) (19). The addition of extra-silver salt leads to a disordered structure made of discrete anions (iodides, molybdates...) in which the doping agents are not discernible (Fig. 1.c.) (20, 21).

Finally, unusual glass formers have also been used to obtain alkali conductive glasses. In that case, ultra-fast quenching techniques are necessary. Glass formation has been observed in systems of tantalates and niobates (22), aluminates, gallates and bismuthates (23), tungstate and molybdates (24), and sulfates (25). Unlike the other glass systems, only small amounts of alkali salts can be incorporated before the occurrence of partial crystallisation (26).

Mobile ion	Ag^+	M^+ (M = Li, Na, K...)	doped by AgX or MX
usual glass former cation	Si, Ge, B, P, Al ...		X = halides and sulfate
unusual glass former cation	Mo, As, Cr, W, Se, Te, V (quenched)	Ta, Nb, W, Mo, La, Ga (ultrafast quenched)	X = Cl and Br
			X = halides and sulfate

TABLE 1 : OXIDE GLASS COMPOSITIONS

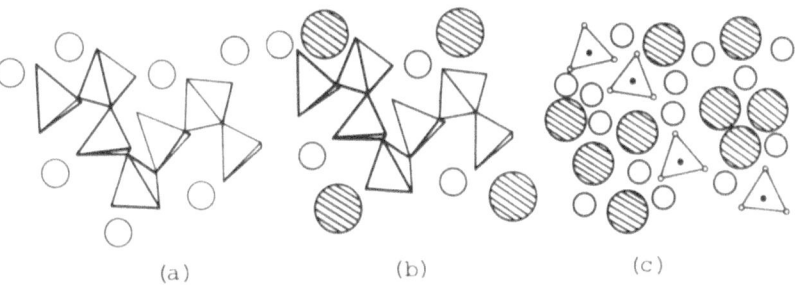

(a)   (b)   (c)

Fig. 1 : Schematic structure representation for (a) $AgPO_3$ glass ; (b) $AgPO_3$ - AgI glass ; (c) $Ag_2 MoO_4$ - AgI glass.

2) Sulfide glasses.

Table 2 gives the different glass compositions investigated in the sulfide systems. Due to their hygroscopicity, the synthesis of sulfide glasses needs special attention. For this reason, the history of sulfide glasses is more recent (27) and the list of studied compositions much less exhaustive than that of oxide glasses.

Nevertheless, the glass forming ability of the common, as well an unusual (excepted for alkali glasses) glass formers has been tested in the sulfide systems, along with silver or alkali halide additions (28-30). Transport number measurements performed on Na_2S-SiS_2 and Na_2S-GeS_2 glass compositions show that the electronic contribution to the conductivity is quite negligible (27) and confirm the ionic character of the conduction in sulfide glasses.

Mobile ion	Ag^+	M^+ (M = Li, Na, K...)	doped by AgX or MX
usual glass former cation	Si, Ge, B, P		X = halides (excepted F)
unsual glass former cation	As, Ga, La, Sb		X = halides (excepted F)

TABLE 2 : SULFIDE GLASS COMPOSITIONS

3) Fluoride glasses.

These glasses are currently under considerable development mainly because of their potential use for making infrared optical components and ultra low-loss optical fibres. Their compositions are given elsewhere in this book. They exhibit relatively high fluorine ion conductivities and have been suggested as likely candidates for solid electrolytes (31). Fluorine conductivity was confirmed on fluorozirconate glasses by the Tubandt test (32) ; the anionic transport in disordered structure provides a new topic for fundamental investigation since only partial interpretations have been presently proposed (33 and 34).

4) Some prospects for new ionic conductive glass compositions.

Since the selenide and telluride-base glasses are known to be electronic conductors, it seems at first inspection that most of the glass compositions showing ion conduction have been investigated. The ability of unusual glass formers (like Ta, Nb, Mo...) to form sulfide glasses through ultra-fast quenching techniques has still to be tested but the size of the obtainable samples will make these materials of questionable interest. Nevertheless, the field of compositions is still wide open for future investigations and some suggestions are proposed here :

- extending the choice of available doping agents to low decomposition or low melting temperature salts through the use of a soft method of glass synthesis.
- preparation of mixed organic-inorganic ion conducting glasses (35).
- investigation of proton conduction in glasses. Although the glass literature is full of references to proton transport, by diffusive exchange with other monovalent cations, electromigration or mixed alkali effect induced by water (36), very little is reported on protonic bulk conductivity in glasses (37).

II <u>CONDUCTIVITY MEASUREMENTS</u>.

For many years, the techniques of conductivity measurements (dc polarization or single frequency techniques) were inappropriate to study apart the different phenomena occuring in solid state ionic conductors under an applied electric field. AC fields are required for measuring bulk conductivities in order to prevent ion/electron charge transfer limitations or space charge building up which always happens at the electrodes of ionic conductive materials tested under dc current flow. The use of impedance spectroscopy is truly the beginning of noticeable progress in the study of conduction and interface phenomena and is now widely used for experimental investigations. Furthermore, in the case of heterogeneous glasses (partially crystallised or phase separated), it can provide valuable information on the different existing phases.

The method consists of measuring the electrical complex impedance of the sample versus frequency. When plotted in the complex plane (i.e imaginary part of Z vs. its real part for different frequencies), the impedances show more or less well defined successive circular arcs. The first arc (high frequency range) goes through the origin and has been proven to be representative of the bulk conduction relaxation phenomena (Fig. 2). The ohmic bulk resistance are given by the intersection of this circular arc with the real axis, which may require some extrapolation when the time constants of the bulk relaxation and of the electrode polarization phenomena do not differ sufficiently. The conductivities σ, are then calculated and the activation energies for conduction, E_σ, are deduced from the slope of the straight lines obtained in the Arrhenius plots.

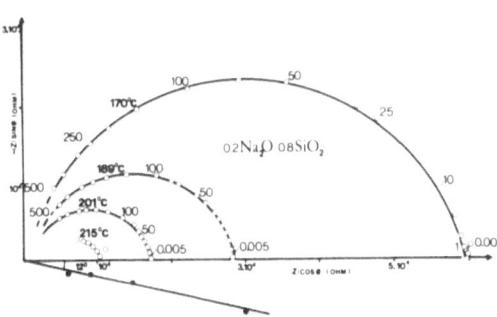

Fig. 2 : Typical complex impedance plots at different temperatures (frequencies are given in kHz).

Semicircular curves are expected when capacitance effects are superimposed on the conduction. This is the case for materials, like ionic conductive glasses, with conductivities ranging intermediately between those of true resistors and of true insulating capacitors. The top frequency of the circular arc ($f_o = \omega_o/2\pi$) is given by:

$$\omega_o\ RC = 1 \qquad (1)$$

where R and C are the ohmic resistance and the electrical capacitance of the sample ; the conductivity is then related to the top frequency by the equation :

$$\sigma = f_o\ 2\pi\ \varepsilon_o\ \varepsilon_r \qquad (2)$$

where $\varepsilon_o\ \varepsilon_r$ is the static dielectric constant. This relation has been experimentally tested for binary oxide glasses (38). Deviations from semicircles are systematically observed : the experimental points are seen to fall along circular arcs whose centers are located below the real axis. A physical significance has been proposed which is based on the dispersion of the local conductivity σ (r) within the sample and hence its microscopic homogeneity (39).

The conductivities are also deducible from the determination of the top frequencies exhibited in complex impedance diagram without knowing the geometrical parameters of the sample (the relative dielectric constant may be chosen close to a value of ten which is a reasonable order of magnitude for many solid electrolytes). This method can be of some utility, for instance, when measuring the conductivity of thin film materials.

Another application is concerned with heteregeneous glass samples for which the geometrical parameters of the existing phases are not discernable. Figure 3 shows the circular arcs obtained on a fluoride ion conducting oxyfluoride glass sample exhibiting a vitreous phase separation. The high frequency curve corresponds to the highly conductive (fluoride rich phase) regions while the low frequency arc corresponds to the much less conductive (oxide rich phase) regions. Values of the conductivities calculated from the top frequencies are then plotted in an Arrhenius diagram (see Fig. 4) for the determination of the activation energies for both vitreous phases.

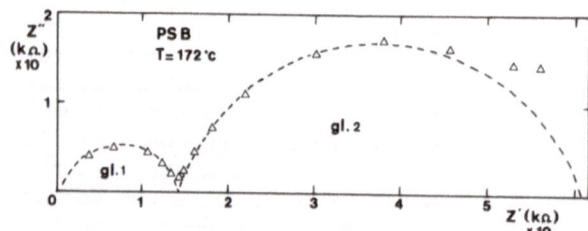

Fig. 3 : Complex impedance diagram for a phase separated glass (34).

Valuable information could also be provided, for instance on the nucleation (or crystal-

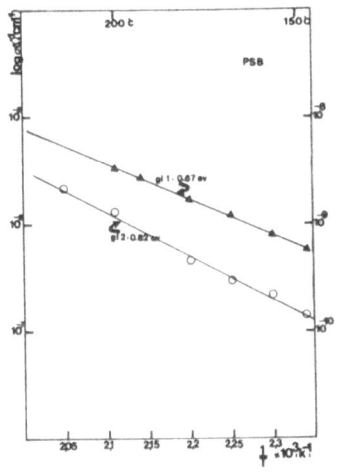

lisation) process during a thermal treatment by using a similar procedure.

In most cases, the ionic character of the conduction is non ambiguous just by considering the chemical composition, the order of magnitude of the conductivity, or the temperature dependance of the conductivity. Otherwise, transference numbers are measured by the usual techniques of solid state electrochemistry (e.m.f., Tubandt's test, Wagner polarization technique...).

Fig. 4 : Arrhenius diagram of conductivities as deduced from top frequencies (for gl. 2 use right hand scale).

III DATA AND STRIKING FEATURES.

Fig. 5 shows an Arrhenius plot of conductivity measurements carried out on lithium and silver amorphous electrolytes. All the data shown come from investigations within the past five years. At room temperature, the conductivity values are spread over more than 10 orders of magnitude though Fig. 5 does not include much less conductive glasses (e.g. simple binary oxide glasses). Silver glasses (and copper glasses to a lesser extent) exhibit higher conductivities than alkali glasses.

The performances of amorphous electrolytes are quite comparable to those of the best ionic conductive crystals. At the present time, silver conductivities in glasses are one order of magnitude below the best conductivity exhibited by the superionic-conductor $RbAg_4I_5$ (Fig. 6). As a result of the research to improve ionic conductivity in glasses, the best lithium solid state conductor is a glass ((30), also in Fig. 5) with a R.T. conductivity value of 10^{-3} $(\Omega cm)^{-1}$. In any case, the search for highly conducting materials will be limited to a R.T. upper limit which can be estimated to be close to 1 $(\Omega cm)^{-1}$ (conductivity value of a molar solution of KCl in water). Doped sulfide glasses have already been successfully tested in solid state cells (41, 42) and they now appear as the most promising materials for solid state high energy batteries (43).

This remarkable improvement in glass conductivities has been obtained in less than ten years. It has been the result of two contributions : the replacement of oxygen by sulphur and the introduction of salts in base-glasses. The improvement in conductivity due

446

to the first effect is always higher than one order of magnitude. As a general trend, it appears that the smaller the difference in electronegativity between the modifier cation and the anion, the higher is the conductivity ; also, the larger the difference in electronegativity between the former cation and the anion, the higher is the conductivity. This trend could also be suggested from considerations on the effectiveness of the non-bridging anions to trap the mobile cations.

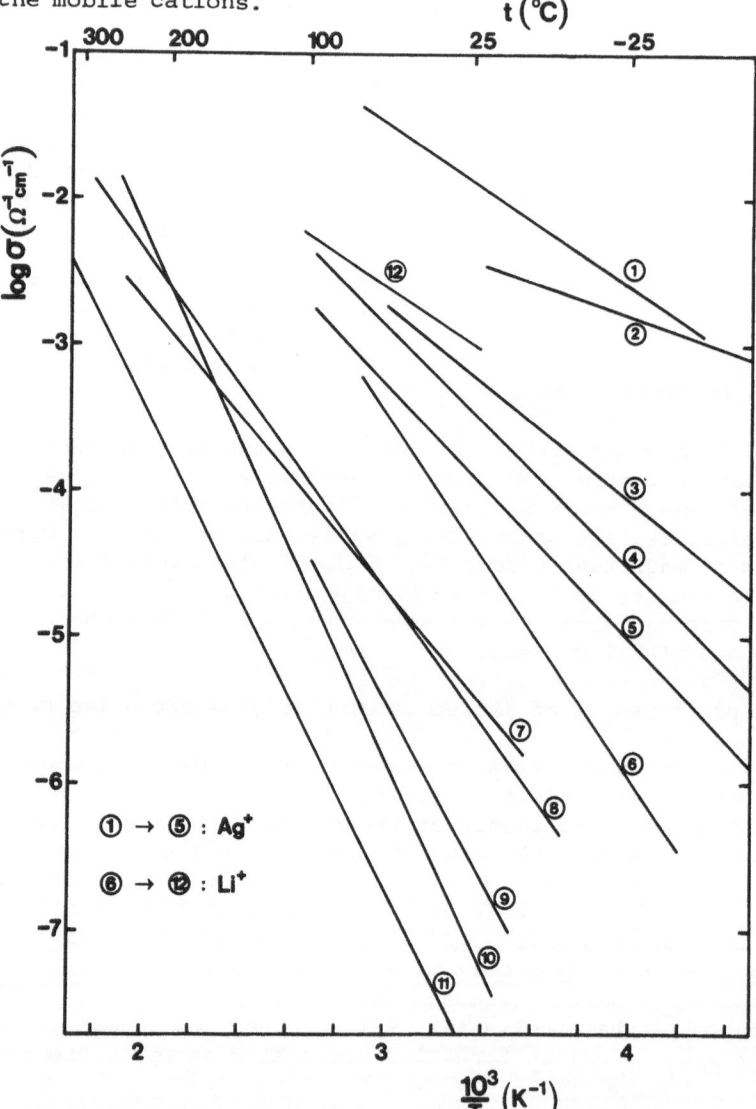

Fig. 5 : Arrhenius diagram of conductivity for various glasses (40).
1. $AgPO_3$-AgI ; 2. Ag_2MoO_4-AgI ; 3. $AgPO_3$-AgBr ; 4. GeS_2-GeS-AgI ;
5. $AgAs_2S_3$-Ag_2S ; 6. GeS_2-Li_2S ; 7. $LiTaO_3$; 8. B_2O_3-Li_2O-LiCl ;
9. $LiPO_3$-LiBr ; 11. SiO_2-Li_2O-Li_2SO_4 ; 12. P_2S_5-Li_2S-LiI.

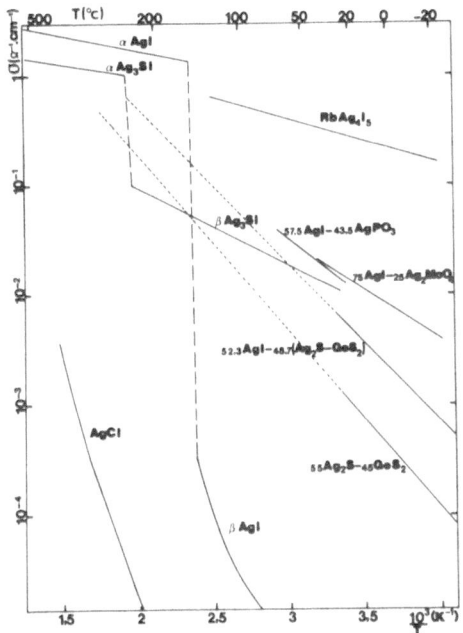

A few examples of doping effects are given in Table 3. Its magnitude depends on the nature of the support glass (higher for borate glasses), the halide size (higher for larger size anion), the doping content (limited by the occurence of partial crystallisation).

The electrical conductivity of a glass is largely dependent on the concentration of its constituents. A simple example is given by the SiO_2-Na_2O system where increasing the concentration per unit volume of sodium atoms by a factor of 5 multiplies the conductivity by a factor 2000. The same behaviour is found in more complex glasses as shown in Fig. 7 for silver conducting glasses. Such large effects of compositions can be considered as an unique feature among solid electrolytes. Considering that the conductivity is given by the product of the mobility and the

Fig. 6 : Arrhenius diagram of conductivity for silver conducting solids.

charge carrier concentrations, only two ways are suitable for interpretation : either the mobility sharply increases, or the mobile ion concentration varies considerably which in turn implies that only part of the existing cations are free to move.

Glass composition	σ at 25°C (Ω^{-1} cm^{-1})	E_σ (eV)
$LiPO_3$	2×10^{-9}	0.70
$0.7\ LiPO_3 + 0.3\ LiCl$	1×10^{-7}	0.60
$0.7\ LiPO_3 + 0.3\ LiBr$	3×10^{-7}	0.55
$0.7\ LiPO_3 + 0.3\ LiI$	1×10^{-6}	0.52
$0.33\ P_2S_5 + 0.66\ Li_2S = A$	1.1×10^{-4}	
$0.29\ A + 0.71\ LiI$	1.0×10^{-3}	0.31
Glass composition	σ at 200°C (Ω^{-1} cm^{-1})	E_σ (eV)
$0.64\ B_2O_3 + 0.36\ Li_2O = B$	2×10^{-6}	0.72
$0.54\ B + 0.46\ LiCl$	2.5×10^{-3}	0.46
$0.62\ B + 0.38\ Li_2SO_4$	4×10^{-4}	

TABLE 3 : CONDUCTIVITIES AND ACTIVATION ENERGIES FOR GLASSES CONTAINING SALT ADDITIVES.

448

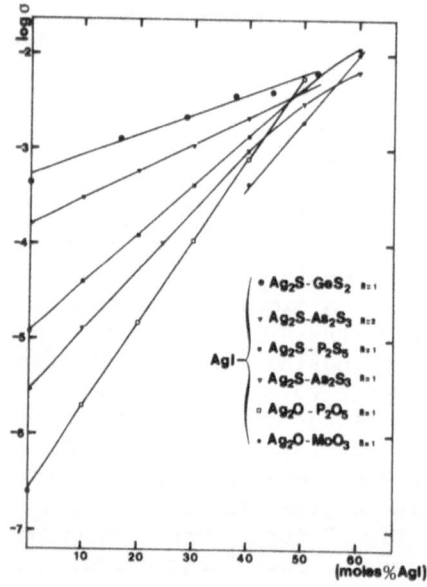

Fig. 7 : Conductivity vs. AgI content for various glass systems (47).

Another feature of glass conductivity is the well known mixed alkali effect which is extensively discussed in the next chapter of this book. It refers to a large deviation from linearity in glass conductivities as a function of composition as one alkali ion is replaced by another one. Although the effect is not unique in glasses, it is always associated with a structure exhibiting a high degree of disorder, like in ß-alumina (44, 45) or in hydrate melts (46).

Another particular point has been observed in doped glasses : at very high concentrations of a given doping salt, experimental conductivity values all approach the same limit whatever the vitreous support (30 and 47) as shown by results collected in Fig. 7 for silver iodide doped systems. Minami et al. (48), then Schiraldi (20), have proposed that the addition of AgI would lead to the formation of metastable microdomains of αAgI (superconducting phase) which would become joined as soon as their volume occupied about 70 % of the total volume. Another interpretation is based on a displacement of silver ions simply by rotating from one anion to another without loss of contact (46). This displacement would happen as soon as the distance between the anions is close to 5 Å.

IV THEORETICAL CONSIDERATIONS ON ION TRANSPORT IN GLASSES.

1) Microscopic approach.

The classical microscopic approach of ionic transport in solids is a hopping mechanism. For a one-dimensional potential profile, the jump frequency ν is given by :

$$\nu = \nu_o \exp \left[- E_m/kT\right] \qquad (3)$$

where ν_o is the attempt frequency corresponding to the oscillation frequency of the cation about an equilibrium position and E_m the barrier height. Under an external electrical field $\vec{E}$, the potential profile is slightly skewed and the jump frequency is split accordingly to the forwards direction, $\underset{\rightarrow}{\nu}$, and the backwards direction, $\underset{\leftarrow}{\nu}$, with :

$$\underset{\rightarrow}{\nu} = \nu_o \exp - [\frac{E_m \mp \frac{a}{2} z e |\vec{E}|}{kT}] \qquad (4)$$

where a is the jump distance and z e the particle charge.

The average velocity, v, is given by :

$$v = a [\underset{\rightarrow}{\nu} - \underset{\leftarrow}{\nu}] \qquad (5)$$

In the linear approximation : $a z e |\vec{E}| << kT$, equation (5) becomes :

$$v = \frac{a^2 z e}{2 kT} |\vec{E}| \nu_o \exp [- \frac{E_m}{kT}] \qquad (6)$$

The conductivity, σ, is given by :

$$\sigma = C z e \frac{v}{|\vec{E}|} \qquad (7)$$

where C is the concentration of mobile charge carriers per unit volume. Combining Eqs. (6) and (7), the conductivity for a three - dimensional network is :

$$\sigma = \frac{a^2 z^2 e^2}{6 kT} C \nu_o \exp [- \frac{E_m}{kT}] \qquad (8)$$

When estimations of the preexponential term are possible (ν_o lies in the vibrational frequency range of $10^{11} - 10^{13}$ Hz (49)), the calculated values always lie a few orders of magnitude below the experimental value (50). The hopping mechanism appears to be over-simplified for the description of the conduction process in glasses. Correlation effects between consecutive jumps, local field considerations or ionic dissociation analogies for the calculation of mobile ion concentration have to be taken into account.

Different attempts have been made to consider the random network of glass structure. Stevels (51), and Taylor (52) proposed a model in which there are potential barriers of various heights. It was assumed that for dc conduction the highest energy barrier had to be overcome, while for ac conduction the migration across the limited distance overcoming the lower energy barrier was responsible.

Charles (53) postulated the existence of a number of equivalent positions for an alkali ion around each non-bridging oxygen. Different processes of alkali migration can take place involving defect formation (two alkali ions on the same NBO) and polarization (rotation of an alkali ion around one NBO). Quantitatively, the defect concentration is calculated using the procedure for estimating the point defect concentration in ionic crystals.

These models imply the existence of dielectric relaxation

effects in glasses. Actually, many papers deal with dielectric phenomena (54), including field distribution in inhomogeneous glass
structure (55), and relaxation processes involving diffusion of defects (56). They all invoke ion migration mechanism over long distances. As mentioned in different papers (57, 58), the low frequency dielectric characteristics are often difficult to obtain because
of the dc conductivity contribution, which is the major portion of
the loss, and which has to be subtracted from the total dielectric
loss. From this point of view, the impedance spectroscopy method to
investigate the a.c. electrical properties of ionically conducting
glasses has brought a major contribution for the understanding of
relaxation losses at low frequencies.

Other techniques have been used to investigate alkali-ion motion in silicate glasses, such as : tracer diffusion coefficient
measurements (59-61) and thermally stimulated currents (62, 63).
The values obtained for the correlation factor and the Haven ratio
have been discussed using analogies with proposed diffusion mechanisms in crystalline solids. The overall picture of alkali diffusion is that, at low temperatures, the alkali ions diffuse by a
2-atom interstitialcy mechanism ; as the temperature increases, a
larger fraction of the alkali ions diffuses individually by a vacancy mechanism which consists of an alkali on a lattice site jum-
ping to a vacancy at one of the 6 nearest-neighbor alkali lattice
sites.

2) Weak electrolyte model.

These microscopic descriptions of conduction mechanisms in
glasses are unable to predict or to explain the large variations
observed for glass conductivities versus composition. Thermodynamic
considerations are much more appropriate to provide an interpretation when large composition ranges are available for investigation.

The conductivity of all electrolytes is given in principle by
an equation of the form :

$$\sigma = \Sigma \; C.ze.u \qquad\qquad\qquad (9)$$

where u is the mobility of ions of charge ze and C the number of
mobile ions per unit volume. If all the alkali (or silver ions) are
equally mobile, then the situation is analogous to the complete
dissociation of strong electrolytes in aqueous solution. If however
the number of mobile ions is less than the stoichiometric concentration, then such glasses can be regarded as weak electrolytes.
Earlier support for the weak electrolyte model in glasses came from
Myuller (64) who postulated the formation of Frenkel defects,
Proctor and Sutton (65) who studied space charge development in
glass, and Haven and Verkerk (60) who discussed evidence for the
interstialcy mechanism in cationic transport.

Let us consider conductive glasses as solid solutions in which the network former components or the doping salts, behave as weakly dissociated electrolytes, we can then write the following dissociation equilibria :

$$M_2O \rightleftharpoons M^+ + OM^-$$

which implies : $[M^+] = K^{1/2} a_{M_2O}^{1/2}$ (10)

or (for doped glasses) :

$$MX \rightleftharpoons M^+ + X^-$$

which in turn implies : $[M^+] = K^{1/2} a_{MX}^{1/2}$ (11)

where $[M^+]$ is the concentration of dissociated ions, K the dissociation constant (independent of concentration) and a_{M_2O}, a_{MX} the thermodynamic activities of the corresponding glass components. Although such an approach does not provide any description of the physical state of these species, M^+ may be regarded as dissociated "free" alkali (or silver) ions, M_2O as trapped entities and OM^- as vacancies in the vicinity of non-bridging oxygens. Since only the first kind of species mentioned are able to move under an applied electric field, it is tempting experimentally to correlate the conductivity variations to those of thermodynamic quantities. This has been done for different silica glasses (9) : the ratios of the thermodynamic activities of two different glass compositions have been obtained from concentration cell emf measurements. Fig. 8 shows a plot of these ratios versus the corresponding ratios of the electrical conductivities for various pairs of glasses. In logarithmic scales, they fit a linear relationship, according to :

$$\sigma_1/\sigma_2 = [a_{M_2O}^{(1)} /a_{M_2O}^{(2)}]^{1/2}$$ (12)

A similar relation has been observed from calorimetric measurements for AgX (X = Cl, Br, I)-doped phosphate glasses (66). As a consequence of eqs. (9) and (10), relation (12) suggests that the mobility of the free ions is independent of concentration for a given glass system.

Comments.

(i) The analogy with the liquid weak electrolyte only holds if there are no structural limiting factors. This implies that eqs. (10) and (11) can be used to calculate the concentration of mobile ions if the number of available interstitial sites

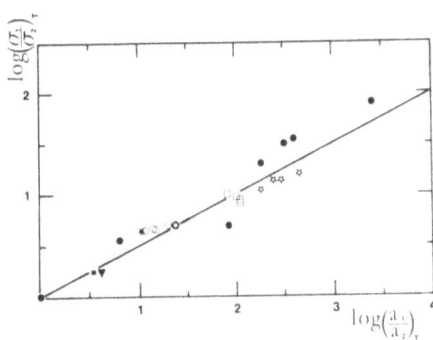

Fig. 8 : Conductivity ratio vs. activity ratio for various pairs of glass (9).

452

for the dissociated ions in the rigid glass network is large. Density measurements suggest the existence of voids in the glass structure where these interstitial sites could be located. Also, it has been recently proved that up to 30 % of sodium atoms per phosphorus atom can be incorporated in a phosphovanadate glass without noticeable change in volume (8). This gives some consistency to the applicability of the weak electrolyte model in glasses.

(ii) Disordering in the glass structure is responsible for the existence of voids and for the availability of intestitial sites. Extended defects also exist in crystalline superionic-conductors, forming sub-regions of highly disordered structure. The weak electrolyte model can then be expected to apply in such circumstances. As an example, a theory of the mixed alkali effect based on the weak electrolyte model has been used in the analysis of conductivity isotherms in β-alumina (44).

3) Using the weak electrolyte model.

a – Temperature dependance for the conductivity.

From eqs. (9) and (10), the activation energy for conduction, E_σ, can be calculated as a linear combination of three terms : E_m (migration energy corresponding to the activation energy for the mobility), ΔH_d (activation of the dissociation constant, independant of the concentration), ΔH_{M_2O} (partial free enthalpy for the component M_2O).

$$E_\sigma = E_m + \frac{\Delta H_d - \Delta H_{M_2O}}{2} \qquad (13)$$

Using a regular solution model for calculating the mixing enthalpy, we can deduce ΔH_{M_2O} and estimate the concentration dependance of E_σ. For sodium and potassium glass systems, it has been shown that they agree well with experimental variations of E_σ (67) (Fig. 9).

b – Observation of conductivity maximum.

Although in most glasses conductivity increases with mole fraction of alkali, a maximum conductivity has been observed in ultra-fast quenched oxide glasses (24) and in boro-aluminate

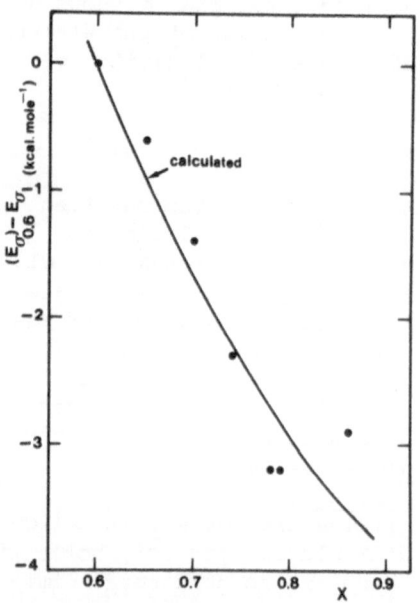

Fig. 9 : Comparison of calculated and experimental values of E_σ for x SiO_2-(1-x) Na_2O glasses (67).

glass systems (68). The structure of tantalate and niobate glasses is made of a random array of corner linked octahedra. The concentration of NBO's is a minimum for a Li/Nb ratio value of 0.5, corresponding to the formula.

$$2 \, [Nb \, O_{6/2} - Li] = Nb_2O_5 - Li_2O$$

At this concentration level, the disappearance of NBO's moves the dissociation equilibrium (eq. 10) to the right hand side, and a maximum in mobile ion concentration and in conductivity may be expected in accordance with the experimental observation (24).

c - Mixed alkali effect.

A weak electrolyte model for the mixed alkali effect on electrical conductivity in glass has been developed for the dilute foreign alkali region (69). It has been proposed that adding small amounts of foreign alkali leads to the formation of alkali-foreign alkali interstitial pairs which, by mass action, suppresses the mobile species concentration.

The weak electrolyte model is shown to be a macroscopic approach to the interpretation of many poorly undestood results in the field of vitreous electrolytes. Physical investigations or computer simulation experiments are nevertheless necessary to provide some support and to confirm ideas concerning mobile and non-mobile ion concentration, description of the interstitial sites, ion distribution on sites of different energy states...

REFERENCES.

(1) Warburg, G. Ann. Phys. 21 (1884) 622.
(2) Morey, G.W. The Properties of Glass (Princeton, Van Nostrand - Reinhold, 1954).
(3) Hughes, K. and Isard, J.O. The Ionic Transport in Glasses, in Physics of Electrolytes (Hladik, J.,New-York, Academic 1972).
(4) Tuller, H.L., Button, D.P. and Uhlmann, D.R. J. Non-Cryst. Solids 40 (1980) 93.
(5) Levasseur, A., Brethous, J.C., Reau, J.M. and Hagenmuller, P. Mater. Res. Bull. 14 (1979) 921.
(6) Barrau, B., Ribes, M., Maurin, M., Kone, A. and Souquet, J.L. J. Non-Cryst. Solids 38 / 39 (1980) 271.
(7) Malugani, J.P. and Robert, G. Solid State Ionics 1 (1980) 519.
(8) Pagnier, T., Fouletier, M., Souquet, J.L. Solid State Ionics, 9 et 10 (1983) 649.
(9) Ravaine, D., Souquet, J.L., Phys. Chem. Glasses 18 (1977) 27.
(10) Day, D.E., J. of Non-Cryst. Solids 21 (1976) 343.
(11) Ingram, M.D., Moynihan, C.T. and Lesikar, A.V., J. of Non-Cryst- Solides, 38 and 39 (1980) 371.

454

(12) See for instance :
   a) Owen, A.E. in Prog. Ceram. Sc. 3 (1963) 77-196.
   b) Ravaine, D., Souquet, J.L., Ionic Conductive Glasses in Solid Electrolytes, 277-290, New-York Academic (1978).
(13) Malugani, J.P. and Robert, G. Mat. Res. Bull. 14 (1979) 1075.
(14) Levasseur, A., Kbala, M., Brethous, J.C., Reau, J.M., Couzi, M. and Hagenmuller, P. Solid State Com. 32 (1979) 839.
(15) Otto, K. Phys. Chem. Glasses 7 (1966) 29.
(16) Malugani, J.P., Wasniewki, A., Doreau, M., Robert, G. and Al Rikabi, A. Mat. Res. Bull. 13 (1978) 427.
(17) Minami, T., Shimizu, T. and Tanaka, M., Solid State Ionics 9/10 (1983) 577.
(18) Bonino, F., Lazzari, M., Lonardi, A., Rivolta, B. and Scrosati, B. Solid State Chem. 20 (1977) 315.
(19) M. Lazzari, B. Scrosati and C.A. Vincent, J. of Am. Ceram. Soc. 61 (1978) 451.
(20) Schiraldi, A., Electrochimica Acta 23 (1978) 1039.
(21) Minami, T. and Tanaka, M., J. of Non-Cryst. Solids 38/39 (1980) 289.
(22) Glass, A.M., Nassau, K. and Negran, T.J., J. Appl. Phys. 49 (1978) 4808.
(23) Glass, A.M. and Nassau, K., J. Appl. Phys. 51 (1980) 3756.
(24) Nassau, K., Glass, A.M., Grasso, M. and Olson, D.H., J. Electrochem. Soc. 127 (1980) 2743.
(25) Nassau, K., Glass, A.M., Grasso, M. and Olson, D.H., J. of Non-Cryst. Solids 46 (1981) 45.
(26) Ravaine, D., Nassau, K. and Glass, A.M., Solid State Ionics 11 (1984).
(27) Barrau, B., Latour, J.M., Ravaine, D. and Ribes, M., Silicates Ind. 12 (1979) 275.
(28) Ribes, M., Ravaine, D., Souquet, J.L. and Maurin, M., Rev. Chim. Min. 16 (1979) 339.
(29) Carette, B., Robinel, E. and Ribes, M., Glass Techn. 24 (1983) 157.
(30) Robert, G. Malugani, J.P. and Saida, A., Solid State Ionics 3/4 (1981) 311.
(31) Leroy, D., Lucas, J., Poulain, M. and Ravaine, D., Mater. Res. Bull. 13 (1978) 1125.
(32) Ravaine, D. and Leroy, D., J. of Non-Cryst. Solids 38 (1980) 353.
(33) Ravaine, D., Perera, W.G. and Minier, M., J. de Physique 43 (1982) C9-407.
(34) Ravaine, D., Perera, W.G. and Poulain, M., Solid State Ionics 9/10 (1983) 631.
(35) Cooper, E.I. and Angell, C.A., Solid State Ionics 9/10 (1983) 617.
(36) For basic references, see for instance : Ernsberger, F.M., J. of Non-Cryst. Solids 38/39 (1980) 557.
(37) Hodge, I.M. and Angell, C.A., J. Chem. Phys. 67 (1977) 1647.
(38) Ravaine, D. and Souquet, J.L., J. Chim. Phys. 5 (1974) 693.

(39) Ravaine, D., J. of Non Cryst. Solids 49 (1982) 507.

(40) Souquet, J.L., Ann. Rev. Mater. Sci. 11 (1981) 211.

(41) Carette, B., Maurin, M., Ribes, M. and Duclot, M., Solid State Ionics 9/10 (1983) 655.

(42) Malugani, J.P., Fahys, B., Mercier, R., Robert, G., Duchange, J.P., Baudry, S., Broussely, M. and Gabano, J.P., Solid State Ionics 9/10 (1983) 659.

(43) Gabano, J.P. This book.

(44) Ingram, M.D. and Moynihan, C.T., Solid State Ionics 6 (1982) 303.

(45) Hunter, C.C., Ingram, M.D. and West, A.R. Solid State Ionics 8 (1983) 55.

(46) Moynihan, C.T., J. Electrochem. Soc. 126 (1979) 2144.

(47) Robinel, E., Carette, B. and Ribes, M., J. of Non-Cryst. Solids 57 (1983) 49.

(48) Minami, T., Nambu, H. and Tanaka, M., J. Am. Ceram. Soc. 60 (1977) 467.

(49) Exarhos, G.J. and Risen, W.M., Solid State. Com. 11 (1972) 755

(50) Doi, A., J. of Non-Cryst. Solids 29 (1978) 131

(51) Stevels, J.M., The electrical properties of glass. In Handbuch der Physik 20 (1957) 350, Berlin, Springer.

(52) Taylor, H.E., J. Soc. Glass Tech. 41 (1957) 350 and 43 (1959) 124.

(53) Charles, R.J., J. of Appl. Physics 32 (1961) 1115.

(54) Tomozawa, M. in Treatise on Mater. Sc. and Tech. 12 (1977) 283, Academic Press. N.Y.

(55) Isard, J.O., Proc. Instn. Electr. 44 (1962) 109B.

(56) Hakim, R.M. and Uhlmann, D.R., Phys. Chem. Glasses 14 (1973) 81

(57) Tomazawa, M. Cordaro, J. and Singh, M., J. of Mater. Sc. 14 (1979) 1945.

(58) Cordaro, J.F. and Tomozawa, M., J. Am. Ceram. Soc. 64 (1981) 713.

(59) Terai, R. and Hayami, R., J. of Non-Cryst. Solids 18 (1975) 217

(60) Haven, Y. and Verkerk, B., Phys. Chem. Glasses 6 (1965) 38

(61) Lim, C. and Day, D.E., J. Am. Ceram. Soc. 60 (1977) 198.

(62) Hong, C.M. and Day, D.E., J. Mater. Sc. 14 (1974) 2493.

(63) Agarwal, A.K. and Day, D.E., J. Am. Ceram. Soc. 65 (1982) 111

(64) Myuller, R.L., Sov. Phys. Solid State 2 (1960) 1213.

(65) Proctor, T.M. and Sutton, P.M., J. Am. Ceram. Soc. 43 (1960) 173.

(66) Reggiani, J.C., Malugani, J.P. and Bernard, J., J. Chim. Phys. 75 (1978) 849.

(67) Ravaine, D. and Souquet, J.L., Phys. Chem. Glasses 19 (1978) 115.

(68) Martin, S.W., Cooper, E.I. and Angell, C.A., Am. Ceram. Soc. Spring Meeting, Abstr. n° 87-6-83 (1982)

(69) Moynihan, C.T. and Lesikar, A.V., J. Am. Ceram. Soc. 64 (1981) 40.

Abstract only

THE MIXED ALKALI EFFECT

C.T. MOYNIHAN

Renseelaer Polytechnic Institute
TROY,   NY   12181   (U.S.A.)

The term "mixed alkali (MA) effect", as originally proposed, refers to a large deviation in glass properties as a function of composition as one alkali ion is replaced by another. For example, in XNa_2O - $(1 - X)$ K_2O - 3 SiO_2 glasses intermediate composition $(X \sim 0.5)$ exhibit electrical conductivities several orders of magnitude lower than do the end compositions $(X=0$ and $1)$. The effect is largely confined to properties determined by mobile ions, e.g. electrical conductivity, ionic diffusion and chemical durability. The effect is not unique to glasses; it is also observed in crystals and high viscosity melts. It is also not unique to mixed alkali systems ; it is also observed in mixed Ag^+/Tl^+ and F^-/Cl^- systems, so long as the ions in question are the main charge carriers in the system. Consequently the phenomenon might be better referred to as "mixed mobile ion effect".

Present indications are that the MA effect is best explained in terms of a defect or weak electrolyte model for ionic transport. Here it is presumed that only a small fraction of the conducting ions -i.e., the defects- are mobile at any instant and that additions of small amounts of a second ion can cause major perturbations in the number of mobile defects. A phenomenological treatment of the weak electrolyte or defect model of the MA effect appropriate to the composition region dilute in one on the alkalis will be discussed. Perhaps the main virtue of this treatment is that it is the first model of the MA effect to make a verifiable experimental prediction. However, a detailed understanding of the MA effect in glasses is currently impeded by our lack of knowledge about the exact nature of mobile ionic defects in vitreous materials.

APPLICATIONS OF GLASSES IN ALL SOLID STATE BATTERIES

A REVIEW

J.P. GABANO, SAFT DEPARTEMENT PILES
Rue Georges Leclanché, 86009 POITIERS

I - INTRODUCTION

Recently solid state batteries have been the subject of renewed interest, because of their potential advantages over those using both aqueous and non aqueous liquid electrolytes such as a long shelf life resulting from a better control of corrosion reactions of the electrode materials, an absence of cell leakage, a wide range of temperature operation and an ability to be miniaturized or shaped like an electronic component.

These batteries are based on polycristalline super ionic conductors mainly using silver or alkali ion conducting solid electrolytes ; nevertheless, only a few papers have reported solid state batteries using glassy superionic conductor in spite of the fact that glassy materials may have higher potentialities than polycrystals when used as electrolyte or electrodes in such solid state batteries.

Up to now most of the existing research has been focused on the electrolyte and a considerable effort has been devoted to the improvement of the alkali ion conducting glasses which represent the most promising materials for constituting solid state high energy batteries.

The purpose of this paper is consequently to review the state of the art of existing or prototype solid state batteries using glassy materials both as electrolyte and/or electrode.

II - <u>REQUIRED CHARACTERISTICS OF GLASS MATERIALS AS ELECTROLYTES</u> <u>FOR SOLID STATE BATTERIES</u>

2.1. <u>Electrical conductivity and chemical composition</u>

The prime function of the electrolyte in a battery system is to serve as a medium for the transport of the appropriate ionic species between the anode and cathode reaction sites. In a solid state battery it also provides a barrier that prevents direct contact of the anode and cathode reactions.

In all solid state batteries, the electrolyte is often a large source of cell polarization and consequently a limiting factor regarding the performances that can be obtained. Although the internal resistance can be decreased by minimizing the interelectrode distance, the electrolyte must have a sufficiently high ionic conductivity for the appropriate ions to be transported at the required rate.

As far as alkali inorganic glasses are concernted, a number of purely Li^+ ion conducting glasses ($t^+ = 1$) were recently investigated as solid electrolytes for Li batteries. These materials were first found to belong to some composition of the following sulfide based binary systems : $GeS_2 - Li_2S$, $B_2S_3 - Li_2S$, $P_2S_5 - Li_2S$; then Li I was added in order to improve ionic conductivity of the Li^+ carrier (1-3). In such compositions consisting of complex anions, incorporating sulfur, an element more polarisable than oxygen, the rigidity of the material is the result of the relative immobility of the macro-molecular anionic sublattice in which Li^+ ion can move more or less freely. Further improvement of the ionic mobility obtained by dissolving a monovalent lithium salt having a large anionic radius such as Li I has been explained by some decrease in the packing densit of the structure through a combination of the macromolecular and discrete anion sublattice and a subsequent reduction of the electrostatic dissociation energy, allowing a better mobility for the Li^+ carrier (4).

In table 1, we have reported the conductivity of some specific compositions of lithium inorganic sulfide based glasses measured at 25° C as compared with various Li^+ polycrystalline solid conductors which have already been proposed for use in lithium battery systems (5-8).

TABLE I - CONDUCTIVITY OF LITHIUM ION CONDUCTING SOLID ELECTROLYTES

Solid Electrolyte	Form	Ionic Conductivity at 25° C, $\Omega^{-1} \cdot cm^{-1}$	Activation energy (ev)
0.26 B_2S_3 - 0.30 Li_2S - 0.44 LiI	glass	1.7×10^{-3}	0.30
0.36 GeS_2 - 0.24 Li_2S - 0.40 LiI	glass	1.0×10^{-4}	0.48
P_2S_5 - 2 Li_2S	glass	1×10^{-4}	0.30
0.18 P_2S_5 - 0.37 Li_2S - 0.45 LiI	glass	2.0×10^{-3}	0.30
LiI	pc (*)	5.5×10^{-7}	0.43
LiI(Al_2O_3) $\simeq$ 40 %	DSES (*)	1.0×10^{-4}	0.40
Li_3N	pc (*)	1.5×10^{-3}	0.26
Li_3N - LiOH - LiI	pc (*)	2.0×10^{-3}	-----

(*) pc, DSES refer respectively to polycrystal and dispersed phase solid electrolyte systems.

These specific compositions have been selected from the study of the corresponding binary or ternary, phase diagrams. Practically, the various vitreous compositions have been prepared by mixing appropriate amounts of the starting materials which have been brought progressively to the liquid state at elevated temperature (800 - 900° C) in a vacuum sealed quartz vessel and then quenched in air.

The conductivities of glasses indicated in Table 1 are those of the bulk materials : they were obtained from mechanically polished thin discs ($\emptyset$ = 10 mm, thickness = 3 - 5 mm) on which gold or platinum electrodes were sputtered for achieving a.c. conductivity measurements ; these values given at room temperature are quite comparable to those of the best polycrystalline lithium ion conductors. Moreover, the conductance of these glassy electrolytes (as for the best poly-cristalline ones) exhibits a minimum temperature dependence (fig. 1) as reflected by the low activation energy determined from the Arrhenius law (log σ vs $\frac{1}{T}$) for temperature below the vitreous transition temperature (Tg). It is now well known that both of these properties, viz a high conductivity and low activation energy are complementary to each other and therefore have a common origin.

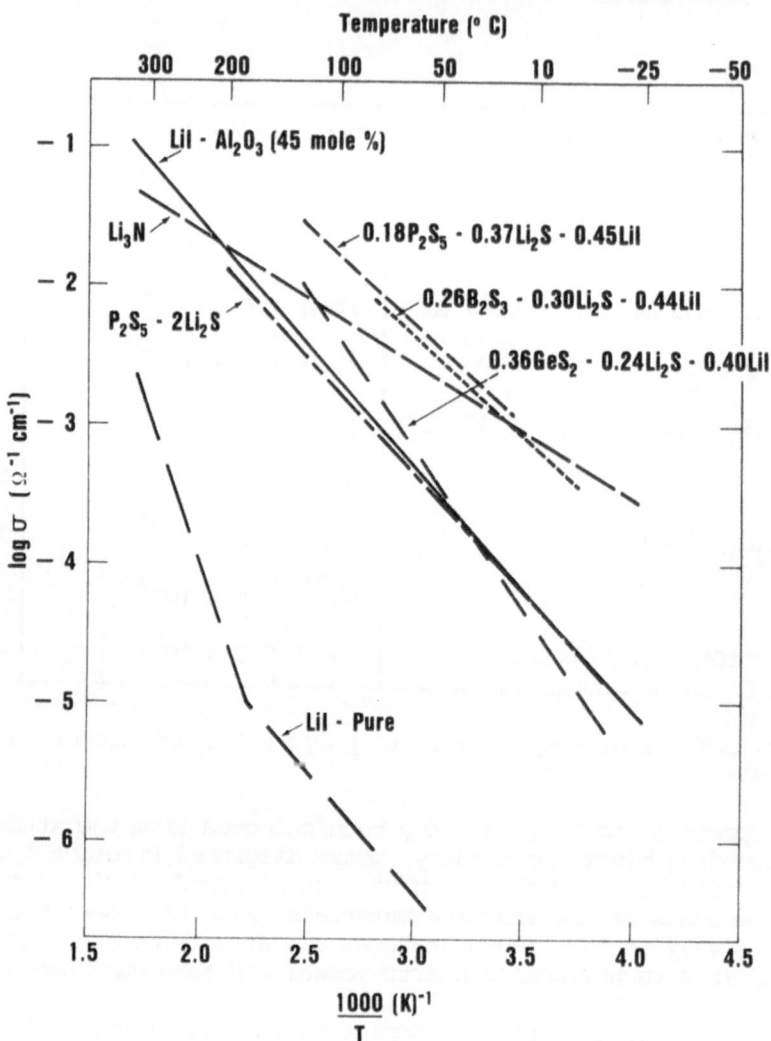

Fig. 1 Electrical conductivity as a function of
inverse temperature (K^{-1}) for a number of Li^+
based electrolytes.

Both properties are best exhibited by solids with disordered open channels, wherein the number of sites outnumber the ions available to occupy them ; in addition they are highly desirable to minimize the effect of temperature change on battery performances especially if a wide operating temperature range is needed.

2.2. Electronic conductivity

It is stated that a solid electrolyte useful for batteries must exhibit an electronic conductivity as low as possible in order to avoid internal short-circuiting consuming electrode materials, resulting in short shelf life.

As far as lithium sulfide based glasses are concerned, a very low level of electronic conductivity was determined using the "blocking" electrode d.c. polarization technique, developed by C. Wagner (9). For instance, with the best conductors, viz. $0.26\ B_2S_3 - 0.30\ Li_2S - 0.44\ LiI$ (2) and $0.18\ P_2S_3 - 0.37\ Li_2S - 0.45\ LiI$ (10) an electronic conductivity of less than $10^{-8}\Omega^{-1}\ cm^{-1}$ was found at room temperature.

Therefore, electrical conductivity may be assumed to be almost exclusively ionic in character, the electronic contribution representing less than 0.01 o/oo of the total.

2.3. Decomposition voltage

A high decomposition potential of the glassy electrolyte is also desirable. Its value should be higher or at least comparable to the E.M.F. of the anode/cathode couple which constitutes the selected battery system.

As far as lithium sulfide based glasses are concerned, the lack of thermodynamic data for these glassy materials prevents the calculation of a theoretical decomposition potential in the same way as it has been carried out for single superionic conductors like Li_3N or LiI. Consequently, the electrochemical stability of these glasses was determined experimentally in two ways :

Firstly by building test cells constituted by an electrolyte sample (same discs as those used for conductance measurements) on which a platinum electrode was sputtered on both sides. By varying the voltage between the electrode and measuring the corresponding current it was possible to determine the voltage decomposition of the electrolyte. Figure 2 represents the curve I = f (E) obtained for the $0.18\ P_2S_5 - 0.37\ Li_2S - 0.45\ LiI$ vitreous composition.

Secondly by constituting non symetrical cells Li/glass/Pt, on which cyclic voltammetry was applied, the lithium electrode being used both as a counter and a reference electrode. Figure 3 represents such a voltammogram for the same glass composition as above. This figure shows that the process which limits the cathodic domain can be assigned to the deposition of lithium metal as demonstrated by the reoxidation peak observed on the return sweep. Moreover, no well defined anodic limit corresponding to the oxidation of some anionic species was found within the voltage range investigated.

Taking into account these various experiments, we can however make a rough estimation of the decomposition voltage. Nevertheless, it will be essential to check of practical oxidizers used as cathode materials and working in this voltage range are compatible with these electrolytes, as this apparent voltage stability determined between platinum electrodes may be due to some kinetic hinderance.

2.4. Transition temperature (Tg)

The vitreous transition temperature (Tg) defines the maximum temperature up to which the material remains indefinitely metastable in the glassy state, that is it behaves as a solid not exhibiting the order of a crystallized phase. Beyond this temperature, it has the behaviour of a supercooled liquid, the properties of which may change with time (i.e. recrystallization).

From a practical point of view, the vitreous transition temperature is determined by D.T.A. In table 2, we have reported the values obtained for both binary and ternary glass sulfide compositions previously mentioned as electrolytes for battery systems.

Table II vitreous transition temperature of Li^+ ion conducting sulfide glass electrolytes (3, 11)

Glass composition	Tg
$0.26\ B_2S_3 - 0.30\ Li_2S - 0.44\ LiI$	$\simeq 85 - 90°\ C$
$0.18\ P_2S_5 - 0.37\ Li_2S - 0.45\ LiI$	$120°\ C$
$P_2S_5 - 2\ Li_2S$	$200°\ C$
$0.36\ GeS_2 - 0.24\ Li_2S - 0.40\ LiI$	$226°\ C$

From this table, it should be noted that the binary system $P_2S_5 - 2\ Li_2S$ exhibits a vitreous transition temperature considerably greater than the corresponding lithium iodide doped one. This is well illustrated by figures 4, on which is reported the influence of LiI addition to such a system.

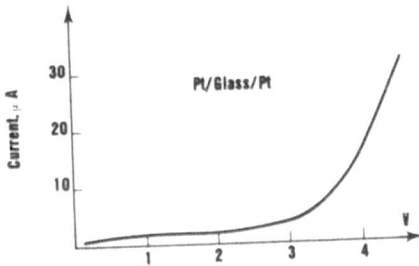

Fig. 2 Current voltage relationship performed on a glass sample having the following composition : $0.18\ P_2S_5 - 0.37\ Li_2S - 0.45\ LiI$ between two platinum blocking electrodes.

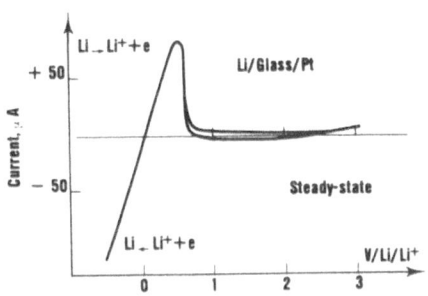

Fig. 3 Voltammogramm performed on a glass sample having the following composition : $0.18\ P_2S_5 - 0.37\ Li_2S - 0.45\ LiI$ between a Li and a platinum electrode

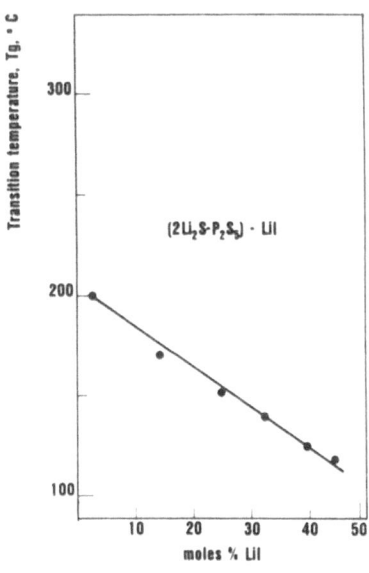

Fig. 4 Effect of electrolyte composition on transition temperature

As far as the temperature range of battery operation is concerned, the knowledge of the vitreous transition temperature allows the proper selection of electrolyte composition, nevertheless a compromise has to be made between a wide temperature range and suitable battery performance. This is shown in the preceeding example, where the ionic conductivity of LiI doped electrolytes was found to increase at the same time that Tg was decreased.

It is however interesting to observe that for a given composition the metastability of the glassy electrolyte was kept unchanged, even when the temperature was raised to a value up to the Tg. This is particularly well shown on figure 5, on which we have demonstrated the stability of samples having the $0.18 P_2S_5 - 0.37 Li_2S - 0.45 LiI$ composition, that we have maintained at $110 \stackrel{\sim}{=} 115$ ° C during more than 10 months, without any appreciable change of the electrical conductivity.

2.5. <u>Shaping ability</u>

Another criterion of importance for the glassy electrolyte is its ability to be vacuum deposited, sputtered on various electrode supports (thin films) or pelleted (thick layers) from the pulverized material, in order to be shaped to a form appropriate to the type of battery technology required.

As far as sulfide based glass electrolytes are concerned, the pelleting process has been presently developed, taking into account its relative simplicity. It was, however, necessary to ensure that this process did not cause too great a loss of electrical conductivity between a sample of bulk glass and a sample of powdered pressed pellet one. As shown on figure 6 the conductivity between bulk and recompacted glass for the aforementioned $0.18 P_2S_5 - 0.37 Li_2S - 0.45 LiI$ composition indicated a loss of electrical conductivity of only one half of a long unit (10), moreover no deterioration of the material was found, as demonstrated by the linearity fit of the activation energy in the Arrhenius plot representation.

III - <u>REQUIRED PROPERTIES OF ELECTRODE MATERIALS FOR USE IN GLASSY ELECTROLYTE BATTERY SYSTEMS</u>

In the previous section, we have discussed the properties of the best available Li[+] vitreous electrolyte for use in high energy Li based battery systems. Although during the last few years, much of the attention has been directed towards finding more highly conductive Li[+] glassy electrolyte, the role of electrode systems and related interface effects (electrode/electrolyte) are perhaps as important as that of electrolyte. The electrolyte conductance can limit the rate capability of the cell through ohmic polarization one well known mechanism of energy loss in such galvanic cells.

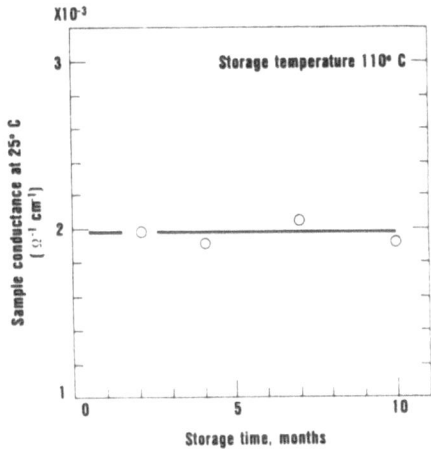

Fig. 5 Effect of electrolyte storage at 110° C on electrical conductivity for a sample having the following composition : 0.18 P_2S_5 - 0.37 Li_2S - 0.45 LiI

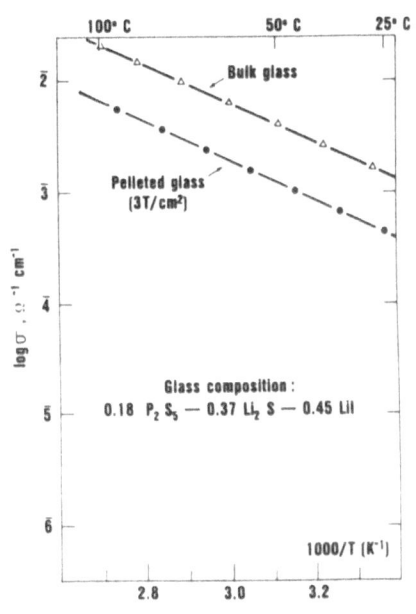

Fig. 6 Electrical conductivity as a function of inverse temperature (K^{-1}) for bulk and pelleted glass samples

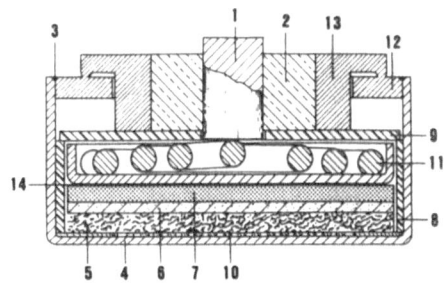

1 Central pin	8 Insulating ring
2 Glass insulator	9 Insulating disc
3 Laser welded seal	10 Stainless steel grid
4 Stainless steel can	11 Spring loader
5 Cathode	12 Cover
6 electrolyte	13 Eyelet
7 Anode	14 Anode current collector

Fig 7 A schematic cross section of the test vehicle cell

Nevertheless other types of polarization involving the electrode/ electrolyte interface may be more severe than the ohmic one due only to the electrolyte ;moreover the anode/cathode couple controls such important features as the open circuit voltage, specific capacity and energy density of cells.

From a practical point of view, the three components of the battery, namely the cathode, the solid electrolyte consisting of the glass Li^+ conducting material and the anode are co-pressed successively according to various methods, in order to produce a cell as represented by figure 7. Where electrodes are obtained from pressed powder materials, both for the anode (Li-alloys) and cathode, it has been found useful to incorporate into each electrode, a certain amount of the glassy electrolyte to facilitate the ionic conducting and related mass transfer properties of the Li^+ carrier to the reaction sites. Eventually addition of a metallic or graphite powder may be made to the cathode in order to improve its electronic conductivity.

3.1. Anode materials

The choice of a Li^+ ion conducting material as electrolyte, leads preferentially to the use of lithium metal, which provides at the same time the highest possible potential for a selected cathode battery system. The functioning of such a system involves the "dissolution" migration of metal (Li) ions through the electrolyte to react with the cathode and form resulting discharge products. Thus this "dissolution" of metal ions may leave empty sites at the anode surface, leading to formation of voids at the anode/electrolyte interface and to the increase in the contact related impedance of the cell.

A possible approach for minimizing this problem would be to improve the surface of contact between the anode and electrolyte. This can be achieved by using a pressed pellet anode made by the method previously described. Because of the powdered character of Li-Al alloys (50/50) and the case be shaped as pressed pellet electrodes when blended with the various vitreous electrolytes, Li-Al anodes containing 30 % of electrolyte (on weight basis) were generally selected instead of Li anodes, for a first evaluation of various cathode systems.

3.2. Cathode materials

The choice of cathode materials is relatively quite large compared to that for anodes. In view of the general requirements that a cathode material useful for miniature batteries should have a low equivalent volume and be as electronegative as possible in order to provide useful voltage compatible with the electrochemical window of the electrolyte, various metal halides, sulfides, oxides and oxysalts already used in lithium organic or solid electrolyte battery systems immediately became potential condidate for the cathode. Table III, reports important features of such cathode materials, some of which have been tested and found to be of interest in vitreous electrolyte solid state batteries.

TABLE III - THEORETICAL VALUES OF SPECIFIC CAPACITY OF SOME CATHODE MATERIALS, EXPERIMENTAL OPEN CIRCUIT VOLTAGE (OCV) AND WORKING VOLTAGE (U) MEASURED EITHER IN ORGANIC OR SOLID ELECTROLYTE LITHIUM BATTERY SYSTEMS

	Cathode material	Capacity density Ah/cm3	Assumed cell reaction product	OCV vs Li (V)	U vs Li (V)
Organic electrolytes	CuO	4.37	$Li_2O + Cu$	2.40	1.50
	$Bi_2Pb_2O_5$	2.64	$Li_2O + Pb + Bi$	2.60	1.60
	$Bi_4B_2O_9$	2.60	$Li_{12}B_2O_9 (?) + Bi$	2.60	1.90
	$Cu_4O (PO_4)_2$	2.00	$Li_2O + Li_3PO_4 + Cu$	2.70	2.20
	TiS_2	0.77	TiS_2Li_x	2.70	2.20
	MnO_2	1.55	MnO_2Li_x	3.50	3.00
	V_2O_5	0.49	$V_2O_5Li_x$	3.50	3.40
Solid Electrolyte	PbI_2	0.79	$LiI + Pb$	1.90	1.45
	PbS	1.68	$Li_2S + Pb$	1.80	1.45

It should be pointed out that these different materials are generally composed of well crystallized structures. As far as amorphous cathode materials are concerned, very little work has been done up to now, in spite of the fact that the latter may have some potential advantages over the crystallized ones, namely in the case of host intercalation material (TiS_2, V_2O_5 ....), the structure of which offers a hope for rechargeable all solid state batteries. Presently existing vitreous cathode materials are based either on phosphovanadates or vanadium boron oxide glasses ($0.6\ V_2O_5 - 0.4\ P_2O_5$; $(V_2O_{4.85})_{0.9} - (B_2O_3)_{0.1}$ (12, 13) they are unfortunately not compatible with the electrochemical window of the Li^+ ion conducting sulfide based glass electrolytes, taking into account the rather strong oxidizing character of the V^{5+}/V^{4+} redox system (~ 3.5 V. vs Li).

3.3. Electrode materials, stability against the electrolyte

The electrodes (anode and cathode) must be quite stable in contact with the electrolyte. This important requirement has to be checked quite carefully for battery systems which have to be stored or used for extended periods of time. Although the decomposition voltage of the electrolyte, determined as previously described, gives a first idea of the possible electrode systems to be used, there is no acceptable technique to determine how such systems would practically behave during long periods of storage, particularly if metathetical reactions in addition to the above may have to occur between cathode and electrolyte materials.

Thus, a new methodology has been developed, based on very sensitive techniques, in order to secure long term compatibility between electrolyte and electrode materials. One such technique, microcalorimetry has been extensively used for detecting any chemical or electrochemical parasitic reactions, through the heat flow periodically measured at a constant temperature (37° C) evolving from samples made of an intimate mixture of electrode/electrolyte powdered material, shaped by pelleting as small discs and put in moisture free hermetically sealed stainless steel containers for practical determinations.

In table IV, we have reported the heat flow so obtained at different periods of time for various electrode materials in the presence of the $0.18 \, P_2S_5 - 0.37 \, Li_2S - 0.17 \, LiI$ glassy electrolyte selected as an example. These determinations have been run on a microcalorimeter designed by TRONAC, which is a twin cell differential heat flow instrument with a peak to peak noise level of less than 0.3 µwatt and a precision better than 1 µwatt on a sample measurement.

TABLE IV - CALORIMETRIC RESULTS FOR ELECTRODE/ELECTROLYTE PELLETED SAMPLES (100 mg) AT 37° C.

SAMPLE	Initial	Heat Flow 3 months	6 months
Glass (Blank)	0 - 4 µW	0	0
Glass + LiAl	0 - 5 µW	1 µW	3 µW
Glass + $Bi_2Pb_2O_5$	0 - 5 µW	1 µW	1 µW
Glass + Ti S_2	-	1 µW	1 µW
Glass + $Cu_4O \, (PO_4)_2$	6 - 9 µW	5 µW	6 µW
Glass + MnO_2	15 - 35 µW	10 µW	10 µW

From this table, it can be seen that there is a quite good compatibility of the glassy electrolyte with cathodic materials such as $Bi_2Pb_2O_5$ and TiS_2. On the other hand, the rather high values of the heat flow observed with MnO_2 reflect the oxidizing power of this material with respect to the electrolyte. Moreover the values recorded for the $Cu_4O(PO_4)_2$ may equally reflect some interactions, in spite of the fact that the latter has the same electronegative character as TiS_2 (see table III).

As far as LiAl is concerned, the heat flow so measured indicates a pretty good compatibility with the electrolyte and thus may justify the use of this material for building complete cells.

As complementary to the above, another technique based on the determination of the open circuit voltage (O.C.V.) of cells using the Li-Al pelleted anode together with the various cathode materials and the glassy electrolyte previously described, has been undertaken in the same isothermal conditions (37° C) at different periods of time. In table V, we have reported the results of such O.C.V. measurements.

TABLE V - O.C.V. MEASUREMENTS FOR Li-Al VITREOUS ELECTROLYTE

BATTERY SYSTEMS (37° C)

CELL	O.C.V., V.		
	Initial	1 month	3 months
LiAl/$Bi_2Pb_2O_5$	2.32	2.32	2.31
LiAl/TiS_2	2.37	2.37	2.34
LiAl/$Cu_4O(PO_4)_2$	2.36	2.30	2.22
LiAl/MnO_2	2.70	2.76	2.55

As compared to the Li-battery systems, the O.C.V. of Li Al cells may be estimated from table III taking into account that the Li-Al anode has a potential of about 0.3 V more positive than Li. The experimental values reported in table V are quite in agreement with those predicted from these estimations, except for the Li Al/MnO_2 cell which demonstrated an initial O.C.V. of about 0.5 V lower than expected. This behaviour results from the formation of a mixed potential at the cathode/electrolyte interface, confirming a reaction between the electrolyte and the oxidizer as already indicated by the microcalorimetrical measurements.

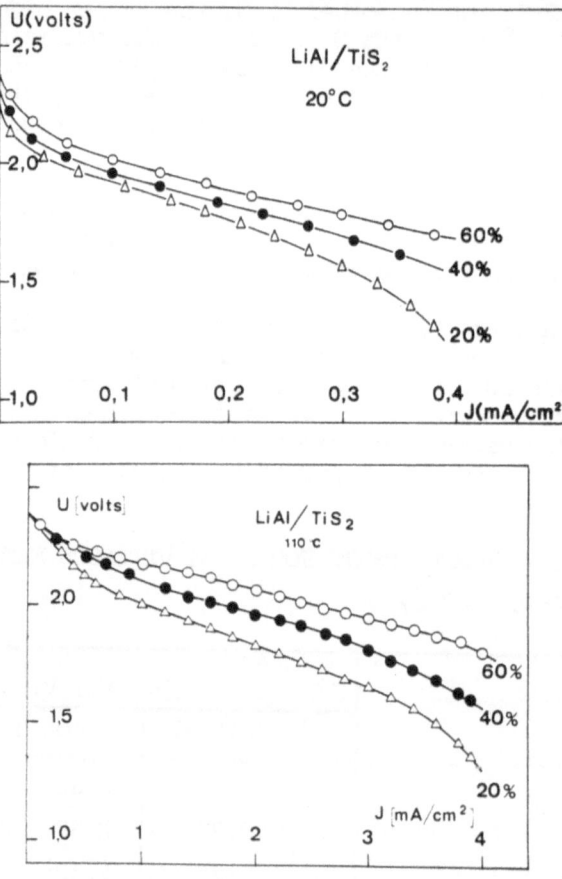

Fig. 8 Cell voltage current density relationships
of LiAl/TiS$_2$ cell as a function of electrolyte
amount in the cathode both for 20° C and 110° C.

As far as the $Cu_4O(PO_4)_2$ cathode material is concerned, a small decrease of the cell O.C.V. seems to appear with increasing storage time. We believe that this phenomena is indirectly linked to a pure chemical process. A metathetical reaction of the following type :

$$Cu_4O(PO_4)_2 \ + \ 4 \ Li_2S \ \longrightarrow \ 4 \ CuS \ + \ 2 \ Li_3PO_4 \ + \ Li_2O$$

involving one of the constituant of the electrolyte may possibly occur explaining the heat flow previously observed.

IV - LiAl AND Li GLASS ELECTROLYTE BATTERY SYSTEMS

LiAl and Li miniature cells using the $0.18 \ P_2S_5$ - $0.37 \ Li_2S$ - $0.47 \ LiI$ glass composition as electrolyte, and some of the selected cathode materials were tested in various environmental conditions.

4.1. LiAl cells

These cells involve a mixture of LiAl powder (50/50 alloy) and electrolyte (30 % in weight ratio) for the anode and a mixture of the selected active positive material with various amount of electrolyte, in which 10 % of graphite has been eventually added to increase the initial electronic conductivity of the cathode. The cathode, the vitreous electrolyte and the anode were sequentially introduced and pressed in a stainless steel die, followed by pressing them together at a somewhat higher pressure (~ 3 t/cm2) than that used for individual pressing. The electrolyte layer is typically 0.5 mm thick and the total cell thickness is 1.2 mm with a diameter of 10.7 mm. The cell capacity is limited by the cathode. The battery is made by placing the cell in a hermetically sealed stainless steel container with a glass to metal seal current feedthrough, moreover the contact between the different battery components during discharge is maintained by spring loading (figure 7).

4.1.1. LiAl/0.18 P_2S_5 - 0.34 Li_1S - 0.47 LiI/TiS$_2$

The first system to be studied was the LiAl/TiS$_2$ Cell, as TiS$_2$ is well researched in both organic and solid electrolyte battery systems (14, 15), moreover this material was found to be entirely compatible with the glassy electrolyte.

LiAl/TiS$_2$ cells were built with various cathode formulations based on a mixture of TiS$_2$ and electrolyte, the amount of which ranged from 20 % to 60 % in weight. Taking into account that TiS$_2$ and its metallic intercalates TiS_2Li_x have a good electronic conductivity, it was not found necessary to add graphite as an electronic conductor.

Figure 8 shows the voltage current density curves obtained for such cells both at ambient temperature and $110°$ C, as a function of the electrolyte amount in the cathode. The improved performance attained at high temperature is evident : about one order of magnitude for investigated current densities is obtained when the temperature is raised from $20°$ C to $110°$ C. Figure 9 shows discharge characteristics

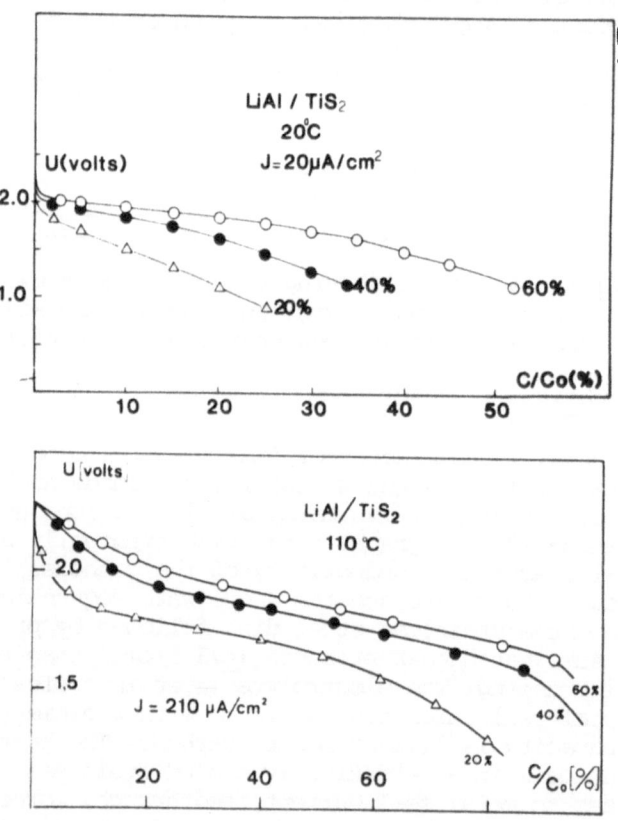

Fig. 9 Cell discharge characteristics of LiAl/TiS$_2$ cell as a function of electrolyte amount in the cathode both for 20° C and 110° C.

expressed by cell voltage as a function of depth of discharge (Co representing the stoichiometric capacity of the cathode, C the capacity delivered at any time of discharge) for the same cells discharged respectively at the 20 $\mu A/cm2$ current density at 20° C and at about ten time more (210 $\mu A/cm2$) at 110° C. As can be seen from this figure, addition of electrolyte to the cathode improves cell performance both in term of cell voltage and discharge efficiency : its benefit seems however more marked at ambient temperature than at 110° C. Moreover, it appears that the Li_xTiS_2 formed as a reaction discharge product, presumably enhances the transport of Li^+ ion in the cathode and helps to reduce the amount of electrolyte to be added : this has been confirmed later through the use of non intercalation cathodes such as $Cu_4O(PO_4)_2$.

4.1.2. LiAl/0.18 P_2S_5 - 0.37 Li_2S - 0.47 $LiI/Cu_4O(PO_4)_2$

In spite of its low reactivity with the electrolyte this cathode material (14) has been investigated, taking into account its higher capacity density than TiS_2 (2 Ah/cm3 as compared to 0.77 Ah/cm3 for TiS_2) for a comparable cell working voltage range.

Figure 10 represents the voltage current density characteristics at 20° C and 110° C for cells incorporating various amount of electrolyte in the cathode (10 - 60 % in weight). Figure 11 shows discharge characteristics obtained for these cells at the same temperatures and corresponding discharge rate as run for LiAl/TiS_2 cells.

In contrast to the cells using the layered structure TiS_2 cathode material, a drastic reduction of cell performance was observed when the amount of electrolyte in the cathode was decreased ; this is particularly well illustrated for cells discharged at room temperature, indicating a serious lack of Li^+ mass transfer properties inside the cathode. In view of this result coupled with the long term instability already demonstrated, no more work has been done with this system.

4.2. Li cells

These cells were built according to the same procedure as above, except that the LiAl anode was replaced by a thin lithium metal disc deposited on the glassy electrolyte according to a suitable technique in order to minimize any interfacial problems at the anode/electrolyte interface.

4.2.1. Li/0.18 P_2S_5 - 0.37 Li_2S - 0.47 $LiI/Bi_4B_2O_3$

A survey of Table III reveals the attractive features of metal oxides as cathode materials : as far as capacity density is concerned the calculated figures for the oxides are in general more impressive than the corresponding halides or sulfides, particularly when their

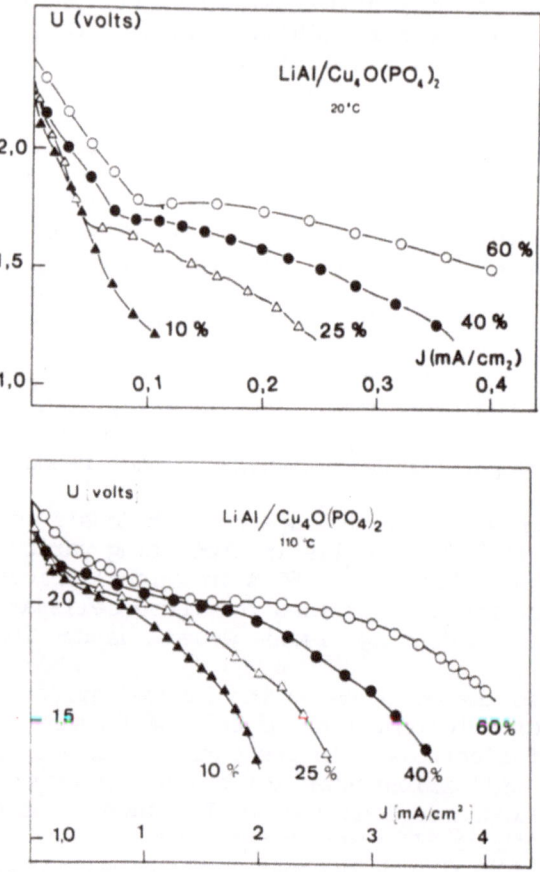

Fig. 10. Cell voltage current density relationships
of LiAl/Cu$_4$O(PO$_4$)$_2$ cell as a function of electrolyte
amount in the cathode both for 20° C and 110° C.

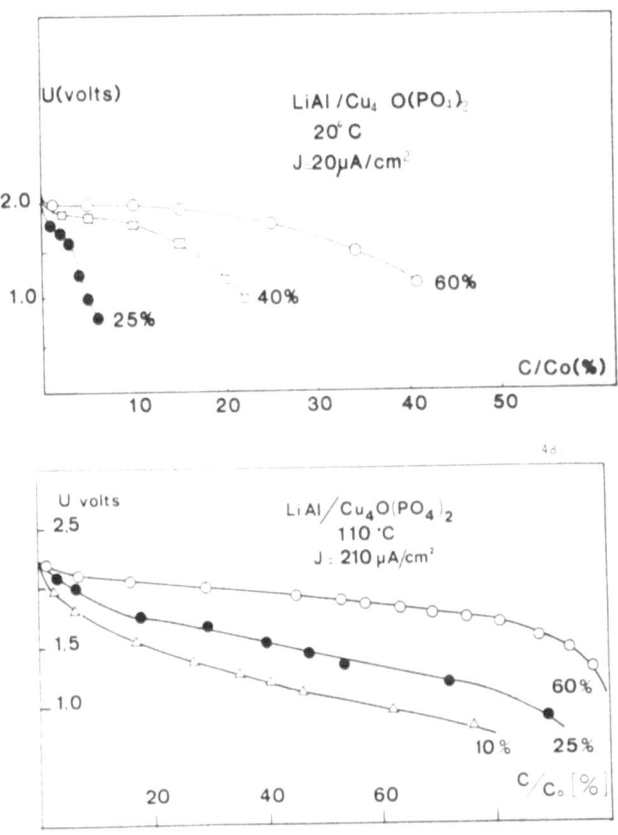

Fig. 11 Cell discharge characteristics of LiAl/
$Cu_4O(PO_4)_2$ cell as a function of electrolyte amount
in the cathode both for 20° C and 110° C.

electrochemical reduction is assumed to occur through the metallic state. Although it should be pointed out that most of these oxides were generally found to react with the vitreous electrolyte through a metathetical reaction to give the corresponding sulfide (i.e. $CuO + Li_2S \longrightarrow CuS + Li_2O$) some of them, like those of bismuth and lead were in contrast found to be quite compatible (i.e. $Bi_2Pb_2O_5$). Also a number of investigations on such similar compounds including either bismuth or lead as a metal were recently investigated and led to the discovery of a new material called bismuth borate $Bi_4B_2O_9$ (17) obtained through a suitable combination of B_2O_3 and Bi_2O_3 at elevated temperature. Such material was found to be entirely stable with the glassy electrolyte, thus $Li/Bi_4B_2O_9$ cells were manufactured.

Figure 12 shows a typical discharge characteristic obtained for cells having the following cathode mass composition :

$Bi_4B_2O_9$	35 %
Glass	60 %
Graphite	5 %

and discharged at 110° C under the 15 KΩ load (127 µA or 140 µA/cm2). It is interesting to note that the depth of discharge exceeds 100 %, when the cell cut off voltage is below 1 V. This phenomenon is believed to be due to the formation of Li - Bi alloys (eg Li_3Bi through an electrochemical alloying process $3 Li + Bi \longrightarrow Li_3Bi$) as it has already been found to occur in organic electrolytes (18).

The storage characteristics of the $Li/Bi_4B_2O_9$ battery system are shown on figure 13. Cells stored at 110° C (near the transition temperature of the vitreous electrolyte) up to 7 months and discharged in the same conditions as before, demonstrated negligable loss of capacity and cell voltage deterioration. This excellent shelf life results from the chemical compatibility of the cell components, the absence of parasitic chemical and electrochemical reactions between the electrodes and the electrolyte and the low electronic conductance of the electrolyte (less than 10^{-7} Ω^{-1} cm^{-1} at 110° C) which minimizes self discharge.

4.2.2. $Li/0.18 P_2S_5 - 0.37 Li_2S - 0.45 LiI/Bi_4B_2O_3$
       v.s. $Li/LiI(Al_2O_3)/PbI_2$, PbS

A comparison of the performance of a $Li/Bi_4B_2O_9$ glassy electrolyte battery with the commercially available Li/PbI_2, PbS solid state battery system using a $LiI(Al_2O_3)$ dispersed phase solid electrolyte (19), has been made on the basis of the pelleted unit cell, using for the $Li/Bi_4B_2O_9$ a cathode formulation of the following composition :

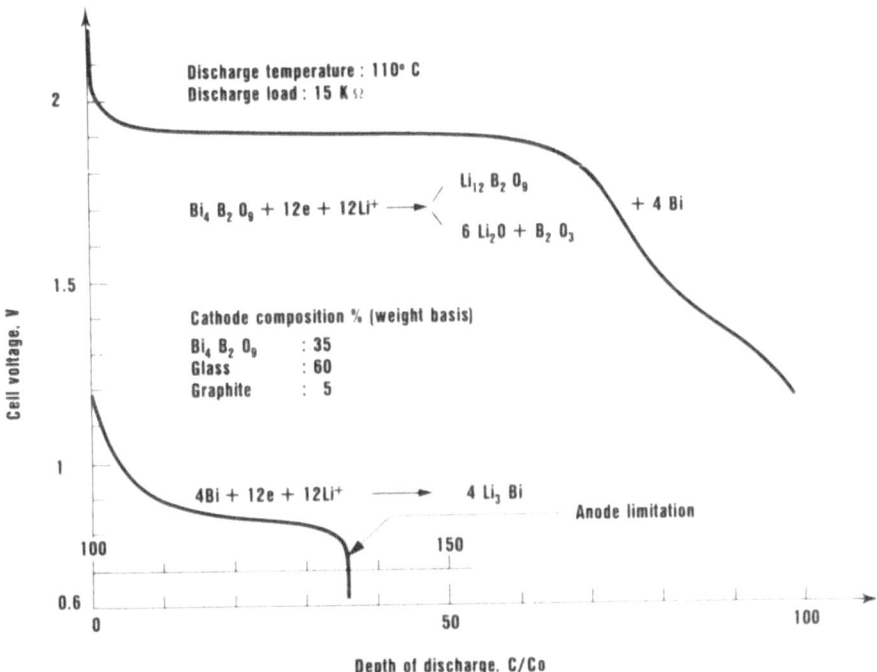

Fig. 12 Discharge behaviour of the Li/Bi$_4$B$_2$O$_9$ cell at 110° C

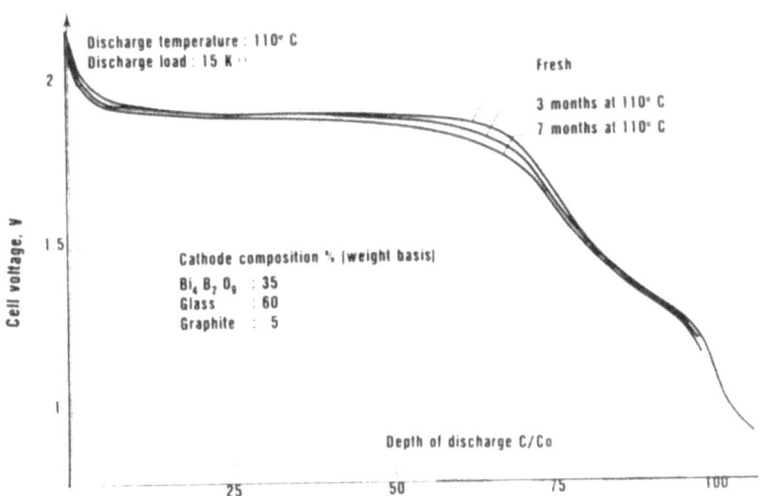

Fig. 13 Effect of cell storage on discharge characteristics of Li/Bi$_4$B$_2$O$_9$ cell

$Bi_4B_2O_9$	65 %
glass	30 %
graphite	5 %

and for the Li/PbI_2, PbS cell, a cathode formulation involving only a mixture of PbI_2, PbS and Pb.

On figure 14, we have reported comparative performance of the two systems at 110° C (accelerated tests) discharged at the same specific rate. The advantage of addition of electrolyte in the cathodic mixture is clearly illustrated on this figure ; the Li/PbI_2, PbS system which does not include any electrolyte in the cathode, taking into account that the ionic path in this electrode is obtained through one of the discharge products (LiI), suffers at the same time a severe ohmic drop inside the cathode, due to the poor ionic conductivity of pure LiI.

CONCLUSIONS

The relevant properties of presently available Li^+ sulfide based vitreous electrolytes and battery electrodes (anode and cathode) were reviewed and examined in view of their practical requirements. Some good Li^+ sulfide based vitreous electrolytes with ionic conductivities in excess of 10^{-3} Ω^{-1} cm^{-1} at room temperature, belonging to some compositions of ternary systems involving either B_2S_3 or P_2S_5 associated with Li_2S and LiI, were found to be of interest for practical application in all solid state batteries.

As for the anode, Li, the most electropositive metal is unquestionably the most practical choice of anode as compared to the Li alloys, the capacity density of which is elsewhere drastically reduced (0.5 Ah/cm3 for the LiAl (50/50) anode vs 2 for Li). There exists a whole spectrum for possible cathode materials : metal sulfides, oxides and oxysalts have been successfully used and found to be of interest, particularly when electrochemically reduced through to the metallic state, because of the high capacity density so recovered. Some materials, that seem to be promising like CuO (4.4. Ah/cm3), $Cu_4O(PO_4)_2$ (2 Ah/cm3) are however limited by chemical processes involving the vitreous electrolyte. Future research is oriented towards bismuth compounds which have been proven to provide both chemical compatibility and performance. New materials exceeding 3 Ah/cm3 have been recently discovered and tested ; they are potential candidates for use in such Li battery systems which are characterized by a high energy density, a long shelf life in various environmental conditions and high reliability.

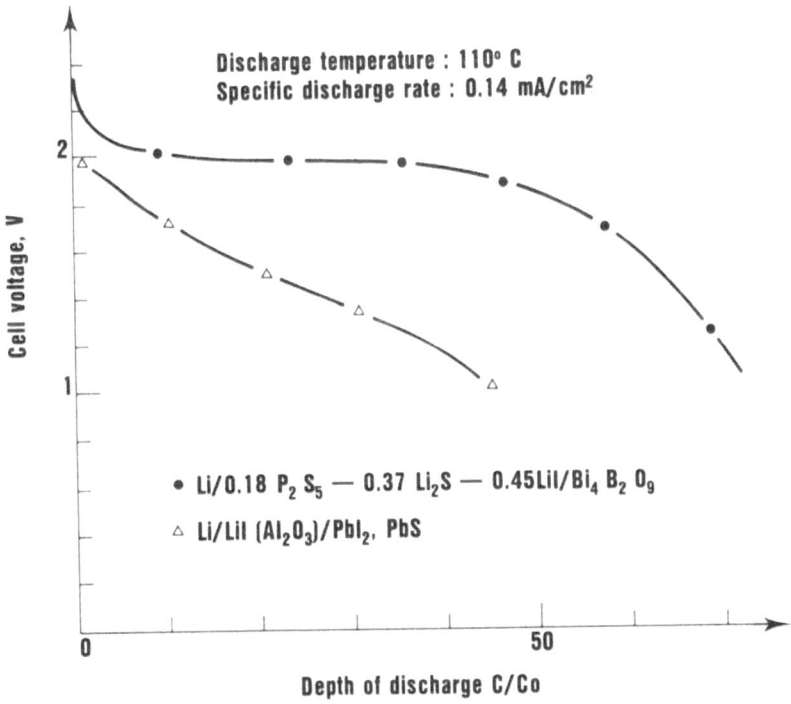

Fig. 14 Comparative performance of Li/Bi$_4$B$_2$O$_9$ test cells vs Li/PbI$_2$, PbS commercial cells.

ACKNOWLEDGEMENTS

The author wishes to express his thanks to Pr ROBERT, Dr MALUGANI from "Laboratoire d'Electrochimie de la Faculté des Sciences de Besançon" ; Pr LECERF from "Laboratoire de Chimie des Matériaux Inorganiques et de Cristallographie de l'Institut National des Sciences Apppliquées de Rennes", Dr BROUSSELY, Dr DUCHANGE and Mrs BAUDRY from "SAFT Departement Piles", for their helpful contribution in the preparation of this review.

REFERENCES

1. B. Barrau, M. Ribes, M. Maurin, A. Kone and J.L. Souquet, J. Non-Cryst. Solids 37, 1 (1980)
2. M. Wada, M. Menetrier, A. Levasseur and P. Hagenmuller, Mat. Res. Bull 18, 189 (1983)
3. J.P. Malugani and G. Robert, Solid State Ionics, 2, 519 (1980)
4. J.L. Souquet, Solid State Ionics, 5, 77, (1981)
5. C. Schkaijer and C.C. Liang, J. of Electrochem. Soc. 118, 14-47 (1971)
6. C.C. Liang, A.V. Joshi and N.E. Hamilton, J. of Applied Electrochem 8, 445 (1978)
7. Von Alpen, and M.F. Bell In "Fast Ion Transport in Solide (P. Vashita, J. Handy and G.K. Shenoyeds), 463 (1979) North-Holland
8. H. Obayashi, R. Nagai, A. Gotoh, S. Mochizufi and T. Kudo, Mat. Res. Bull. 16, 587 (1981)
9. C. Wagner, 7th CITCE, 381 (1955)
10. J.P. Malugani, B. Fahys, R. Mercier, G. Robert, J.P. Duchange, S. Baudry, M. Broussely and J.P. Gabano, Solid State Ionics 9-10, 659 (1983)
11. B. Carette, These Montpellier (1982)
12. T. Pagnier, M. Fouletier and J.L. Souquet, Solid State Ionic, 9-10, 649 (1983)
13. A.C. Leech, J.R. Owen and B.C. Steele, Solid State Ionic, 9-10, 645 (1983)
14. M.S. Whittingham, Ann. Chim. Fr. 7, 204 (1982)
15. J.R. Rea, L.H. Barnette, C.C. Liang and A.V. Joshi in Proceedings of the Symposium on Power Sources for Biomedical Implantable Applications and Ambient Temperature Lithium Batteries (E.E. Owens and N. Margalit Ed), The Electrochem. Soc. Princeton N.J. (1980)
16. M. Broussely, A. Lecerf, J.P. Gabano, 13th Power Sources n° 9, 451, (1983) Acad. Press London
17. M. Broussely, A. Lecerf D.B.F. 8312809
18. J.O. Besenhard and H.P. Fritz, Electrochem. Acta, 5, 513 (1975)
19. Duracell "Designers guide to Lithium Systems", page 61, 1982.

THE ORIGIN OF THE GLASS ELECTRODE RESPONSE

Friedrich G.K. Baucke

Schott Glaswerke, Mainz, Germany

ABSTRACT

After a summary of the operational basis of pH and pM glass electrodes the current state of understanding their response is presented and discussed. As proposed earlier, the equilibria of functional groups at the glass surface with cations of contacting solutions, which determine the ionic species entering the glass by interdiffusion and in electric fields, are thought to be responsible also for the formation of potential differences between membrane glass and solution. Application of electrode kinetics yields equations which describe the experimental results also at pure pH and pM response of the glass. The functioning of glass electrodes with protonated membranes excludes mechanisms based on an exchange of different ionic species between glass and solution. The equilibrium surface coverage was determined by electrolyses of membranes in contact with appropriate anodic solutions and subsequent concentration profiling the glass by IBSCA. Three different kinds of sodium error are expected and verified experimentally.

1 INTRODUCTION

Since the discovery of the potential difference between glasses and electrolyte solutions by Cremer (1) in 1906 and its characterization by Haber and Klemensiewicz (2) in 1909 the construction and application of glass electrodes have developed at a high rate supported by the simultaneous development of electronic measuring equipment. The understanding of the glass electrode mechanism, however, has been much slower and leaves some unanswered questions.

The reason has been the lack of experimental methods sufficiently sensitive to analyse glass components and their depth dependence in glass surfaces with only some to some hundred nanometers
thickness. The history of the glass electrode can thus roughly be
devided into three periods:

1. the thermodynamic period from 1906 to about 1955, during which
emf measurements, analyses of solutions, and few glass electrolyses
gave information on the thermodynamics of the electrode response
and which was characterized by four simultaneously existing theories, (3)

2. the period of measuring elemental distribution, from about 1955
to 1968, during which the availability of isotopes resulted in information on the gross distribution of certain elements between
the glass an the solution, (4, 5) and

3. the dynamic period commencing with Boksay's ingenious method of
measuring sodium concentration profiles in glass surfaces by frac-
tional glass dissolution in 1968 (6) and characterized by modern
surface-analytical methods as SIMS, IBSCA, AES, SEM/EDXA, and NRS
(for a discussion, see e.g. (7)) and by extremely sensitive analyses of solutions.

This paper is not intended to be a literature review as excellent reviews are available (8-1o) but is to inform on the present
state of understanding the mechanism of pH and pM glass electrodes.
Since the author has been involved in this field for almost
15 years, his own work will be the guide-line of this report. After
a summary of the operational elements of glass electrodes, the
interaction between ion-sensitive glasses and aqueous solutions
will be presented as they generate the basis of potential formation
between the two phases. From the properties of the glass surface
layers generated by the corrosion the mechanism of glass electrodes
will then be deduced according to the present understanding.

2 THE OPERATIONAL BASIS OF GLASS ELECTRODES

The functioning of glass electrodes is based on the property
of special, and to some extent of all, oxide glasses to develop
Galvani voltages at interfaces they share with contacting liquid
phases. The magnitude of these interfacial voltages is a function
of the activities of certain cations, mainly hydronium, sodium, and
potassium, in the solution, and it depends on the glass composition
to which ion the glass is more sensitive. For making use of this
effect the system consisting of a liquid and a solid electrolyte is
extended by additional phases in order to obtain an electrochemical
cell with metallic terminals,

Reference Electrode	Reference Electrolyte	Measuring Solution	Glas Membran	Internal Buffer	Internal Ref. El.	(1)
ε_1	ε_j	ε_2	ε_3	ε_4		

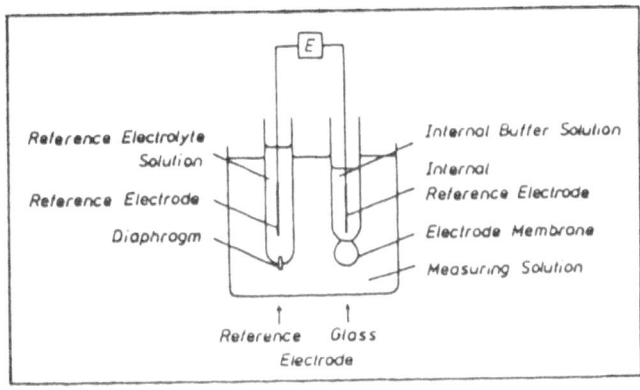

Fig. 1 Schematic of glass electrode cell.

The condition is a defined Galvani voltage at each interface introduced. Fig. 1 shows a practical arrangement. Galvani voltages $\mathcal{E}_4$ and $\mathcal{E}_1$ are kept constant by employing electrodes of the second kind, e.g. Ag/AgCl (11, 12), Hg/Hg_2Cl_2 (calomel, (11) and Tl(Hg)/TlCl (Thalamid^R) (13-15) and $_3$ by an internal buffer solution with high buffer capacity. Liquid junction potential $\mathcal{E}_j$, although slightly dependent on the solution composition, can be regarded sufficiently constant (16). The electromotive force, E, of the cell is thus a function only of the Galvani voltage, $\mathcal{E}_2$, between the membrane glass and the measuring solution and, consequently, of the activity of that cation to which the glass is sensitive,

$$E = const + f(a_i) . \qquad (2)$$

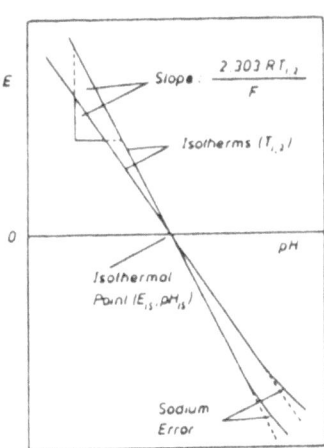

Fig. 2. Response of pH glass electrode cell (schematic).

Fig. 2 shows schematically the response of a pH glass electrode. Functions E(pH) are linear within $\triangle pH \simeq 10-11$ with a nearly theoretical slope, (dE/dpH) = 2.3o3 RT/F. The intersection of the isotherms depends on the internal and external reference electrodes and is positioned close to the abscissa if the same type is used for both electrodes (symmetrical cell). At high pH and alkali, e.g. sodium, ion concentration, positive deviations from the straight lines are observed (sodium or alkaline error) and indicate participation of sodium ions in the potential formation.

The electrode response is described by Nicolsky's equation (7).

$$E = E^O + (RT/F) \ln (a_{H^+} + K_{HNa} \, a_{Na^+}) \, , \tag{3a}$$

or by Eisenman's general equation (18),

$$E = E^O + (nRT/F) \ln (a_i^{1/n} + K_{ij}^{1/n} \, a_j^{1/n}) \tag{3b}$$

where E^O = standard potential of the glass electrode cell, a_{H^+} and a_{Na^+} = activities of hydrogen and sodium ions, respectively, a_i and a_j = activities of ions i and j, respectively, K_{HNa} and K_{ij} = selectivity constants of the membrane glass, and R, T, and F have their usual meanings. pH glass electrodes function equally well in deuterium oxide solutions (16) without the necessity of preconditioning the glass in this solvent if conditioned in H_2O.

Glass electrode cells are generally standardized by two or more standard buffer (16, 19, 2o) or technical buffer solutions (21)

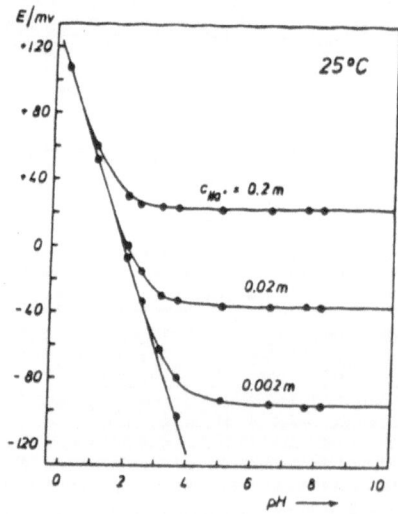

Fig. 3. Response of pNa glass electrode cell.

since the pH response of glass electrodes may be slightly below theoretical. Also available are pD standard buffers in D_2O solutions (16). In most cases, however, pD can be obtained from cells standardized in pH buffers by

$$pD = pH \text{ (meter reading)} + C \tag{4}$$

where C = o.45 and o.41 on the molal and molar scale, respectively (16, 22).

While silicate glasses exhibit a pH function over a wide pH range, alumosilicate, borosilicate, etc., glasses are distinguished by a wide range of alkali, e.g. sodium, selectivity and by only a narrow range of pH response (8, 18). Fig. 3 gives a typical example demonstrating nearly theoretical response to sodium according to eqs. (3a) and (3b).

3 INTERACTION OF GLASSES WITH SOLUTIONS

3.1 Surface Equilibria

The reaction of glass with contacting solutions can be best understood by the electrochemical definition of ion-conducting oxide glasses:
oxide glasses are solid electrolytes consisting of an atomic network with fixed, i.e. immobile, anionic groups and containing cations with finite mobilities. The sum of the transport numbers of all mobile cations in glasses is unity.
The mobile cations of a glass, consequently, respond to concentration (or activity), potential, and temperature gradients and may also leave the network if charge neutrality is maintained, e.g. by replacement by other cations (23-25). Glasses, however, are not ion exchangers as they are not transparent to solvent molecules, although they may take up some water in surface-near regions.

Most important for electrode glasses is that anionic groups positioned at the glass surface, although bound to the network, are subject to chemical reactions with contacting phases (26), e.g. to the mass action law governing ionic equilibria in a solution (23, 27-31). The exchange of different cations between the surface groups and a liquid is driven by the free enthalpy of the surface equilibria rather than by the high concentration gradients at the interface. Surface concentrations thus established differ considerably from those in the glass and in the solution and play a key role for the ion exchange between the phases.

Assume a glass in contact with a solution containing M^+, besides H_3O^+ ions and the anions necessary to balance their charge. The glass surface contains functional groups, e.g. $\equiv SiOM'$,

$[\equiv AlOSi\equiv]$ M', $[\equiv GaOSi\equiv]$ M', (M' = alkali of the glass), of which the siloxy group may be taken as an example for the following derivation. On contact with the solution the $\equiv SiOM'$ group dissociates completely since the liquid contains no M'^{+} ions (27). The siloxy group, $\equiv SiO^{-}$, thus formed is subject to competing equilibria with hydrogen and alkali ions (3o). The equilibrium with hydrogen ions, written as dissociation,

$$\equiv SiOH(s) + H_2O(soln) \rightleftharpoons \equiv SiO^{-}(s) + H_3O^{+}(soln), \qquad (5)$$

is characterized by the dissociation constant,

$$K_D = (a'_{SiO^{-}} \, a_{H_3O^{+}})(a'_{SiOH} \, a_{H_2O})^{-1} \exp(-F\mathcal{E}/RT) \qquad (6a)$$

or

$$K''_D = (c'_{SiO^{-}} \, a_{H_3O^{+}}) \, c'^{-1}_{SiOH} \exp(-F\mathcal{E}/RT) \qquad (6b)$$

if surface concentrations are introduced by defining

$$K''_D = K_D \, \gamma'_{SiOH} \, a_{H_2O} / \gamma'_{SiO^{-}} \qquad (7)$$

(a, a' and c, c' = activities and concentrations on solution (soln) and at the glass surface (s), respectively, γ' = activity coefficients at the glass surface). The "association equilibrium",

$$\equiv SiO^{-}(s) + M^{+}(soln) \rightleftharpoons \equiv \overset{(-)}{SiO} \cdots \overset{(+)}{M}(s) , \qquad (8)$$

is fixed by the "association constant",

$$K_A = a'_{SiOM}(a'_{SiO^{-}} \, a_{M^{+}})^{-1} \exp(F\mathcal{E}/RT) \qquad (9a)$$

or

$$K'_A = c'_{SiOM}(c'_{SiO^{-}} \, a_{M^{+}})^{-1} \exp(F\mathcal{E}/RT) \qquad (9b)$$

if K'_A is defined by

$$K'_A = K_A \, \gamma'_{SiO^{-}} / \gamma'_{SiOM} \qquad (1o)$$

Combination of eqs. (5) and (8) yields the overall surface equilibrium

$$\equiv SiOH(s) + M^{+}(soln) + H_2O(soln) \rightleftharpoons \equiv \overset{(-)}{SiO} \cdots \overset{(+)}{M}(s) + H_3O^{+}(soln) \qquad (11)$$

whose equilibrium constant is equal to the product of dissociation and association constants,

$$K_D K_A = (a'_{SiOM} \, a_{H_3O^{+}})/(a'_{SiOH} \, a_{M^{+}} \, a_{H_2O}), \qquad (12a)$$

Table 1

Glass, State	T/$^{\circ}$C	Ions	pK_D'' + pK_A'	pH_{tr} = - log ($K_D'' K_A'$ a_M+) at a_M+ = o.o2 mole/kg
pNa, Rate-Cooled	25	H^+,Li^+	2.44	4.2
	5o	H^+,Li^+	2.24	4.o
	25	H^+,Na^+	o.94	2.7
	5o	H^+,Na^+	o.76	2.5
pH, Rate-Cooled	25	H^+,Li^+	11.3	13.o
	5o	H^+,Li^+	1o.6	12.3
	75	H^+,Li^+	9.9	11.6
pH, Quenched	25	H^+,Li^+	12.1	13.8
	5o	H^+,Li^+	11.3	13.o
	75	H^+,Li^+	1o.6	12.3

or

$$K_D'' K_A' = (c_{SiOM}' \, a_{H_3O^+})/(c_{SiOH}' \, a_{M^+}) \cdot \tag{12b}$$

Table 1 gives numerical values for a rate-cooled lithium alumo-silicate pNa glass and for rate-cooled and quenched membranes of a lithium silicate pH glass.

It is advantageous to express surface concentrations by mole fractions, x',

$$x_{SiOM}' = c_{SiOM}' \, / \, \Sigma c' \quad , \tag{13}$$

where $\Sigma c'$ is the sum of concentrations of all species at the surface

$$\Sigma c' = c_{SiOH}' + c_{SiOM}' + c_{SiO^-}' \tag{14a}$$

or

$$\Sigma c' = c_{SiOH}' + c_{SiOM}' \tag{14b}$$

since the siloxy concentration, although never zero, may be neglected in most practical cases. Combining eqs. (13) and (14b) with (6b) and (9b) results in the relationship

$$x_{SiOM}' = 1 - x_{SiOH}' = (\frac{1}{K_D'' K_A'} \cdot \frac{a_{H_3O^+}}{a_{M^+}} + 1)^{-1} \quad , \tag{15}$$

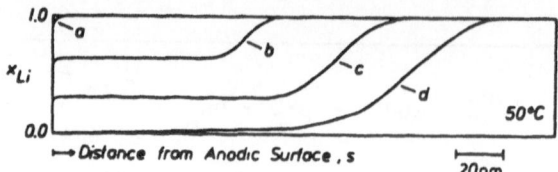

Fig. 4. Li concentration profiles below anodic surface of pH (Li silicate) glass after transferring various Li^+ and H^+ mole fractions from solution into glass in an electric field: $x_{Li} = 1$ (a); $x_H = 0.35$ (b); 0.66 (c); 1 (d).

which yields mole fractions of silanol and metal siloxy groups at the glass surface as a function of ionic activities in solution and specific glass properties. The transition between surfaces covered predominantly by one species ($x^{,}_{SiOM} = x^{,}_{SiOH} = 0.5$) is given by

$$(a_{H_3O^+}/a_{M^+})_{x^,=0.5} = K_D'' K_A^? \tag{16}$$

or by the transition pH

$$pH_{tr} = -\log (K_D'' K_A^? c_{M^+}) = pK_D'' + pK_A^? + pM . \tag{17}$$

Some data are given in Table 1.

The surface coverage was measured in quantitative agreement with eq. (15) by applying electric fields transporting various cations into surface-near glass regions and by subsequent concentration profiling the elements by IBSCA (23, 25, 3o, 32). Fig. 4 gives concentration profiles of lithium at the surface of a lithium silicate pH glass membrane measured after electrolyses with anodic lithium-containing solutions with high pH. Fig. 5 shows the results plotted according to eq. (15). The plot yields the product ($K_D'' K_A^?$),

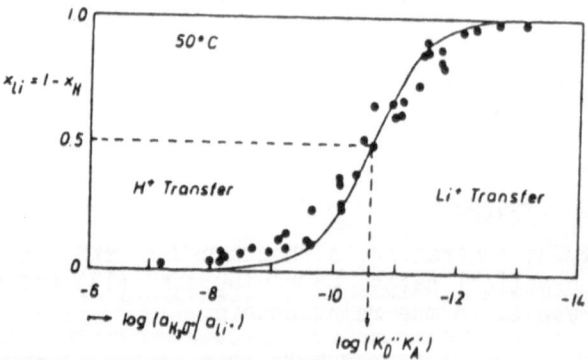

Fig. 5. Dependence of Li mole fraction, $x_{Li} = 1 - x_H$, transferred into Li silicate glass by electric field on ionic activities a_{Li^+} and $a_{H_3O^+}$ in anodic solution.

Table 1, which is equal the ion exchange constant K_{HLi} of Nicolsky's equation according to eq. (3) as expected.

Transfer and emf measurements yielded also relative dissociation constants of the silanol and the deutero silanol groups of several glasses, $K_D(\equiv SiOH)/K_D(\equiv SiOD) = 2.5$ to 2.7 (lithium silicate pH glasses, 25 °C) and 2.o - 2.2 (lithium alumosilicate pNa glasses, 25 °C 1). These values are based on the convention underlying the pH and pD scales (16). The relative numbers, however, are slightly dependent on the glass composition within each group of glasses and distinctly different for pH and pNa glasses. This suggests that the values cited for constant C of eq. (4) are only average values for pH glasses and not valid for the relative pH and pD response of pNa glasses.

3.2 Interdiffusion

Functional groups of the glass surface are also subject to interaction with ions in the glass, fig. 6. Due to extremely small ionic diffusivities in the solid, however, the surface equilibria with ions of the solution are always maintained, independent of the glass composition (23). Any ions diffusing through the surface in either direction under the influence of concentration gradients do not disturb these equilibria. The surface is thus a constant source of ions for the glass and determines the interdiffusing species in the solid (31). While the driving force of the surface equilibria

	Glass	Surface	Solution
Concentrations	c_g , (dc_g/ds)	c'	c
Activities	a_g , (da_g/ds)	a'	a
	SiOH $\underset{}{\overset{K_{as}}{\rightleftharpoons}}$ SiOH $\underset{}{\overset{K_{gs}}{\rightleftharpoons}}$ SiOH $\underset{}{\overset{K_s}{\rightleftharpoons}}$ H$^+$·H$_2$O		
	SiO$^-$ $\underset{}{\overset{K_{vs}}{\rightleftharpoons}}$ SiOM $\underset{}{\overset{K_{gs}}{\rightleftharpoons}}$ SiO$^-$ $\underset{}{\overset{K_v}{\rightleftharpoons}}$ M$^+$		
	SiOM		SiO...M
Mobility / Diffusivity	low		high
Exchange Rate		low	high

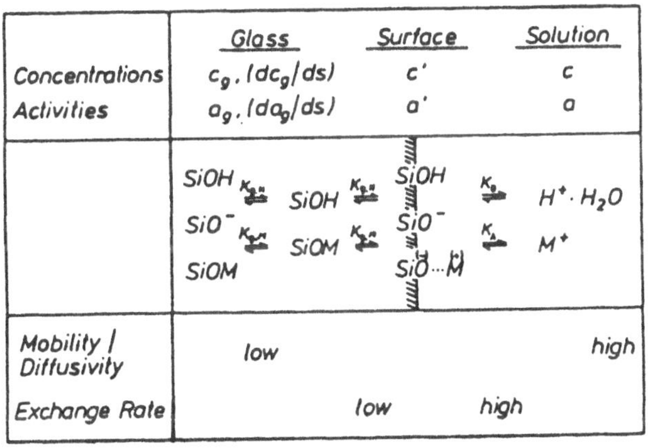

Fig. 6. Schematic of equilibrium of functional groups at glass surface with ions of contacting solution determining interdiffusion within glass.

1. Baucke, F.G.K., unpublished results

is given by the free enthalpies, e.g. for the overall surface equilibrium, eq. (11), by

$$\Delta G_{overall} = - RT \ln K_D'' K_A^{\circ} + RT \ln (c_{SiOM}^{\circ} a_{H_3O^+})/(c_{SiOH}^{\circ} a_{M^+}), \quad (18)$$

the subsequent interdiffusion is driven by ionic concentration
gradients in the glass with the boundary condition of constant
surface concentrations at all times. Three basically different kinds
of interdiffusion must thus be distinguished (31):
1. at pH $<$ (pH$_{tr}$ - 1.5) 2), i.e. at x_{SiOH}° = 1, hydrogen ions of
the solution interdiffuse with alkali ions of the glass,
2. at pH $>$ (pH$_{tr}$ + 1.5) 2), i.e. at x_{SiOM}° = 1, alkali ions of the
solution are exchanged for alkali of the glass, and
3. at (pH$_{tr}$ - 1.5) $<$ pH $<$(pH$_{tr}$ + 1.5) 2), i.e. at x_{SiOH}° $<$ 1 and
x_{SiOM}° $<$ 1, both alkali and hydrogen ions of the solution are exchanged for alkali ions of the solid.

Interdiffusion is complicated by hydrolysis and subsequent
dissolution of the glass network whose rate depends on the solution
pH and the ionic species at the surface. Silicate glasses, for
instance, are subject to slow hydrolysis catalyzed by hydrogen ions
at pH $<$ 9 and to increasingly fast dissolution catalyzed by hydroxyl
ions at higher pH values (16). Network dissolution has to be taken
into account when ionic interdiffusion is studied.

3.2.1 Interdiffusion at pH $<$ (pH$_{tr}$ - 1.5) - Glass Leaching. The
product (K$_D''$K$_A^{\circ}$) is a small number for pH sensitive silicate glasses,
Table 1. Most solutions which are not extremely alkaline thus have
pH $<$ (pH$_{tr}$ - 1.5) causing an exchange of hydrogen and alkali ions,
called leaching. This reaction has been investigated most extensively in the past, e.g. see (6, 33-38). Alumosilicate pNa glasses
exhibit leaching within a much more narrow pH range due to the
larger product (K$_D''$K$_A^{\circ}$) (31), Table 1.

Fig. 7A presents the time-dependent lithium concentration
profiles at the surface of an untreated lithium silicate glass
membrane during the initial contact with a solution having
pH $<$ (pH$_{tr}$ - 1.5) (28). Lithium ions are replaced by hydrogen ions
as confirmed by infrared spectra. The ion exchange is diffusion-
controlled in the first stage as seen by the linear relationship
between lithium loss and square root of time, fig. 7B. The process
soon becomes more complicated since hydrogen ions catalyze the
hydrolytic attack of the glass network (39) leading to an uptake of
water and to the dissolution of the leached layer. This makes the
concentration-dependent interdiffusion coefficient (4o) dependent
also on the glass layer structure. The hydrolysis may change the
concentration of functional groups at the surface, but this does

2. The transition pH range covers approximately ΔpH = 3 to 4

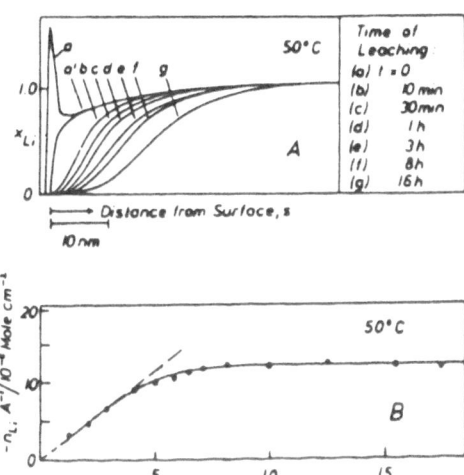

Fig. 7. Development of a leached layer at the surface of a Li silicate glass at pH < (pH$_{tr}$ - 1.5). (A) Li concentration profiles
after storing at the atmosphere (a); after wiping (a'); after contact with solution (b-g); time-independent profile (g). (B) Li
deficit of surface as a function of the square root of time.

not influence the surface equilibrium as only groups of the same
kind are generated. The final, constant concentration profile g,
fig. 7A, and the corresponding time - independent lithium loss after
about two days, fig. 7B, demonstrate a steady state characterized
by equal rates of layer formation and dissolution.

Comparable steady state concentration profiles are found with
lithium silicate and alumosilicate glasses as long as
pH < (pH$_{tr}$ - 1.5). Alkali ions of the solution are excluded from
entering the glass under this condition although neither lithium
nor, in some cases, sodium ions are sterically hindered.

The ion exchange rate between the layer surface and the bulk
glass is largely determined by the low proton mobility in the nearly
unchanged glass network of the transition range between leached
layer and glass, fig. 7A. Measurements on protonated glass layers
yielded substantially smaller proton than lithium ion mobilities in
the glass within the entire concentration range, fig. 8 (25, 41).
This causes high local resistivities, ρ_{Tr}, within the transition
layer with maxima exceeding the resistivity of the bulk glass,
ρ_{gl}, by more than two orders of magnitude: ρ_{Tr}/ρ_{gl} = 34o (25 °C),
2oo (5o °C) (silicate glass) (28) and 38o (25 °C) (alumosilicate
glass) (31). Wikby reported a factor of 1ooo for a pH glass (42).

492

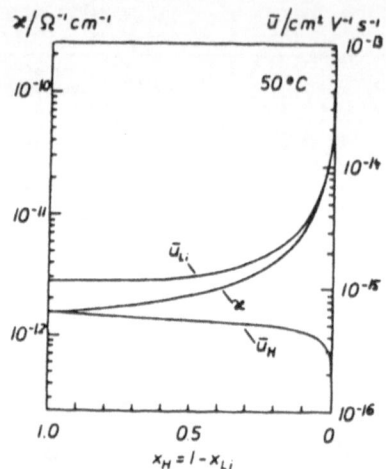

Fig. 8. Transport data of protonated Li silicate glass as a function
of proton mole fraction, x_H .
$\bar{u}_{Li}$, $\bar{u}_H$ = average mobilities of lithium ions and guest protons
introduced by electrolysis and $\varkappa$ = conductivity.

Thickness and maximum concentration gradient of steady state
leached layers depend on the dissolution rate of the layer and on
the interdiffusion coefficient determining its formation rate.
While the temperature controls both, only the dissolution rate is
also dependent on solution parameters, e.g. pH, hydrodynamics, or
HF content (29). Maximum concentration gradients, $(dx/ds)_{max}$,
within steady state layers may be estimated according to eq. (19),

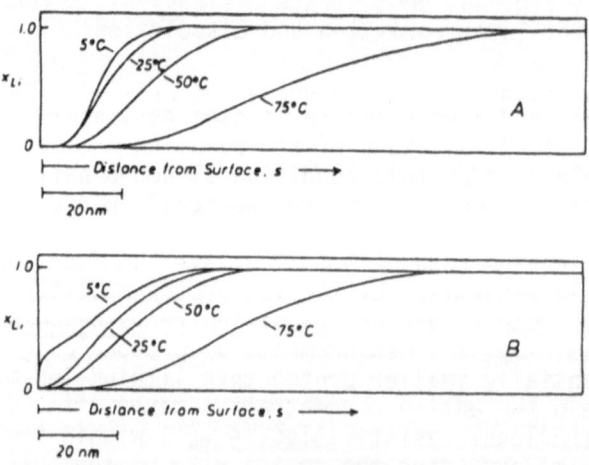

Fig. 9. Steady state lithium concentration profiles at the surface
of a lithium silicate (A) and a lithium alumosilicate (B) glass
generated by leaching at pH $<$ (pH_{tr} - 1.5).

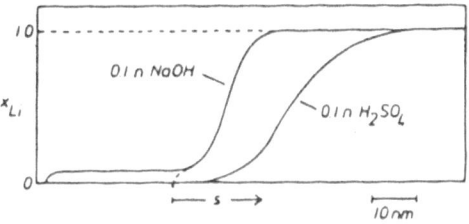

Fig. 1o. Steady state Li concentration profiles generated by leaching a lithium silicate glass, 5o °C. The layer deposited in NaOH could be dissolved in dilute acid.

$$(dx/ds)_{max} = \bar{D}_{min}^{-1} \; (ds/dt) \quad , \tag{19}$$

where x = mole fraction of interdiffusing ions, s = depth below surface, $\bar{D}_{min}$ = concentrations-dependent minimum interdiffusion co-efficient in the range of maximum concentration gradient, (ds/dt) = temperature - and solution parameter - dependent dissolution rate. Fig. 9 presents steady state lithium concentration profiles at the surface of a pH and a pNa glass at various temperatures and demonstrates a similar character of leaching (31). According to eq. (19) the profiles indicate a larger temperature dependence of the inter-diffusion coefficient than of the dissolution rate for both glasses. Fig. 1o presents lithium concentration profiles at the surface of

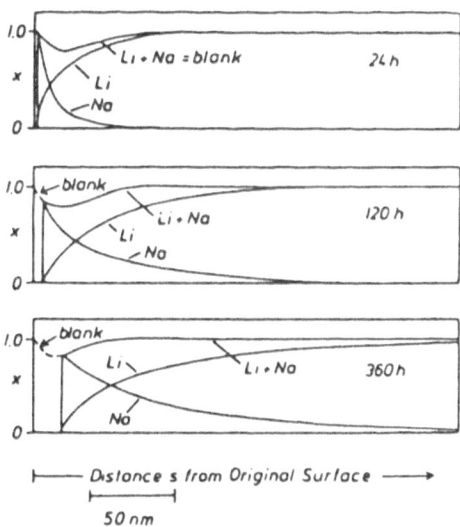

Fig. 11. Concentration profiles at the surface of Li alumosilicate glass after contact with Na^+-containing solution with $pH > (pH_{tr} + 1.5)$. The sum of lithium and sodium mole fractions is identical to the internal part of the profile of the untreated glass.

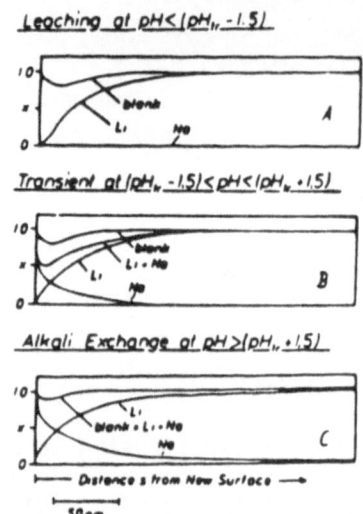

Fig. 12. Leaching at pH = 1.1, alkali interdiffusion at pH = 6.5, and transient at pH = 2.4 of a lithium alumosilicate glass in Na^+ containing solutions with pH_{tr} = 2.7, 16o h, 25 °C.

the same glass at low and high pH at 5o °C and indicates a larger glass dissolution rate at large pH.

Changing the rate-determining factors causes new steady states within times comparable to the primary leaching period, also after a temperature increase. Upon decreasing the temperature, however, drastically longer times are observed, e.g. 3 years after a temperature change of $\triangle T$ = - 7o K (29), since the thick layer formed at the high temperature must be reduced in thickness at the low temperature and thus at small dissolution rate.

3.2.2 Interdiffusion at $pH > (pH_{tr} + 1.5)$. A literature survey may suggest that leaching is the only kind of glass solution-interaction. Eisenman's report (5) on an exchange of alkali ions between pNa glasses and solutions is rarely cited, probably because alkali exchange is often obscured and thus rarely observed.

The concentration profiles of fig. 11 demonstrate an exchange of alkali ions, and fig. 12 shows that the condition is $pH > (pH_{tr} + 1.5)$. Sodium ions are not sterically hindered as seen from the linear dependence of lithium loss and sodium uptake on the square root of time, fig. 13, stating also diffusion control of the exchange. These results are in agreement with Eisenman's data (5) from radiochemical experiments on ion exchange between pNa glasses and solutions, whose pH values were estimated from his

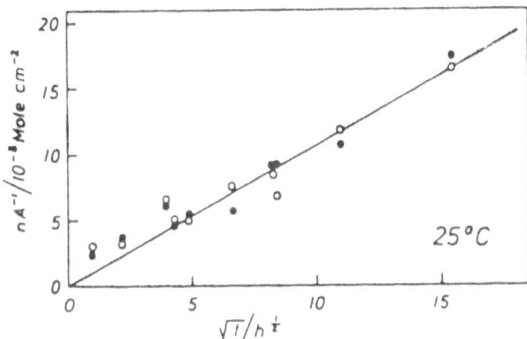

Fig. 13. Li loss (o) and Na uptake (●) by Li alumosilicate glass at pH > (pH_{tr} + 1.5) as a function of square root of time.

data and found to be pH > (pH_{tr} + 1.5) when alkali exchange had been observed (31).

The respective equal amounts of exchanged lithium and sodium, fig. 13, suggest that hydrogen ions do not substantially take part in the interdiffusion. This was confirmed by ir spectra of thin membranes after long periods of contact with this solution (35 days) and by electrolyses with the solution as anodic medium during which negligible amounts of hydrogen entered the glass. Further, the sum of lithium and sodium concentration profiles are identical to respective parts of the blank profile. Due to the absence of hydrogen ions and water in the ion-exchanged layer, the dissolution cannot be "prepared" by hydrolysis of internal parts of the network as during leaching. The dissolution process at pH > (pH_{tr} + 1.5) must therefore be different from that at pH < (pH_{tr} - 1.5) although the pH may be equal in both cases.

If, under the condition pH > (pH_{tr} + 1.5), the alkali ions at the surface are sterically hindered to enter glass or surface layer, an exchange of hydrogen for alkali ions is observed. Thus, Li concentration profiles similar to those formed at pH < (pH_{tr} - 1.5) are found when the glass is leached in dilute KOH, the potassium ion being known from electrolyses to be sterically hindered (3o). The exchange is obviously caused by the large concentration gradients at the interface and by circumventing the surface equilibria and may be called a "forbidden" ion exchange.

3.2.3 Interdiffusion at (pH_{tr} - 1.5) ≤ pH < (pH_{tr} + 1.5).
Fig. 12 compares leaching (A), Li^+/Na^+ ion exchange (C), and transition behaviour (B) after equal periods of glass solution-contact. Both sodium and hydrogen are exchanged for lithium ions in case (B), and the presence of hydrogen seems to retard the Li^+/Na^+ exchange to some extent. Different profiles observed in this pH range demonstrate complicated interdiffusion and probably interfering network hydrolysis. This range, although representing the range of

sodium error has not yet been investigated further.

4 THE GLASS ELECTRODE RESPONSE

4.1 Phase Boundary Potentials; Dissociation Mechanism

The response of glass electrodes is fast, reversible, and thermodynamically correct although neither membrane glass nor surface layer are in equilibrium with the solution and even a steady state of the layer is not maintained during practical measurements. The equilibria of functional groups at the glass surface with cations of the solution were thus proposed to be responsible also for the potential formation (27, 28, 29):
1. the surface groups belong to both the glass and the solution,
2. obviously form the only reversible equilibria, and,
3. due to their position at the layer solution-interface, can cause the fast electrode response; besides,
4. the equilibria are independent of the surface layer structure, and
5. are neither sensitive to ionic interdiffusion nor to network dissolution.

The pH response of glass electrodes is thus described by eq. (6b) or, after rearrangement, by

$$\mathcal{E} = - (RT/F) \ln K_D'' + (RT/F) \ln \left[(c'_{SiO^-} \, a_{H_3O^+})/c'_{SiOH} \right] , \tag{2o}$$

and the pNa response by eq. (9b) or by

$$\mathcal{E} = (RT/F) \ln K_A' + (RT/F) \ln \left[(c'_{SiO^-} \, a_{M^+})/c'_{SiOM} \right] . \tag{21}$$

Equating the electrochemical potentials of hydronium ion and silanol group and observing eqs. (6b), (9b), and (14a) yield the Nernst equation of the glass electrode,

$$\mathcal{E} = \mathcal{E}^0 + (RT/F) \ln (a_{H_3O^+} + K_D'' K_A' \, a_{M^+} + K_D'' \, e^{F\mathcal{E}/RT}) , \tag{22}$$

which is actually a more detailed form of Nicolsky's and Eisenman's equation, eqs. (3a) and (3b), respectively, and shows that the ion exchange constant represents the product of dissociation and association constants of eq. (12b), $K_{HM} = K_D'' K_A'$.

The derivation raises two questions:
1. The second term on the right side of eqs. (2o) and (21) contains equilibrium activities and surface concentrations. It is thus striking that the experimental electrode response to hydronium and alkali ion activities according to

$$\mathcal{E} = \text{const} + (RT/F) \ln a_{H_3O^+} \tag{23}$$

and

$$\mathcal{E} = \text{const} + (RT/F) \ln a_{M^+} \tag{24}$$

suggests constant ratios (c'_{SiO^-}/c'_{SiOH}) and (c'_{SiO^-}/c'_{SiOM}), respectively, although the surface concentration, expected to participate in the equilibrium, eq. (2o) and (21), should be changed by ionic activity changes and should contribute to the magnitude of $\mathcal{E}$. The electrode response should thus be different from that observed experimentally, eqs. (23) and (24).

2. Eqs. (12a) and (12b) cannot be rearranged, as eqs. (6b) and (9b), to give an equation for the Galvani voltage in the transition region of pH since the overall surface equilibrium, eq. (11), is purely chemical in nature and not directly connected to any potential difference.

These obviously mechanistic questions can only be answered by electrode kinetics. Thus, at equilibrium (and e.g. pure pH response) there is a continuous exchange of protons between locally fixed silanol groups at the surface and water molecules in the solution at an equal anodic and cathodic rate. The exchange is enabled by the small concentration of siloxy groups which serve as the necessary vacancies. The quantity describing the exchange is the ionic equilibrium exchange current density,

$$i_{0,H} = \overrightarrow{i_H} = F \overrightarrow{k}_{a,H} \, a_{H_3O^+} \exp(-\beta F\mathcal{E}/RT) = \tag{25}$$

$$= \overleftarrow{i_H} = F \overleftarrow{k}_{c',H} \, c'_{SiOH} \exp\left[(1-\beta)F\mathcal{E}/RT\right]$$

where $\overrightarrow{i_H}$, $\overleftarrow{i_H}$ = cathodic and anodic current densities, respectively; $\overrightarrow{k}_{a,H}$, $\overleftarrow{k}_{c',H}$ = cathodic and anodic rate constants, respectively, and β = symmetry factor. Eq. (25) yields the relationship

$$\mathcal{E} = (RT/F) \ln (\overrightarrow{k}_{a,H}/\overleftarrow{k}_{c',H}) + (RT/F) \ln (a_{H_3O^+}/c'_{SiOH}) \quad , \tag{26}$$

which may be written

$$\mathcal{E} = (RT/F) \ln (\overrightarrow{k}_{a,H}/\overleftarrow{k}_{c',H} \, c'_{SiOH}) + (RT/F) \ln a_{H_3O^+} = \tag{27}$$

$$= \text{const} + (RT/F) \ln a_{H_3O^+}$$

since $c'_{SiO^-} \ll \Sigma c'$ because of electrostatic reasons and, consequently,

$$c'_{SiOH} \cong \Sigma c' = \text{const}.$$

Eq. (27) demonstrates that $\mathcal{E}$ is indeed the function of the hydronium ion activity observed experimentally, eq. (23). A corresponding derivation yields

$$\mathcal{E} = (RT/F) \ln (\overrightarrow{k}_{a,M} \overleftarrow{k}_{c',M} \, c^2_{SiOM}) + (RT/F) \ln a_M^+ =$$

(28)

$$= const + (RT/F) \ln a_M^+$$

for the alkali ion sensitivity. This treatment also justifies the derivation of the Nernst equation, eq. (22), by equating the electrochemical potentials of H_3O^+ in solution and $\equiv$SiOH at the glass surface and neglecting that of the $\equiv$SiO$^-$ groups.

The electrode response in the transition range is characterized by the simultaneous flow of an equilibrium exchange current of each ionic species, $i_{0,H}$ and $i_{0,M}$. The overall surface equilibrium, eqs. (11), (12a,b), merely controls the surface concentrations, c^2_{SiOH} and c^2_{SiOM}, which determine the magnitude of the ionic exchange current densities via the potential difference $\mathcal{E}$ between glass and solution according to eq. (25) and to a corresponding equation for the alkali ion.

The proposed mechanism is supported experimentally by pH and pNa glass electrodes with membranes completely protonated up to at least 5 μm thickness in electric fields (27-29). These electrodes exhibit ideal response to hydrogen, lithium, and sodium ions within the entire pH range as before the protonation. This proves that the ionic species within the glass surface region are neither essential for the pH nor for the pM function of glass electrodes and excludes mechanisms based on an exchange of different ions between glass and solution, e.g. as assumed for the ion exchange theory (17).

4.2 Diffusion Potentials

4.2.1 Diffusion potentials caused by hydrogen/alkali ion interdiffusion. The interdiffusion of ions with different diffusivities causes an (inter-) diffusion potential within the surface layers of the membrane glass (8). Diffusion potentials caused by hydrogen and alkali ions within leached layers are most common since they are observed not only with surfaces completely covered with $\equiv$SiOH groups but also at higher pH if the alkali ion at the surface is sterically hindered to enter the glass. Since, according to Eisenman (8, 13), the diffusion potential is independent of time and concentration profile once the concentrations at the surface are fixed, it must also be independent of pH and pM within the ranges of pure pH and pM function, respectively. Within the transition range, however, the diffusion potential depends on the ionic concentration ratio at the glass surface.

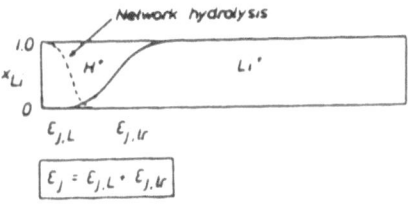

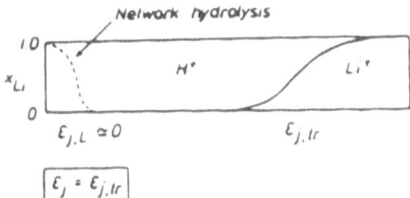

Fig. 14. Diffusion potentials in untreated and protonated glass membrane (schematic).

This could be shown experimentally by pH electrodes with protonated membranes (27). Protonation separates temporally ion exchange from glass hydrolysis and dissolution. The transition layer is displaced to a deeper region in the bulk glass, fig. 14, so that the exchange of lithium ions and protons in the unchanged network is separated from surface-near processes. The difference of electrode potentials, $\Delta\varepsilon$, measured before and after the protonation represents, as a first approximation, the diffusion potential, $\varepsilon_{j,L}$, within the outer part of the leached layer of the nonprotonated membrane. As shown by fig. 15, this difference is indeed constant at pH $<$ (pH$_{tr}$ - 1.5) but changes, and even changes sign, at higher pH values, i.e. in the transition pH region. The linear part of these curves yields

$$\Delta\varepsilon = \varepsilon_{before} - \varepsilon_{after} = \varepsilon_{j,L} = -14 \text{ mV}.$$

The bulk glass is negative with respect to the layer surface, i.e. the electric field is directed into the glass, as is the field within the transition layer, which was concluded earlier (27).

4.2.2 Time-dependent, irreversible diffusion potentials. As shown by figs. 11 and 12, an exchange of alkali ions between glass and solution is observed at pH $>$ (pH$_{tr}$ - 1.5) if the alkali ions are not sterically hindered to enter the glass. This exchange also results in a diffusion potential, which, however, increases during contact with the solution (31). The rate of increase, which slowly decreases with time, is independent of the solution composition as

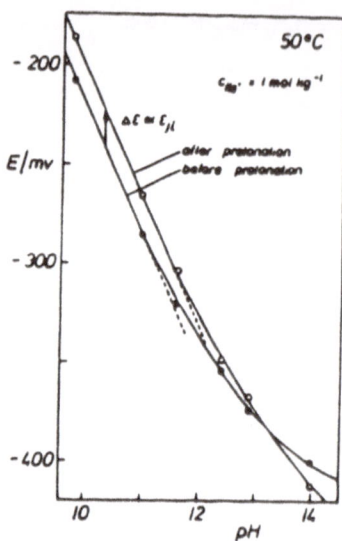

Fig. 15. pH response and sodium error of glass electrode before and after protonation of outer membrane surface.

long as pH $>$ (pH$_{tr}$ + 1.5). At lower pH a reversal of the drift is observed. The original electrode potential, however, is never resumed even if the alkali ion of the glass is offered to the surface for reexchange. Two experimentally important examples were discovered.

1. Fig. 16 presents the pH and pNa response of the pNa electrode glass shown in figs. 11 to 13 and Table 1. As long as the unchanged glass is in contact only with solutions with pH $<$ (pH$_{tr}$ - 1.5), the electrode exhibits pH response (a). On contact with solutions with pH $>$ (pH$_{tr}$ - 1.5), however, a drift to more positive potentials is observed (e.g. + 25 mV within 3 days) whose time course is indepent of the solution composition if pH $>$ (pH$_{tr}$ + 1.5). After the main drift the electrode response to pNa and to pH is displaced, and it is evident from fig. 16 that the diffusion potential was added to the phase boundary potential.

It is concluded that alkali ion-sensitive glass electrodes must be "conditioned" and stored in solutions with the alkali ion to be measured in order to anticipate potential changes caused by interdiffusion and to minimize, at best, emf changes during application. Aqueous o.1 mol kg^{-1} alkali chloride solution recommended (44, 45) has pH $>$ (pH$_{tr}$ + 1.5), which, however, is not the generally known reason for its use. Contact of "conditioned" electrodes with solutions with lower pH should be avoided in order to maintain the appropriate state of the membrane glass.

501

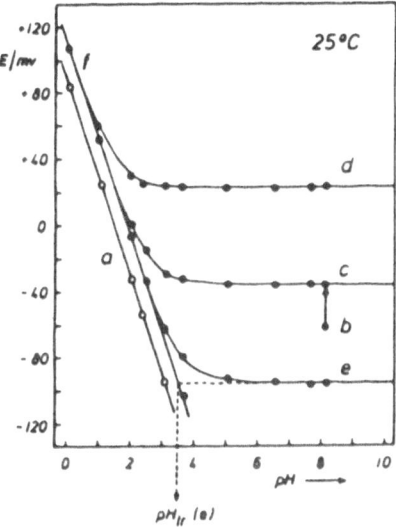

Fig. 16. Effect of sodium uptake at pH > (pH$_{tr}$ - 1.5) upon electrode response of pNa electrode (lithium alumosilicate glass): pH response after leaching fresh membrane at pH < (pH$_{tr}$ - 1.5) (a), potential change during sodium uptake at pH = 8 (b→c), response after Na uptake (f-e; o.oo2 m Na$^+$), (f-d; o.o2 m Na$^+$), (f-c; o.2 m Na$^+$).

  2. The sodium error of pH glass electrodes is reversible only if the alkali ions involved are sterically hindered to enter the glass. This was particularly obvious during measurements on electrodes with two pH membrane glasses (No. 6 and 7) recently patented (46). These glasses impose no sterical hindrance on lithium and sodium ions as shown by concentration profile measurements, and even small amounts of potassium can penetrate the glass. The entire electrode response was shifted to more positive potentials, e.g. by more than + 15o mV by sodium ions within four days at 5o ^{o}C. In solutions outside the alkali error range the potentials drifted to more negative values. These glasses can be used only if solutions with pH > (pH$_{tr}$ - 1.5) are strictly avoided, which, however, is unrealistic for practical measurements.

4.3 The Sodium Error of pH Glass Electrodes

  It is concluded that three different kinds of sodium error of pH glass electrodes can be observed:
1. The reversible sodium error is caused by the participation of alkali ions in surface equilibria. The formation rate is comparable to that of pH response, and the position is symmetrical with respect to pH$_{tr}$.
2. The influence of the reversible sodium error after 1. upon the diffusion potential within the leached layer is observed if the

alkali ions are sterically hindered to penetrate the glass. Quantitative treatment is difficult because of complicated interdiffusion. This effect is unsymmetrical with respect to pH_{tr} , e.g. see fig. 3. 3. The time-dependent, irreversible alkali error is caused by partial or complete coverage of the glass surface with alkali ions which are not sterically hindered to diffuse into the glass. The entire electrode response is shifted, and the original potentials may never be resumed. Acid solutions reverse the potential drift, although not completely. Practical measurements with electrodes showing this effect are strictly confined to $pH < (pH_{tr} - 1.5)$.

REFERENCES

1. Cremer, M., Über die Ursache der elektromotorischen Eigenschaften der Gewebe, Zugleich ein Beitrag zur Lehre von den polyphasi-
   schen Elektrolytketten, Z. Biol., 47 (19o6) 562
2. Haber, F. and Z. Klemensiewicz, Electrical Forces at Phase
   Boundaries, Z. Phys. Chem., 67 (19o9) 385
3. Schwabe, K. and H.D. Suschke, Theorie der Glaselektrode, Angew.
   Chem., 76 (1964) 39-49
4. Schwabe, K. and H. Dahms, Untersuchung des Ionenaustauschs an
   Glaselektroden mit Radioindikatoren, Isotopentechnik,
   1 (196o) 34-39
5. Eisenman, G., The Origin of the Glass-Electrode Potential, in:
   G. Eisenman, ed., Glass Electrodes for Hydrogen and Other
   Cations, Principles and Practice (New York: Dekker, 1967)
   pp. 133-173
6. Boksay, Z., G. Bouquet, and S. Dobos, Diffusion Processes in
   the Surface Layer of Glass, Phys. Chem. Glasses, 8 (1967)
   14o-144; 9 (1968) 69-71
7. Bach, H., Oberflächen- und Dünnschichtanalysen an Glasober-
   flächen und Oberflächenbelägen, Glastechn. Ber., 56 (1983)
   1-18 (Part I), 56 (1983) 29-46 (Part II), 56 (1983) 55-62
   (Part III)
8. Eisenman, G., ed.: Glass Electrodes for Hydrogen and Other
   Cations, Principles and Practice (New York: Dekker, 1967)
9. Johansson, G., Bo Karlberg, and A. Wikby, The Hydrogen-Ion
   Selective Electrode, Talanta, 22 (1975) 953-966
lo. Schwabe, K., pH-Meßtechnik (Dresden: Steinkopff, 1976)
11. Ives, D.J.G., and G.J. Janz, eds.: Reference Electrodes,
    Theorie and Practice (New York, London: Academic Press,
    1961) pp. 179-23o, pp. 127-178
12. Baucke, F.G.K., Thermodynamics of Solid-State Connected Ion-
    Sensitive Membrane Electrodes: The Silver-Silver Chloride
    System, J. Electroanal. Chem., 67 (1976) 277-289 (Part I),
    291-299 (Part II)
13. Fricke, H.K., Zum Problem der pH-Messung bei höheren Tempe-
    raturen, in: E. Schott, ed., Beiträge zur angewandten Glas-
    forschung (Stuttgart: Wissensch. Verl. Ges., 1959) pp. 175-198
14. Fricke, H.K., Eine neue Bezugs- und Ableitelektrode für Glas-
    elektroden-Meßketten, in: DECHEMA Monogr., Vol. 43 (Weinheim:
    Verlag Chemie, 1962) pp. 1-12
15. Baucke, F.G.K., Standard Potentials $(\mathcal{E}^{o'} + \mathcal{E}_j)$ of the
    Thalamid^R Reference Electrode, Hg,Tl(4o wt %)/TlCl(s)/KCl
    (s)//..., in Aqueous Solutions Between 5 and 9o °C, J. Electro-
    anal. Chem., 33 (1971) 135-144
16. Bates, R.G., Determination of pH, Theory and Practice
    (New York, London, Sydney, Toronto: Wiley, 2nd edn., 1973)
    pp. 32-58, 85-86, pp. 251-253, 375-376, pp. 59-133
17. Nicolsky, B.P., Theory of the Glass Electrode, Acta Physico-
    chim. (USSR), 7 (1937) 597-6o2

504

18. Eisenman, G., Cation Selective Glass Electrodes and Their Mode of Operation, Biophys. J., 2 (1962) 259-323
19. British Standard Institute, British Standard 1647, Spezifi- cation for pH Scale, 1961)
2o. DIN 19 266, pH-Messung; Standardpufferlösungen (Berlin: Beuth-Verl., 1979)
21. DIN 19 267, pH-Messung; Technische Pufferlösungen vorzugsweise zur Eichung von technischen pH-Meßanlagen (Berlin: Beuth-Verl., 1978)
22. Covington, A.K., M. Paabo, R.A. Robinson, and R.G. Bates, Use of the Glass Electrode in Deuterium Oxide and the Relation Between the Standardized pD (p_{aD}) Scale and the Operational pH in Heavy Water, Anal. Chem., 4o (1968) 7oo-7o6
23. Baucke, F.G.K., Cation Migration in Electrode Glasses, in: A.R. Cooper and A.H. Heuer, eds., Mass Transport Phenomena in Ceramics (New York: Plenum, 1975) pp. 337-354
24. Bach, H. and F.G.K. Baucke, Investigation of Glasses Using Surface Profiling by Spectrochemical Analysis of Sputter- Induced Radiation: I, Surface Profiling Technique with High- In-Depth Resolution, J. Amer. Ceram. Soc., 65 (1982) 527-533
25. Baucke, F.G.K. and H. Bach, Investigation of Glasses Using Surface Profiling by Spectrochemical Analysis of Sputter- Induced Radiation: II, Field-Driven Formation and Electro- chemical Properties of Protonated Glasses Containing Various Proton Concentrations, J. Amer. Ceram. Soc., 65 (1982) 534-539
26. Bach, H. and F.G.K. Baucke, Investigations of Reactions Between Glasses and Gaseous Phases by Means of Photon Emission Induced During Ion Beam Etching, Phys. Chem. Glasses, 15 (1974) 123-129
27. Baucke, F.G.K., Contribution to the Electrochemistry of pH Glass Electrode Membranes, in: E. Pungor, ed., Ion-Selective Electrodes (Amsterdam, Oxford, New York: Elsevier, 1978) pp. 215-234
28. Baucke, F.G.K., Investigation of Surface Layers, Formed on Glass Electrode Membranes in Aqueous Solutions, by Means of an Ion Sputtering Method, J. Noncryst. Solids, 14 (1974) 13-31
29. Baucke, F.G.K., Investigation of Electrode Glass Membranes: Proposal of a Dissociation Mechanism for pH-Glass Electrodes, J. Noncryst. Solids, 19 (1975) 75-86
3o. Baucke, F.G.K., Simultaneous Transfer of Different Cations Across Anodic Electrolyte Solution-Glass Interfaces in Elec- tric Fields, in: G.H. Frischat, ed., The Physics of Non- Crystalline Solids (Aedermannsdorf, Switzerland: Trans Tech Publications, 1977) pp. 5o3-5o8
31. Baucke, F.G.K., Equilibria of Functional Groups of Glass Surfaces With Cations in Contacting Solutions, in: The Glassy State, Proc. 7th All-Union-Conference, Leningrad, Oct. 13-15, 1981 (Leningrad: Academy of Sciences USSR, 1983), pp. 96-1o8
32. Baucke, F.G.K., Field-Driven Redistribution of "Guest Protons" Within Protonated Silicate Glasses, J. Noncryst. Solids, 4o (198o) 159-169

33. Wikby, A., The Surface Resistance of Glass Electrodes in Acid Solutions, J. Electroanal. Chem., 33 (1971) 145-159

34. Wikby, A., The Surface Resistance of Glass Electrodes in Neutral Solutions, J. Electroanal. Chem., 38 (1972) 429-443

35. Hench, L.L. and D.E. Clark, Physical Chemistry of Glass Surfaces, J. Noncryst. Solids, 28 (1978) 83-1o5

36. Scholze, H., D. Helmreich, and J. Barkardjiev, Untersuchung über das Verhalten von Kalk-Natrongläsern in verdünnten Säuren, Glastechn. Ber., 48 (1975) 237-247

37. Jvanovskaya, J.S., A.A. Belijustin, M.M. Schultz, and T.P. Worabjewa, Sodium Distribution in Surface Layers of Sodium Silicate Glasses by Interaction with Aqueous Solutions, Phys. Chem. Glasses (USSR) 1 (1975) 156-161

38. Smets, B.M.J. and T.P.A. Lommen, SIMS and XPS Investigation of the Leaching of Glasses, Verres Refract., 35 (1981) 84-9o

39. Boksay, Z. and G. Bouquet, The pH-Dependence and an Electrochemical Interpretation of the Dissolution Rate of a Silicate Glass Network, Phys. Chem. Glasses, 21 (198o) 11o-113

4o. Doremus, R.H., Interdiffusion of Hydrogen and Alkali Ions in a Glass Surface, J. Noncryst. Solids, 19 (1975) 137-144

41. Baucke, F.G.K., Transport Properties of Glasses Containing Mobile Original and Anodically Introduced "Guest" Ions, in: XIth International Congr. on Glass (Prague: CVTS-DÛm techniky Praha, 1977) pp. 347-356

42. Wikby, A., The Resistance of the Surface Layers of Glass Electrodes, Phys. Chem. Glasses, 15 (1974) 37-41

43. Eisenman, G., The Electrochemistry of Cation-Sensitive Glass Electrodes, Advan. Anal. Chem. Instr., 4 (1965) 213-369

44. Dutz, H., Eigenschaften einer selektiven Natriumelektrode, Glastechn. Ber., 39 (1966) 139-14o

45. Cammann, K., Das Arbeiten mit ionen-selektiven Elektroden (Berlin, Heidelberg, New York: Springer, 2nd edn., 1977) p. 57

46. U.S. Patent No. 4,o28,196; June 7, 1977

DEVELOPMENT OF AN ELECTROCHROMIC MIRROR

Friedrich G.K. Baucke

Schott Glaswerke, Mainz, Germany

ABSTRACT

Laboratory samples of up to 16o cm^2 large, unstructured, reflecting, all-solid-state, electrochromic devices have been developed.
These mirrors allow the continuous change of the reflectance between
about 5 % and 6o %, the chosen reflectance being constant without
energy supply. A possible future application is an inside and outside
automotive rear view mirror with adjustable reflectance. The basic
electrochromic reaction, different modes of operation, and optical
and electrical properties of e.c. mirrors according to the present
state of development are reported and discussed.

1 INTRODUCTION

Layer systems with variable optical properties, especially with
controllable light absorption, are of interest for displays and for
transparent and reflecting optical devices, and much effort has been
put into the development of new systems alternative to liquid crystals. Most of them are based on electrochemical processes. Thus,
inorganic (1) and organic (2) electrochromic, electrophoretic (3), and
electrolytic (4) systems as well as devices based on electrochemi-
luminescence (5) and on electrocapillarity (6) have been described.
The most promising are inorganic electrochromic (e.c.) layer systems since they are distinguished by a number of favourable properties not shared by other devices. They have a sharp and nearly
angle-independent contrast. Visibility is excellent, and the blue
colour of WO_3-based electrochromic systems mostly investigated is
pleasant. The intensity of coloration is continuously variable, and
supply of energy is needed only for changing the coloration inten-

sity. The operating voltage is below 3 volts, and a threshold voltage is exhibited by certain configurations (7). The response time is sufficiently short for most applications, i.e. some tenths of a second to some seconds, depending on the active area. The operating temperature ranges from approximately -3o ^{o}C to +1oo ^{o}C. Besides, all-solid-state electrochromic devices offer sandwiching between glass plates and thus protection from the environment.

Schott has been involved in research on the inorganic electrochromic effect for some years and has concentrated on reflecting systems (8,9), which could, for instance, be applied as automotive rear view mirrors with adjustable reflectance (1o). This paper presents the basis of transparent and reflective electrochromic devices and reports about e.c. mirrors with adjustable reflectance according to the present state of development. It is noted, however, that the work is still performed on a laboratory scale.

2 THE ELECTROCHROMIC REACTION

Electrochromic optical devices are characterized by a layer between light source and receiver, or observer, whose transparency can be changed by an external electric voltage. The basic "electrochromic reaction" causing the change is the formation of "bronzes" from oxides of polyvalent metals, e.g. tungsten, molybdenum, vanadium, indium, or titanium, with an alkali metal or hydrogen (11-14). Bronzes have been known for as long as 15o years and were investigated in detail by Glemser (15,16). The reaction is represented by the equation

$$WO_3 + x \, Me \longrightarrow Me_x WO_3 \, , \qquad (1)$$

if tungsten trioxide is taken as an example 1). (Me = $\frac{1}{2}$ H$_2$, Li, Na, K, Rb, Cs and o $<$ x $\leq$ 1). As the reaction proceedes and x increases, the yellowish WO$_3$ is transformed into increasingly dark blue tungsten bronze, Me$_x$WO$_3$.

3 APPLICATION TO OPTICAL DEVICES

For applying the reaction, eq. (1), and the intense colour change to optical purposes, WO$_3$ is made a layer of a thin-layer cell such that the cell reaction caused by an external voltage either provides hydrogen to the WO$_3$ or consumes and thus removes it from the oxide. Hydrogen is preferably used as reactant since it guarantees complete reversibility of the reaction and yields the

---

1. Different reaction mechanisms have also been discussed, e.g. reduction of WO$_3$ resulting in WO$_{3-x}$ (17) and a colour centre mechanism (1).

largest reaction rate among the elements applicable. The electrochromic process, eq. (1), can be brought about electrochemically in
two different ways leading to two functionally different constructions of electrochromically variable optical systems.

3.1 Diffusion-Controlled Electrochromic Systems

The amorphous WO_3 layer forms a phase adjacent to an inert
electrode, which generates or consumes the hydrogen. It is, however,
not positioned within the cell and is not subjected to electric
fields created between the electrodes of the thin layer cell by an
external voltage. The WO_3 layer thus participates only indirectly in
the cell reaction in that it stores the hydrogen supplied by the
contacting inert electrode by forming blue hydrogen tungsten bronze.
On reversing the polarity of the external voltage, the electrode consumes the stored hydrogen, and the layer is gradually bleached.
The cell reaction,

$$WO_3 + \frac{x}{2} H_2 \rightleftharpoons H_x WO_3 ,$$
(2)

corresponds to eq. (1), but is restricted by the condition $0 < x \leq 0.3$

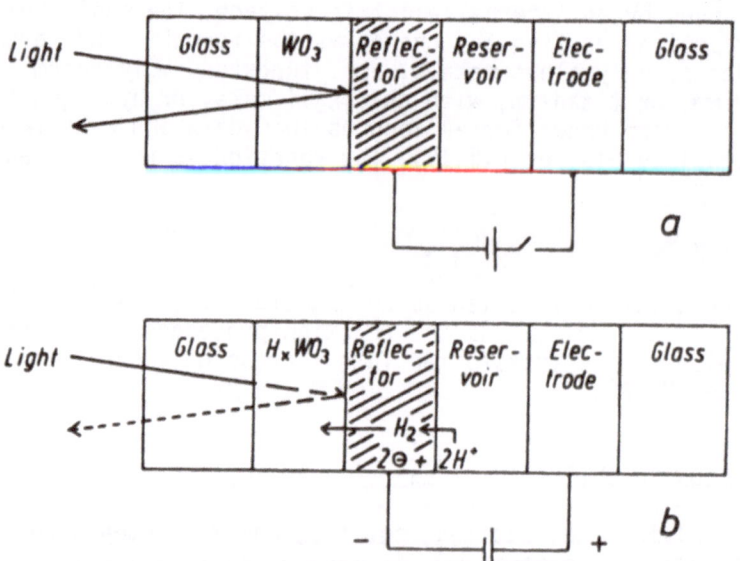

Fig. 1. Scheme of diffusion-controlled, reflective, electrochromic
system. The front electrode functions as reflector. (a) The hydrogen
is stored within the reservoir behind the reflector, reflectance is
high. (b) Hydrogen generated at front electrode diffuses into WO_3
layer and forms blue hydrogen tungsten bronze which reduces the
reflectivity. The system is sandwiched between two glass plates and
protected from the environment.

since the structure of hydrogen tungsten bronze changes at larger hydrogen contents thus damaging the WO_3 layer. Since the uncharged hydrogen enters the WO_3 layer by diffusion, the rate of bronze formation, or removal, can be controlled only indirectly, i.e. by controlling the hydrogen concentration by the rate of its formation or consumption, respectively, at the inert electrode. Such electrochromic layer cells are called "diffusion-controlled" cells.

As an example fig. 1 presents the scheme of a diffusion-controlled, reflective, e.c. cell. In the fully reflecting state the hydrogen content is stored as some compound, e.g. water, in the storing layer (reservoir) behind the reflector, which serves also as the front electrode of the cell, fig. 1a. When the reflecting electrode is made the cathode by applying an external voltage, fig. 1b, hydrogen is generated, diffuses through the metal (grain boundary diffusion) into the WO_3 layer, and causes the formation of tungsten bronze whose colour reduces the reflectance of the system. Changing the polarity of the external voltage reverses the reaction and increases the reflectance of the mirror.

3.2 Field-Controlled Electrochromic Systems

Other than in diffusion-controlled systems, the amorphous WO_3 layer is positioned between a transparent front electrode 2) and an electrolyte layer, thus constitutes a phase of the cell, is subjected to electric fields generated by external voltages, and takes part directly in the cell reaction. Fig. 2a presents the cell scheme of a field-controlled, electrochromic layer system in its fully reflecting state. The hydrogen of the system is stored as a compound, e.g. water, within the storing layer (reservoir) behind the reflector, which does not serve as an electrode but floats electrically. An external voltage, which makes the transparent ITO electrode the cathode, fig. 2b, generates hydrogen ions at the positive rear electrode, which, driven by the electric field, migrate through the reflector (probably grain boundary diffusion of intermediately generated hydrogen atoms or molecules) and the electrolyte into the WO_3 layer. Simultaneously, electrons are supplied to the WO_3 layer by the ITO cathode. The formation of hydrogen tungsten bronze thus arises from the separate introduction of hydrogen ions and electrons through opposite surfaces of the WO_3 layer according to

$$x\ e^- + WO_3 + x\ H^+ \rightleftharpoons H_xWO_3 \ . \tag{3}$$

The rate of the reaction can be controlled, within certain limits, directly by the external voltage. Such systems are thus termed "field-driven" or "field-controlled" e.c. systems.

---

2. Thin transparent layers of semiconducting oxides, e.g. SnO_2, or double oxides, e.g. $(In_2O_3)_x(SnO_2)_{1-x}$ = indium tin oxide (ITO).

510

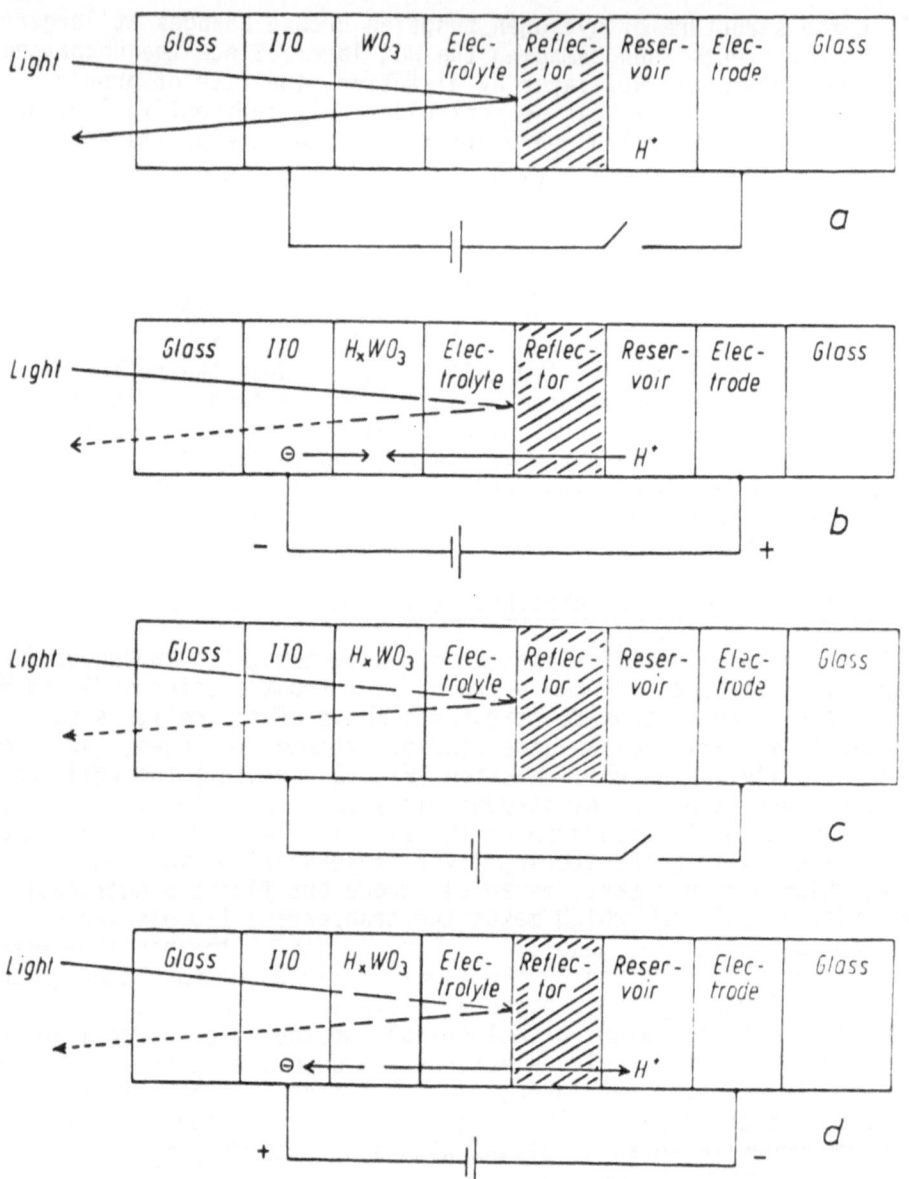

Fig. 2. Scheme of a field-controlled, electrochromic mirror. (a) Reflectance is high while the hydrogen is stored within reservoir behind reflector. (b) Reflectance decreases while hydrogen ions from reservoir and electrons from transparent ITO electrode are driven into the WO_3 layer by external voltage and form blue hydrogen tungsten bronze. (c) Reflectance remains unchanged after external voltage has been turned off. (d) Increase of reflectance on reversing polarity of external voltage, see also (b).

4 THE ELECTROCHEMICAL BATTERY

The total cell reaction, which includes the electrochromic reaction, eq. (2) or eq. (3), is either forced by an external voltage in the direction in which the cell gains energy or proceedes spontaneously under loss of energy when the cell is short-circuited. (An external voltage accelerating the spontaneous reaction, however, is applied in most practical cells). It depends on the specific cell construction which direction is distinguished by energy gain or loss. In either case, however, the reaction can be halted simply by disconnecting the external voltage or by interrupting the short circuit. Electrochromic cells can thus be characterized as secondary thin-layer batteries whose energetic state, or charge, is indicated quantitatively by light absorption. Since the electrochromic reaction, eqs. (2) and (3), is nonstoichiometric and x can attain any number between o and o.3, it follows that 1. the change of colour intensity is continuous, 2. any colour intensity can be selected, and 3. the intensity of coloration is constant in an open circuit cell. This means that any selected colour intensity and thus information are stored without energy supply, an important property of e.c. systems not shared by most other displays, e.g. liquid crystal devices.

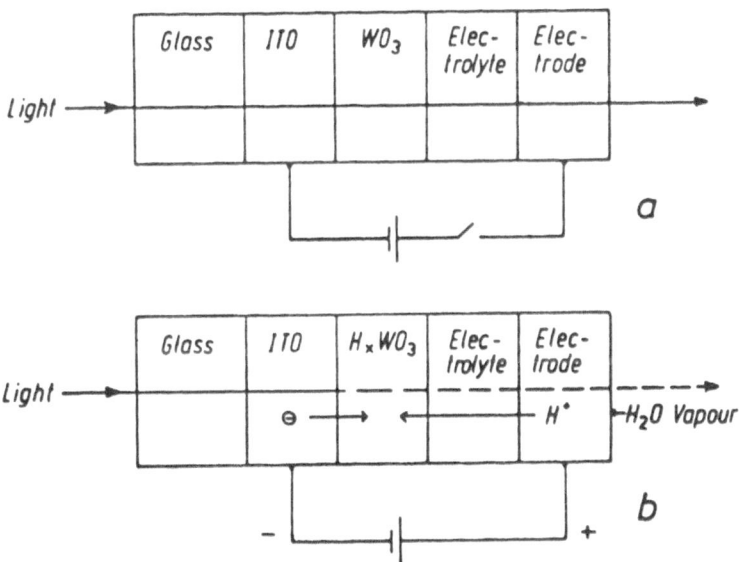

Fig. 3. Scheme of unprotected ("open"), field-driven, transparent, electrochromic system. (a) All layers are transparent. (b) An external voltage generates blue tungsten bronze by injecting electrons from ITO electrode and hydrogen ions generated by electrolysis of water adsorbed at rear electrode into WO_3 layer. The transmittance decreases correspondingly.

512

5 THE ELECTROCHROMIC CELL; DETAILS

5.1 Open Cells

The essential electrode reactions of e.c. cells are the generation, or consumption, of hydrogen in diffusion-controlled systems
and the supply, or removal, of hydrogen ions in field-controlled
devices. The reaction at the counter electrode depends on the cell
construction. In "open" cells (18) whose counter electrode is in
direct contact with the atmosphere adsorbed water is electrolysed.
This results in the formation of oxygen or oxide layers at the
surface of the rear electrode. As an example, fig. 3 shows the cell
scheme of an open, transparent, e.c. system in which water vapour
takes part in the cell reaction. The rear electrode consists of a
thin transparent gold layer. Fig. 4 presents the current-voltage
curve, $i(V)$, and the corresponding transmission-voltage curve, $T(V)$,
measured simultaneously at 5 mv s^{-1} on this device. The minimum
voltage of approximately +1.2 volts for coloration indicates electrolysis of water while bleaching of the system involving regeneration
of water starts at about +o.2 volts. This behaviour is important
for the construction of electrochromic displays.

5.2 Enclosed Cells

Open systems are subject to mechanical and chemical impairment
and require a certain minimum content of atmospheric humidity. Enclosed electrochromic cells have, therefore, been developed. The
devices shown in figs. 1 and 2, for instance, are hermetically
sealed by sandwiching the e.c. systems between two glass plates.
If the cell reaction of enclosed e.c. systems is also based on the

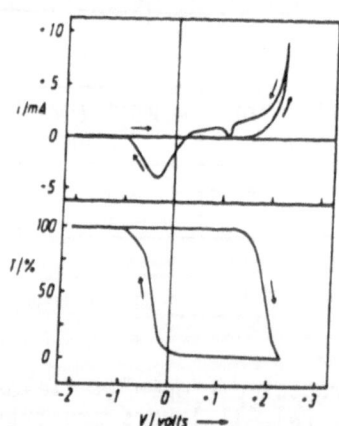

Fig. 4. Current-voltage and transmittance-voltage curves of unprotected ("open") electrochromic system according to fig. 3.
Frequency: 1o^{-3} cps, active area: 6.6 cm^2.

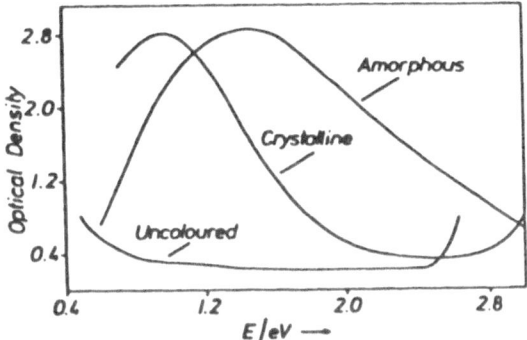

Fig. 5. Absorption spectra of amorphous and crystalline hydrogen tungsten bronzes and of uncolored WO_3.

electrolysis of water, the water must be stored within the system. The devices given in figs. 1 and 2, for instance, contain a reservoir behind the reflector. Oxygen generated during electrolysis must be stored as an oxide layer at the counter electrode since OH^- ions can lead to corrosion of the reservoir. Besides, complete regeneration of water must be guaranteed during the reverse electrolysis, i.e. during bleaching of the e.c. layer (19). Current-voltage and transmission-voltage curves of enclosed e.c. systems involving water electrolysis are similar to those of open systems, fig. 4, and show that also enclosed systems are suitable for constructing displays.

5.3 Hydrogen-Storing Layers (Reservoirs)

WO_3 layers can also be utilized as reservoirs for hydrogen (1o). The necessary hydrogen content of the devices must be supplied before the cells are sealed. In reflecting devices coloration of these layers during hydrogen storage does not interfere with the e.c. function as long as the layers are behind the reflector. In transparent systems crystalline WO_3 layers can serve as hydrogen-storing layers since crystalline WO_3 absorbs in the infrared region of the spectrum (2o), fig. 5. The cell reaction of cells with WO_3 reservoir consists of a displacement of hydrogen ions between the optically active and the storing WO_3 layer, and vice versa, and of a corresponding electron shift through the external voltage supply or wire. Such systems represent hydrogen-concentration cells whose current-voltage curves are continuous and show considerable overlapping of the branches for coloration and bleaching, fig. 6.

5.4 Electrolyte Layers

Electrochromic systems can be constructed as all-solid-state devices by introducing solid-electrolyte layers (1o). This is advantageous as solid electrolytes eliminate problems encountered with

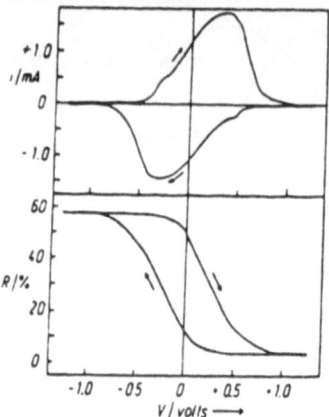

Fig. 6. Current-voltage and reflectance-voltage curves of protected ("enclosed") electrochromic mirror with hydrogen-storing WO_3 reservoir, see fig. 2. External voltage shifts hydrogen ions between the two WO_3 layers. Frequency: 10^{-4} cps; active area $10 \ cm^2$.

cells containing liquid (mixtures of glycerol and concentrated sulfuric acid) or polymer electrolytes (organic polymers with acid groups). Thus, dissolution and recrystallization of dissolved WO_3, caused by temperature changes, as well as spilling of the acid on device damage are excluded. All-solid-state e.c. devices are manufactured completely by vapour deposition or sputtering and have thicknesses between some tenths of a μm and some μm. They can thus be sandwiched between the glass substrate and a cover glass plate and are protected from mechanical and chemical damage. Such glued devices are also non-splintering because of their laminated structure offering a high degree of safety during application to automotive purposes.

6 THE ELECTROCHROMIC LAYER

6.1 Absorbance

The absorptivity of tungsten bronzes is extremely large (5). A WO_3 layer with a thickness of about 500 nm yields a sufficiently low transmittance or reflectance in the coloured state. On maximum coloration this device shows primarily the reflectance of the front glass plate although the bronze content of the layer in the darkened state corresponds only to the formula $H_{0.3}WO_3$; see above.

6.2 Mechanism of Light Absorption

Several mechanisms of light absorption by tungsten bronzes have been discussed. Most probable is intervalence transfer absorp-

tion (21) also found in Prussian blue. It is characterized by the
transfer of an electron localized at a certain metal ion, e.g.
tungsten(VI), to a neighbouring metal ion by photon absorption,
$h\nu$, which enables the crossing of the energy barrier between the
ions,

$$e^- \cdot W(I)^{6+} + W(II)^{6+} + h\nu \longrightarrow W(I)^{6+} + e^- \cdot W(II)^{6+} + E \, . \tag{4}$$

The absorbed photon energy is given off as radiationless energy, E,
after the transfer process (phonon emission).

6.3 The Role of the Mobile Cations

Cations introduced into WO_3 layers during coloration, eqs. (1)
to (3), only balance the charge of the electrons introduced simultaneously. Correspondingly, the spectrum of tungsten bronzes is independent of the cation species present. The charge-balancing alkali
ions can thus be exchanged for other alkali ions without changing
the optical properties of the bronze layer. Because of this possibility, vapour-deposited WO_3 layers have been characterized as
"ion exchangers whose capacity can be electrically controlled" (22).
The mobility of alkali and hydrogen ions in WO_3, however, determines
largely the kinetics of coloration and decoloration. The response
rate of e.c. devices is thus basically limited since the transport
of charge causing the optical effect is linked to the transport of
charge-balancing matter.

7 DEVELOPMENT AT SCHOTT, MAINZ; ELECTROCHROMIC MIRRORS

Two properties, continuously adjustable colour intensity and
storage of a chosen intensity without energy supply, make electrochromic layer systems preferably applicable in transparent and
reflective optical devices. Schott, Mainz, has concentrated on the
development of mirrors since an application as automotive rear view
mirrors with continuously adjustable reflectance can be envisaged.
They would be an entirely new product as no experience concerning
the effect, mode of application, etc. is yet available, except for
the conventional dipping mirror, which, however, offers only extreme
reflectances (8o % and 4 %) and is available only as interior
mirror. Electrochromic rear view mirrors could be adjusted to individual requirements and specific lighting conditions between about
5 % and 6o % reflectance, fig. 7. It is up to future experience
whether the driver chooses the appropriate reflectance in a continuous or discontinuous mode or whether the reflectance is adjusted
automatically.

Enclosed all-solid-state e.c. system (1o) were chosen because of
the advantages described above. The driving voltage is $\pm$ 1.5 volts

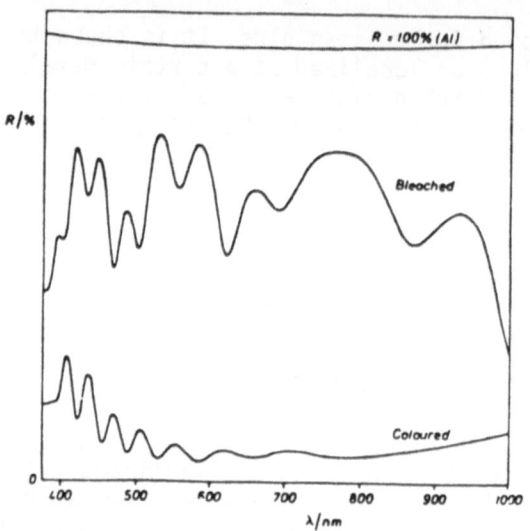

Fig. 7. Spectral reflectance of field-controlled e.c. mirror in bleached (WO_3) and maximum colored ($H_{0.3}WO_3$) state. Interferences are caused by thin layers of the system.

and can be supplied by the battery of the car. The maximum current density, 4 mA cm^{-2}, corresponds to a short maximum current peak of 32o mA for a mirror with 8o cm^2 active area, which lasts less than one second. The charge density for a change of reflectance between 6o % and 1o % and vice versa amounts to approximately 17 mC cm^{-2} corresponding to 1.36 C for a mirror of the size mentioned.

Tests run at Schott indicated that complete dimming is seldom or never necessary and that a reduction to $R \simeq 15$ % or even to $R \simeq 2o$ % is sufficient in almost all traffic situations. The time required for a change from maximum to minimum operating reflectance

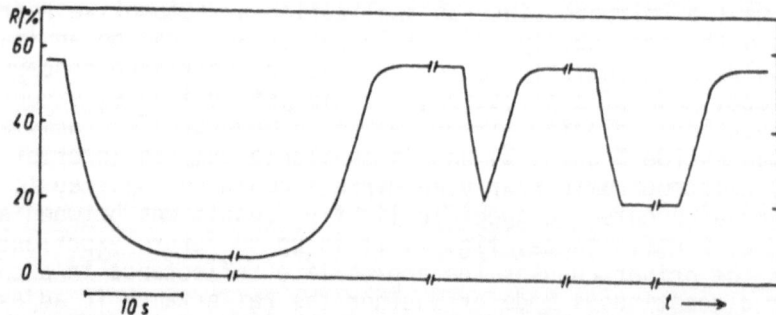

Fig. 8. Reflectance-time plot, R(t), of protected ("enclosed") electrochromic mirror with hydrogen-storing WO_3 reservoir, see fig. 2. Active area: 8o cm^2; driving voltage: 1.5 volts; 25 °C.

and vice versa would thus be about 5 s, according to the present state of development. Fig. 8 shows the time dependence of the reflectance exhibited by a mirror with 8o cm^2 active area driven by $\pm$ 1.5 volts at 25 ^oC. Reaction rates are unproportionately smaller at low reflectances, where, however, different colorations have almost no effect. Coloration and decoloration are faster at high temperatures, e.g. by a factor 2 to 3 at 9o ^oC, compared to 25 ^oC. Slow response at low temperatures can be overcome only by heating. Tests were conducted at -3o ^oC.

All e.c. mirrors tested for cycle life time reached 10^4 cycles without impairment; those tested longer reached also 10^5 cycles without damage. 10^4 cycles were also achieved by mirrors tested at high (+9o ^oC) and low (-3o ^oC) temperatures.

The development of electrochromic mirrors at Schott, Mainz, has not been completed. The results obtained, however, have shown that the choice of enclosed all-solid-state systems was obviously a right step. Such systems also enable the development of other devices, e.g. of large size electrochromic displays.

518

REFERENCES

1. Deb, S.K., Optical and Photoelectric Properties and Colour
   Centres in Thin Films of Tungsten Oxide, Phil. Mag.,
   27 (1973) 8o1-823
2. US Pat. 3,451,741, June 24, 1969
3. Chiang,A., A Matrix-Addressable Electrophoretic Display,
   Electrodisplay, 1981, 1o7-11o
4. Duchene, J., Meyer, R., and Delappiere, G., Electrolytic
   Display, IEEE Trans. in Electron. Devices, ED 26 (1979) 1243
5. Kabayama, M.A., Pighin, A., and Coderre, W.M., SID Symp.
   Digest, 1973
6. Beni, G. and Hackwood, S., Electrowetting Displays, Appl. Phys.
   Lett., 38 (1981) 2o7-2o9
7. Beni, G. and Schiavone, L.M., Matrix-Addressable Electrochromic
   Display Cell, Appl. Phys. Lett., 38 (1981) 593-595
8. Baucke, F.G.K., Beat the Dazzlers, Schott information,
   1 (1983) 11-17
9. DE PS 28 28 332, 9.7.1981
1o. DE OS 3o o8 768.4, Published: 7.3.198o
11. Hersh, H.N., Kramer, W.E., and McGee, J.H., Mechanism of Electro-
    chromism in WO_3, Appl. Phys. Lett., 27 (1975) 646-648
12. Faughnan, B.W., Crandall, R.S., and Lampert, M.A., Model for the
    Bleaching of WO_3 Electrochromic Films by an Electric Field,
    Appl. Phys. Lett., 27 (1975) 275-277
13. Crandall, R.S. and Faughnan, B.W., Dynamics of Coloration of
    Amorphous Electrochromic Films of WO_3 at Low Voltages, Appl.
    Phys. Lett., 28 (1976) 95-97
14. Arnoldussen, T.C., Electrochromism and Photochromism in MoO_3
    Films, J. Electrochem. Soc., 123 (1976) 527-531
15. Glemser, O. and Sauer, H., Über Wolframoxide, Z. anorg. allg.
    Chem., 252 (1943) 144-159
16. Glemser, O. and Naumann, C., Kristallisierte Wolframblauverbin-
    dungen; Wasserstoffanaloga der Wolframbronzen H_xWO_3, Z. anorg.
    allg. Chem., 265 (1951) 288-3o2
17. Chang, I.F., Gilbert, B.I., and Sun, T.I., Electrochromic
    Systems for Display Applications, J. Electrochem. Soc.,
    122 (1975) 955-962
18. US Pat. 3,521,941; July 28, 197o
19. DE OS 31 15 894 A1, Published: 21.o4.1981
2o. Schirmer, O.F., Wittwer, V., Baur, G., and Brandt, G., Depen-
    dence of WO_3 Electrochromic Absorption on Crystallinity,
    J. Electrochem. Soc., 124 (1977) 749-753
21. Faughnan, B.W., Crandall, R.S., and Heyman, P.M., Electro-
    chromism in WO_3 Amorphous Films, RCA Review, 36 (1975) 177-197
22. Pickelmann, L. and Schlotter, P., The Xerogel Structure of
    Thermally Evaporated Tungsten Oxide Layers and its Consequences
    for ECD Application, Eurodisplay, 1981, 87-9o

Round Table Discussion

Ion Transport: Applications and Theories

Leader: Malcolm D. Ingram (University of Aberdeen, Scotland)

Contributors: A.J. Dianoux, P.H. Gaskell, P.G. LeComber,
               S.W. Martin, C.T. Moynihan and D. Ravaine.

APPLICATIONS

The main uses of glasses in electrochemistry are as ion
selective electrodes, as discussed elsewhere by Baucke.   However,
with the discovery in recent years of more highly conductive
glasses, there could be a wider range of applications.

One possible application was highlighted as long ago as 1968
when patents were issued[1,2] for the sodium-sulphur battery which
operates in the range $300^\circ - 350^\circ C$.   Most batteries of this type
now employ the β-alumina solid electrolyte, but one of the original
ideas was to employ a Na^+-ion conducting glass.   This glass would
have to be inert towards moisture and atmospheric CO_2, as well as
to molten Na and S at high temperatures.   It is interesting to see
how a vitreous electrolyte of sufficient conductivity might be
selected for this purpose.

It is widely known that the conductivity of $Na_2O - SiO_2$ glasses
increases with Na_2O content.   But how can the conductivities of
glasses of different systems, e.g. $Na_2O - B_2O_3$ and $Na_2O - Al_2O_3 - SiO_2$ be compared?

In Fig (1) are shown the conductivities of a variety of
glasses, with log σ ($150^\circ C$) being plotted against the calculated
"optical basicity", Λ.   ($\Lambda = X_1 \Lambda_1 + X_2 \Lambda_2 \ldots$, where X_1, $X_2 \ldots$
are equivalent fractions of oxide components, Λ_1, $\Lambda_2 \ldots$ are
basicities assigned to individual oxides from spectroscopic data.[3]
Λ expresses the *electron donor power* of oxygen atoms in the glass.)
Clearly σ ($150^\circ C$) increases with increase in basicity, in fact from
10^{-14} to 10^{-4} S cm^{-1}.   Correspondingly, the activation energy falls

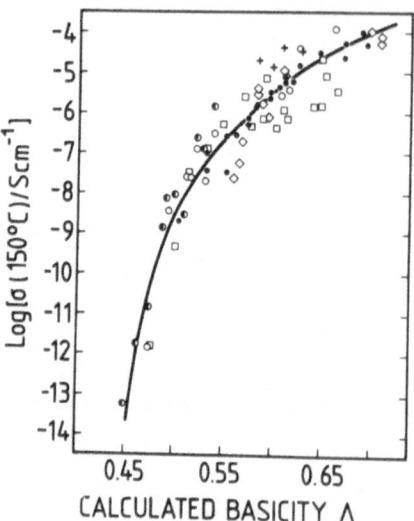

Fig. (1)   Variation of Na⁺-ion conductivity with glass basicity
in various network glasses.   Conductivity data taken
from reference (2);   ◑ , borates; ▱ , aluminoborates;
● , silicates; ○ , borosilicates;
◔ , aluminosilicates;   and ✛ , NASIGLAS compositions.

from ca. 120 kJ mol⁻¹ to 50 kJ mol⁻¹.   Glasses could become
serious rivals to β-alumina solid electrolytes if the activation
energy could be brought down to less than 40 kJ mol⁻¹.

Recent studies[4] of compositional effects over a wide range
of Na₂O contents in sodium aluminoborate glasses are especially
interesting in this context.   At first the conductivity increases
with increasing Na₂O (as expected), but beyond 50 m/o Na₂O the
trend is reversed.   The maximum in conductivity is associated with
a minimum activation energy (55 kJ mol⁻¹).   Probably the
conductivity decrease is associated with an increase in the "salt
like" character of the glass.

So far a really attractive glass for use in Na/S cells has not
been developed but these researches are pointing the way towards a

better understanding of what controls ionic conductivity in glass.
Thus the basicity parameter, Λ, has been shown to correlate with
the molar refractivity of O^{2-} ions.[3]   The increase in
conductivity with increasing Λ is part therefore of a general trend
in solid electrolytes, where cation mobility is strongly influenced
by anion polarisability.[5]   Another way of increasing the anion
polarisability, and therefore the conductivity, is to replace O^{2-}
by S^{2-} ions.   Gabano has shown elsewhere in these proceedings how
this effect may be exploited in solid state batteries employing
vitreous Li_2S . P_2S_5 . LiI electrolytes.   These batteries are
intended for application at lower (i.e. ambient) temperatures, e.g.
in the memory boards of computers, and are at the stage of
commercial development.

THEORIES

These attempts to develop vitreous fast ion conductors have
gone hand in hand with fresh thinking about the mechanisms of ionic
conduction in glass.   The conductivity can be written formally as:

$$\sigma = n . e_o . z . u \qquad (1)$$

where n is the number of charge carriers (mobile ions) per unit
volume, e_o is the electronic charge, z is the valency of the ion,
and u is the corresponding electrical mobility.   But if this
equation is to be useful, we must be able to identify the charge
carriers, and to discover the factors which influence their
mobility.

In this context, two different versions of the "weak
electrolyte theory" deserve mention.   In the first,[6]
thermodynamic arguments are applied to dissociation equilibria such
as:

$$Na_2O \rightleftharpoons ONa^- + Na^+ \qquad (2)$$

where the implication is that the concentration of "mobile" Na^+
ions is much less than the total Na^+ content.   Compositional trends
in conductivity (such as the increase in σ with glass basicity) are
explained by varying concentrations of mobile ions, and the mobility
is assumed to be constant.

In the second approach,[7] comparisons are made with
crystalline electrolytes and the role of *defects* in glass is
emphasised.   It is assumed that glasses contain large numbers of
empty interstitial sites, and equilibrium (2) is interpreted as the
promotion of cations into these positions.   These "point defects"
migrate *via* some kind of interstitialcy mechanism written
schematically as:

$$M_i^{+*} + M \rightarrow M^* + M_i^+ \qquad (3)$$

where only the virtual or "excess" charges associated with interstitial cations are shown for clarity, and the asterisk (*) identifies a particular atom. According to this latter theory, increases in conductivity could reflect increases either in the numbers of these defects or their mobilities.

However, the special case of the mixed alkali effect (where e.g. substituting Na into a K silicate glass causes a sharp, initial fall in conductivity) is understood to arise from a decrease in the concentration of mobile ions. The foreign cations (Na) are immobilised in interstitial positions, thus suppressing the concentration of mobile K_i^+ cations. Moynihan and Lesikar[7] showed how the mixed alkali conductivity isotherms can be analysed to yield the concentrations of mobile ions in the *single alkali* glass. They found that for $K_2O . 3SiO_2$ glass the concentrations of mobile K^+ defects (expressed as fractions of the total K^+ concentration) increased from about 0.5% at room temperature to about 5% at the glass transition.

But what are these defects in glass? Is the point defect model, which has been developed for crystalline systems, appropriate to the vitreous state? To take an extreme position, if glass is said to be highly disordered and "full of defects", why are some defects mobile and not others? At present there are no firm answers to these questions, and there is plenty of scope for speculation. One idea is that alkali silicate and other "network" glasses might contain small ordered domains (> 10 Å across), where the local structure is not very different from that of the corresponding crystal. Conducting defects could arise quite naturally at the interfacial boundaries – which could be considered as analogous to the conduction planes in β-alumina. The adsorption of foreign cations into these interfacial regions could then be the "underlying cause" of the mixed alkali effect.

To develop these ideas further there is clearly a need for independent values either for the mobilities or the concentrations of mobile ions. Progress in amorphous semiconductors (see for example the papers of Owen and LeComber) has relied heavily on data from thermoelectric power, Hall effect, and drift mobility experiments. Even the mobilities of cations in liquid Ne were determined from drift mobility experiments.[8] So far these techniques have not been widely applied to *ionic* glasses, but a variety of spectroscopic methods are now being investigated. For example the diffusive motions of Ag^+ ions in $AgI - AgPO_3$ (1:1) glasses over the range $-143°$ to $+95°C$ are being measured in Grenoble[9] by neutron scattering experiments. The high diffusion coefficients and low activation energy (≈ 9 kJ mol^{-1}) suggest that Ag^+ ions are not hopping from site to site within the glass but that there is a "quasiliquid motion", much as in the "superionic" conductor, α-AgI.

Clearly there is no such thing as a single, clearly perceived conductivity mechanism in glass. The consensus of this discussion is that the field is wide open to new ideas and to the development of new experimental methods.

REFERENCES

1. J.T. Kummer and N. Weber, U.S. Patent No. 3,404,035, 1 Oct. (1968); W.E. Brown, G. Heitz and C.A. Levine, U.S. Patent No. 3,476,602, 4 Nov. (1968).
2. C.C. Hunter and M.D. Ingram, Solid State Ionics (in press).
3. J.A. Duffy and M.D. Ingram, J. Non-Cryst. Solids 21, 373 (1976); J. Chem. Soc. Farad. Trans. I, 74 (1978) 1410.
4. S.W. Martin and C.A. Angell, J. Amer. Ceram. Soc. (in press).
5. R.D. Armstrong, R.S. Bulmer and T. Dickinson, J. Solid State Chem. 8, 219 (1973).
6. D. Ravaine and J.L. Souquet, Phys. Chem. Glasses 18, 27 (1977).
7. C.T. Moynihan and A.V. Lesikar, J. Amer. Ceram. Soc. 64, 40 (1981).
8. R.J. Loveland, P.G. LeComber and W.E. Spear, Phys. Lett. 39A, 225 (1972).
9. A.J. Dianoux, A. Tachez, R. Mercier and J.P. Malugani (unpublished work).

CHAPTER VI: COMPOSITE MATERIALS

THE LONG TERM BEHAVIOUR OF GLASSFIBRE REINFORCED COMPOSITES

B. A. Proctor

Pilkington Brothers p.l.c., R&D Laboratories, Lathom,
Nr. Ormskirk, Lancashire, England.

Synopsis

The major factors governing the strength of glasses are briefly
reviewed in relation to the behaviour of reinforcement fibres.
Examples are given of the strength loss of glass fibres in a range
of environments and attention drawn to the effects of glass
composition, solution pH, and applied stress. The need for careful
and relevant test procedures is emphasised.

The ageing of glass fibre reinforced plastics (GRP) is briefly
discussed and an accelerated ageing procedure for predicting the
long term strength of glass fibre reinforced cement (GRC) materials
is described.

A. F. Wright and J. Dupuy (editors), Glass ... Current Issues. ISBN 978-94-010-8758-2
© 1985, Martinus Nijhoff Publishers, Dordrecht.

1. INTRODUCTION

The most widely held view of the structure of silicate glasses is based on the fundamental unit of a silicon-oxygen tetrahedron in which a central silicon ion is surrounded by four oxygen ions. A continuous but irregular three-dimensional spatial network is built up from these tetrahedra joined corner to corner via a common oxygen ion. Other oxide cations fit into (and create additional) holes in the structure and cause a number of broken linkages in the network which may itself be modified in a few places by the presence of small amounts of other network forming oxides, such as boric oxide.

This rather simplified picture of structure and components of commercial glasses is essentially applicable to all glass fibres used for reinforcement in composite materials.

The -Si-O-Si- chemical bonds are strong and have a significant covalent, directional, character. This, together with the complex irregular structure, gives a mechanical behaviour similar to that of oxide crystals - that is plastic deformation is extremely limited and only occurs at very high stresses (references 1, 2, 3). Glasses are thus brittle materials.

The strength of a brittle materials is governed by the presence, nature and severity of stress raising flaws (Refs.4, 5, 6, 7). This is particularly true of single crystals which lack major structural faults such as grain boundaries, and silicate glasses with their continuous network structures of chemical bonds. In a well made commercial glass or glass fibre internal flaws (cracks, voids, foreign inclusions, devitrification) are rare and the strength is controlled by the condition of the surface.

Common experience of glass articles - windows, tubing, household and laboratory ware - is that glass is both brittle and weak, but also that its (poor) strength is relatively insensitive to quite severe corrosive environments, to heating, to prolonged weathering, or to changes in the composition of the glass. The weakness is due to the presence of a multitude of minute cracks in the glass surface caused by cutting damage at edges of window plates or tubes, and by abrasive scratching damage on the surfaces of other articles. This is most readily demonstrated by the very large increase in strength which can be achieved by hydrofluoric acid etching which rounds out the cracks and reveals their locations (Refs.8, 9). The weakening effect of mechanical abrasion is normally much more severe than the effects of corrosion and weathering - and effectively obscures any effects which these and other conditions may have on the strength.

In the manufacture of continuous filament glass fibres for the reinforcement of composites, great care is taken to protect the surface of the fibre from undue damage during both the production process and prior to incorporation in the composite. Before any mechanical contact with other fibres or guide surfaces, the fibre is coated with a multi-component "size" (Ref.10) which serves to

(i)   protect filaments from abrasive damage by lubrication.
(ii)   bind individual filaments together into convenient bundles or "strands" for subsequent processing and use.
(iii) improve adhesion between the glass fibres and resin matrices in the final composite.

The pristine glass surface is extremely sensitive to abrasion and there is inevitably some reduction in strength from the absolutely untouched value of about 3500 MPa to around 1500 to 2000 MPa (Ref.11). However this still represents a relatively "perfect" glass surface which can be extremely sensitive to corrosive attack leading to roughening and strength loss - which would be quite undetected in low strength glass articles.

This paper surveys the strength loss of different composition glass fibres in various corrosive conditions and relates it to the long term behaviour of fibre reinforced ccomposites.

2. GLASS FIBRE BEHAVIOUR

In studying the strength loss behaviour of glass fibres in corrosive conditions and the subsequent relationship to composite behaviour, it is important to distinguish between two different regimes of exposure under which behaviour may be quite different. These are

1. Exposure in an unstressed state - subsequently referred to as "ageing". Possible deterioration is frequently assessed by a strength measurement after removal of the sample from the test environment where it has been stored for a given period.
and
2. Exposure under stress until failure at a stress lower than the initial strength value - subsequently referred to as "static fatigue" or "stress-corrosion". Here the glass attack may be accelerated or even initiated by the applied stress and its rate is markedly stress level dependant. Degradation in these conditions is common to a number of brittle solids as well as glasses and is assessed by exposure of samples, under stress, to corrosive conditions and observing the time to failure whilst still under load in the test environment.

2.1 Ageing of Glass Fibres

A borosilicate glass composition known as E-glass (Table 1) is now almost universally used for the reinforcement of plastics.

TABLE 1

GLASS FIBRE COMPOSITIONS (% WT)

Composition	E-glass (range)	A.R. glass (range)	C-glass (range)
SiO_2	52 – 56	60 – 70	59 – 64
Al_2O_3	12 – 16	0 – 5	3.5 – 5.5
Fe_2O_3	0 – 0.5		0.1 – 0.3
B_2O_3	8 – 13		6.5 – 7
ZrO_2		15 – 20	
MgO	0 – 6		2.5 – 3.5
CaO	16 – 25	0 – 10	13.5 – 14.5
Na_2O	)		
K_2O	) 1		0.4 – 0.7
Li_2O	)		
TiO_2		0 – 5	
F_2	0 – 1.5		

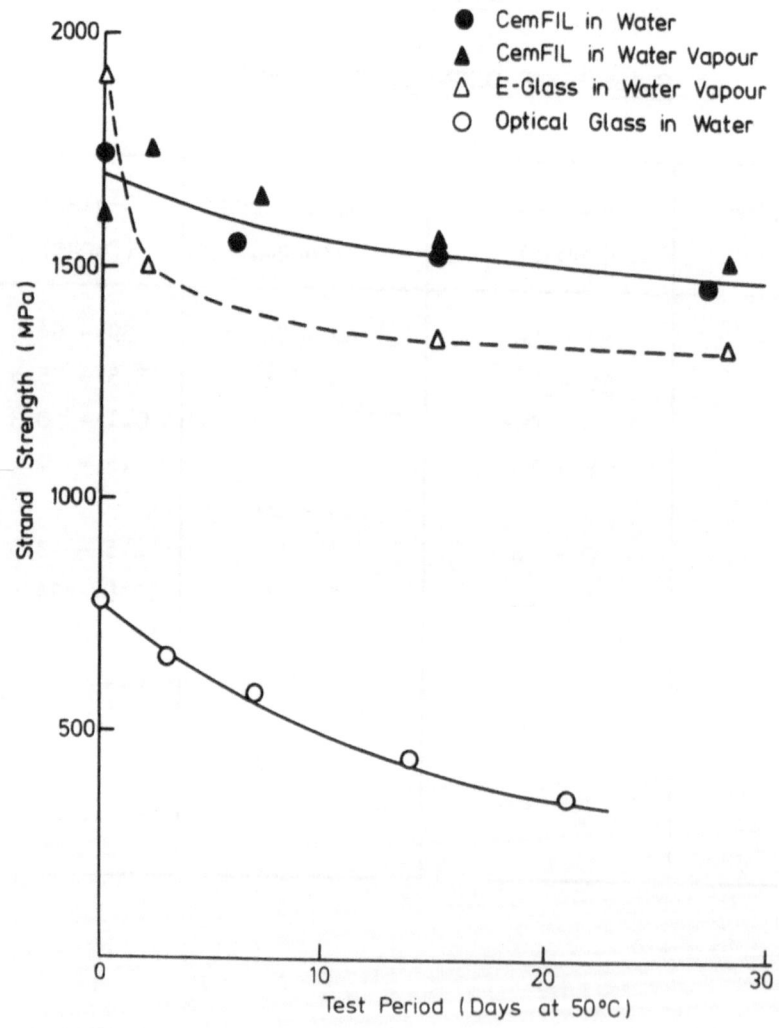

FIGURE 1 STRENGTHS OF GLASSFIBRE STRANDS IN
WATER OR WATER VAPOUR AT 50°C

This was chosen for its low electrical conductivity (an important factor in many early GRP uses) and for its excellent resistance to the most common corrosive element – water. Both Thomas (Ref.12) and Cameron (Ref.13) have reported a drop of some 20% to 25% in strength over about 4 months in 100% RH at room temperature – but this is from the extremely high "untouched" fibre strength level of 3500 GPa. At the slightly lower practical levels of commercial reinforcing strands E-glass shows little strength loss even in accelerated ageing at 50°C (Fig.1). Also shown in Fig.1 are results for the newer Cem-FIL alkali resistant glass (Table 1) developed for cement reinforcement, which again shows excellent strength retention in water.

Not all silicate or borosilicate glasses are so resistant to water however and Fig.1 also contains results for an early crucible drawn optical fibre glass of approximate composition SiO_2 52% (wt), B_2O_3 14%, PbO 16%, $R_2O(R = Na + K)$ 15%. For these experiments this glass was prepared as multifilament strands, similar to the reinforcement fibres, by drawing from a single tip bushing. In this procedure the fibres suffered some surface damage and the initial, unaged, strand strength was rather low at 785 MPa. Thus it may have been expected to have been less sensitive to water attack than the E-glass or Cem-FIL strands with their start strengths of 1600-1800 MPa. It can be seen however that the crucible drawn optical glass suffered further significant strength loss in this test whereas a different type of optical fibre with essentially a pure silica surface, prepared by the Molecular Stuffing Process, showed virtually no strength loss over 3 months in water at 50°C. These results indicate the sensitivity of some glasses to water attack and the wide variation in behaviour between different glass compositions.

Silicate glasses have long been known to be subject to dissolution and/or severe and progressive surface corrosion in strong alkali solutions, the basic silicon–oxygen–silicon network structure suffering disruption from hydroxyl ion attack according to

$$-Si-O-Si- + OH^- \longrightarrow -Si-OH + SiO^- \text{ (in solution)}$$

This results in drastic strength loss for E-glass fibres when aged in strong alkali solutions as shown in Fig.2, and prevented the early development of glass fibre reinforced cement composites based on E-glass. The use of alkali resistant (A.R.) glass compositions containing significant proportions of zirconia (Refs.14, 15 – Table 1) led to considerable improvements in strength retention as shown in Figure 2 for Cem-FIL glass fibre and hence to the development of cement based composites with useful long term strengths (see Section 3.2 below).

530

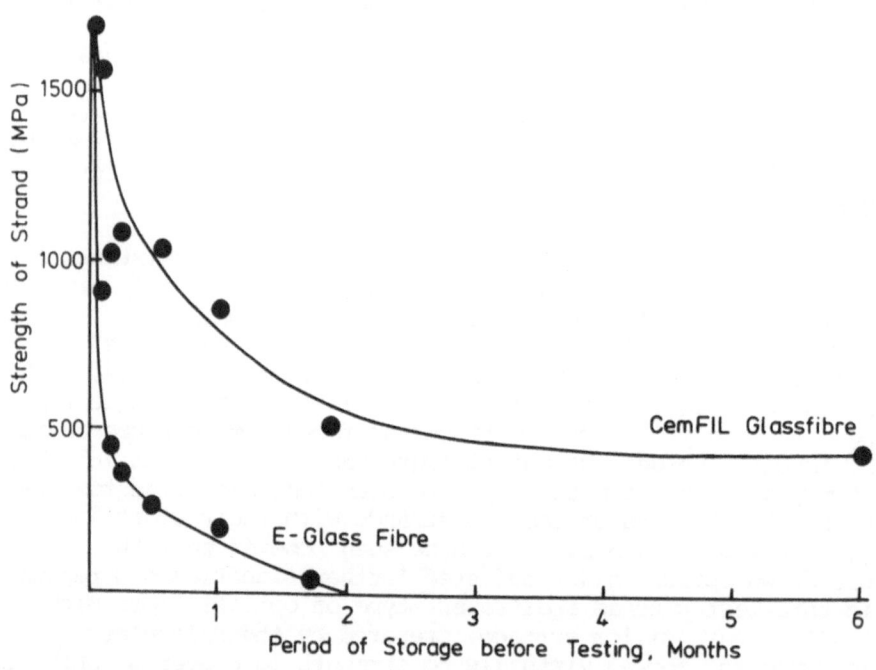

FIGURE 2 STRENGTHS OF GLASS FIBRE STRANDS IN CEMENT EXTRACT SOLUTION AT 50°C

At the other end of the pH range, Metcalfe et al (Ref.16) drew attention to severe strength loss of E-glass fibres in strong acids and attributed this to ion exchange between hydrogren ions in solution and alkali metal ions in the glass which leads to contraction, and hence tensile stresses, in the surface of the fibre.

$$SiONa + H^+ \longrightarrow SiOH + Na^+ \text{ (in solution)}$$

Recent results of strand strength tests after 16 hours exposure at room temperature to 2N H_2SO_4 for standard E-glass fibres and an experimental "alkali free" E-glass were:-

Standard "E"   - Strand strength 63 MPa.
Alkali free "E"- Strand strength 941 MPa.

which would tend to confirm Metcalfe et al's hypothesis. However C-glass tissue (Table 1) has long been used as a surfacing mat for GRP exposed to acid conditions because of its known acid resistance. C glass contains significant amounts of alkali and should be more

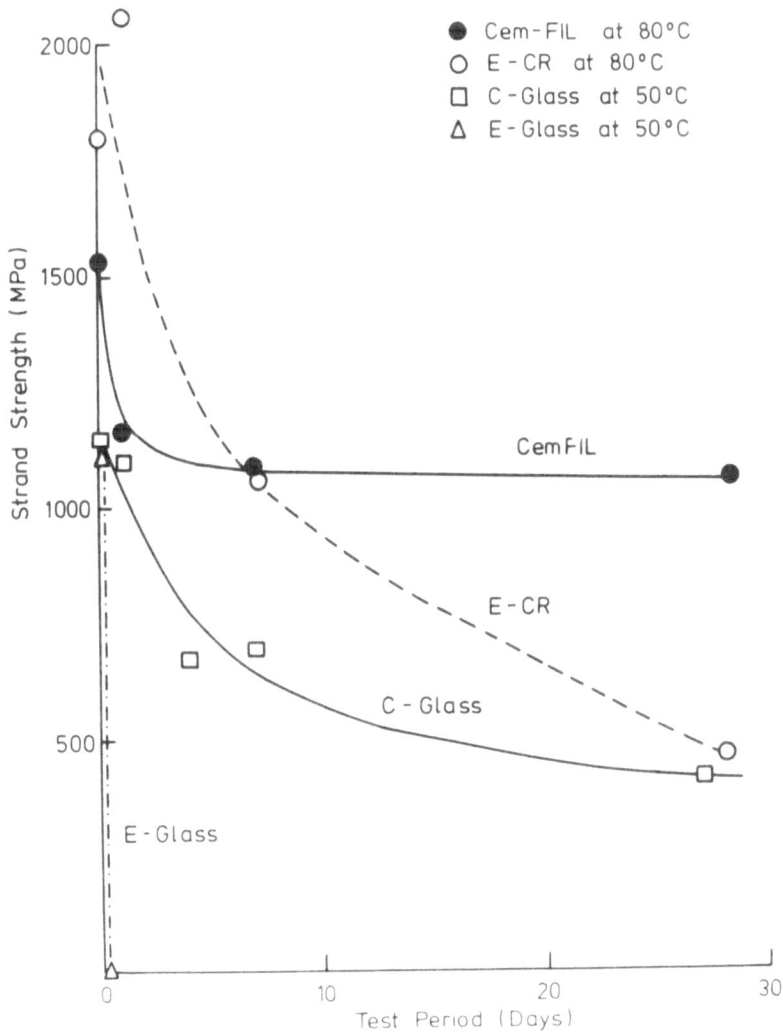

FIGURE 3 STRENGTHS OF GLASSFIBRE STRANDS IN N−H₂SO₄ AT 80°C AND 50°C

sensitive to ion—exchange on the Metcalfe model than E—glass. Its strength retention is in fact very much better as shown in Figure 3 so the mechanism of weakening must be rather different. Also shown in Fig.3 are results for Cem—FIL A.R. glassfibre and a new "boron—free E—glass" fibre (E—CR, E—chemically resistant) which is now being sold for use in acid conditions. Both Cem—FIL and E—CR show noticeably better strength retention than C glass in hot acid. Cem—FIL contains significantly more alkali than E—glass (Table 1) and E—CR contains a similar amount to standard E—glass.

It is clear that strength loss of glass fibres in corrosive environments is critically dependant on both the glass composition and the nature of the corrosive environment. In particular Cockram (Ref.17) has drawn attention to the need to test across the full pH range and Scrimshaw (Ref.18) has carried out a detailed investigation of affects of HCl and H_2SO_4 in the range N – 20N, on E glass strength. There is a relatively sharp maximum in the degradation in the range 1 – 10N, with the position of the maximum depending somewhat on acid type and exposure temperature. Fig.4 gives results for general surveys across the pH range for C—glass, E—glass, E—CR glass and Cem—FIL fibre strands showing the much better strength retention of Cem—FIL, E—CR and C glass (in that order) in the acidic region, the susceptibility of all glasses to attack in strong alkali, but nevertheless the overall better strength retention of Cem—FIL glass above about pH 4.

Because of the complexity of strength loss behaviour it is important to assess the durability of glass fibres by means of direct measurements of the relevant property (i.e. strength) in the appropriate environment. Cockram et al (Ref.19) have drawn attention to some of the anomolies which can arise in using general chemical stability tests such as weight loss, or chemical extraction, to rank glasses in order of strength retention. It should also be clear from the above that the relative behaviour of (say) two glasses may change simply by altering the pH (or some other factor) in the ageing environment. Evaluation should be carried out in conditions as close to the practical use environment as it is possible to achieve in laboratory experiments.

2.2 Static Fatigue Behaviour

Any possible corrosive reaction between a glass and its environment may be accelerated or even initiated by an applied stress. The extent of acceleration and/or the initiation (or not) of this reaction are critically dependent on stress level. Classical papers, describing this behaviour in some detail for glasses, have been written by Mould and Southwick (Refs.20—23) but the phenomena may occur with almost any brittle solid and has been reported for a lithium alumino silicate glass ceramic, alumina and silicon carbide (Ref.24).

533

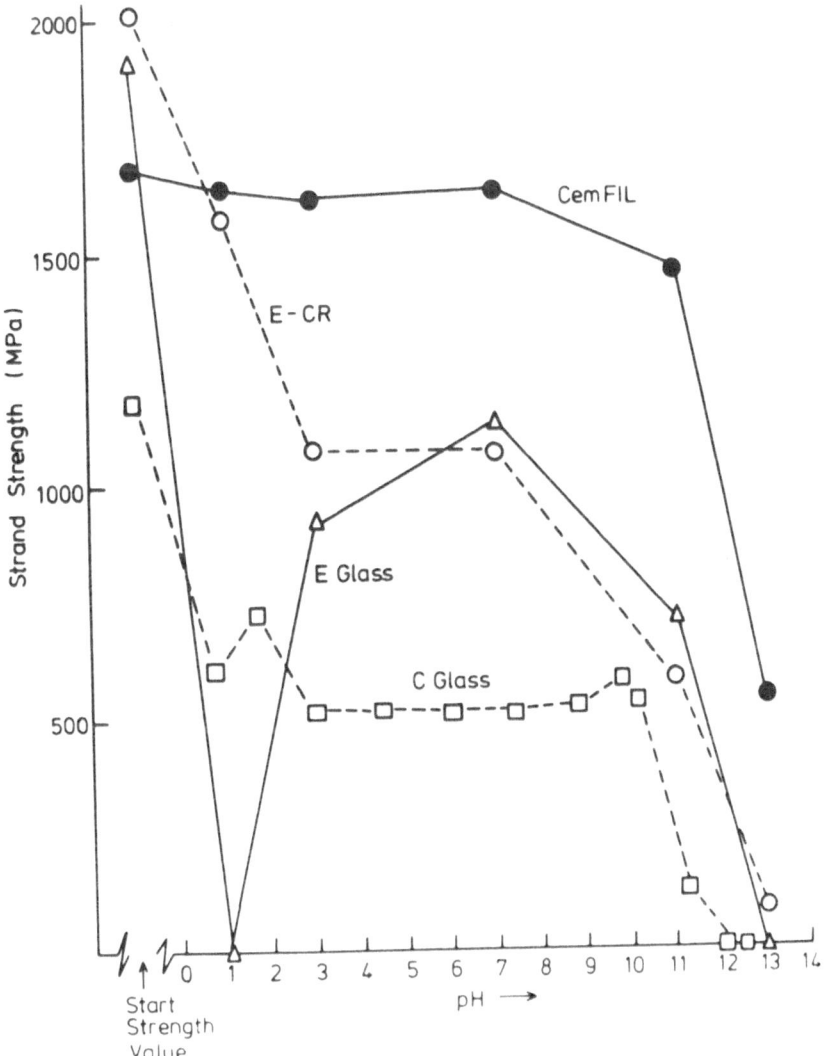

FIGURE 4 STRENGTHS OF GLASSFIBRE STRANDS IN SOLUTIONS OF DIFFERENT pH AFTER IMMERSION FOR 2 DAYS AT 80°C

Static fatigue is normally attributed to stress enhanced corrosion in the highly stressed region at the tip of a crack - lengthening and/or sharpening the crack to make it a more severe stress-raiser - a model first propounded by Charles (Ref.25). Since static fatigue occurs with very similar relative stress/failure time behaviour for high strength fibres (Ref.26) as for badly flawed glasses (Refs.20-23) the exact details of the Charles model may be questioned, but the broad principle of stress enhanced flaw growth/sharpening at even a minor surface blemish is probably correct (Ref.26).

In recent years considerable progress has been made in using fracture mechanics to relate direct observations of the growth of macroscopic cracks (in for example a double cantilever beam specimen) to the static fatigue behaviour of high strength fibres (Refs.29, 30, 31, 32). Weidman and Holloway (Ref.33) and Dabbs et al (Ref.34) have also considered the effects of residual stresses due to plastic deformation on crack growth which may ultimately explain the actual increase in fatigue life observed by Proctor et al (Ref.26) when a high stress fatigue period was immediately followed by the application of a lower stress until failure. This could possibly be related to localised stress relaxation/yield behaviour which may eventually need to be incorporated in the Charles model.

Phenomenologically static fatigue is observed as delayed failure at continually applied stresses below the normal breaking strength and "time to failure" at a given stress level is measured. In contrast with unstressed "ageing" it is not really possible to detect any significant weakening if samples are unloaded prior to failure and then tested normally. The reduction in strength seen when samples are tested at slower rates of loading is also an aspect of static fatigue behaviour - often referred to as dynamic fatigue - and may also be quantitatively related to static fatigue results via fracture mechanics and other approaches (Refs.31, 32, 26).

Published data on the static fatigue behaviour of reinforcement fibres is sparse although it has been quite widely studied and reported in the optical fibre field (Ref.35). The general behaviour is shown in Figure 5 for single high strength silica fibres in ambient conditions (Ref.26). Delayed failure after 1 year under stress occurred at about one half of the short term fracture stress value yet no strength loss on "ageing" without stress was detected over that time period.

Aveston et al (Ref.36) have studied static fatigue using multifilament strands of E-glass and Cem-FIL fibres in air and distilled water at room temperature. In air the failure loads fall from about 6 kg to 3 kg over 1 month, similar behaviour to the

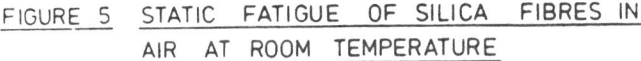

FIGURE 5 STATIC FATIGUE OF SILICA FIBRES IN
AIR AT ROOM TEMPERATURE

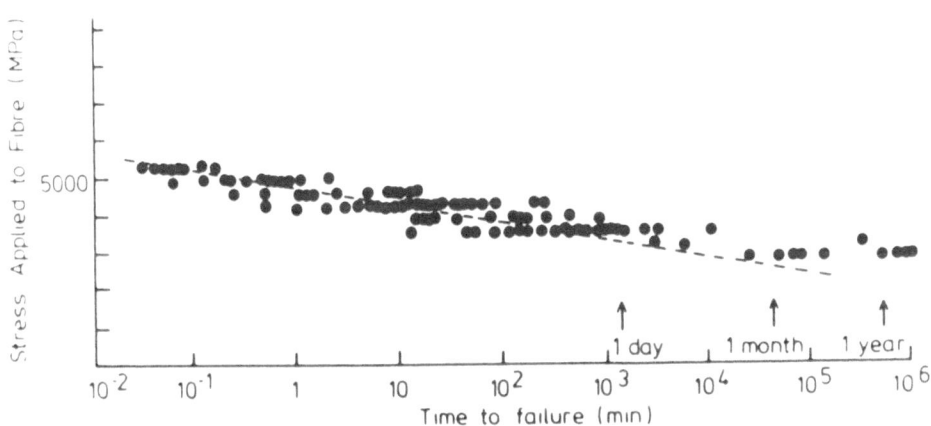

silica fibres shown in Fig.5. The trend lines for results in
distilled water for E-glass and Cem-FIL have been replotted in
Fig.6. The E-glass results are only a little lower than those in
ambient air - the Cem-FIL fibre is seen to sustain a higher
proportion of its initial strength for longer periods.

Hillig (Ref.37) showed that static fatigue was much increased
with fused silica when carried out in an Hydrofluoric Acid
environment. HF is known to attack fused silica. Scrimshaw
(Ref.18) reported static fatigue results for E-glass strands in
$4N$ H_2SO_4 at room temperature and the trend line for these
results has been replotted on Fig.6 which also shows trend lines for
recent preliminary results for Cem-FIL and E-CR strands in
N H_2SO_4. It is difficult to make absolute numerical comparisons
because of unknown differences in start strength and gauge length
for the tests carried out by different workers but it is clear that
both Cem-FIL and E-CR show much better strength retention under
stress in acid than normal E-glass, with Cem-FIL possibly superior
to E-CR. Further evidence of the dependance of static fatigue on
the chemical nature of the environment is given by the results of
tests (Table 2) with a crucible drawn optical fibre glass similar to
that described previously.

The much longer times to failure in alkali are suprising since
this environment would be expected to attack the glass and cause
strength reduction in the absence of stress. Whilst it is generally
true that static fatigue/stress corrosion, i.e. ageing under a
significant applied stress, normally accelerates expected glass

536

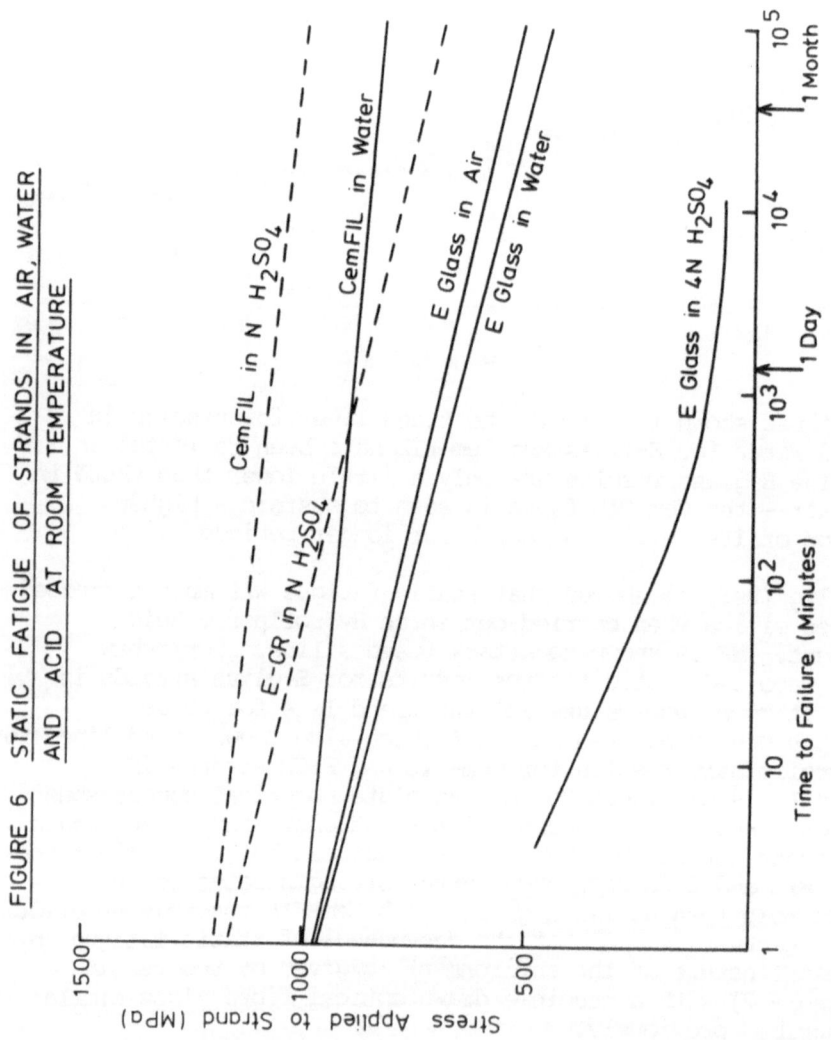

FIGURE 6   STATIC FATIGUE OF STRANDS IN AIR, WATER
AND ACID AT ROOM TEMPERATURE

TABLE 2

Failure times of optical fibres in different pH solutions

Applied Stress Level (MPa)	Mean Time to Failure (minutes) in			
	pH 1.0	pH 5.5	pH 8.5	pH 13
1676	–	1	15	390
1310	48	45		1200
786		1560		21,500

attack mechanisms, there are some occasions when <u>unexpected</u> differences do occur between ageing and fatigue behaviour – and the whole field would repay detailed comparative study as a function of glass composition/environment.

3. COMPOSITE BEHAVIOUR

3.1 Glassfibre Reinforced Plastics (GRP)

The strength and stiffness of fibre reinforced composites are influenced by many factors (Ref.38) in addition to the properties of the reinforcing fibre itself. The strength and stiffness of the resin matrix and the shear strength of the interfacial bond between fibre and resin are particularly important.

Resins may suffer chain scission in U.V. radiation and oxidative damage on heating. Water permeates most plastics relatively easily, resins may swell and hence reduce the bond to the fibre. Ashbee and Wyatt (Ref.39) reported shrinking of resin films and they, Pritchard and Tanega (Ref.40) and Apicella et al (Ref.41) reported loss of low molecular weight components of the resin in water while Ward and Proctor (Ref.42) reported swelling of samples and explained changes in stiffness of composites almost entirely on the basis of changes in resin modulus.

The first glass fibre reinforced plastic materials suffered severe loss of strength on storage in wet conditions due to debonding of the resin from the glass due to diffusion of water to the interface. The development of coupling agents, particularly silanes (Ref.43), which chemically link the glass surface to the resin led to enormous improvements in wet strength retention and enabled GRP to become a practical material for boat construction, water tanks, pipes, etc.

Silanes are included in the size applied to the glass fibre during manufacture (Ref.10). They have the general form RMX$_3$ where R is a reactive organic group, selected for specific reactivity to the type of resin used in the composite e.g. epoxy or polyester, M is a central complex containing silicon, and X's are hydrolyzable groups which link to the glass, first as –Si–OH.HO–Si(glass) linkages in the aqueous application conditions and subsequently (at least in theory) condensing on heating/drying to   –Si–O–Si(glass)   bonds.

Depending on the ageing conditions, nature of resin and type of glass there will still generally be some deterioration of interfacial bonding due to water penetration. Ashbee and Wyatt (Ref.39) suggested an osmotic mechanism. Kasturiarachchi and Pritchard (Ref.44) pointed out that the interfacial environment could contain corrosive material leached from the resin – it should also be realised that leaching can occur from the glass so that even in simple water immersion, conditions at the fibre interface may be more aggressive than first thought and approach the acidic or alkaline environments discussed in Section 2 above.

Because of the complexity of these possible degradation mechanisms there is no simple way of predicting the long term strength of GRP Composites. It is common to make comparative assessment of resins, fibres and composites for long term use in normal damp/wet environments by means of a short hot water or boiling water exposure (Ref.42, 46) and Algra and Van der Beck (Ref.45) suggested that 1 day in boiling water was equivalent to the order of 1000 days weather exposure. However many workers question the validity of such high temperature testing which exceeds, for example the glass transition temperature of most resins.

In long term use of GRP in wet conditions the reinforcing fibre will be operating (at least) in a damp environment and static fatigue must be allowed for as well as possible fibre strength loss due to ageing. Long term design stresses are therefore normally taken as a fraction ($\frac{1}{5}$ to $\frac{1}{15}$) of the short term material strength. Despite the degradation mechanisms discussed above, GRP materials have proved to be valuable additions to the range of corrosion resistant materials and have been increasingly used in aggressive conditions. It is partly because of the extension of use that a new practical problem, that of failure of a number of GRP tanks and pipes when used in acidic conditions, has become apparent (Refs.18, 47, 48, 49).

In a number of instances brittle failures have occurred in E–glass GRP laminates under load in acidic conditions despite the fact that the operating stresses were (at least nominally) well within the design range. This phenomena is still being extensively studied but it appears that cracks or flaws in the resin allowed acid to penetrate to the E–glass fibres with consequent rapid loss in strength as discussed in Section 2. The resin crack then spreads allowing acid to contact further fibres, until catastrophic failure

of the whole laminate results. More complete knowledge of the
strength behaviour of glass reinforcing fibres in a range of
conditions, as discussed in Section 2 above, should allow the
development and use of better composites and better durability
fibres – and lead to the solution of this (and similar future)
problems.

3.2 Glassfibre Reinforced Cement (GRC)

Glass fibre reinforced cement came into existence as a
practical construction material some 13 years ago with the first
commercial production of an alkali resistant glassfibre based on
high zirconia glass compositions (Ref.14, 15). Materials in the
construction industry are expected to perform satisfactorily for
many tens of years. To enable a new material to be used in building
it is often necessary to obtain approval from some "Authority", it
is frequently necessary to demonstrate adequate performance to
prospective owners and it is essential to be able to provide
property data to designers and architects. All this is almost
impossible with a new material and in order to obtain the widespread
use of GRC that has occurred over the last 10–13 years it was almost
essential to develop some means of accelerating normal ageing
procedures and hence be able to predict the long term strength of
GRC materials.

Fortunately the problems of satisfactorily simulating ageing
were in many ways less complex than for GRP. In any damp condition,
and this covers exposure to most climates, the overall environment
inside the GRC is highly alkaline, dictated by the nature of
Portland cement, and not affected by conditions outside the
composite. The strength of the matrix, the cement, tends to
increase somewhat with time, as does the fibre/matrix bonding. The
long term strength of the composite therefore becomes dependant on
the effect a continuing slow alkali attack has on the strength of
the fibre. It has been found possible to treat this as a simple
chemical reaction, to accelerate it in hot wet conditions and to
predict composite strengths for very long times ahead in different
climates (Refs.50,51,52). The development and justification of
improved durability fibres and composites has also depended heavily
on this accelerated testing (Refs.51, 52).

The accelerated ageing procedure itself, and the establishing
of a quantitative correlation with real weathering behaviour has
been described in some detail in Reference 50. It was based on 3
main areas of work:

(i)   studies of the extent and kinetics of glass fibre strength
        reduction in a cement environment in hot/wet accelerated
        conditions.

(ii)   studies of the strength loss in GRC composites in similar
        accelerated conditions.

(iii)   comparison of the above results with strength loss in GRC
        composites after weathering over several years in different
        climates.

The glass fibre strength changes were studied by means of a
specially developed test (The Strand-in-Cement or SIC test) in which
a short length of typical glassfibre reinforcement strand is
embedded in a small cement block (Fig.7). The direct tensile
strength of the glassfibre strand, after ageing in a water bath, was
measured by pulling on the ends of the strand (Refs.53, 50, 52).
The time taken for the strand strength to fall to a given value
(e.g. 1000, 900, 800 MN/m^2 etc) was taken as a measure of the rate
of the strength controlling cement glass interaction and plotting
the logarithm of this time against the inverse of the absolute
temperature gave a family of parallel straight lines (Fig.8). This
indicated that one reaction mechanism, or certainly only one
activation energy, was applicable across the whole temperature and
strength ranges - suggesting the validity of extrapolating long term
behaviour at low temperatures from data obtained from hot, short
term, tests.

The strength changes for GRC composites in both hot wet
accelerated tests and real weathering exposure is summarised in
Fig.9. The accelerated ageing results from tests after exposure in
water at 50°C, 60°C and 80°C, on the left hand side of Fig.9, show a
consistent pattern of falling strength, where the composite strength
is directly proportional to the SIC strength (Ref.51), followed by
long term stability. The weathering results on the right hand side
of Fig.9, which extend over 10 years in the UK and from 2 to 5 years
in other climates show a parallel initial fall in strength, with no
unexpected changes in pattern due to changes of drying out, wetting,
temperature, freeze-thaw, etc, experienced in real climates.

Analysis of the rate of strength loss in the composites, by the
method of Arrhenius plots used for the SIC test, indicated that the
same activation energy applied to strand strength changes in hot wet
immersion, composite strength changes in similar hot wet accelerated
conditions and, if the mean annual temperature of the climate was
used to define the weathering condition, to composite strength
changes in different climates. This is illustrated in Fig.10 by
means of a normalised type of Arrhenius plot in which the relative
rate of strength loss, across a wide range of strength levels for
both strands and composites, is plotted against 1/T°K. (Ref.50).

From the relative time displacements shown in Fig.10 it is
possible to derive acceleration factors which relate a period of
ageing in an accelerated test to a corrosponding lifetime, in any

FIGURE 7   STRAND - IN - CEMENT (SIC) TEST SPECIMEN

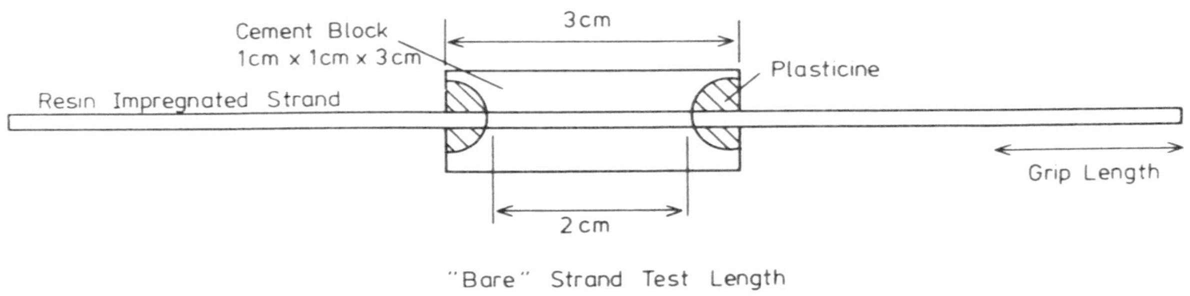

FIGURE 8 RATE OF STRENGTH LOSS (TO
DIFFERENT STRENGTH VALUES)
AS A FUNCTION OF TEMPERATURE

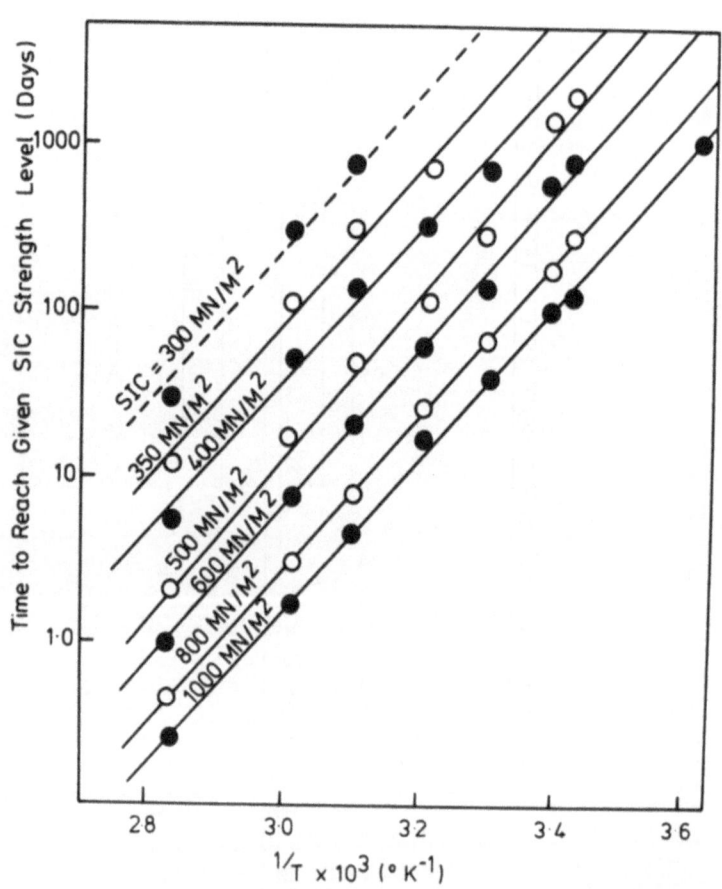

543

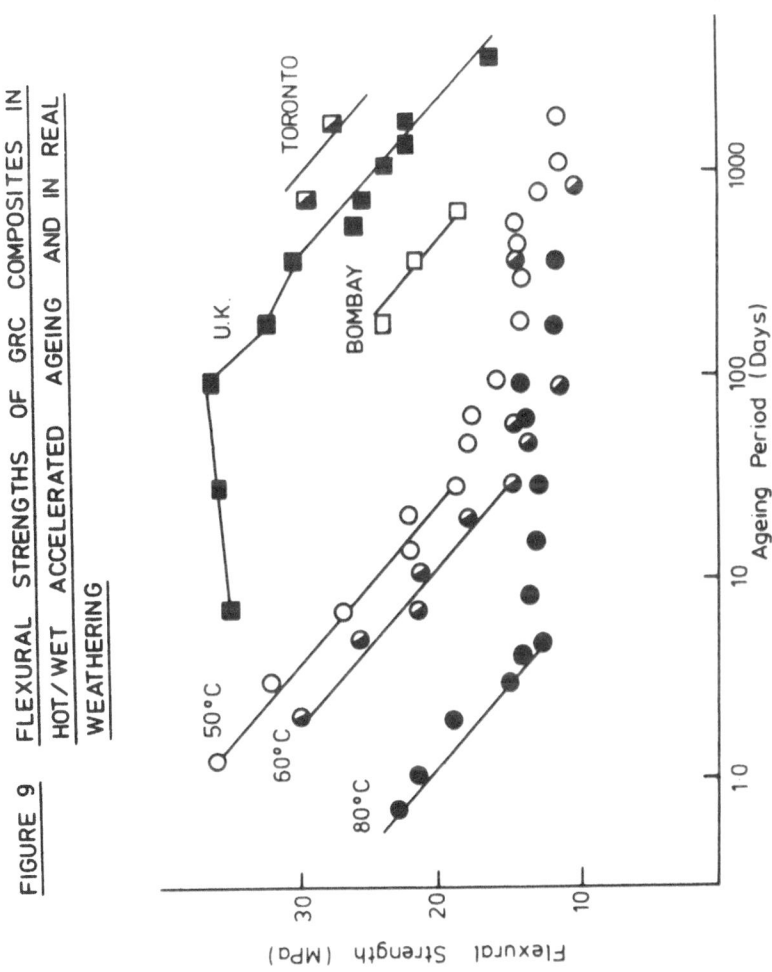

FIGURE 9   FLEXURAL STRENGTHS OF GRC COMPOSITES IN
HOT/WET ACCELERATED AGEING AND IN REAL
WEATHERING

FIGURE 10 <u>NORMALISED RATES OF STRENGTH</u>
<u>LOSS IN ACCELERATED HOT/WET</u>
<u>AGEING CONDITIONS AND IN REAL</u>
<u>WEATHERING FOR SIC's AND</u>
<u>GRC COMPOSITES</u>

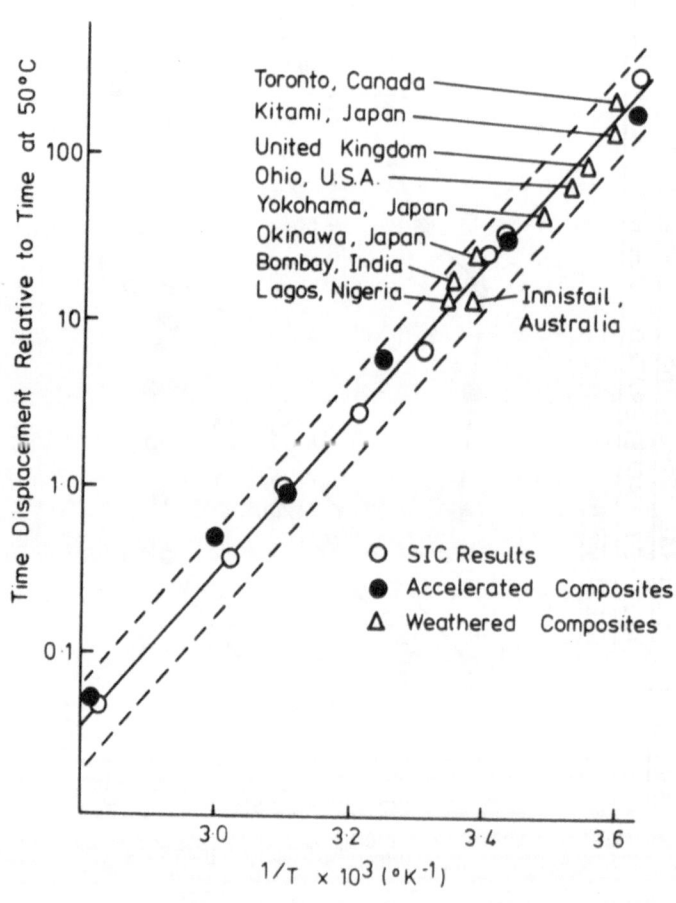

climate, for which the glass fibre strength in the composite will be reduced to the same value (Ref.52). Bearing in mind the fact that neither fibre/matrix bonding nor cement matrix strength are expected to deteriorate, it is thus possible to make a conservative estimate of the strength of the composite over a long period in any climate from the results of accelerated tests. Predictions made in this way for the UK climate are compared with observed results out to 10 years in Fig.11 and the agreement is seen to be very close.

A very considerable concerted effort has been made to establish this accelerated ageing procedure for a standard type of GRC composite. Extensive laboratory testing has been correlated with results from weathering stations set up in many climates world wide. It is now believed that a satisfactory methodology has been developed which can be, and indeed is being, applied to predict the long term performance of many different types of current GRC materials using improved fibres, different fibre contents and orientations, etc. As indicated above the problems of accelerated ageing are somewhat eased for GRC by the simplifications of bond and matrix strength retention, etc. Nonetheless similar concerted effort would probably be repaid by progress in developing accelerated ageing methodologies for other materials.

4. SUMMARY AND CONCLUSIONS

In commonly used glassware the sensitivity of the strength to corrosive conditions is usually hidden by the effects of surface abrasive damage. Reinforcement glass fibres have relatively undamaged surfaces and often lose strength on exposure to acid, alkalis or water. This has led to the development of different fibre types based on compositions optimised for particular use conditions.

Changes in glass fibre strength may be sensitive to quite small changes in composition, or vary from one environment to another for the same glass. It is not sufficient to use one general "chemical stability" test to rank the performance of glasses. Their strength retention should be measured specifically in relevant test conditions. It is also important to distinguish between behaviour when exposed under zero stress ("ageing") and under sustained loads ("static fatigue").

Long term behaviour of glass reinforced plastics (GRP) is extremely complex with possible deterioration in the resin (dependant on resin type) and resin fibre bond as well as the fibre. Water will generally diffuse to the fibre but may also leach materials from the resin or the fibre so that the interfacial conditions are difficult to define. Flaws in the composite, or

546

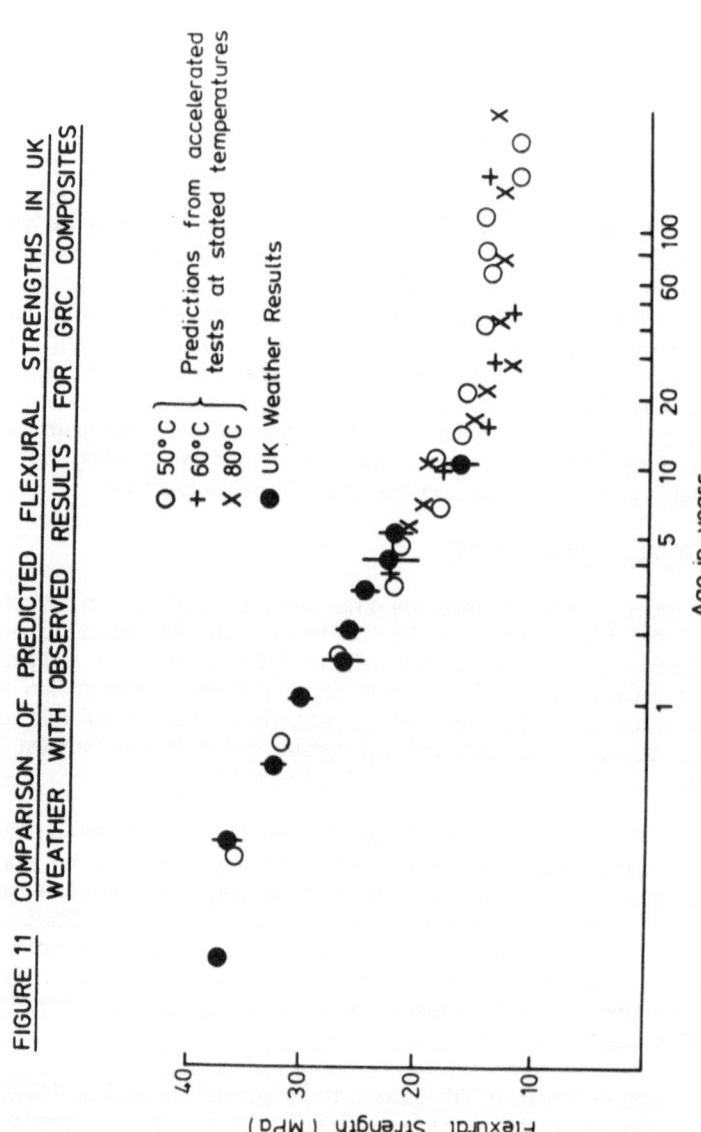

FIGURE 11   COMPARISON OF PREDICTED FLEXURAL STRENGTHS IN UK
WEATHER WITH OBSERVED RESULTS FOR GRC COMPOSITES

O 50°C ⎫ Predictions from accelerated
+ 60°C ⎬ tests at stated temperatures
X 80°C ⎭

● UK Weather Results

Flexural Strength (MPa)

40   30   20   10

Age in years

1   5   10   20   60 100

Figure from: B.A. Proctor:"Properties and Performance of GRC"
In: Fibrous Concrete Proceedings, Concrete International 1980,
published by the Construction Press Ltd. for the Concrete Society

cracks due to stress, may allow the outside environment direct access to the fibre. Accelerated tests are normally comparative and cannot be used to predict long term strengths quantitatively.

In glass fibre reinforced cements (GRC) the internal environment is controlled by the highly alkaline nature of Portland cement. Even alkali resistant fibres lose strength in these conditions but extensive laboratory testing, correlated with observations of GRC behaviour over several years in many climates, has led to the development of a methodology of accelerated ageing by which the expected long term strength in any climate may be confidently predicted.

ACKNOWLEDGEMENT

I am indebted to many colleagues for helpful discussions and previously unpublished information, in particular to D. R. Cockram, K. L. Litherland, D. R. Oakley, B. J. Norman, P. Maguire and R. J. Parry.

This paper is published with the permission of the Directors of Pilkington Brothers P.L.C. and Mr. A. S. Robinson, Director of Group Research and Development.

REFERENCES

1. Marsh D. M., Proc. Roy. Soc. A279, (1964), pp.420-435.

2. Marsh D. M., Proc. Roy. Soc. A282, (1964), pp.33-43.

3. Westbrook J. H. and Jorgensen P. J., Transactions Met. Soc. AIME, 233 (1965), p.425.

4. Dash W. C. in "Growth and Perfection of Crystals", edited by Doremus R. H. and Turnbull D, Wiley, New York, 1958, p.189.

5. Marsh D. M. in "Fracture of Solids" edited by Drucker D. C. and Gilman J. J., Interscience, New York, 1963, pp.119-142.

6. Proctor B. A. in "The Physical Basis of Yield and Fracture", Institute of Physics Conference Series No.1, 1966, pp.218-224.

7. Proctor B. A., Composites, June 1971, pp.85-92.

8. Proctor B. A., Physics and Chemistry of Glasses, 3, (1962), pp.3-27.

9. Wilkinson Betty and Proctor B. A., Physics and Chemistry of Glasses, 3, (1962), pp.203-207.

10. Burns R, Hankin A. G. and Johnson A. E., Eight International Reinforced Plastics Conference, B.P.F., Brighton, October 1972 - Proceedings pp.69-83.

11. Thomas W. F., Glass Technology, 12, No.3 (June 1971), pp.60-64.

12. Thomas W. F., Phys. Chem. Glasses, 1, No.1 (Feb. 1960), pp.4-18.

13. Cameron N. M., Glass Technology, 9, No.5 (Oct. 1968), pp.121-130.

14. Majumdar A. J. and Ryder J. F., Glass Technology, 9, No.3, (1968), pp.78-84.

15. Proctor B. A. and Yale B., Phil. Trans. Roy. Soc., London, A294, (1980), pp.427-436.

16. Metcalfe A. G., Gulden Mary E, and Schmitz G. K., Glass Technology 12, No.1, (February 1971), pp.15-23.

17. Cockram D. R., Glass Technology, 22, No.5, (1981), pp.211-214.

18. Scrimshaw George, Paper 6 in PipeCon - Proceedings Large Diameter Glass Reinforced Plastic Pipes, London, June 1980, published by Fibreglass Ltd. and Amoco Chemicals S.A.

19. Cockram D. R., Litherland K. L., Proctor B. A. and Yale B., "Assessing the Durability of Glass Compositions", XIII International Congress on Glass, Hamburg, July 1983.

20. Mould R. E., J. American Ceram. Soc. 43 (1960), p.160.

21. Mould R. E., J. American Ceram. Soc. 44 (1961), p.481.

22. Mould R. E. and Southwick R. D., J. American Ceram. Soc. 42 (1959), p.542.

23. Mould R. E. and Southwick R. D., J. American Ceram. Soc. 42 (1959), p.582.

24. Multhopp H, Cook R. F. and Lawn B. R., Jnl. Matls. Science, Letters, 2, (1983), pp.683-684.

25. Charles R. J., Jnl. Appl. Physics, 29, (1958), pp.1554-60.

26. Proctor B. A., Whitney I. and Johnson J. W., Proc. Roy. Soc. A297, (1967), pp.534-557.

549

27. Ritter J. E., Phys. Chem. Glasses 11, No.1, (Feb. 1970) pp.16 and 17.

28. Doremus R. H., Jnl. Appl. Phys., 47, No.2, (Feb. 1976), pp.540-543.

29. Evans A. G., Int. Jnl. Fracture, 10, No.2, (June 1974), pp.251-259.

30. Wiederhorn S. M., Fracture Mechanics of Ceramics, Vol.2, edited Bradt R. C., Hasselman D. P. H. and Lange F. F., Plenum Press, New York, 1974, pp.613-646.

31. Kalish D. and Tariyal B. K., J. American Ceram. Soc. 61 No.11-12, (Nov-Dec. 1978), pp.518-523.

32. Ritter J. E., Fiber and Integrated Optics, 1, No.4, (1978), pp.387-399.

33. Weidmann G. W. and Holloway D. G., Phys. Chem. Glasses, 15, No.3, (June 1974), pp.68-75.

34. Dabbs T. P., Lawn B. R. and Kelly P. L., Phys. Chem. Glasses 23, No.2 (April 1982), pp.58-66.

35. Ritter J. E., Jakus K., Cook D. S. - "Predicting Failure of Optical Glass Fibres" - paper to Amer. Ceram. Soc. Annual Meeting, May 1981.

36. Aveston J., Kelly A. and Sillwood J., Proc. 3rd Int. Conference on Composite Materials, Paris, 1980, pp.556-568.

37. Hillig W. B., Compte Rendu, Symposium sur la resistance mecanique du verre et les moyens d'ameliorer, Union Scientifique Continentale du Verre, 1962, p.206.

38. Proctor B. A., Faraday Special Discussions of The Chemical Society, No.2, (1972), pp.63-76.

39. Ashbee K. H. G. and Wyatt R. C., Proc. Roy. Soc., A312 (1969), pp.553-564.

40. Pritchard G. and Taneja N., Composites (Sept. 1973), pp.199-202.

41. Apicella A., Migliaresi C., Nicodemo L., Nicolais L., Iaccorico L. and Roccotell S., Composites (Oct. 1982), pp.406-410.

42. Ward D. and Proctor B. A., Kunstoffe 67 (1977), pp.408-412.

43. Plueddemann E. P., Int. Jnl. Adhesion and Adhesives, (October 1981), pp.305-310.

44. Kasturiarachchi K. A. and Pritchard G., Composites $\underline{4}$, No.3, (July 1983), pp.244-249.

45. Algra E. A. H. and Van der Beek M. H. B., Plastica, $\underline{23}$ (1970), pp.45-55.

46. e.g. BS 3496:1973, BS 3691:1969, BS 3396:1970, etc.

47. Aveston J. and Sillwood J. M., Jnl. Materials Science $\underline{17}$ (1982), pp.3491-3498.

48. Jones F. R., Roch J. W. and Wheatley A. R., Composites $\underline{14}$, No.3, (1983), pp.262-269.

49. Hogg P. J. and Hull D., Metals Science, (1980), pp.441-449.

50. Litherland K. L., Oakley D. R. and Proctor B. A., Cement and Concrete Research, $\underline{11}$, No.3, (1981), pp.455-466.

51. Proctor B. A., Proceedings Int. Congress Glassfibre Reinforced Cement, Paris, November 1981, edited V. Blake, pp.50-67, (The Glass fibre Reinforced Cement Association).

52. Proctor B. A., Oakley D. R. and Litherland K. L., Composites (April 1982), pp.173-179.

53. Litherland K. L., Maguire P. and Proctor B. A., The International Journal of Cement Composites - To be published in 1984.

GLASS CORROSION

L. L. Hench

Department of Materials Science and Engineering
University of Florida, Gainesville, Florida 32611

Studies of soda-lime-silicate (NCS) glasses have led to the identification of at least ten types of corrosion mechanisms (Table I) (1,2). A glass may undergo one or more of the corrosion mechanisms simultaneously or sequentially. The relative contribution of each mechanism to the overall surface change for a given glass is determined by the glass composition and by environmental factors.

When glasses of a given type but different composition behave differently under identical environmental conditions it may be attributed to two possible causes: 1) different mechanisms are occurring and/or 2) the same mechanisms are occurring, but at different rates. The final product of the overall corrosion process is determined by the environment and the reactions' thermodynamics, but the pathways by which the end product is reached are a result of the corrosion mechanisms and the reaction kinetics. The situation is analogous for compositional variation from family to family of glasses, for example NCS to alkali borosilicate (ABS) glasses. However, because the solution chemistry is quite different between different types of glasses, the relative importance of the roles of the various corrosion mechanisms is likely to be altered.

At least eight of the corrosion mechanisms for NCS glasses also have been observed for ABS glasses: ion exchange, congruent dissolution, precipitation, solution concentration, surface layer exfoliation, stable film formation, pitting, and weathering. Closer examination of these mechanisms using data

Table I. Mechanisms of Glass Corrosion

1. Ion Exchange or Selective Leaching
2. Network Dissolution (Congrent and Surface Film Dissolution)
3. Pitting
4. Solution Concentration
5. Precipitation
6. Stable Film Formation
7. Surface Layer Exfoliation
8. Weathering
9. Stress Corrosion
10. Erosion Corrosion

from corrosion experiments with ABS glasses yields a phenomenological basis for a partial theory of ABS glass behavior as
well.

One of the most commonly observed mechanisms for NCS and
ABS glasses is ion exchange (leaching), where selected constituents are extracted from the glass when it is exposed to
water. It is generally accepted that this involves an exchange
between hydrogen or hydronium ions from the solution with mobile
ions such as Na^+ and Ca^{+2} in NCS glasses and Na^+ and B^{3+} from
ABS glasses with low silica content. This reaction results in
the development of a surface layer which is compositionally and
structurally different from the unreacted glass.

Figure 3 in the lecture on Leaching of Nuclear Waste
Glasses in this proceedings is an example of the surface compositional profile resulting from corrosion of an ABS glass with
water. This type of surface is described in the above lecture
as a Type IIIB surface. Three regions are indicated on the
profile. The outer surface corresponding to Region I is
enriched in many of the glass constituents, primarily due to the
precipitation of insoluble compounds. Region I is not expected
to occur on simple ABS glasses corroded in infinitely dilute
solutions (i.e., very low glass surface to solution volume
ratio, SA/V), or in flowing solutions. Region II corresponds to
the depth of ion exchange and many species are depleted with
respect to the unaltered glass (Region III). Selective removal
of the mobile glass constitutents results in the development of
a layer rich in the immobile network formers such as silica
(Region II). Corrosive environments conductive to ion exchange
are low pH (<8), low SA/V ratios, and flowing solutions with
small residence times (i.e., time in which a specific volume of
solution is in contact with the glass). At sufficiently low pH
(<3) all of the species can be leached from the glass surface
with the exception of SiO_2 (1,2).

Scanning electron micrographs (SEM) of corroded glass surfaces often show precipitate formation, surface layer exfoliation and pitting (1). Although there may be several causes for surface layer exfoliation the most common is surface dehydration occurring upon removal of the glass from solution. Dehydration occurs when water is driven out of the glass surface either during drying or upon exposure to vacuum for analysis. The characteristic dehydration cracks extend perpendicularly into the glass surface and run parallel to the surface layer/unaltered glass interface. Exfoliation may occur at this interface or anywhere within the layers that form on the glass surface. The extent of surface cracking and exfoliation appears to be dependent on the extent and depth of ion exchange. Cyclic wetting and drying provide favorable conditions for exfoliation. Pitting, which is due to locallized network dissolution, is seen in the area where the surface layer is missing and is common on all glass surfaces with defects such as polishing scratches or heterogeneities and in phase separated glasses.

Network dissolution is due to the attack of the glass structural bonds by OH^-. Network dissolution is congruent when all constituents are extracted into solution in the same ratio as they are present in the glass. This results in a Type V surface (3) which has the same composition as the bulk glass.

Stable surfaces (stable film formation) can form when additives such as CaO, MgO, and Al_2O_3 are incorporated into the glass either during melting or through solution passivating porcesses (4,5). Stable surfaces can form when ABS glasses are exposed to brine solutions (6). Auger electron spectroscopy (AES) and scanning electron microscopy coupled with energy dispersive spectroscopy (SEM-EDS) have indicated that corroded ABS glass surfaces are rich in Al, suggesting the formation of an alumino-silicate surface layer. Surfaces containing Al_2O_3 are stable in the pH range of 3.0-10.7 (2,7,8).

The extent to which a glass corrodes is dependent on the concentrtion of glass constituents in the solution (solution concentration). As the solution approaches saturation the rate of corrosion decreases. Consequently there is a critical flow rate where corrosion rates reach a maximum due to the absence of solution concentration effects. For ABS nuclear waste glasses this is in the range of 10 ml/hr.

When glass is exposed to an environment containing water vapor and reactive gases such as CO_2 the resulting surface alterations, termed weathering, may be different from those observed when the glass is immersed in water (9). The alterations are usually not homogeneous over the surface. Often water

554

droplets form resulting in localized aqueous corrosion with a very high SA/V ratio. Outside of this area the precipitates are often much richer in Na, indicating that this part of the surface was in contact with water vapor only (2). Since this part of the surface was not in contact with liquid there was no means for the Na precipitate to be removed. Similar behavior has been observed for the NCS glasses (2). The rate of weathering appears to be controlled by the availability of water vapor and reactive gases capable of forming stable salts.

REFERENCES

1. Clark, D.E. and L. L. Hench, Mats. Res. Soc. Symp. Vol. 15 (Elsevier Sci. Pub. Co., New York, 1983).
2. Clark, D.E., C. G. Pantano, Jr. and L. L. Hench, Corrosion of Glass (Magazines for Industry, Inc., New York, 1979).
3. Hench, L.L. and D. E. Clark, J. Non-Crystalline Solids 28 (1978) 83-105.
4. Dilmore, M.F., D. E. Clark and L. L. Hench, Am. Ceram. Soc. Bull. 58 (1979) 1111-1124.
5. Paul, A. and M. S. Zaman, J. Mater. Sci. 12 (1978) 1499-1502.
6. Hermansson, H-P., H. Christensen, D.E. Clark and L. Werme, Effects of Solutions Chemistry and Atmosphere on Leaching of Alkali Borosilicate Glass, in D.G. Brookins, ed., Scientific Basis for Radioactive Waste Management, (Elsevier Sci. Pub. Co., New York, 1983) 143-150.
7. Paul, A. J. Mater. Sic. 12 (1977) 2246-2268.
8. Clark, D.E., C.A. Maurer, A.R. Jurgensen and L. Urwongse, Effects of Waste Composition and Loading on the Chemical Durability of a Borosilicate Glasse, in W. Lutze, ed., Scientific Basis for Nuclear Waste Management, V (Elsevier Science Pub. Co., New York, 1982) 1-14.
9. Chao, Y. and D. E. Clark, Weathering of Binary Alkali Silicate Glasses and Glass Ceramics, proceedings 6th Annual Conference on Composites and Advanced Materials, Am. Ceram. Soc., July 1982.

ALKALI RESISTANT GLASS FIBRES FOR REINFORCEMENT OF CEMENT

B. A. Proctor

Pilkington Brothers p.l.c., R&D Laboratories, Lathom,
Nr. Ormskirk, Lancashire, England.

Synopsis

The potential for glass fibre reinforced cement (GRC) materials was recognised at least as early as the beginnings of glass fibre reinforced plastics (GRP). The fact that practical realisation of cement reinforcement had to wait until about 1970 was due to the fact that silicate and borosilicate glass compositions used for commercial glass fibre manufacture were severely attacked in the highly alkaline environments of common hydraulic cements.

This paper describes the development of glass fibre compositions with a sufficiently high degree of alkali resistance to provide useful reinforcement in Portland cement. The sometimes competing requirements between alkali resistance and the capability for economic large scale fiberising are discussed, together with the various techniques used for evaluating and assessing the performance of alkali resistant fibres in cement.

Some recent developments in glass compositions, described in the patent literature, are reviewed and their effects on alkali resistance considered together with some of the possible mechanisms of their action. Other factors affecting strength retention in cement are also briefly reviewed.

A. F. Wright and J. Dupuy (editors), Glass ... Current Issues. ISBN 978-94-010-8758-2
© *1985, Martinus Nijhoff Publishers, Dordrecht.*

1. INTRODUCTION

The usefulness of, and the large potential market for, glassfibre reinforced cement (GRC) materials had been recognised since the early days of glassfibre reinforced plastics (GRP). The long delay in practical realisation of this concept was due to the fact that the silicon-oxygen-silicon network structure, which forms the skeleton of all conventional silicate and borosilicate glassfibre compositions, is severely attacked in highly alkaline solutions.

$$-Si-O-Si- + OH^- \longrightarrow -Si-OH + SiO^- \text{ (in solution)}$$

All hydraulic cements provide a highly alkaline environment both through the curing/hydration period and when moist thereafter (Ref.1), the commonest, Portland cement, being the most alkaline. Somewhat less alkaline cements, such as High Alumina and Super Sulphate, are not wholly satisfactory as cements and are not widely used. Attempts were made, however, notably by Russian workers in the 1950's (e.g. Ref.2), to use conventional borosilicate E-glass fibres with high alumina cement but these efforts failed and we now know that even this cement attacks E-glass fibres quite drastically (Ref.3).

The real development of practical glassfibre reinforced cement materials began with the work of Nurse, Majumdar and colleagues at the U.K. Building Research Establishment in the 1960's. They realised that a new glass composition would be required to give to glassfibres a high degree of alkali resistance, and hence sufficient strength retention in cement, to provide adequate long term strengths in a GRC composite. The target of an alkali resistant (AR) glassfibre had been identified.

2. DEVELOPMENT OF A PRACTICAL A.R. GLASSFIBRE

Majumdar and co-workers (Refs.1, 4, 5) produced small quantities of continuous filament glass fibre from a single tip fibre-drawing bushing. Diameter changes in alkali solution, single fibre strength measurements and chemical tests were used to evaluate a range of compositions. It had been known since the work of Dimbleby and Turner (Ref.6) that the addition of zirconia to sodium silicate compositions used for bulk glasses conferred alkali resistance, but it was widely believed that such compositions could not be fiberised successfully. Majumdar et al succeeded in producing experimental quantities of a high zirconia content glassfibre (G.20, Table 1) and their fibre strength tests, after immersion in a simulated cement aqueous extract solution at 80°C,

TABLE 1 GLASS COMPOSITIONS (WT%)

Oxide	A-glass (Typical)	E-glass (Typical)	G.20	Cem-FIL	Rockwool (Range of 9 UK Compositions)
SiO_2	73	54	71	62	43.9 – 52.3
Na_2O	13	–	11	14.8	1.5 – 9.9 (includes K_2O)
CaO	8	22	–	5.6	4.3 – 11.6
MgO	4	0.5	–	–	2.1 – 11.1
K_2O	0.5	0.8	–	–	–
Li_2O	–	–	1	–	–
Al_2O_3	1	15	1	0.8	13.3 – 27.7
Fe_2O_3	0.1	0.3	–	–	2.5 – 12.6 (includes FeO)
B_2O_3	–	7	–	–	–
ZrO_2	–	–	16	16.7	–
TiO_2	–	–	–	0.1	0.7 – 2.8
MnO	–	–	–	–	0.05 – 1.0

558

FIGURE 1B G20 A.R. Glass fibres in Portland cement after storage for 10 years in U.K. weather

├─────┤ 10 μm

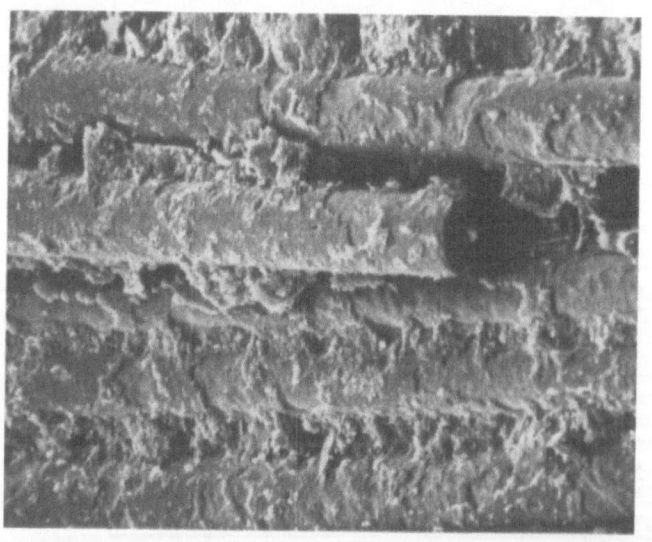

FIGURE 1A E-glass fibres in Portland cement after storage for 2 years in U.K. weather

FIGURE 2 RELATIONSHIP BETWEEN IMPORTANT FIBRE
DRAWING PARAMETERS

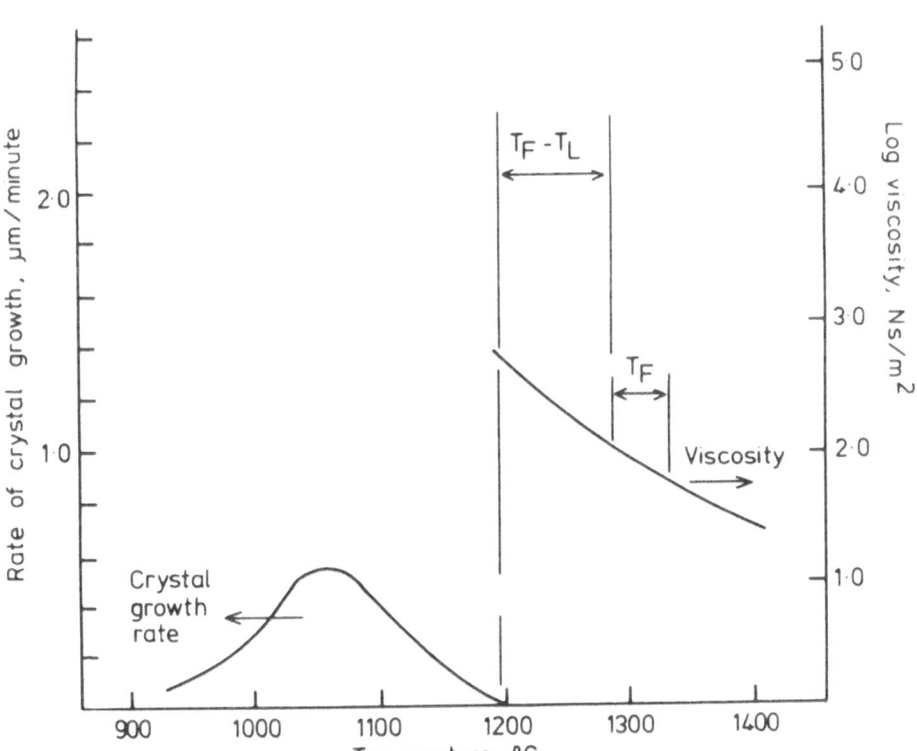

showed this fibre to have significantly better strength retention
than Pyrex which was, in turn, better than E-glass, A-glass
(Table 1) and a group of calcium-magnesium-aluminium-silicate
glasses.

Further evaluation, and development of GRC composite
technology, was concentrated on G.20. It turned out to have
sufficient durability to provide useful reinforcement in Portland
cement. An indication of the very significant improvement in alkali
resistance achieved with G.20 glass is seen in the scanning electron
micrographs of Fig.1.

The requirements for practical, economic, commercial scale
production of glass fibre are very much more severe than those for
small scale laboratory experiments. In large scale production
several hundred individual filaments are drawn simultaneously and

continuously from specially shaped and closely spaced tips in the base of a platinum-rhodium alloy bushing. Breakage of one single filament will cause a drip of glass across the remaining filaments leading to breakdown of the drawing operation for the wholebushing - with drastic effects on production cost. Fibre breakage can be due to mechanical causes - abrasion, snagging; to hydrodynamic instability; and to particles in the fibre from poor melting or devitrification.

To achieve stable fibre drawing the glass viscosity must be in a narrow range close to 100 Nsm^{-2}, which imposes one limitation on the fiberising temperature. To avoid devitrification problems the liquidus temperature (T_L) must be at least 40°C (and preferably 80°C) below the fiberising temperature (T_F).

In addition to these constraints, bushing life decreases rapidly as the fiberising temperature (T_F) exceeds 1300°C, 1320°C to 1350°C being taken as the absolute limit. The interaction of these factors is illustrated in Figure 2. The combination of the required characteristics together with a high degree of alkali resistance is extremely difficult to achieve in a fibre of acceptable final cost for use in a cheap basic construction material such as cement.

G.20 composition had an unacceptably high fiberising temperature of about 1400°C. Thus when Pilkington considered taking a licence on the early BRE work via NRDC the immediate target was to develop the glass composition to give an economically viable full scale production characteristic with minimal compositional change from G.20 and with no loss of alkali durability (Ref.6). This was achieved with the Cem-FIL composition given in Table 1 (Refs.3, 7) as indicated by the strand-in-cement test results (see Section 3 below) given in Figure 3 for test specimens aged at 50°C and 20°C. These results were subsequently confirmed by composite tests which showed that, after weathering exposure in the UK for 10 years, G.20 glass composites had flexural strengths of 18.5 MPa (sd=2.7) whereas similar Cem-FIL glass composites had strengths of 20.4 (1.3), after ageing in water at 18°C-20°C for 10 years composite strengths were: G.20, 18.3 (1.6); Cem-FIL, 19.6 (0.6). These and other test results led to the conclusion that there was no difference in the durability behaviour of the two glass compositions.

3. METHODS OF EVALUATING ALKALI RESISTANCE

Whilst the ultimate criterion for the selection of a glass composition for cement reinforcement was, and still is, satisfactory long term performance in actual cement composites, it was clear that glass development work required a small scale, relatively rapid, test procedure (Refs.3, 8). Majumdar and his colleagues at BRE had used chemical tests, microscopic examination and some single fibre

561

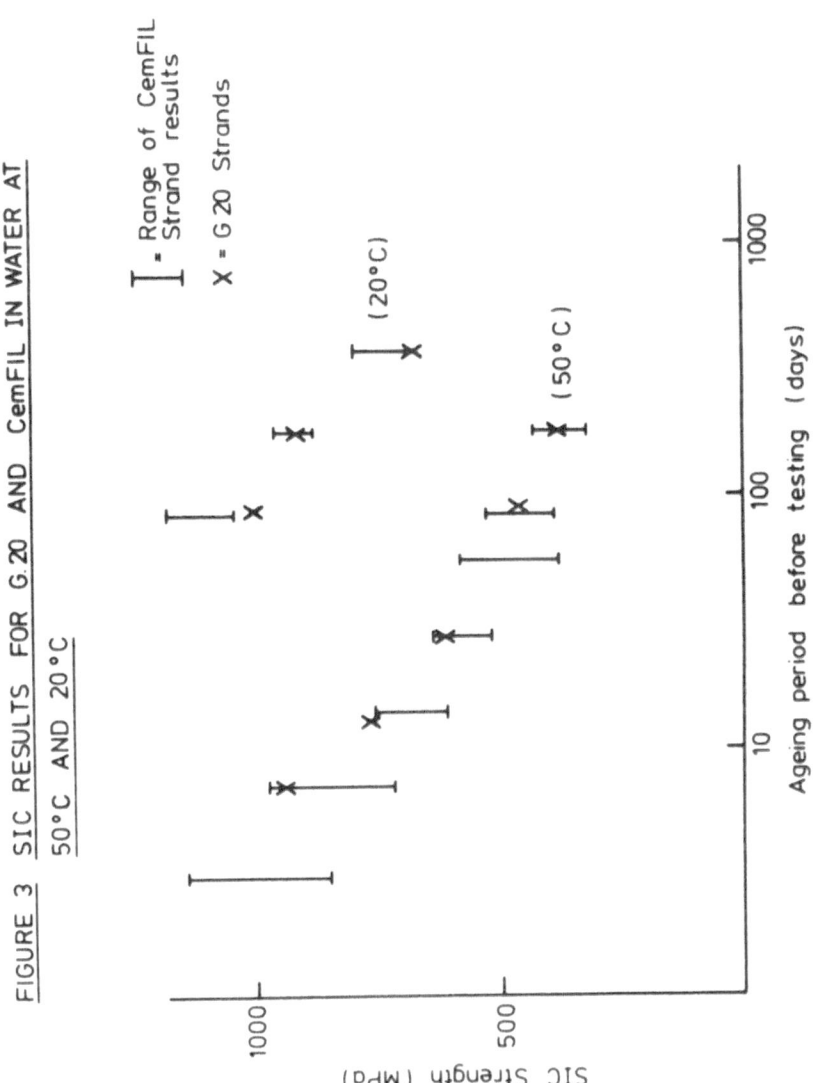

FIGURE 3 SIC RESULTS FOR G20 AND CemFIL IN WATER AT 50°C AND 20°C

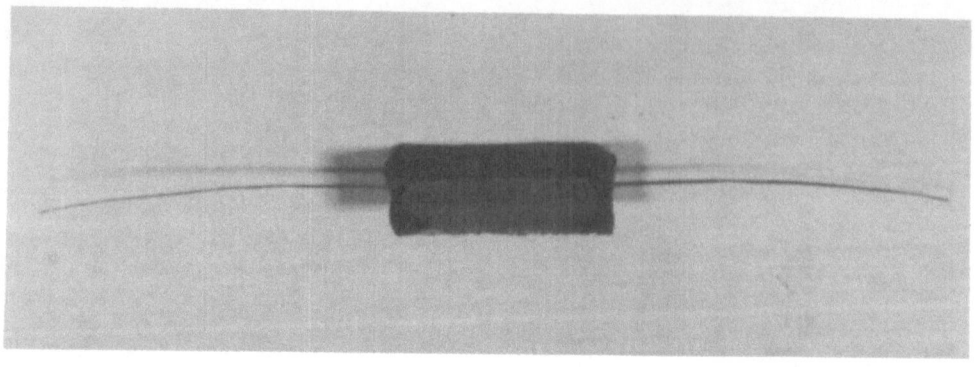

FIGURE 4· Strand of Cem-FIL fibre set into small
cement block for strand-in-cement (SIC) test   1cm

strength measurements (Refs.1, 5, 9) and early Pilkington work was
also based on chemical extraction tests (Ref.3).

However the mechanisms of strength loss in glasses are
extremely complex. They are critically dependant on the existence
of stress concentrations on the glass surface - i.e. on the degree
of smoothness or roughness of the surface (Refs.3, 10) - and hence
are sensitive to the nature, as well as the extent, of surface
attack. Strength changes do not correlate well with the results of
weight loss and chemical extraction tests (Ref.11). Some of the
disparate claims in the literature regarding suitability of glasses
for cement reinforcement may have resulted from reliance on purely
chemical solution type tests.

In addition to these fundamental problems it was desirable that
the test procedure could be conveniently applied to glassfibres in
the normally supplied commercial form of strands or rovings. The
strength retention of glass fibre strands in hot alkaline solution
(Ref.3) provided some useful information but anomalies occurred even
in this test (Refs.3, 11) and the most direct and convenient
evaluation method has proved to be the Strand-in-Cement (SIC) Test
(Ref.12). This test measures the direct tensile load bearing
capacity of a glassfibre strand in an actual hardened cement or
mortar environment (Fig.4) and the effect of changes in the nature
of the cement environment may be assessed as well as differences in
the glassfibre reinforcement itself. The strength controlling
interactions between glass and cement may be accelerated by
immersing the SIC specimens in hot water, changes in SIC strength
correlate directly with changes in the strength of cement composites
in similar accelerated ageing conditions (Fig.5, Ref.13) and the
temperature dependance of these changes also correlates with the

FIGURE 5 <u>RELATIONSHIP BETWEEN FLEXURAL STRENGTH AND SIC STRENGTH IN ACCELERATED TESTS - SPRAY DEWATERED GRC</u> (5% Wt. CemFIL Fibre) (From Reference 13)

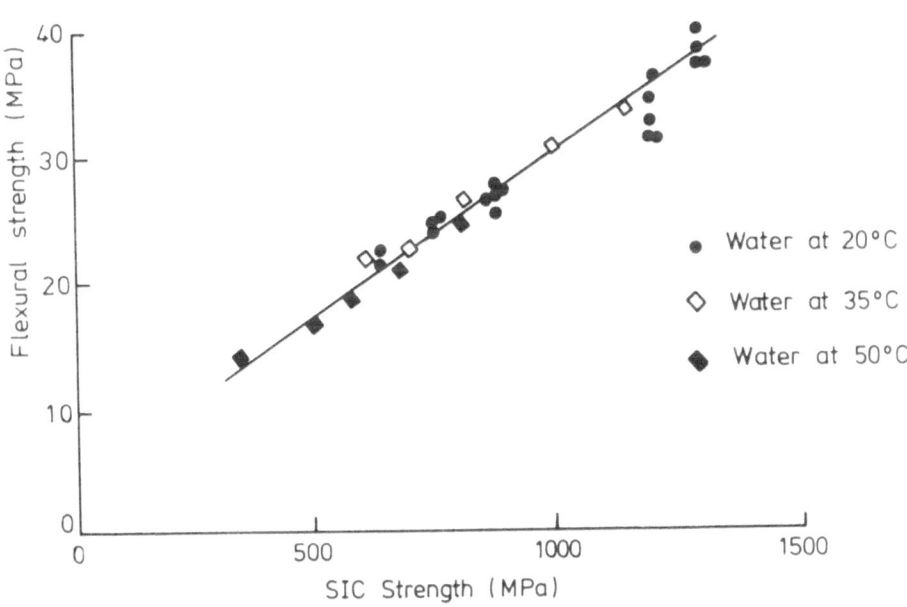

behaviour of composites when exposed to a variety of climates over many years (Refs.10, 13, 14).

This test method has now become our accepted procedure for evaluating and comparing glassfibres for cement reinforcement. Because of the good correlation with actual composite performance accelerated tests carried out over a few days or months enable a prediction to be made of the expected strength of the glassfibre reinforcement in a composite after very many years in a given climate (Refs.10, 13, 14).

4. EFFECTS OF GLASS COMPOSITION

The initial G.20 alkali resistant glass was basically a $Na_2O-ZrO_2-SiO_2$ composition (Table 1). Achievement of practical, commercial scale fiberising characteristics, with no detriment to alkali resistance, was obtained by the addition of some calcia for silica and an increase in soda content to give a four component glass in the system $Na_2O-CaO-ZrO_2-SiO_2$ (Ref.3). These zirconia alkali-resistant glasses have remained, so far, the basis of all successful commercially available cement reinforcement

564

FIGURE 6   EFFECT OF ZIRCONIA CONTENT ON SIC
STRENGTHS IN WATER AT 50°C

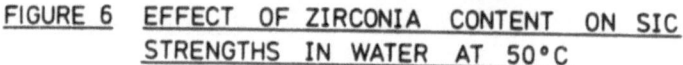

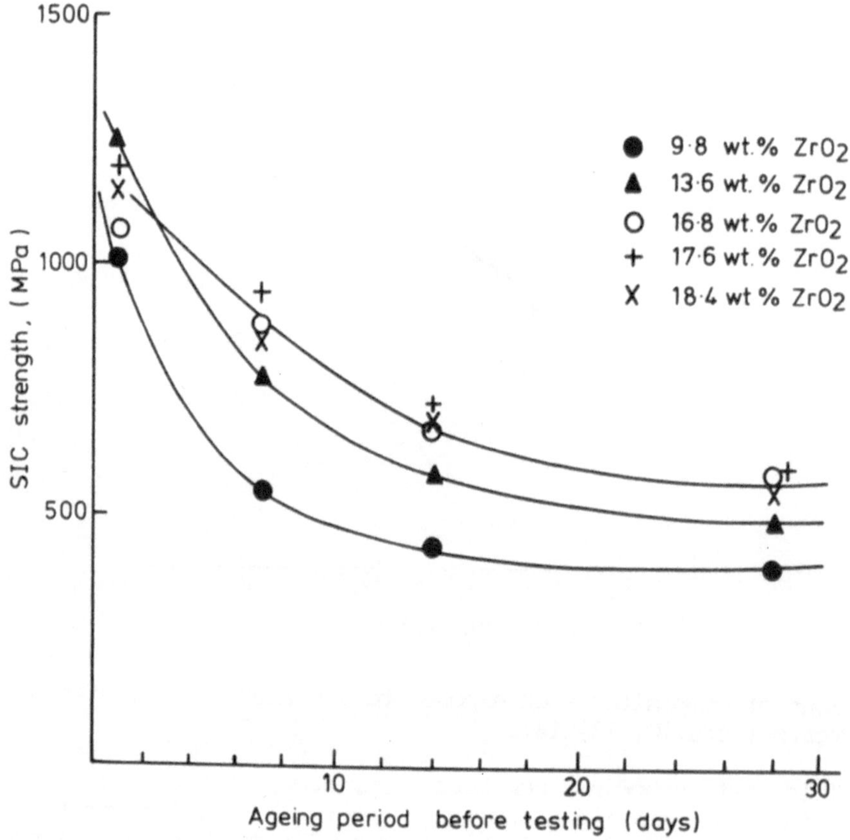

fibres. The alkali resistance and strength retention is essentially
controlled by the zirconia content, as shown in Fig.6, up to about
17% (wt). Above this level there is however less improvement and
the consequent increase in fiberising temperature (T_F) and
liquidus temperature (T_L) lead to significantly higher costs of
fibre. Thus significant further increases in zirconia content do
not appear to be a practical way forward to improved fibre
durability.

Attempts to avoid the fiberising difficulties of higher
zirconia glasses by adding phosphorous and boron actually more than
countered the improvement due to the higher zirconia contents and
gave slightly poorer strength retention (Fig.7) — so even minor
compositional changes must be made with care.

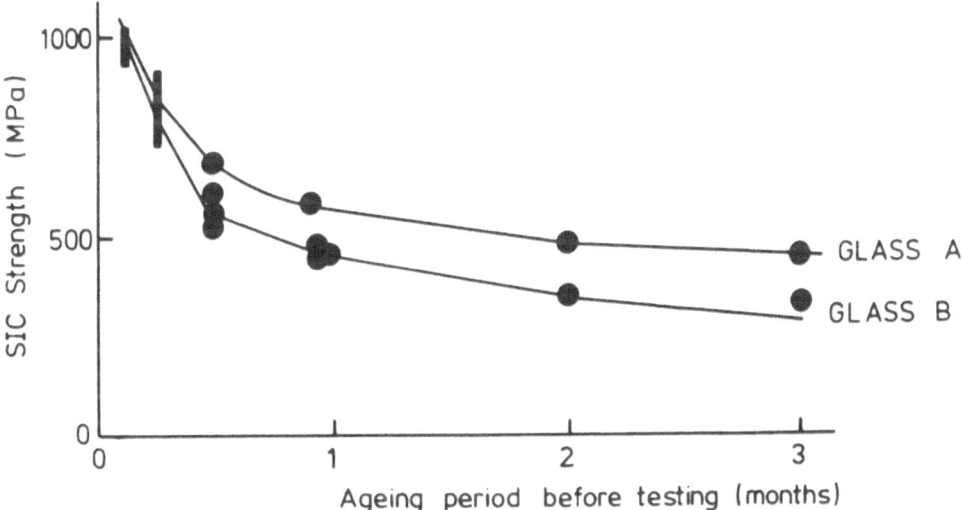

FIGURE 7 SIC STRENGTHS OF TWO ZIRCONIA GLASSES IN WATER AT 50°C; GLASS A 62% SiO_2, 14·8% Na_2O, 5·6% CaO, 16·7% ZrO_2, etc; GLASS B 53% SiO_2, 18% Na_2O, 20% ZrO_2, 5·8% P_2O_5, 1·9% B_2O_3, etc.

During the investigation of the basic four component zirconia containing AR glasses, additions of a number of other oxides were evaluated including Li_2O, K_2O, MgO, SrO, BaO, ZnO, MnO, Fe_2O_3, B_2O_3, La_2O_3, SnO_2 and TiO_2 (Ref.3). None of these gave any marked improvement in durability which might justify the added complexity of a fifth, generally more expensive, constituent. Any attempt to replace significant quantities of ZrO_2 with any of them led to a noticeable reduction in alkali resistance.

Over the last 10-15 years many glass and glassfibre manufacturers have attempted to imitate or improve upon the zirconia alkali resistant glasses G.20 and Cem-FIL. Some measure of the level of activity can be gauged from the fact that over 100 patents have been filed in this area. It is impossible to assess all the suggested compositions but a reasonably comprehensive evaluation suggests that all satisfactory developments are still based upon high zirconia content glasses. Table 2 summarieses the main areas.

TABLE 2   MAIN GLASS COMPOSITION PATENT AREAS

COMPOSITION CHANGE	PATENT EXAMPLE
Rare Earths + Titania	UK 1,460,042
Chromium + Tin Oxides	UK 1,498,917
Thoria	UK 2,046,726A
Trivalent chromium + Titania + Rare Earths	UK 2,071,081A

Additions of rare earth oxides in combination with titania give glasses of good alkali resistance - but without significant improvement over Cem-FIL, and probably at extra cost. Similar conclusions can probably be drawn from the work on addition of chromium plus tin oxides. Significant improvements in strength retention have been obtained from quite low additions of thoria - but problems arising from the level of radioactivity from the thoria make this an unacceptable route for practical application. Perhaps the most promising avenue so far investigated involves the specific use of chromium in the trivalent state in combination with titanium and rare earth oxides. Strand-in-Cement tests indicate that such glasses may give a significant improvement, in the range 2x to 3x for the time taken for strength to decay to any given value, as compared with the original Cem-FIL composition.

However it should be remembered that all the ways of improving durability which have been identified so far carry cost penalties in terms of raw materials, production methods or in a combination of these factors. After some 15 years of alkali resistant glass composition work the basic question is no longer how to make a more durable glass - but rather that of obtaining significant and useful improvements at acceptable economic levels.

In view of the fundamentally higher costs associated with the zirconia based alkali resistant glasses, a number of workers (e.g. Refs.15, 16, 17) have been tempted to investigate glasses in other composition fields. In particular it has been suggested (Refs.16, 17) that glasses of normal rockwool, slate or basalt composition (Table 1) have alkali resistance equivalent to or better than Cem-FIL. Our own assessment of these glasses by means of SIC testing has always shown them to be considerably inferior to Cem-FIL (Fig.8) and, whilst performance may vary somewhat with the different naturally occurring compositions, it is unlikely that any of these compositions will give satisfactory strength retention in cement without significant additions of zirconia.

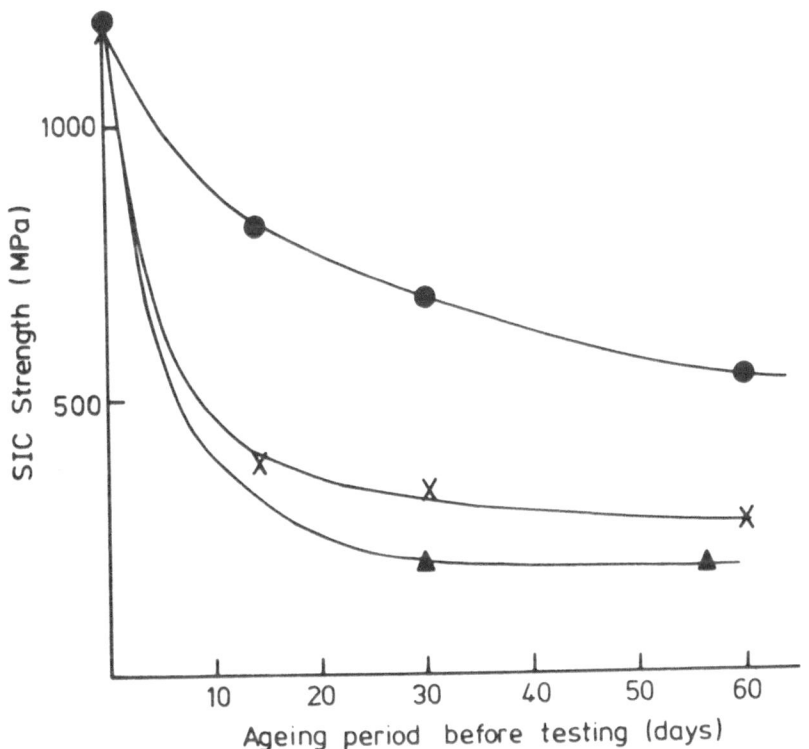

FIGURE 8   SIC COMPARISON IN WATER AT 50°C OF EXPERIMENTAL
CemFIL (●), ROCKWOOL (X), AND
IRON/MANGANESE (▲), GLASSES

Based on an exhaustive survey of the chemical literature, Paul and Youssefi (Ref.15) also sought alkali resistance in potentially cheaper manganese/iron oxide glasses - but again strand in cement strength retention tests showed these glasses to be much inferior to Cem-FIL (Fig.8) confirming the previous comment that high zirconia content glass compositions appear to offer the only presently identified route to satisfactory alkali resistant fibres.

5. OTHER FACTORS AFFECTING DURABILITY

Early workers attempted to avoid the problems of alkali attack on the then available borosilicate E-glass fibres by making composites with high alumina cement. Supersulphate cement is even less alkaline than high alumina, but both are too aggressive for satisfactory use of E-glass borosilicate fibres as well as having limitations as cements in themselves (e.g. Ref.18). However, for

568

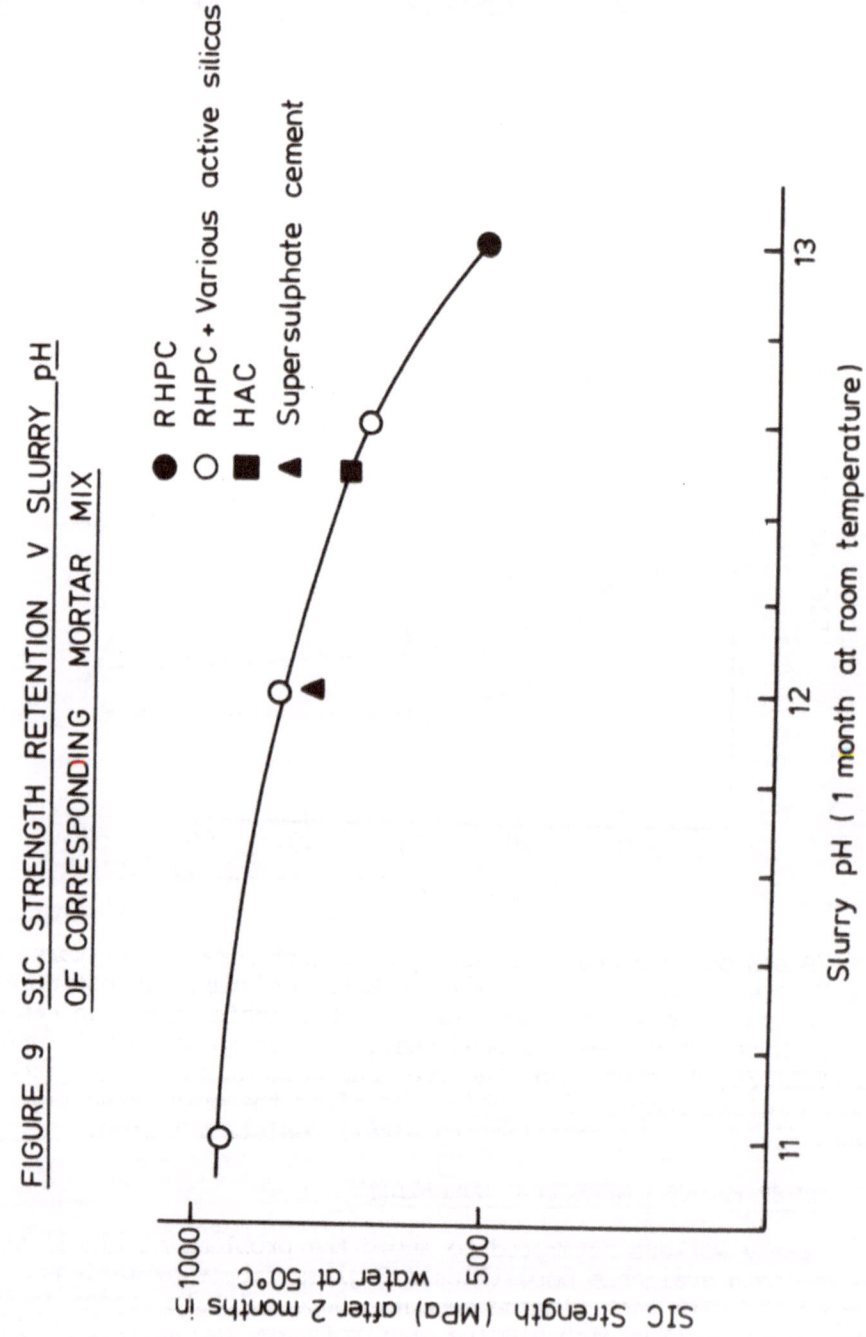

FIGURE 9 SIC STRENGTH RETENTION V SLURRY pH
OF CORRESPONDING MORTAR MIX

● RHPC
○ RHPC + Various active silicas
■ HAC
▲ Supersulphate cement

SIC Strength (MPa) after 2 months in
water at 50°C

Slurry pH (1 month at room temperature)

A.R. glassfibres, in the limited situations in which these cements can be used, the reduced pH does lead to better strength retention as shown in Fig.9.

The pH of a set cement could not be obtained directly so in this figure the SIC strength retention (in a small block of set cement) was compared with the pH measured in a high (1:1) water cement ratio slurry which was kept fluid by stirring. It was felt this would be reasonably representative of the value in a wet, set cement.

The pH of a Portland cement mortar may also be reduced by reaction with active silica materials, and as might be expected, this also leads to an improvement in strength retention in line with the pH changes accompanying alteration of the basic cement. Results for some fine silica/Portland cement mixes are also shown in Fig.9. The behaviour of pfa's (pulverised fuel ashes) is rather more complex and cannot be discussed here - they do not alter the pH of Portland cement slurries, but in some circumstances additions of pfa can lead to improved strength retention in GRC composites.

Whilst the use of protective resin or plastic films does not seem practical or reliable - uneconomically high coating levels are needed to produce significant effects and the presence of pinholes and cracking often renders the coating ineffective - one other approach to durability improvement has proved to be useful and reliable. This is the incorporation of a chemical inhibitor in the size coating which, in the alkaline environment of the set cement, is slowly released around the glass fibre, significantly reducing the rate of alkali/glass reaction and leading to marked improvements in the strength retention of the fibre, and other composite properties (Fig.10, Refs.8, 13). Organic compounds of the poly-hydroxy-phenol family have been found to be effective (Ref.19) and fibres of significantly improved durability are commercially available, known as Cem-FIL 2.

6. DISCUSSION AND CONCLUSIONS

From their studies of alkali resistance of the early zircono-silicate glass fibres Larner et al (Ref.5) concluded that the corrosion of these glasses proceded at a much slower rate than for soda lime or borosilicate glasses because preferential depletion of silica from the glass surface allowed the building up of a (implied protective and insoluble) zirconia rich layer. They did not however positively identify zirconia in the reaction products. Cockram (Ref.20), from radioactive tracer studies, has concluded rather similarly that insoluble reaction products on alkali resistant glasses impose diffusion controlled kinetics on the attack process and hence provide a protective layer. Further work in these

570

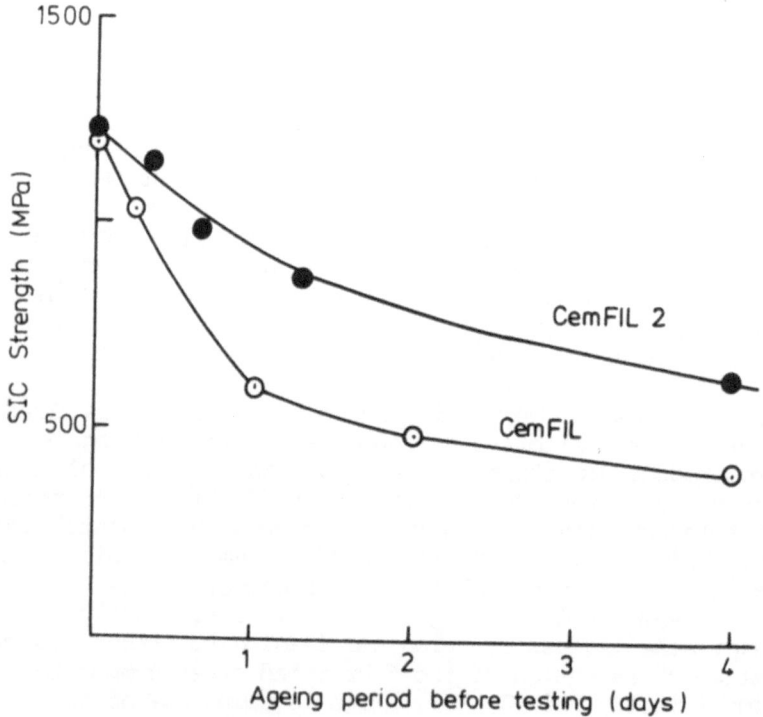

laboratories has indicated that a surface layer on attacked zircono-silicate glass fibres is zirconia rich, implying that alkali resistance is associated with insoluble zirconia containing reaction products.

Shin (Ref.21) took a similar view, but specifically associated alkali resistance with the solubility of the hydroxide. From such solubility data as a function of pH, he went on to deduce that magnesia containing glasses would have better alkali resistance than zirconia glasses. His own test work on single filaments in solution appeared to confirm his claims but practical fibres have not yet appeared, and whilst we have not directly tested his compositions, our general results would lead us to question whether the $SiO_2-Al_2O_3-MgO-P_2O_5$ or $SiO_2-TiO_2-MgO-SnO_2$ compositions which he described would in fact have adequate strength retention in cement.

The true situation is probably much more complex in terms of both the nature of the protective reaction products and the role of other factors. Proctor and Yale (discussion following Ref.3) suggested, in very general terms, that the stability of a complex glass was related to the sum of the stabilities of its constituent oxides - zirconia is insoluble in acids and alkalis and it seems reasonable therefore that it provides a general increase in chemical stability to the glass - in addition to insoluble reaction product layer protection.

Paul and Youssefi (Ref.15) expressed similar views about the required durability of individual oxide constituents in their choice of glasses containing Fe_2O_3 and MnO. Their selection of these oxides appeared to be confirmed by their own testing in solutions up to pH 12. However as shown in Fig.8, these glasses have poor strength retention in the rather more alkaline (pH 13 or so) environment of cement. The conclusion is that zirconia has a particular ability to confer significant alkali resistance on silicate glasses - probably by a combination of "structural strengthening" and insoluble reaction product layer protection.

As discussed in Section 3 the mechanism of strength loss in glasses subject to corrosive attack is extremely complex. It is well known that silicate glasses are attacked and even dissolved by hydrofluoric acid - however etching in HF is also used to strengthen glass articles by rounding out surface cracks and thus reducing their stress raising effect (Ref.22). However in the very same solutions, if accumulations of reaction products are allowed to occur on the glass surface, these may lead to uneven etching, rough surfaces, and very much weaker regions; indicating the importance of the physical, topographical nature of the attack as well as the basic chemistry and extent of interaction. Discrepancies between direct measurements of strength retention and indirect chemical assessment of alkali resistance have been pointed to in previous publications (Refs.3, 8, 11, 13) and serve to emphasize the second conclusion that assessment of the suitability of glasses for cement reinforcement should be based on direct tensile strength measurements, made on fibres, in the form used in the final composite, in a cement environment.

Finally it has been suggested (Ref.23) that strength loss in glass fibre reinforced cement composites is not due to chemical attack of alkali, but to abrasion of glass fibres by growing crystals of calcium hydroxide in the hydrating cement. Apart from the fact that calcium hydroxide crystals are in any case softer than glasses, it is clear that glass fibres loose strength in alkaline solutions in the absence of any crystal formation in a broadly similar manner to the behaviour in cement (Ref.4). The appearance of fibre surfaces after ageing in cement is consistent with chemical attack rather than mechanial damage (as seen in Fig.1) and the very big changes in fibre strength retention brought about by changes in glass composition discussed above reinforce the view that chemical corrosion is the mechanism of strength loss in cement. Experiments in which polymer is mixed in with cement in order to protect fibres from the assumed mechanical damage still show severe fibre strength loss and surface appearances consistent with corrosive attack (Ref.10). All this evidence leads to the third and final conclusion – that glass fibre strength loss in cement environments is due to alkaline chemical attack, and consequently that the avenues to improvement in GRC durability performance lie in further modifying the glass composition, controlling the immediate fibre environment in the cement, or/and reducing the overall aggressiveness of the cement itself.

ACKNOWLEDGEMENT

I am indebted to many colleagues for helpful discussion and use of previously unpublished information – in particular to D. R. Cockram, K. L. Litherland, K. M. Fyles, B. Yale, P. Maguire and D. R. Oakley.

This paper is published with the permission of the Directors of Pilkington Brothers P.L.C. and Mr. A. S. Robinson, Director of Group Research and Development.

REFERENCES

1.   Majumdar A. J. and Ryder J. F., Glass Technology, 9, No.3 (June 1983), pp.78-84.

2.   Biryukovitch K. L. and Biryukovitch Yu L., Stroit Mater. 1, No.11 (1961), pp.18-20.

3.   Proctor B. A. and Yale B., Phil. Trans. R. Soc. London, A294 (1980), pp.427-436.

4.   Majumdar A. J., Proc. Roy. Soc. London, A319, (1970), pp.69-78.

5.   Larner L. J., Speakman K. and Majumdar A. J., J. Non-Crystalline Solids, 20, (1976), pp.43-74.

6.   Dimbleby V. and Turner W.E.S., J. Soc. Glass Technology, 10, (1926), pp.304-358.

7.   U.K. Patent 1,290,528.

8.   Proctor B. A., Oakley D. R. and Litherland K. L., Composites, April 1982, pp.173-179.

9. Majumdar A. J., West J. M. and Larner L. J., Jnl. Material Science, 12, (5, 1977), pp.927-936.

10. Proctor B. A., Composites, June 1971, pp.85-92.

11. Cockram D. R., Litherland K. L., Proctor B. A. and Yale B., "Assessing the Durability of Glass Compositions", Proceedings XIII International Congress on Glass, Hamburg, July 1983.

12. Litherland K. L., Maguire P. and Proctor B. A., The International Journal of Cement Composites and Lightweight Concrete, To Be Published in 6 (No.2) May 1984.

13. Proctor B. A., Proceedings Int. Congress Glassfibre Reinforced Cement, Paris 1981, edited Vincent Blake, The Glassfibre Reinforced Cement Association, pp.50-67.

14. Litherland K. L., Oakley D. R. and Proctor B. A., Cement and Concrete Research, 11, (3, May 1981), pp.455-466.

15. Paul A. and Youssefi A., Jnl. Materials Science, 13 (1978) pp.97-107.

16. Mackenzie J. D., US Bureau of Mines, Materials Research Contract Report JO199056, 1981, University of California.

17. Ramachandran B. E., Velpari V. and Balasubramanian N., Jnl. Materials Science, 16 (1981), pp.3393-3397.

18. UK Building Research Establishment Curent Paper CP34/75, "High Alumina Cement in Buildings.

19. UK Patents 1,465,059; 1,524,232; 1,519,041; 1,565,823.

20. Union Scientific Continentale du Verre, Conference Proceedings 1984, to be published.

21. Shin, Woo Seung, Proceedings Int. Congress Glassfibre Reinforced Cement, Paris 1981, edited Vincent Blake, The Glassfibre Reinforced Cement Association, pp.359-374.

22. Proctor B. A., Phys. Chem. Glasses, 3 (1962), pp.7-27.

23. Bjien J., Proceedings Int. Congress Glassfibre Reinforced Cement, London, 1979, edited edited Hayden Jeffery, The Glassfibre Reinforced Cement Association, pp.62-67.

24. Proctor B. A., ibid pp.85-86.

COMPOSITES AS BIOMATERIALS

June Wilson

Department of Surgery
J. Hillis Miller Health Center
University of Florida
Gainesville, Florida 32610

The search for new biomaterials concerns many different workers, including clinicians who are able to define their needs in terms of what a device will do and for how long. This is generally derived from the nature and behavior of the tissue or organ which is in need of augmentation or replacement. However, a very long and complex process begins once a need is identified, a process which involves materials scientists who can prescribe materials with properties which match the requirements of the clinicians, engineers who must design and fabricate a functioning device from such materials and toxicologists, biochemists, bacteriologists, pathologists who must ensure that the introduction of devices made from these materials will not produce effects in the patient other than those intended. Investigations are concerned mainly with the effect of the implant material on local and distant tissues but must also examine the effect of body fluids and the physiological environment on materials, which may be unexpected.

The primary requirement of a successful biomaterial is that it reproduces exactly the function of the tissue which it replaces and since all replaceable tissues are themselves composites it would seem reasonable to assume a place for composite materials in the range of successful implants. Unfortuntely this is not so - or not yet so.

A tissue which frequently needs repair and augmentation is bone. Bone is a composite of two main components - collagen fibers and mineral crystals. The mechanical properties of bone derive from the interactions between the properties of these two. A traditional demonstration of this interaction requires two identical long bones, from one the mineral phase is chemically removed, from the other the collagen is removed by digestion. The bones look identical after treatment but the first may be tied in a knot and the second crumbles when struck (1,2). However in life the two components interact in a way which allows them to function as one. To produce a composite to mimic this function is simple only in theory. The components must each be non-toxic, biocompatible and resistant to degradation by either dissolution or wear. The composite must have the bonds between the components also resistant to degradation by dissolution or breakdown at the interface. The implant made from the composite must also interface with the adjacent bone so that stresses are passed across that interface in a natural and acceptible fashion. This last, positive interaction with natural tissues has been an enormous barrier to the use of new materials in surgery. Until recently the only acceptable biomaterials were those designed to be inert and to rely on the scar tissue which comes to surround such implants to keep them in place. Such a material is silicone rubber, which is widely used for all kinds of augmentation. Over the past two decades inert materials have been introduced in a porous form, the critical pore size which allows control over the type of tissue which will grow is now well defined (3). However since stress is not transferred across voids such materials, although they consist of two phases, are not considered to be composites and will not be included here.

The critical event which showed that a chemical and not merely a mechanical bond could be achieved between tissues occurred with the discovery and development of the various Bioglass™ systems by Dr. Hench and his co-workers at the University of Florida since 1969. These will be the subject of talks in the next session. The important difference between these and inert glasses is that the active surface layers which develop when Bioglasses™ are exposed to tissue fluid encourages the integration of collagen fibers or bone mineral with them, allowing transfer of stress across the bonded interface.

Combinations of various materials with glasses have been suggested as useful biomaterials. These are listed in Table 1. All of these materials have been tested in animal models, almost all unsuccessfully (4,5). The components of the composite in each case have been found individually to be non-toxic. However when combined as a composite and exposed in a physiological environment adverse tissue reactions have occurred due to the generation of wear particles which produce an effect merely by

Matrix	Dispersed Phase
Pyrex glass	Carbon fiber (4)
Pyrex glass	Alumina fiber (4)
Pyrex glass	Silicon carbide fiber (4)
High Density Polyethylene	Bioglass™ particles (4)
High Density Polyethylene	Bioglass™ on alumina particles (4)
Bioglass™	Stainless steel fibers (6)
Polypropylene	Borosilicate glass fibers (5)

Table 1. Examples of Composites with Glass as One Phase.

their shape and presence in the tissues. There is a whole field of research which is concerned with the effect of shape and size on tissue response which it is inappropriate to introduce at this stage, however it should be remembered that the effect of particulates, especially fibrous particulates, depends on the aspect ratio as well as on the chemistry of the material. So-called inert materials which are indigestible can seriously perturb the cells in which they are contained. When composites are placed into a soft or hard tissue bed, such as subcutis or bone the prevention of micromovement at the interface is essential and for many materials impossible. The movement of the implant against tissue is sufficient to damage both tissue and implant causing release of particles and dissolution of the interphase bonds at the surface of the material. This was the mode of deterioration of all those composites which included inert glass and fibers (see Table 1).

After implantation in soft tissue for four weeks all of the implants were surrounded by a thick cellular capsule with many phagocytes within the capsule (4). The capsules surrounding the implants which had carbon as the fibrous component were indeed grey in color to the naked eye. The combination of inert glass with fiber reinforcement has not proved to be useful as a bio-material. The first step in the deterioration of these materials however is a mechanical one rather than chemical, abrasion at the surface causing an influx of phagocytic cells which provide high concentrations of liposomal enzymes which may then chemically attack the material. If micromovement at the inteface can be prevented, these cells do not accumulate and the progressive deterioration of the material should not follow. To test this hypothesis several combinations of particulate Bioglass™ materials in a polyethylene matrix have been made (4). The modulus of this type of composite may be varied by varying the thickness of the slice of material tested and a positive bond with tissue achieved by having a sufficient amount

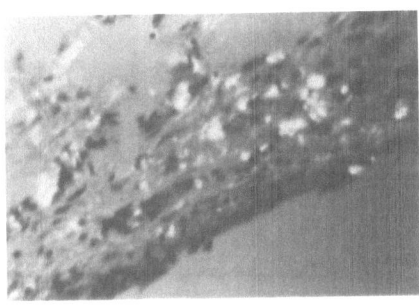

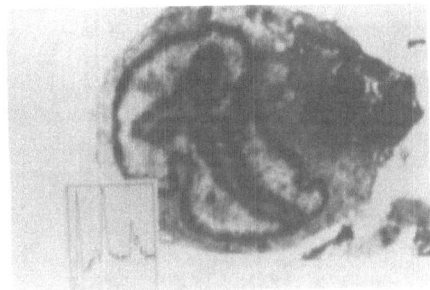

Fig. 1.   Fig. 2

of Bioglass™ exposed at the interface with tissue. With a
proper modulus match and enough glass to bond the interface the
micromovement should be eliminated.

A series of materials were made and tested in this experiment but the interphase bonds were insufficient to prevent the
release of fragments of polymer into the adjacent tissues (Fig.
1). Fragments of released Bioglass™ were not seen adjacent to
these implants since earlier studies have shown that particulate
Bioglass™ does not adversely affect the cells which take it
up (7) (Fig. 2). When tested in bone these materials were
bonded in place when the Bioglass™ component was high enough to
get a properly bonded interface. Further work will be done to
develop these materials since they do show promise in certain
applications.

The final, and to date most successful composite material
using glass is that developed by Ducheyne and Hench,
incidentally funded by NATO (6). They developed a material in
which stainless steel A1S1 316L fibers are used to reinforce
Bioglass™. Thermal shock, tensile tests and three point bending
all show a material with a marked increase of strength over the
parent glass and a substantial improvement in toughness. The
material has elastic properties which closely match those of
bone, an important factor in the maintenance of healthy bone
after implantation. The glass facilitates bonding to adjacent
bone and the interphase bonds between metal and glass surface do
not separate from each other under stress. This is due to the
fact that this is a graded interface as shown from EDX and EMP
analyses. Next to the fibers the layer is rich in Cr and
in Fe and next to the glass the reverse is true across the
interface from metal to glass Ca and Si are gradually altered
also (Fig. 3).

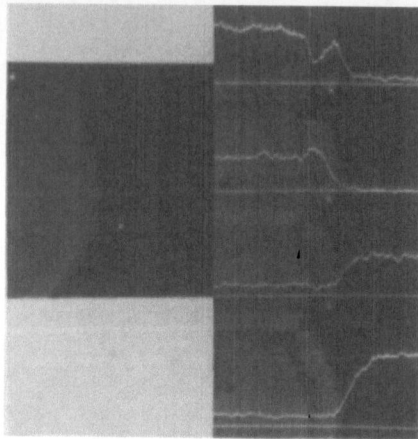
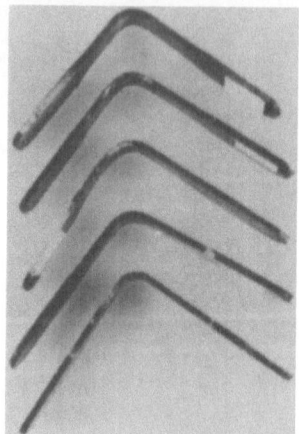

Fig. 3   Fig. 4.

This metal fiber reinforcement has successfully transformed a brittle glass into a tough composite and these composite materials withstand thermal shock of much greater value than could bulk glass or glass as a coating on metal. Before the introduction of the metal the glass is not ductile, the composite is, may be deformed by bending through 90° (Fig. 4). All of these properties allied to the bone bonding property of the glass make this Bioglass™/metal composite very attractive as an implant material for orthopedic and dental use. Implantation in bone has shown that these composite materials do bond in place under different loading techniques and do not cause any adverse tissue reactions. The immobilization by bonding, the careful optimization of the interphase bonds do not allow the micromovement and breakup seen in other cases, the mechanical properties of the composite can accurately reproduce those of the tissue being replaced and the result is potentially a valuable composite biomaterial with a wide range of application in orthopedics and surgery.

REFERENCES

1. Park, Joon B., Biomaterials, An Introduction (Plenum Press, New York, 1979) 104-110.
2. Ham, Arthur W., Histology, 7th Ed. (J. B. Lipencott Co., Philadelphia, Pennsylvania, 1969) 382.
3. Hench, L.L. and E.C. Ethridge, Biomaterials: An Interfacial Approach (Academic Press, New York, New York, 1982) 68.
4. Wilson, J. and L.L. Hench, unpublished data.
5. Christel et al., in Hastings and Williams, eds. (John Wiley & Sons, New York, 1980) 367-377.

6. Ducheyne, P. and L.L. Hench, J. Mat. Science 17 (1982) 595-606.
7. Wilson, J, et al., J. Biomed. Mat. Res. 15 (1981) 805-817.

INORGANIC/ORGANIC POLYMERS FOR CONTACT LENSES BY THE SOL-GEL
PROCESS

H. SCHMIDT and G. PHILIPP

Fraunhofer-Institut für Silicatforschung, Neunerplatz 2, D-8700
Würzburg, F.R.G.

ABSTRACT

The use of organoalkoxysilanes as network forming materials
allows the introduction of organic groups into an inorganic network by the sol-gel process. This possibility, in addition to the
inorganic network modification, opens a wide variability of material synthesis. Facing the desire for high performance materials
for contact lenses, an attempt is made to meet the requirements
of contact lens materials, as oxygen permeability and wetting.
The paper shows the basic principles which can be carried out in
order to taylor material for specific properties.

1 INTRODUCTION

In the field of material development an increasing demand
for materials with special properties is obvious. Considering the
properties of different types of materials like ceramics, metals,
or organic polymers, one started very early to think about the
combination of different properties in order to get materials for
special applications which unify the desired different properties
in one material. This led to the conception of composites. Examples of composites might be laminates (fig.1a) which are successfully used e.g. for windshields of cars, for skis, or to create
protective surfaces; other examples are filled organic polymers
(fig.1b), as they are used for tires of cars. From this point of
view a ceramic material with a crystalline phase and a glassy
phase as binder may be considered as a composite material, too;
as a last example fibre reinforced materials may be cited
(fig.1c). This type of composite is used e.g. with organic poly-

mers or inorganic materials like concretes. These examples represent composites where the combination of two materials is per-

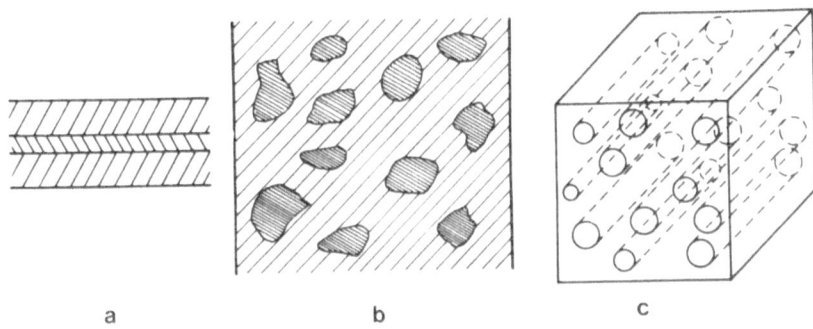

a   b   c

Fig.1a-c. Principle possibilities of composites.
(a) laminates; (b) filled organic polymers;
(c) fibre reinforced material

formed on a macroscopic scale. In order to process such materials, with a few exception, combination techniques have to be developed (mixing, sealing etc.). In the most cases this needs a more-step processing.

From this the question arises, if the principle of composites may be extended to the microscopic scale. In the field of material development one advantage might be (if such material are produced by the rules of chemistry) to apply an one-step processing. Another question is which technical means may be used in order to perform this.

A simple way could be to combine components on a molecular scale from which it is known that they cause different material properties. Examples might be components which lead to non-metallic inorganic networks on the one hand and organic substituents or organic polymers on the other hand; organic radicals might be functional groups or polymerizable components. In order to get a desirable combination of these materials, it seems to be necessary to develop a method where these two types of components may be linked together by well-defined chemical bonds like covalent ones. There might be considered two general possibilities. One is to build up an organic backbone and to link inorganic components to this backbone; this principle is realized in some organic polymers, so-called ionomers, as they are shown in fig.2.

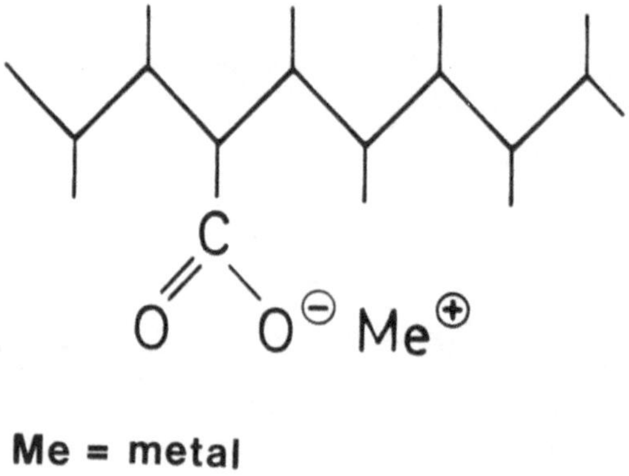

Me = metal

Fig.2. Scheme of an organic ionomer

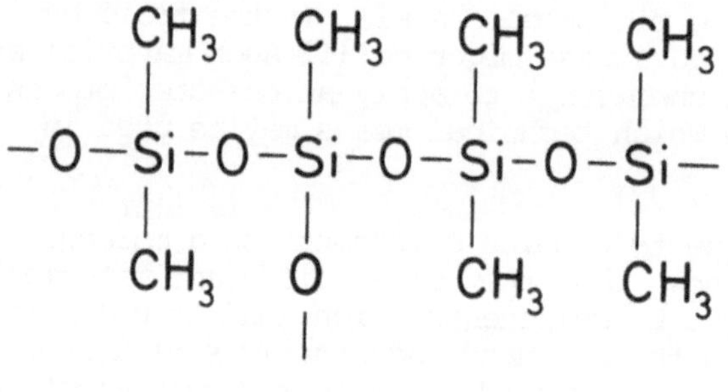

Fig.3. Scheme of a silicone polymer

The second possibility is to build up an inorganic (metal oxide)
backbone which is partially realized in silicones (fig.3.).

   Silicones normally are produced by hydrolsis and condensation of the proper chlorosilanes. In principle the preparation of
this type of material is possible, too, by the means of the sol-
gel process i.e. by the use of alkoxysilanes. The advantage of
this route is that the steps of hydrolysis and condensation are

easier to be controlled because of the lower reactivity of the
alkoxysilanes compared to the chlorosilanes [1-3]. The sol-gel
process moreover enables the introduction of different other compounds like network forming elements into the material as it is
described by Gulledge, Andrianov, and Noll [4-6]. An additional
principle of modification is the use of organoalkoxysilanes,
where the Si-C bond can be used to introduce organofunctional
groups which may have special chemical properties or which can
carry polymerizable ligands [7-11] according to equation (1).

$$Me^{IV}(OR)_4 + Si(OR)_4 + (RO)_m Si(\diagup\!\!=\!\!\diagdown)_{4-m} + (RO)_n Si(R'Y)_{4-n} \longrightarrow$$

$$
\begin{array}{cccc}
| & | & | & | \\
O & O & O & O \\
| & | & | & | \\
-O-Me^{IV}-O & -Si-O- & Si-O- & Si-O- \\
| & | & & R' \\
O & O & & Y \\
| & | & &
\end{array}
\tag{1}
$$

$m = 3;\ n = 2$

Me = metal e.g. Ti, Zr
R = alkyl, aryl
R' = alkylen, arylen
$\diagup\!\!=\!\!\diagdown$ = polymerizable double bond
Y = organofunctional group e.g. $-NH_2$, $-CHO$

This conception leads to a very flexible system of material development. It was of high interest to find out, if this conception
may be advantageously used in order to prepare special materials
for contact lenses.

2 CONCEPTION OF MATERIAL DEVELOPMENT

2.1 Requirements for Contact Lenses

As a result of the special type of application (i.e. the
material is in contact with the cornea) it is pretty well-known
which requirements materials for contact lenses should fulfill
[12]. The main requirements are listed in table 1. The wettability is necessary in order to maintain the lachrymal film between
the material and the cornea; this wettability should not be lost,
if the maerial is processed e.g. shaped by mechanical procedures.
The oxygen permeability is necessary in order to supply the cornea with oxygen, since the cornea has no other supplying system.
The mechanical properties have to be sufficient for a secure handling and the tendency of the surface not to deposit components of

Table 1. Main requirements for hard contact lens material

Wettability	contact angle with water 30° in the hydrated state,
O_2-permeability	permeability coefficient P = 10×10^{-11} ml $O_2.cm^2.ml^{-1}.s^{-1}.mm$ Hg^{-1},
Physical properties	density: 1.1–1.2 g cm^{-3}, sufficient flexibility, high hardness, good scratch resistance,
Optical properties	refractive index n^{20}_D = 1.43, transmission ≥98%,
Chemical properties	water absorbing capacity < 10 wt.%, low tendency of deposition of components of lachrymal fluid, chemical stability (no unlinked components of contact lens material; resistance to acids, bases, organic solvents, microorganisms, and UV-light).

---

the lachrymal film is necessary in order to avoid microbial contamination. The common polymers in the couple of special developments [12] are not able to unify all these requirements in one
and the same material.

2.2 Possibilities of Realization

It seems to be useful in order to have a high oxygen permeability to use structure elements of silicones, since silicones
have an oxygen permeability of P = 79×10^{-11} ml $O_2.cm^2.ml^{-1}.s^{-1}.mm$
Hg^{-1} [12]. But silicones are extremely hydrophobic. That means,
the idea should be to build up a silicone-like network with hydrophilizing components incorporated. The hydrophilizing components have to be non-ionic in order to avoid deposition of components of the lachrymal fluid. Therefore alcoholic groupings are
to be preferred.

In principle compounds of the type $(RO)_n Si(alkylen OH)_{4-n}$
should be useful in combination with another network forming Si
compound like $Si(OR)_4$ [10]. Then the condensation reaction should
follow equation (2), but there appears one complication, since,
as proved by Voronkov et al. [14], transesterification reaction
occurs very easily with mixtures of alkoxysilanes. Reaction according equation (3) may take place and lead to a decrease of hydro-

$$\text{Si(OR)}_4 + \text{(RO)}_3\text{Si-alkylene-OH} \xrightarrow[\text{- ROH}]{\text{+ H}_2\text{O}} \begin{array}{cc} \text{O} & \text{O} \\ | & | \\ \text{-O-Si-O-\!\!-\!\!Si-alkylene-OH} \\ | & | \\ \text{O} & \text{O} \end{array} \qquad (2)$$

$$\text{I}$$

R = alkyl

philicity. Since equation (3) easily may occur under average condensation conditions, especially if one uses catalysts like protons, the complication has to be avoided. Measurements of alcoholic solutions of compound I, equation (2), show that if the solvent is evaporated to a high extent type II (equation (3)) oligomers are formed. This was proved by mass spectrometric analysis.

$$\equiv\!\text{Si-alkylene-OH} + \text{RO-Si}\!\equiv \xrightarrow{\text{-ROH}} \equiv\!\text{Si-alkylene-O-Si}\!\equiv \qquad (3)$$

$$\text{II}$$

R = alkyl

In order to avoid this complication a better route should be to use latent hydrophilic organic groupings which develop the real hydrophilicity after the condensation process has taken place. This principle should be performed according to equation (4).

$$\text{Si(OR)}_4 + \text{(RO)}_3\text{Si-alkylen-Y} \longrightarrow \begin{array}{cc} \text{O} & \text{O} \\ | & | \\ \text{-O-Si-O-Si-alkylen-Y} \\ | & | \\ \text{O} & \text{O} \\ | & | \end{array}$$

$$\qquad (4)$$

$$\begin{array}{cc} \text{O} & \text{O} \\ | & | \\ \text{-O-Si-O-Si-alkylen-Y} \\ | & | \\ \text{O} & \text{O} \\ | & \end{array} \longrightarrow \begin{array}{cc} \text{O} & \text{O} \\ | & | \\ \text{-O-Si-O-Si-alkylen-OH} \\ | & | \\ \text{O} & \text{O} \\ | & | \end{array}$$

R = alkyl; Y = latent hydrophilic grouping

A technical possibility in order to realize this idea could be to use the commercially available epoxysilane $\text{(RO)}_3\text{Si}\!\sim\!\overline{\text{CHCH}_2\text{O}}\!\!\!\!\rfloor$ (III). A cocondensation of Si(OR)_4 with (III) should lead to highly crosslinked and therefore to brittle materials. The influence of the organic grouping on the network modification was unknown, and therefore it was necessary to take care for building up sufficient elasticity. If not, it will become necessary to incorporate structure elements which lead to a more linear crosslinking. The possibility for this is demonstrated in equation (5).

$$\equiv Si-alkylen-\!\!/\!\!/ \quad + \quad n \overset{\overset{\displaystyle R}{|}}{/\!\!/} \quad + \quad /\!\!\backslash alkylen-Si\equiv$$

$$\downarrow$$

$$\equiv Si-alkylen-\!\!\Big(\!\!\wedge\!\!\overset{R}{\wedge}\!\!\wedge\!\!\Big)_{\!\!n}\!\!\wedge alkylen-Si\equiv$$

$$R = alkyl, -COOR'$$

(5)

3 RESULTS

3.1 Hydrophilicity

If water is added to the epoxysilane (III), the epoxide ring opens to the glycol grouping (IV) according to equation (6).

$$(CH_3O)_3SiC_3H_6OCH_2-\overset{\displaystyle O}{\overset{\diagup\diagdown}{CH-CH_2}} \quad \xrightarrow[-CH_3OH]{H^+ \text{ or } OH^-/H_2O} \quad -O-\overset{\displaystyle \overset{O}{|}}{\underset{\displaystyle \overset{|}{O}}{Si}}C_3H_6OCH_2-\overset{\displaystyle \overset{OH}{|}}{C}H-\overset{\displaystyle \overset{OH}{|}}{C}H_2$$

III $\qquad\qquad\qquad\qquad\qquad\qquad\qquad\qquad$ IV $\qquad$ (6)

The reaction conditions under which this reaction can run are basically the same where hydrolysis and condensation may take place (alcoholic solutions, protons as catalysts, addition of water). The overall reaction according to equation (6) does not exclude a reaction according to equation (3), leading to a decrease of the hydrophilicity. The proof that the ring opening reaction takes place at the same time as hydrolysis and condensation was given by IR- and NIR-spectroscopy (fig.4). Investigations to which extent the reaction according to equation (6) decreases the hydrophilicity were not carried out, but another way

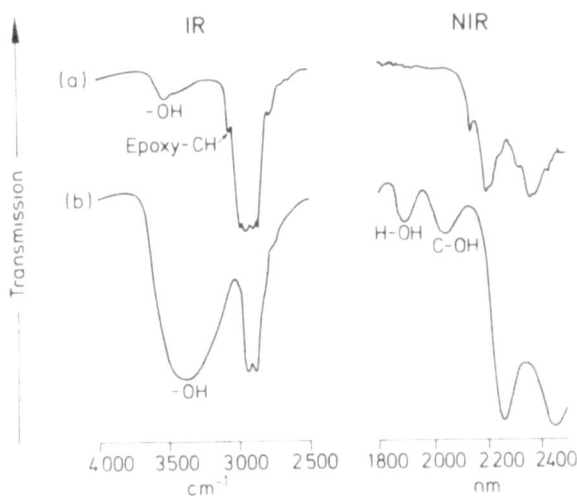

Fig.4. IR- and NIR-spectra of the epoxysilane III (a) before and
(b) after hydrolysis and condensation

which shows the good possibility to avoid this is given by Voronkov [12] as shown in equations (7) and (8). This "water-free"
condensation reaction was proved by IR- and NIR-spectroscopy
which showed that the epoxide ring was not opened and by mass
spectroscopy which proved the appearance of RX. By this, reaction
can be run to a viscous material and then by the addition of water the epoxide ring will be opened to the glycol grouping. After
this, the viscous material can be cured at temperatures of 100-
130°C to a solid material. But as a function of the high crosslinking, brittle material were produced and during the drying
procedure the desaggregation to a granulate occurred.

$$2 \equiv SiOR + HX \longrightarrow \equiv Si-O-Si\equiv \ + ROH + RX \tag{7}$$

R = alkyl; X = halogen

$$2 \ R'OH + 2 \ HX \longrightarrow 2 \ R'X + 2 \ H_2O$$
$$2 \ H_2O + 2 \equiv SiOR \longrightarrow 2 \equiv SiOH + ROH \tag{8}$$
$$2 \equiv SiOH \longrightarrow \equiv Si-O-Si\equiv \ + H_2O$$

R, R' = alkyl; X = halogen

In order to run the condensation reaction to a higher condensed material and to decrease the shrinkage problem, in addition to $Si(OR)_4$ $Ti(OR)_4$ was used. This is easily possible because the water-free precondensation process avoids the precipitation of TiO_2 and leads to a homogeneous Ti-containing precondensate. The use of water in the presence of $Ti(OR)_4$ very often leads to the precipitation of TiO_2, since the hydrolysis and condensation rates are far higher than that of $Si(OR)_4$ containing identical organic substituents. The overall reaction demonstrates equation (9). By addition of water to the precondensate, the epoxyring is opened and the glycol grouping is formed. The viscous is cured to compact rods from which contact lenses can be shaped by a mechanical processing. The materials have an oxygen permeability of $(13+1).10^{-11}$ ml $O_2.cm^2.ml^{-1}.s^{-1}.mm$ Hg^{-1}, a contact angle with water in the hydrated state of $(25+5)°$, a tensile strength of $2.1-3.6$ $MN.m^{-2}$ and a modulus of elasticity of $(29-34).10^2$ $MN.m^{-2}$. The handling of contact lenses from these materials is very difficult, since their mechanical strength is not sufficient.

$$Ti(OR)_4 + (CH_3O)_3SiC_3H_6OCH_2-\overset{O}{\overset{/\backslash}{CH}}-CH_2 \xrightarrow[R'OH]{HX} -O-\overset{OR}{\underset{OR}{\overset{|}{\underset{|}{Ti}}}}-O-\overset{OCH_3}{\underset{OCH_3}{\overset{|}{\underset{|}{Si}}}}C_3H_6OCH_2-\overset{O}{\overset{/\backslash}{CH}}-CH_2$$

$$(9)$$

$$-O-\overset{OR}{\underset{OR}{\overset{|}{\underset{|}{Ti}}}}-O-\overset{OCH_3}{\underset{OCH_3}{\overset{|}{\underset{|}{Si}}}}C_3H_6OCH_2-\overset{O}{\overset{/\backslash}{CH}}-CH_2 \rightarrow -O-\overset{O}{\underset{O}{\overset{|}{\underset{|}{Ti}}}}-O-\overset{O}{\underset{O}{\overset{|}{\underset{|}{Si}}}}C_3H_6OCH_2-\overset{OH}{\overset{|}{CH}}-\overset{OH}{\overset{|}{CH_2}}$$

R = alkyl

In order to reduce brittleness, the incorporation of linear polymerizing components should be helpful. Therefore the polymethacrylate system was chosen by reason of their mechanical properties and their very low tendency of deposition of components of the lachrymal fluid.

The realization of the incorporation of PMMA chains should be possible according to equation (10) by the use of methacryloxysilane $(RO)_3SiC_3H_6OOCC(CH_3)=CH_2$ and methylmethacrylate (MMA). Again the water-free precondensation seems to be useful as well as the use of titanium. The precondensation process (equations 7 and 8) is performed by the epoxysilane (III) methacryloxysilane

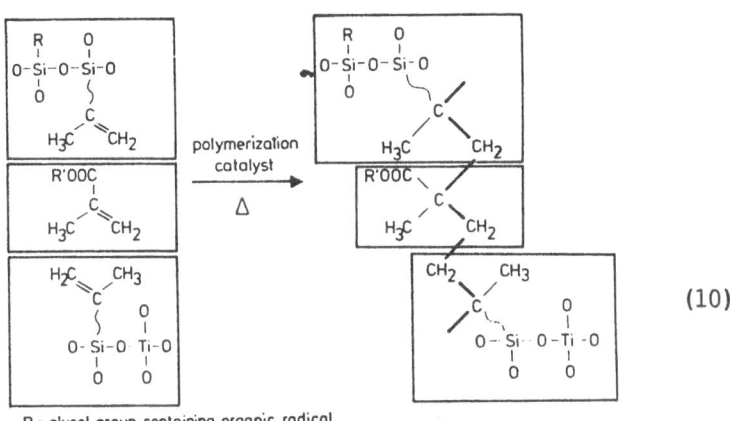

R = glycol group containing organic radical

R' = CH$_3$; C$_2$H$_4$OH

$\sim$ = $-$ C$_3$H$_6$OOC $-$

(10)

and Ti(OR)$_4$ until a viscous liquid is received. Then this viscous liquid is mixed with MMA or hydroxymethylmethacrylate (HEMA) and a polymerization catalyst and cured at temperatures of 100-130°C. This curing reaction leads to compact rods which easily can be cut into blanks which are processed to contact lenses. The properties of these materials are given in table 2. These materials unify in one and the same product all necessary mechanical, optical and surface properties for a contact lens material since the deposition test with a lachrymal simulation fluid shows that the tendency of deposition of components lachrymal fluid is very low.

Table 2. Main properties of copolymers[a] containing 20-30 mol.% methacrylates[b], cured at 130°C

Tensile strength[c] (MN m^{-2})	4,85	$-$ 5,15
Modulus of elasticity x 10 (MN m^{-2})	33	$-$ 34
Mohs'hardness	3	
Shore D hardness	67	$-$ 73
Refractive index n$_D^{20}$ [d]	1,499	$-$ 1,503
Contact angle with water[e] (°)	25	5
O$_2$-permeability coefficient P = 10x10^{-11} ml.O$_2$.cm^2.ml^{-1}.s^{-1}.mm Hg^{-1})	11.5	$-$ 13.3

a) Silicate composition: 60-70 mol% epoxysilane III, 5 mol.% Ti(OR)$_4$ and 5 mol.% methacryloxysilane.

b) Methacrylates R'OOCC(CH$_3$) = CH$_2$ with R' = CH$_3$, C$_2$H$_4$OH.

590

Figure 5 shows a photograph of rods processed by the procedure described above.

Fig.5. Examples of monolithic condensates

4 CONLUSIONS

The investigations, carried out in order to combine inorganic and organic networks, show that it is possible to prepare a molecular scale composite material which unifies properties of non-metallic inorganic components (e.g. preparation by sol-gel processing, structure) with that of organic polymers (oxygen permeability elasticity). This leads to the conclusion that this way of material development can be successfully used for the preparation of very special materials.

5 ACKNOWLEDGEMENT

The authors thank Professor Dr.H.Scholze for valuable discussions. They gratefully acknowledge the assistance of their coworkers, especially of Mrs.E.Popp with the experiments, and the industrial companies for the financial support.

6 REFERENCES

[1] H.Dislich, Angew.Chem. 83 (1971) 428.

[2] R.Roy, J.Amer.Ceram.Soc. 52 (1969) 344.

[3] K.S.Mazdiyasni, R.T.Dollof and J.S.Smith, II, J.Amer. Ceram.Soc. 52 (1969) 523.

[4] H.C.Gulledge, Titanated organo-silcon-oxy compounds, US 2 512 058 (1950).

[5] K.A.Andrianov, Organic silicon compounds (State Sci.Publ. House for Chemical Literature, Moscow, 1955).

[6] W.Noll, Chemie und Technogie der Silicone. 2.Aufl. (Verlag Chemie, Weinheim, 1968).

[7] H.Schmidt and H.Scholze, J.Non-Crystalline Solids 63 (1984) 1.

[8] A.Kaiser and H.Schmidt, J.Non-Crystalline Solids 63 (1984) 261.

[9] G.Philipp and H.Schmidt, J.Non-Crystalline Solids 63 (1984) 283.

[10] H.Schmidt, O.v.Stetten, G.Kellermann, H.Patzelt, and W.Naegele, Proc.Radioimmunoassay and Related Procedures in Medicine 1982, 111-121, Vienna 1982.

[11] H.Schmidt, Organically Modified Silicates by the Sol-Gel Process. Proc.Materials Research Society 1984 Spring Meeting, Albuquerque, N.M., USA (in print).

[12] C.F.Kreiner, Kontaktlinsenchemie (Median-Verlag, Heidelberg, 1980).

[13] M.G.Voronkov.V.P.Mileshkevich and Y.A.Yuzhelevskii, The Siloxane Band (Consultants Bureau, New York and London, 1978).

Abstract only

GLASS FIBER REINFORCED PLASTIC IN THE AUTOMOTIVE INDUSTRY

Y. APPELL

S.E.P. - LE HAILLAN
B.P. 37 - 44165 St MEDARD EN JALLES

The objective of this paper is to give an idea of what is the interest of Glass Fibers Reinforced Plastic (GFRP) in the Automotive Industry and what will be their future.

Some data are given as an indication of how GFRP can be advantageously used in the car.

The main objective of Automotive Engineering functions is the energy factor and reduction in fuel consumption.

Some examples of achievements using composites on vehicles from the Peugeot Group are given.

Mention is made of the advantages and restrictions relative to use of these materials in the Automotive Industry.

Potential applications of GFRP will greatly differ between structural parts, body parts and mechanical parts. Certain technical orientations are given.

The potential of composites in automobiles will depend, to a large extent, on the development of new adapted resins and new processes permitting high production rates.

We must not expect a revolution but rather a gradual evolution leading to a rather extensive use of GFRP in the 90's.

Abstract only

COMPOSITE MATERIALS - AN OVERVIEW

D. Hull

University of Liverpool, England.

The essence of composite materials technology, particularly
in the field of fibre composites, is the ability to put strong,
stiff reinforcing elements (fibres) in a matrix in the right place,
in the right orientation, with the right volume fraction. Implicit
in this approach is the concept that in making the composite
material one is also making the final product. This means that
there must be very close collaboration between those who design
composite materials at the micro scale and those who design and
manufacture the final engineering component. The importance of this
concept is illustrated by reference to the structure of wood. The
primary structure consists of fibre tracheids which are hollow
tubes. These are bonded together and are oriented in the axial and
radial growth directions. The walls of the tracheids are themselves composite materials and consist of many layers of well-
oriented cellulose fibrils held together with lignin. The orientation of the fibre tracheids and cellulose fibrils is related to
the load bearing requirements of the tree.

Composite materials can be classified in many different ways.
Two classifications will be given. The first a broad classification
into natural composite materials, microcomposite materials and
macrocomposites, and the second a more detailed classification of
microcomposite materials.

Man-made microcomposite materials usually have the following
characteristics (a) they consist of two or more physically distinct
and mechanically separable materials, (b) they are made by mixing
separate materials in such a way that the dispersion of one material in the other can be done in a controlled way to achieve

optimum properties and (c) their properties are superior, and possibly unique in some specific respects, to the properties of the individual components. It is the last point which provides the main impetus for the development of composite materials. Many manufacturing routes are available for the manufacture of components from composite materials and these will be reviewed briefly.

The multiplicity of possible composite materials prevents a detailed overview. Specific reference will be made to the most important group of materials which consist of fibres embedded in a polymeric matrix. There are three important classes of fibres : carbon, glass and organic. Their properties are closely related to their molecular structure. A knowledge of this structure provides an insight into the limitations in properties which are achievable. A particularly important feature of glass fibres is the variability of strengths which arises because of the existence of surface flaws. The properties of the polymer matrix and the fibre matrix interface have a profound effect on the properties of the composite and some aspects of this problem will be outlined.

The properties of fibrous composite materials are strongly dependent on microstructural parameters such as fibre diameter, fibre length distribution, volume fraction of fibres and the alignment and packing arrangement of fibres. The effect of each parameter varies from one property to another. An insight into the range of possibilities will be given by reference to unidirectional laminae, woven rovings, in-plane random arrays with long and short fibres.

An example of the interaction between fibre, matrix and interface properties on the one hand and microstructural parameters on the other will be given by reference to the stiffness and strength properties of an unidirectional lamina tested in tension at different angles to the fibre direction. This example will be used to illustrate the composite principle and provide an insight into the complexity of engineering design using composite materials.

Reference

D. Hull, 'An Introduction to Composite Materials', Cambridge University Press.

Abstract only

STRESS CORROSION OF GLASS REINFORCED COMPOSITES

D. Hull

University of Liverpool, England.

The performance of glass reinforced composites in chemical plant and other corrosive environment is usually predicted on the basis of a simple immersion test. However, under the combined influence of a stress and a corrosive environment, a completely different mode of failure occurs which can lead to catastrophic failure of GRP vessels. This is due to a phenomenon called stress corrosion cracking.

The main features of stress corrosion will be described by reference to (1) a ring compression test in which the corrosive liquid is contained inside a circular section of the composite material which is compressed diametrically and, (2) a Mode I opening fracture mechanics test which involves measuring the rate of crack growth in flat sheet specimens immersed in the environment. The most important characteristic is the formation of a very flat fracture surface. Scanning electron microscopy has been used to study the mechanism of crack growth.

Stress corrosion is primarily due to the effect of an acidic environment on the strength of glass fibres. A model for stress corrosion provides an explanation for the experimentally observed effects of fibre type, chemical resistance and mechanical properties of the resin, orientation of fibres in relation to applied stress and stress intensity on the rate of crack growth and appearance of the fracture surfaces. The relevance of this model is model is discussed with respect to the design of vessels.

596

References

P.J. Hogg and D. Hull, 'Corrosion and environmental deterioration of GRP' in Developments in GRP Technology - Ed. B. Harris, Applied Science Publ., 1983, p. 37-90.

J.N. Price and D. Hull, 'Propagation of stress corrosion cracks in aligned glass fibre composite materials", J. Mater. Sci., 18, 1983, p. 2798-2810.

P.J. Hogg, J.N. Price and D. Hull, 'Stress Corrosion of GRP', 39th Annual Conference, RP/PI, Soc. Plastics Industry, New York, 1984, Session 5C, p. 1 - 6.

CHAPTER VII: LONG-RANGE APPLICATIONS OF GLASS

ATOMIC PROCESSES IN CHARGED PARTICLE IRRADIATION OF GLASSES

Paolo Mazzoldi

Department of Physics of the University, Via Marzolo n.8,
35100 Padova (Italy)

ABSTRACT

Glass modifications induced by ion implantation will be discus_sed with particular reference to alkali silicate glasses.

Charged particle irradiation introduces network damage and alters the chemical composition. Compositional changes are due to internal electrical field formation or to sputtering processes, connected to different stopping power (electronic or nuclear) regimes of incident particles. A comparison between the alkali profile modi_fications induced by light or heavy particle irradiations is presen_ted and discussed on the basis of phenomenological models.

H decoration of defects and mechanical modifications are repor_ted.

1 INTRODUCTION

Glasses are the most promising candidate materials for high le_vel waste solidification, due to their chemical stability, high solubilities for the waste elements, because of their amorphous network structure, well-known and economic fabrication procedures.

The efforts to optimize glasses are proceeded on the last 20 years for a number of demands as high crystallization and low melting temperatures, good mechanical and leaching stability, etc.

Problems that need further study are the response to radiation, applied stresses and the presence of water, including synergistic effects.

The investigation of long-term stability of glasses under irra_diation requires the use of accelerated tests, so as to simulate in a short time what would occur in many years. The validity of such experiments is critically dependent on a clear identification of ba sic damage mechanisms and related defects.

In this paper I will discuss the results of charged particle irradiation experiments in glasses and the correlated phenomenological models.

2 RESULTS

When glasses are irradiated with electrons(as in Auger Electron Spectroscopy AES) or ion beams (as in Secondary Ion Mass Spec troscopy) their surface composition undergoes marked modifications (see for example refs. 1-5). Such a modification, mainly connected to alkali migration, together with network damage, causes a variation in the refractive index, in the near surface hardness and in the tensile surface stress. The alkali migration can be correlated to different mechanisms, clearly connected to the different stopping power regimes of incident particles (6). Particularly in the electronic stopping power regime (low-mass particle irradiation), the alkali profile evolution appears to be governed by the ordinary and field assisted diffusion, with an electric field which is a function of the depth (7,8). In the nuclear stopping power regime (heavy-ion irradiation) the observed alkali depletion at the surfa ce (9-11) can be interpreted on the basis of a phenomenological model, which takes into account a preferential ejection of alkali atoms from the surface and of an enhanced diffusion over the range of implanted ions (12-14).

2.1 Low Mass Ion or Electron Irradiation (Electronic Stopping Power Regime)

In figure 1 we report the depth concentration profiles after 2.5 keV electron irradiations, at two currents, of soda lime glasses (15,16). The Na tends to accumulate towards the sample interior, at a depth comparable to the maximum electron range. The glasses were irradiated in a "XPS-Auger" apparatus and a decay of the Na surface concentration was observed during irradiation, in agree ment with data reported in the current literature. The Na depth profile were obtained by using the nuclear reaction $^{23}Na(p,\alpha)^{20}Ne$(17). The comparison between the Na profiles obtained after irradiation at the same energy but for different currents clearly indicates an effect of the beam power deposition in the sample. The Na surface depletion is accompanied by a Ca surface accumulation, detected by analyzing the same sample with the Rutherford backscattering technique (18) (see inset in fig.1).

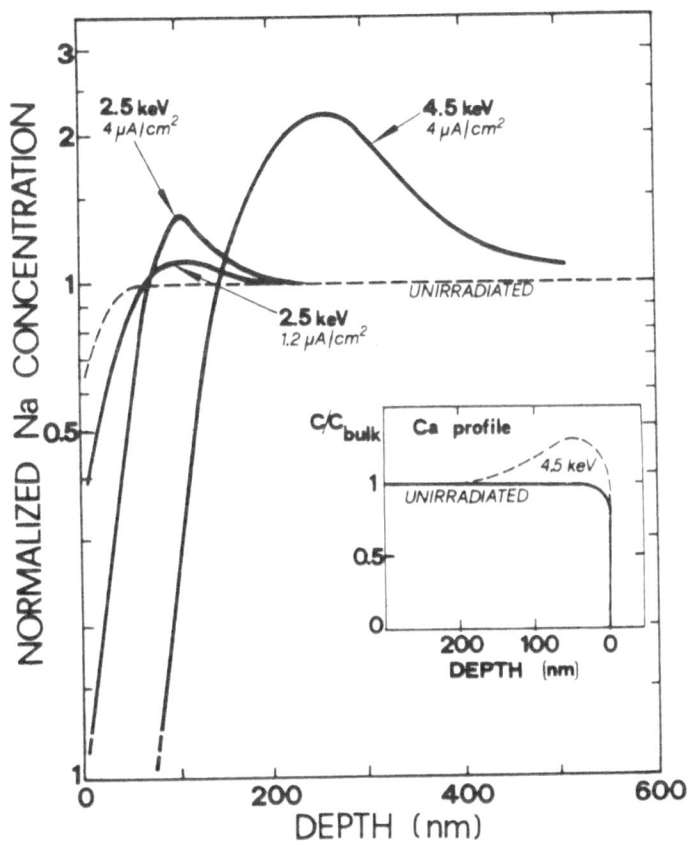

Fig. 1 - Sodium depth profiles, normalized to the unirradiated bulk value, for irradiation of 2.5 keV, at two current densities (I_1 = = 1.2 µA cm^{-2}, I_2= 4 µA cm^{-2}) and 4.5 keV electrons at a current of 4 µA cm^{-2}. The normalized calcium profiles for unirradiated and 4.5 keV electron irradiated glass are shown in the inset (ref.16).

In figure 2 we show the depth concentration profiles (full points), for unirradiated and 600 keV-proton irradiated soda lime glasses (15). We observe a Na migration towards the surface accompa̲ nied by a Ca surface depletion (see the inset).

Figs. 3a and b report the Na distribution (full dots) after 600 keV proton irradiation at two current density values in a soda lime glass, which showed an initial Na depletion, due to sample pre̲ paration (open dots) (8). The results are consistent with those re-

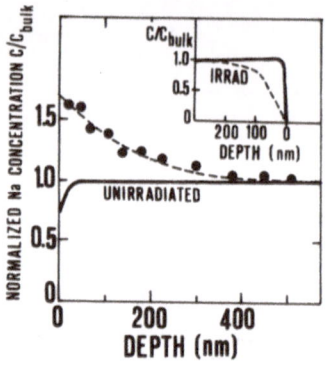

Fig.2-Normalized Na depth profile for unirradiated (-) and 600 keV proton irradiated (•) soda-lime glasses. Calculated profiles (---) are also reported with the indication of diffusion coefficient D and surface field strength F(0) values used in the calculations. The proton current density was 7.5 µA cm^{-2}. The normalized calcium profiles for unirradiated (full curve) and irradiated (broken curve) samples are also reported in the inset (ref. 15).

ported in figure 2. The influence of the beam power deposition is evident, as observed for electron irradiation, from the comparison between the two final Na profiles, obtained for equal implant duration, but at different current densities.

A study of the time evolution of the alkali migration process

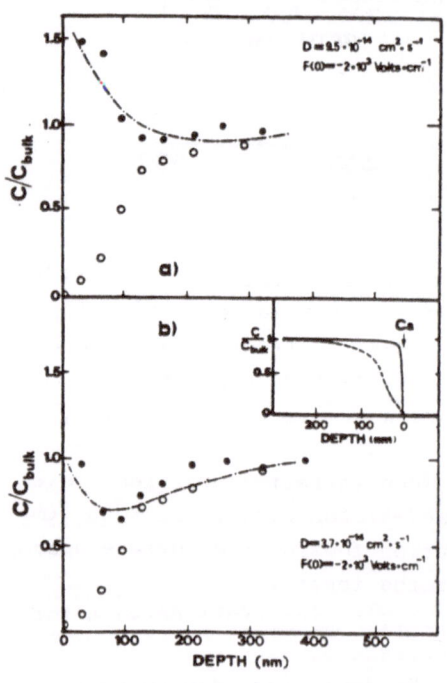

Fig.3 - Normalized sodium depth profile for unirradiated (o) and irradiated (•) soda-lime glasses for two proton current densities (a) 11.2 µA cm^{-2} and (b) 7 µA cm^{-2}. The irradiation energy was 600 keV and the irradiation time t = 36 min. Calculated profiles (---) are also reported. The normalized calcium profiles for unirradiated (full curve) and irradiated (broken curve) samples are shown in the inset (ref. 8).

confirms that the atomic processes, which govern the alkali profile evolution, are the same for low mass ion or electron irradiation (electronic stopping power regime).

Figure 4 shows a plot of the peak-to-peak heights, normalized to the peak-to-peak height at zero time, for the Na signal from a soda-lime glass as a function of the time of exposure to the exciting electron beams and for several values of beam current, in AES experiments (19). The Na signal decay rate is accomplished by a delay time, called residence time, t_r, before the Na signal starts to decay. t_r value increases as the beam current decreases.

AES experiments (19), performed at different temperatures, showed that t_r is strongly temperature dependent, with an activation energy of some hundredths of eV. Such a value appears very low in comparison to the ordinary activation energy for alkali diffusion in glasses. The time evolution of the Na surface accumulation during proton bombardment has been measured, for different beam currents and target temperatures (20).

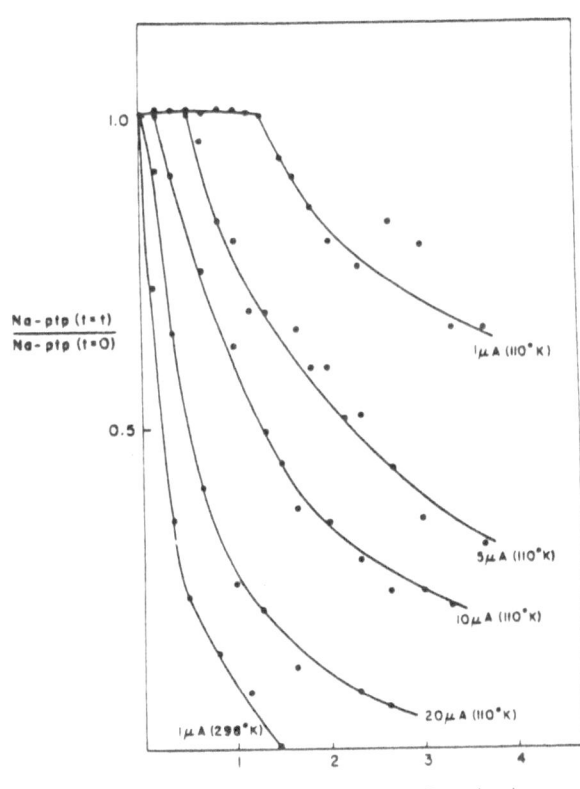

Fig. 4 – The decay at 110 K of the Na peak-to-peak height, normalized to the peak height at zero time, as a function of the time of electron irradiation for beam currents of 1, 5, 10, and 20 µA. A decay curve measured at room temperature and 1 µA is also shown (ref.19).

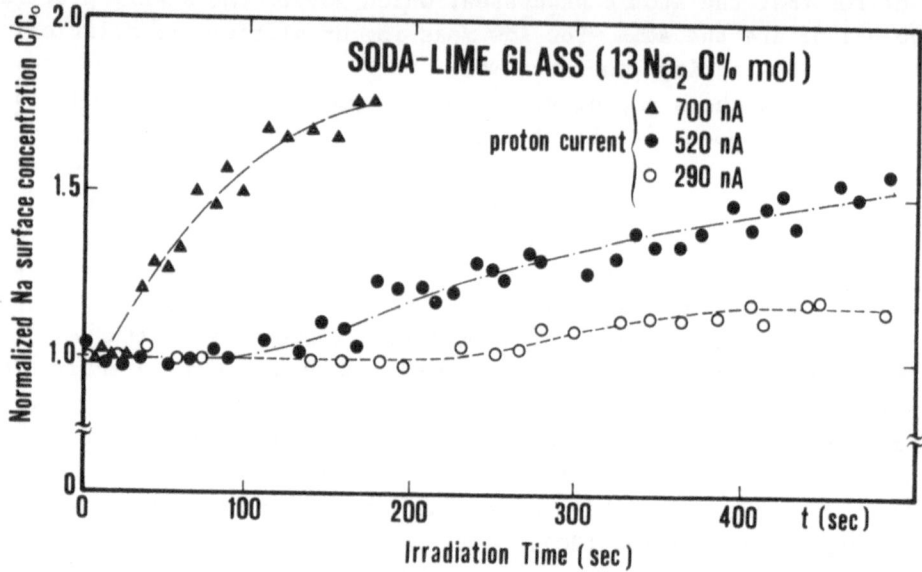

Fig. 5 - Surface Na concentration, normalized to the initial one, as function of 600 keV proton irradiation time at room temperature, for a soda-lime glass irradiated at three proton currents (ref.20).

Figure 5 reports the ratio between the Na surface concentration c, and the starting concentration, c_o, for a soda-lime glass, as a function of the irradiation time for three different current values of the 600 keV proton beam, at room temperature (20). A residence time, which is clearly beam current dependent, is evidenced.

The room temperature current dependence of t_r is shown in figure 6 (20). Moreover, the t_r dependence on beam current for different target temperature is reported in figure 7 (20).

From the above reported results, the t_r dependence from the total beam current, the lack of a critical proton dose for Na mobilization and the measured t_r values greater than those expected for the establishment of an electric field in the glasses (evaluated about some hundredths of a second (21)) or necessary to reach a steady value of the temperature under proton bombardment (22), make the role played by the electric field in the first stage of the alkali migration phenomenon questionable. Moreover the target temperature strongly influences t_r and if the assumption is made that, at constant current, the t_r temperature dependence is connected to a thermally activated processes, an activation energy of some hundredths of eV may be caluclated, consistent with the one reported

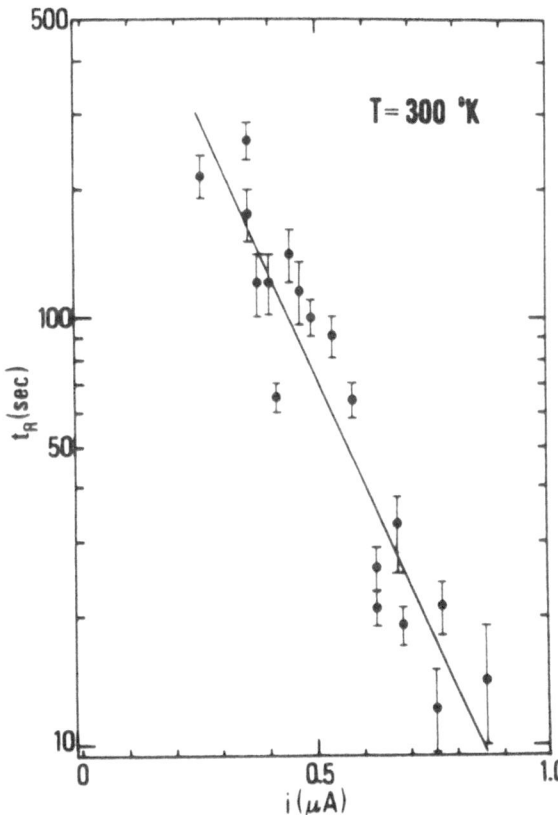

Fig. 6 - Residence time, t_r, as function of proton current, for room temperature irradiations of soda-lime glass. The full line is the best-fit of experimental data (20).

by Pantano et al (19) in electron bombardment experiments. This phenomenon, which is certainly threshold, may be connected to the production of defects, which from a certain moment onward make the Na ions move within the glass. Models, based on the percolation theory, which take into account both defect distribution and their time evolution, could be investigated.

The results on the alkali migration during low mass particle irradiation (electrons or protons) have been analyzed by assuming a field assisted Na diffusion due to an electric field constant in time but function of depth (7), going to zero at a depth X_0, roughly corresponding to the incident particle range.

The alkali flux due to a concentration gradient in the presence of an electric F(X) (23) is

$$J = - D \frac{\partial C(X,t)}{\partial X} + C(X,t) \, \mu F(X), \qquad (1)$$

where D and μ are the Na diffusion and mobility constants (connected by the Einstein relation) and C(X,t) the Na concentration at

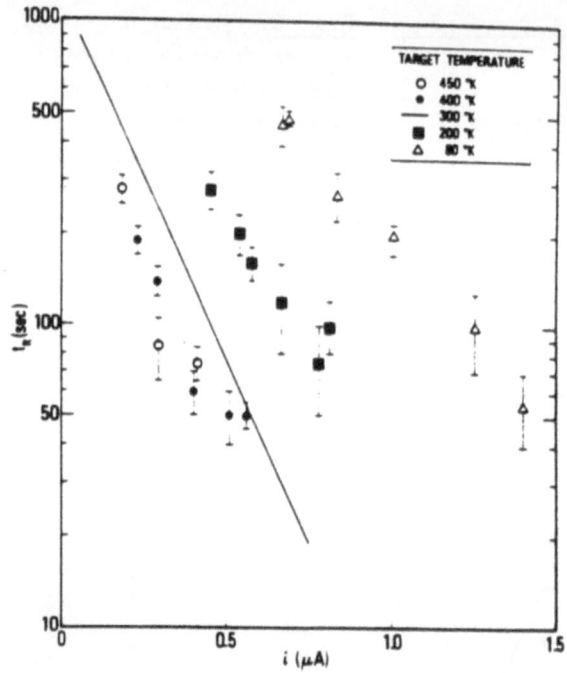

Fig. 7 – Residence time, t_r, as function of proton current, for irradiations at different temperatures. The full line is the best-fit of experimental data at room temperature (Figure 6) (ref.20).

depth X and time t. By the continuity equation we obtain:

$$\frac{\partial C(X,t)}{\partial t} = D\frac{\partial^2 C(X,t)}{\partial X^2} - \mu\frac{\partial}{\partial X}\left|C(X,t)\,F(X)\right|, \qquad (2)$$

assuming D to be constant.

Eq.(2) may be integrated numerically (19) starting from the initial alkali concentration, determined by the sample analysis before irradiation, and with the boundary condition:

$$-D\frac{\partial C(X,t)}{\partial X} + \mu F(X)\,C(X,t) = 0 \qquad (3)$$

at X=0 (surface position) and X=L, i.e. the maximum depth at which the Na diffusion (ordinary and field assisted diffusion) plays a ro le. In addition if we consider the electric field acting only up to the depth X_o<L, then

$$F(X_o)\,C(X_o,t) = 0 \qquad (4)$$

and the second term on the right hand side of eq.(2) may be omitted for X_o<X<L. The theoretical curves, reported in figs.2-3 for proton and in fig. 8 for 2.5 keV electron irradiations, are obtained by a fitting procedure. The calculated electric fields for positive ion irradiation are opposite in sign to those evaluated for the electron

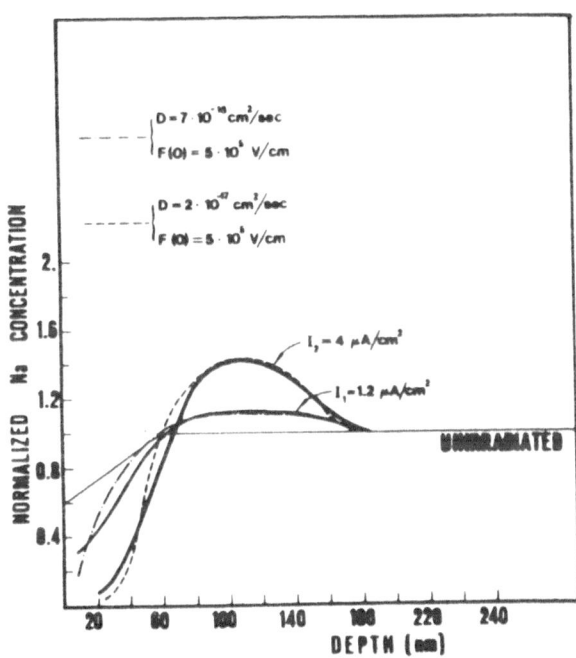

Fig. 8 - Calculated sodium depth profiles for 2.5 keV electron irradiations at two current densities. The full curves report the experimental results, shown in fig.1 (ref.7).

irradiated glasses. This fact suggests that the incident particle charge deposition mainly determines the electric field formation.

The migration process may be clearly ascribed to the fact that when low-mass particles strike the glass, electron bonds of alkali are broken according to Lineawever (25) and Puglisi et al. (26) in the incident particle range region, and so alkali ions are more susceptible to diffuse, not only at the surface (6), but also in the bulk where defects may be created by the irradiation process. As a consequence Miotello and Mazzoldi (27) introduced a physical idea of an alkali enhanced diffusion process, during electron irradiation of glasses, which may explain the results of Na Auger signal decay at liquid Nitrogen temperature reported by Ohuchi and Holloway (28).

The electric field formation cannot be determined only by the trapping of charged particles, indeed a very simple calculation, based on Maxwell equations, indicates in this case, the occurrence of the breakdown phenomena in very short times, of the order of 10^{-2} - 10^{-4} seconds (21). The electric field value, evaluated of the order of 10^5 V/cm in the case of keV-AES experiments (7) is controlled on the contrary by the electron mobility, which is related to the incident particle energy deposition.

Desorption of alkali ions, detected by a quadrupole mass filter, from glasses, exposed to a scanning electron beam has been reported by Ohuchi and Holloway (28). The desorption originates only from the surface monolayer (29). Such effect becomes important at current densities higher than 5 10^{-3} A/cm^2 and don't appear relevant to explain the decay rate of alkali signal in AES experiments (30).

I want to underline two problems connected with the above phenomenological analysis of the alkali profiles after low-mass particle irradiation. The first is related to the observation that the alkali migration is accompanied by a partial Ca/alkali ion exchange. These structural and compositional changes in the near-surface region of the sample may further contribute to a change of the activation energy of the self-diffusion coefficient and ionic exchange processes. The second problem is whether the Einstein relation between the mobility and diffusion coefficients is valid for the present experimental conditions (i.e. during irradiation processes): in fact, they are very different from simpler ones (23), for which this relation is shown to be appropriate.

2.2 Heavy Ion Irradiation (Nuclear Stopping Power Regime)

Soda-lime - silica glasses have been irradiated with Ar ions at different doses and energies. The effects of 50 keV-Ar$^+$ irradiation as a function of dose are reported in fig. 9 (11, 14, 30).

A Na depletion increasing with the irradiation dose is observed. Steady-state profiles are obtained for implantation doses higher than $4 \cdot 10^{16}$ ions/cm^2. The study has been extended by varying the current density between 0.5 to 2 A/cm^2. The Na profiles appeared independent from the implantation current (43).

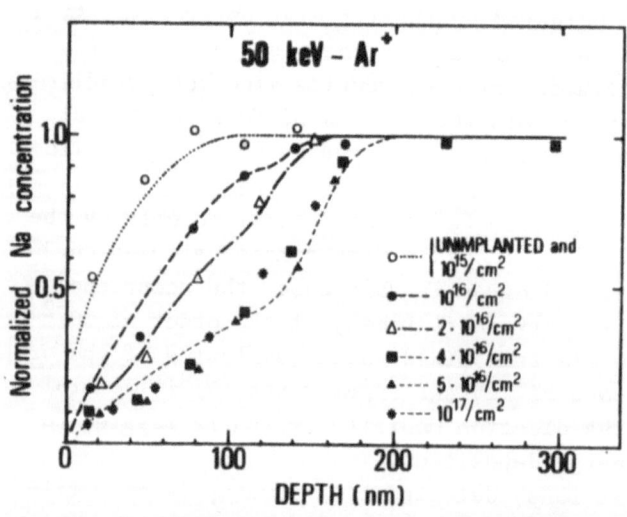

Fig. 9 - Experimental (see refs. 8,11,13) and theoretical Sodium profiles (curve) after 50 keV - Ar irradiation at different doses. For the calculation, a D*-value, $2 \cdot 10^{-14}$ cm^2 s^{-1}, greater than the D value evaluated for a temperature activated mechanism, has been assumed (ref. 43).

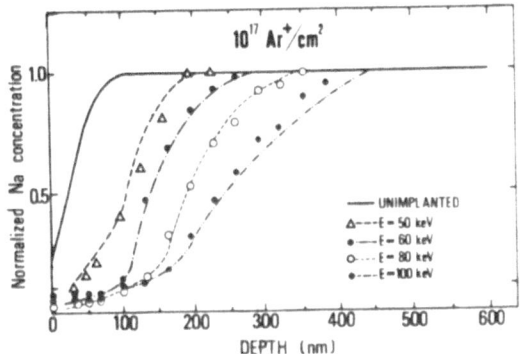

Fig. 10 - Experimental and theoretical sodium profiles (curves) after irradiation at different energies(43).

Fig. 10 shows the Na concentration modifications as a function of implantation energy between 50 and 100 keV, for a dose at which certainly steady-state profile was attained (10^{17} ions/cm^2) (43).

From the results reported in figs. 9-10 we can underline the following points:

1) the thickness of the alkali depleted layer is a function of the implantation dose, reaching a steady-state value.

2) The alkali profile, for a fixed irradiation dose, is independent of the incident ion current density.

3) The depleted layer thickness increases with the implantation energy.

The reported experimental results may be qualitatively described as follows (13): the alkali atom remotion rate larger than for other matrix components, preferential sputtering (31), and the subsequent migration towards the $\Delta \ell$ surface layer, involved in the sputtering processes ($\Delta \ell \leqslant 10$ Å) (32) induce the observed alkali depletion up to a large depth. The alkali migration towards the $\Delta \ell$ surface layer may be described (12,13,33) from a mathematical point of view by the transport equation, in a one-dimensional form

$$\frac{\partial c(X,t)}{\partial t} = D^* \frac{\partial^2 c(X,t)}{\partial X^2} + U \frac{\partial c(X,t)}{\partial X} \qquad (5)$$

where $c(X,t)$ is the alkali concentration at a time t and a depth X and D^* is the effective diffusion coeffcient. D^* accounts for the enhanced transport of alkali atoms in the matrix region affected by defect production, due to the local energy deposition in nuclear collisions processes of incoming ions.

The last term in eq.(5) takes a matrix erosion process, through matrix erosion speed, U. For the implantation energies used in the present experiments, the last term in eq.(1) can be neglected (34).

The use of the diffusion transport instead of the Boltzmann transport equation is justified under conditions of high energy deposition densities (35).

We account for the preferential sputtering of alkali atoms from the surface by means of the condition

$$D^* \frac{\partial c(X,t)}{\partial X} = H c(X,t), \quad X = 0 \tag{6}$$

assuming that the alkali loss is proportional to the concentration itself at the surface (12,33).

In fig. 9 the calculated sodium profiles (curves) are compared to the experimental results. The D^* value is $2 \cdot 10^{-14}$ $cm^2 \cdot sec^{-1}$ for an irradiation density of 2 $\mu A/cm^2$. For different irradiation currents $i(0.5 - 2 \ \mu A/cm^2)$ a linear relation between D^* and i values is observed.

In fig. 10 we report the sodium distribution (curves) evaluated following the previous calculation model. D^* appears a function of irradiation energy, reaching a value of $8 \cdot 10^{-14}$ cm^2/sec for 100 keV irradiation energy. Such a result can be explained considering the increase of the total number of vacancies, created by incident ions, as function of irradiation energy. The mechanism by which ion irradiation influences atomic transport have been discussed, including enhanced diffusion via mobile point defects, by Myers (36).The increased vacancy concentration due to the irradiation causes a proportional increase in diffusion of atoms at low temperature by the vacancy mechanism. In addition substitutional atoms are ejected into interstitial sites, from which they diffuse rapid. An analytical expression for D^* has been introduced by Myers (36) on the basis of the formalism of Dienes and Damask (37).

For high irradiation flux, vacancy agglomeration processes near the surface are expected due to lower vacancy distance. Such mechanism determine partial trapping of certain solutes and formation of blistering phenomena. Such trapping phenomena do not avoid the depletion in depth (43).

Enhanced diffusion processes govern the alkali migration during low or heavy mass ion irradiation in connection to the local deposited energy. These phenomena can modify the ionic exchange mechanisms and influence the glass chemical durability. Such observations assume a particular importance for the nuclear waste disposal problems.

2.3 Defect evaluation – Thermal stability

Recently (38,39) the damage caused by the implantation has

been characterized by measuring the enhanced of the penetration of water into irradiated silica glasses.

The enhanced penetration is due to creation of chemically reac tive defects which greatly increase the reaction between water and the silica network.

Figure 11 reports the H depth profiles determined by means of the nuclear reaction H $(^{15}N,\alpha\gamma)^{12}C$, in a 207 keV - Pb implanted sili ca glass leached during 3 hours in deionized water at 100°C in sta tic conditions. The implantation dose ranged from 10^{12} to $10^{15}Pb/cm^2$ (39).

Figure 12 shows the H profiles in silica implanted at the same dose, 10^{15} Pb/cm^2, and leached for increasing time (39).

There are a number of features, as discussed in ref. 39, in the reported results, useful to establish the defect distribution during heavy ion implantation: a) the damage induced by ion implant ation increases the rate of water (Hydrogen) penetration into the silica. b) The effects of ion implantation penetrate much farther into the glass than the range of the implanted ions (≈600 Å for 207 keV-Pb ions). This fact could indicate the tail of the damage distribution. c) The thickness of the hydrated layer increases with the leach time. For 3 hours leach times the H concentration profile shows a nearly gaussian shape with a maximum at a depth of 600 Å, which reflects the defect distribution. For very long leaching ti mes the profile extends to 2000 Å, loosing the specific effect of ion induced defects. Such H profiles are observed in leached alka li silicate glasses and are attributed to the formation of a hy drated surface gel layer (39). For 13 days leaching a decrease of hydrated layer thickness is due to the dissolution of the sur

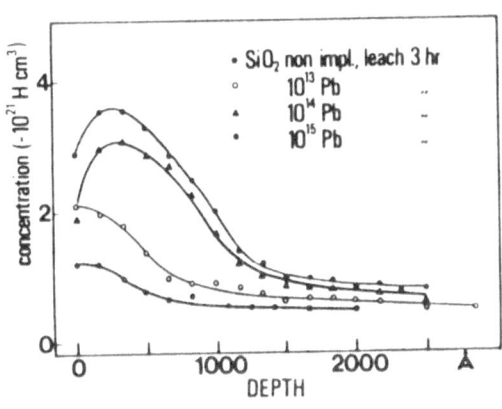

Fig. 11 - H depth pro files for silica im planted at increasing doses and leached at a fixed time (ref.39).

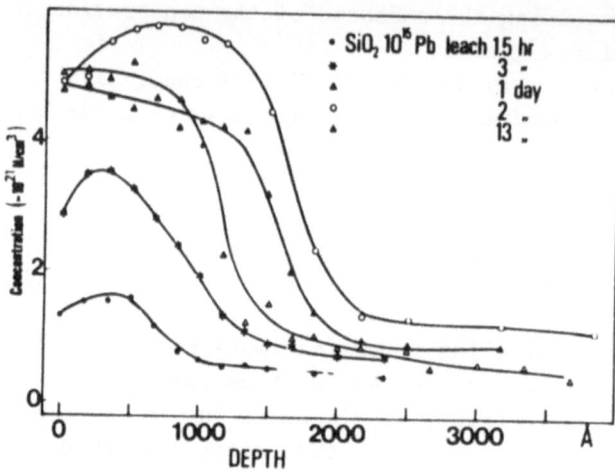

Fig. 12 - H depth profiles for silica implanted at 10^{15} Pb·$\cdot cm^{-2}$ and leached at increasing times (ref.39).

face damaged layer. d) The maximum H concentration in the implanted layer is largely lower than the total number of the evaluated atomic displacements, produced by Pb implantation. This result could be interpreted on the basis of two different hypothesis: first, the H reacts only with a given type of defects, second, most of defects anneal or recombine during the implantation.

Figures 13 and 14 report the H concentration in silica, implanted at low or high dose respectively, annealed at increasing temperatures and then leached at a fixed time (40).

We observe an annealing of the defects in the 10^{13} Pb/cm^2 implanted silica, at a temperature lower than that for the higher im

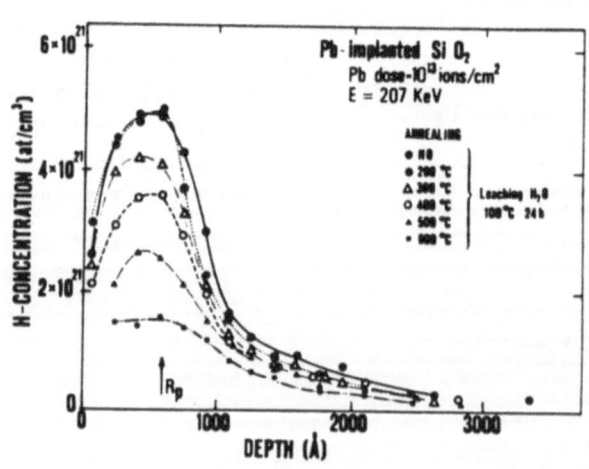

Fig. 13 - H depth profiles for silica implanted at 10^{13} Pb·$\cdot cm^{-2}$, annealed and leached at fixed time (ref.40).

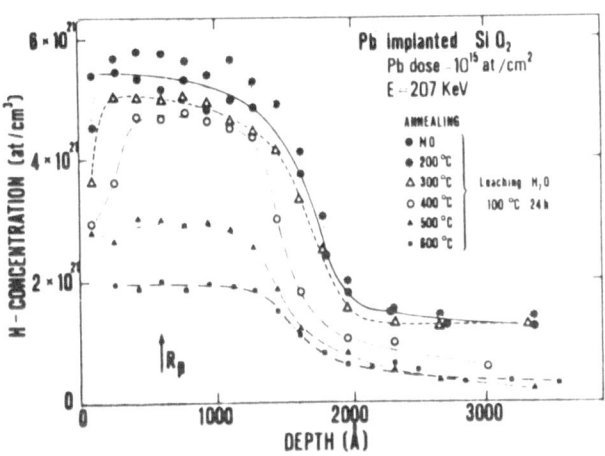

Fig. 14 - H depth profiles for silica implanted at 10^{15} Pb·cm^{-2}, annealed and leached at fixed time (ref. 40).

plantation dose. This fact indicates that the implanted layer is constituted by point defects, defect clusters and extended defects which anneal at different temperatures. The formation of interstitial clusters has been assumed to increase the fracture thoughness (41,42), observed in Ar implanted soda-lime glasses (42).

Figure 15 shows the percentage of cracks,which develop after application of a loaded square pyramidal indented on 100 keV-Ar implanted soda-lime glasses, as function of implantation dose for three applied loads (41). A reduction of crack development is observed for all implantation fluences. Figure 16 reports the diffe_rence, Δc, of the average crack length, c, between the implanted and the unimplanted glasses, at the same indented load, as a func_tion of the ion dose. A reduction in the crack length, the maximum of which occurred for implantation doses of the order of $5 \ 10^{15}$ ions/cm^2, is evident. A reduction of crack formation is ac_

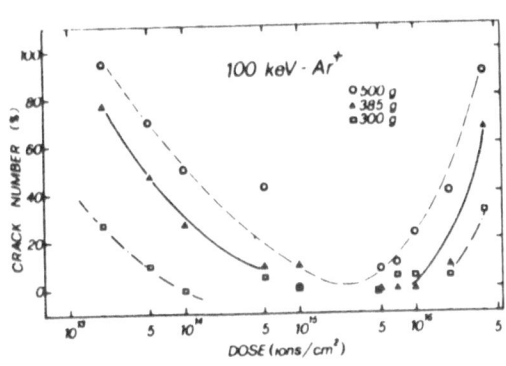

Fig. 15 - Percentage of cracks versus implantation dose for three applied loads (ref.42).

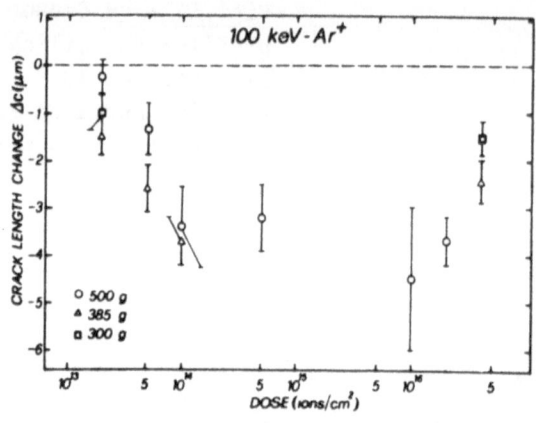

Fig. 16 - Change of radial crack length as function of implantation dose for three applied loads (ref. 42).

companied by a shorter length (surface compression). The thermal stability of the surface mechanical resistance of implanted glass has been studied by performing indentation tests on sample annealed at different temperatures and for a constant time (42). The annealing results have been analyzed by studying the temperature dependence of the P^X load, at which the crack percentage reaches the value 50%. Figure 17 shows the parameter P^X using a logarithm scale, as a function of $1/T$, where T is the annealing temperature. An activation energy of 0.07 eV, characteristic of point defect annihilation, is obtained.

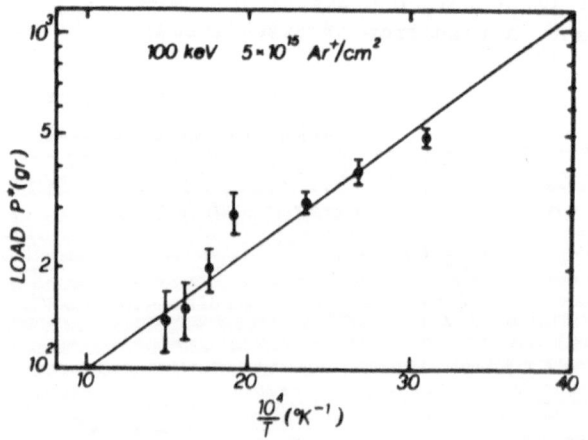

Fig. 17 - Applied load at which the percentage of cracks reaches a value of 50% versus the annealing temperature (ref.42).

3 CONCLUSIONS

Alkali migration observed during particle irradiation of glasses, has been explained in terms of two different mechanisms, connected to the stopping power regimes of incident particles. Leaching experiments, performed on Pb implanted silica glasses, show a defect decoration by the Hydrogen, indicating that the implanted layer is constituted by point defects and defect clusters. Annealing experiments support this interpretation. The formation of interstitial clusters causes an increase in fracture thoughness, a very beneficial effect.

The author is grateful to Drs. G. Battaglin, G. Della Mea, G. De Marchi, S. Lo Russo (Dipartimento di Fisica, Università di Padova); R. Del Maschio, V. Gottardi, M. Guglielmi (Istituto di Chimica Industriale - Facoltà di Ingegneria, Università di Padova); A. Miotello (IRST Povo, Trento), for their contribution to the reported results and for stimulating discussions.

REFERENCES

1. R.G. Gossink, M. Van Doveren and J.A.T. Verhoeven. J. Non-Cryst. Sol. 37 (1980) 111.
2. F. Ohuchi, M. Odino, P.M. Halloway and C.G. Pantano Jr. Surface and Interface Analysis 2 (1980) 85.
3. P. Champion and P. Lacharm. Verres Réfract. 35 (1981) 65.
4. O.S. Heavens and D.M. Usher. Verres Réfract. 35 (1981) 76.
5. D.M. Usher. J.Phys.C 14 (1981) 2039.
6. P. Mazzoldi. Nucl.Instr.Meth. 209/210 (1983) 1089.
7. A. Miotello, P. Mazzoldi. J.Phys.C 15 (1982) 5615.
8. G. Battaglin, G. Della Mea, G. De Marchi, P. Mazzoldi, A. Miotello and M. Guglielmi, J. Phys. C 15 (1982) 5623.
9. G.W. Arnold and P.S. Peercy. J. Non-Cryst.Sol. 41 (1980) 359.
10. G.W. Arnold, P.S. Peercy and B.L. Doyle. Nucl.Instr. Meth.182/183 (1981) 733.
11. V. Chinellato, V. Gottardi, S. Lo Russo, P. Mazzoldi, F. Nicoletti and P. Polato. Rad.Eff. 65 (1982) 31.
12. Z.L. Liau, J.W. Mayer, W.L. Brown and J.M. Poate. Appl.Phys. 49 (1978) 5295.
13. A. Miotello and P. Mazzoldi, J. Phys.C 16 (1983) 221.
14. G. Battaglin, G. Della Mea, G. De Marchi, P. Mazzoldi, A. Miotello and M. Guglielmi. Journ.de Phys. 9 (1982) 645.
15. G. Battaglin, G. Della Mea, G. De Marchi, P. Mazzoldi and O. Puglisi, Rad.Eff. 64 (1982) 99.

16. G. Battaglin, G. Della Mea, G. De Marchi, P. Mazzoldi and O. Puglisi. J.Non-Cryst.Sol. 50 (1982) 119.
17. A. Carnera, G. Della Mea, A.V. Drigo, S. Lo Russo and P. Mazzoldi. J.Non-Cryst.Sol. 23 (1977) 123.
18. P. Mazzoldi and G. Della Mea. Thin Sol.Films 77 (1981) 181.
19. C.G. Pantano, D.B. Dove and G.Y. Onoda Jr. J.Vac.Sci.Tech. 13 (1976) 414.
20. G. Battaglin, G. Della Mea, G. De Marchi, P. Mazzoldi. To be published.
21. A. Miotello. To be published.
22. G. Della Mea, G. De Marchi, E. Grinzato, A.Mazzoldi, P. Mazzoldi and A. Miotello. Journ.Phys.C 16 (1983) 6329.
23. Y. Adda and J. Philibert. "La diffusion dans les solides", Press Universitaires de France, Institut National des Sciences et Tech niques Nucléaires, ch. XVI (1966).
24. M.S. Carslaw and J.C. Jaeger. "Conduction of heat in solids",Oxford University Press, London, ch.XVIII (1959).
25. J.L. Lineweaver. J.Appl.Phys. 34 (1963) 1786.
26. O. Puglisi, G. Marletta, A. Torrisi. Journ.Non Cryst.Sol. 55 (1983) 433.
27. A. Miotello and P. Mazzoldi. J.Phys.C, in press (1984).
28. F. Ohuchi and P.M. Holloway. Journ.Vac.Sci.Tech. 20 (1982) 863.
29. D. Menzel. "Topics in Applied Physics" ed.by Gomer (Springer, Berlin 1975).
30. A. Miotello and P. Mazzoldi. Rad.Eff., in press (1984).
31. L. Holland. Brit.J.Appl.Phys. 9 (1958) 410.
32. R. Behrisch, ed. "Sputtering by particle bombardment I, Topics in Applied Physics", Springer Verlag, Berlin-Heidelberg-New York (1981).
33. A. Miotello and P. Mazzoldi. J.Appl.Phys. 54 (1983) 4235.
34. R.L. Hines. J.Appl.Phys. 28 (1957) 587.
35. R. Collins and G. Carter. Rad.Eff. 54 (1981) 235.
36. S.M. Myers. Nucl.Instr.Meth. 168 (1980) 265.
37. G.J. Diemes and A.C. Damask. J.Appl.Phys. 29 (1958) 1713.
38. C. Burman and W.A. Lanford. J. Appl.Phys. 54 (1983) 2312.
39. G. Della Mea, J.C. Dran, J.C. Petit, G. Bezzon and C. Rossi-Alvarez. Mat.Res.Soc.Conference, Boston 1983.
40. G. Della Mea, J.C. Dran, J.C. Petit, G. Bezzon and C.Rossi-Alvarez. To be published.
41. Hj.Matzke and G. Linker, Int.Conf.Rad.Eff.Insulators 1983.
42. G. Battaglin, R.Del Maschio, G.Della Mea, G. De Marchi, V.Gottardi, M. Guglielmi, P. Mazzoldi and A. Paccagnella, in ref.41.

RADIATION DAMAGE OF GLASSES FOR NUCLEAR WASTE STORAGE:
OPTICAL AND MICROSTRUCTURAL ASPECTS

M. Antonini[1], P. Camagni[2], A. Manara[2] and M. Sacchi[1]

1 GNSM and Istituto di Fisica dell'Università,
  41100 Modena - Italy

2 Joint Research Centre - Ispra Establishment,
  21020 Ispra (Va) - Italy

ABSTRACT

A possible way to achieve a stable nuclear waste form consists of incorporating the different radionuclides in borosilicate glasses. As a consequence of the long meanlife of some active species, e.g. actinides, radiation damage is produced in the storage matrix, which may potentially affect its long-term stability.

Glasses based on SiO_2 have a mixed ionic-covalent bonding, hence both ionization and atomic displacements are of relevance in determining the damage. Precursor atomic defects, present before irradiation, as well as added modifiers, including sodium, interact with the bombarding particles: γ-rays, electrons, light and heavy ions. Under certain circumstances, individual defects tend to coalesce and defect clusters, usually in the form of gas bubbles, nucleate and grow.

The overall damage thus consists of relatively strong modifications of the atomic bonding and of the local microstructure. Such modifications have been examined. Optical absorption spectra of pure SiO_2 provide the basis of interpretation for individual defects. Results of High Voltage Electron Microscopy, where irradiations and "in situ" observations take place simultaneously, allow to discuss the agglomeration effects.

Both aspects of radiation damage are discussed in connection with the vitrification of nuclear waste, with special regard to possible implications upon the leaching of active species.

1. INTRODUCTION

Among the factors influencing the choice of a suitable glass for long-term nuclear waste storage, radiation stability has deserved an interest which can be compared only to leach resistance (1). In the earlier period, most of the research was devoted to screening tests or to assess changes caused by radiation to such quantities, e.g. density or stored energy, characterizing the bulk behaviour of the glass or its thermal response (2-5). The observed enhancement of the leaching rate under irradiation, originally reported by Dran et al. (6), has successively concentrated the attention upon the presence or absence of this effect under potentially reproducible long-term storage conditions (7). While radiation enhanced leaching is certainly a subject of great technological interest as a major potential cause of damage under long-term irradiation, the full evidence for long-term stability and survivability of nuclear waste forms requires an understanding of the basic chemical and physical processes associated with radiation damage of borosilicate glasses. This is particularly necessary in those cases, such as accelerated tests, where results obtained in a relatively short period of time, say from a few minutes to a few years, are interpreted as representative of processes taking place over the mean half-decay life of actinides stored in the vitrified waste form ($\sim 10^5$ years).

Over the past few years, we have undertaken a program aimed at identifying the basic phenomena taking place in silica based glasses irradiated with particles of different mass and energy. As is required for most other materials, these studies have involved the identification of the atomic defects (8,9) and the observation of the kinetics of their formation and annealing (10). Besides the formation of new defects during irradiation, the modifications of precursor defects, already present before particle bombardment, have also been followed.

More recently, we have extended our investigations to explore defect clustering effects with special regard to the nucleation and growth of oxygen gas bubbles in electron irradiated borosilicate glasses (11-14). In the following, we shall describe some of the most relevant obtained experimental results which will further be discussed within the limits of a first attempt to unify into a single frame of interpretation the different phenomena taking place both at the atomistic and at the microstructural scale.

2. BASIC DEFECTS AND BASIC DAMAGE MECHANISMS

The relatively open atomic network of silica – based oxide glasses, together with the mixed ionic-covalent character of the Si-O bond (15) represents a source of complexity which often hinders a clear interpretation of the observed effects in terms of electronic and atomic defects. Here, we neither describe all types of detected defects nor discuss the several problems still to be solved in order to obtain a complete picture of the defect states in vitreous SiO_2, since they have been described both experimentally and theoretically by Griscom (16), Greaves (17) and Mott (18). The obtained results have had a direct impact not only in the field of nuclear waste vitrification, but also in other glass technologies including fiber-optics, metal-oxide semiconductors and materials for fusion reactors.

We shall instead focus our attention upon those defects which involve oxygen atoms, since they are probably present in bubbles. The best known effect relating oxygen defects to optical absorption is the so-called E_1' centre, i.e. a narrow optical absorption band located at 5.85 eV in the U.V. region (Fig.1). An equivalent signal is observed in the electron spin resonance spectrum. E_1' centres are associated with triply coordinated silicons having a "dangling" bond which contains only one unpaired electron (Fig.2).

Different types of precursor defects, formed during irradiation or even present prior to irradiation, may however produce such a final configuration. If the precursor defect is an oxygen vacancy, the E_1' centre is produced after cleavage of the Si-Si bond which generates two SiO_3 asymmetric groups, one of which captures the unpaired electron during ionization. This is the case during massive particle irradiation (light or heavy ions), or with bombardment with high doses of high energy electrons, e.g. those produced in a high voltage electron microscope. For γ irradiation or electron irradiations at lower doses, the E_1' centres are instead generated from isolated triply coordinated silicons whose concentration directly corresponds to that of Non-Bridging Oxygen-atoms (NBOs). The two types of defects, namely oxygen vacancies and NBOs, have however different concentration in SiO_2, since the first one can be directly produced during irradiation and attains a saturation level of about $5 \cdot 10^{19}$ centres/cm^3, while those observed after γ or low dose electron irradiation have to be considered as structural imperfections already present in the network prior to irradiation, which become observable after ionization generated by the impinging particles. The concentration of such E_1' centres is much lower than that achievable by oxygen vacancies, being about 10^{17} centres/cm^3, as reported in one of our previous papers (9). Moreover, this number refers to pure "dry" silica. In "wet" commercial silica samples, the high concentration of OH groups is likely to block the formation of isolated triply coordinated silicons leading to a reduced amount of E_1' centres (19).

In a conventional annealing experiment, the kinetics of recombination of oxygen vacancies is hidden by thermal bleaching of the electrons producing the observable centre, which takes place at a lower temperature. A suitable way of observing the annealing kinetics of the precursor defect is to re-establish the ionization field at each annealing temperature by means of a purely ionizing radiation. Fig.3 shows the result of such a procedure. The excess E_1' precursors, originally produced by bombardment with massive, displacing particles, are readily coloured again after low dose

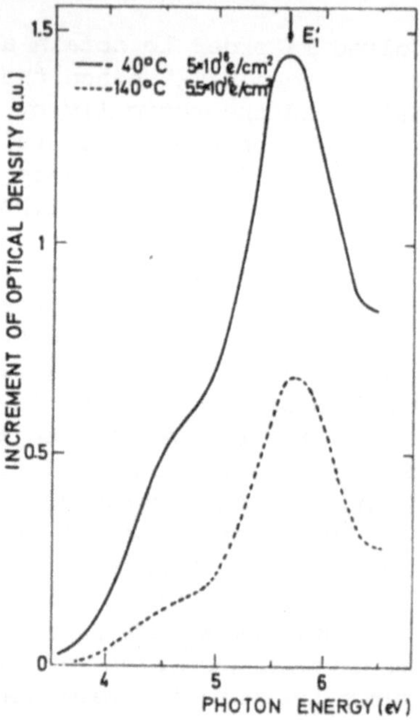

618

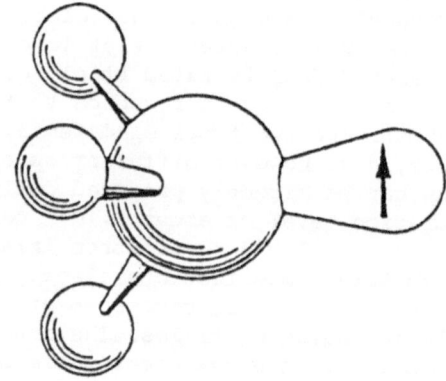

Fig. 1 - Increment of optical density due to E'_1 centres after electron irradiation at two different temperatures.

Fig. 2 - Simplified model of an E'_1 centre. Small spheres represent oxygen atoms (from Griscom, ref. 14).

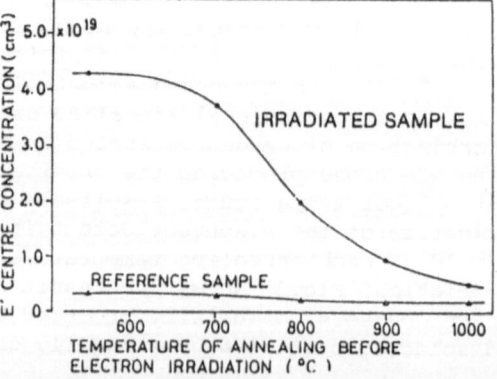

Fig. 3 - Annealing of excess E'_1 precursors in heavy-ion damaged v-SiO$_2$, compared to an undamaged sample (ref. 10 b).

electron irradiation, and start to disappear only at relatively high temperatures, around 900°C (10b) indicating a great stability of this type of defect. The network oxygen diffusion coefficient in amorphous SiO_2 has recently been measured by Mikkelsen (20). Oxygen self diffusivities of the order of 10^{-15} cm^2s^{-1} were found, together with an activation energy of 2 eV. These values can be compared with analogous results obtained in more complex soda-lime silicate glasses by Terai and Oishi (21), where in the same temperature range higher diffusion coefficients (D $\sim$ 10^{-11} cm^2s^{-1}) and lower activation energies (E = 1.54 eV) are reported. It seems therefore that a higher mobility of oxygen atoms produced during irradiation of more complex silicate glasses has to be expected.

A second type of oxygen-related centre is the so-called per-oxy-radical which produces a pronounced optical absorption band at 7.6 eV (The E-band) and was originally discovered by Friebele et al. (22) following ionizing irradiation. The precursors of these defects are envisaged to be $\equiv$ Si-O-O-Si $\equiv$ structures, and although pre-existent in the silica, they are formed in greater number during neutron bombardment. When observing that the annealing behaviour of this centre is close to that of the B2 band, which is only formed after massive particle irradiation (19), we attribute also this precursor to displaced atoms.

The atomic damage mechanisms operating in glasses more complex than pure silica, such as alkali borosilicate glasses, are difficult to explore, owing to their high optical absorption in the u.v. range. One could infer that the efficiency of oxygen vacancy production is increasing with the number of NBOs, since to this type of bond corresponds a lower energy. On the other hand, the presence of additional network formers, like boron, may favour saturation of NBOs, similarly to what has been observed above in connection with wet silica. We hope that future experiments will elucidate better this important point.

3. NUCLEATION AND GROWING OF BUBBLES UNDER ELECTRON IRRADIATION

In the aim to clarify the damage mechanisms underlying the effects of local, but intense beam irradiation, and in order to achieve an overall picture of the evolution of defects in silica-based glasses, we have extended our previous studies on colour centre production and annealing to include microstructural changes. The most suitable tool for these studies is High Voltage Electron Microscopy (HVEM), since it allows us both to produce and observe damage in small portions of the sample at various temperatures simultaneously. Moreover, the high energy electrons (E = 1 MeV) produced by the HVEM are transmitted without excessive absorption through relatively thick samples ($\sim$ 1 µm). In these conditions bulk damage may be observed without escape of defects from the sample surface. On the other hand, only relatively high damage rates, both in terms of displacement and ionization, may be applied to the sample from the microscope intense electron source, creating

620

nominal local defect concentration of about 0.01 dpa/s*. Therefore, the obtained results may be considered as corresponding to extremely high local damage conditions which could be conceived in nuclear waste glasses. The first observations, performed in pure vitreous silica, were rather deceiving, since no defect agglomeration was detected, even with fully focussed beam at each of the adopted irradiation temperatures from RT to about 800 K (11). Such absence of microstructural changes has also been confirmed by other authors (23) and it is an important factor which needs justification in all models of interpretation of those phenomena which conversely take place readily in more complex glasses. In fact, HVEM irradiation of alkali borosilicate glass produces, under appropriate conditions, large spherical regions where electrons are scattered much less than in the unirradiated area. These prevalent microstructures have shapes similar to those of gas filled bubbles as previously found in metals (24) and other insulators (25), and are unaffected by annealing at high temperature, contrary to what would be expected for voids nucleation. The interfaces between pairs of different sized cavities are curved in the direction expected from an inverse relationship between diameter and internal pressure. For these reasons we believe that they are indeed gas filled bubbles although a direct evidence for the presence of gas in the pores has not yet been achieved. Fig.4 is indicative of the high rate at which bubble growing can take place under appropriate temperature, dose and dose rate conditions in alkali borosilicate glasses having typical compositions.

The rapid growing of the microstructure is better followed with the help of an image intensifier TV camera and video tape. Quantitative measurements of the bubble radius, their number density in the irradiated volume and their evolution with time and temperature are made using an image analyser connected to a computer on-line, where data are stored and processed. It seems advisable to begin to describe our results from a macroscopic approach, similar to that adopted to analyse the growth of voids or bubbles in irradiated metal (26). We define a "swelling" parameter:

$$S = \frac{\Delta V}{V} \qquad (1)$$

which is the ratio between the total volume ΔV occupied by bubbles and the original volume V in which they were produced. A plot of the dose dependence of the swelling parameter is given in Fig.5,

---

* 1 dpa corresponds statistically to one displacement for each atom present in the irradiated volume. The calculation is performed assuming an average value of $\sim$ 25 eV for the displacement energy.

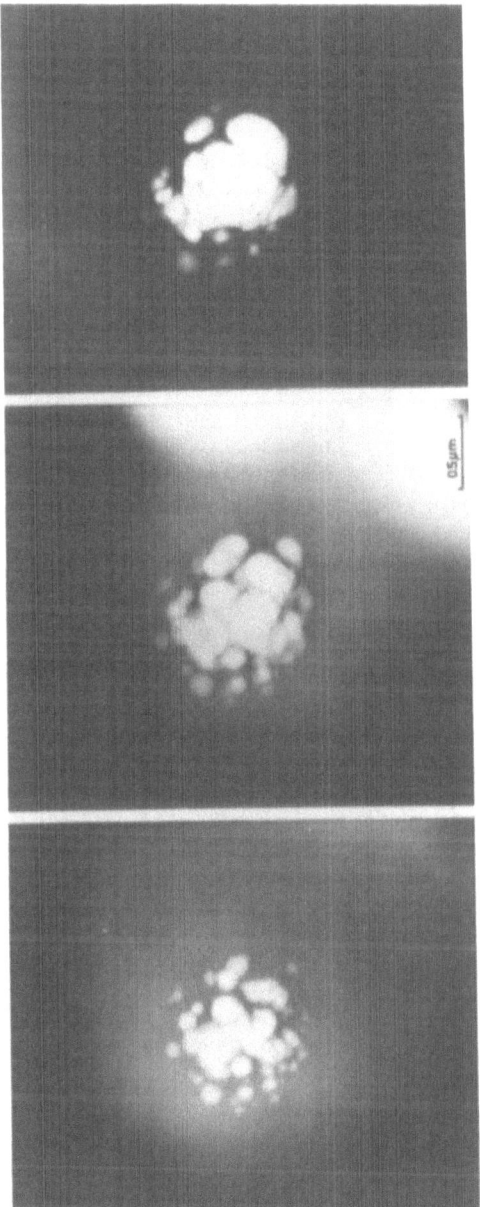

Fig. 4 - Bubbles in electron irradiated alkali borosilicate glasses. Irradiation times are 10, 30 and 90 sec for figures (a), (b) and (c), respectively. The dose rate is $1.7 \cdot 10^{22}$ e m^2 s^{-1} (ref. 11). Temperature is about 200°C.

for a typical glass composition, dose rate and irradiation temperature. Three characteristic stages present in all our observations are detectable from the figure:
- an incubation dose, which in this case is $6 \cdot 10^{25}$ e $m^{-2}s^{-1}$, required before any swelling is observable (bubble radius less than about 10^2 Å);
- a short, almost parabolic nucleation stage, immediately followed by a linear increase;
- a tendency towards saturation at high doses, over 10^{26} e $m^{-2}s^{-1}$.

At higher doses quantitative measurement of swelling becomes increasingly difficult, since adjacent bubbles superimpose each other.
    Other general features of the phenomenon, which do not appear from observation of Fig.5 are:
- the presence of a minimum dose rate below which swelling does not take place;
- a strong dependence of both quantities, i.e. incubation dose and critical dose rate, on temperature and sample composition. The temperature dependence of the swelling rate shows a maximum at a temperature value, which decreases with the increase of alkali concentration (Fig.6).

It is interesting to compare the swelling curves for the glass and a gas doped metal (24) (Fig.7). The two curves have a quite similar behaviour. The dose unit for the case of the metal is expressed in dpa, since atomic displacement is considered as the prevalent driving mechanism in the case of metal swelling. Also other features of the metal processes, such as the temperature dependence of the swelling rate and the presence of a denuded zone at the surface where bubbles do not nucleate, are reproduced in glasses. On the other hand, bubble growing takes place even in pure metals, on condition that extra gas atoms are implanted prior to or during irradiation. In glasses, the presence of additional components, i.e. alkali oxides, determines both the onset and the continuation of the phenomenon. However, no extra gas is necessary.
    We now discuss, with the help of previously unpublished data, a possible damage mechanism which is capable of explaining many of the observed quantitative aspects.
    Three major questions need an answer. First, what is the condition determining the onset of nucleation? Second, why bubbles grow linearly and how can the observed temperature dependence of the swelling efficiency be explained with a kinetic model? Finally, what is the composition of the observed bubbles?
    We start from some considerations relative to the last point, which, however, throws a considerable light on the entire process.
    The average concentration of gas atoms present in a bubble is readily calculated from the measurement of their size distribution and number density. A bubble of radius R in a material with surface energy σ at a temperature T contains approximately

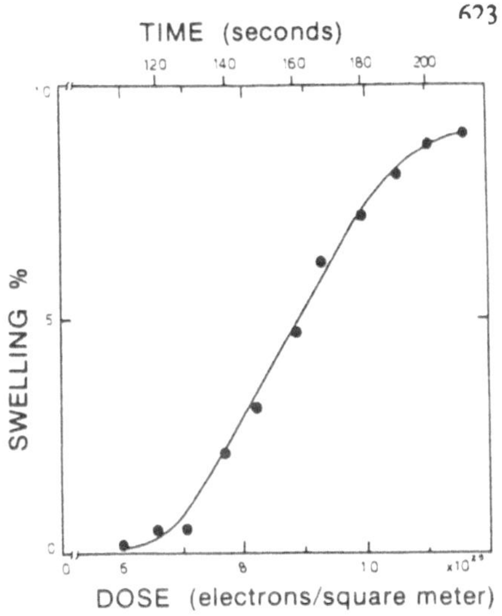

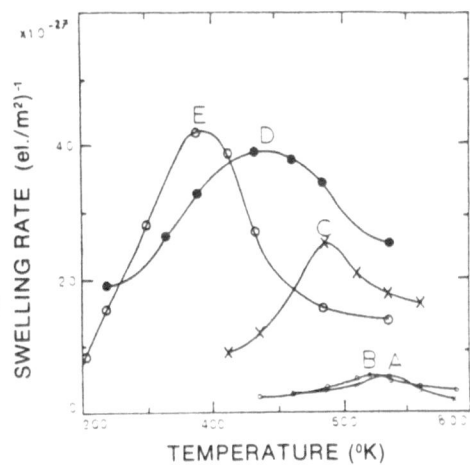

Fig. 5 - Dose dependence of swelling in an alkali borosilicate glass with composition given by column B of Table 1. The dose rate was $5.5 \cdot 10^{23}$ e m^{-2}s^{-1} and the irradiation temperature 523 K.

Fig. 6 - Temperature dependence of the swelling at various glass compositions. The letters refer to the corresponding columns of Table 1.

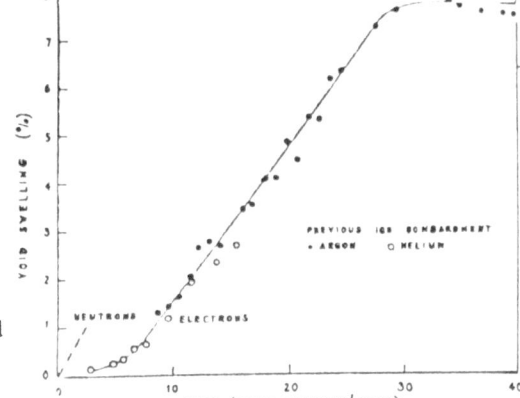

Fig. 7 - Typical swelling curve for a metal. The plotted data are obtained by Norris (24) after electron irradiation of nickel previously implanted with argon and helium ions.

624

$$n = \frac{8\pi\sigma R^2}{3KT}$$

(2)

gas atoms. This expression is a good approximation for large bubbles (R ≥ 1000 Å). For smaller bubbles the equation of state of a perfect gas used in eq.(2) should be replaced by the Van Der Waals equation. Setting

$$m = \left(\frac{R}{r}\right)^3$$

(3)

as the number of vacancies of radius r filling the bubble, the ratio

$$\frac{m}{n} = \frac{3\pi KT}{8\sigma} \frac{R}{r^3}$$

(4)

represents the relative amount of vacant sites to gas atoms. As can be seen from eq.(4), this number increases proportionally with R and it attains values of over 10^3 when our data and typical experimental conditions (R ∿ 1000 Å, r ∿ 1 Å, σ = 250 ergs/cm^2 (25), T ∿ 500 K) are inserted in eq.(4). Higher values of σ are also reported in the literature (28) from fracture analysis. However, these values are much exaggerated upper limits since they contain also the contribution from plastic deformation. Thus, as many as about 1000 vacancies are required for each additional gas atom entering a bubble. Although gas atoms may be required during the nucleation stage to stabilize the void, atomic or ionic depletion seems therefore as the driving mechanism to bubble growth. Since no bubbles grow in pure silica, such a depletion must be due to a fast diffusing species which does not enter the silica network. Ion depletion in alkali silicate glass is not a new phenomenon (29), although it was not so far observed in direct association with bubble formation during bulk electron irradiation. Moreover, alkali migration following irradiation has already been reported by Primak (30), as a specific effect of the non-network ions of the glass. We have therefore examined the atomic composition of the irradiated volume by means of X-ray Energy Dispersive Spectroscopy (XEDS). The comparison between spectra obtained before and after irradiation, reported in Fig.8, confirms our expectations. Sodium atoms move from the centre of the irradiated volume towards the periphery, where they probably concentrate in a colloidal form, as is observed from Fig.9. The escape takes place immediately after the beginning of irradiation, before any bubble is observed. Parallel measurements of the transmitted electron current have confirmed this result, as is seen in Fig.10.

Starting from such a discovery, a more systematic study of the compositional dependence of swelling has been performed at

625

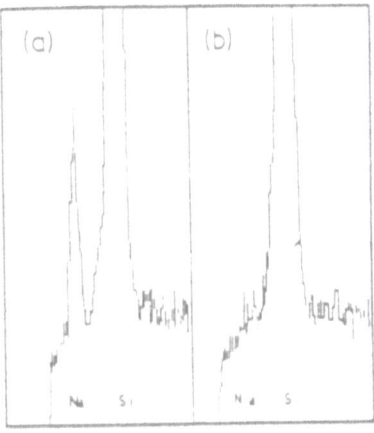

Fig. 8 - XEDS spectra recorded before (left side) and after (right side) two minutes of irradiation from a sample of type D (Table 1) at a temperature of 300 K using an electron beam of 300 KeV.

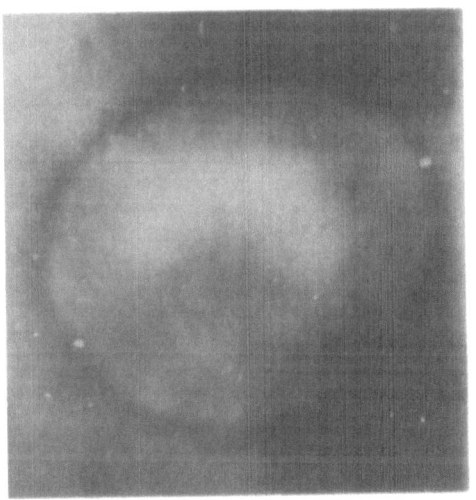

Fig. 9 - Micrograph of the electron-irradiated region in a borosilicate glass. The micrograph has been taken after 1.5 minute of irradiation with a dose rate of $5.1\cdot10^{23}$ e m^{2}s^{-1} at room remperature. The alkali borosilicate glass has 15.5%wt concentration of Na_2O.

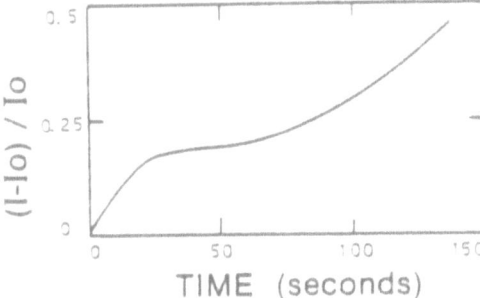

Fig. 10 - Current variation vs time of the primary electrom beam transmitted through the irradiated region.

molar concentrations of Na_2O, given in Table 1. The measured values of the maximum swelling at various contents of Na_2O are reproduced in Fig.11. These results confirm the general statement that micro-structural damage increases with the amount of alkali oxides present in the glass. In a parallel work on a similar subject, De Natale and Howitt (31) interpret their results on bubble formation in terms of a mechanism involving the fragmentation of oxide bonds, followed by oxygen stabilization by a cation migration process. This general statement is in agreement with our findings. However, these authors attribute to ionization induced reduction and charge neutralization the preferential migration of the cationic species. The results of our experiments on sodium migration do not confirm such interpretation, but the presence of a driving force accelerating the thermal motion appears to be required to justify the observed high rate of Na diffusion. This effect of acceleration is evident from Fig.12, where our Na diffusion coefficient during irradiation is compared with literature values (32,33) measured from thermal processes. Following a suggestion by Buckley (34), we attribute the existence of such a driving force to the generation of a strong electric field ($\sim 10^5$ V/m) associated with secondary electrons generated by ionization.

TABLE 1 -   Glass compositions

	A	B	C	D	E
SiO_2	80	75	70	65	53
B_2O_3	12.9	15	15	15	16.3
Na_2O_3	3.5	10	15	20	25
Al_2O_3	2.1				5.7
K_2O	1.5				

We now turn to the last question, the atomic composition of the bubbles. In agreement with De Natale and Howitt (23,31), we believe that the gas atoms present in bubbles are oxygen atoms. Arguments supporting this assumption are:
- oxygen is the major component of a silicate glass (~ 60 atomic %) and it can segregate into a gaseous phase;
- many of the atomic defects generated during irradiation are related to oxygen atoms, as described in section 2 above;
- being a relatively light atom, oxygen is difficult to observe in bubbles by those techniques such as XEDS or EELS. This circumstance may explain, together with the difficulty of separating oxygen in bubbles from that present in the glass matrix, the unsuccessful attempts to detect the presence of gas atoms.

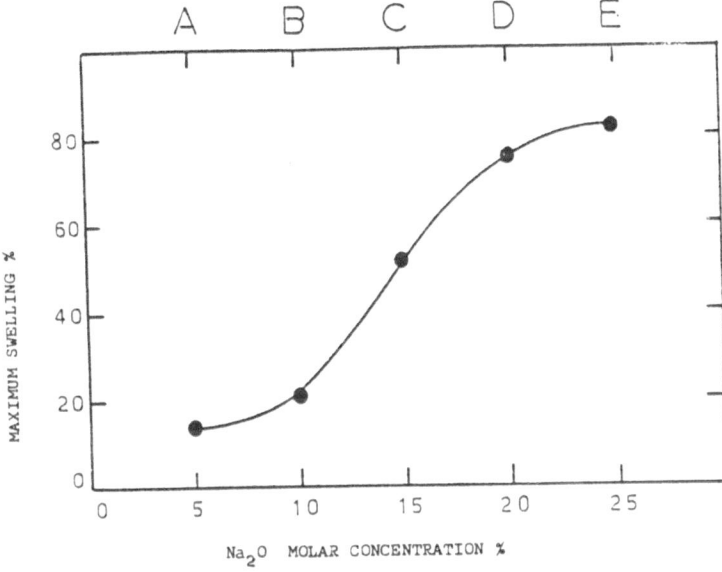

Fig. 11 - Compositional dependence of the swelling parameter from Na_2O molar concentration. The data refer to the temperature of maximum swelling rate, as shown in Fig. 6.

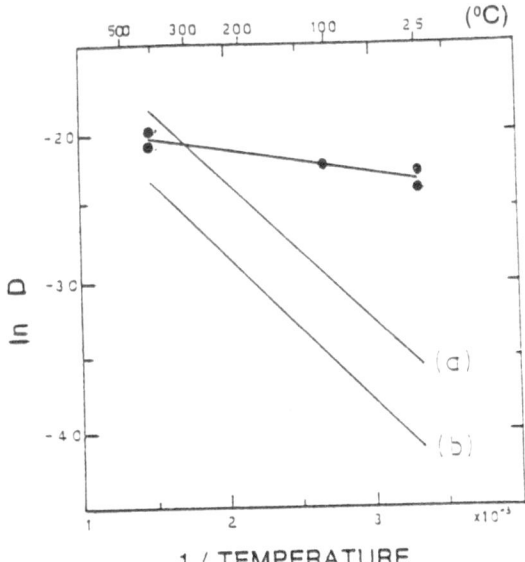

Fig. 12 - Arrhenius plot of Na diffusion coefficient, as obtained from out data on Na migration during irradiation (Glass D of Table 1, dose rate $= 2.7 \cdot 10^{23}$ e $m^{-2}s^{-1}$). The underlying straight lines refer to literature values for thermal diffusion of sodium in $Na_2O\text{-}3SiO_2$ (curve a, ref. 32) and $Na_2O\text{-}3B_2O_3$ (curve b, ref. 33).

It remains to establish whether thermal or field stimulated diffusion of oxygen atoms displaced during irradiation may be responsible for the gas present in bubbles or whether, as suggested by De Natale and Howitt, the local concentration of ionization fragmented oxide bonds unrelated to the silica network represents the main source for gas filling into bubbles. By taking our suggested value of the surface energy σ (see eq.2), we find that the number of gas atoms present in bubbles ranges from $3.8 \cdot 10^{18}/cm^3$ to $3.8 \cdot 10^{19}/cm^3$. These values are roughly comparable with those reported in Fig.3, as the concentration of E_1' centres measured at saturation in amorphous silica, after heavy ion irradiation ($\sim 4.5 \cdot 10^{19}$ E_1' centres/cm^3). However, the same number is much higher than the concentration of NBOs, which have the already quoted value of about $10^{17}/cm^3$. When considering that, at low ($\sim 5\%$) alkali concentration, bubble growing still persists from our observations* we are prone to consider the neutral displaced oxygens as providing most of the gas atoms in bubbles. Moreover, on the basis of the calculated number of vacant sites, we suggest that Na vacancies are the most probable candidates to form the second component.

4. CONCLUSIONS

From the results reported and discussed in the previous sections, it is evident that the study of fundamental radiation damage in glasses has still to be pursued in order to obtain an acceptable confidence in data resulting from accelerated irradiation tests in nuclear waste vitrified products. It also appears that close connections are present between the radiation enhanced leaching observed by some authors and the enhanced alkali diffusion and dissolution observed during our HVEM observations. Surface exfoliation and ionic exchanges are probably, among leaching mechanisms, those which are most susceptible of being affected by these processes. We believe that a more systematic investigation of such connections, already started by us with etching experiments (13), may provide a significant contribution to the present research activities on glasses for nuclear waste storage.

ACKNOWLEDGEMENTS

Dr. S.N. Buckley and Mr. S.A. Manphorpe, AERE Harwell, U.K., have shared during both experimental and discussion sessions of our work at the Harwell HVEM. We wish here to thank them for their invaluable assistance.

---

* At such low alkali concentration values only a few NBOs are probably added to those already present in the slica network (35).

REFERENCES

1. For updated critical reviews concerning radiation effects on borosilicate glasses for nuclear waste storage, see:
   a) W.J. Weber, R.P. Turcotte, F.P. Roberts, Radioactive Waste Management, 2, 295 (1982).
   b) F. Lanza, A. Manara, M. Antonini, F. Van Rutten, Proc. of Intern. Seminar on Chemistry and Process Engineering for High Level Liquid Waste Solidification, June 1981, Ed. R. Odoj, E. Merz, p.733.

2. M. Antonini, P. Camagni, F. Lanza and A. Manara, Scientific basis for nuclear waste management, Vol.2 (1980) Northrup Ed., p.127.

3. A.R. Hall, J.T. Dalton, B. Hudson, J.A.C. Marples, Management of radioactive wastes from the nuclear fuel cycle, Vol.2 (1976) p.3.

4. F.P. Roberts, G.H. Jenks and C.D. Bopp, Radiation effects in solidified high level waste, BWL-1944, UC-70 (1976).

5. G. Malow, J.A.C. Marples, C. Sombret, Radioactive waste management and disposal, Proc. of the 1st European Community Conf., Luxembourg, May 1980, Ed. R. Simon and S. Orlowski, p.341.

6. J.C. Dran, K. Langevin, M. Maurette, J.C. Petit, The scientific basis for nuclear waste management (Northrup, Jr., C.J.M., Ed.) Vol.2, p.136, Plenum Press, N.Y. (1980). See also Science 209, 1518 (1980).

7. Updated papers in this field will be published in the Proc. of the Int. Conf. on The Scientific Basis for Nuclear Waste Storage, Boston (Mas.), November 14-17, 1983, McVay G.L., Ed., Plenum Press, N.Y., in press. See also: W.G. Burns, A.E. Hughes, J.A.C. Marples, R.S. Nelson and A.M. Stoneham, J. of Nuclear Materials, 107 (1982) 245, North-Holland Publishing Company.

8. M. Antonini, P. Camagni, A. Manara, Recent developments in condensed matter physics, Vol.2, Ed. J.T. DeVreese, L.F. Lemmens, V.E. Van Doren and J. VAn Royen (Plenum Publ. Corp., 1981) p.217.

9. M. Antonini, P. Camagni, A. Manara and L. Moro, J. Non-Cryst. Solids, 44 (1981) 321.

10. a) M. Antonini, P. Camagni, P.N. Gibson and A. Manara, Rad. Effects, 65 (1982) 41.
    b) M. Antonini, P. Camagni, P.N. Gibson and A. Manara, ibidem.

11. M. Antonini and A. Manara, Rad. Effects, 65 (1982) 55.

12. M. Antonini, S.N. Buckley, P. Camagni, P.N. Gibson and A. Manara, The scientific basis for nuclear waste management, Stephen V. Topp, Ed., Elsevier Publ. Co. (1982) 709.

13. A. Manara, M. Antonini, P. Camagni and P.N. Gibson, Proc. 2nd Int. Conf. on Radiation Effects in Insulators, May 30 - June 3, 1983, Albuquerque, New Mexico, to be published in Nucl. Instr. Methods.

14. M. Antonini, A. Manara, S.N. Buckley, S.A. Manthorpe, submitted for publication.
15. G.J. Dienes, J. Phys. Chem. Solids, 13 (1960) 272.
16. D.L. Griscom, J. Non-Cryst. Solids, 40 (1980) 211.
17. G.N. Greaves, Phil. Mag. B 37 (1978) 447.
18. N.F. Mott, J. of Non-Chryst. Solids, 40 (1980) 1.
19. P.N. Gibson, Radiation damage in amorphous silica, Ph.D. Thesis (unpublished).
20. J.C. Mikkelsen, see ref.7 above.
21. R. Terai and Y. Oishi, Glastechn. Ber. 50 (1977).
22. E.J. Friebele, D.L. Griscom and M. Stapelbroek and R.A. Weeks, Phys. Rev. Lett. 42 (1979) 1346.
23. J.F. De Natale and D.B. Howitt, Proc. of the 2nd Conf. on Radiation Effects in Insulators, Albuquerque, New Mexico (1983) to be published in Nucl. Instr. Methods.
24. D.I.R. Norris, Rad. Effects, 14 (1972) 1.
25. J.M. Bunch, J.G. Hoffman and A.H. Zeltmann, J. Nucl. Mat. 73 (1978) 65.
26. M.J. Makin, Electron microscopy in material science, Academic Press, N.Y. (1971).
27. G.W. Morey, The properties of glass, 2nd ed., p.200, Reinhold Publ. Co., N.Y. (1960).
28. J. Nakajama, J. Am. Ceram. Soc. 48 (1969) 583.
29. D.E. Carlson, J. Am. Ceram. Soc. 57 (1974) 291.
30. W. Primak, Nucl. Technology, 60 (1983) 199.
31. J.F. De Natale and D.G. Howitt, Proc. of the 7th Int. Conf. on High Voltage Electron Microscopy, Berkeley, CA (1983).
32. R.H. Doremus, J. Elett. Soc. Sol. St. Sci. 115 (2) (1962) 181.
33. C. Lim, D.E. Day, J. Cer. Soc. 5 (6) (1977) 198.
34. S.N. Buckley, to be published.
35. W.J. Dell, P.J. Bray, S.Z. Xiao, J. Non-Cryst. Solids, 58 (1983) 1.

LEACHING OF NUCLEAR WASTE GLASSES

L. L. Hench

Department of Materials Science and Engineering
University of Florida, Gainesville, Florida 32611

INTRODUCTION

Resistance to aqueous corrosion is the most important requirement of glasses designed to immobilize high level radioactive wastes. Obtaining a highly durable nuclear waste glass is complicated by the requirement that melting, homogenization, and casting be rapid and capable of remote operations and occur at temperatures generally around 1150°C in order to reduce volatilization of Cs and Ru radionuclides. Consequently, addition of large quantities of Al_2O_3, ZrO_2 or use of high $SiO_2/MO+M_2O$ ratios to improve durability are not possible for nuclear waste glasses since such compositions increase glass viscosity and melting temperatures. In spite of these difficulties, highly corrosion resistant nuclear waste glasses have been developed with SiO_2 contents of less than 50 w/o.

Previous efforts to generalize the surface behavior of silicate glasses proposed five types of glass surfaces to represent a broad range of glass-environment interactions, Fig. 1 (1). It was suggested that any silicate glass could be described in terms of one of the five surface types at any particular instant in its processing and environmental history. Recent studies of the leaching of nuclear waste glasses have shown that another type of surface, III B, must be added to this scheme (2). Fig. 1 incorporates this new development.

Studies of the kinetics of the above processes show that Type II surfaces are present when

632

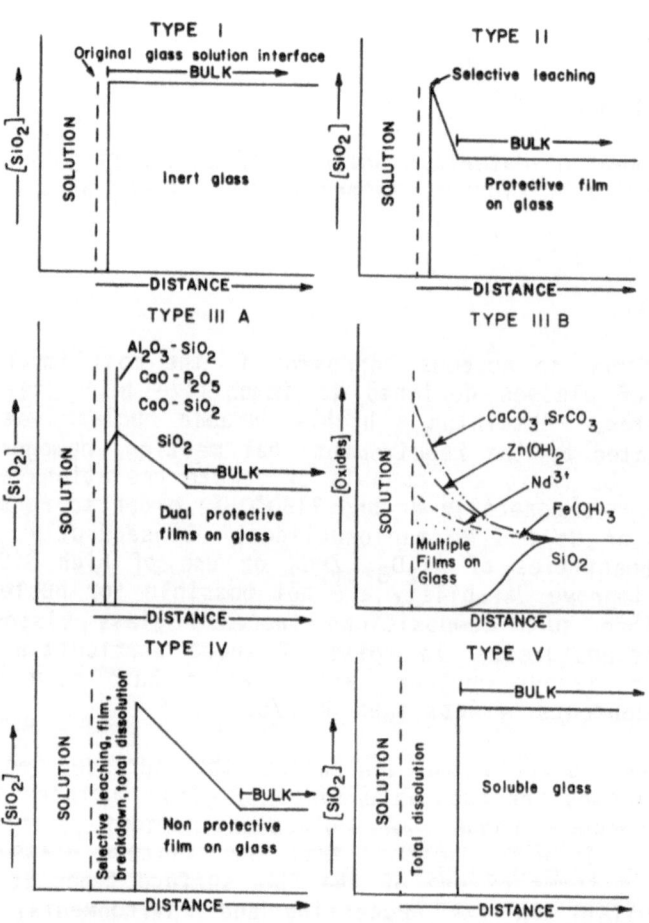

Fig. 1. 6 types of glass surfaces.

$$C_1 = k_1 t^{1/2} \tag{1}$$

where C_1 is the concentration of species in solution, t is reaction time and k_1 is the rate constant for the diffusion controlled process. Type V surfaces correspond to a regime of kinetics where

$$C_2 = k_2 t^1 \tag{2}$$

where C_2 and t have the same meaning as in Equation 1 and k_2 is the rate constant for the interface controlled network dissolution process. Type IV surfaces result when both processes occur together; e.g.

$$C_3 = k_1 t^{1/2} + k_2 t \tag{3}$$

Type III Glass Surfaces

Type III surfaces result when the dissolved species from reactions (1) or (2) reach saturation in the reaction solution and subsequently precipitate or back react with the glass surface. Thus, Eq. 4 describes the overall reaction kinetics for Type III surfaces.

$$C_4 = k_1 t^{1/2} + k_2 t - k_3 t^x \tag{4}$$

Previous surface studies of biologically active invert silicate glasses containing CaO and P_2O_5 (Bioglass™) (3) and certain compositonal ranges of Li_2O-Al_2O_3-SiO_2 glasses (4) showed that dual films can develop on glass. These glasses with a dual layer are termed Type III A. Certain compositional ranges of alkali borosilicate nuclear waste glasses develop multiple layers of oxides or hydroxides on their surface when exposed to water. This behavior is designated Type III B (Fig. 1). The most leach resistant Type III B nuclear waste glasses occur within a small compositional range. This is illustrated in Fig. 2 which compares the compositional dependence of the leach rates of 27 nuclear waste glasses studied worldwide (5). All glasses were leached in DI water using MCC-1 static leach methods (6) at 90°C for 28 days. The oxides of Si, B, and Na (the major constituents of the non-radioactive glass frit) are grouped along one binary axis. The oxides of Al, Fe, and all other constituents in the frit or in the waste, labeled WP, are added together and comprise the third axis. A narrow field of compositions, but still extensive, possess leach rates of 0.1 to 0.2 g/m^2/day. However, only a few compositions exhibit the very low range of 0.02 g/m^2/day and these all contain nearly equivalent 51-53 w/o SiO_2 and ±2% of the other constituents.

A theoretical basis for Type III B glasses is Grambow's work which predicts the formation of a series of insoluble reaction products on glass surfaces (7). He shows that the strong pH dependence of nuclear waste glass leaching can be explained in terms of the pH dependent solubility limits of simple reaction compounds of many constituents in the glass. In low and high pH leachants glass dissolution is controlled by congruent dissolution (eq. 2) but at intermediate pH values the solubility of reaction species such as $Fe(OH)_3$, $Zn(OH)_2$, $Nd(OH)_3$, $SrCO_3$, $CaCO_3$, or $CaSiO_3$ control the surface reactions. Thus k_3 in Eq. 4 involves a summation of the solution-precipitation reactions of all of the above constituents.

Although a short (several days) period of alkali-hydrogen ion exchange occurs for III B glasses, the dominant, long term mechanism is matrix dissolution followed by incongruent dissolution and solution/precipitation reactions. The extent of matrix dissolution and onset of surface precipitation depends on the time for various species in the glass to reach saturation in solution. Saturation of species (i) is a function of the initial solution pH, amount of alkali in the glass and rate of alkali release, T, initial concentration of species (i) in the solution, SA/V or flow rate which influences solution concentration. Until saturation of some species in solution is reached, the glass dissolves congruently at a rate proportional to kt^1.

When solution saturation of species (i) is reached there is no longer any driving force for that species to leave the glass surface. Consequently species (i) accumulates at the glass-solution interface as the matrix dissolves, leaving species (i) behind in the glass. If the matrix dissolution releases alkali ions, as will be the case for most glasses, there is a concomitant rise in pH proportional to the flow rate or SA/V of the system. An increase in pH can have several simultaneous effects on the glass, the solution, and the glass-solution interface. At the new pH, a second species (j) may reach solution saturation and subsequently be retained in the glass surface along with species (i). The extent of incongruent dissolution of the glass is thereby increased. In addition, the pH can have either one of three effects on species (i), previously in saturation; 1) it remains saturated but at a higher concentration; 2) it becomes supersaturated and precipitates either on the glass or other surface or as a colloid; or 3) it becomes undersaturated and species (i) in the glass surface once again begins to be released. The sequence of events that occurs is predictable based upon the solubility limits of each species at a given pH. Thus, we conclude (3) that the solubility limits that establish equilibrium ionic concentrations for ground waters should also establish the multiple barrier films to protect nuclear waste glasses in contact with those ground waters.

REPOSITORY EFFECTS

Relating the leaching rates of nuclear waste glasses to long term stability in a geologic repository is essential. However, this is complicated by the presence of many components in a storage system, such as canisters, overpacks, backfill minerals, and differing geological constituents. A series of in-situ burial studies have been conducted in deep (350 m) bore holes in Swedish granite (STRIPA Mine) to test the interactions of nuclear waste glasses with various components of the storage system (8-11). Results show that Type III B surfaces form under conditions expected in long term geologic storage. Figure 3 summarizes surface compositional gradients formed on a nuclear waste glass during one year 90°C burial in STRIPA. The depth of Na and B depletion is 0.35 μm, the maximum depth observed for this glass composition after one year in contact with wet 90°C granite. The Si concentration is increased by more than 20% throughout the reaction layer. Ion exchange of Ca and Mg from the ground water for Na and B in the glass is one of the most striking features of the profiles, especially in the highly alkaline earth enriched outer 0.1 μm. High valence species such as Fe also tend to concentrate in the surface layer giving rise to the multiple layer Type III B surface.

Profiles shown in Fig. 3 are characteristic of glass/glass, glass/granite, and glass/bentonite interfaces. The only major differences are a 3-5x greater depletion depth for glass/bentonite and ~2x greater depth for glass/glass interfaces. These differences are established during the first 3 mos. of burial. After 3 mos. the rate of attack is nearly the same, $<10^{-7}$ μm/yr.

ACKNOWLEDGMENT

The author acknowledges support of the U.S. Nuclear Reg. Com. Contract # NRC-04-78-252 and Alex Lodding, Chalmers Institute of Technology, Gothenburg, Sweden for the SIMS analysis shown (11).

NUCLEAR WASTE GLASS CONSTITUENT OXIDES

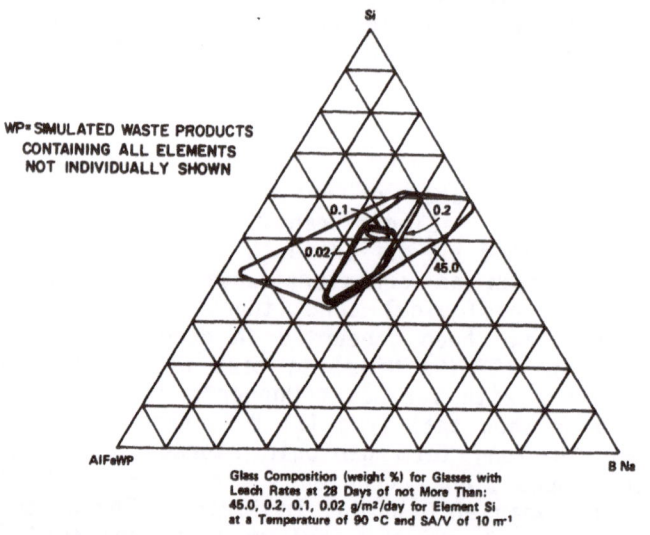

Fig. 2. Compositional field for nuclear waste glasses.

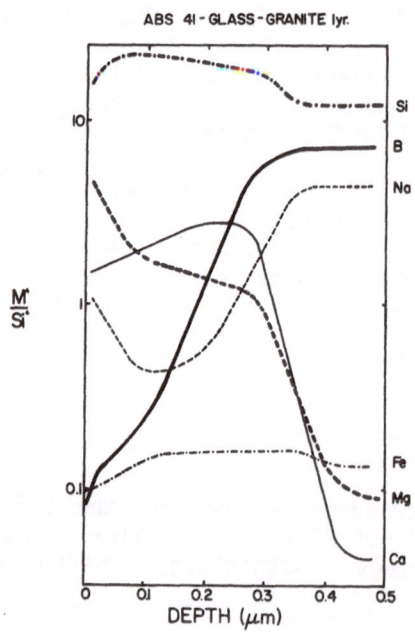

Fig. 3. Compositional profiles for nuclear waste glass ABS 41 glass/granite after one year 90°C burial.

REFERENCES

1. Hench, L. L. and D. E. Clark, J. Noncrystl. Sol, 28, (1978) 83-105.
2. Hench, L. L., "Glass Surfaces - 1982," Proceedings of Int. Conf. on Physics of Amorphous Solids, Montpellier, France, July 1982, J. Zarzycki, ed.
3. Hench, L. L. and E. C. Ethridge, Biomaterials: An Interfacial Approach, Academic Press, N.Y. (1983).
4. Dilmore, M. F., D. E. Clark, and L. L. Hench, Amer. Ceram. Soc. Bull., 57, (1978) 1040-1044.
5. Hench, A. A. and L. L. Hench, Computer Analysis of Compositional Effects in Nuclear Waste Glass Leaching, Nuclear and Chemical Waste Management (in press).
6. Materials Characterization Center (MCC) Test Methods, Preliminary Version, (PNL 3940) Pacific Northwest Laboratory, Richland, WA.
7. Grambow, B., "The Role of Metal Ion Solubility in Leaching of Nuclear Waste Glasses," in Scientific Basis for Nuclear Waste Management, V, V. W. Lutze, ed., Elsevier Science Pub. Co., (1982) 93-102.
8. Hench, L. L., L. Werme and A. Lodding, "Burial Effects on Nuclear Waste Glass," in Scientific Basis for Nuclear Waste Management-V, W. Lutze, ed., Elsevier Science Pub. Co., (1982) 153-162.
9. Hench, L. L., A. Lodding and L. Werme, "Analysis of One Year In Situ Burial of Nuclear Waste Glasses in STRIPA," presented at the Second International Symposium on Ceramics in Nuclear Waste Management, April 24-27, 1983, Chicago, IL.
10. Clark, D. E., B. F. Zhu, R. S. Robinson and G. G. Wicks, "Preliminary Report on a Glass Burial Experiment in Granite," presented at the Second International Symposium on Ceramics in Nuclear Waste Mangement, April 24-27, 1983, Chicago, IL.
11. Hench, L. L., A. Lodding, L. Werme, "Nuclear Waste Glass Interfaces After One Year Burial in STRIPA, Part 1 Glass/Glass, Part 2 Glass/Bentonite, Part 3 Glass/Granite, Part 4 Surface Profiling", J. Nuclear Materials (1984) in press.

FRENCH EXPERIENCE IN VITRIFICATION OF RADIOACTIVE WASTE SOLUTIONS

F.L. LAUDE
Service des Déchets de Haute Activité, C.E.A. Valrho,
Site de Marcoule BP 171, 30200 Bagnols sur Cèze, France.

1 GENERAL

The residual solutions produced by spent fuel reprocessing are the most important liquid waste in the nuclear fuel cycle, not by their volume but by their radioactivity level. These solutions contain 98 % of the radioactivity in the reprocessing plant.
After concentration, these "fission product solutions" are temporarily stored in liquid form in stainless steel tanks for at least one year. In a nitric acid medium, they contain most of the uranium fission products and the actinides formed in the reactor (except for the plutonium which is recovered), corrosion products such as Fe, Ni, Cr and Mo, fuel impurities or additives, and chemical compounds introduced during reprocessing or formed by solvent radiolysis.
The total solution volume is relatively small : depending on the type of fuel, its specific burnup and the cooling time before reprocessing, the volume ranges from 300 liters per metric ton for light water reactor fuel to 100 $l.mt^{-1}$ for graphite-gas reactor fuel. The solution activity, however, is very high : on the order of a thousand curies per liter. A typical 1000 MW civil PWR produces about 10 m^3 of such solutions annually.
A number of observations can be made regarding the radioactive decay and residual power curves (Figures 1 et 2).
During the first 300-500 years, most of the radioactivity is attributable to two fission products : ^{137}Cs and ^{90}Sr. The activity due to α emitters and to $\beta\gamma$ emitters is similar up to 300 years, but the α emitters

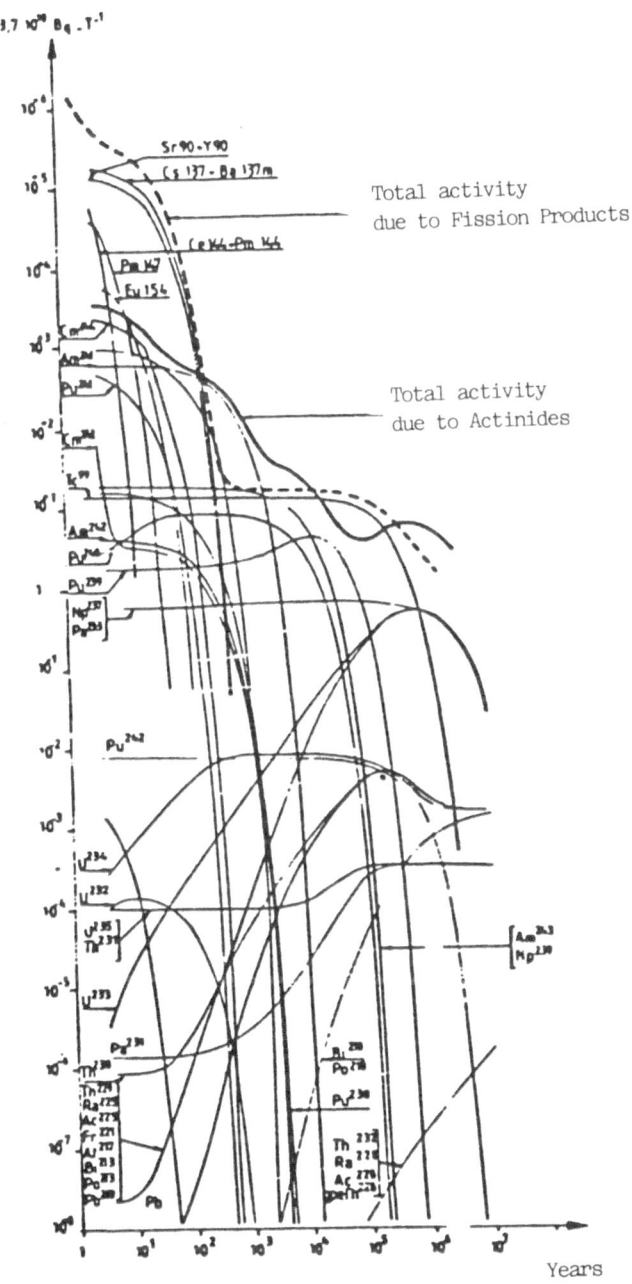

≈ 3,7 10^{10} Bq . T^{-1}

Sr 90 - Y 90

Cs 137 - Ba 137m

Total activity
due to Fission Products

Ce 144 - Pm 144

Pm 147

Eu 154

Total activity
due to Actinides

Am 241

Pu 241

Cm 244

Cs 90

Am 242

Pu 240

Pu 239

Np 237
Pu 233

Pu 242

U 234

U 232

U 235
Th 231

Am 243
Np 239

U 233

Pu 234

Th 230
Th 229
Ra 225
Ac 225
Fr 221
At 217
Bi 213
Po 213
Pb 209
Pb

Bi 210
Po 210

Pu 238

Th 232
Ra 228
Ac 228
etc Th

Years

Figure 1 : VERSUS TIME OF THE ACTIVITY
OF THE FISSION PRODUCT SOLUTION

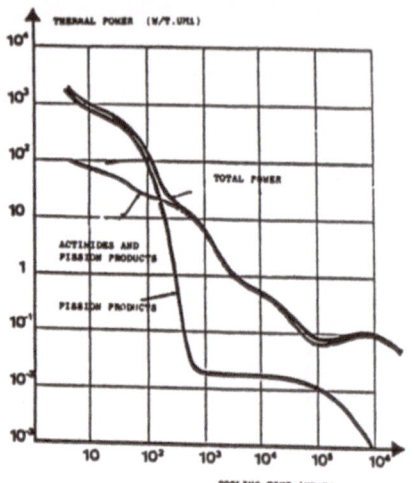

THERMAL POWER (W/T.UM1)

TOTAL POWER

ACTINIDES AND
FISSION PRODUCTS

FISSION PRODUCTS

COOLING TIME (YEAR)

Figure 2 : VERSUS TIME OF THE THERMAL POWER
OF THE FISSION PRODUCT SOLUTIONS

dominate thereafter. For LWR fuel reprocessed after a
3-year cooling period, the α emitters account for 52 %
of the activity after 300 years and 91,9 % after 500
years. The only fission products remaining after 1000
years are ^{99}Tc and the nuclide pair ^{93}Zr-Nb. After
10^6 years, the total α activity is 5.6 Ci.mt^{-1}.
The radiotoxicity of the waste solutions after 1000
years is almost exclusively due to α emitters.
The thermal power released from the solution due to
radioactivity energy absorption varies in proportion
with the radioactive decay curve (Figure 2), and is thus
relatively high during the first hundred hears (approx.
2 to 0.2 W.l^{-1}. After 1000 years, most of the residual
power is attribuable to the actinides (0.01 W.l^{-1}).
Before reprocessing, additives may be included in the
fission product solutions. Current plans call for the
addition of alkaline concentrates from extraction
solvent rinsing, containing plutonium and suspensions of
dissolution fines. As a result, the reprocessing volume
will be increased to about 600 l.mt^{-1}.
Although the experience to date has been satisfactory in
storing liquid solutions in a nitric acid medium cooled
by water coils over a period of several decades, it is
not possible to maintain liquid storage conditions much
longer. Storage tank maintenance costs are high, and
leakage is a permanent hazard.
The solidification of these liquid wastes will permit
continued development of nuclear power with fuel
reprocessing.

A further advantage of the solidification process is a significant waste volume reduction, which is appreciable from a storage standpoint.
Among potential solidification processes, vitrification has received the most attention and is the only process implemented on an industrial scale. Glass has a number of advantages :
- It is capable of dissolving all of the chemically different fission product elements at high temperatures, and can be implemented with considerable flexibility.
- It is homogeneous, isotropic materiel that is also impermeable and thus presents only a limited exchange surface area.
- It is highly resistant to heat and radiation.
- It is easy to manufacture with a minimum of intermediate process steps liable to generate secondary wastes.
The general vitrification process can be summarized as follows. The fission product solutions are evaporated, and the resulting nitrates are calcined at temperatures ranging from 130° to 600°C, converting most of them to oxides. After the addition of glass frit network formers, the mixture is melted at about 1150°C, refined and cast into a stainless steel container. These operations can be carried out in one or more steps, depending on the technology used.
After welding the cover to provide a leaktight unit, the container is decontaminated. It is then immediately transferred to a surface storage yard for an initial about 10-year cooling period, following which it can be moved to a long-term geological repository.
Fission product glass must meet a number of criteria related to the manufacturing process (the solution composition, the melting technology, etc...) and must be sufficiently stable to act as a primary barrier throughout the interim and ultimate storage periods.
Borosilicate glass formulation are widely used. The presence of boron lowers the melting point, improves retention of the transition element oxides and has a buffering action in limiting the pH of leaching water in the event of leaching with slight water renewal.
The glass compositions are matched with the solutions to be vitrified. The concentration of fission products and elements form the fission product solution is limited to maintain a uniform network and to limit the thermal power of the glass. The content is not excessively reduced, however, since many fission products (especially the transition elements) improve materiel strength as a result of their low solubility.

In the current state of technology, the glass is manufactured at about 1150°C. Elements not found in the fission product solutions, notably the network formers, are added in the form of preprocessed glass frit to accelerate the vitrification kinetics.

2 HISTORICAL BACKGROUND

The idea of vitrification dates to about 25 years ago, and the first application experiments were implemented in CANADA at CHALK RIVER (1).
The first French studies began during the same period, followed by work in other countries (2), (3).
After the first radioactive glass compositions were developed, a demonstration pilot facility (PIVER) began active operation in 1969 at MARCOULE, using a melting pot batch process with glass casting. At the same time, the initial glass blocks were stored in air-cooled vertical shafts in a concrete-covered storage pit.
Meanwhile, R and D work focused on countinuous glass manufacturing, and the continuous vitrification process with separate calcining and vitrification operations was adopted for industrial application in 1972. This process allows a compact facility to achieve major production throughout capability, and permits considerable flexibility in operation.
A continuous vitrification facility began operation under radioactive conditions for COGEMA in 1978 at MARCOULE (4). Since that time, the unit has satisfactorily vitrified 750 m³ of graphite-gas Fission Product solutions. The one-thousandth container was filled in March 1984. The total weight of glass produced is 342 tons ; 1000 containers are stored. The total activity incorporated in this glass is 140.10^6 curies.
A second reprocessing plant exists at LA HAGUE, in Normandy, for oxide fuel from commercial PWR power plants. In order to meet French and foreign throughput requirements, the decision was made to design and build two additional plants on the same site : UP2 and UP3, each with an annual capacity of 800 metric tons. UP2 is already partially in use, and should be in full operation by 1989. UP3 is scheduled to be ready for operation in 1987. Vitrification facilities will be provided for each reprocessing plant, together with interim storage dumps.
Both plants include 3 vitrification lines using the 2-step continuous processes. A full-scale prototype is now in operation at MARCOULE with simulated fission

product solutions. The first vitrification facility is scheduled to begin operation in 1987.

3 COMPOSITION OF THE VITRIFICATION WASTE SOLUTIONS

The waste solutions are generated by reprocessing spent fuels irradiated in commercial, defense and material testing reactors. These nitric acid solutions have radioactive and chemical compositions that vary according to a number of factors :

- the type or nuclear fuel,
- the specific burnup,
- the composition of the cladding hulls if they have not been removed,
- reprocessing additives,
- possible neutron poisons,
- Pu and U partitioning efficiency,
- the final concentration,
- the interim storage duration (corrosion products).

Typical solutions compositions are indicated in Table I.

TYPE OF REACTOR	TYPE OF FUEL	BURN-UP	CONCENTRATION	ACIDITY (N)	CHEMICAL COMPOSITION $(g.l^{-1})$									
					Al	Na	Mg	Fe	Ni	Cr	P	F	Gd	ACT. + PF
MTR	U Al/PuAl	500 MWj.kg^{-1}	12 m³.t^{-1}	~1,8	81	2/3		1/2				10/12		LOW
GRAPHITE GAZ	SICRAL	1000 MWj.t^{-1}	30 l.t^{-1}	1,5/2	30/35	19/23	4/6	16/17	1/2	1/2	8	1		23
		4000 MWj.t^{-1}	100 l.t^{-1}	1,5	10	3	3	8	0,3	0,4		1		40
LWR	UO$_2$	33000 MWj.t^{-1}	660 l.t^{-1}	1,0/1,5	<1	15		9	1,5	1,5			0,6	45/50
FBR	UO$_2$/PuO$_2$	100000 à 120000 MWj.t^{-1}	1700 l.t^{-1}	1/1,5		20/30		8/10	1/2	2/3			25 ou 0	30/33

TABLE 1.: COMPOSITION OF SOME FISSION PRODUCT SOLUTIONS

4 FISSION PRODUCT GLASS

Glass is a material with a disorderly structure constituting the vitreous state, which can be compared with the crystalline state. This state is conducive to special properties resulting from the irregular, isotropic, non-stoechiometric and mobile internal structure in which each atom does not occupy a well defined position (4), (5).

While a vitreous network is not well-ordered like a crystal lattice, it is characterized by a degree of organization over a short but highly variable distance. It is thus difficult to apply general rules, even restricted to the very limited category of silica-based glasses. However, although any chemical element can enter this lattice in theory, its concentration must be limited in order to maintain a uniform vitreous lattice. Moreover, the vitreous state is metastable, and at a given temperature any glass may change to a lower internal energy state from which it is separated only by a potential barrier.

Existing fission product glasses represent a relatively limited range dictated by the chemical properties of the materials to be vitrified and by technological limitation. The properties and behaviour of these glass formulations are similar in many respects (Figure 3). Fission product glass is subjected to a number of thermal, chemical, mechanical and nuclear stresses.

The glass composition selected for solidification of a given radioactive waste is determined by three factors : the radioactive liquid constituents, the vitrification process, and the desired waste containment properties (i.e. the safety-related physical and chemical characteristic required for interim storage and long-term disposal).

4.1 Glass compositions for Different Fission Product Solutions

The structure or silica-based glasses is generally flexible enough to lodge the various waste consituents, so that a looser network (e.g. boric oxide glass) is unnecessary. Nevertheless, B_2O_3 is often used to lower the melting point ; boric acid is in fact preferred over sodium oxide, which induces the same effect but is less favorable from a chemical resistance standpoint.

The major requirement of a vitrification process involving casting is that the glass viscosity must not exceed a specified value at the process temperature.

TWO DIMENSIONAL REPRESENTATION OF A FP-Na SILICATE GLASS

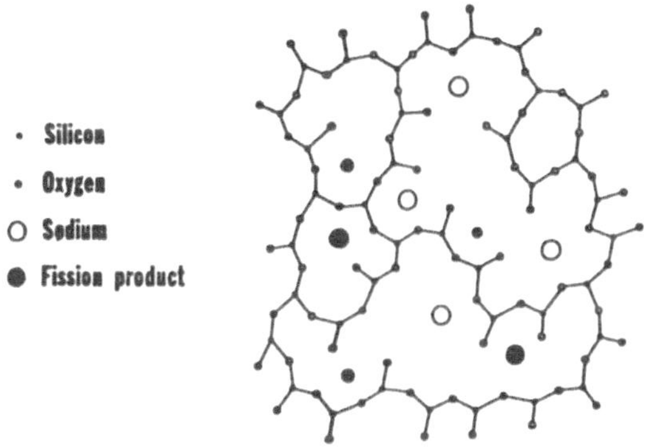

- Silicon
- Oxygen
- O Sodium
- ● Fission product

Figure 3

This is ensured by the use of B_2O_3 or, to a lesser extent, by Na_2O. Both static and flowing processes require low corrosion of equipement surfaces in contact with the molten glass : consequently work has been discontinued with phosphate glasses, although the P_2O_5 network is very flexible and permits a low melting point. Other process considerations include the volatility of certain elements, melt homogeneity and (when Joule heating techniques are used) electrical conductivity versus temperature.

A typical glass composition for vitrifying LWR liquid wastes is shown in Table II.

4.2 Glass Properties Related to the Manufacturing Process

Three important aspects must be considered with respect to the fabrication process described above :
- the corrosiveness of the molten glass with regard to metals,
- the viscosity of the molten glass,
- volatilization during glass fabrication.

The corrosiveness of the glass is of considerable importance, since it determines the lifetime of the melting furnace and thus affects the quantity of the secondary solid wastes produced. It is difficult to modify this corrosiveness (e.g. by reducing the Na_2O content) without significantly affecting other properties, and viscosity in particular. It is preferable to select a materiel with a favorable

SiO$_2$	45.2
B$_2$O$_3$	13.9
Al$_2$O$_3$	4.9
Na$_2$O	9.8
CaO	4.0
Fe$_2$O$_3$	2.9
NiO	0.4
Cr$_2$O$_3$	0.5
P$_2$O$_5$	0.3
ZrO$_2$ (fiings)	1.0
Li$_2$O	2.0
ZnO	2.5
Actinide oxides	0.9
+ Metallic particles	0.7
Fission product oxides	11,0

Table II : Possible Glass Composition for Solidification
of High Level LWR Waste (values in weight %)

cost/lifetime ratio ; laboratory corrosion studies have shown the advantages from this standpoint of using Inconel 601.
Suitable viscosity is essential in any process involving molten glass transfer, as for casting. The maximum viscosity at the casting temperature has been set at about 400-500 poises. Once again, trade-offs are necessary among the Na$_2$O, B$_2$O$_3$ or Al$_2$O$_3$ content values, since a lower viscosity (obtained by increasing the Na$_2$O and B$_2$O$_3$ content or reducing the Al$_2$O$_3$ content) is incompatible with the chemical resistance of the glass. The volatility of certain non-radioactive elements can raise problems, but above all ruthenium is capable of clogging outgassing lines by RuO$_2$ buildup. In order to assess this, volatilization experiments must be conducted with radioactive samples using the same equipment setup. These experiments were carried out at MARCOULE using tracers in the ATLAS one-half scale continuous vitrification mackup. A reduction in ruthenium volatility could be achieved either by denitrating or by sugaring the process solution. Sugar hydrolyzes in a acid medium, yielding levulose and glucose, which is a reducing agent. An adequate amount

of sugar in the calciner ensures a sufficiently reducing environment to prevent the formation of ruthenium oxides complexes, forming RuO_2 directly in the unit. The results shoved that a process solution sugar content of 30 $g.l^{-1}$ reduces the volatilization to an acceptable value or less than 2 %, compared with almost 30 % in an unsugared solution in the same process equipment.

4.3 Glass Properties Related to Interim Storage and Long-Term Disposal

In addition to its fission product and actinide oxide retention properties and specifically technological properties (viscosity, volatility, etc...), the glass must maintain the necessary containment integrity for a time that depends on the radionuclide half-time (4),(5). During this period, which exceeds 10 000 years for the actinides, the glass will be subjected to temperature rises caused by radiation self-absorption heat, to irradiation by the trapped β, γ and α emitters, and to chemical aggression primarily due to the leaching action of water in storage dumps and repositories.
The glasses are therefore selected by a parametric sensitivity study involving the following major properties :
- nuclear stability   : low self-irradiation sensitivity ;
- thermal stability   : devitrification, crystallization ;
- chemical stability   : resistance to water leaching action ;
- mechanical stability : limited fracturing which would increase the potential leaching area.

Irradiation stability : Owing to the considerable differences in their half-lives and properties, it is convenient to discriminate between glass behavior with regard to fission products ($\beta\gamma$ emitters) and actinides (α emitters).
- Fission Product Sensivity : Glass samples were subjected to electron bombardment in accelerators reaching doses of 1.2×10^{11} Rads, corresponding to the doses integrated by actuel glasses in 500 years of storage. No significant alterations of the glass properties were observed.
- Actinide Sensivity : This is the major safety-related problem involving long-term storage. Not only is the

α particle emission energy higher (about 5 MeV), but electron capture leads to helium formation in the glass. The most widely used method for simulating long-term irradiation by α emitters is to dope the test glasses with ^{241}Am, ^{238}Pu, ^{244}Cm or ^{242}Cm in order to simulate a few thousand years in only a few years. This is limited by the maximum actinide content in the glass, which must be low enough not to modify the glass properties (the upper limit is generally on the order of 3 wt.% of actinide oxides). This method, considered suitable if the dopant is evenly distributed in the glass, is used in France, the UK and the USA. In France, a glass sample doped with ^{244}Cm has already integrated 4×10^{18} αdisintegrations per gram, corresponding to over 20 000 years of actual glass storage.

It is unlikely that irradiation induces any devitrification in a glass medium. No crystallization has been detected in samples investigated to date (France, UK, USA) at does of up to 10^{18} – 10^{19} disintegrations per gram. Morever, the opposite effect, i.e. crystal amorphization, was noted.

Furthermore, if an irradiated glass was submitted to heat, any possible damage would be cured between 200 and 400°C, a temperature range below the incipient crystallization temperature ($> 500°C$).

Density charges have occasionnally been observed in doped glasses. The cause and mechanism of the alteration has not been fully elucidated to date. The most likely assumption is the tendency of glasses under irradiation to approach their theorical density, although some nuclear glasses behave differently.

Residual and diffused helium measurements in doped glasses were used to calculate the diffusion coefficient at various temperatures.

In general, the variation is governed by an Arrhenius type law, but a change in the diffusion coefficient as a function of the cumulative dose has also been observed.

The helium diffusion coefficient was calculated to be 2×10^{-11} $cm^2.s^{-1}$ in French Pu or Am doped glasses at room temperature. In that case (borosilicates containing 2,8 % PuO_2 or 2.8 % AmO_2), 10 % of the daily helium production was released by diffusion after a 4-year storage period. In curium-doped borosilicate (1.3 % CmO_2) steady-state conditions were reached at 3×10^{18} disintegrations per gram : the remaining helium quantity no longer increased, and the daily He yield completely diffused.

It should be noted that the steady-state helium concentration (34×10^{-3} $ml.g^{-1}$) depends on the sample shape and the specific activity.

Helium appears to have a minor effect on the mechanical properties of nuclear glasses , but longer storage durations will be required to confirm this.

Chemical stability : The major hazard during interim and long-term storage arises from the action of water. Studies of glass stability when subjected to leaching cover two inherent material properties : chemical alterability (material damage resulting from the action of water), and containability with regard to the activity of various radionuclides, fission products and, above all, actinides.

Suitable leaching methods are designed accordingly. In general, leaching alterability tests are conducted in Soxhlet devices and the alteration rate is measured by the weight loss. the containability is measured by the rate of activity transfer to the water, or "leach rate" (6).

Similar alterability results were obtained on a 100 kg industrial size block of PWR glass cooled under simulated realistic conditions.

For containability tests, the leach rates are measured on radioactive glass samples prepared by a pot vitrification process. In general, the water action is not congruent, i.e. different elements are leached at different rates. The following leach rates were measured for PWR glasses :

^{137}Cs approx. $1.5 \times 10-7$ $g.cm^{-2}.d^{-1}$
^{90}Sr " 5.0×10^{-8} "
$^{106}Ru-Rh$ " $5.0 \times 10-8$ "
^{144}Ce " 3.0×10^{-8} "
actinides   approx. 10^{-7} to 10^{-9} "

The actinide containment is only slightly affected by the simulated α dose. In the temperature range compatible with storage conditions, temperature variations have no effect.

Thermal stability. It is evaluated in the following ways :
- Glass Crystallization Property Measurements : Glass samples are subjected to a 15-20 hours heat treatment at 30°C temperature intervals up to 1200°C to induce microcrystallization phenomena for which the following points are measured :
. occurrence temperature range
. crystalline density
. maximum crystal growth rate
. maximum growth temperatures

. upper and lower crystallization temperatures.
The crystalline phases are identified where possible, although this is made difficult by the small crystal fraction (well below 1 %). - Crystalline Phase Development Tests : The Glass blocks are held for 100 hours at the previously determined maximum crystallization temperatures in order to confirm the presence of crystals in massive samples, to obtain, crystal growth, to quantify the crystalline density by X-ray diffraction or image analysis, and to identify the crystalline phases formed.
In most cases, complex molybdic microcrystallization occurred, containing primarily sodium, rare earth or strontium molybdates, spinels (mixed Cr, Ni, Fe oxides), cesium and uranium oxides.
Crystallization generally remains very low, never exceeding a few percent.
- Examination after Extended Isothermal Treatment : Samples are held for 6 months or 1 year at a temperature 50°C above the transformation point, and at 450°C (the maximum currently permissible glass core temperature for interim storage). Crystalline phases are again identified and quantified.
- Induced Crystallization Effects on Mechanical Properties and Alterability : In the selected glasses with their low crystallization rates, the appearance of crystalline phases generally did not modify their mechanical properties. With regard to their fission product containment power, crystallyzation raises the cesium leaching rate by about a factor of 5, slightly increases, the strontium leaching rate and does not affect or even reduces the leaching rates for other fission products.

Mechanical Properties : It is essential to prevent the glass from fracturing into small particles during transport, cooling and container stocking in the storage area.
Certain physical properties are also measured, including the following :
. specific heat
. thermal conductivity versus temperature
. expansion coefficient versus temperature
. compression modules
. microhardness
. Young's modulus
. biaxal flexure strength
. stress intensity (KIC).
Moreover, controlled cooling of glass blocks immediately

after casting is a promising method for limiting their fracturation (See Table III some physical properties of LWR glass).

Specific gravity	2.75
Viscosity at 1100°C	8 Pa.s (80 poises)
Transformation temperature	502°C
Man expansion coefficient	$8.3 \times 10^{-6} \cdot C^{-1}$
Thermal conductivity at room temperature	$1.1 \text{ W.m}^{-1} \cdot C^{-1}$
Young's modulus at room temperature	8.4×10^{10} N.m
Stress intensity factor	0.95 MPa.m$^{1/2}$

Table III : Some Physical Properties of the LWR Glass

5 VITRIFICATION PROCESS AND FACILITIES

The vitrification technique is based on a two-stage process. In the initial calcination step the feed solution is continuously supplied to a rotary kiln where it is solidified. In the next step, the calcined products are blended with suitable raw materials in an induction-heated melter to produce the desired glass at a temperature in the 1100°-1250°C range (Figure 4).
The molten glass is cast every eight hours into stainless steel canisters that are introduced into the vitrification cell.
Gas generated in the melter and the calciner is extracted through the rotary kiln. The primary components are steam, nitrogen compounds and dust particles. Most of the latter are soluble in nitric acid and are trapped in a dust remover and then fed back into the calciner. The gas stream then flows across a condenser and a standard treatment system comprising an absorption column, a washing column, an absolute filter and an extraction blower.

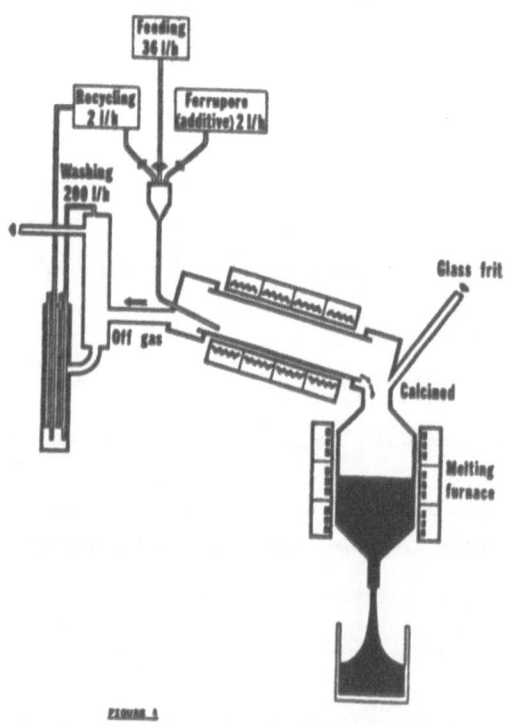

FIGURE 5

5.1 MARCOULE Reprocessing Plant (UP1)

The maximum plant capacity is 36 l.h^{-1}, and the maximum glass output is about 16-18 kg.h^{-1} (Figure 5).
Each canister is filled with 360 kg of glass (150 l) corresponding to three castings. Several hours after filling, the canister is transferred to another station in the vitrification cell where the lid is welded on using a plasma torch.
After welding, the canisters are cleaned by high pressure water jetting in a separate cell, from which they are then transferred to an air-cooled storage facility where they are stocked on metal racks. Forced-air cooling is initially provided but subsequently switched off to permit natural convection cooling. The storage facility was designed to prevent the glass temperature from exceeding 450°C (Figure 6).

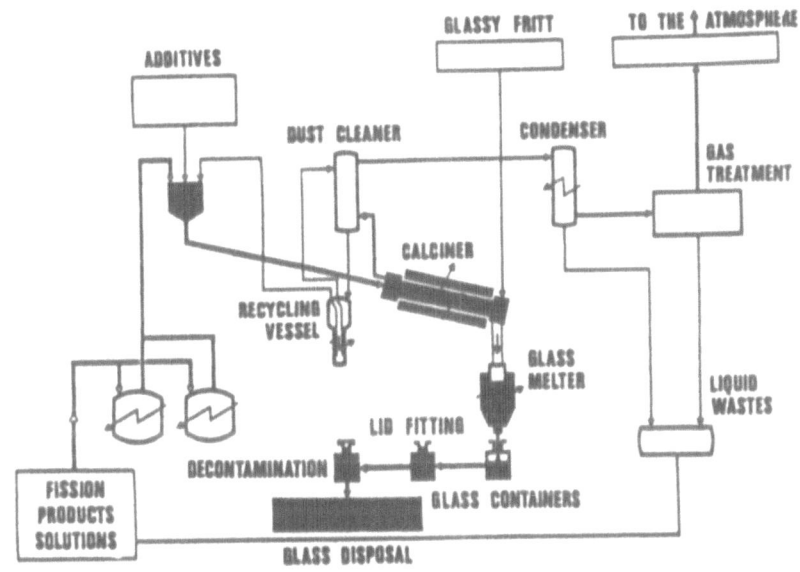

Figure 5 : MARCOULE UP1 CONTINUOUS VITRIFICATION PROCESS SCHEMA

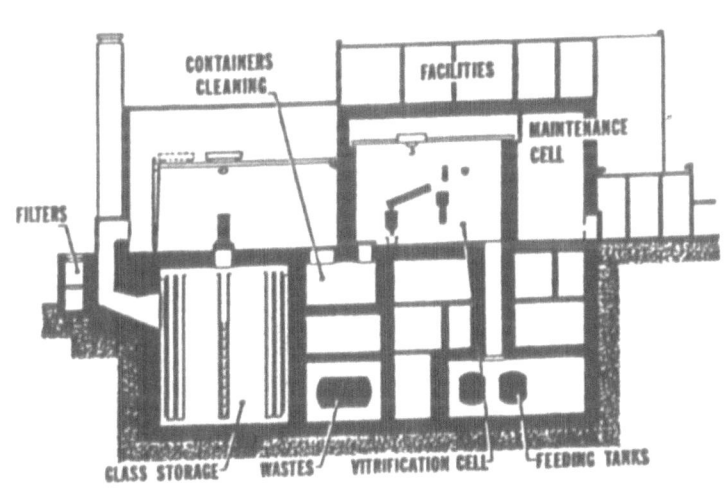

Figure 6 : AVM VERTICAL SECTION

5.2 LA HAGUE Reprocessing Plant (UP2-UP3)

Two identical vitrification facilities are planned for the LA HAGUE site : R7, to process the UP2 throughput, and T7 for the UP3 Plant. Both will implement the same process used at MARCOULE. R7 is now under construction and is scheduled to become operational late in 1986 or early in 1987. Each planned unit will include three identical continuous vitrification lines.
A full-scale prototype was built for equipment test purposes at MARCOULE, and is now operating with non-radioactive simulated solutions. This prototype has already logged 4000 hours of operation, including 500 hours of uninterrupted continuous vitrification.

The LA HAGUE facility differs from the AVM facility in a number of fundamental respects. To begin with, the liquid waste material itself is different : as mentioned earlier, it consists of an LWR solution to which are added fine metallid particles in suspension and solvent rinsing solutions, representing a total of about 660 liters per metric ton of uranium.
A significant design difference in R7 is the process layout with three independent vitrification lines, each with the same equipment as the MARCOULE prototype. Each line includes a dismantling cell which, unlike the AVM facility, is separate from the vitrification cell. Another important difference is that the glass is not cast in the vitrification cell but in a separate cell. After two castings (instead of 3 as in the AVM facility) totaling 150 liters, the canisters exits via a passageway accessible from all three vitrification cells. All the canisters are wipe-tested for possible contamination before they are transfered to the storage area. If evidence of contamination is found, the canisters are decontamined using high pressure water jets.
The containers inside their transport casks are transfered to the storage area by a traveling crane, where they are deposited inside tubular steel storage pits suspended from a metal framework beneath the concrete slab. Forced-air cooling is provided for a maximum specific power of 25 W per liter of glass to limit the glass core temperature to 450°C. A quality control program is implemented throughout the fabrication and storage operations.

6 OUTLOOK

France has now acquired industrial expertise with the
vitrification facilities now in operation or under
construction.
Work is now oriented towards producing glasses at higher
melting temperatures (1350-1400°C) to obtain higher
performance characteristics using a new generation of
melting furnace.

REFERENCES

(1) Watson L.C., Paper 135 - 2nd United nation Conf. on
    Peaceful use of Atom Energy - Geneva 1958 -
(2) Bonniaud R., Laude F., Sombret C., Processings of
    Management of Radioactive Waste from Fuel
    Reprocessing - DECD - Paris 1972 -
(3) Laude F., Management of radioactive Waste from the
    Nuclear Fuel Cycle - Symposium IAEA - Vienne 1976 -
(4) Laude F., Le verre comme première barrière pour le
    stockage à long terme des déchets de haute activité
    OECD - CEC - Ispra - Mai 1977 -
(5) Jacquet-Francillon N., Bonniaud R., Sombret C.,
    Glass as a Material for the final Disposal of
    Fission Products - Radiochimice Acta 25 -
    p. 235-240 - 1978 -
(6) Laude F., Etat des Etudes de la lixiviation des
    verres en France - KWU - Karlstein/Mein - Nov 1983 -
(7) Bonniaud R., Jouan A., Sombret C., Nuclear and
    Chemical Waste Management - Vol. 1, n° 1 - 1980 -
(8) Sombret C., Waste Management in Nuclear Facilities
    Cours IAEA - Karlsrhue - Sept. 1983 -
(9) Sombret C., Vitrification of high level wastes in
    France - 11° Annual Energy Conference - Knoxville -
    USA - Fev. 1983 -

COMPOSITION AND BONDING MECHANISMS IN BIOGLASS™ IMPLANTS

L. L. Hench and D. B. Spilman

Department of Materials Science and Engineering
University of Florida, Gainesville, Florida   32611

INTRODUCTION

Controlled surface active glasses and glass-ceramics were developed to achieve a direct chemical bond of an implant with living tissues (1-4). Devices made from a specific compositional range,(see Fig. 1) that bond to tissues are termed Bioglass™ implants. When this same compositional range is partially or fully crystallized the materials are termed Bioglass™-Ceramic implants. Surface active glass-ceramics made from another specific range of oxide compositions are termed Ceravital® implants (5,6). Other compositional ranges of surface active glasses and glass-ceramics are still being developed (7,8) and the limitations on glass or glass-ceramic compositions that will bond to living tissues are still unspecified.

A large number of in vitro and in vivo experiments (1-4,9-15) have established that Bioglass™ implants develop, when exposed to body fluids, a silica-rich layer, and an outer, biologically active, calcium phosphate layer which bonds to collagen, mucopolysaccharides, and affects interfacial cellular responses. A consequence of this controlled bioactivity is the formation of a strong bond to bone when osteogenesis is stimulated under functional loads (4,9). When exposed to soft tissues, Bioglass™ implants bond via attachment of a very thin layer of collagen fibrils to the active apatite layer (12-14). This layer does not significantly increase in thickness in various animal models during long term experiments (12,13). In some otological applications such as partial or total ossicular replacement, Bioglass™ prostheses bond via the collagen

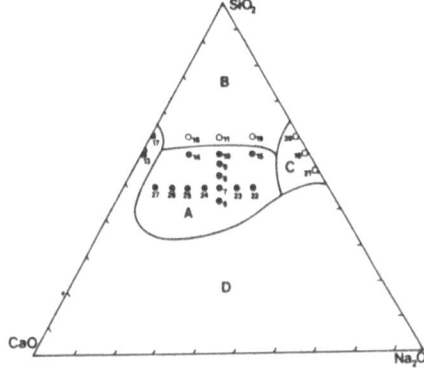

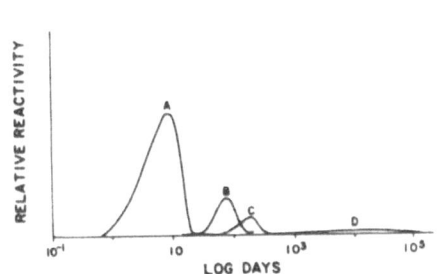

Fig. 1. Bioglass™ bone bonding composition diagram (with constant 6.0 wt% P_2O_5). Region (A) bonding at 30 days or less.

Fig. 2. Relative reactivity of biomaterials. (A) resorbable bioceramics; (B) moderate-surface-activity; (C) low-surface-activity; (D) nearly inert bioceramics.

mechanism (12,16). Other applications that involve significant replacement of bone, such as fixation of orthopedic devices or maxillo-facial reconstruction, result in a bone-Bioglass™ bond (2,4,13). Bioglass™ dental implants can produce both hard and soft tissue bonding (14).

COMPOSITIONAL EFFECTS

Bioglass™ and Bioglass™-Ceramic materials constitute only a certain range of compositions within the field of surface active or bioreactive glasses and glass-ceramics. Figure 1 shows the compositional range for bonding Bioglass™ implants to bone within 30 days (in rat) (3). Compositions outside area (A) do not bond due to too little reactivity (B) or resorb (C) or cannot be made as devices (D). The compositional dependence of Bioglass™ implants shown in Fig. 1 depends on the relative bioreactivity of these materials. This is shown in Fig. 2.

There are three primary types of biomaterials. (See Ref. 3 for discussion of all three types and examples).

Nearly inert biomaterials (curve D) show little chemical change during long term exposure to physiological solutions. Tissue response to this class of biomaterials involves a very thin - several μm or less - fibrous membrane surrounding but unbonded to the implant materials. Because fibrous tissues do not chemically bond to nearly inert biomaterials, fixation within the body must be established by a mechanical interlock with tissues achieved by use of controlled porosity in the implant, as with biomaterials such as Plastipore™ and Proplast™. Relative advantages and disadvantages of this method of tissue attachment have been described previously (12).

Consider again Fig. 2; at the other extreme of the biomaterials reactivity spectrum (curve A), are totally resorbable biomaterials. Such materials must have compositions that contain only elements that are easily processed through normal metabolic pathways such as Ca, P, H_2O, and CO_2. With time, such reative biomaterials are totally resorbed by the body and replaced with tissues. Consequently, the function of totally resorbable biomaterials is to serve as a scaffolding or filler of space permitting tissue infiltration and replacement.This function is similar to that provided by bone grafts from the host. The major advantages of the use of resorbable ceramics over host bone grafts are a ready supply, controlled variations in size, and elimination of a second surgical procedure. However, a disadvantage is the significant strength reduction that occurs during resorption. Consequently, mechanical designs are difficult to eliminate fracturing of the tissue and resorbable ceramic structure during the intermediate stages of healing if the implant is loadbearing.

In Fig. 2, the biomaterials in the middle of the reactivity spectrum (curves B and C) are those with controlled surface reactivity. In this class of biomaterials the composition is designed such that the surface undergoes a selected chemical reaction with the physiological system establishing a chemical bond between tissues and the implant surface (1-4). The chemical reactions are such that, ideally, the bonded interface protects the implant material from further change with time. Thus this approach combines the high strength of nearly inert biomaterials with surface chemical reactivity favorable to tissue bonding. Since stabilization of controlled surface reactive implants is not restricted to mechanical interlocking with tissues, more flexibility in device design and fabrication is possible.

Various Bioglass™ and Bioglass-Ceramic™ formulations as well as Ceravital™ surface active glass-ceramics (5,6) have been found to behave biologically with a range of bioreactivities, depicted as curves B and C in Fig. 2. For example, as a Bioglass™ composition nears the bone-bonding-boundary in Fig. 1 its bioreativity decreases. This control over rate of bioreactivity is important with respect to the nature of interfacial cellular reactions in vivo.

NATURE OF THE BIOGLASS™ IMPLANT-TISSUE BOND

Since 1969, in vitro and in vivo investigations have been pursued in order to understand the bonding mechanism between Bioglass™ and host tissues (1-4,7-16). These findings are summarized for the 45S5 Bioglass™ composition since this is the composition used for the dental implants and middle ear devices now in clinical trial (15). For details of bonding mechanisms the reader is referred to Refs. 3, 9 and 10.

Upon implantation of Bioglass™ devices there is a rapid exchange of alkali ions from the glass surface with H^+ or H_3O^+ ions from body fluids. This exchange results in the formation of a highly biologically active silica-rich layer on the Bioglass™ surface while at the same time maintaining an alkaline pH at the implant interface. The consequence of high bioreactivity and alkaline pH is development of a $CaO-P_2O_5$ rich layer on the implant within mins. to hrs. after implantation. Nucleation of the $CaO-P_2O_5$ layer occurs on the Bioglass™ surface but growth proceeds by incorporation of Ca and P ions from body fluids as well as from the glass network. Consequently an anionic balance of OH^- and CO_3^{2-} groups characteristic of natural tissues is achieved within the growing $CaO-P_2O_5$ layer. Within a 1 to 2 week period of time crystallization of the calcium phosphate layer into a hydroxyl-carbonate-apatite proceeds to completion.

A consequence of the growth of the apatite layer in vivo is that organic metabolic constituents such as collagen, mucopolysaccharides, and glyco-proteins are incorporated intimately within the inorganic layer, satisfying mutual molecular bonding sites. A second consequence of the growing bioactive interface is its effect on cellular populations, and perhaps, differentiation (3). Cells with small concentrations of fibronectin such as fibroblasts, when exposed to a Bioglass™ substrate are maintained in a resting state for many days (11). In contrast, highly specialized cells are able to attach, divide, and multiply rapidly, within hours, on Bioglass™ surfaces (3).

IN VIVO RESPONSE TO BIOGLASS™ BONDING

The net effect in vivo of the cell-specific response to Bioglass™ implants when put into a bony defect is a rapid replacement of post-surgical debris and soft tissue with bone. The rate of formation of bonded bone at the interface of surface reactive glasses and glasses is a function of the bioreactivity of the materials, as depicted in Fig. 2. Refer to Ref. 3, pp.

279-288, for more detail discussion of bone bonding mechanisms in vivo.

Both in vitro and in vivo studies show that a Bioglass™ interface forms a stable bond even when osteogenic stem cells are not present. Attachment of 45S5 Bioglass™ implants to connective tissue occurs in rats via collagen fibers (13).

Clinical experience suggests that non-porous implants are preferable to those fixed by ingrowth, especially in the infected ear (17). Bioglass™ implants in the middle ear are not porous, do not bond to the bone, promote a thin and stable bonding layer, and show normal tissue response to the implant in animals experimentally infected with haemophilus influenzae streptococcus-pneumoniae. These attributes make Bioglass™ implants ideal for a wide variety of otological applications. Clinical trials, which are now in progress are described in following presentations.

CONCLUSION

The controlled surface active Bioglass™ system provides a solution to stabilization of the interface between implants and tissues. However, the need for further basic research into the mechanisms of interfacial reactions is clear. The interface between tissue and implant surface and between phases in a composite, such as a coating and substrate, are potential weak links in long-term reliability. Only by a thorough study of the mechanisms and kinetics of interfacial reactions will it be possible to determine why failures occur and thereby design better devices with predicted lifetimes of 20-30 years.

REFERENCES

1. Hench, L.L., R.J. Splinter, W.C. Allen and T. K. Greenlee, Bonding Mechanisms at the Interface of Ceramic Prosthetic Materials, J. Biomed. Mater. Res. Symp., No. 2, Part 1 (1972) 117-141.
2. Hench, L.L. and H. A. Paschall, Direct Bone of Bioactive Glass-Ceramic Materials to Bone and Muscle, J. Biomed. Maters. Res. Symp., No. 4 (1973) 25-42.
3. Hench, L.L. and E.C. Ethridge, Biomaterials: An Interfacial Approach (Academic Press, New York, 1982).
4. Hench, L.L., H.A. Paschall, W.C. Allen and G. Piotrowski, Interfacial Behavior of Ceramic Implants, NBS Special Publications (US) 415 (1974) 19-35.

5. Blencke, B.A., H. Bromer and K.K. Dutscher, Compatibility of Long-Term Stability of Glass Ceramic Implants, J. Biomed. Mater. Res. 12 (1978) 307-316.
6. Gross, U.M. and V. Strunz, The Anchoring of Glass Ceramics of Different Solubility in the Femur of the Rat, J. Biomed. Mater. Res. 14 (1980) 606-618.
7. Onoda, G.Y., Jr., D.B. Dove and C.G. Pantano, Jr., Mater. Sci. Res. 7 (1974) 39-56.
8. Clark, A.E., L.L. Hench and H.A. Paschall, The Influence of Surface Chemistry on Implant Interface Histology: A Theoretical Basis for Implant Material Selection, J. Biomed. Mater. Res. 10 (1976) 161-174.
9. Hench, L.L. and A.E. Clark, Adhesion to Bone in D.F. Williams, ed., Biocompatibility of Orthopedic Implants, Vol. II, Chapt. 6 (CRC Press, Boca Raton, Florida, 1982)
10. Hench, L.L. Stability of Ceramics in the Physiological Environment in D.F. Williams, ed., Fundamental Aspects of Biocompatibility, Vol. I, Chapt. 4 (CRC Press, Boca Raton, Florida, 1982).
11. Seitz, T.L., K.D. Noonan, L.L. Hench and N.E. Noonan, Effect of Fibronectin on the Adhesion of an Established Cell Line to a Surface Reactive Biomaterial, J. Biomed. Mater. Res. 16 (1982) 195-207.
12. Merwin, G.E., J.S. Atkins, J. Wilson and L.L. Hench, Comparison of Ossicular Replacement Materials in a Mouse Ear Model, Otolaryngol Head Neck Surg. 90 (1982) 461-469.
13. Wilson, J., G.H. Pigott, F.J. Schoen and L.L. Hench, Toxicology and Biocompatibility of Bioglasses™, J. Biomed. Mater. Res. 15 (1981) 805-817.
14. Wilson, J., H.R. Stanley and L.L. Hench, Chapt. 50 in James W. Clark, ed., Dental Appliations of Bioglass™ Clinical Dentistry (Harper and Row, Philadelphia, Pennsylvania, 1981).
15. Ogino, M., F. Ohuchi and L.L. Hench, Compositional Dependence of the Formation of Calcium Phosphate Films on Bioglass™, J. Biomed. Maters. Res. 14 (1980) 55.
16. Merwin, G.E., J. Wilson and L.L. Hench, Current Status of the Development of Bioglass™ Ossicular Replacement Implants in J.J. Grote, ed., Biomaterials in Otology (The Hauge: Martinus Nijhoff Publishers, 1983).
17. Austin, D.F., Avoiding Failures in the Restoration of Hearing with Ossiculoplasty and Biocompatible Implants, Otolaryngologic Clinics of North America 15 (1982) 763-771.

CLINICAL APPLICATIONS OF BIOGLASS™

June Wilson

Department of Surgery
J. Hillis Miller Health Center
University of Florida
Gainesville, FL 32610

An ideal implant material would have many potential applications in surgery. Such a material, as outlined by Garrington in 1972 (1) and reiterated regularly since then, should be biocompatible, inert, nonallergenic, nontoxic, and noncarcinogenic and should restore to normal function the part it replaces. Bioglasses™ in several different forms and compositions conform to most of these requirements and are being used in clinical applications. The materials have been tested in bone in many different species: subcutaneously in rats, intramuscularly in rabbits, intravenously in mice, intraperitoneally in powder and solid form in rats, and in vitro by exposure to several different cell culture systems. This work has been reviewed, and in no case has any toxicity been detected (2). Bioglasses™ can not however be described as "inert" since they possess a unique surface activity. The final requirement, that the normal function of the part is reproduced, is limited by the mechanical properties of the bulk glass. Techniques have been developed to use the Bioglass™ as a coating on substrates to provide the properties needed for orthodontic and dental devices. Another area in which devices made from Bioglass™ will be used is in ear surgery.

Clinical trials have already begun at the University of Florida on tooth root replacement and ossicular chain replacement. Preclinical experiments are nearing completion in several other areas, periodontal packing material, ridge augmentation material, orthodontic anchors, many orthopedic devices, including hip devices which are fixed without cement, and spinal fusion. Other areas in which preclinical experiments have begun or are planned include joint resurfacing, finger joint anchor. It is not possible to cover all these applications here and only the experimental work which has led to the applications in or near clinical trial will be described.

DENTAL APPLICATIONS

Loss of teeth, for many reasons provides an enormous number of people, indeed the majority of aging adults in the population, who could advantageously receive replacement tooth implants. The functional and aesthetic advantages of a replicate tooth form implant are self-evident, and the successful reproduction of crowns has been successfully achieved. However, the anchoring of the tooth into the jaw bone is not always easily accomplished, particularly in the presence of periodontal disease. In orthodontic applications, a material providing a strong biocompatible bond with the bony tissues in the jaw would allow both tooth and bone movement.

Cotton showed that unwanted tooth movement occurred when teeth were used as anchors for a device for palatal expansion for restoration of arch size in cleft palate cases (3). Stable bonded anchorage points, such as can be achieved with a successful implant, could avoid the use of teeth as anchors and the subsequent problems related to their movement. The use of Bioglass™ as skeletal augmentation could be valuable in maxillofacial surgery.

Various formulations of Bioglass™ have been used in the studies of Stanley and co-workers (4,5). In both studies, replicate tooth forms replaced extracted incisor teeth in adult female baboons. Extracted teeth were cleaned, impressions made and then replaced in the sockets and splinted in place. In this way the sockets were maintained during the time taken to carve and implant the replacement teeth. Healing was allowed for 3 months. After which all restraints were removed and the animals were kept a further 3 months.

At the end of 6 months only teeth from certain Bioglass™ formulations were in place, others and control inert teeth had been lost. The glass known as 45S5 and a formula which had some of the CaO replaced by CaF produced the most satisfactory histological response, that is one which most closely approximated to normal healing. These were the glasses used in the 2-year study. In this study, only two formulations were used together with control teeth of cobalt chrome or of solid polymethyl methacrylate. Replacement teeth, on the whole, did not withstand normal activity, and many broke off at the gingiva, leaving the roots firmly in situ. The initial retention rate at 3 months of around 65% for Bioglass™ and 25% for cobalt chrome implants was, by the end of 2 years, 55% for Bioglass™ and 20% for cobalt chrome.

Histologically, the Bioglass™ formulations had induced ankylosis at 2 years, as after 6 months. It was clear that the success of splinting and initial immobilization of the implant against bone is essential to the bond. A successful bond depends on absence of micromovement at the interface. If total immobilization is satisfactorily achieved, only then can the chemical bonding properties of the Bioglass™ be exploited.

Stanley and associates have shown that, given optimal clinical condition, proper postoperative periodontal care combined with an appropriate amount of restraint replacement teeth of Bioglass™ can be used to replace extracted teeth (5).

In these baboon studies a very large number of the replacement teeth broke at the gum line leaving the root portion buried. Where this occurred normal bone grew over the fresh glass surface leaving the glass within the jawbone. This, combined with the observations of many clinicians that broken roots of natural teeth could be left in place suggested the possibility that cone shaped pieces of Bioglass™ could be put into extraction sockets and would bond in place there, successfully maintaining the integrity of the jawbones. It is the loss of this integrity, causing alteration to the normal mechanical stimuli on the jawbone, which is responsible for the progressive loss of bone found after multiple extractions. It is this application which is currently in a clinical trial at the University of Florida under the direction of Dr. Harold Stanley. Patients with multiple extractions have had cone shaped devices, known as ERMI'S (Fig. 1) placed tightly in fresh reamed sockets, and kept in place by suturing of the overlying gingiva. It is too early to report clinical results since long term maintenence of the alveolar ridge height will require regular monitoring. To date however reports auger well.

When a tooth is stressed either accidentally or intentionally, it may be moved in the direction of the load. The movement is the consequence of the combined activity of osteoclasts, which remove bone under the influence of pressure., and of osteoblasts, which lay down new bone in the area of decreased pressure. Such is the basis of orthodontic movement of teeth. However, ankylosed teeth cannot be moved in this way (6,7).

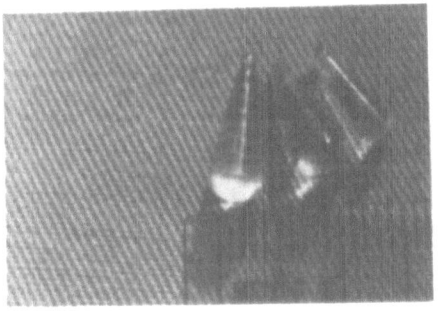

Fig. 1.

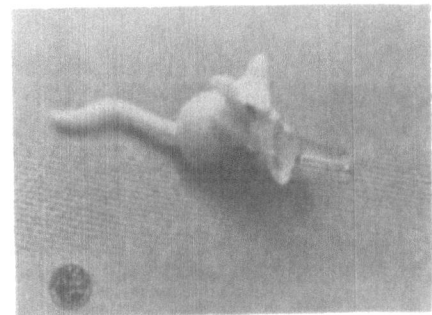

Fig. 2.

Smith, used 45S5 glass to coat alumina implants of a blade shape that were inserted into the alveolar ridge of monkeys (8). The implants were allowed to heal with a short rod protruding through the gingiva. These rods were removed after 9 weeks and replaced by the posts that would hold a lingual arch appliance. Forces of 425 gm were applied to these appliances and the palatal expansion measured radiographically. In one experiment, forces of 950 gm were applied for several weeks.

Smith observed the fusion of implants to the surrounding bone and that the bond remained intact after several weeks of continuous loading. There was no movement of the implants in the bone and no resorption of bone adjacent to the implant as might be found if natural teeth were so loaded.

Paige recognized that, to be really clinically useful, orthodontic implants should be small and preferably simple in shape (9). Much previous work had used implants that were irregularly shaped and relatively large, providing a large surface area to the tissues for reaction. In his experiment, Paige showed that vitallium wires of 1 mm diameter coated with 0.5 mm Bioglass™ would bond to bone, and then take a continuous load of 180 gm applied to protruding ends for 4 weeks with no measurable movement. This work has been further extended by Grey et al. (10) and it is their designs which will be used in clinical trials of orthodontic anchors.

In all these studies it was clear that those implants that were stable were those in which initial surgical techniques, animal health, and selection of implant and site gave optimal conditions for bone bonding. The implants were more stable than adjacent teeth even when only parts of the implant bonded to bone.

MIDDLE EAR APPLICATIONS

Of the many causes of complete or partial deafness otitis media, a middle ear infection is probably the most common. The ear consists of outer, middle and inner parts and the middle ear is an aerated cavity connected via the eustachian tube to the throat. Sound waves collected by the outer ear are relayed as vibrations by the ear drum. The vibrations are transmitted by the ossicles of the middle ear to the inner ear fluids, which result in the stimulation of nerve endings and the perception of sound.

In order to relay vibrations faithfully atmospheric pressure on each side of the drum must be equal and this is effected by the eustachian tube. At the same time the integrity of the ossicular chain must be maintained. Unfortunately in chronic otitis media the eustachian tube can provide regular access of infective organisms to the middle ear and recurrent infection can destroy part or all of the ossicular chain. Many alloplastic materials have been used to restore the continuity of the chain where this has happened. These include polymeric materials, porous materials and inert materials.

In a recent series of experiments the mouse incus was replaced by Silastic (an inert polymer) PlastiPore (an inert, porous polymer) Proplast (an inert porous composite) and Bioglass™ (an active solid glass). Particular note was taken of the state of the middle ear with regard to inflammation, exudate and adhesions. Also noted was the position of the implants at the end of the experimental period, their stability and ability to transmit motion. Full data are given in reference 11. In every case Bioglass™ implants showed improvement over the other materials, all of which are in current clinical use. Macroscopically and microscopically differences were even more clearly demonstrable.

Gross inspection of ears revealed a clear middle ear in 10 of 10 Bioglass™, clear adhesions in 8 of 9 Silastic, fluid or opaque and clear adhesions in 6 of 17 Plasti-Pore, and fibrosis with adhesions or excess bone in 6 of 9 Proplast implanted mice. Tympanic membranes were intact in all mice.

Bioglass™

After one and two months in the middle ear, Bioglass™ implants are surrounded by a very thin capsule, over the surface of the implant. The collagen fibers are usually adherent to the surface of the implant. Normal-appearing respiratory epithelium of the middle ear formed a covering on the outer surface of this thin capsule. Some small capillaries were present in the capsule. No polymorphs or giant cells were seen in any sample.

Silastic

In several animals, scraps of refractile material (presumed to be remnants of polymer) were seen embedded in fibrous scar tissue that formed adhesions within the middle ear cavity. Giant cells and macrophages , indicating breakup of the material, were invariably seen, as were polymorphs and plasma cells, both in the scar tissue and free in middle ear cavity exudate.

Plasti-Pore

After one month the implant had a thick capsule with many inflammatory cells, including giant cells. However, after two months this reaction had subsided in one animal, although the capsule did not contain mature collagen. Another animal had a continuing acute inflammatory response.

Proplast

After one month the implant was attached by fibrous adhesions to ossicles and the walls of the middle ear cavity. Bone had grown into the pores of the material, forcing the fibrous reinforcement and polymer matrix apart. There was no inflammation or macrophage and giant cell activity; respiratory epithelium covered one surface of the implant. After two months the bone ingrowth had further separated the fibers and polymer in the scar tissue, and osteoneogenesis had provided an attachment between the ossicles and the implant. There was no inflammation.

The results of this study indicate that Bioglass™ compares favorably with alloplastic materials currently used in human ear surgery. It maintains the position of initial placement and develops a tenacious bond with the tissues with which it is in contact.

The histologic picture of Bioglass™ in the middle ear shows that a thin fibrous capsule develops, covered by a normal respiratory epithelium with no apparent scarring, giant cells, or inflammation. Proplast implants showed invasion of thick fibrous tissue and new bone growth into the pores with no inflammation. Plasti-Pore also showed thick cellular scar tissue with continued inflammation. The Silastic implants revealed a scar tissue capsule with inflammation. The Bioglass™ implants showed the least reaction, inflammation and minimal

668

encapsulation; Proplast induced a more reactive response but appeared to be compatible. Plasti-Pore and Silastic, in this model, demonstrated persistent inflammation. Longer term, up to 12 months, experiments have confirmed these short term findings.

INFECTION STUDIES

An ear which has been subject to repeated chronic infection may be repaired with a suitable device but may continue to suffer infection. Porous materials have been critized since they are particularly prone to harbour infections and the effect of bacterial contamination on the materials used for devices must be assessed. Karlan et. al. (12) have studied the effect of interactions between Teflon, silicone rubber, and Bioglass™ and various bacteria in a subcutaneous model and Merwin et. al (13) have further investigated interactions between Bioglass™ and organisms which cause o. media in the mouse ear model. In both sets of experiments there was no evidence of potentiation of infection by Bioglass™ implants nor was the implant affected by the infection.

Extrusion through the ear drum is the reason for long term failures of some middle ear protheses . Where there is no posi-tive bond between implant and drum, micromovement causes abrasion, increased cellularity, thickening and enventual break-down at the interface allowing perforation of the drum. In experiments in which the ear drums of mice were stretched as much as possible over the end of an implant, Bioglass™ implants have shown none of these effects. Other materials have produced varying degrees of inflammation and thickening, although none to date have perforated. These experiments are continuing.

All the data support previous findings of early adhesion or stabilization by soft tissue bonding of Bioglass™ in a mouse ear model. Using the same model, Bioglass™ compares favorably with other materials in current clinical use. Modes of failure seen clinically can be mimicked experimentally in this model as can the effect of infectious with those organisms responsible for chronic otitis media. All the preclinical data suggest that middle ear prosthetic devices made from Bioglass™ will provide longer and more reliable alleviation of deafness than do those currently available. Fopr these reasons they are now being used in a clinical trial at the University of Florida under the direction of Dr. Gerald Merwin. The implants (known as Merwin Struts) (Fig. 2) are placed so that the flat plate bonds, in this case by a soft-tissue bond to the ear drum. The stem is then contoured and grooved so that it fits the remains of the patients ossicular chain which interfaces with the inner ear.

As with the dental implants immediate restoration of function occurs but it is in the long term that an advantage over other treatments is anticipated.

REFERENCES

1. Garrington, G.E. and P. M. Lighthouse , J. Biomed. Mat. Res. Symp., 2 (1982) 333.
2. Wilson, J. et al., J. Biomed. Mat. Res., 15 (1981) 805-817.
3. Cotton, L.A. Am. J. Orthod., 73 (1978) 1.
4. Stanley, H., et al., Inter. J. Oral Surg., 42 (1980) 339.
5 Stanley, H., et al., Intern J. Oral Implant, 2 (1981) 26-36.
6. Mitchell, D.L. and J. D. West, Am. J. Orthod., 68 (1975) 404.
7. Utley, K.R. Am. J. Orthod., 54 (1968) 167.
8. Smith, J.R. Am. J. Orthod., 76, (1979) 618.
9. Paige, S.F., U.F. Thesis in Orthodontics (1980).
10. Grey, J.B., et al., Am. J. Orhtod., 82 (1983) 312-317.
11. Merwin, G.E., et al., Otol. Head and Neck surg., 90 (1982) 461-469.
12. Karlan, M., et al., Otol. Head and Neck Surg., 89 (1981) 528-534.
13. Merwin, G., et al, to be published.

MACHINABLE BIOACTIVE GLASS-CERAMIC

W. HÖLAND, W. VOGEL, K. NAUMANN
Friedrich-Schiller-University Jena, Otto Schott-Institut,
6900 Jena, DDR

J. GUMMEL
Humboldt University Berlin, Orthopedic Clinic (Charité)
1040 Berlin, DDR.

INTRODUCTION

The development of glassy-crystalline materials, the glass-ceramics, showed that materials can be formed which have new properties that any other inorganic material. So called "machinable glass-ceramics" are important for industrial appplications[1,2]. These materials consist of a glassy phase and a mica crystal phase. The mica crystal, e.g.,the fluorophlogopite $(Na,K)Mg_3/AlSi_3O_{10}/F_2$ is responsible for the excellent machinability. This means that the material is machinable with standard metal-working tools or instruments for the machining of ceramic materials.

On the basis of the investigations of machinable glass ceramics, it was possible to develop new glass-ceramics with a combination of different properties. Machinable glass-ceramics with magnetic properties were developed [3] and now a new bio-material for human bone substitution in medicine, a "machinable bioactive glass-ceramic", has been prepared and will be reported in detail in this paragraph [20].

SYSTEMS OF BIOACTIVE GLASS-CERAMICS

Biomaterials for bone substitution are bioactive, if a stable bond to the bone is formed. Hench [4] could show that a bioglass (R) can bond to bone in animals. The interface reactions between bioglass (R) and bone, the formation of different Ca, P-rich and SiO_2-layers depending on the composition of the implant, the environment and reaction time were studied (5-9).

It was shown that a preferred intergrowth of bone and bio-material (glass-ceramic or sintered ceramic) can be achieved if the biomaterial contains apatite crystals already in the basic material (10-19). The sintered bioactive ceramics of Osborn et al. (10), Ducheyne, de Groot (11), Acki et al. (12), Jarcho et al. (13) are based on hydroxylapatite and have bioactivity because of their similar composition (30 vol-%"calciumphosphate" /70 vol-% hydroxy-lapatite) to the human bone. The materials contain hydroxylapatite crystals, whitlockite (h-ca$_3$(PO$_4$)$_2$ and the bending strength is about 115 MPa. Bioactive glass ceramics can have higher bending strength than glasses and sintered ceramics and it is possible to combine different properties in one material. The first bioactive glass-ceramics were also developed by Hench (14) in the same SiO$_2$-CaO-Na$_2$O-P$_2$O$_5$ glass system as the bioglass (R). The main crystal phase is devitrite Na$_2$Ca$_3$Si$_6$O$_{16}$), but the bending strength is not much higher than that of the glass.

Glass-ceramics of Blencke, Brömer et al. (18,16) are based on the system SiO$_2$-CaO-MgO-P$_2$O$_5$-Na$_2$O-K$_2$O. This material is known as ceravital (R) (Leitz/Wetzlar, FRG). Some variations in Na$^+$/K$^+$ content (Käs (17)) and additions of Al$_2$O$_3$, Ta$_2$O$_5$, TiO$_2$ are possible. The crystal phases in the material are hydroxylapatite and devi-trite; thus bioactivity results. But it could be shown by in vivo investigations that a leaching region of about 100 μm between material and bone exists.
New materials by Deutscher and Brömer (18) do not show such a great interface region. These new materials were developed in the SiO$_2$-CaO-P$_2$O$_5$ system. Glass-ceramics by Wihsmann (19) show reactions similar to those of Ceravital (R). Trojer (26) could develop glass-ceramics in a system similar to that of Brömer (15). Additions of La$_2$O$_3$ (and other oxides) to glasses of the SiO$_2$-CaO-P$_2$O$_5$-Na$_2$O system result in crystallization of La^{3+}-containing apatites, which are isomorph with hydroxylapatite.

Bioglass ceramics by Kokubo (27) were produced in the alkali-free SiO$_2$-MgO-CaO-P$_2$O$_5$-CaF$_2$ system in a sintering process.

Therefore a material which should be machinable and bioactive has to contain both mica and apatite crystals.

THE PREPARATION OF BIOACTIVE MACHINABLE GLASS-CERAMICS

It was possible to develop such a new biomaterial in the glass system 19-52 SiO$_2$, 12-33 Al$_2$O$_3$, 5-15 MgO, 9-30 CaO, 3-10 K$_2$O/Na$_2$O, 0,5-7 F$^-$ and 4-24 wt.-% P$_2$O$_5$(20).
The preparation of the machinable bioactive glass-ceramic took

place in two processes. The first processes are the melting of the basic glass at about 1500°C and cooling to room temperature. The basic glass contains three glassy phases which are the basis for the nucleation and crystallization to the final glass ceramic (Fig. 1).

The next process, heat treatment of the basic glass at a temperature range between 610°C and 1050°C allows an <u>in situ</u> crystallization to a machinable bioactive glass-ceramic. The material contains two crystal phases (apatite and mica) and a glassy phase (Fig. 2). To achieve these properties, this material can also be produced in different types. E.g. a high phlogopite containing glass-ceramic with very good machinability or a glass-ceramic, which consists of a high concentration of apatite crystals can be formed. The crystal phases in the glass-ceramic were analysed by by X-ray diffraction. A new type of phlogopite crystals was discovered (20,21) and the apatite could be identified as $Ca_{10}(PO_4)_6(OH)_xF_{1-x}$ with a 937,2 ± 0,8 pm and 0 = 687,4 ± 2,1 but an incorporation or other ions from the glassy phase into the apatite lattice should also be possible (22,23).

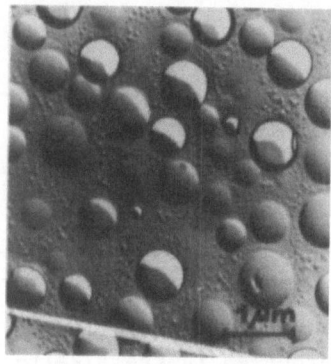

Fig. 1. Replica electron micrograph of a basic glass specimen for a machinable bioactive glass-ceramic. The glass contains a large P_2O_5-rich droplet phase, a silicon droplet phase and a SiO_2-rich glass matrix phase. - etched sample -

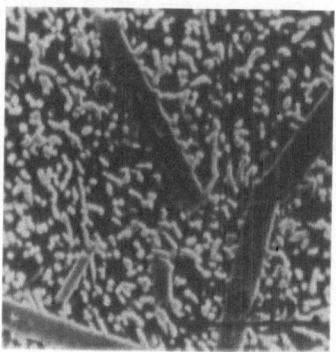

Fig. 2. Scanning electron micrograph of a machinable bioactive glass-ceramic specimen. Phlogopite and apatite crystals grow in the basic glass after heat treatment.

BASIC PROPERTIES OF BIOACTIVE MACHINABLE GLASS-CERAMICS

Implantation of the glass ceramic was carried out to study the reactions between biomaterial and bone (tibia) of guinea pig. The surface of the implant (cube-shape sample) terminates the surface of the corticalis (Fig. 5). This method allowed the implant to have contact with corticalis and spongiosa.

A biocompatible but bioinactive reference implant did not show a direct bond to bone. There was a microscopically visible gap. A direct bonding was achieved between machinable bioactive glass-ceramic and bone (Fig. 3). The X-ray intensity profiles (measured from glass-ceramic to bone) for the elements Si and K in comparison to Ca_K and P_K show bonding formation 16 weeks post implantation (Fig. 4). Fig. 4 indicates a Ca-phosphate rich interface with a thickness of about 5-10 μm between the bone and the implant. This means that a Ca-phosphate rich surface on the glass-ceramic was formed after a reaction time of 16 weeks. The properties of the three phases of the glass ceramic (phlogopite and apatite crystals, glassy phase) are the driving forces for the bonding formation. The glassy phase is slightly soluble but the apatite crystals are stable. The slight solubility of the glass-ceramic is necessary because of a stimulated mineralization process and the activity of bone functions (24).

A better estimation of the solubility of the glassy phase in glass ceramics was obtained from in-vitro investigations. A glass ceramic which was cooked one week in Ringer's solution

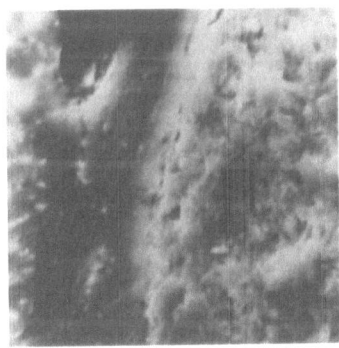

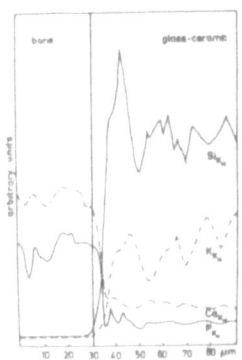

Fig. 3.   Scanning electron micrograph of a bone – implant interface.

Fig. 4.   X-ray intensity profile for the elements Si, K, P, Ca along a line from implant (machinable bioactive glass-ceramic to bone.

shows a slight solubility of the glassy phase. Similar solubility of about 0.5 m was observed at a glass-ceramic implant of 16 weeks post implantation.

CONCLUSIONS AND APPLICATIONS

A Ca-phosphate rich interface layer with apatite crystals and a thickness of about 5-10 μm grows as the result of a solid state reaction between machinable bioactive glass-ceramic and bone. The new biomaterial has the advantage that it is workable by the surgeon (if necessary during operation) and that the material has a good bioactivity (20,25). The bending strength of the material is about 200 MPa and the shear strength between bone and bio-material is about 4MPa (in comparison : Al_2O_3-ceramic 0.1-0.5 MPa). The material might be therefore applicable for bone substitution in the human body.

REFERENCES

1. Vogel, W., E. Heidenreich, E.D. Maschinell bearbeitbare Glaskeramik, DDR, WP Nr. 113 885 (1973).
2. Vogel, W., W. Höland, K. Naumann, Fluorophlogopithaltige, Glaskeramik mit bester maschineller Bearbeitbarkeit, DDR-WP Nr. 0153 108 (1980).
3. Höland, W., P. Nguyen Anh, E. Heidenreich, E. Tkalcec, W. Vogel, Einfluß von Eisenoxiden auf Kristallisationskinetik und Eigenschaften glimmerhaltiger maschinell bearbeitbarer Glaskeramiken, Teil 1 und 2, Glastechn. Berichte, 55 (1982), 41-49 und 70-74.
4. Hench, L.L., R.J. Splinter, W.C. Allen, T.K. Greenlee Jr., Bonding mechanism at the interface of ceramic prosthetic materials, J. Biomed. Mater. Res. Symp., 2 (1972) 117-141.
5. Pantano C.G. Jr., A.E. Clark, Jr., L.L. Hench. Multilayer corrosion films on bioglass (R) surfaces, J. Am. Ceram. Soc., 57 (1974) 412-413.
6. Ogino M., F. Ohuchi, L.L. Hench. Compositional dependence of the formation of calcium phosphate films on bioglass (R), J. Biomed. Matr. Res., 14 (1989) 55-64.
7. Ogino, M., L.L. Hench, Formation of calcium phosphate films on silicate glasses, J. Non-Cryst. Sol., 38 (1980), 673-678.
8. Ducheyne P., L.L. Hench, A. Kagan II, M. Martens, J.C. Mulier, Short term bonding behaviour of bioglass (R) coatings on metal substrate, Arch. Orthop. Traumat. Surg. 94 (1979) 155-160.
9. Griss P., P.C. Greenspan, G. Heimke, B. Krempien, R. Buchinger, L.L. Hench, G. Jentschura. Evaluation of a bioglass coated Al_2O_3 total hip prosthesis in sheep, J. Biomed. Mater. Res., 10 (1976) 511-518.
10. Osborn J.F., E. Kovacs, A. Kallenberger, Hydroxylapatit-Entwicklung eines neuen Biowerkstoffes und erste tierexperi-mentelle Ergebnisse, Dr. Zahnärztl. Z., 35 (1980), 54-56.

11. Ducheyne, P., K. de Groot, In vivo surface activity of a hydroxyapatite alveclar bone substitute, J. Biomed. Mater. Res., 15 (1981), 441-445.

12. Aoki, H., K. Kato, T. Kabata, M. Ogiso, Orthopedic and dental implant ceramic composition and process for preparing same, Us-Ps Nr. 4 149 893 (1979).

13. Jarcho, M., C.H. Bolen, M.B. Thomas, J. Bobick, J.P. Kay, Doremus, Hydroxylapatite synthesis and characterization in dense polycrystalline form, J. Mat. Sci. 11 (1976) 2027-2035.

14. Hench, L.L., H.A. Paschall, Direct chemical bond of bioactive glass-ceramic materials to bone and muscle, J. Biomed. Mat. Res. Symp. 4 (1973) 25.

15. Blencke B., H. Brömer, K. Deutscher, E. Pfeil, Glaskeramiken für Osteoplastik und Osteosynthese, Bundesministerium für Forschung und Technologie, Forschungsbericht T 77-91 (1977).

16. Strunz V., M. Bunte, U.M. Gross, K. Männer, H. Brömer, K. Deutscher, Beschichtung von Metallimplantaten mit bioaktiver Glaskeramik Ceravital, Dtsch, Zahnärztl. Z., 33 (1978) 862-865.

17. Käs, H.H., Glaskeramisches Material mit Apatit-Kristallphase, insbesondere zur Verwendung für prothetische Zwecke sowie Verfahren zu seiner Herstellung, BRD-PS Nr. 2 349 859 (1973).

18. Deutscher H., New glass-ceramic materials paper at ISIR-conference, Berlin (West),1983.

19. Wihsmann, F., G. Berger, "Bioaktive Implantate auf der Basis von Vitrokerammaterialien, Wiss. Zeitschr. der Friedrich-Schiller-Universität Jena, Math.-nat.-Reihe 32 (1983) 553-569.

20. Höland W., K. Naumann, W. Vogel, J. Gummel, Maschinell bearbeitbare bioaktive Glaskeramiken, Wiss. Zeitschr. der Friedrich-Schiller-Universität Jena, Math.-Nat.-Reihe 32 (1983) 571-580.

21. Höland W., W. Vogel, W. Mortier, P.H. Duvigneaud, G. Naessens, E. Plumat, "A new type of phlogopite crystals in machinable glass-ceramics, Phys. Chem. Glasses.

22. Chiranjeevirao, S.V., J. Hemmerle, J.C. Voegel, R.M. Frank, A method of preparation and characterization of magnesium-apatites, Inorg. Chim. Acta 67 (1982) 183-187.

23. Okazaki, M. J. Tahakashi, H. Kimura, T. Aoba, "Crystallinity, solubility and dissolution rate behaviour of fluoridated Co_3-apatites, J. Biomed. Mater. Res., 16 (1982) 851-860.

24. Gross U.M., V. Strunz, The anchoring of glass ceramics of different solubility in the femur of the rat, J. Biomed. Mater. Res., 14 (1980), 607-618.

25. Gummel, J., W. Höland, K. Naumann, W. Vogel, M. Timm, "Ergebnisse zytotoxikologischer und histologischer Untersuchungen zur Eignung einer maschinell bearbeitbaren bioaktiven Glaskeramik als Knochenersatz", VII. Symposium Extrakorporale Systeme und Jahrestag der Sekt. Biomed. Mechanik, 2.-5.3.1983, Rostock, DDR.

Abstract only

THE APPLICATION OF GLASSES IN AGRICULTURE AND MEDICINE

P. Knott

Department of Ceramics, University of Leeds, England.

The solubility of low network former glass compositions can be varied to give controlled release of ions over a period of months or years in a specific environment. Such glasses have been developed for application to both animals and humans as a local source of trace elements. Field trials of soluble glass capsules to provide trace elements for ruminant animals grazing on deficient pastures have proved the success of this application. Since the capsules are swallowed, but remain in the gut of the animal, the glass composition must be matched to the corrosivity of the digestive juices, and in particular to the pH. The product is now commercialised for a number of specific ion deficiencies.

Other applications include controlled release fertilisers, corrosion inhibiting glassy paints, ion leachable glasses for dental fillings and insoluble glasses for the immobilisation of enzymes.

SHORT COMMUNICATIONS

A DETAILED STUDY OF THE DEVITRIFICATION PROCESSES OF
THE ELECTROLYTE SYSTEMS LiCl nD_2O AND LiCl nH_2O

A. AOUIZERAT-ELARBY[*], P. CHIEUX[**], P. CLAUDY[***],
J.M. LETOFFE[***], J. DUPUY[*], A.F. WRIGHT[**]

[*] Département de Physique des Matériaux, LA 172
Université Claude Bernard, Villeurbanne (France)

[**] I.L.L., 156 X 38042 Grenoble Cédex (France)

[***] I.N.S.A. de Lyon, Thermochimie Minérale
Villeurbanne (France)

The aqueous electrolytes $LiClnD_2O$ can be prepared in the glassy state over a very large range of concentration. The devitrification processes have been studied by using Differential Scanning Calorimetry (D.S.C.) and neutron scattering (Diffraction and Small Angle Scattering (S.A.N.S.)). We identify different behaviour for the two concentration ranges with x above and below 0.15 (x = LiCl/LiCl + D_2O).

Concentration range : x < 0.15.

In this range the first hydration ion spheres are full and some free H_2O molecules remain. We observe the crystallization of the free water as a function of concentration :

For x < 0.08 hexagonal ice I_h precipitates during cooling at a rate of 3K.sec^{-1}.

For 0.08 < x < 0.14, a vitreous phase is easily formed and heat treatments allow a controlled crystallization of metastable cubic ice I_c. I_c transforms to hexagonal ice I_h at 150 K for x = 0.1.

For 0.14 < x < 0.15 (LiCl $6H_2O$) no crystallization occurs.

Concentration range : x > 0.15.

In this range the Li^+ and Cl^- ions influence over the H_2O molecules is more important. Here the stoichiometric compounds $LiCl\ 5H_2O$, $LiCl\ 3H_2O$ and $LiCl\ 2H_2O$ are crystallized. This may be preceded by the crystallization of metastable phases.

For x=0.2 a "window" occurs where no crystallization is observed.

For this system, it is easy to control the glassy stage as well as metastable and stable crystallization. We can analyse the processes of nucleation and crystal growth with respect to the time and temperature of thermal treatment. The ranges where no crystallization occurs are specially interesting. They may be explained in terms of frustration between completing hydration spheres.

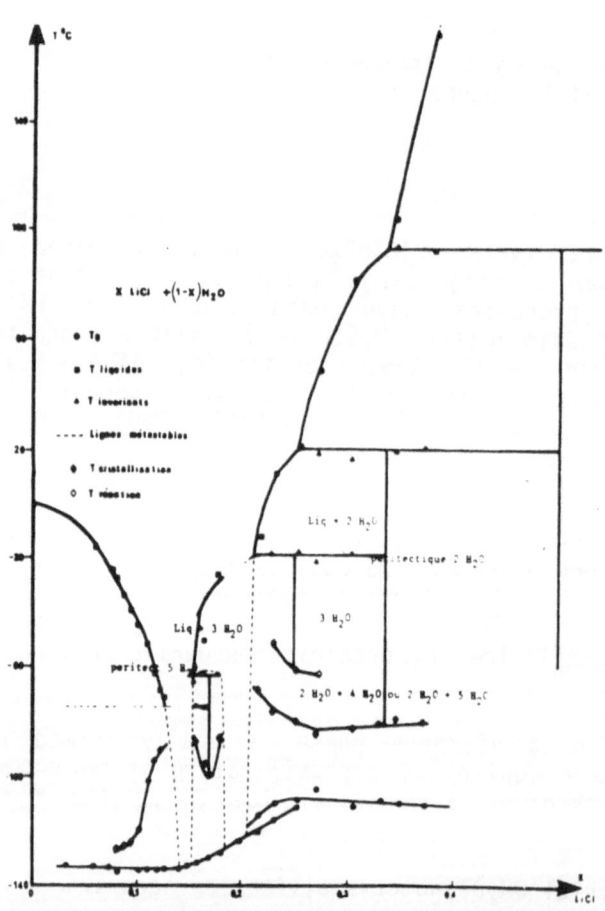

LASER SPECTROSCOPY

B. CHAMPAGNON

Laboratoire de Physico-Chimie des Matériaux Luminescents
Université Lyon I 43, boulevard du 11 Novembre 1918
69622 VILLEURBANNE Cédex (France)

Coworkers : A. BOUKENTER, G. BOULON, F. DURVILLE and E. DUVAL

Laser spectroscopy is a powerfull tool to study the environment of a luminescent ion. In glasses the structure, usually hidden in the wide distribution of sites, can be observed by site selective laser excitation (Fluorescence Line Narrowing). Time Resolved Spectroscopy allows to distinguish emission with different time constants even if these emission take plase at the same wavelength.

Example of the use of the laser spectroscopy are illustrated in order to study the earlier stages of nucleation in a cordierite glass containing chromium oxide as nucleating agent (52% SiO_2, 34.7 % Al_2O_3, 12.5 % MgO, 0.8 % CrO_3). Characterization of the nucleus (the size of which is known form Small Angle Neutron Scattering) is achieved. Fluorescence Line Narrowing in the $^2E \longrightarrow {}^4A_2$ emission band heat treated 10 mn at 950 °C allows to identify the microcrystals formed as being $MgAl_2O_4$: Cr^{3+} spinel (fig. 1) ; the time resolved spectroscopy in a glass heat-treated 4 h at 875 °C and 2 h at 900 °C shows that in a earlier stage of nucleation a chromium concentrated phase develops : a $(^2E, {}^4A_2) \longrightarrow ({}^4A_2, {}^4A_2)$ broad band with a long life time appears and can be attributed to chromium couples ions (fig. 2).

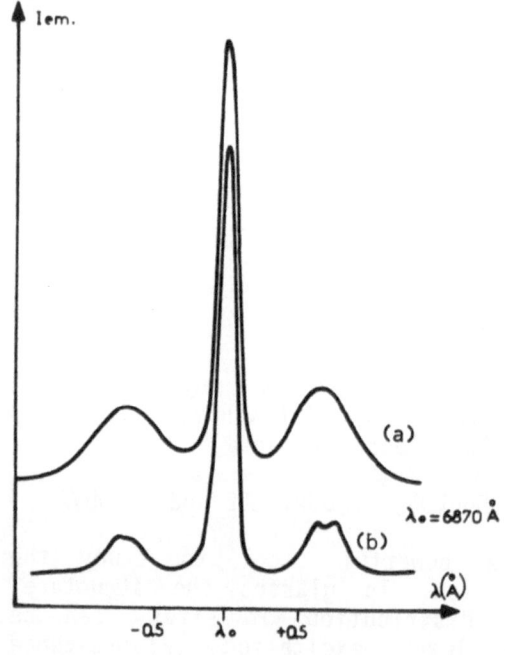

Figure 1 :

Fluorescence Line Narrowing at 4.2 K

a) Glass heat treated 10 mn at 950 °C

b) Synthetic spinel powder $MgAl_2O_4 : Cr^{3+}$.

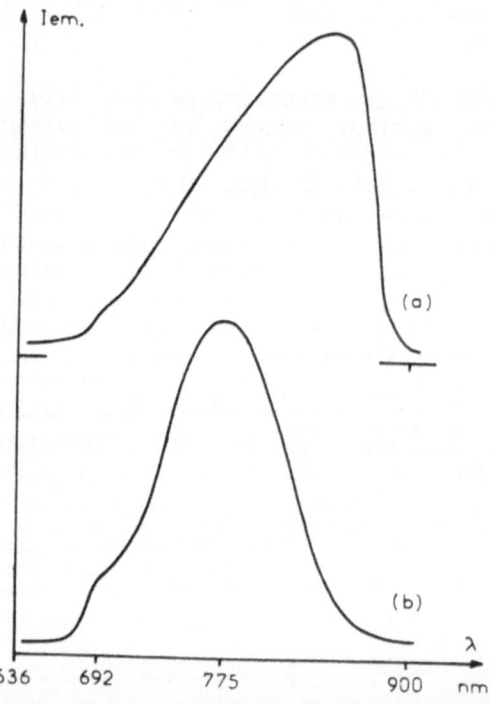

Figure 2 :

Time Resolved Spectroscopy at 4.2 K excitation 6000 Å for a sample heat-treated 4h 875°C + 2h 900°C

a) Normal spectrum

b) With a delay of 1ms and a gate of 3 ms.

GLASS-CERAMICS

W. HÖLAND, W. VOGEL

Friedrich-Schiller - Universität Jena
Sektion Chemie - Otto-Schott-Institut
6900 JENA (G.D.R.)

Four types of glass-ceramics are described, which were developed at the Otto Schott Institute of the Friedrich Schiller University Jena. The basic glass system of all types is the SiO_2-Al_2O_3-MgO system.

A glass-ceramic with good mechanical properties can be formed from organic gels if the final product contains TiO_2 (or Ti (III)- Ti (IV)).

A new type of machinable glass-ceramics with new, curved fluorophlogopite crystals can be formed, if the basic glass contains additions of Na_2O, K_2O, F^-. Electron and scanning electron micrographs were shown.

Machinable glass-ceramics with ferrimagnetic properties contain FeO, Fe_2O_3, K_2O, F^- and the basic glass components SiO_2, Al_2O_3, MgO. After melting and cooling mica-type crystals can grow in the glass together with ferrospinels, e.g. magnesium-ferrite. The material is magnetic soft and machinable.

Machinable bioactive glass-ceramics contain as main crystal phases phlogopite and apatite. The materials bonds to bone, it is bioactive and machinable by standard metal working tools and the material might be applicable for bone substitution.

STRUCTURE OF AMORPHOUS BIMETALLIC CHAINS MM'EDTA nH_2O

E. CORONADO[a], D. BELTRAN[a] and M. DRILLON[b]

(a) Dep. Quimica Inorganica VALENCIA (Spain)
(b) Dep. Science des Matériaux E.N.S.C.S.
STRASBOURG (France)

The local structure of MM'EDTA n H_2O amorphous compounds (with [MM'] = (NiNi), (CoCo) and (CoCu)) is discussed from its magnetic properties. This behavior is interpreted in terms of a chain arrangement with metal ions couples magnetically by an exchange interaction randomly distributed along the chain. This result agrees with LAXS investigations which suggest the presence in the chains of a random distribution in the MM' distances and $\widehat{MM'M}$ angles.

The present study permets to describe the local structure as "flexible" or "deformable" chains and evidences the essential role of the magnetic properties on the better knowledge of the structure in disordered materials.

STRUCTURE OF METALLIC GLASSES

J.M. DUBOIS

Laboratoire de Métallurgie, LA 159
Parc de Saurupt
54042   NANCY   (France)

Strong experimental evidence is now in favour of the existence of a metallic prismatic coordination of the metalloid in transition metal M - metalloid X glasses. This type of coordination is quite common in the crystalline counterparts when the atomic radii ratio lies in the range $0.65 < r_X/r_M < 0.85$. These compounds present a large variety of structures which can all be simply described in terms of structural operations generating (and correlating) the prismatic units by periodically twinning -at unit cell level- close packed structures. Such operations on the one hand preserve the high packing fraction of the basically metallic M network whereas on the other hand they offer enough room to the metalloid X.

Models of $a-Ni_{80}B_{20}$, $a-Co_{80}B_{20}$ and $a-Pd_{80}Si_{20}$ were computer built by using the structural operation which generates the cementite-like compounds Ni_3B, Co_3B and Pd_3Si. The disorder was introduced by limiting the spatial extent of the structural operation. Domains of average size (1-2 nm) were generated by operations acting in different directions in adjacent domains. The coordinates of the atoms linking neighbouring domains were also generated by the same structural operation which was restricted to be invariant for a local rotation. Thus, the glass structure had overall homogeneity and might fulfill the requirement of gauge invariance introduced by Rivier.

Evidence was given by the model coordinates that the metalloids belong to prismatic units. Excellent agreement was obtained between experimental and calculated partial and total structure factors and correlation functions. The extent of medium range order in the model was investigated and compared favourably with the available experimental data.

Some of the consequences of such a model might be of interest from both the fundamental and applied point of views. First, there are two distinct sources of disorder originating from i - frustration effects and ii - concentrations fluctuations. Frustration arises from the invariance of the connection at the domains boundaries. The periodicity of the chemical twinning imposes the stoichiometry. Therefore, concentration fluctuations involves different periodicities which also contribute to the disorder. Different types of defects result from these two effects which may play a role in the dynamical properties of the glasses. Secondly, the filling of the space by structural units cannot be achieved when the X concentration is below a limit concentration x_ℓ which depends on the nature of the M and X elements. Thus, two types of M environments are expected in the $x < x_\ell$ range. Several binary glasses exhibit slope changes in their physical properties - concentration relationships which may be related to this effect. Finally, the application of the different possible structural operations allows a classification of the metallic amorphous structures. Specifically in the $x = 0.20-0.25$ range, two different kinds of connection are distinguishable, namely connection by tetrahedra and by half-octahedra. This point is supported by some experimental results and was developed with the aim of predicting new glass forming compositions.

Work done in collaboration with P.H. Gaskell and G. Le Caer to be published as :

- A model for metallic glasses generated by chemical twinning
  J.M. Dubois, P.H. Gaskell, G. Le Caer   RQ5 Conference
- Modelling the structure of metallic glasses by chemical twinning
  J.M. Dubois, P.H. Gaskell, G. Le Caer   Proc. Roy. Soc.
- Investigation of the medium range order in an amorphous structure model
  J.M. Dubois, W.O. Saxton, P.H. Gaskell, G. Le Caer, Phil. Mag. B
- Ordre local et propriétés physiques des verres métalliques riches en fer
  J.M. Dubois, G. Le Caer   Acta Met.
- The chemical twinning model as a possible guide to the choice of new glass forming compositions
  J.M. Dubois, G. Le Caer, K. Dehghan   RQ5 Conference

CRYSTALLISATION IN SOME TRANSITION METAL-METALLOID GLASSES

Joanne E. ROUT

Department of Metallurgy and Materials Science
University of Cambridge   Pembroke Street
CAMBRIDGE CB2 3QZ   (ENGLAND)

The aim of the current study[1] is to investigate the crystallisation of well-characterised iron-boron and iron-nickel-boron metallic glasses containing a few atomic percent of aluminium. The glasses are produced by the melt spinning technique and the amorphous nature of the as-spun material is confirmed by X-ray diffraction. The effect of production variables and composition on the crystallisation process is being investigated by Differential Scanning Calorimetry (DSC) and X-ray diffraction techniques.

The isothermal and non-isothermal DSC thermograms of $Fe_{40}Ni_{40}B_{20}$ both exhibit one crystallisation exotherm. However, Johnson-Mehl-Avrami (JMA) analysis of the isothermal crystallisation below 663 K indicates a two stage process. The values of the JMA exponents, which increase discontinuously during the course of the reaction, suggest that diffusion controlled crystal growth initially occurs on pre-existing nuclei or from surface sites followed by interface controlled growth. This is consistent with primary crystallisation of metallic Fe-Ni ffollowed precipitation of metal borides.

Initial results indicate that as boron is replaces by aluminium in $Fe_{40}Ni_{40}B_{20}$, the non-isothermal crystallisation begins at lower temperatures and a second crystallisation exotherm is observed at higher temperatures. The changes are approximatively 5 and 10 k/atomic percent aluminium respectively. Work is in progress to investigate these reactions in more detail, in particular the nucleation stage, and to identify the crystalline phase formes using X-ray diffraction and TEM.

1) J.E. Rout and J.A. Leake to be presented at Rapidly Quenched
   Metals 5, Wurzburg, Germany, Septembre 1984

STUDY OF DISORDERED MATERIALS BY MÖSBAUER SPECTROSCOPY

J.M. GRENECHE

Laboratoire de Spectroscopie Mössbauer
Faculté des Sciences - Route de Laval - B.P. 535
72017 LE MANS Cédex   (France)

Some soft metallic glasses ($Mt_{80}Me_{20}$ with MT = Fe, Cr and Me = P, C, Si) elaborated by Saint-Gobain Company using melt-spinning technique (1), have been studied by Mössbauer spectroscopy which is a well-adapted method to investigate the magnetic properties. Some development of spin-texture studies are presented here.

The spin texture wich is the prefered orientation of ferromagnetic domains can be describes by an angular distribution function D ($\cdot$, :) (2) and we showed that is any case, by a convenient rotation of reference axes, there is an orthogonal set of principal axes for which texture respects the D_{2h} symmetry. So the number of independant parameters to describe the texture can be reduced from 5 to 2 (3).

Considering the ^{57}Fe Mössbauer Spectroscopy, we developped a method to obtain a random powder spectrum from any sample, with any hyperfine interaction in the case of an unpolarized radiation : the spectrum is recorded by rotating the sample at a given angle (the "magic angle") to the radiation direction (4).

On the order hand, the Mössbauer spectra of disordered systems exhibit very broad and overlapping lines and the line intensities can be accurately measured by using a linear combination of "standard spectra" whose line intensities are well-known (5), (6), (7).

We also showed that the spin texture can be described in an equivalent way through 3 populations of spins parallel to the principal axes. By use of appropriate shields, we established a schematic map of the spin-texture ; external stresses have been applied and the results are interpreted in terms of magnetostriction and internal stresses due to the elaboration process (8), (9).

References

(1) - This work has received financial support from the french Ministry of research and Industry and Saint-Gobain Company

(2) - H.D. PFANNES, M. FISCHER, Applied Physics 13 (1977) 317

(3) - J.M. GRENECHE, F. VARRET, J. Phys. C. 15 , 5333 (1982)

(4) - J.M. GRENECHE, F. VARRET, J. Physique 43, L-233 (1982)

(5) - J.M. GRENECHE, M. HENRY, F. VARRET
      Proc. Int. Conf. Appl. Mössbauer Effect (Jaipur, 1981) p.900

(6) - F. VARRET, J. Physics E, to be submitted

(7) - F. VARRET, J.M. GRENECHE, J. TEILLET
      Workshop of Mössbauer Spectroscopy (Seehiem, 1983)

(8) - J.M. GRENECHE, M. HENRY, F. VARRET, J. Magn. Mat. (1982)
      26, 153

(9) - M. BOURROUS, M. HENRY, J.M. GRENECHE, F. VARRET
      to be presented RQM 5

ZERO CHROMATIC DISPERSION - LOW LOSS
SINGLE MODE OPTICAL FIBRES IN THE 1.3 - 1.6 m RANGE

J.F. BAYON

LAB/MER/FOG
C.N.E.T. route de Trégastel B.P. 40
22301 LANNION   (France)

The future long band and high bit rate optical telecommunication systems will need single mode fibres with the actual following characteristics :

- for those operating at 1.3 μm, loss of about 0.40 DB/km and zero chromatic dispersion ;

- for those operating at 1.55 μm, loss close to 0.2 DB/km and a chromatic dispersion near 15 ρs/lm.nm.

But, as actual lasers working at 1.55 μm are not rigourously monochromatic, it would be useful to achieve zero chromatic dispersion at such wavelength without increasing the optical fibre loss.

Moreover, is zero chromatic dispersion cancels within a long wavelength range (1.3 - 1.6 μm) the two optical windows (1.3 - 1.55 μm) become usable and multiplexing is feasible. This can be solved by using the quadruple clad fibres, so called "Q fibers" first reported by the Bell Labs.

After presenting briefly the apparatus and C.V.D. method, the structure of a "Q fiber" is defined and the experimental results are given.

- First we compared the refractive index profile on both the bulk preform (interfatial method) and the fibre (near refracted field technique) and found a good agreement.

- Then we studied the opto-geometrical parameters (core and

cladding diameters, Δn) resulting in different dispersion curves versus such parameters).

- The influence of the only geometrical parameters by drawing one preform at five different diameters is also shown as well as the evolution of the dispersion curve.

- The homogeneity of the fibre was controlled by dispersion measurements at both fibre ends (using an interferometric method) and on the whole fibre. These three measurements are in a very good agreement about 1 PS/lm-nm on all the scale.

- Finally we present a "Q fiber" of 900 m length with a chromatic dispersion lower than 2 PS/km.nm in the 1.45 μm - 1.7 μm range, a cut off wavelength of 0.95 μm. The loss is 0.3DB/km at 1.55μm and the mode size (ωo) 4.84 μm.

The loss value at 1.7μm was found to be 0.5 DB/km, indicating that the fibre was quite well resistant to bending (not high loss increase between 1.55 and 1.7μm).

INFRARED ZrF$_4$ BASED FLUORIDE GLASS OPTICAL FIBRES

H. POIGNANT

LAB/MER/FOG
C.N.E.T. route de Trégastel B.P. 40
22301 LANNION (France)

Fluoride glasses are nowaday of great interest owing to their projected ultralow loss ($\sim 10^{-2}$ db/km at $\sim$ 2.5 - 3 µm), making them very useful for the achievement of long haul repeaterless optical fibre links. In this paper, theoretical and experimental results are reported which confirm the interest for such materials :

a) - Material dispersion $M(\lambda) = - d^2/d\lambda^2 |n(\lambda)|$: it was found to be lower than for silica and to vary gently with wavelength. The zero dispersion wavelength is between 1.6 and 1.8 µm.

b) - Rayleigh scattering loss are calculated to be about twice lower than for vitreous SiO$_2$, which mean: α_{RS} = 2.5 x 10^{-2} db/km at 2.5 m. The experimental results lead to α_{RS} = 6 db/km at 0.75m on a bulk glass rod (8 mm diameter, 150 mm long).

c) - The multiphonon adsorption occuring in the IR edge was estimated to few 10^{-4} db/km at 2.5µm wavelength, which is negligible compared to intrinsic scattering loss.

The glass casting technique was used to prepare bulk fluoride glass preforms. The influence of the purity of the starting materials on the optical loss (both adsorption and scattering) was pointed out by microcalorimetric measurements in the 0.65 - 0.75 µm range.

Viscosity measurements were also performed on two different glass composition samples versus the temperature, allowing the energy for viscous flow to be calculated.

Finally, fibre drawing was carried out ; however, the observed loss are still high. So, further improvements in the glassmaking and drawing experiment are needed.

GRADIENT-INDEX ROD LENS, PLANAR LENS AND PLANAR WAVEGUIDE

JoJI SUZUKI (TOKYO)
Toshiaki MIZUNO (MONTPELLIER)

NIPPON SHEET GLASS Co, LTD

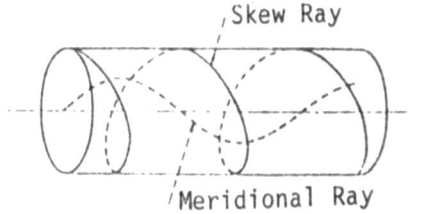

Fig. 1 : Light path in rod lens

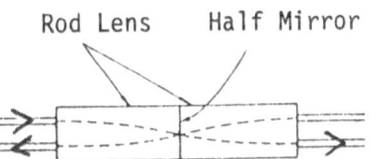

Fig. 2 : Beam splitter using rod lens

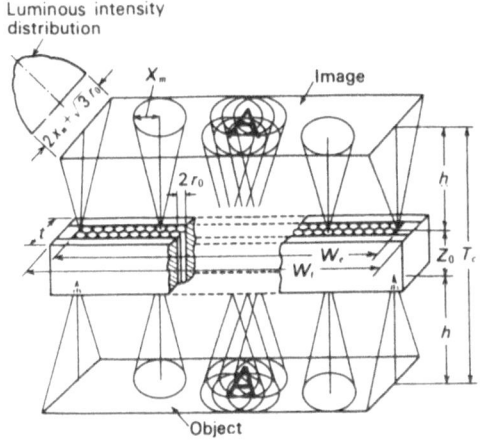

Fig. 3 :
Copier using lod lens array

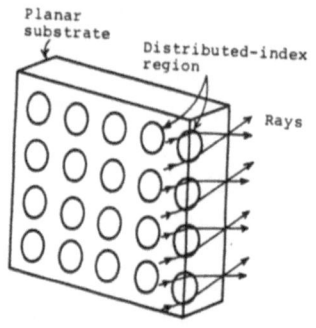

Fig. 4 :
Planar lens configuration

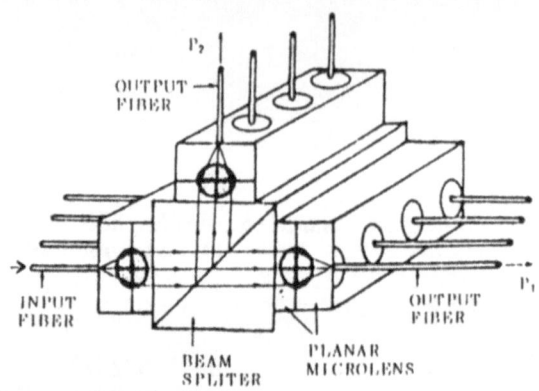

Fig. 5 :   Beam splitter using
planar lens

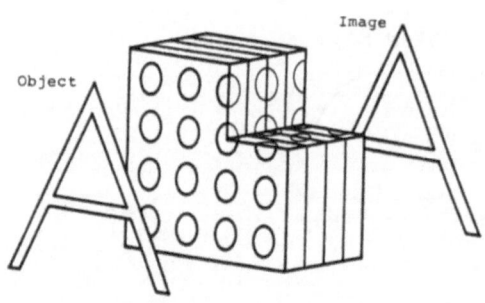

Fig. 6 :   Imaging system using planar lens

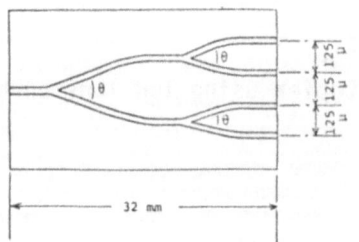

Fig. 6 :
Beam splitter using planar
waveguide

NEW GLASS FORMING SYSTEMS

C.A. ANGELL[*] and A.M. ELIAS[**]

[*] Purdue University, W-Lafayette, IN 47907 (U.S.A.)
[**] Lisbon University, 1294 LISBON Codex (Portugal)

Mass transport properties and cohesion physical properties have been measured for the binary and ternary systems. Complex ionic species in the glass forming systems, $|CoCl_4|^{2-}$, $|NiCl_4|^{2-}$, $|NiF_2Cl_4|^{4-}$, $|CoF_2Cl_4|^{4-}$, are formed and stable. This statility occurs since the cation of the organic compounds, α-picH$^+$, has a small broken effect on the complex species due to its low polarizing power. the formation of the complexes shown by the phase diagram is also clear in the isotherms of the conductivity for the systems, which show maxima and minima at the complex stoichiometry. The glass transition temperatures T_g and T_0 are remarkably affected by the formation of the complex, since it converts configurational degrees of freedom of free chloride and fluoride ions to internal degrees of freedom of chloro- and fluoro-complexes. In addition, the breakdown of the hydrogen bonding structure of the α-picolinium-Cl and α-picolinium-F, as the halogen ion is replaced by the less polarizable ionic form, $|MCl_4|^{2-}$ or $|MF_2Cl_4|^{4-}$, determines the dependence of T_0 and T_g on composition. Similarly, melts fluidity shows characteristic changes in all the measured systems.

The UV-Vis. spectrum have supported the structural evidence of six and four coordination related to complexes formed on the glasses.

STRUCTURE AND PROPERTIES OF a-Si : H : F FILMS OBTAINED BY R.F. GLOW-DISCHARGE IN SiH$_4$-F$_2$ MIXTURES

Rui M. ALMEIDA and Antonio G. DIAS

Centro de Fisica Molecular - Instituto Superior Tecnico
av. Rovisco Pais
1000 LISBOA   (Portugal)

Amorphous Si : H : F films have been prepared by R.F. flow discharge in a SiH$_4$-F$_2$ mixture. A substantial fluorine incorporation was achieved, as found from Rutherford back-scattering and infrared absorption spectra. New vibrational bands have been found at 1070 cm^{-1} and 1170 cm^{-1}, which have been interpreted mainly in terms of SiH$_2$F and SiHF$_2$ complexes, whose frequencies were calculated. Dark and photoconductivities have also been measured for films with and without fluorine. Both the incorporation of fluorine and an increase of the substrate temperature led to lower conductivity values and to a reduction in the density of gap states. The ratio between the photoconductivity and the dark conductivity was found to increase substantially with the incorporation of fluorine in the amorphous alloy.

POROUS GLASS MATRIX FOR NUCLEAR WASTE STORAGE - PART I :
PREPARATION AND CHARACTERISATION OF SPINODAL PHASE SEPARATION

P.C. SANTOS VENTURA, D.I. DOS SANTOS, M.A. AEGERTER

Institute of Physics and Chemistry of Sao Carlos
University of São Paulo
13560   SÃO CARLOS (SP)   (Brasil)

We have studied the phase separation behavior by spinodal decomposition of two sodium borosilicate glasses of the following composition (weight %) $8Na_2O-32B_2O_3-60SiO_2$ (glass A) and $8Na_2O-27B_2O_3-65SiO_2$ (glass B). From photographs of glass samples obtained with a Scanning Electron Microscope, we have determined the growth of the mean size $\bar{r}$ of the microstructures of the least phase as a function of the time of the thermal treatment (0-100 h) and for different temperatures (580, 600, 620 and 640 °C). The results are in agreement with the theory of Lifshitz-Slyozov-Ardell which predict a growth of the microstructures of the least phase through the insoluble phase by a diffusion controlled mechanism such that $\bar{r}^3 = A_0 t \exp(-\Delta E/RT)$. The activation energy ΔE and the pre-exponential term A_0 of the diffusion controlled mechanism are : $\Delta E = 70$ Kcal/mol and $A_0 = 2.2 \ 10^{24}$ Å^3/h (glass A) and $\Delta E = 92$ Kcal/mol and $A_0 = 2.2 \ 10^{29}$ Å^3/h (glass B). The curves of the sizes distribution of the microstructures allow us to suggest for the glass presently under study the most adequate thermal treatments for the preparation of porous glass matrix to store nuclear liquid waste materials.

POROUS GLASS MATRIX FOR NUCLEAR WASTE STORAGE
PART II : SOLIDIFICATION, CHARACTERIZATION AND LEACHING

D.I. DOS SANTOS, P.C. SANTOS VENTURA, M.A. AEGERTER
Institute of Physics and Chemistry of São Carlos
University of Sao Paulo
13560 SÃO CARLOS (SP)   (Brasil)

We present a study of the solidification and the leaching of high silica porous glass matrices which contain 20 weight % of simulated liquid waste solution of high radioactivity level of the type Savanah River Laboratory. The porous glass matrices have been prepared from a $60\%SIO_2-32\%B_2O_3-8\%Na_2O$ (glass A) and $65\%SiO_2-27MB_2O_3-8\%Na_2O$ (glass B) after heat treatment at 4, 10 and 20 h at 560 °C and leaching in HCl-3N at 90 °C. The size of the pores determined by BET techniques were typically 100-250 Å diameter. After sintering at 1300°C in air, the samples were physically and chemically characterized with the standards tests MCC1, Soxhlet (MCC5) and a Static Test during 28 days. We have determined the loss of the total weight the pH and the integral and differential leaching rates of Si, B, Ca, Mn, Al, Fe and Ni using ICP techniques and flame spectroscopy for Na. The results are compared with those obtained with borosilicate reference glasses made with conventional fusion technique (SRL 131, PNL 76-68, MCC 76-68, SRL TDS 131, AVM-M1 to M7), glasses made with the sol-gel techniques (TDS 211), glasses of high silica content (CU PGM), Synroc-D, ceramics concrete FUETAP and metalic matrices. Our results are analogous to those obtained with the best borosilicate glasses presently used.

CONCLUDING REMARKS

JEAN-PIERRE CAUSSE

Vice President for Research & Development
Compagnie de Saint-Gobain, Paris, France

I would like, on behalf of all of you, to say that this was a
successful meeting and to present our very sincere and warm
thanks to the organizing team, Dr. Albert Wright from the
Institute Laue-Langevin, Dr. Josette Dupuy from the CNRS, and
her husband, Professor Claude Dupuy, Président de l'Université
de Lyon. They had the initiative of holding this conference,
they selected the place, and I think it was a good choice. You
imagine the miriad of details that they had to solve to get a
two-week meeting like ours to work as smoothly as it did. So
congratulations again. I would like to add my own personal
thanks to them for having invited me to join in the preparation
of this meeting.

Let me turn now to the main purpose of this evening, which is to
review the conference, to have a discussion on what we tried to
accomplish, and to see what lessons we can learn for the future.

I will start with a few words about the NATO Science Programme.
As you know, NATO was created shortly after the War, and from
its very inception it was clear that the defense of the Occident
was not to be based only on military means, but also on
scientific and cultural actions. The United States appointed as
the first Science Directors of NATO top quality scientists like
Frederick Seitz and Norman Ramsey. Their policy of exchanges
between scientists helped to improve the level of science in
Europe. More recently, however, some people thought it was not
so important to increase the scientific level as to make sure
that this level had an impact on the economy and the industry of
our countries. Among them, Robert Chabbal, a well known French

physicist, who was in 1980 the NATO Assistant Secretary General for Scientific Affairs, had his council approve the idea of the so-called "double jump programme", which meant that besides helping people to go from one country to another, NATO should also help them go from the academic world to applied science, or even to industry. This was originally done by offering fellowships for people from industry to go not only to universities, but also in another country; and Chabbal also suggested that meetings take place which would be attended not only by people from the academic world but also from industry. This idea was followed by Josette Dupuy when she organized this conference, which was approved by NATO, who funded 50% of the cost provided that it be part of the "double jump" programme.

The aims were clear and I summarize them in three sentences - to cover new developments in the field of glass; to present the complementary roles of the academic and the industrial worlds; and to promote long-term collaboration between these groups. I think we have achieved these goals to a large extent. Certainly one key aspect was to bring together people from the academic and the industrial worlds. Of the 120 people who attended the seminar, I counted 23 from industry and 12 from what I call the traditional glass industry, i.e. 10%. The number of talks totals 53, and industry had its share of the burden, with 30% (and 20% coming from the traditional glass industry). If we count the number of people from each company in the traditional glass industry, Saint-Gobain comes first with four people, including one from our Spanish subsidiary ; then we find two from Corning, Nippon Sheet Glass and Philips; and one each from Pilkington and Schott. Very large companies like PPG, LOF, Owens-Illinois, Owens Corning Fiberglas, Asahi or Guardian, have not sent any representatives. If I have gone as far as giving names of companies who had no participants, it is not to put any blame on them, but to remind you that the 12 people from the traditional glass industry represent only a small part of the glass industrial world. It is not obvious for everyone that the glass industry needs research! It is one of the questions I put to the audience.

I come now to the question of our Institute. My own view is that we should be proud of it. All speakers were at least good, some of them were excellent, and a few were memorable.

Having a meeting lasting two weeks is obviously a problem. Some of the people left at the end of the first week, others came at the beginning of the second week. We all agree it would be much better if we could all stay together for the whole seminar, and I think a future Institute should be a little shorter, and consequently either a little more dense, or not so broad. I think there is value in having people live and work together, as we have seen. I think we all enjoyed it, and we are sorry for the people who joined us late or left early.

I would like now to discuss the contents of our programme. If an extraterrestrial had attended the conference, coming from a flying saucer from another planet, he would have heard about a new material called glass with many interesting applications : it is an electrical conductor, has magnetic properties, is made of metal or at low temperatures, and it even has some useful optical properties, primarily in the infra-red.

What about traditional glass ? I do not want to leave the impression that silicate glass is not a current issue. So I present here a plea from the traditional glass industry. Of course it is the glass which is traditional, not the industry. Our industry is lively, or at least trying to remain lively, and I would like to devote the remainder of my speech to two ideas : glass is in competition, and glass needs research.

There is a general phenomenon in materials. Some people have called it the materials revolution, which is the acceleration of the fact that there is constant creation of new materials, and materials are fighting one against the other. This has happened since the stone age, with the bronze age, the iron age, etc. but it took...ages. Now things happen at a rapid pace. I have made a small list here of only a few of the problems, but it is already very impressive.

Take flat glass for building. Roughly half of the flat glass goes into the building industry. We don't really have a competing material, but a window appears as a heat loss in a building, and the competition to glass is that we have less windows. After the 1973 energy crisis there was a trend to reduce the window area, which hurt the glass industry.

The reply of the glass industry is multiple glazings, and energy-efficient coatings on glass. I will not elaborate this point because Jean Blétry's thin-film presentation covered this point and gave examples. This is a very important challenge and will lead to significant changes for the years to come : coatings rejecting excess solar energy, and coatings keeping the inside of the home warm. In the future perhaps these effects will be combined, and functional glasses may have different performances during summer and winter and during day and night.

Solar collectors may be a new function for glass, but at the moment nobody really knows if this is going to be a significant factor. As probably you all know, the original enthusiasm has lost some of its momentum, but it remains certainly a possibility that collectors can be used to provide heat or photovoltaic electric power, relying in both cases on the unique properties of glass.

Flat glass in the automotive industry faces serious problems, both the traditional problem of glass - it is fragile - but also new ones : weight, safety, shape. New cars have a more aerodynamic shape than before, again primarily to save energy. Here the competition of plastics is becoming very active. Glass has to evolve, to modify its performance, and this may be done by making composites. We, especially in our Company, are convinced that windshields in the future will be a glass-plastic composite, not only laminated glass with two sheets of glass and plastic in the middle, but with more layers, performing other functions like safety ; our Securiflex, which was recently approved for use in the United States, has two layers of glass and three layers of plastic. This is an example of a trend.

We also have to use thinner glass, and because it is thinner it is not as strong. Therefore it must be strengthened, so the subjects of glass toughening and glass strengthening must be reviewed again.

In container glass, where the situation is felt by some to be desperate, the competition comes from plastics. PVC has already taken most of the bottled water business, and PET is used with carbonated drinks. Competition comes also from cans in steel, aluminium, and paper. The tetrapak process is very successful, at least in Europe. It has replaced all milk bottling, and now is used for some fruit juices. Fortunately wine remains a good customer for glass! Other packagings also tend to reduce the role of glass, such as the "blister" presentation for pills, which traditionally have been sold in small bottles.

The reply of container companies must obviously be to make lighter and stronger bottles, and then reinforcement is needed - either by plastic coating (a composite solution) or by ion exchange. Of course, this must be done at a reasonable cost.

Another way glass can fight back involves the concept of quality, often linked to the image of glass. By design things can be done with glass that cannot be done with other materials.

Also opthalmic glass is in competition with plastics. Again we believe composites may be the solution. Our Company has developed, together with Corning, spectacle lenses which are composites : glass is on the outside - it is hard, and can withstand the environment ; and plastic is on the inside - the same plastic layer we had developed for safety windshields, and which now provides safety to the eye.

Fibers for insulation

I do not think this word was used during our meeting, and I regret it. Our Company alone, together with its licensees, made one million tons of fiberglass in 1983, so it is a big business. The competition there is primarily with plastics, and the reply of glass is the same: we need a thinner product, less glass, but with the same performance. We also have to emphasize the qualities of glass ; for instance, it is fireproof. Importantly, only fiber glass has acoustic as well as thermal properties. Plastic foams have reasonable thermal properties but do not have good acoustic properties, and we know the importance of soundproofing as an element of the "quality of life". We also follow the route of composites by bringing together materials which can complement each other, for instance plaster - glass wool sandwiches in products which are ready for use.

Fibers for reinforcement

I will be brief, since we had a session on this topic and the competition was explained. There are new fibers (aramides, carbon). But there is also the reaction of the existing products. For instance steel is fighting back. The penetration of the glass fiber-reinforced products in the automobile industry is not what we composite people thought it would be, because the steel of today is not what it was ten years ago. Plastics themselves have been improved. For instance the bumper of the Renault R5 was the first mass-produced composite with glass fiber, but the bumper of the recent R9 has no glass in it -- it is made only with plastic. Obviously this plastic is better than the one we used ten years ago when the R5 appeared. In this very active field glass should not only fight back, but also be on the attack and try to find new applications. One example lies with the new ZMC forming process which is used successfully for complex automotive body parts, and in our opinion opens new areas.

Significant market developments can also be expected for new glass fibers if the bridgeheads opened by alkali or corrosion resistant glasses prove to be successful.

At this point I suppose I have convinced you that there is a lot to be done, but some of you may think that the problems require only efforts of engineering or technology. This is undoubtedly true, but also I am convinced that we need more science. The understanding of the properties of glass and the physics and chemistry of the process involved must be deepened before further technological progress can be achieved. This applies to melt properties, the melting and fining processes (with the impact of all-electric furnaces), corrosion phenomena of refactories, fracture mechanics and related strengthening processes, forming techniques (with the new possibilities opened by modeling) etc. ; last but not least, research on new glass compositions can lead to a very fruitful harvest of exciting properties for new applications.

I sincerely hope that the better understanding of the amorphous state of the matter, which was rightly the cornerstone of this meeting, will give a new momentum to all glass research, including the so-called traditional glass.

DISCUSSION (summary)

Mr. Suzuki, from Nippon Sheet Glass Company (NSG), a major producer of flat glass in Japan, indicated that he came for the purpose of looking for new advanced products and technologies for his Company's diversification. He appreciated the information received and thanked the organizers and the participants. He also offered copies of publications on recent developments in gradient index lenses and optical systems at NSG and of a Japanese Government report entitled "Japan Science Outlook".

Dr. Proctor, from Pilkington Brothers, introduced three practical "constructive criticisms" for future meetings.
- The choice of Tenerife, a holiday resort, was a mistake from an industrial standpoint.
- A fortnight was too long for most people to be able to attend the whole meeting.
- The programme did not concentrate enough on common problems of glass ; however, the nucleus of interesting cross fertilization was present.

Dr. Hench, from the University of Florida, expressed the view that the primary purpose of a school is to stimulate ideas, and not to answer questions. This Institute had done that admirably. He himself is leaving with several ideas he didn't have before he came, a very positive result of the meeting.

He also defended the choice of Tenerife, which had been an enticement for him to come. He would not have accepted to attend such a meeting in an industrial region of Europe. Being a "captive" on the island benefited greatly attendance at the school.

Dr. Hench said he thinks that it is a great challenge for the traditional glass industry to be able to assimilate the fast-moving progress in fields that previously would have seemed totally foreign to them and apply them to their own industry. Academics are not going to do that for them. If there is a failing in the meeting, it is the failing of some traditional glass industries to recognize the importance of participating in the expanding knowledge which is occurring.

He described his past experience as an officer with the American Ceramic Society, where he had observed over the last decade a disappointing trend in the traditional glass industry, which tends to concentrate on short term problems and on the bottom line of the company's balance sheet. Some of the better scientists had not been able to present their work openly or even in some cases had not been allowed to come to ACS meetings. Yet the list of problems to be solved, which was presented today, was not very different fifteen years ago.

Finally Dr. Hench introduced the idea, now widely discussed at least in the United States, that whereas the stone age, the bronze age, etc. have marked civilizations, we are already making the transition from a materials age to an information age. This meeting has shown that glass is a critical component in this information age and that glass has a leadership role to play.

Dr. Kreidl commented also on the absence of research people from the classical glass industry. Some such industries have practically closed their research and development departments, and acquired other companies making competitive materials, or even spent the money thus saved in completely different fields. As an optimist, he believes that this is only a wave, and that the situation will correct itself. A similar situation existed years ago in materials, but now materials research and development is a flourishing field.

Dr. Kreidl's concluding remarks were that as a man who has attended many meetings, the arrangements and the contents for this meeting were among the best (although he thinks that the duration of two weeks is not easily acceptable). He thanked the chairman of this meeting for his coherent description and the frank discussion on what had been accomplished in the school.

List of participants

Aegerter, M.A.	Institute of Physics and Chemistry of São Carlos, University of Sao Paulo, 13560 Sao Paulo (SP) Brasil.
Allemand, J.P.	Institut National des Sciences Appliquées, G.E.N.P.P.M., Bât.502 20, Avenue Albert Einstein 69622 Villeurbanne Cédex, France
Almeida, R.M.	Centro de Fisica Molecular, Complexo I. I. S. T. 1000 Lisboa, Portugal
Anderson, I.	Institut Laue – Langevin, B.P.156 Centre de Tri 38042 Grenoble Cédex, France
Angell, C.A.	Purdue University, West Lafayette Indiana 47907, U.S.A.
Antonini, M.	Istituto di Fisica, University of Modena, Via Campi 313-A Modena, Italy
Aouizerat, A.	Université Claude Bernard Lyon I Département de Physique des Matériaux, 43 Bd. du 11 Novembre 1918, 69622 Villeurbanne Cédex, France
Appell, Y.	Chef de production, S.E.P. Le Haillon, B.P. 37 33165 St. Médard-en-Jalles, France
Baro, M.D.	Facultad Sciencias, Universidad Autonomo Bellaterra, Barcelona, Spain

Battaglin, G.

Dipartimento di Fisica 'Galileo Galilei', Via Marzolo 8 35131 Padova, Italy

Baucke, F.G.

Schott Glaswerke, Hattenberstr. 10 6500 Mainz, R.F.A.

Bayon, J.F.

C.N.E.T., Route de Tregastel, B.P. 40, 22301 Lannion, France

Beall, G.

Corning Glass Works, FR 5 Sullivan Park, Corning, N.Y. 14832 U.S.A.

Beatrice, C.

Corso Massimo D'Azeglio n° 42, Istituto Elettrotechnico Nazion. G. Ferraris, 10125 Torino, Italy

Beck, W.R.

3M Center BLD 6201-1W-31, St. Paul, Minn. 55144, U.S.A.

Bhat, P.K.

Department of Physics, University of Sheffield, U.K.

Blétry, J.

St. Gobain Recherches, 39 Quai Lefranc, B.P. 135 93304 Aubervilliers Cédex, France

Cachard, A.

U.E.R. Sciences, 23 Rue du Docteur Michelon, 42100 St. Etienne, France

Calas, G.

Laboratoire Minéralogique-Cristallographique, Universite Paris VI, 4 Place Jussieu 75230 Paris Cédex 05, France

Campbell, A.G.

University of Edinburgh, Department of Electrical Engineering, School of Engineering, King's Building, Edinburgh EH9 3JL, UK

Cases-Andrea, R.

Departamento de Optica, Facultad de Ciencias, Universidad de Zaragosa, Zaragosa, Spain

Causse, J.P.	Directeur Général Adjoint chargé de la Recherche, Compagnie Saint-Gobain, Les miroirs, 18 Avenue d'Alsace, Cédex 27 92096 Paris La Défense, France
Champagnon, B.	Université Claude Bernard Lyon I Bâtiment 205, 43 Bd. du 11 Novembre 1918 69622 Villeurbanne Cédex, France
Chieux, P.	Institut Laue – Langevin, 156X 38042 Grenoble Cédex, France
Coronado, M.	Dept. Science des Matériaux, Ecole Normale Supérieure de chemie de Strasbourg, 1 Rue Blaise Pascal, 67008 Strasbourg, France
Davies, L.A.	Allied Corporation, Corporate Technology, P.O. Box 1021 R, Morristown, N.J. 07960, U.S.A.
De Laat, F.	Engineering Specialist, Litton Glass, 5500 Canoga Avenue, Woodland Hills CA 91367, U.S.A.
Della Mea, G.	Dipartimento di Fisica, Università di Padova, Via Marzolo 8 35100 Padova, Italy
De Waal, H.	Technich Physiche Dient TNO-TH Postbus 155, 2600 AD Delft, The Netherlands
Dianoux, A.J.	Institut Laue – Langevin, 156X 38042 Grenoble Cédex, France
Drexhage, M.G.	U.S. Air Force, RADC/ESM, Solid State Sciences Division, Rome Air Development Center, Hanscom AFB, MA 01731, U.S.A.
Dubois, J.M.	Laboratoire de Métallurgie, E.N.S.M.I.M., Parc de Saurupt, 54042 Nancy Cédex, France

Durand, J.	Université de Nancy I, Faculté des Sciences, B.P. 239 59506 Vandoeuvre-les-Nancy Cédex France
Dupuy, J.	Université Claude Bernard Lyon I Département de Physique des Matériaux, 43 Bd. du 11 Novembre 1918, 69622 Villeurbanne Cédex France
Dupuy, C.H.S.	Université Claude Bernard Lyon I Département de Physique des Matériaux, 43 Bd. du 11 Novembre 1918, 69622 Villeurbanne Cédex France
El Bayoumi, O.H.	U.S. Air Force RADC/ESM, Solid State Sciences Division, Rome Air Development Center, Hanscom AFB, MA P131, U.S.A.
Elias, M.	Lisbon University, 1294 Lisboa Cédex, Portugal
Elliott, S.R.	Department of Physical Chemistry, University of Cambridge, Lensfield Road, Cambridge, U.K.
Fares, V.	I.T.S.E., Asea Ricerca Roma, Via Salario km 295 00016 Monterotondo / Roma, Italy
Fiorani, D.	I.T.S.E., Asea Ricerca Roma, Via Salario km 295 00016 Monterotondo / Roma, Italy
Gabano, J.P.	S.A.F.T., 11 Rue Georges Leclanché, 86000 Poitiers, France
Gambino, R.	I.B.M. Research, T.J. Watson Research Center, Yorktown Heights N.Y. 10598, U.S.A.
Gannon, J.R.	Corning Glass Works, Research and Development Division Fiber Optics Division, Corning N.Y. 14821, U.S.A.

Gaskell, P.H.	Cavendish Laboratory, University of Cambridge, Cambridge, U.K.
Gottardi, V.	Università di Padova, 35000 Padova, Italy
Gowan, J.G.	European Office of Aerospace Research and Development EOARD 223 Old Marylebone Road London, U.K.
Grenèche, J.M.	Laboratoire de Spectrométrie Mösbauer, Faculté des Sciences ERA 682, B.P. 535, Route de Laval 72017 Le Mans Cédex, France
Griscom, D.	Naval Research Laboratory, Code 6570, Washington D.C. 20375, USA
Hench, L.	Department Materials Sciences and Bio-Engineering, University of Florida, Gaines Ville FL 32611 Florida, U.S.A.
Henry, M.	Spectrochimie du Solide, Université Paris VI, 4 Place Jussieu 75230 Paris Cédex, France
Hoeland, W.	Otto-Schott-Institut der Sektion Chemie der Friedrich-Schiller-Universität, Steiger S H III 6900 Jena, DDR
Hull, D.	Department of Metallurgy and Materials Science, P.O. Box 147 Liverpool L 69 3BX, U.K.
Ingram, M.D.	Department of Chemistry, University of Aberdeen, Aberdeen AB4 2VE, Scotland, U.K.
Joglar-Tamargo, A.	Cristaleria Espagnola SA, Apartado 88, Aviles, Spain
Johnson, W.L.	W.M. Keck Laboratory of Engineering, California Institute of Technology, Pasadena CA 91125, U.S.A.

Knott, P.	Department of Ceramics, University of Leeds, LS2 9JK Leeds, U.K.
Köster, U.	ABT. Chemietechnik, Universität Dortmund 4600 Dortmund 50, R.F.A.
Kreidl, N.J.	1433 Canyon Road, Santa Fe New Mexico, U.S.A.
Laude, D.	SDHA, CEA, B.P. 171 Marcoule, 30200 Bagnoles/Cèze France
Lecomber, P.G.	Carnegie Physics Laboratory, University of Dundee, Dundee DD2 4HN, Scotland, U.K.
Lespade, P.	Aérospatiale, Laboratoire Matériaux et Procédés Nouveaux, B.P. 11, 33160 St. Médard-en-Jalles, France
Livage, L.	Université Pierre et Marie Curie, Spectrochimie du Solide, 4 Place Jussieu 75230 Paris Cédex 05, France
Luborsky, F.	General Electric Research and Development Center, P.O. Box 8 Schenectady N.Y. 12301, U.S.A.
Lucas, J.	Université de Rennes, Laboratoire de Chimie Minérale D, Avenue du Général Leclerc 35031 Rennes Cédex, France
Magini, M.	Divisione Chimica/TIB-ENEA C.R.E. - Casaccia, Via Anguillarese, Roma, Italy
Maret, M.	Institut Laue - Langevin, 156X 38042 Grenoble Cédex, France
Martin, S.W.	Purdue University, Department of Chemistry, Chemistry Building Box 610, West Lafayette, Indiana 47907, U.S.A.

Mazzoldi, P.   Istituto di Fisica Galileo Galilei
                                Università di Padova, Via Marzolo
                                8, 35131 Padova, Italy

Mendiratta, S.K.   Universidade de Aveiro, Depart-
                                ment of Physics, 3800 Aveiro,
                                Portugal

Mizuno, T.   Laboratoire des Verres du CNRS
                                Université de Montpellier, Place
                                Eugène Bataillon
                                34060 Montpellier, France

Montenero, A.   Istituto di Strutturistica Chimica
                                Università di Parma, Via Massimo
                                d'Azeglio 85, 41100 Parma, Italy

Moynihan, C.T.   Materials Engineering Department
                                Rensselaer Polytechnic Institute
                                Troy, N.Y. 12181, U.S.A.

Mukkerjee, S.P.   Jet Propulsion Laboratory
                                48000 Oak Grove Drive, Pasadena
                                CA 91109, U.S.A.

Murawski, L.   Technical University of Gdansk,
                                Politichniko Gdonslo, Institute
                                of Physics, 80-952 Gdansk, Poland

O'Neill, M.A.   The Cavendish Laboratory, Metal
                                Physics Section, Madgley Road,
                                Cambridge CB3 OHE, U.K.

Oversluizen, G.   Philips Research Laboratories,
                                P.O. 80 000, 5600 JA Eindhoven,
                                The Netherlands

Owen, A.E.   University of Edinburgh, Depart-
                                ment of Electrical Engineering,
                                School of Engineering, King's
                                building, Edinburgh EH9 3JL, U.K.

Perez, J.   I.N.S.A., Département de Métal-
                                lurgie, 20 Avenue Albert Einstein
                                69100 Villeurbanne, France

Phalippou, J.   Université des Sciences et Tech-
                                niques du Languedoc, Place Eugène
                                Bataillon, 34060 Montpellier,
                                France

Pinna, G.	Istituto di Chimica Fisica Via Ospedale 72, 09100 Cagliari, Italy
Poignant, M.	C.N.E.T., Route de Tregastel, B.P. 40, 22301 Lannion, France
Potkay, E.	AT & T., Engineering Research Center, P.O. Box 900, Princeton NJ 08540, U.S.A.
Proctor, B.A.	Pilkington PLC, R & D Lathom, Ormskirk, Lancashire, L40 SUF, UK.
Ravaine, D.	E.N.S.E.E.G., B.P. 75 38400 St.-Martin-d'Hères, France
Rodmacq, B.	D.R.F. Physique du Solide, CENG B.P. 85, 38041 Grenoble Cédex, France
Rout, J.	University of Cambridge, Department of Metallurgy and Material Sciences, Penbrooke Street, Cambridge CB2 3QZ, U.K.
Savage, J.A.	Royal Signals and Radar Establishment, St. Andrews Road, Malvern, U.K.
Schmidt, H.	Fraunhofer Institut für Silicatforschung, Neunerplatz 2 8700 Würzburg, F.R.G.
Scott, M.G.	S.T.L. Ltd., Optical Waveguide Division, Harlow Essex
Seddon, A.	Department of Ceramics, Glasses and Polymers "Elmfield", Northumberland Road, Sheffield, U.K.
Simonis, F.	Institute of Applied Physics TNO-TH, Stieltjesweg 1, P.O. Box 155, 2600 AD Delft, The Netherlands
Smith, C.	Allied Corporation, Corporate Technology, P.O. Box 1021 R, Morriston, N.J. I7960, U.S.A.

Spierings, G.

Nederlandse Philips, Bedrybenglass
Central Development, Building TYI,
5600 MD Eindhoven, The Netherlands

Steeb, S.

Max-Planck-Institut für Metallforschung, Serstrasse 92
7000 Stuttgart 1, F.R.G.

Surinach-Cornett, S.

Facultad Ciencias, Universitad
Autonomo Bellaterra, Barcelona,
Spain

Suzuki, J.

Nippon Sheet Glass Co. Ltd., Corporate Planning Department,
5-11-3 Shimbashi, Minato-Ku
Tokyo 105, Japan

Torell, L.

Department of Physics, Chalmers
University of Technology,
41296 Göteborg, Sweden

Tran, D.C.

Optical Techniques Branch, Code
6570, Naval Research Laboratory,
Washington D.C. 20375, U.S.A.

Uhlmann, D.R.

Department of Materials Science
and Engineering, Massachussetts
Institute of Technology, 13-4005
Cambridge Mass. 02139, U.S.A.

Wright A.C.

J.J. Thomson Physical Laboratory,
University of Reading, Whiteknights
Reading, RG6, 2AF, U.K.

Wright, A.F.

Institut Laue - Langevin,
156X
38042 Grenoble Cédex, France

Wenzel, J.

Compagnie Saint-Gobain, Direction
de la Recherche, Cédex 27,
92096 Paris La Défense, France

Wilson-Hench, J.

Department of Materials Sciences
Bioengineering, University of
Florida, Gainesville, FL 32611,
U.S.A.

Zarzycki, J.

Directeur du Laboratoire des
Verres du CNRS, Lab. de Sciences
des Matériaux, Université de Montpellier, 34060 Montpellier, France

INDEX

718